Springer Earth System Sciences

The Springer Earth System Sciences series focuses on interdisciplinary research linking the lithosphere (geosphere), atmosphere, biosphere, cryosphere, and hydrosphere that build the system earth. The series seeks to publish a broad portfolio of scientific books, aiming at researchers, students, and everyone interested in this extremely interdisciplinary field. It covers the entire research area of earth system sciences including, but not limited to, Earth System Modeling, Glaciology, Climatology, and Human-Environment/Earth interactions. Springer Earth System Sciences includes peer-reviewed monographs, edited volumes, textbooks, and conference proceedings.

More information about this series at https://link.springer.com/bookseries/10178

Alejandro Otero · José L. Carballido · Diego Pol
Editors

South American Sauropodomorph Dinosaurs

Record, Diversity and Evolution

Springer

Editors
Alejandro Otero
Facultad de Ciencias Naturales y Museo
CONICET—Consejo Nacional de
Investigaciones Científicas y Técnicas
La Plata, Argentina

José L. Carballido
Museo Paleontológico Egidio Feruglio
CONICET—Consejo Nacional de
Investigaciones Científicas y Técnicas
Trelew, Argentina

Diego Pol
Museo Paleontológico Egidio Feruglio
CONICET—Consejo Nacional de
Investigaciones Científicas y Técnicas
Trelew, Argentina

ISSN 2197-9596 ISSN 2197-960X (electronic)
Springer Earth System Sciences
ISBN 978-3-030-95958-6 ISBN 978-3-030-95959-3 (eBook)
https://doi.org/10.1007/978-3-030-95959-3

This Springer imprint is published by the registered company Springer Nature Switzerland AG
The registered company address is: Gewerbestrasse 11, 6330 Cham, Switzerland

To Leo Salgado, for his great contribution to the understanding of the South American sauropods and for being the mentor or worked with most of the contributors of this book

To Zulma Gasparini, for her constant support of the younger generations

To José Bonaparte, who laid the modern foundations and brought South American dinosaurs to the forefront of vertebrate paleontology

Foreword

Sauropodomorpha Huene 1932 is one of the most successful groups of dinosaurs that includes the most abundant and diverse herbivorous forms with a worldwide record, extending from the Late Triassic to the Late Cretaceous. Sauropodomorphs include early forms (the formerly traditionally called 'prosauropods') and the iconic sauropods. With more than 150 valid species and a worldwide distribution, Sauropoda Marsh 1878 comprises the dominant herbivorous dinosaurs from the Middle Jurassic to the Late Cretaceous. The sauropodan body plan, characterized by gigantic size, graviportal locomotion, long necks and tails, and a comparatively small skull, made this group of sauropodomorphs part of the popular culture since the late nineteenth century.

Worldwide, the current knowledge about sauropods, their classification, anatomical and paleobiological knowledge, is the result of the joint work of numerous researchers, and I could not name some of them without taking the risk of leaving out others. However, abusing a bit of the generosity of Alejandro, José Luis, and Diego, the editors of this book who invited me to preface it, I will recall two that profoundly influenced many of the sauropod specialists of my generation, including myself: Jack McIntosh (1923–2015), Professor at Wesleyan University (Connecticut, USA), and José Bonaparte (1928–2020), for years Head of the Vertebrate Paleontology Section of the Museo Argentino de Ciencias Naturales (at Buenos Aires city, Argentina). Particularly Bonaparte is, to date and without doubts, the greatest exponent of Mesozoic Vertebrate Paleontology in the entire South American continent.

This book deals with South American sauropodomorphs, although South America did not exist as a separate continent when sauropodomorphs lived. Actually, for most of the evolutionary history of this group of dinosaurs, South America was integrated to other landmasses, at first to the rest of Gondwana and later to West Gondwana. Beyond this, the geographical demarcation thought for this book is totally justified based on the superlative record of sauropodomorphs in what is today South America, with forms ranging from the Carnian—when they make their appearance in the paleontological record—up to the Maastrichtian—when they became extinct—. During all that long period of time (more than 160 million years), sauropodomorphs experienced important evolutionary events: diversifications and global or local extinctions, modifying their body plans with the acquisition of unique evolutionary novelties, many of which are directly linked to their herbivory and huge sizes.

Many of the events that marked the evolution of the Sauropodomorpha can be better studied from the South American record, and several of these extreme evolutionary transformations have excellent examples in sauropodomorphs recorded in this subcontinent; to mention just two, the earliest giant sauropodomorph (the lessemsaurid *Ingentia*) and the largest sauropods so far recorded (the lognkosaurs *Argentinosaurus* and *Patagotitan*) lived in what is now South America. In this way, the first and last steps toward the gigantism are recorded in this subcontinent; it can be said that, in some way, the evolutionary history of sauropodomorphs begins and ends in South America.

Among all the forms that integrate the list of South American sauropodomorphs, titanosaurs undoubtedly occupy a central place, to the point that they have deserved several chapters of the book (8–10). Here, I will again abuse the kindness of the editors bringing to mind to a third sauropodologist that laid the foundations of the current knowledge of the South American representatives of this group of macronarians: Jaime Powell (1953–2016), from the Instituto 'Miguel Lillo' at Tucumán, Argentina. Among some of his discoveries, Jimmy was the one who recognized the existence of osteoderms in some titanosaurs, a characteristic unique to this group of huge reptiles.

In this book, there are 37 participant contributors. Although there is a majority of Argentinians, many of them are from other countries in South America and from other continents. After all, South American sauropodomorphs have always raised international interest and are not few of the researchers that, year after year, arrive in the region with the purpose of consulting old collections and seeing new materials: new materials that are increasing at an accelerated rate.

This book contains 15 chapters, including systematics, covering the main lineages within Sauropodomorpha (Chapters 1 to 10), paleobiology (Chapters 11 to 14), and taphonomy (Chapter 15), a broad approach that accounts the breadth of interests of its authors.

Beyond the notably increase in anatomical and systematic knowledge achieved, especially in the last 30 years, new lines of research in sauropodomorph paleontology have been opened with the incorporation of young researchers; among these are paleosteohistology, reproductive biology, and ichnology, which have also been enhanced with the incorporation of new and not so new technologies, now widely used, such as computed tomography, scanning electron microscopy, or photogrammetry. Hopefully, in the coming years, these new lines of investigation will allow us to learn more about these dinosaurs.

Sauropodomorphs went extinct along with the rest of the non-avian dinosaurs, but, needless to say, the fascination they still produce has not diminished. Ninety years after the group was recognized, books like this continue to prove it.

Leonardo Salgado
Consejo Nacional de Investigaciones
Científicas y Técnicas, Instituto de
Investigación en Paleobiología y
Geología
Universidad Nacional de Río Negro
Río Negro, Argentina
lsalgado@unrn.edu.ar

References

Huene F von (1932) Die fossile Reptil-Ordnung Saurischia, ihre Entwicklung und Geschichte. Monographien zur Geologie und Paläontologie, 4:1–361
Marsh OC (1878) Principal characters of American Jurassic dinosaurs. Part 1. Am J Sci. Series 3, 16:411–416

Acknowledgements

We would like to express our gratitude to Jorge Rabassa for his kind assistance since the beginning of this book.

Special thanks to all contributors who made this volume possible. Each one of them put time, effort, and expertise to make an incredible work. Here, they are (in alphabetical order):

Cecilia Apaldetti
Sebastián Apesteguía
Flavio Bellardini
Jonathas S. Bittencourt
Mario Bronzati
Jorge O. Calvo
José I. Canudo
Gabriel A. Casal
Ignacio Cerda
Martín D. Ezcurra
Claire Peyre de Fabrègues
Mariela S. Fernández
Leonardo S. Filippi
Anthony R. Fiorillo
Pablo Gallina
Juan P. Garderes
Kevin Gomez
Bernardo J. González Riga
Femke M. Holwerda
John R. Hutchinson
Max C. Langer
Lucas Lerzo
Philip D. Mannion
Júlio C. A. Marsola
Ricardo N. Martínez

Miguel Moreno-Azanza
Rodrigo T. Müller
Leonardo D. Ortiz David
Agustín Pérez Moreno
Oliver W. M. Rauhut
Leonardo Salgado
Rodrigo M. Santucci
María B. Tomaselli
Bernat Vila

We would also like to express our gratitude to all the reviewers who kindly reviewed each chapter of this book, highly improving its quality (in alphabetical order): Daniel Barta (Oklahoma State University, USA), Jennifer Botha (National Museum, Bloemfontein, South Africa), Mario Bronzati (Universidade de São Paulo, Brazil), Ignacio Canudo (Universidad de Zaragoza, Spain), Kimberly Chapelle (American Museum of Natural History, USA), Michael D'Emic (Adelphi University, USA), Verónica Díez Díaz (Museum für Naturkunde, Germany), Peter Falkingham (Liverpool John Moores University, United Kingdom), Pablo Gallina (Fundación de Historia Natural Félix de Azara, Argentina), Lucio Ibiricu (Centro Nacional Patagónico, Argentina), Ashu Khosla (Panjab University, India), Verónica Krapovickas (Universidad de Buenos Aires, Argentina), Susannah Maidment (Natural History Museum, United Kingdom), Phillip Mannion (University College London, United Kingdom), Adam Marsh (Petrified Forest National Park, USA), Blair McPhee (blair.mcphee@gmail.com), Rodrigo T. Müller (Universidade Federal de Santa Maria, Brazil), Pedro Mocho (Natural History Museum, USA), Stephen Poropat (Australian Age of Dinosaurs Museum of Natural History, Australia), Flavio Pretto (Universidade Federal de Santa Maria, Brazil), Leonardo Salgado Universidad de Río Negro, Argentina), Vanda Santos (Faculty of Sciences of the University of Lisbon, Portugal), Rodrigo Santucci (University of Brasília, Brazil), Rodrigo Tomassini (Instituto Geológico del Sur, Argenitna), Fidel Torcida (Museo de Dinosaurios, Spain), Emanuel Tschopp (Universität Hamburg, Germany), John Whitlock (Carnegie Museum of Natural History, USA).

Contents

The Early Radiation of Sauropodomorphs in the Carnian (Late Triassic) of South America 1
Max C. Langer, Júlio C. A. Marsola, Rodrigo T. Müller, Mario Bronzati, Jonathas S. Bittencourt, Cecilia Apaldetti, and Martín D. Ezcurra

Non-sauropodiform Plateosaurians: Milestones Through the "Prosauropod" Bauplan 51
Alejandro Otero and Claire Peyre de Fabrègues

South American Non-Gravisaurian Sauropodiformes and the Early Trend Towards Gigantism 93
Cecilia Apaldetti and Ricardo N. Martínez

Sauropods from the Early Jurassic of South America and the Radiation of Eusauropoda 131
Diego Pol, Kevin Gomez, Femke M. Holwerda, Oliver W. M. Rauhut, and José L. Carballido

Highly Specialized Diplodocoids: The Rebbachisauridae 165
Leonardo Salgado, Pablo A. Gallina, Lucas Nicolás Lerzo, and José Ignacio Canudo

Southernmost Spiny Backs and Whiplash Tails: Flagellicaudatans from South America 209
Pablo A. Gallina, Sebastián Apesteguía, José L. Carballido, and Juan P. Garderes

The Rise of Non-Titanosaur Macronarians in South America 237
Jose L. Carballido, Flavio Bellardini, and Leonardo Salgado

Titanosauria: A Critical Reappraisal of Its Systematics and the Relevance of the South American Record 269
José L. Carballido, Alejandro Otero, Philip D. Mannion, Leonardo Salgado, and Agustín Pérez Moreno

Time for Giants: Titanosaurs from the Berriasian–Santonian Age 299
Pablo A. Gallina, Bernardo J. González Riga, and Leonardo D. Ortiz David

Last Titans: Titanosaurs From the Campanian–Maastrichtian Age 341
Rodrigo M. Santucci and Leonardo S. Filippi

Eggs, Nests, and Reproductive Biology of Sauropodomorph Dinosaurs from South America 393
Mariela Soledad Fernández, Bernat Vila, and Miguel Moreno-Azanza

Body Size Evolution and Locomotion in Sauropodomorpha: What the South American Record Tells Us 443
Alejandro Otero and John R. Hutchinson

South American Sauropodomorphs: What Their Bone Histology Has Revealed to Us 473
Ignacio A. Cerda

Sauropod Ichnology: Overview and New Research Lines from a South American Perspective 503
Jorge Orlando Calvo, Bernardo J. González Riga, Sebastián Apesteguía, and María Belén Tomaselli

Taphonomy: Overview and New Perspectives Related to the Paleobiology of Giants 541
Bernardo J. González Riga, Gabriel A. Casal, Anthony R. Fiorillo, and Leonardo D. Ortiz David

The Early Radiation of Sauropodomorphs in the Carnian (Late Triassic) of South America

Max C. Langer, Júlio C. A. Marsola, Rodrigo T. Müller, Mario Bronzati, Jonathas S. Bittencourt, Cecilia Apaldetti, and Martín D. Ezcurra

Abstract Carnian (Late Triassic) deposits of South America provide the oldest unequivocal dinosaur records worldwide, most of which has been assigned to the sauropodomorph lineage. This includes *Eoraptor lunensis*, *Panphagia protos*, and *Chromogisaurus novasi*, from the Ischigualasto Formation, Argentina, and *Saturnalia tupiniquim*, *Pampadromaeus barberenai*, *Buriolestes schultzi*, and *Bagualosaurus agudoensis*, from the Santa Maria Formation, Brazil. Here, we

---

Electronic supplementary material The online version contains supplementary material available at (10.1007/978-3-030-95959-3_1).

M. C. Langer (✉) · M. Bronzati
Departamento de Biologia, FFCLRP, Universidade de São Paulo, Av. Bandeirantes 3900
Ribeirão Preto 14040-190, Brazil
e-mail: mclanger@ffclrp.usp.br

J. C. A. Marsola
PPG Biologia Animal, IBILCE, Universidade Estadual Paulista, R. Cristóvão Colombo 2265
São José do Rio Preto 15054-000, Brazil

Departamento de Biologia e Zootecnia, FEIS, Universidade Estadual Paulista, R. Monção 226, Ilha Solteira 15385-000, Brazil

R. T. Müller
Centro de Apoio à Pesquisa Paleontológica da Quarta Colônia, Universidade Federal de Santa Maria, R. Maximiliano Vizzotto 598, São João do Polêsine 97230-000, Brazil

J. S. Bittencourt
Departamento de Geologia, Universidade Federal de Minas Gerais, Av. Antônio Carlos 6627, Belo Horizonte 31270-901, Brazil
e-mail: jsbittencourt@ufmg.br

C. Apaldetti
Consejo Nacional de Investigaciones Científicas y Técnicas (CONICET), Instituto y Museo de Ciencias Naturales, Universidad Nacional de San Juan, Avenida España 400 Norte, San Juan 5400, Argentina

M. D. Ezcurra
Consejo Nacional de Investigaciones Científicas y Técnicas (CONICET), Museo Argentino de Ciencias Naturales 'Bernardino Rivadavia', Av. Ángel Gallardo 470, Buenos Aires C1405DJR, Argentina

A. Otero et al. (eds.), *South American Sauropodomorph Dinosaurs*, Springer Earth System Sciences, https://doi.org/10.1007/978-3-030-95959-3_1

demonstrate that their holotypes anatomically differ from one another, supporting the taxonomic validity of the species. In addition, a morphological disparity analysis, with significant statistical support, clustered some of the better-known specimens of *E. lunensis*, *Sat. tupiniquim*, and *Bu. schultzi*, with the respective holotypes. For the latter two taxa, this was corroborated by a specimen-level phylogenetic analysis that also found *Ba. agudoensis* as the sister taxon to post-Carnian sauropodomorphs. Our results also suggest that *Bu. schultzi* and *E. lunensis* are the earliest branching sauropodomorphs and that *Sa. tupiniquim* and *Pam. barberenai* are closer to Bagualosauria. A species-level phylogenetic analysis further suggests that *Bu. schultzi* and *E. lunensis* form a clade, that *Sa. tupiniquim* is the sister taxon to Bagualosauria, and that *Pan. protos* and *Ch. novasi* are also more highly nested, forming a clade with *Pam. barberenai.*

Keywords Dinosauria · Sauropodomorpha · Bagualosauria · Ischigualasto formation · Santa Maria formation

1 Introduction

Research on Carnian (early Late Triassic) sauropodomorphs started about twenty years ago with the description of *Saturnalia tupiniquim* from south Brazil (Langer et al. 1999). Because coeval dinosaurs known at the time were either assigned to Ornithischia (*Pisanosaurus mertii*) or Theropoda (*Eoraptor lunensis*), or had unclear affinities (herrerasaurids), Sauropodomorpha was until then the only of the three major dinosaur lineages lacking an Ischigualastian (≈Carnian; Langer 2005; Langer et al. 2018) record. Funny enough, the present knowledge reveals that the most abundant Carnian dinosaurs were sauropodomorphs (including the 'ex-theropod' *E. lunensis*), whereas the record of coeval ornithischians and theropods is meagre. In fact, the Carnian diversity of the latter clades may have been even reduced, as neither group has currently undisputed representatives of that age (see Novas et al. 2021). This is because the putative theropod affinity of herrerasaurs continues under debate (e.g. Pacheco et al. 2019), as it is also the case for the affinities of *Nhandumirim waldsangae* and *Eodromaeus murphi* to that group (e.g. Langer et al. 2017; Pacheco et al. 2019) and the ornithischian affinity of *Pi. mertii* (e.g. Agnolin and Rosadilla 2018; Baron et al. 2017a).

Carnian sauropodomorphs recognised after *Sat. tupiniquim* (Fig. 1) were described in the last ten years or so (Martínez and Alcober 2009; Ezcurra 2010; Cabreira et al. 2011, 2016; Pretto et al. 2019), namely *Panphagia protos*, *Chromogisaurus novasi*, *Buriolestes schultzi*, *Pampadromaeus barberenai*, and *Bagualosaurus agudoensis*. The latter three were found in the Alemoa Member of the Santa Maria Formation, in south Brazil, which also yielded *Sat. tupiniquim*. The former two came from the Ischigualasto Formation, north-western Argentina, along with *E. lunensis*, a taxon that since the proposal of Martínez et al. (2011) has been most frequently accepted as belonging to Sauropodomorpha. Both stratigraphic units

Fig. 1 Sauropodmorph diversity on the Carnian of South America. **a** *Buriolestes schultzi* (modified from artwork of MS Garcia following Cabreira et al. 2016). **b** *Panphagia protos* (modified from Martínez and Alcober 2009). **c** *Saturnalia tupiniquim* (modified from artwork of MS Garcia following Langer 2003). **d** *Eoraptor lunensis* (modified from artwork of MS Garcia following Sereno et al. 2012). **e** *Pampadromaeus barberenai* (modified from artwork of MS Garcia following Langer et al. 2019). **f** *Chromogisaurus novasi* (modified from artwork of DH Heman following Müller et al. 2020). **g** *Bagualosaurus agudoensis* (modified from artwork of MS Garcia following Pretto et al. 2019)

have been dated based on radioisotopic studies (Rogers et al. 1993; Martínez et al. 2011, 2012a; Langer et al. 2018; Desojo et al. 2020; Colombi et al. 2021), all of which agree on a late Carnian age for their main dinosaur-bearing beds; 231–229 Ma for the lower Ischigualasto Formation and ca. 233 Ma for the upper Santa Maria Formation.

Following an original proposal by Salgado et al. (1997), some authors (e.g. Langer 2003; Langer et al. 2010; see also Sereno 1998) employed node-based definitions for Sauropodomorpha that, based on most phylogenetic arrangements so far proposed, would exclude the Carnian members of the lineage. The alternative maximal-clade (i.e. stem-based) definitions (Upchurch 1997; Galton and Upchurch 2004) better fit the most common usage of the term (Sereno et al. 2005), including such early branching Carnian taxa, and this was fixed by Fabbri et al. (2020) in *Phylonyms*. Indeed, as current phylogenetic studies mostly concur in placing the seven taxa that form the core of this revision closer to *Saltasaurus loricatus* than to either *Allosaurus fragilis* or *Iguanodon bernissartensis*, they should, by definition, be referred to as sauropodomorphs. In fact, the understanding that such Carnian taxa belong to Sauropodomorpha broke some paradigms about the paleobiology of the early representatives of the group, hitherto inferred based on 'prosauropod-grade' taxa, as relatively large-sized, small-headed, long-necked, omnivore/herbivore, and facultatively quadruped animals. The Carnian forms revealed that none of those typically sauropodomorph traits was present in the earliest radiation of the group, which was represented by small, lightly build animals that had larger heads and shorter necks, and were most probably faunivorous and fully biped (Bronzati et al. 2017).

After an original suggestion by Bonaparte et al. (1999), Ezcurra (2010) proposed that the Norian dinosaur *Guaibasaurus candelariensis* nested, along with some Carnian taxa, into a clade of early sauropodomorphs named Guaibasauridae. In addition, Ezcurra (2010) proposed that *Sat. tupiniquim* and *Ch. novasi* formed a minimal clade named Saturnaliinae. These suggestions were followed by some authors (e.g. Novas et al. 2011; Baron et al. 2017b; Cau 2018), whereas others allied *Gu. candelariensis* to theropods (Yates 2017a, b; Langer et al. 2011; Marsh et al. 2019). This gave rise to the notion that at least some Carnian sauropodomorphs form a clade, either including *Gu. candelariensis* or not, exclusive of most younger members of the group (e.g. Martínez et al. 2011; Langer et al. 2017; Baron et al. 2017b; Müller et al. 2018a). Instead, other studies recover a more pectinate phylogenetic pattern for early sauropodomorph radiation (e.g. Martínez et al. 2012b; Cabreira et al. 2016; Pretto et al. 2017; Müller et al. 2018a), and intermediate arrangements have also been proposed. For example, a clade including part of the Carnian sauropodomorph diversity was termed Saturnaliidae by Langer et al. (2019), translated from Saturnaliinae Ezcurra (2010), whereas its sister clade was named Bagualosauria. The latter clade includes a single Carnian taxon, the name-bearing *Ba. agudoensis* (Pretto et al. 2019). The phylogenetic study proposed here will tackle the relations of several 'guaibasaurids', but will not investigate the possible sauropodomorph affinity of *Gu. candelariensis*. A broader sample of saurischians are required to properly evaluate that possibility, which is beyond the scope of this work. For the same reasons, the recent proposals that (1) most Carnian 'sauropodomorphs' nest outside of

Eusaurischia (Pretto et al. 2019), (2) the putative theropod *Nh. waldsangae* (Marsola et al. 2018) may belong to Saturnaliidae (Pacheco et al. 2019), and (3) the still often suggested non-sauropodomorph affinity of *E. lunensis* will not be tested here.

So far, Carnian sauropodomorphs were positively recognised only in South America. A possible exception corresponds to a partial femur from the Pebbly Arkose Formation of Zimbabwe (Raath 1996), which Langer et al. (1999) suggested to be closely allied to *Sat. tupiniquim* and Ezcurra (2012a) considered an indeterminate saurischian. More recently, further dinosaur material coming from that stratigraphic unit was reported by Griffin et al. (2018), including a partial skeleton with sauropodomorph affinities. The confirmation of this find would highlight the similarities of that African paleofauna to those of 'Ischigualastian' deposits of South America, as already inferred (Langer et al. 2018) by the presence of the rhynchosaur *Hyperodapedon*. Another likely coeval stratigraphic unit (Langer 2005), the lower Maleri Formation of India, yielded the controversial dinosaur *Alwalkeria maleriensis* (Chatterjee 1987). As previously discussed (Langer 2004; Remes and Rauhut 2005; Novas et al. 2011; Ezcurra 2012a), this taxon shares several traits with early sauropodomorphs, but its fragmentary and chimeric nature hampers a proper evaluation of its affinities. Also from India, but from the younger (possibly Norian) upper Maleri Formation, Novas et al. (2011) described a fragmentary specimen (ISI R277) that may belong to Guaibasauridae. Finally, the only proposed non-Gondwanan record of the group corresponds to *Agnosphitys cromhallensis* (Fraser et al. 2002; Ezcurra 2010). Yet, this hypothesis was mostly abandoned lately, given the composite nature of the taxon and its position in more recent phylogenetic studies (e.g. Baron et al. 2017a, b; but see Chapelle et al. 2019). Hence, except for the still undescribed Zimbabwean form, all other non-South American putative guaibasaurids and/or Carnian sauropodomorphs are very controversial, residing outside the scope of the present work, which is to evaluate in detail the alpha-taxonomy and relations of the better-known South American members of the group.

Recently, Baron et al. (2017a, b) assigned the putative dinosaur *Nyasasaurus parringtoni* from the Middle Triassic Lifua Member of the Manda beds of Tanzania (Nesbitt et al. 2013) to Sauropodomorpha. Yet, as fully discussed by various authors (Langer et al. 2017; Ezcurra et al. 2017; Novas et al. 2021), both the affinities of *Ny. parringtoni* and age of the Manda beds are controversial, and this taxon is not discussed further here.

The description of five new taxa in a ten-year interval raised questions about possible synonymies among the Carnian sauropodomorphs of Brazil and Argentina (Langer et al. 2019; Müller and Garcia 2019): are they really different from one another, or could this be a case of taxonomic inflation? In an attempt to tackle this and other questions, we present below a brief review of the status of each of the seven South American Carnian sauropodomorphs, followed by phylogenetic and Principal Coordinates Analysis, aimed to better understand their morphological diversity. In the end, we hope to integrate these data to address the above-proposed question.

Institutional Abbreviations CAPPA/UFSM: Centro de Apoio à Pesquisa Paleontológica da Quarta Colônia, Universidade Federal de Santa Maria, São João do Polêsine, Brazil; **MACN**: Museo Argentino de Ciencias Naturales 'Bernardino Rivadavia', Buenos Aires, Argentina; **MCP**: Museu de Ciências e Tecnologia, Pontifícia Universidade Católica do Rio Grande do Sul, Porto Alegre, Brazil; **PUC/RS**: Pontifícia Universidade Católica do Rio Grande do Sul, Porto Alegre, Brazil; **PVSJ**: Museo de Ciencias Naturales, Universidad Nacional de San Juan, San Juan, Argentina; **UFRGS**: Universidade Federal do Rio Grande do Sul, Porto Alegre, Brazil; **ULBRA**: Universidade Luterana do Brasil, Canoas, Brazil.

2 Methods

2.1 Protocol for Building the Taxon-Character Matrix

Because we intended to design a phylogenetic study specifically to tackle the issue of Carnian sauropodomorph relationships, we conducted a simple protocol to extract information from the phylogenetic literature about the group. First, starting with Langer et al. (1999) and ending in June 2020, we identified all numerical phylogenetic studies that included, as terminal taxa, at least two of the seven early dinosaurs that form the core of this study. There were some exceptions, however, such as studies focused on pseudosuchians that employed modified versions of the Nesbitt (2011) data-matrix, which includes *E. lunensis* and *Sat. tupiniquim*. For these studies, we inferred that the part of the data that is of interest to the present revision was not modified in more recent iterations (at least not substantially), as the authors did not aim at investigating early dinosaur relations. Likewise, studies on theropods that employed modified versions of the data-matrix of Smith et al. (2007), which also includes *E. lunensis* and *Sat. tupiniquim*, were not selected for the second step of the protocol (see below). This first step resulted in the identification of 147 phylogenetic studies (see Supplementary Material) with suitable data-matrices.

The data-matrices of those 147 studies were then subject to a manual search for characters with variable scoring among the seven taxa discussed here, i.e. characters were selected if not scored equally (with the same state) for those taxa. In that search, missing entries were not considered different states; otherwise, the high number of such entries for incomplete taxa, such as *Ch. novasi*, would result in the selection of almost all characters in those matrices. Also, considering that characters that do not vary within those Carnian taxa could still be phylogenetically informative, because they could support their nesting in a single clade, sister to younger sauropodomorphs, e.g. Müller et al. (2018a), we expanded the search for variable characters to other early branching members of the group, including the genera *Pantydraco*, *Thecodontosaurus*, *Efraasia*, *Macrocollum*, and *Plateosaurus*, regardless of their specific assignments. This expanded search resulted in a selection of over 3,500 variable characters, which were then manually compared in search for the multiple expected overlapping/duplication among them. Purged of the duplications, a list of nearly 800 non-overlapping characters was built. Their definitions were standardised, mostly following the grammar of Sereno (2007b), when the characters themselves and their compartmentations in different states were revised for clarity, avoiding ambiguous statements. In addition, new characters gathered from the comparisons conducted in Sect. 3.1, were included. All characters were then scored de novo based on first-hand observations of all terminals. During the scoring process, improvements to character definition and state compartmentation were identified and incorporated into the character list without further ado. In addition, despite being originally scored as variable for the selected taxa in data-matrices gathered for this revision, some characters patently refer to anatomical traits that are unseen among early sauropodomorphs and were excluded from the final list. This included, for example, the presence of nasal crests, of an otic incisure, a double-headed ectopterygoyd, a caudodorsal process in the lacrimal, a transverse ridge along the basioccipital/parabasisphenoid articulation, caniniform teeth, and pneumatised nasal, articular, and ectopterygoid, among others. We also excluded characters with very ambiguous definitions, in particular those dealing with traits of serially homologous elements (e.g. vertebrae, teeth) with variable conditions, but lacking more precise indication about which individual elements were under evaluation. This all resulted in a data-matrix with 771 characters (see Supplementary Material), which was employed in the analysis.

The selection of terminals followed the taxonomic revision provided in the following sections, where the uniqueness of all Carnian sauropodomorph holotypes was corroborated, confirming the validity of the taxa typified by them. These were included in the data-matrix following two strategies. Firstly, the seven holotypes, as well as five other specimens of *E. lunensis*, *Sat. tupiniquim*, and *Bu. Schultzi,* were scored separately, but composite terminals of these three species were then built based on the conjoined scoring of the specimens originally attributed to them. Such composite terminals were scored multistate when two or more states were positively identified for the individual specimens it represents. On the other hand, if one of the specimens was scored multistate and a single state was given to the others, that state was assigned to the composite terminal.

Another premise of the protocol was that the seven taxa that form the core of the study are members of the sauropodomorph branch of dinosaurs. Clearly, that was not the case for *E. lunensis* in the first studies to include the taxon, and it is fair to say that this is still not a completely settled issue (see below). Yet, we opted to follow this premise so that we could focus our efforts on investigating the relations among those putative early sauropodomorphs. Likewise, we assumed that such Carnian taxa are all external to the minimal clade formed by all known post-Carnian sauropodomorphs, which are represented in the data-matrix by *Pantydraco caducus*, *Efraasia minor*, *Plateosaurus engelhardti*, and *Macrocollum itaquii*. On the other hand, the outgroup taxa include the herrerasaurids *Herrerasaurus ischigualastensis* and *Gnathovorax cabreirai*, the possible herrerasaurian *Tawa hallae*, and the neotheropod *Coelophysis bauri*, as well as the non-dinosaur dinosauromorph *Lewisuchus admixtus*, which was used to root the topologies. The final taxon-character matrix, with 771 characters scored for 24 terminals—five outgroup taxa, twelve Carnian specimens, three composite Carnian taxa, and four Norian taxa—can be seen in the Supplementary Material.

2.2 Tree Search Strategy and Branch Support/Stability

The data-matrix was analysed using both the composite scorings for *E. lunensis*, *Sat. tupiniquim*, and *Bu. schultzi* ('combined' analysis) and their type and referred specimens as independent terminals ('specimen-based' analysis). The analyses were conducted under equally weighted parsimony using TNT 1.5 (Goloboff et al. 2008; Goloboff and Catalano 2016). Heuristic search of 1,000 replications of Wagner trees (with random addition sequence) followed by TBR branch swapping (holding ten trees per replicate) was performed. Branches with a maximum possible length of zero among any of the recovered most parsimonious trees (MPTs) were collapsed (rule 3 of Swofford and Begle 1993; Coddington and Scharff 1994). Based on the two premises outlined in the previous section, we applied constrains using an a priori built tree that forced the monophyly of post-Carnian sauropodomorphs and Sauropodomorpha. The following multistate characters were ordered because they represent nested sets of character states: 1, 13, 14, 23, 27, 43, 49, 56, 63, 71, 72, 73, 89, 91, 94, 97, 109, 120, 135, 137, 163, 165, 173, 174, 176, 177, 190, 195, 197, 214, 219, 221, 224, 237, 269, 271, 274, 275, 276, 282, 284, 299, 300, 302, 314, 341, 343, 344, 345, 352, 358, 370, 379, 382, 383, 384, 385, 393, 394, 398, 407, 415, 429, 439, 446, 454, 455, 461, 462, 463, 472, 477, 478, 486, 501, 504, 509, 518, 520, 524, 552, 557, 562, 564, 587, 588, 593, 596, 601, 606, 609, 612, 613, 616, 618, 623, 640, 643, 659, 660, 668, 676, 681, 690, 692, 693, 695, 701, 718, 719, 731, 744, 762, 766, 767, and 768. Consistency and retention indices were calculated considering only those terminals active during the tree search (using the 'maxstepsact' function), in a modified version of the STATS.RUN script. After the tree searches, the possible occurrence of topologically unstable terminals was tested using the iterPCR protocol (Pol and Escapa 2009). As a measure of branch support, decay indices (=Bremer support) were calculated (Bremer 1988, 1994) and, as a measure of branch stability, a bootstrap resampling analysis (Felsenstein 1985) was conducted, with 10,000 pseudo replications. Both absolute and GC (i.e. difference between the frequency whereby the original group and the most frequent contradictory group are recovered in the pseudo replications; Goloboff et al. 2003a, b) bootstrap frequencies were reported. Analyses forcing topological constraints were conducted to find the minimum number of steps necessary to force alternative suboptimal positions for the Carnian sauropodomorph specimens or species.

2.3 Morphological Disparity Analysis

The morphological diversity (disparity) of the Carnian sauropodomorph specimens was quantified based on the matrix of discrete characters described above (excluding the combined terminals for *E. luensis*, *Sat. tupiniquim*, and *Bu. schultzi*). A distance matrix was generated from the taxon-character matrix using the Maximum Observable Rescaled Distance (MORD) (Lloyd 2016; see Lehmann et al. 2019) with the R package Claddis v0.6.1 (Lloyd 2016). An ordination of the distance matrix was performed using a Principal Coordinates Analysis (PCoA) without the necessity of trimming specimens before the analysis. We conducted the PCoA using the Lingoes correction because of the presence of negative eigenvalues. Subsequent Permutational Multivariate Analysis of Variance (PERMANOVA) and Linear Discriminant Analysis (LDA) based on the results of the PCoA used the first three coordinates (58.46% of accumulated variance), which were chosen after detecting the first major break of slope in the scree plot of explained variances. These two analyses were conducted in order to determine if the hypodigms of *E. luensis*, *Sat. tupiniquim*, and *Bu. schultzi* could be statistically differentiated from one another. In order to test if the morphospace distribution was significantly driven by body size, we conducted a generalised least squares regression between the values of the first three PCos and the logarithm of femoral length (as a proxy of body size) of each specimen.

3 Systematic Palaeontology

Dinosauria Owen 1842 [Langer et al. 2020].

Saurischia Seeley 1888 [Gauthier et al. 2020].

Sauropodomorpha Huene 1932 [Fabbri et al. 2020].

Eoraptor Sereno, Forster, Rogers and Moneta 1993

E. lunensis Sereno, Forster, Rogers and Moneta 1993

Holotype The holotype of *E. lunensis* (PVSJ 512) corresponds to a fairly complete skeleton of a probable young adult approaching skeletally maturity (Sereno et al. 2012). This is one of the most complete Carnian dinosaur skeletons known to date, lacking only most of the scapula, the coracoid, and manual phalanges from the left side and caudal vertebrae distal to the 17th position (Sereno et al. 1993, 2012).

Referred Specimens The referred specimens of *E. lunensis* (Sereno et al. 2012) include rather incomplete partial skeletons (PVSJ 559, 745, 860, 862), as well as isolated bones (PVSJ 852, 855, 876), from inferred adult (PVSJ 559, 855, 860, 876) and subadult (PVSJ 745, 852, 862) individuals (Table 1).

Table 1 List of specimens attributed to the Carnian sauropodomorphs of South America

Taxon/specimens	Parts preserved and maturity
Eoraptor lunensis	
PVSJ 512 (holotype)	Articulated skeleton including the skull and most of the postcranium, lacking most of left scapulocoracoid, most of left manual phalanges, and vertebral elements distal to caudal vertebra 17 (Sereno et al. 2012)
PVSJ 559 (referred material)	Two cranial trunk vertebrae, rib shafts, partial right hind limb, including a femur lacking the head, tibia, distal half of the fibula, astragalus, calcaneum, and metatarsal fragments (Sereno et al. 2012)
PVSJ 745 (referred material)	Partial basicranium, including basioccipital and parabasisphenoid, postaxial cervical vertebrae, multiple trunk, sacral, and caudal vertebrae, sacral rib, partial ilia, ischia, femora, and fibulae, distal end of tibia, and proximal portions of metatarsal II–IV (modified from Sereno et al. 2012; MDE pers. obs.)
PVSJ 852 (referred material)	Right femur (Sereno et al. 2012)
PVSJ 855 (referred material)	Right femur (Sereno et al. 2012)
PVSJ 860 (referred material)	Proximal and distal ends of left femur, distal end of right femur, proximal and distal ends of right tibia, proximal end of left tibia, and proximal end of right fibula (Sereno et al. 2012)
PVSJ 862 (referred material)	Proximal end of right humerus, distal ends of both femora, distal end of right tibia, proximal end of right fibula, and right astragalus (Sereno et al. 2012)
PVSJ 876 (referred material)	Right femur lacking midsection (Sereno et al. 2012)
Saturnalia tupiniquim	
MCP 3844-PV (holotype)	Articulated postcranial skeleton including most of the presacral vertebral series, both sides of the pectoral girdle, right humerus, partial right ulna, right radius, both sides of the pelvic girdle with the sacral series, left femur and most of the right limb (Langer 2003)
MCP 3845-PV (paratype)	Partial skeleton including part of the skull with braincase, the natural cast of a mandibular ramus bearing teeth, presacral series including caudal cervical and cranial trunk vertebrae, both sides of the pectoral girdle, right humerus, right side of the pelvic girdle and most of the right hind limb (Langer 2003)
MCP 3846-PV (paratype)	Incompletely prepared skeleton, from which a partial tibia and foot, as well as some trunk vertebrae, are visible (Langer 2003)

(continued)

Table 1 (continued)

Taxon/specimens	Parts preserved and maturity
Panphagia protos	
PVSJ 874 (holotype)	Partial disarticulated skeleton including right nasal, prefrontal and prootic, left frontal, both parietals and quadrates, supraoccipital, rostral half of the left hemimandible, right hemimandible lacking the rostral tip of the dentary, one cranial and two caudal cervical vertebrae, four caudal trunk neural arches, one trunk centrum, first primordial sacral vertebra, two proximal, one proximal-middle, and 15 distal caudal vertebrae, left scapula, ilium, pubic apron, ischium and proximal half of probable metatarsal 4, right tibia, astragalus, metatarsal 3, and four pedal phalanges of uncertain position, one of which is an ungual (Martínez and Alcober 2009)
Chromogisaurus novasi	
PVSJ 845 (holotype)	Partial postcranial skeleton including one proximal and two middle caudal vertebrae, proximal haemal arch, glenoid region of left scapulocoracoid, proximal end of right ulna (sensu Ezcurra 2010; interpreted as the caudal end of a rhynchosaur right hemimandible by Martínez et al. 2012b), partial ilia and femora, right tibia, proximal end of left tibia, partial right fibula and proximal end of left fibula, partial left metatarsal II (sensu Ezcurra 2010; interpreted as right by Martínez et al. 2012b), articulated phalanges of—possibly left—pedal digit II (sensu Ezcurra 2010; alternatively interpreted as belonging to the right pedal digit III by Martínez et al. 2012b), and unidentified bone fragments (Ezcurra 2010; Martínez et al. 2012b)
Pampadromaeus barberenai	
ULBRA-PVT 016 (holotype)	Disarticulated partial skeleton including a semi-articulated cranium set with right premaxilla, maxilla, lacrimal, left palatine, and an indeterminate partial palatal bone; skull bones including right frontal, prefrontal, postorbital, and pterygoid, left nasal, parietal, jugal, squamosal, quadrate, and pterygoid; nearly complete left dentary, with possible portions of the angular and surangular; partial right dentary; semi-articulated set of postdentary bones of the right lower jaw including, angular, surangular, articular, and prearticular; left prearticular; vertebrae including atlas/axis complex, third neck vertebra, eleven trunk vertebrae, articulated pair of sacral vertebrae and ribs, and 17 tail vertebrae; various neck and trunk ribs and haemal arches; partial left scapula; right scapula, humerus, and ulna; partial ilia; proximal portion of the left ischium; femora, tibiae, and fibulae; left metatarsals I and II, partial right metatarsal II, partial metatarsals III, and right metatarsal IV; two phalanges (Langer et al. 2018)

(continued)

Table 1 (continued)

Taxon/specimens	Parts preserved and maturity
CAPPA/UFSM 0027 (referred material)	Left femur (Müller et al. 2016a)
CAPPA/UFSM 0028 (referred material)	Left femur (Müller et al. 2017a)
Buriolestes schultzi	
ULBRA-PVT280 (holotype)	Articulated skeleton including partial skull and both lower jaws and partial postcranium, including few presacral, three sacral and 42 tail vertebrae; left scapula and forelimb lacking most of the manus; paired ilia and ischia; partial left pubis; and a nearly complete left hindlimb (Cabreira et al. 2016)
CAPPA/UFSM 0035 (referred material)	Articulated skeleton including a nearly complete skull and lower jaws and partial postcranium, including complete vertebral column lacking the last sacral vertebra and the caudal series; partial left scapula and coracoid, a fragmentary left humerus, both ilia, the proximal portion of both pubes, the proximal portion of the right ischium, an almost complete right femur, a fragmentary left femur and partial right tibia and fibula, and some phalanges from the right pedal digits III and IV. (Müller et al. 2018a)
ULBRA-PVT289 (referred material)	Isolated right femur (Müller et al. 2018a)
ULVRA-PVT056 (referred material)	Associated elements including two cervical vertebrae, ilium, proximal portion of the pubis, and femur from the right side, plus some phalanges (Müler et al. 2018a)
CAPPA/UFSM 0179 (referred material)	An isolated axis (Müller et al. 2017b)
Bagualosaurus agudoensis	
UFRGS-PV-1099-T (holotype)	Semi-articulated skeleton, including partial skull and mandible, trunk vertebrae, sacrum and isolated caudal vertebrae, fragmented ribs, gastralia, isolated haemal arches, both ilia, right pubis, femora, tibiae, fibulae and partial left pes (Pretto et al. 2019)

Geographic and Stratigraphic Provenance The holotype and all specimens referred to *E. lunensis* were collected from the Cancha de Bochas and Valle Pintado sites, Hoyada de Ischigualasto, Ischigualasto Provincial Park, San Juan, Argentina (Fig. 2; Sereno et al. 2012). They were found in rocks corresponding to the La Peña, Cancha de Bochas, and Agua de la Peña members of the Ischigualasto Formation, belonging to the *Hyperodapedon-Exaeretodon-Herrerasaurus* biozone (Martínez et al. 2012a; Colombi et al. 2021). The maximum age of *E. lunensis* is constrained by a radioisotopic date of 231.4 ± 0.3 Ma close to the base of the Ischigualasto

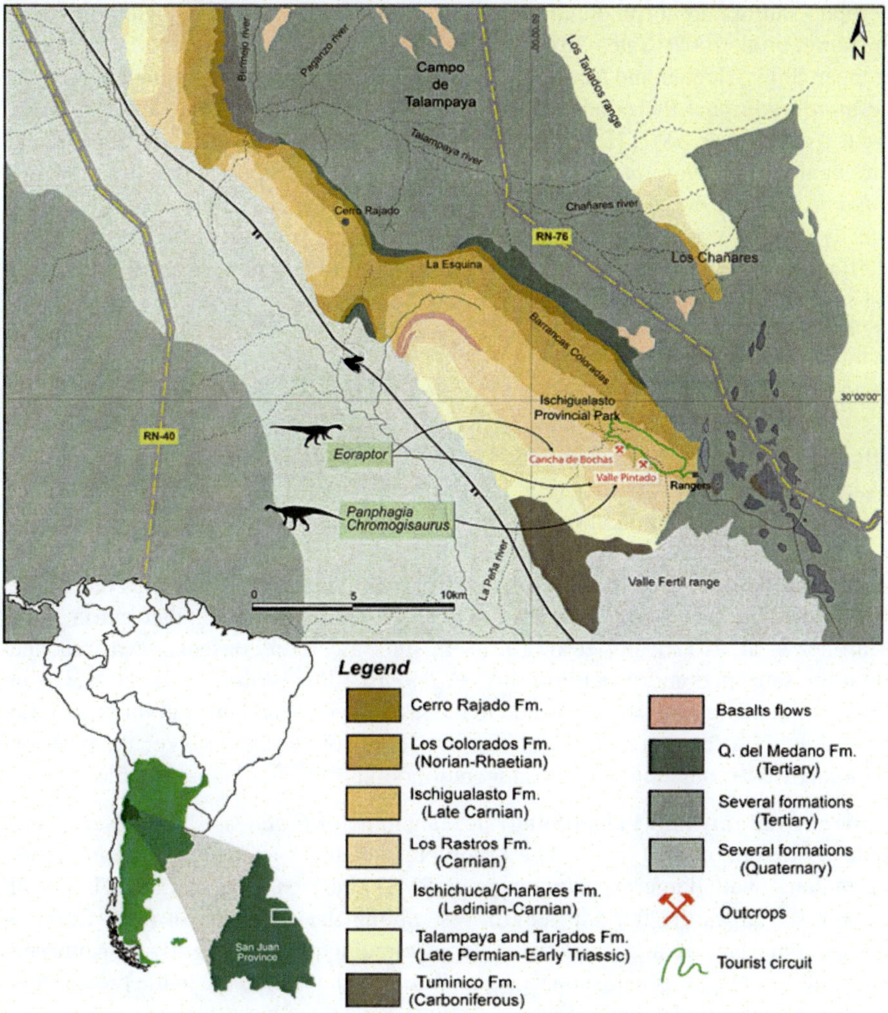

Fig. 2 Geological surface distribution map of the area around the Ischigualasto Provincial Park, San Juan, Argentina, indicating the sites where the specimens of *Eoraptor lunensis*, *Panphagia protos*, and *Chromogisaurus novasi* were found

Formation (Rogers et al. 1993). The upper stratigraphic range of that biozone, hence of *E. lunensis*, in the Provincial Park was constrained by an absolute age of 228.91 ± 0.14 Ma.

Proposed Phylogenetic Relations The first phylogenetic analysis to test the affinities of *E. lunensis* found the taxon as the earliest branching theropod, sister taxon to herrerasaurids plus neotheropods (Sereno et al. 1993). The theropod affinity of *E. lunensis* was supported by subsequent studies during the 1990s (e.g. Novas 1996; Sereno 1999). However, this interpretation started to be contradicted by some phylogenetic analyses in the beginning of this century, which found the species outside the theropod-sauropodomorph dichotomy (e.g. Langer 2004; Langer and Benton 2006; Upchurch et al. 2007; Yates 2017a, b; Nesbitt and Chatterjee 2008; Martínez and Alcober 2009; Alcober and Martínez 2010). Yet, other analyses continued to recover the more traditional theropodan position of *E. lunensis* (e.g. Ezcurra 2006, 2010; Ezcurra and Novas 2007; Nesbitt et al. 2009; Langer et al. 2011; Novas et al. 2011; Sues et al. 2011). More recently, a third alternative was proposed, this time placing *E. lunensis* as one of the earliest branching sauropodomorphs (Martínez et al. 2011). This hypothesis was supported by multiple subsequent studies (e.g. Ezcurra 2012b; Martínez et al. 2012b; Nesbitt and Ezcurra 2015; Cabreira et al. 2016; Baron et al. 2017a; Langer et al. 2017; Müller et al. 2018a; Marsola et al. 2018; Langer et al. 2019; Marsh et al. 2019; Pacheco et al. 2019; Ezcurra et al. 2020a; Müller and Garcia 2020; Pol et al. 2021), although a number of others carried on recovering theropodan affinities for *E. lunensis* (e.g. Martínez et al. 2012a; Baron and Barrett 2017; Baron et al. 2017b; Baron and Williams 2018). *Eoraptor lunesis* has been recovered in different positions among the earliest known sauropodomorphs, such as the earliest branching member of the group (e.g. Langer et al. 2019; Marsh et al. 2019), the sister taxon of all other sauropodomorphs with the exception of *Bu. schultzi* (e.g. Cabreira et al. 2016, 2018b; Pacheco et al. 2019; Müller and Garcia 2020; Garcia et al. 2021), within Saturnaliidae (e.g. Martínez et al. 2011; Müller et al. 2018a), or as one of the most immediate successive sister taxa to the Saturnaliidae plus Bagualosauria clade (Martínez et al. 2012b; Pol et al. 2021). In sum, a certain consensus was reached about the sauropodomorph affinities of *E. lunensis*; this position was recovered in most phylogenetic analyses published in the last ten years, and most of those that did not are based on datasets built during the first decade of this century. Yet, the position of *E. lunensis* among non-bagualosaur sauropodomorphs has been unstable.

General Anatomy and Paleobiology The anatomy of *E. lunensis* was described in detail by Sereno et al. (2012). The holotypic skeleton was estimated to be about 1.2 m long, with larger specimens (e.g. PVSJ 559) ca. 10% larger. The skull is about 0.8 times the femoral length, resembling the condition in most Carnian sauropodomorphs. It has a relatively large circular orbit, a slightly downturned premaxilla, but lacks a clear subnarial gap. The skull has four teeth in the premaxilla, 17 in the maxilla, and at least 20 in the dentary. Palatal teeth are present in the pterygoid. The marginal tooth crowns have a slight basal constriction and most are distally recurved, with the exception of those in the caudal part of the maxilla. 'Cheek-teeth' crowns have a rounded labial eminence and mesial and distal denticles/serrations. The cervical vertebrae are moderately elongated, with a centrum length twice its height. The sacrum is composed of three vertebrae, one here reinterpreted as incorporated from the caudal series. The scapula has a relatively short and narrow blade with a moderately craniocaudal expanded end. The forearm is shorter than the relatively robust humerus, and the manus has five metacarpals, but only the first three digits have phalanges; although the fourth digit may have had or not a phalanx (Sereno et al. 2012). The manual ungual phalanges are slightly recurved. The pelvis has a partially opened acetabulum, an ilium with conspicuously developed pre- and postacetabular alae, and a pubis longer than the ischium. Femur and tibia are subequal in length, the former bearing an asymmetric fourth trochanter. The tibia has a short, laterally curved cnemial crest and the distal end is sub-squared, with a lateral groove separating a poorly developed caudolateral process from the facet for reception of the ascending process of the astragalus. Metatarsal III is the longest and metatarsal V the shortest, the latter lacking phalanges. Metatarsal I is slightly longer than half the length of metatarsal III and reaches the proximal end of the metatarsus.

Sereno et al. (2012) reviewed several palaeobiological aspects of *E. lunensis*, inferring that the antorbital fossa was occupied by an air sac emanating from the nasal cavity, but with no evidence of accessory diverticuli from the antorbital sinus. The rostrum was inferred to be akinetic, given the long suborbital ramus of the premaxilla and the premaxilla-maxilla contact lacking a significant diastema, but with a subnarial foramen. On the contrary, Sereno et al. (2012) suggested the presence of an intramandibular joint, with both dorsal and ventral articulations, allowing limited flexure on the vertical plane. Tooth crown anatomy—distal margins straight or only slightly concave in labial/lingual views, mesial margin with large (six per millimetre) and obliquely set denticles—was used to infer a pulping function suitable for plant matter, rather than a meat-cutting function. Also, Sereno et al. (2012) suggested the presence of a small keratinous beak at the rostral tip of the lower jaw, as indicated by the presence of neurovascular foramina and a retracted first dentary tooth. They also identified an extreme hollowing in some vertebrae, formed by internal cavities lacking pneumatic external communications. Traits of the cervical centra suggest that the neck formed a sigmoid curve, elevating the skull above the level of the trunk, with the ribs forming a flexible rod, ventrolateral and parallel to the centra. A forelimb shorter than half the hind limb length suggests that *E. lunensis* was biped, but the interosseous gap between the forearm bones, the long metacarpals 4 and 5, and the twisted phalanx 1 of the pollex precludes raptorial functions for the arm. The proportions among the hind limb parts (femur, epipodium, metatarsus), suggest more cursorial habits compared to Norian sauropodomorphs, but less than early ornithischians and theropods.

Saturnalia tupiniquim Langer et al. 1999

Holotype The holotype of *Sat. tupiniquim* (MCP 3844-PV) corresponds to an articulated partial skeleton (Table 1).

Reffered Specimens The two paratypes of *Sat. tupiniquim* (MCP 3845-PV and 3846-PV) are the only other specimens so far referred to that species (Langer et al.

1999). MCP 3846-PV is the less complete of them and MCP 3845-PV the only with cranial remains (Table 1).

Geographic and Stratigraphic Provenance The type-series of *Sat. tupiniquim* was collected in the site known as 'Cerro da Alemoa' or 'Waldsanga' (Langer 2005). This is located in the eastern outskirts of Santa Maria (Fig. 3), south of RS-509 road (coordinates: 29° 41′ 51.86″ S, 53° 46′ 26.56″ W). The site exposes the red mudstones of the Alemoa Member, Santa Maria Formation (Da Rosa 2015), overlaid by a basal fluvial conglomerate. MCP 3844-PV and 3845-PV were excavated about three metres below the conglomerate, whereas MCP 3846-PV was found three metres further down (Langer 2005). Between them (five metres below the conglomerate)

Fig. 3 Geological surface distribution map of the central part of Rio Grande do Sul, Brazil, indicating the sites where the specimens of *Saturnalia tupiniquim*, *Buriolestes schultzi*, *Pampadromaeus barberenai*, and *Bagualosaurus agudoensis* were found

rock samples were radioisotopically dated as 233.23 ± 0.73 Ma (Langer et al. 2018). The entire mudstone package at 'Cerro da Alemoa' corresponds to the upper portions of the Alemoa Member in the area (Da Rosa 2015), which in turn belong to the lower part of the Candelária Sequence, within the Santa Maria Supersequence (Horn et al. 2018; Schultz et al. 2020). In biostratigraphic terms, the fauna of the site fits the *Hyperodapedon* Acme-Zone (Langer et al. 2007), within the eponymous Assemblage-Zone (Schultz et al. 2020).

Proposed Phylogenetic Relations Because it was the first Carnian sauropodomorph recognised as such, at a time when *E. lunensis* was not assigned to the group, *Sat. tupiniquim* was depicted as sister to all other sauropodomorphs in the first phylogenetic studies dealing with it (Langer et al. 1999; Yates 2003, 2004, 2017a, b; Yates and Kitching 2003; Langer 2004; Pol 2004; Langer and Benton 2006; Ezcurra 2006; Sereno 2007a; Ezcurra and Novas 2007; Upchurch et al. 2007; Irmis et al. 2007; Martínez 2009; Nesbitt et al. 2009; but see Galton and Upchurch 2004; Barrett et al. 2007). With the description of other Carnian members of the group, the position of *Sat. tupiniquim* shifted among phylogenetic proposals, although most frequently forming a minimal clade with *Ch. novasi* (Ezcurra 2010; Novas et al. 2011; Apaldetti et al. 2011; Martínez et al. 2012b; Otero and Pol 2013; McPhee et al. 2014, 2015; Otero et al. 2015; Müller et al. 2016a, b, 2017b, 2018a; Cabreira et al. 2016; Cerda et al. 2017; Wang et al. 2017; Bronzati et al. 2018, 2019a; Bronzati and Rauhut 2018; Zhang et al. 2018; McPhee and Choiniere 2018; Marsola et al. 2018; Chapelle and Choiniere 2018; Marsh and Rowe 2018; Chapelle et al. 2019; McPhee et al. 2018, 2020; Langer et al. 2019; Garcia et al. 2019; Pacheco et al. 2019; Pretto et al. 2019; Pol et al. 2021). When all non-bagualosaur sauropodomorphs are joined in a clade (e.g. Ezcurra 2010; Novas et al. 2011; Martínez et al. 2011; Langer et al. 2017), *Sat. tupiniquim* is sometimes allied (apart from *Ch. novasi*) with *Pan. protos* and/or *Pam. barberenai* (Baron et al. 2017a; Müller et al. 2018a; McPhee et al. 2020). When early sauropodomorph phylogeny is arranged in a more pectinate fashion, *Sat. tupiniquim* most frequently nests closer to bagualosaurs than to other Carnian forms (Martínez and Alcober 2009; Alcober and Martínez 2010; Cabreira et al. 2011, 2016; Martínez et al. 2012b; Bittencourt et al. 2015; McPhee et al. 2015; Müller et al. 2016a, b, 2017a, b, 2018a; Wang et al. 2017; Agnolín and Rozadilla 2018; Pretto et al. 2017; Bronzati et al. 2017, 2018, 2019a; Zhang et al. 2018; Marsola et al. 2018; Bronzati and Rauhut 2018; Dal Sasso et al. 2018; Garcia et al. 2019; Pacheco et al. 2019; but see Cabreira et al. 2011; Baron and Barrett 2017; Baron et al. 2017b; Parry et al. 2017; Chapelle and Choiniere 2018; 2018b; Chapelle et al. 2019; Pretto et al. 2019; Pol et al. 2021), notably *E. lunensis* and/or *Bu. schultzi* (Marsh and Rowe 2018; McPhee and Choiniere 2018; Cau 2018; Baron and Williams 2018; McPhee et al. 2018; Marsh et al. 2019), sometimes forming a clade with *Ch. novasi* and other taxa (2018c; Langer et al. 2019; Müller 2020). One investigative analysis of Pretto et al. (2019) was the only so far not to find *Sat. tupiniquim* as a sauropodomorph, but instead forming a Guaibasauridae clade outside Eusaurischia. More recently, Müller and Garcia (2020) found *Sat. tupiniquim* forming, together with *Nh. waldsangae*, *Ba. agudoensis*, and *Ch. novasi*, the sister clade (= Saturnaliidae) to post-Carnian sauropodomorphs, within which Garcia et al. (2021) found a sister taxon relation between *Sat. tupiniquim* and *Nh. waldsangae*.

General Anatomy and Paleobiology The sacral and pelvic (girdle and limb) anatomy and scapular skeleton of *Sat. tupiniquim* were respectively described by Langer (2003) and Langer et al. (2007). Later, a series of studies described the cranial (Bronzati et al. 2019a) and endocranial (Bronzati et al. 2017, 2019b) anatomy of this taxon. Femoral circumference allowed estimating its body mass in about 6.5 to 11 kg (Delcourt et al. 2012; Benson et al. 2014), showing that *Sat. tupiniquim* was a light-weighted, gracile dinosaur. Its neck is as long as ca. 0.6 of the trunk, which is slightly above the length seen in other early dinosaurs like *E. lunensis* (Bronzati et al. 2017). Hence, although the femur and tibia of all specimens are subequal in length (ca. 15 cm) to those of the *E. lunensis* holotype, *Sat. tupiniquim* was most probably somewhat longer, reaching about 1.5 *m* in length. Its hindlimbs are about 1.5 times longer than the forelimbs, and comparisons between its humeral and femoral circumferences, as well as the position of the humerus in relation to the shoulder girdle, indicate that *Sat. tupiniquim* was most likely a bipedal animal (Delcourt et al. 2012; McPhee et al. 2018). Yet, humeral traits as the large deltopectoral crest and expanded distal articulation suggest that *Sat. tupiniquim* was somehow intermediary, in terms of employing the forelimb for locomotion, between the condition seen in coeval dinosaurs, such as *E. lunensis*, and their Norian relatives (Langer et al. 2007).

The skull length of MCP 3845 was estimated at less than 10 cm, based on the lengths of the frontals and dentaries. This is about two-thirds the femoral length, as seen in younger sauropodomorphs, and unlike all other Carnian members of the group, except for *Ba. agudoensis*, for which skull and femora are known (i.e. *Bu. schultzi*, *E. lunensis*, *Pam. barberenai*, and *Pan. protos*; Bronzati et al. 2017, 2019a). The teeth of *Sat. tupiniquim* have small serrations, and some are gently curved distally, fitting better a faunivorous diet. Some studies suggested that it was either an herbivore (Langer et al. 1999) or an omnivore (Barrett and Upchurch 2007), but the lack of both coarse denticles and overlap between adjacent crowns contradicts this inference. On the other hand, the brain of *Sat. tupiniquim* exhibits a relatively large cerebellar flocculus, a structure related to the control of head and neck movements and it is also involved in gaze stabilisation. Predatory birds generally have larger flocculi in relation to their non-predatory relatives (Ferreira-Cardoso et al. 2017). In this sense, the large cerebellar flocculus of *Sat. tupiniquim* could potentially be related to a feeding habit involving the capture of small and elusive prey (Bronzati et al. 2017).

Panphagia protos Martínez and Alcober 2009

Holotype The holotype and only known specimen of *Pan. protos* (PVSJ 874) corresponds to a skeletally immature individual represented by a partial skull and postcranium (Martínez and Alcober 2009; Table 1).

Geographic and Stratigraphic Provenance The holotype of *Pan. protos* was found at the Valle Pintado locality, Hoyada de Ischigualasto, Ischigualasto Provincial Park, San Juan, Argentina (Fig. 2). It came from the lower levels of the Cancha de

Bochas Member, 40 m above the base of the unit (*Hyperodapedon-Exaeretodon-Herrerasaurus* biozone), near the beds dated in 231.4 ± 0.3 Ma (Martínez and Alcober 2009; Martínez et al. 2012b; Colombi et al. 2021).

Proposed Phylogenetic Relations *Panphagia protos* was originally recovered as the earliest branching member of Sauropodomorpha (Martínez and Alcober 2009), and this position was also found in some subsequent studies (e.g. Alcober and Martínez 2010; Martínez et al. 2012b). Other phylogenetic analyses have recovered *Pan. protos* within Saturnaliidae (e.g. Ezcurra 2010; Martínez et al. 2011; Novas et al. 2011; Müller et al. 2018a, b; Langer et al. 2019) or as one of the sister taxa to the Saturnaliidae + Bagualosauria clade (e.g. Pacheco et al. 2019). Müller and Garcia (2020) found *Pan. protos* forming a clade with *Pam. barberenai* sister to a Bagualosauria including *Sat. tupiniquim* and *Ch. novasi*. Pol et al. (2021) found *Pan. protos*, along with *Bu. schultzi*, as one of the earliest branching sauropodomorphs. Beyond these alternative positions, the non-bagualosaur sauropodomorph affinities of *Pan. protos* have been stable among the published phylogenetic analyses.

General Anatomy and Paleobiology The general anatomy of *Pan. protos* has been described by Martínez and Alcober (2009), whereas its cranial elements were described in more detail by Martínez et al. (2012c). The tibia of PVSJ 874 is subequal in length to those of *E. lunensis* and *Sat. tupiniquim* (i.e. ca. 15 cm), matching the total skeletal length reconstructed as 1.3 m by Martínez and Alcober (2009). The preserved lower jaw indicates that *Pan. protos* has a relatively long skull, as in *E. lunensis* and proportionally longer than those of *Sat. tupiniquim* and bagualosaurs (Martínez and Alcober 2009). The floccular fossa of the prootic and supraoccipital is proportionally large, as those of several other Carnian dinosauriforms (e.g. *Lewisuchus admixtus*: Ezcurra et al. 2020b; *Sat. tupiniquim*: Bronzati et al. 2017; *Gnathovorax cabreirai*: Pacheco et al. 2019; *Bu. schultzi*:2020), indicating enhanced gaze stabilisation and coordination of eye, head, and neck movements (Bronzati et al. 2017). The dentary tooth crowns are somewhat expanded at the base, with labial and lingual eminences, and relatively small and obliquely set denticles on the mesial and distal margins. This dental morphology has been interpreted as evidence of an omnivorous diet (Martínez and Alcober 2009). The cervical vertebrae are moderately elongated, similar to those of other Carnian sauropodomorphs. The scapula has a fan-shaped blade. The ilium has a partially opened acetabulum and the ischium has a conspicuously craniocaudally expanded distal end. The tibia has a sub-squared distal end, with an extensive facet for reception of the ascending process of the astragalus. The astragalus is similar to that of other early saurischians in the presence of a cranially prominent craniomedial corner of the body and a transversely reduced fibular facet.

Chromogisaurus novasi Ezcurra 2010

Holotype The holotype and only known specimen of *Ch. novasi* (PVSJ 845) corresponds to the partial skeleton of a probable adult individual (Table 1), but some of its elements have been subject to different interpretations. For example, the proximal end of the right ulna described by Ezcurra (2010) was alternatively interpreted as the caudal end of a rhynchosaur right hemimandible (Martínez et al. 2012b). Also, a partial metatarsal II was assigned either to the left (Ezcurra 2010) or to the right (Martínez et al. 2012b) side, and articulated phalanges were assigned to the left pedal digit II by Ezcurra (2010) and to the right pedal digit III by Martínez et al. (2012b).

Geographic and Stratigraphic Provenance The holotype of *Ch. novasi* was collected at the Valle Pintado locality, Hoyada de Ischigualasto, Ischigualasto Provincial Park, San Juan, Argentina (Fig. 2). PVSJ 845 was found in the lower levels of the Cancha de Bochas Member, 40 m above the base of the unit (*Hyperodapedon-Exaeretodon-Herrerasaurus* biozone), at about the same level dated in 231.4 ± 0.3 Ma (Ezcurra 2010; Martínez et al. 2012a; Colombi et al. 2021).

Proposed Phylogenetic Relations In the phylogenetic analysis that accompanied its first description, *Ch. novasi* was recovered as the sister taxon of *Sat. tupiniquim*, forming a clade of early sauropodomorphs now recognised as Saturnaliidae (Ezcurra 2010). All subsequent analyses found a sister taxon relationship between *Ch. novasi* and *Sat. tupiniquim* among non-bagualosaur sauropodomorphs (e.g. Novas et al. 2011; Martínez et al. 2012b; Cabreira et al. 2016; Langer et al. 2019; Pacheco et al. 2019; Pol et al. 2021). More recently, Müller and Garcia (2020) found *Ch. novasi* forming, together with *Sat. tupiniquim Nh. waldsangae*, and *Ba. agudoensis*, the sister clade (=Saturnaliidae) to post-Carnian sauropodomorphs.

General Anatomy and Paleobiology The holotype of *Ch. novasi* was described in detail by both Ezcurra (2010) and Martínez et al. (2012b). It corresponds to a small-sized dinosaur with relatively gracile hindlimbs. Yet, its ca. 17.5 cm long tibia suggests that it was somewhat larger than *E. lunensis*, *Sat. tupiniquim*, and *Pan. protos*, likely surpassing 1.5 m in total body length. Unfortunately, the specimen lacks cranial bones and this precludes assessing several aspects of its palaeobiology. The presence of closed neurocentral sutures in the caudal vertebrae and fusion between the scapula and coracoid indicates that the holotype was probably a skeletally mature individual (Ezcurra 2010; Martínez et al. 2012b). The only possible preserved forelimb bone is the proximal end of the right ulna (Ezcurra 2010), which was also interpreted as a partial rhynchosaur hemimandible (Martínez et al. 2012b). Yet, the glenoid region of a rhynchosaur hemimandible (e.g. *Hyperodapedon sanjuanensis*; MACN-Pv 18,185) differs from this element in the presence of a transversely broad ventral surface, with a distinct longitudinal change of slope on the lateral surface of the surangular; an upturned caudal end of the articular; a transversely broader glenoid fossa; and a smooth lateral surface of the hemimandible. By contrast, the bone of PVSJ 845 closely resembles the proximal end of the ulna of *Sat. tupiniquim*, including the presence of a long olecranon process and a strongly striated lateral surface of the bone, and may indeed represent a right ulna (MDE pers. obs.). The size of the olecranon process and its striated surface indicate an extensive insertion area for the *M. triceps* and probably strong forearm extension. The ilium has a partially closed acetabular wall and a relatively long postacetabular process. The preserved femora are incomplete, but their length was probably subequal to that of the tibia. The tibia has a long and laterally curved cnemial crest, and the distal end is sub-squared, with an extensive facet for reception of the ascending process of the astragalus. Metatarsal II is the only preserved metatarsal, but interpreted as from either the left (Ezcurra 2010) or the right (Martínez et al. 2012b) side. Although the profile of the distal end of the bone resembles that of the right metatarsal II of other early dinosaurs (Martínez et al. 2012b), the curvature of the shaft would result in an unusual bowing towards metatarsal III and not inwards. Thus, although there is conflicting evidence for the interpretation of this bone, the bowing of the shaft favours the interpretation as a left side element (MDE pers. obs.). Similarly, the articulated pedal digit was interpreted as either a left digit II (Ezcurra 2010) or a right digit III (Martínez et al. 2012b). This digit has two non-ungual and one ungual phalanges; thus, it would be complete if interpreted as a digit II but would lack its proximal most phalanx if interpreted as digit III. The proximal articular surface of the most proximally preserved phalanx is continuously concave, indicating that it is articulated with a metatarsal. By contrast, if it articulated with a missing proximal phalanx, it would have had a median vertical ridge for articulation with the ginglymus of that phalanx. Thus, it seems more likely that this digit represents a complete left digit II (MDE pers. obs.).

Pampadromaeus barberenai Cabreira, Schultz, Bittencourt, Soares, Fortier, Roberto da Silva and Langer 2011

Holotype The holotype of *Pam. barberenai* (ULBRA-PVT016) corresponds to a partial skeleton (Table 1), with most elements preserved disarticulated over an area of less than half square metres, within a single block of sediment. Few other bones assigned to the holotype were collected from around that block. The assignment of these elements to a single individual is possible due to the lack of duplicated parts, similar taphonomic signatures, and matching morphology and size.

Reffered Specimens and Discoveries Specimens referred to *Pam. barberenai* include two fairly complete and isolated left femora: CAPPA/UFSM 0027 (2015) and 0028 (2017a).

Geographic and Stratigraphic Provenance The holotype and both referred specimens of *Pam. barberenai* were collected in the site known as 'Sítio Janner' or 'Várzea do Agudo' (Fig. 3; Cabreira et al. 2011; 2015, 2017a; Da Rosa 2015; Pretto et al. 2015, 2019) that is located about two kilometres to the west of the town of Agudo (coordinates: 53° 17′ 34.20″ W, 29° 39′ 10.89″ S). In the site, fossils are concentrated in the upper half of the massive to laminate, red mudstones interpreted to have accumulated in a distal floodplain palaeoenvironment, overlaid in erosive contact by a light-coloured, cross-bedded sandstone that represents a river channel (Pretto et al. 2015; Da Rosa 2015). The holotype (ULBRA-PVT-016) was collected at the base of the fossiliferous layer, about eight metres below the sandstone (Pretto et al. 2015), which also yielded the two referred femora (2017a). The mudstones at 'Sítio Janner' correspond to the upper portions of the Alemoa Member in the area (Zerfass 2007; Da Rosa 2015; Godoy et al. 2018), which in turn belongs to the lower part of the Candelária Sequence, Santa Maria Supersequence (Horn et al. 2018; Schultz et al. 2020). In biostratigraphic terms, the record of *Hyperodapedon* places the site in the eponymous Assemblage Zone (Schultz et al. 2020). However, rhynchosaurs are scarce in the site, which is dominated by the cynodont *Exaeretodon*, suggesting a placement above the *Hyperodapedon* Acme-Zone of Langer et al. (2007). Thus, the dinosaur-bearing beds of 'Sítio Janner' are probably slightly younger than those of 'Cerro da Alemoa', dated as 233.23 ± 0.73 Ma (Langer et al. 2018).

Proposed Phylogenetic Relations *Pampadromaeus barberenai* was first considered as an early diverging sauropodomorph, with variable positions depending on the phylogenetic dataset employed by Cabreira et al. (2011): i.e. closer to *Pan. protos*, in a polytomy with *Sat. tupiniquim* + *Ch. novasi*, *Gu. candelariensis* and bagualosaurs (Ezcurra 2010); forming a polytomy with *Sat. tupiniquim*, *Pan. protos*, and bagualosaurs (Martinez and Alcober 2009); sister taxon to the clade formed by *Sat. tupiniquim* + bagualosaurs (Nesbitt et al. 2010); forming a clade with bagualosaurs, which is sister to the clade formed by *E. lunensis*, *Pan. protos*, and *Sat. tupiniquim* (Martínez et al. 2011). A similarly floating positioning was also found by other subsequent phylogenetic studies. When Saturnaliidae is recovered, *Pam. barberenai* can be found both as one of the early diverging members of the clade (Langer et al. 2017, 2019; 2018c; Müller 2020) or more deeply nested, close to *Sat. tupiniquim* (Baron et al. 2017a; Dal Sasso et al. 2018; Müller et al. 2018a; Pretto et al. 2019). Alternatively, *Pam. barberenai* is also found outside Saturnaliidae, as sister to the clade formed by that group and bagualosaurs (Martínez et al. 2012b, 2015; Pretto et al. 2019; Müller and Garcia 2020), which frequently also includes *Pan. protos* in a polytomy (Cabreira et al. 2016; Müller et al. 2017b, 2018a; Baron and Williams 2018; Garcia et al. 2019; Marsola et al. 2018). Other hypotheses place *Pam. barberenai* as sister to the clade formed by Guaibasauridae and post-Carnian sauropodomorphs (Baron and Barrett 2017; Parry et al. 2017; 2018b), sister to post-Carnian sauropodomorphs (Baron and Williams 2018), or even to all other sauropodomorphs (2017a). Besides, the taxon sometimes forms a minimal clade with either *Pan. protos* (Müller et al. 2018a; Bronzati et al. 2019a; Pacheco et al. 2019; Müller and Garcia 2020) or *Bu. schultzi* (Pretto et al. 2019).

General Anatomy and Paleobiology The holotype of *Pam. barberenai* was described in a preliminary fashion by Cabreira et al. (2011), and more comprehensively by Langer et al. (2019). Its humerus and femur have been reconstructed from fairly complete bones to ca. 8.5 and 14 cm long, respectively, and a complete fibula is 15 cm long. These measurements broadly match those of *E. lunensis*, suggesting a body length somewhat below 1.5 *m*, and its partially fused sacral zygapophyses suggest that ULBRA-PVT-016 was reaching osteological maturity. The skull of *Pam. barberenai* is plesiomorphicaly long, compared to the shorter skulls of *Sat. tupiniquim* and bagualosaurs. The pectoral limb is characterised by a transversely expanded distal end of the humerus and a long ulna with a relatively short olecranon process (although the association of this bone to the holotype is not beyond uncertainty). The iliac acetabulum is semi-perforated, and the pelvic epipodium is significantly longer than the femur, although this could be the result of proximodistal compression of the femora. Recent optimisations of early dinosaurs feeding behaviour have suggested faunivory as the ancestral sauropodomorph condition, including *Pam. barberenai* (Cabreira et al. 2016). However, most of its teeth are lanceolate with coarse denticles along the carinae, more closely resembling those of bagualosaurs than those of other Carnian forms such as *Sat. tupniquim*, *E. lunensis*, and *Bu. schultzi*. This dental pattern is more broadly accepted as related to an omnivorous diet, rather than with pure faunivory. One of the isolated femora referred to *Pam. barberenai* (CAPPA/ UFSM 0028) is about 80% the length of the other, corresponding to an animal also relatively smaller that the holotype (2017a). It bears some traces related to osteological immaturity, probably representing a juvenile individual.

Buriolestes schultzi Cabreira, Kellner, Dias-da-Silva, Roberto da Silva, Bronzati, Marsola, Müller, Bittencourt, Batista, Raugust, Carrilho, Brodt and Langer 2016

Holotype The holotype of *Bu. schultzi* (ULBRA-PVT280) is composed of a partial skull (lacking most of the roof, palate, and braincase), complete lower jaw, and partial postcranial skeleton (Table 1; Cabreira et al. 2016).

Reffered Specimens Müller et al. (2017b, 2018a) referred further specimens to *Bu. schultzi* (Table 1), the most complete of which (CAPPA/UFSM 0035) preserves the entire skull and a partial postcranial skeleton, lacking forearm, manus, and tail. Other, less complete specimens referable to *Bu. schultzi* include a partial axis (CAPPA/UFSM 0179) and a complete right femur (ULBRA-PVT289). Müller et al. (2018a) also assigned the slightly more complete ULBRA-PVT056—preserving some neck vertebrae, a partial pelvic girdle, a right femur, and pedal phalanges—to *Bu. schultzi*. This corresponds to a significantly smaller specimen, about 2/3 the linear size of the holotype, which may indeed represent a less mature individual (Müller et al. 2018a). In fact, some of its notable anatomical differences relative to other specimens of *Bu. schultzi* can be explained by ontogeny, but a more comprehensive study of that specimen is needed to fully endorse that proposal.

Geographic and Stratigraphic Provenance All specimens mentioned in the previous section were collected in the site known as 'Buriol ravine' (Cabreira et al. 2016; Müller et al. 2018a), which is located about five kilometres south of São João do Polêsine (Fig. 3), in the eponymous municipality (coordinates: 29° 39′ 34.2″ S, 53° 25′ 47.4″ W). The site exposes the mudstones of the upper part of the Alemoa Member, Santa Maria Formation (Zerfass 2007; Godoy et al. 2018), which corresponds to the lower portion of the Candelária Sequence (Horn et al. 2018) of the Santa Maria Supersequence (Zerfass et al. 2003). The abundance of hyperodapedontine rhynchosaurs and absence of the cynodont *Exaeretodon* suggest that this site belongs to the *Hyperodapedon* Acme-Zone (Langer et al. 2007), within the eponymous Assemblage-Zone (Schultz et al. 2020). As such, it may have a similar age to the 'Cerro da Alemoa' site, which was radioisotopically dated as 233.23 ± 0.73 Ma (Langer et al. 2018).

Proposed Phylogenetic Relations In its original description, *Bu. schultzi* was placed as the sister taxon to all other sauropodomorphs (Cabreira et al. 2016). This hypothesis has been corroborated by following studies that employed modified versions of that dataset (Müller et al. 2017b, 2018a; Bronzati et al. 2019a; Garcia et al. 2019, 2021; Marsola et al. 2018; Pretto et al. 2019; Pacheco et al. 2019;

Müller and Garcia 2020), as well as different datasets (Cau 2018; 2018c; Baron and Williams 2018; Ezcurra et al. 2020a; Müller 2020). In the case of Pol et al. (2021), that position is shared in a polytomy by *P. protos*. Instead, other phylogenetic hypotheses, also showing a pectinate arrangement of early sauropodomorphs, found *E. lunensis* in such earliest branching position, with *Bu. schultzi* grouped with all other sauropodomorphs, either as their sister taxon (Bronzati et al. 2019a; Langer et al. 2019) or more highly nested (Pretto et al. 2019). Studies that found a clade of Carnian sauropodomorphs are mostly derived from the study of Langer et al. (2017). In these cases, *Bu. schultzi* is never the sister taxon of all other members of the clade, but more highly nested instead, although usually external to the clade formed by *Sat. tupiniquim* and *Pan. protos* and/or *Pam. barberenai* (Parry et al. 2017; Baron et al. 2017a; McPhee et al. 2020; Baron 2019; Garcia et al. 2019). Dal Sasso et al. (2018) and one of the investigative analyses of Pretto et al. (2019) were the only studies not to find *Bu. schultzi* as a sauropodomorph. Instead, the taxon was positioned, respectively, as a theropod closer to neotheropods and forming a Guaibasauridae clade of non-eusaurischian saurischians.

General Anatomy and Paleobiology The holotype of *Bu. schultzi* (ULBRA-PVT280) was described in a preliminary fashion by Cabreira et al. (2016), but CAPPA/UFSM 0035 was fully described by (Müller et al. 2018a) and the axis CAPPA/UFSM 0179 by Müller et al. (2017b). Together, ULBRA-PVT280 and CAPPA/UFSM 0035 reveal details of the almost entire skeleton of this dinosaur, lacking only parts of the tarsus and manus. They correspond to individuals of about the same size, with femora of ca. 13 cm of length. Compared to other coeval sauropodomorphs, this suggests a total body length slightly below that of *E. lunensis*, whereas the isolated axis CAPPA/UFSM 0179 reveals a larger individual and the possible sub adult ULBRA-PVT056 is below 1.0 m of total body length (Müller et al. 2018a). The general body plan of *Bu. schultzi* resembles that of other coeval sauropodomorphs, suggesting a fully bipedal posture. The skull is relatively long, the external nares are reduced, and the long rostrum is about half the skull length. There is a marked subnarial gap separating the alveolar margins of the premaxilla and maxilla, the former of which is downturned. The teeth are blade-like, distally recurved, with fine serrations that form right angles with the crown margins, a condition associated with a faunivorous feeding behaviour (Cabreira et al. 2016). The neck is about two-thirds of the trunk length and two vertebrae form the bulk of the sacral articulation. The humerus is gracile, lacking a strongly expanded distal end, and the ulna lacks a pronounced olecranon process. The acetalubum is almost fully closed and the tibia is longer than the femur. An almost complete cranial endocast was reconstructed from the skull of CAPPA/UFSM 0035 (Müller et al. 2020). It resembles that of *Sat. tupiniquim*, with a relatively large cerebellar flocculus (Bronzati et al. 2017). In adition, the *Bu. schultzi* endocast allowed the reconstruction of the entire forebrain, revealing olfactory bulbs significantly larger than those predicted for dinosaurs of similar body mass (Müller 2021). Therefore, in addition to advanced coordination of eye, head, and neck movements, *Bu. schultzi* would have had an enhanced sense of smell.

Bagualosaurus agudoensis Pretto, Langer and Schultz 2019

Holotype The holotype and only known specimen of *Ba. agudoensis* (UFRGS-PV-1099-T) comprises a partial skull associated with a partial postcranial skeleton, including some vertebrae, the pelvic girdle, and hindlimbs (Table 1; Pretto et al. 2019).

Geographic and Stratigraphic Provenance The holotype was unearthed from 'Sítio Janner' (see *Pam. barberenai* above; Fig. 3), about three metres below the sandstone layer that tops the outcrop (Pretto et al. 2019).

Proposed Phylogenetic Relations In the original description, *Ba. agudoensis* was found as the sister taxon to all other bagualosaurs (Pretto et al. 2019), a clade that circumscribes the entire diversity of unambiguous post-Carnian sauropodomorphs (Langer et al. 2019). Such an affinity was corroborated by most later analyses (Müller et al. 2018c; Bronzati et al. 2019a; Langer et al. 2019; Müller 2020; Pol et al. 2021), but Pacheco et al. (2019) found the taxon in a polytomy with other bagualosaurs. Müller and Garcia (2020) and Garcia et al. (2021) were so far the only studies to positively question the placement of *Ba. agudoensis* as the most immediate sister taxon to all post-Carnian bagualosaurs, suggesting instead that it forms a Carnian clade with only *Sat. tupiniquim*, *Ch. novasi*, and *Nh. waldsangae*.

General Anatomy and Paleobiology *Bagualosaurus agudoensis* is unique among Carnian sauropodomorphs for its relatively large size. It is about 2.5 m long, whereas coeval sauropodomorphs (see above) were animals of ca. 1.5 m. An estimation of its body mass, applying the Campione et al. (2014) equation to the femoral circumference of the holotype (= 83 mm) results in 40 kg. In addition, *Ba. agudoensis* shares with post-Carnian sauropodomorphs (and *Sat. tupiniquim*) a proportionally reduced head (Bronzati et al. 2019b; Pretto et al. 2019). The alveolar margin of the premaxilla and maxilla forms a straight line, unlike the ventrally sloped alveolar margin of the premaxilla of *Bu. schultzi*, *Pam. barberenai*, and *E. lunensis*. The teeth have large denticles, forming oblique angles to the crown margins (Pretto et al. 2019). The pelvic girdle and hind limb of *Ba. agudoensis* are plesiomorphic in comparison to those of younger bagualosaurs. The medial wall of the ilium is well developed ventrally, so that the acetabulum is not fully perforated; the femur is sigmoid, has a trochanteric shelf, and the fourth trochanter is within the proximal half of the bone. The foot is gracile, with elongated phalanges, when compared to most bagualosaurs. The jaw/tooth anatomy reveals a probably omnivorous animal, but more dependent on plant intake that other Carnian sauropodomorphs. The gracile hindlimb, with femur and epipodium of about the same length, indicates that *Ba. agudoensis* was most probably not an obligate quadruped.

4 Alpha-Taxonomy of the Carnian Sauropodomorphs

4.1 Uniqueness of the Holotypes

One of the questions raised about the alpha-taxonomy of the Carnian sauropodomorphs from South America is the possible synonymy between some of the named taxa. In order to tackle this question, we identify below a minimal set of anatomical traits that allow differentiating each of them from one another. Only after such a procedure, we can validate their inclusion as unique terminals in the phylogenetic analyses.

4.1.1 Eoraptor lunensis (PVSJ 512)

Sereno et al. (2012) provided a revised diagnosis of *E. lunensis* based on the following autapomorphies: dorsomedial ramus of the caudal process of premaxilla slender with tongue-shaped distal expansion; nasal with transversely broad, horizontal shelf with a convex lateral margin that overhangs the antorbital fossa; pterygoid process on caudal palate margin that articulates laterally in a synovial socket in the ectopterygoid; narrow premaxilla-maxilla diastema approximately one crown in width; maxillary crowns with a prominent lateral eminence; accessory articular process on the medial aspect of mid-cervical prezygapophyses; extreme hollowing of dorsal centra and neural arches. Yet, several of these features cannot be accessed in the holotypes of other Carnian sauropodomorphs, so further comparison is required to establish the uniqueness of PVSJ 512 and, henceforth, *E. lunensis*.

We concur with some of the traits used by Sereno et al. (2012) to differentiate PVSJ 512 from PVSJ 874 (holotype of *Pan. protos*), namely: a shallower lateral neurovascular groove on the dentary; a less pronounced ridge on the lateral aspect of the surangular; a less distally expanded scapular blade, with a distal margin broadly perpendicular to its long axis; a more elongate pubic blade (more than four times the distal width). In addition, we agree with some other differential traits of *E. lunensis*, relative to *Pan. protos*, mentioned by Martínez and Alcober (2009), i.e.: nasal with a more convex lateral margin and lacking an elongated rostral fossa; a transversally narrower lateral flange of the quadrate, with a smaller and more medially paced quadrate foramen; a straight caudal half of the ventral border of the dentary (rather than concave) in lateral view; maxillary and dentary 'cheek' tooth-crowns with concave distal margins; stouter (craniocaudally) mid-cranial neck vertebrae; pubic peduncle of the ilium with a sharp (rather than rounded) dorsal margin; cranially arched pubic shaft; ischium with a less expanded distal end and subtriangular (rather than rounded) mid-shaft section and distal outline.

Likewise, we endorse some traits identified by Ezcurra (2010), Sereno et al. (2012), and Martínez et al. (2012b), which differentiate PVSJ 512 from PVSJ 845 (holotype of *Ch. novasi*), namely: a non-hypertrophied olecranon process of the ulna; a more prominent iliac supra-acetabular crest; a more medially expanded femoral head; a markedly asymmetrical fourth trochanter; a more transversely expanded distal end of the tibia (although its transverse compression may be a taphonomic artefact in *Ch. novasi*); a less concave distal margin of the tibia in medial view.

In addition to its distinction relative to the holotypes of *Pan. protos* and *Ch. novasi*, PVSJ 512 differs from MCP 3844-PV (holotype of *Sat. tupiniquim*) by a less concave caudal margin of the scapular blade, a less pointed (not 'hook-like') distal corner of the humeral deltopectoral crest, a less lateromedially expanded distal articulation of the humerus, a not enlarged olecranon process of the ulna, and a less dorsoventrally expanded distal end of the ischium. We also concur with Langer et al. (2018) that PVSJ 512 differs from ULBRA-PVT280 (holotype of *Pam. barberenai*) in that the premaxilla has a longer dorsomedial ramus of the caudal process, the base of the dorsal ramus of the maxilla lacks a large rostrally opening lateral foramen, the antorbital fossa is not excavated by a promaxillary fossa, the ventral margin of the antorbital fossa is marked by a rounded ridge, a web of bone spans rostroventrally from the junction between rostral and ventral rami of lacrimal, there is a raised lip forming the prearticular margin of the internal mandibular fenestra, there is a set of rostral foramina at the lateral surface of the dentary, the first premaxillary tooth bears denticles, the maxilla has less than 20 teeth, the maxillary and dentary 'cheek' tooth-crowns have concave distal margins, denticles set perpendicular to the tooth margins, and not restricted to their apical part, the pterygoid bears a transverse row of palatal teeth, the first dentary tooth is inset from the rostral margin of the bone, the scapular blade is short relative to its minimal craniomedial breadth, the dorsal margin of the acromion process forms a lower angle to the cranial margin of the scapular blade, the distal end of the humerus is less transversally expanded, the brevis shelf is connected to the supra-acetabular crest, and metatarsal IV has a broader than deep distal outline.

PVSJ 512 also differs from ULBRA-PVT280 (holotype of *Bu. schultzi*) in that the preorbital region of the skull is shorter, the dorsomedial ramus of the caudal premaxillary process is longer, the maxilla-premaxilla contact bears a subnarial foramen, the forked part of the caudal ramus of the jugal is more rostrally located along the ventral margin of the lower temporal fenestra, the deltopectoral crest of the humerus is proximodistally longer, and the pubic pair lacks a bevel on its distal margin. Finally, PVSJ 512 differs from UFRGS-PV-1099-T (holotype of *Ba. agudoensis*) for its much smaller size, proportionally longer head, first tooth not inset from the rostral margin of the premaxilla, concave ventral margin of the premaxilla-maxilla contact, more ventrally placed subnarial foramen, maxillary and dentary 'cheek' tooth crowns with concave distal margins and smaller denticles along the carinae, and distal end of the tibia lacking a caudomedial notch.

4.1.2 Saturnalia tupiniquim (MCP 3844-PV)

Langer et al. (2007) provided the last emended diagnosis of *Sat. tupiniquim*, but this was based only on the pectoral skeleton and elaborated when no other Carnian sauropodomorph, except for *E. lunensis* (not assigned to the group at the time), was known. As such, it obviously fails to differentiate *Sat. tupiniquim* from more recently described taxa, and this is attempted below.

In addition to the differences relative to the holotype of *E. lunensis* (provided in the previous section), MCP 3844-PV differs from PVSJ 874 (holotype of *Pan. protos*) based on—partially as reviewed by Martínez and Alcober (2009)—a markedly concave caudal margin of the scapular blade, an ilium with an incipient postacetabular embayment and a concave (rather than straight, in dorsal view) caudal margin of the postacetabular ala, a subtriangular (rather than hemispherical) mid-shaft section and distal outline of the ischium, a more cranially placed lateral condyle of the tibial proximal articulation, and a triangular (rather than parallelogram shaped) outline of the proximal end of metatarsal III. Likewise, we concur with Ezcurra (2010) and Martínez et al. (2012b) that MCP 3844-PV differs from PVSJ 845 (holotype of *Ch. novasi*) by an ilium with a less dorsoventrally extensive blade, a more expanded supra-acetabular crest, and a straighter ventral margin of the acetabular wall, a larger fibular condyle of the femur, a more cranially located lateral condyle of the tibia, a cnemial crest more cranially expanded close to the proximal margin of the tibia, and a not ventrally expanded lateral condyle of the distal end of metatarsal II.

Partially as reviewed by Langer et al. (2018), MCP 3844-PV differs from ULBRA-PVT016 (holotype of *Pam. barberenai*) based on a more concave caudal margin of the scapular blade, a greatly enlarged olecranon process of the ulna, the incorporation of a caudal vertebra to the sacrum, a supra-acetabular crest that reaches the distal end of the pubic peduncle, a straighter ventral margin of the iliac acetabular wall, a not hypertrophied fibular condyle of the femur, and a lateromedially broader metatarsal I distal articulation. MCP 3844-PV also differs from ULBRA-PVT280 (holotype of *Bu. schultzi*) by a longer deltopectoral crest of the humerus, with a hook-like distal corner, a lateromedially broader distal articulation of the humerus, a marked fossa olecrani on the caudal surface of the distal end of the humerus, a greatly enlarged olecranon process of the ulna, a distal end of the ischium more caudodorsally expanded and with a triangular distal outline, a cranially convex distal femur outline, and no caudal knob medial to the intercondylar notch of the tibia. Finally, MCP 3844-PV differ from UFRGS-PV-1099-T (holotype of *Ba. agudoensis*) by its overall smaller body size, as well as by a more ventrally expanded brevis shelf in the caudal end of the postacetabular ala, the lack of a groove excavating the ambiens process of the pubis, a 'semi-pendant' fourth trochanter on the femur, and tibia lacking a caudomedial notch in the distal end.

4.1.3 Panphagia protos (PVSJ 874)

Martínez and Alcober (2009) proposed that the holotype of *Pan. protos* differs from the only other Carnian sauropodomorphs known at the time (i.e. *E. lunensis* and *Sat. tupiniquim*) by the presence of a rostrocaudally elongated fossa on the base of the rostroventral process of the nasal, a wide lateral flange on the quadrate, with a large foramen located far from the shaft, a deep groove on the lateral surface of the lower jaw surrounded by prominent dorsal and ventral ridges, extending from the position of ninth tooth to the surangular foramen, a caudoventral process of the dentary bifurcated in two slender rami that overlap the lateral surface of the angular, a long retroarticular process of the articular transversely wider than the articular area for the quadrate, an oval scar on the lateral surface of the caudal border of the cervical centra, distinct prominences located caudodorsally to the diapophyses on the neural arch of the cranial cervical vertebra, a dorsal end of the scapular blade nearly three times wider than the neck, a scapular blade with an expanded caudodistal corner limited by a wedged caudal border, and a medial lamina of brevis fossa twice wider than the iliac spine. Yet, not all of these features are preserved in PVSJ 512 and/or MCP 3844-PV, so that this differential diagnosis should be cross-checked with those given above for *Sat. tupiniquim* and *E. lunensis*.

In addition, PVSJ 874 differs from PVSJ 845 (holotype of *Ch. novasi*)—partially as reviewed by Martínez et al. (2012b)—by proximal tail vertebrae with less transversely compressed centra and leaf-shaped (rather than subtriangular) transverse processes, ilium with a transversely broader caudomedial shelf, a supra-acetabular crest not reaching the distal end of the pubic peduncle, a less concave ventral margin of the iliac acetabuluar wall, and a more prominent antitrochanter, and tibia with cnemial crest more cranially expanded closer to the proximal margin of the bone.

Partially as reviewed by Langer et al. (2018), PVSJ 874 differs from ULBRA-PVT016 (holotype of *Pam. barberenai*) in that the quadrate foramen is larger, the first dentary tooth is inset from the rostral margin of the bone, maxillary and dentary 'cheek' tooth crowns have smaller denticles, the scapular blade is shorter relative to its minimal craniocaudal breadth, forms a lower angle to the acromion process, and is more craniocaudally expanded towards its dorsal end, the pubic peduncle of the ilium has a rounded dorsal margin, and the caudal end of the brevis shelf is not so ventrally projected. PVSJ 874 also differs from ULBRA-PVT280 (holotype of *Bu. schultzi*) by tooth serrations forming oblique (rather than right) angles to the crown margin, less dorsoventrally expanded iliac lamina and preacetabular ala, a transversely broader caudal end of the postacetabular area (=brevis plus 'caudomedial' shelves), a pubis lacking a bevel on its distal margin, and a proximal articulation of the tibia that lacks a caudal knob, medial to the intercondylar notch. Finally, PVSJ 874 differs from UFRGS-PV-1099-T (holotype of *Ba. agudoensis*) in its smaller body size, proportionally longer head, and distal end of the tibia lacking a caudomedial notch.

4.1.4 Chromogisaurus novasi (PVSJ 845)

Martínez et al. (2012b) diagnosed *Ch. novasi* by the general combination of an ilium with a marked caudal projection of the postacetabular ala, an incipient perforation of the acetabular wall, and a supra-acetabular crest with a strongly concave acetabular surface, but not well projected laterally, a reduced fibular condyle in the femur, a medial surface of the proximal end of the fibula with an elongate rugosity adjacent to the cranial margin of the shaft, and metatarsal II with strongly dorsoventrally asymmetric distal condyles (the latter two traits considered autapomorphic). In addition to the differences relative to the holotypes of *E. lunensis*, *Sat. tupiniquim*, and *Pan.*

protos (provided in the previous sections), more specific comparisons for PVSJ 845 are given below.

Partially as reviewed by Langer et al. (2018), PVSJ 845 differs from ULBRA-PVT016 (holotype of *Pam. barberenai*) by an enlarged olecranon process of the ulna, a supra-acetabular crest of the ilium that reaches the distal end of the pubic peduncle, a femur with a symmetrical fourth trochanter and a not hypertrophied fibular condyle, and a fibula that bears a rugose cranial ridge on the medial surface of its proximal end and lacks a more distal rugose iliofibularis muscle insertion. It also differs from ULBRA-PVT280 (holotype of *Bu. schultzi*) in that the ulna has an expanded olecranon process, the ilium has a concave ventral margin of the acetabular wall and a more caudally facing distal facet of the ischial peduncle, the femur has a symmetrical fourth trochanter and a smaller fibular condyle, the distal end of the tibia is lateromedially compressed, and the fibula lacks a rugose *M. iliofibularis* insertion. Finally, PVSJ 845 differs from UFRGS-PV-1099-T (holotype of *Ba. agudoensis*) in its smaller body size, concave ventral margin of the iliac acetabular wall, and tibia with a well-developed fibular crest and lacking a caudomedial notch in the distal end.

4.1.5 Pampadromaeus barberenai (ULBRA-PVT016)

Langer et al. (2019) differentiated *Pam. barberenai* from other Carnian sauropodomorphs by the unique combination of partially fused zygapophyses in the primordial sacral pair, ulna longer than 80 per cent the humeral length, intercondylar groove of the femur broader lateromedially than the lateral and medial condyles, and metatarsal I with an L-shaped proximal outline, including a lateral expansion that covers part of the cranial surface of metatarsal II. Also, in addition to the differences relative to the holotypes of *E. lunensis*, *Sat. tupiniquim*, *Pan. protos*, and *Ch. novasi* (provided in the previous sections), more specific comparisons for ULBRA-PVT016 are given below.

Partially as reviewed by Langer et al. (2018), ULBRA-PVT016 differs from ULBRA-PVT 280 (holotype of *Bu. schultzi*) in that the premaxilla lacks a second foramen above the premaxillary foramen and has a not downturned ventromedial ramus of the caudal process, the first maxillary tooth is directed strictly ventrally, a promaxillary fossa is seen within the antorbital fossa, the forked portion of the caudal ramus of the jugal reaches base of the dorsal ramus, maxillary and dentary 'cheek' tooth crowns have sigmoid distal margins with large denticles set oblique to their margins, the second primordial sacral vertebra has a dorsally (rather than dorsocaudally) directed neural spine, the distal end of the humerus is transversely broader, the supra-acetabular crest is less laterally expanded, the pubic peduncle of the ilium has a sharp dorsal margin, and the femoral head has a less expanded medial tubercle. Also, ULBRA-PVT016 differs from UFRGS-PV-1099-T (holotype of *Ba. agudoensis*) for its smaller size, proportionally longer head, premaxillary and dentary with the first tooth not retreated, a premaxilla/maxilla contact lacking a subnarial foramen and a straight buccal margin, an antorbital fossa that does not reach the caudal portion of the maxilla, a more slender dentary lacking a ventrally sloped rostral tip, a more concave acetabular roof, and an epipodium longer that the femur.

4.1.6 Buriolestes schultzi (ULBRA-PVT280)

ULBRA-PVT280 can be differentiated from the holotypes of *E. lunensis*, *Sat. tupiniquim*, *Pan. protos*, *Ch. novasi*, and *Pam. barberenai* based on the comparisons provided in the previous sections. Some of such futures were highlighted by Müller et al. (2018a), who listed a general combination of traits unique to *Bu. schultzi* among coeval sauropodomorphs, namely: skull slightly shorter than the femur; short dorsomedial ramus of the caudal premaxillary process; no premaxillary fossa on the medial maxillary wall; marked subnarial gap; forking part of the caudal process of the jugal projected from a pedicel; zyphodont dentition; craniocaudally short, raised rugose process on the dorsocaudal margin of the iliac blade; marked protuberance between the craniomedial crest and the dorsolateral trochanter of the femur; ovoid striated tuberosity on the craniomedial margin of the proximal third of the fibula. In addition to that, ULBRA-PVT280 differs from UFRGS-PV-1099-T (holotype of *Ba. agudoensis*) in its smaller body size, proportionally longer head, a premaxilla/maxilla contact lacking a subnarial foramen and a straight buccal margin, maxillary and dentary 'cheek' tooth crowns distally concave and with smaller denticles forming right angles to the tooth margin, epipodium slightly longer than the femur, and distal end of the tibia lacking a caudomedial notch.

4.1.7 Bagualosaurus agudoensis (UFRGS-PV-1099-T)

As revised above, if not only for its significantly larger size, UFRGS-PV-1099-T also differs from the holotypes of all the other Carnian sauropodomorphs from South America based on the series of traits previously mentioned for these taxa. Also, Pretto et al. (2019) diagnosed the taxon based on a short skull, less than two-thirds of femoral length, premaxillary and dentary teeth retracted from the rostral margin of the bones, first premaxillary tooth at least as high as the highest maxillary tooth, no subnarial gap or diastema, most teeth lanceolate with coarse serrations along the carinae, straight ventral margin of iliac acetabulat wall, straight dorsal surface of the iliac acetabulum, lateromedially widened pubic peduncle, with no dorsal crest, pubic tubercle with a distinct longitudinal sulcus, femoral subequal in length to tibia and/or fibula, tibia lacking a marked fibular crest and with a conspicuous caudomedial notch in the distal end, and gracile metatarsals.

4.2 Referred Specimens

Panphagia protos, *Ch. novasi*, and *Ba. agudoensis* are known based only on their holotypes, whereas other Carnian sauropodomorphs have assigned paratypes (*Sat. tupiniquim*) or referred specimens (*E. lunensis*, *Pam. barberenai*, *Bu. schultzi*). Among these, one of the specimens referred to *Bu. schultzi* (CAPPA/UFSM 0035), one of those referred to *E. lunensis* (PVSJ 559), and one of the *Sat. tupiniquim* paratypes (MCP-PV 3845) have been extensively used to complement the holotypes when it came to scoring the respective taxon in phylogenetic analyses (Martínez et al. 2012b; Bronzati et al. 2017; Müller et al. 2018a). Instead, other referred specimens are more incomplete, including isolated femora referred to *E. lunensis* (PVSJ 852, 855 876), *Pam. barberenai* (CAPPA/UFSM 0027, 0028), and *Bu. schultzi* (ULBRA-PVT289), and an isolated axis (CAPPA/UFSM 0179) referred to the latter taxon. These assignments were mostly based on topotypy (i.e. the specimens come from the type-localities), although this does not actually apply to PVSJ 855 and 860, which came from the Valle Pintado site (see above). In fact, we agree with Sereno et al. (2012) that the referral of the isolated femora to *E. lunensis* is very tentative because this bone is unknown in *Pan. protos* and those of *Ch. novasi* are only partially preserved and crushed. This is somehow also the case for the isolated femora ascribed to *Pam. barberenai* and *Bu. schultzi*, which mostly agree in anatomy with the respective taxon, but cannot be unambiguously differentiated from all coeval taxa, from both Argentina and Brazil (but see Müller et al. 2018a). Hence, because they (1) do not significantly add to the understanding of the respective taxon with anatomical data that are not already available from more complete specimens and/or (2) are not demonstrably closer in anatomy to those taxa than to other Carnian sauropodomorphs of South America, those isolated bones are no further discussed here.

The original assignment of the syntypes of *Sat. tupiniquim* (Langer et al. 1999, 2007; Langer 2003) was also based on their general anatomical resemblances and close association (Langer 2005), as well as on their 'early sauropodomorph' phylogenetic signal. Indeed, as no coeval sauropodomorphs have been identified as such at the time, no overlap was recognised between the morphological variation within the type-series and that now recognised for the entire diversity of Carnian sauropodomorphs. This halted justifications to taxonomically split the type-material, a situation that no longer stands given the recent discovery of several coeval/similar forms. Moreover, although most skeletal parts of the type-series have been studied in detail (except for the vertebrae), a diagnosis based on apomorphies shared by the three type-specimens could not be built so far. This is in part because skeletal parts with key anatomical features (notably the skull) are not available in all specimens and also because there are indeed conspicuous differences among them.

Langer et al. (2007) noticed several similarities in the scapular girdle and forelimb of the holotype of *Sat. tupiniquim* and paratype MCP-PV 3845. In fact, they proposed an amended diagnosis based on this part of the skeleton, emphasising that the diagnostic traits are also seen in other early dinosaurs, thus not representing autapomorphies. Despite the similarities, Langer et al. (2007) listed several differences that could be related to the more robust construction of the holotype; i.e. thicker long bone walls, broader elements such as the deltopectoral crest, the shaft and distal end of the humerus, the distal part of the femur, and the proximal portions of ulna, tibia, fibula, and metatarsals. Additional traits of the holotype pectoral skeleton, unlike that of MCP-PV 3845, include a less expanded scapular prominence, the acromion forming a lower angle to the long axis of the scapular blade, a more conspicuous coracoid tuber, a broader preglenoid ridge, and a subglenoid buttress reaching the medial margin of the coracoid (with the subglenoid fossa facing caudodorsally, and not laterally as in the paratype).

A preliminary account of the vertebral column shows that the variation in length of the presacral centra is similar in both the holotype of *Sat. tupiniquim* and paratype MCP-PV 3845. The cranial postaxial neck centra are shorter than those of the mid-cervical vertebrae, but longer than those of cranial trunk vertebrae. Yet, marked differences are seen in the sacral series; the holotype showing a caudal element incorporated into the sacrum (Langer 2003), whereas a trunk vertebra is incorporated instead in the paratype. In both specimens, the primordial sacral vertebrae are similar in shape, including their attachment to the medial surface of the ilium (Langer 2003; Marsola et al. 2018) and the incorporated vertebra, either from the trunk or tail series, bears robust transverse processes articulating with the ilium.

Other key-elements for comparison among the type-specimens of *Sat. tupiniquim* are the ilium, femur, and tibia, partially preserved in all of them. The femora are very similar, except for the absence of a trochanteric shelf in MCN-PV 3846. Otherwise, the ilia have conspicuous differences regarding the length of the postacetabular alae and the shape of the supra-acetabular crest. Indeed, the postacetabular ala of the holotype and MCN-PV 3845 are about 1.3 times longer than the space between the pre- and postacetabular embayments of the ilium (Langer and Benton 2006), a ratio that is significantly lower (slightly above. 1.0) in MCN-PV 3846. As for the supra-acetabular crest, it has a more rounded lateral profile in the holotype, whereas it is straighter and caudodorsally to cranioventrally oriented in both paratypes. Also, the fibular condyle of the tibia is more caudally placed in the paratypes, whereas the distal end of that bone is more lateromedially expanded in MCN-PV 3846 than in the holotype (this portion of the tibia is very deformed in MCN-PV 3845).

The assignment (or not) of the two more complete specimens referred to *Bu. schultzi* is more straightforward. As mentioned above, the small-sized ULBRA-PVT056 requires a more detailed analysis, but it would represent a juvenile if assigned to *Bu. schultzi*. As such, its inclusion along with ULBRA-PVT056 and CAPPA/UFSM 0035 in the phylogenetic study (see below) would patently violate the principle of not comparing different semaphoronts of the same species (Hennig 1966), hampering to elucidate the relationships of the Carnian sauropodomorphs. As for CAPPA/UFSM 0035, its assignment to *Bu. schultzi* is justified further than on topotypical principles and overall similarity with the holotype. In fact, the two specimens share a suite of traits unseen in the holotypes of other Carnian sauropododomorphs from both Brazil and Argentina (Müller et al. 2018a), including:

longer head (unlike *Ba. agudoensis*); short dorsomedial ramus of the caudal premaxillary process (unlike *E. lunensis*); lack of promaxillary fossa (unlike *Pam. barberenai*); forked part of the caudal ramus of jugal more caudally located (unlike *Pam. barberenai*); fully ziphodont maxillary and dentary 'cheek' teeth (minimally unlike *Pam. barberenai, Pan. protos*, and *Ba. agudoensis*); base of the scapular blade with a straight caudal margin (unlike *Sat. tupiniquim*); caudal end of the brevis shelf not projecting much more ventrally that the 'caudomedial shelf' (unlike *Pam. barberenai* and *Sat. tupiniquim*); and a rugose iliofibularis muscle insertion on the craniomedial margin of the fibula (unlike *Ch. novasi*).

A comprehensive revaluation of the more complete specimens assigned to *E. lunensis* (PVSJ 559, 745, 860, and 862; see Table 1) is a much more complex task, which is beyond the scope of the present work. PVSJ 559 was found in the same site as the holotype, and its referral to *E. lunensis* could be based on topotypy (although we endorse that such referrals should be always based on anatomy). Indeed, PVSJ 559 was assigned to *E. lunensis* by Sereno et al. (2012) also partially based on the broad proportions of its tibia and astragalus compared to those of *Pan. protos*. In fact, the astragalus and the distal end of the tibia are more transversely expanded in PVSJ 559 than in the holotype of *Pan. protos* (PVSJ 874), but this cannot be confirmed in PVSJ 512 (hootype of *E. lunensis*), because the distal end of its tibia is not fully exposed and the caudal portion of its astragalus is missing. In any case, as the distal articulation of the tibia is transversely compressed in the holotype of *Ch. novasi* (PVSJ 845), so that the opposite condition may differentiate PVSJ 559 from that taxon. Also, we agree with Sereno et al. (2012) that the ascending process and caudal fossa of the astragalus are lateromedially broader in PVSJ 559 than in PVSJ 874, and this is also the case for the incomplete astragalus of PVSJ 512. Hence, given the current diversity of the Ischigualasto Formation sauropodmormorphs, the assignment of PVSJ 559 to *E. lunensis* seems the most likely, but far from certain option.

PVSJ 745 was also referred to *E. lunensis* by Sereno et al. (2012), but with no further discussions. This specimen was not collected from the type-locality of *E. lunensis*, but from that of *Pan. protos* and *Ch. Novasi*, i.e. Valle Pintado site. It is very similar to PVSJ 512 in the shape of the basal tubera of the basioccipital, the proportions of the cervical vertebrae, femoral head and fourth trochanter anatomy (MDE pers. obs.). As for PVSJ 860 and 862, they were also collected in the Valle Pintado site. Accordingly, their referral to *E. lunensis* should be backed-up by a very detailed differentiation relative to those two other taxa. For Sereno et al. (2012), the ascending process of the astragalus of PVSJ 862 corresponds to about one-third of the width of the bone, as in PVSJ 559 and PVSJ 512, but unlike the narrower structure of PVSJ 874. In the context of the Ischigualasto Formation sauropodmormorph diversity, this could point to the affinity of PVSJ 862 to *E. lunensis*, but a more detailed evaluation of the two more complete Valle Pintado specimens assigned to that taxon is needed. This is beyond the scope of this work, so that PVSJ 860 and 862 will not be further discussed.

As briefly reviewed above, among the referred specimens of South American Carnian sauropodomorphs, the attribution of CAPPA/UFSM 0035 to *Bu. schultzi* is relatively well supported. Yet, this is not the case of the paratypes of *Sat. tupiniquim*—which accumulate several differences relative to the holotype—neither of the specimens referred to *E. lunensis*—which lack strong anatomical evidence for their association with the holotype. Hence, their unjustified employment to complement the scoring of the respective taxon in phylogenetic datasets could lead to misleading results. Accordingly, in order to investigate the possibility that MCN-PV 3845 and 384 and PVSJ 559 and 745 do not respectively belong to *Sat. tupiniquim* and *E. lunensis*, we will run phylogenetic analyses having them as individual terminals. We will also conduct analyses with a more traditional arrangement, in which two of the more complete '*E. lunensis*' specimens (PVSJ 559, 745) and the type-series of *Sat. tupiniquim* are integrated into composite terminals. For consistency, we will also include a composite *Bu. schultzi* formed of its holotype and CAPPA/UFSM 0035, as well as with those specimens as individual terminals.

5 Phylogenetic Study

5.1 Parsimony Analyses Results

The 'specimen-based' analysis found ten most parsimonious trees (MPTs) of 1,544 steps, with a consistency index (CI) of 0.45142 and a retention index (RI) of 0.40268 (best score hit 778 times of the 1000 replicates). The strict consensus tree (Fig. 4a) shows a massive polytomy composed of *Pam. barberenai*, *Pan. protos*, *Ch. novasi*, the three specimens of *E. lunensis*, Bagualosauria, and monophyletic sets of the *Bu. schultzi* and *Sat. tupiniquim* specimens. In the alternative MPTs, *Pan. protos* may form clades with the *Eoraptor* specimens (sister to PVSJ 559), with *Sat. tupiniquim* plus Bagualosauria, or with *Pam. barberenai* plus *Ch. novasi* (as sister to the *Sat. tupiniquim* plus Bagualosauria clade). The other unstable taxon—*Ch. novasi*—is found either with *Pam. barberenai* and *Pan. protos* in a clade sister to that including all other sauropodomorphs except for the *E. lunensis* and *Bu. schultzi* sets of specimens, or in a clade with *Pam. barberenai* and the *Sat. tupiniquim* syntypes.

The a posteriori pruning of *Pan. protos* slightly improves the resolution, with the three *E. lunensis* specimens forming a clade, but with the interspecific relations among Carnian sauropodomorphs still unresolved. The additional *a posteriori* pruning of *Ch. novasi* results in a trichotomy composed of monophyletic sets of *E. lunensis* and *Bu. schultzi* specimens, plus a clade of more deeply nested sauropodomorphs (Fig. 4b). The latter includes a trichotomy formed by *Pam. barberenai*, the *Sat. tupiniquim* clade, and Bagualosauria. Bremer supports for the *E. lunensis* and *Bu. schultzi* clades are minimal, as it is also the case for the clade that includes *Pam. barberenai*, the *Sat. tupiniquim* clade, and Bagualosauria. Similarly, the bootstrap frequencies of these branches are below 50%, with exception of those of the *Bu. schultzi* clade. Bremer supports and absolute bootstrap frequencies

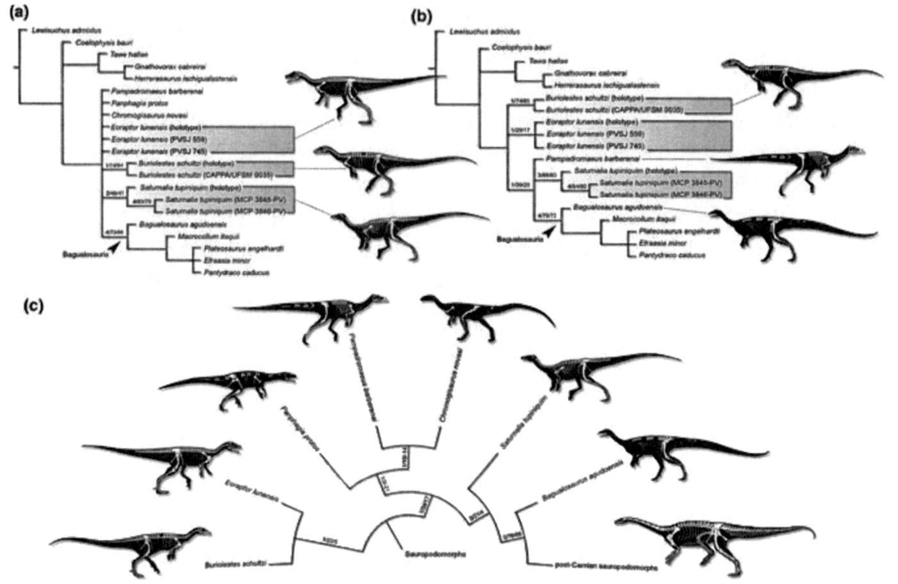

Fig. 4 Phylogenetic relations of the Carnian sauropodomorphs. **a** Strict consensus tree of the 'specimen-based' analysis. **b** Strict reduced consensus tree of the 'specimen-based' analysis with the a posteriori pruning of *Panphagia protos* and *Chromogisaurus novasi*. **c** single MPT of the 'combined' analysis

of Bagualosauria and the *Sat. tupiniquim* clade are higher: 3/66% and 4/75%, respectively. Finally, it is interesting to note that the branch support of the clade composed of the two paratypes of *Sat. tupiniquim*—MCP 3845-PV and MCP 3846-PV—is very high, with a Bremer support of 4 and an absolute bootstrap frequency of 84%.

The 'combined' analysis found a single MPT of 1,453 steps with a consistency index (CI) of 0.47970 and a retention index (RI) of 0.38286 (best score hit 713 times of the 1000 replicates). In the fully resolved optimal tree (Fig. 4c), *Bu. schultzi* and *E. lunensis* are joined in a clade sister to all other Sauropodomorpha. *Pan. protos* is found as a sister to *Pam. barberenai* plus *Ch. novasi*, in a clade sister to that including *Sat. tupiniquim* plus Bagualosauria. Bremer supports are usually minimal along the part of the tree that includes Carnian sauropodomorphs, with the exception of Bagualosauria and the clade it forms with *Sat. tupiniquim*, which have decay indices of 3 and 2, respectively. Similarly, bootstrap frequencies are all lower than 50%, except for those of Bagualosauria, which has absolute and GC frequencies of 76% and 66%, respectively.

We found the following results under constrained topologies: one additional step is necessary to force the position of *Bu. schultzi* as the earliest-diverging sauropodomorph; two extra steps are required to force the sister taxon relationships between *Sat. tupiniquim* and *Ch. novasi*, between *Sat. tupiniquim* and *Pam. barberenai*, and between *E. lunensis* and *Pan. protos*, as well as to form a clade composed of the three Ischigualastian species; nine extra steps are needed to recover

Pam. barberenai as the sister taxon to *Ba. agudoensis*. Finally, only Bagualosauria and its clade with *Sat. tupiniquim* are found in the strict consensus tree generated from suboptimal trees one step longer than the MPTs, with all other interrelations among Carnian sauropodomorphs being unresolved.

5.2 *Carnian Sauropodomorph Relationships*

The above results reveal that there is no disagreement between phylogenetic hypotheses when the three taxa with multiple specimens—*Sat. tupiniquim, E. lunensis*, and *Bu. Schultzi*—are analysed based on either their assigned specimens or combined scorings. Indeed, although the strict consensus tree of Fig. 4a is much less resolved than that of Fig. 4c, they show no conflict and at least concur in those three taxa (as well as *Pam. barberenai, Pan. protos*, and *Ch. novasi*) are external to Bagualosauria. Moreover, the specimens of *Bu. schultzi* and *Sat. tupiniquim* form clades, so that their assignment to the respective taxon is backed up by this phylogenetic study (see below for *E. lunensis*). Hence, our more conservative result—i.e. the strict consensus tree of the 'specimen-based' analysis—answers three of our proposed questions, supporting the monophylies of *Sat. tupiniquim* and *Bu. schultzi* (as composed of their specimens included in this analysis), as well as that of Bagualosauria (as composed of post-Carnian sauropodomorphs plus *Ba. agudoensis*, but no other Carnian taxon). The latter hypothesis agrees with most phylogenetic arrangements proposed so far (but see Müller and Garcia 2020) and may be considered a settled issue based on the currently available evidence.

The sister-group relation between *Ba. agudoensis* and post-Carnian sauropodomorphs, forming Bagualosauria, is supported in the present study by a series of synapomorphies, the most noteworthy of which are (see complete list in the Supplementary Material): larger size; inset first dentary tooth (also seen in *Pan. protos* and *E. lunensis*); ventrally curved rostral end of dentary; 'cheek tooth' crowns with enlarged denticles (also seen in *Pam. barberenai* and *E. lunensis*); smooth medial surface of the proximal portion of fibula (also seen in *E. lunensis*). As for the grouping of the two *Bu. schultzi* specimens into a clade, this is supported by a single trait in the analysis: concave area above supra-acetabular crest restricted to the dorsal half of the iliac blade. Yet, the status of that feature as autapomorphic for the species is jeopardised by its presence also in other Carnian sauropodomorphs.

Regarding *Sat. tupiniquim*, the apomorphies that group its syntypes are: dorsoventrally shallow sacral ribs; ventral surface of proximal caudal centra keeled or strongly constricted lateromedially; crested craniolateral margin of femoral shaft proximal portion. The latter two traits are also seen in other Carnian sauropodomorphs, so that our study also did not reveal a convincing set of autapomorphies for *Sat. tupiniquim*. Interestingly, the grouping of the two *Sat. tupiniquim* paratypes into a minimal clade is supported by a broader array of features, including a caudoventrally oriented ischiadic peduncle of the ilium, a more craniocaudaly elongated lateral condyle of the femur, a fibular shaft with a more marked insertion of m. iliofibularis, a cranially straight cnemial crest of the tibia, and a distal end of the tibia that is lateromedially broader, has an acute craniomedial corner, and lacks a proximodistaly elongated ridge on the medial portion of its caudal surface.

Pruned of the more volatile *Pan. protos* and *Ch. novasi*, the topology of Fig. 4b agrees with that of the 'combined' analysis (Fig. 4c) in that *E. lunensis* and *Bu. schultzi* are external to a clade that includes *Sat. tupiniquim*, *Pam. barberenai*, and bagualosaurs. In addition, the specimens assigned to *E. lunensis* now form a clade. This answers two further questions, supporting the monophyly of *E. lunensis* (as composed of the three specimens analysed here) and a more 'pectinate' phylogenetic arrangement for early sauropodomorphs. Indeed, some studies suggested that all Carnian sauropodomorphs (except for *Ba. agudoensis*) form a clade exclusive of other members of the group (Martínez et al. 2011; Langer et al. 2017; Baron et al. 2017a; Müller et al. 2018a). Even if this is still a possibility based on the topology of Fig. 4a, those of Fig. 4b, c indicate otherwise. As such, we understand that the present study provides reasonable evidence against a clade grouping all non-bagualosaur sauropodomorphs, suggesting instead the 'higher-nesting' of some taxa in the direction of Bagualosauria.

The nesting of *Sat. tupiniquim* and *Pam. barberenai* in a clade with bagualosaurs to the exclusion of *E. lunensis* and *Bu. schultzi* is supported in the present studies by a series of synapomorphies (see Supplementary Material), including portion of the lacrimal lateral lamina covering the antorbital fossa positioned at the mid-length of its caudal margin, tooth crowns labio lingually and mesiodistally expanded at base, and 'cheek tooth' crowns with a convex basal half of the distal margin. Besides, the grouping of the three specimens of *E. lunensis* analysed here is supported by epipophyses limited to more cranial postaxial cervical vertebrae, trunk vertebrae lacking prezygoparapophyseal laminae, and ilium with the lateral tip of the supra-acetabular crest closer to ischiadic peduncle and smooth origins for mm. flexor tibialis and iliotibialis. Interestingly, the latter three features are shared only by *Pan. protos* among Carnian sauropodomorphs, matching the possible affinity of those two Ischigualasto taxa as seen in some MPTs of the 'specimen-based' analysis.

The best resolution of the phylogenetic hypotheses presented here is that resulting from the 'combined' analysis. Apart from agreeing with the arrangements seen in the 'specimen-based' results (Fig. 4a, b), that topology reveals further hypotheses of relationships (Fig. 4c), including the sister-group relations between *E. lunensis* and *Bu. schultzi*. This has never been previously proposed, but is supported in our study by several synapomorphies, the most noteworthy of which are (see complete list in the Supplementary Material): a longer dorsolateral process of the premaxillary caudal ramus; a maxilla not significantly contributing to the external naris; a rugose ridge on the laterodorsal corner of the lacrimal rostral ramus; lacrimal with ventral ramus broader that the rostral; lacrimal with a lateral lamina covering part of the internal antorbital fenestra; ventral ramus of postorbital with a rostrally deflected end; a caudal vertebra incorporated into the sacrum; cranial margin of scapular blade not markedly concave; a stouter pubic peduncle of the ilium.

The topology of Fig. 4b already placed *Sat. tupiniquim* and *Pam. barberenai* in a clade with bagualosaurs, exclusive of *E. lunensis* and *Bu. schultzi*. Yet, the 'combined' analysis also includes *Pan. protos* and *Ch. novasi* into that clade, which is supported by the following main synapomorphies (see full list in the Supplementary Material): no diastema between last premaxillary and first maxillary alveoli, large and rostrally opened foramen perforating the lateral surface of the base of the maxillary ascending process, portion of the lacrimal lateral lamina covering the antorbital fossa positioned at the mid-length of its caudal margin, more laterally positioned paraquadratic foramen, deeper postdentary portion of lower jaw, tooth crowns labio lingually and mesiodistally expanded at base, higher tooth crowns in the rostral quarter of the tooth series, first premaxillary tooth crown with smooth carena, 'cheek tooth' crowns with a reduced distal concavity of the long axis and a convex basal half of the distal margin, pointed or right angled apex of deltopectoral crest, pubic shaft with nearly straight outline in lateral/medial views, distal end of tibia with more oblique facet for reception of ascending process of astragalus, and pedal digit II with ungual phalanx longer than second phalanx. Indeed, the placement of *E. lunensis* and *Bu. schultzi* outside a clade including all other sauropodomorphs is replicated in several previous phylogenetic analyses and is the arrangement favoured by the present investigation.

The 'combined' analysis (Fig. 4c) also identified a sister-group relation between *Sat. tupiniquim* and Bagualosauria. In most previous studies in which *Sat. tupiniquim* appears closely related to bagualosaurs, it forms a minimal clade with *Ch. novasi*. In fact, very few phylogenetic analyses (e.g. Bittencourt et al. 2015) found *Sat. tupiniquim* closer to Norian sauropodomorphs than to *Ch. novasi*. Here, this relation was supported by the following main synapomorphies (see the Supplementary Material for a complete list): smaller head; prefrontal lacking a bone sheet expanding rostroventrally from the intersection of rostral and ventral processes; broader interorbital portion of the frontal; stouter dentary; diapophysis and parapophysis nearly touching in cervical vertebrae 3–7; proximal surface of metatarsal II with a non-concave lateral margin; metatarsal III narrower caudally than cranially in proximal outline. Yet, it is important to mention that *Ch. novasi* does not preserve the skeletal parts related to any of those features, so that its closer relation to *Sat. tupiniquim* cannot be fully discarded based only on the results of this phylogenetic study.

Sister to *Sat. tupiniquim* plus Bagualosauria, the combined analysis recovered a clade composed of the 'lesser-known' Carnian sauropodomorphs, i.e. *Pan. protos*, *Ch. novasi*, and *Pam. barberenai*. This is supported by a paraquadratic foramen almost fully enclosed within the quadrate, an everted caudolateral margin of the quadrate creating a caudally facing fossa, and a relatively stouter tibia, the distal end of which has a concave caudal margin. In addition, some synapomorphies place *Pam. barberenai* and *Ch. novasi* more closely related to one another, including a relatively deeper ilium and acetabulum, a sharp dorsal margin of the iliac pubic peduncle, and a straight ventral margin of iliac acetabular wall. A close relation between *Pan. protos* and *Pam. barberenai* was already proposed by some authors (Müller et al. 2018a; Bronzati et al. 2019a; Pacheco et al. 2019; Müller and Garcia 2020), but never along with *Ch. novasi*. In fact, the overall weak support of the 'combined' analysis results shows that these relations have to be considered with care. Indeed, alternative arrangements revealed by some MPTs of the 'specimen-based' analysis include the position of *Pan. protos* both within the set of *E. lunensis* specimens or more highly nested, close to the Bagualosauria plus *Sat. tupiniquim* clade. The first of those hypotheses would indicate that either these two taxa are synonymous or that some *E. lunensis* specimens may actually belong to *Pan. protos*. As for *Ch. novasi*, it may form a clade with *Pam. barberenai* and *Sat. tupiniquim*, but was not found closer to the latter taxon in any of the MPTs, as it has been frequently suggested in previous studies.

6 Morphological Disparity Analysis

The morphospace generated from the first and second PCos (22.39% and 19.16% of variance, respectively) separated the hypodigms of *E. luensis*, *Sat. tupiniquim*, and *Bu. schultzi* into distinct clusters (Fig. 5a). *Eoraptor lunensis* specimens are positioned on the upper left quadrant, *Bu. schultzi* specimens on the lower left quadrant, and those of *Sat. tupiniquim* on the lower right quadrant of the morphospace. *Chromogisaurus novasi* is found closer to the *Bu. schultzi* cluster than to other taxa in the first two PCos, whereas *Pan. protos* and *Pam. barberenai* are positioned very close to one another and well separated from all other species in the same axes. *Bagualosaurus agudoensis* is the closest species to the latter two, within the upper right corner of the morphospace.

The generalised least squares regression between the values of the first three PCos and the logarithm of femoral length (as a proxy of body size) did not recover a significant regression for any of the PCos ($p > 0.14$), indicating that body size does not explain the morphospacial structure (Fig. 5b). The PERMANOVA found a

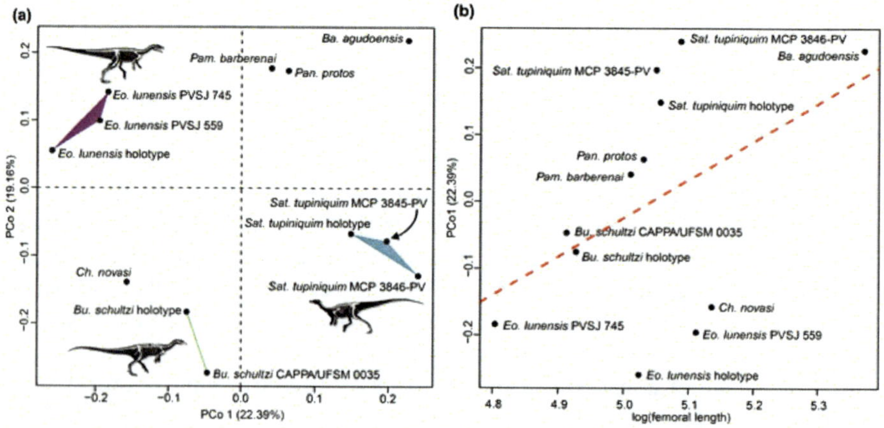

Fig. 5 Morphospace of the Carnian sauropodomorphs. **a** Morphospace represented by PCo1 and PCo2, hypodigms composed of more than one specimen highlighted by coloured convex hulls. **b** log(femoral length), as proxy of body size, versus PCo1 showing the non-significant linear regression (red dotted line) between both variables

significant difference ($p = 0.04107$) between the three hypodigms of *E. luensis*, *Sat. tupiniquim*, and *Bu. schultzi*. Similarly, the LDA predicted correctly the assignment of each specimen to its respective species with posterior probabilities ≥ 0.99.

The morphospace and statistical analyses derived from it provide strong support for the assignment of the evaluated specimens of *E. luensis*, *Sat. tupiniquim*, and *Bu. schultzi* to the respective species. This agrees with the results of the 'specimen-based' phylogeny for the two latter taxa (Fig. 4a) and helps support the assignment of PVSJ 559 and 745 to *E. lunensis*, which was only phylogenetically supported when *Pan. protos* was excluded from the analysis (Fig. 4b). The morphological disparity analysis takes into account overall dissimilarity and not only apomorphic conditions, contrasting with the phylogenetic analysis. Thus, they are complementary, distinguishing species based on unique combinations of character states (disparity analysis) and autapomorphies (phylogenetic analysis). In the end, this distance matrix-based analysis could be also potentially useful to explore the alpha-taxonomy of Carnian sauropodomorphs when new specimens are available.

7 Conclusions

- The holotypes of the seven Carnian sauropodomorphs of South America—*Ba. agudoensis*, *Bu. schultzi*, *Ch. novasi*, *E. lunensis*, *Pam. barberenai*, *Pan. protos*, and *Sa. tupiniquim*—can be anatomically differentiated from one another, hence supporting the taxonomic validity of the species they represent.
- A specimen-based phylogenetic analysis supports the referral of the *Sa. tupiniquim* paratypes and the best-preserved specimen referred to *Bu. schultzi* to the respective species. This is also supported by topotypy, anatomical congruence (especially for *Bu. schultzi*), and their statistically significant groupings in the distance matrix-based morphospace generated from the same dataset as the phylogenetic analysis.
- The referral of the various specimens previously assigned to *E. lunensis* is not supported by strong anatomical congruence and neither (for some specimens) on topotypy. Some of these specimens share putative autapomorphies with the holotype, as well as unique features relative to other Ischigualasto dinosaurs. Two of them nested close to the holotype in our morphospace analysis, but the specimen-based phylogenetic analysis failed to strongly support their affinity. In some resulting MPTs, *Pan. protos* was positioned within the clade of *E. lunensis* specimens, calling for a much-needed taxonomic revision of the specimens referred to that taxon.
- All phylogenetic analyses conducted here support the sister-group relation between *Ba. agudoensis* and post-Carnian sauropodomorphs (forming Bagualosauria). They less strongly support the hypotheses that *Bu. schultzi* and *E. lunensis* represent the earliest branches of Sauropodomorpha and that *Sa. tupiniquim* and *Pam. barberenai* are more highly nested in the direction of Bagualosauria.

- A species-level phylogenetic analysis further indicates that *Bu. schultzi* and *E. lunensis* form a clade, that *Sa. tupiniquim* is the sister taxon to Bagualosauria, and that *Pan. protos*, *Ch. novasi*, and *Pam. barberenai*, also form a clade. These clades are, however, not strongly supported by robustness measurements in the phylogenetic tree, warranting that more research is needed to untangle their relations.
- Alternative relations emerging from subsets of MPTs include the proximity of *Pan. protos* to either *E. lunensis* or Bagualosauria and a possible clade formed by *Sat. tupiniquim*, *Ch. novasi*, and *Pam. barberenai*, also requiring further investigation.

Acknowledgements This study was partially funded by São Paulo Research Foundation (FAPESP 2020/07997-4 to MCL and 2018/18145-9 to M.B.), Agencia Nacional de Promoción Científica y Técnica (PICT 2018-01186 to MDE), Coordenação de Aperfeiçoamento de Pessoal de Nível Superior (CAPES 88887.572782/2020-00 to J.C.A.M.), Fundação de Amparo à Pesquisa do Estado do Rio Grande do Sul (FAPERGS 21/2551-0000680-3 to R.T.M.), and Concelho Nacional de Desenvolvimento Científico e Tecnológico and Fundação de Amparo à Pesquisa do Estado de Minas Gerais (FAPEMIG PPM-00304-18 to J.S.B.). Access to the free version of TNT 1.5 was possible due to the Willi Henning Society.

Supplementary Information Available at: https://osf.io/h8qs3/?view_only=3100c6a2c8d54e9aa91fbf6655f9a2df

References

Agnolín FL, Rozadilla S (2018) Phylogenetic reassessment of *Pisanosaurus mertii* Casamiquela 1967 a basal dinosauriform from the Late Triassic of Argentina. J Sys Palaeontol 16(10):853–879

Alcober OA, Martínez RN (2010) A new herrerasaurid (Dinosauria Saurischia) from the Upper Triassic Ischigualasto formation of northwestern Argentina. ZooKeys 63:55–81

Apaldetti C, Martínez RN, Alcober OA, Pol D (2011) A new basal sauropodomorph (Dinosauria: Saurischia) from Quebrada del Barro Formation (Marayes–El Carrizal Basin) northwestern Argentina. PLoS One 6(11):e26964

Baron MG (2019) *Pisanosaurus mertii* and the Triassic ornithischian crisis: could phylogeny offer a solution? Hist Biol 31(8):967–981

Baron MG, Barrett PM (2017) A dinosaur missing–link? *Chilesaurus* and the early evolution of ornithischian dinosaurs. Biol Lett 13(8):20170220

Baron MG, Williams ME (2018) A re–evaluation of the enigmatic dinosauriform *Caseosaurus crosbyensis* from the Late Triassic of Texas USA and its implications for early dinosaur evolution. Acta Palaeontol Pol 63:129–145

Baron MG, Norman DB, Barrett PM (2017a) Baron et al reply Nature 551(7678):E4–E5

Baron MG, Norman DB, Barrett PM (2017b) A new hypothesis of dinosaur relationships and early dinosaur evolution. Nature 543(7646):501–506

Barrett PM, Upchurch P (2007) The evolution of feeding mechanisms in early sauropodomorph dinosaurs. Spec Pap Palaeontol 77:91–112

Barrett PM, Upchurch P, Zhou XD, Wang XL (2007) The skull of *Yunnanosaurus huangi* Young 1942 (Dinosauria: Prosauropoda) from the Lower Lufeng Formation (Lower Jurassic) of Yunnan China. Zool J Linn Soc 150(2):319–341

Benson RBJ, Campione NE, Carrano MT, Mannion PD, Sullivan C, Upchurch P, Evans DC (2014) Rates of Dinosaur Body Mass Evolution Indicate 170 Million Years of Sustained Ecological Innovation on the Avian Stem Lineage. PLoS Biol 12(5):e1001853

Bittencourt JS, Arcucci AB, Marsicano CA, Langer MC (2015) Osteology of the Middle Triassic archosaur *Lewisuchus admixtus* Romer (Chañares Formation Argentina) its inclusivity and relationships amongst early dinosauromorphs. J Sys Palaeontol 13(3):189–219

Bonaparte JF, Ferigolo J, Ribeiro AM (1999) A new early Late Triassic saurischian dinosaur from Rio Grande do Sul state, Brazil. Nat Sci Mus Monogr 15:89–109

Bremer K (1988) The limits of amino acid sequence data in angiosperm phylogenetic reconstruction. Evolution 42:795–803

Bremer K (1994) Branch support and tree stability. Cladistics 10:295–304

Bronzati M, Rauhut OW, Bittencourt JS, Langer MC (2017) Endocast of the Late Triassic (Carnian) dinosaur *Saturnalia tupiniquim*: implications for the evolution of brain tissue in Sauropodomorpha. Sci Rep 7(1):1–7

Bronzati M, Rauhut OW (2018) Braincase redescription of *Efraasia minor* Huene 1908 (Dinosauria: Sauropodomorpha) from the Late Triassic of Germany with comments on the evolution of the sauropodomorph braincase. Zool J Linn Soc 182(1):173–224

Bronzati M, Benson RB, Rauhut OW (2018) Rapid transformation in the braincase of sauropod dinosaurs: integrated evolution of the braincase and neck in early sauropods? Palaeontology 61(2):289–302

Bronzati M, Müller RT, Langer MC (2019a) Skull remains of the dinosaur *Saturnalia tupiniquim* (Late Triassic, Brazil): With comments on the early evolution of sauropodomorph feeding behaviour. PLoS OONE 221:e0221387.

Bronzati M, Langer MC, Rauhut OWM (2019b) Braincase anatomy of the early sauropodomorph *Saturnalia tupiniquim* (Late Triassic, Brazil). J Vert Paleontol 38(5): https://doi.org/10.1080/02724634.2018.1559173

Cabreira SF, Schultz CL, Bittencourt JS, Soares MB, Fortier DC, Silva LR, Langer MC (2011) New stem-sauropodomorph (Dinosauria Saurischia) from the Triassic of Brazil. Naturwissenschaften 98(12):1035–1040

Cabreira SF, Kellner AWA, Dias-da-Silva S, da Silva LR, Bronzati M, Marsola JC, Müller RT, Bittencourt JS, Batista B, Raugust T, Carrilho R, Brodt A, Langer MC (2016) A unique Late Triassic dinosauromorph assemblage reveals dinosaur ancestral anatomy and diet. Curr Biol 26(22):3090–3095

Campione NE, Evans DC, Brown CM, Carrano MT (2014) Body mass estimation in non-avian bipeds using a theoretical conversion to quadruped stylopodial proportions. Methods Ecol Evol 5:913–923

Cau A (2018) The assembly of the avian body plan: a 160–million–year long process. Boll Soc Paleontol Ital 57(1):1–25

Cerda IA, Chinsamy A, Pol D, Apaldetti C, Otero A, Powell JE, Martínez RN (2017) Novel insight into the origin of the growth dynamics of sauropod dinosaurs. PLoS ONE 12(6): e0179707

Chapelle KE, Choiniere JN (2018) A revised cranial description of *Massospondylus carinatus* Owen (Dinosauria: Sauropodomorpha) based on computed tomographic scans and a review of cranial characters for basal Sauropodomorpha. PeerJ 6:e4224

Chapelle KE, Barrett PM, Botha J, Choiniere JN (2019) Ngwevu intloko: a new early sauropodomorph dinosaur from the Lower Jurassic Elliot Formation of South Africa and comments on cranial ontogeny in *Massospondylus carinatus*. PeerJ 7:e7240

Chatterjee S (1987) A new theropod dinosaur from India with remarks on the Gondwana-Laurasia connection in the Late Cretaceous. In McKenzie GD (ed.). Gondwana 6: Stratigraphy, Sedimentology, and Paleontology. Geophy Monogr 41:183–189

Coddington JA, Scharff N (1994) Problems with zero-length branches. Cladistics 10:415–423

Colombi C, Martínez RN, Césari SN, Alcober O, Limarino CO, Montañez I. (2021). A high-precision U–Pb zircon age constraints the timing of the faunistic and palynofloristic events of the Carnian Ischigualasto Formation, San Juan, Argentina. J South Am Earth Sci 111: 103433

Da Rosa AAS (2015) Geological context of the dinosauriform-bearing outcrops from the Triassic of Southern Brazil. J South Am Earth Sci 61:108–119
Dal Sasso C, Maganuco S, Cau A (2018) The oldest ceratosaurian (Dinosauria: Theropoda) from the Lower Jurassic of Italy sheds light on the evolution of the three–fingered hand of birds. PeerJ 6:e5976
Delcourt R, Azevedo SAK, Grillo ON, Deantoni FO (2012) Biomechanical comments about Triassic dinosaurs from Brazil. Pap Avulsos Zool 52:341–347
Desojo JB, Fiorelli LE, Ezcurra MD, Martinelli AG, Ramezani J, Da Rosa AAS, Baczko MB, Trotteyn MJ, Montefeltro FC, Ezpeleta M, Langer MC (2020) The Late Triassic Ischigualasto Formation at Cerro Las Lajas (La Rioja, Argentina): fossil tetrapods, high-resolution chronostratigraphy, and faunal correlations. Sci Rep 10:12782
Ezcurra MD (2006) A review of the systematic position of the dinosauriform archosaur *Eucoelophysis baldwini* Sullivan Lucas 1999 from the Upper Triassic of New Mexico USA. Geodiversitas 28(4):649–684
Ezcurra MD (2010) A new early dinosaur (Saurischia: Sauropodomorpha) from the Late Triassic of Argentina: a reassessment of dinosaur origin and phylogeny. J Sys Palaeontol 8(3):371–425
Ezcurra MD (2012a) Comments on the taxonomic diversity and paleobiogeography of the earliest known dinosaur assemblages (late Carnian-earliest Norian). Hist Nat 2:49–71
Ezcurra MD (2012b) Sistemática, biogeografía y patrones macroevolutivos de los dinosaurios terópodos del Triásico Tardío y Jurásico Temprano. Universidad de Buenos Aires, Buenos Aires, Tesis de Licenciatura, p 512
Ezcurra MD, Novas FE (2007) Phylogenetic relationships of the Triassic theropod *Zupaysaurus rougieri* from NW Argentina. Hist Biol 19(1):35–72
Ezcurra MD, Fiorelli LE, Martinelli AG, Rocher S, von Baczko MB, Ezpeleta M, Taborda JRA, Hechenleitner EM, Trotteyn MJ, Desojo JB (2017) Deep faunistic turnovers preceded the rise of dinosaurs in southwestern Pangaea. Nat Ecol Evol 1:1477–1483
Ezcurra MD, Nesbitt SJ, Bronzati M, Dalla Vecchia FM, Agnolin FL, Benson RBJ, Egli FB, Cabreira SF, Evers SW, Gentil AR, Irmis RB, Martinelli AG, Novas FE, Roberto da Silva L, Smith ND, Stocker MR, Turner AH, Langer MC (2020a) Enigmatic dinosaur precursors bridge the gap to the origin of Pterosauria. Nature 588:445–449
Ezcurra MD, Nesbitt SJ, Fiorelli LE, Desojo JB (2020b) New specimen sheds light on the anatomy and taxonomy of the early Late Triassic dinosauriforms from the Chañares Formation, NW Argentina. Anat Rec 303:1393–1438
Fabbri M, Tschopp E, McPhee B, Nesbitt S, Pol D, Langer MC (2020) Sauropodomorpha F. R. von Huene. In de Queiroz K, Cantino PD, Gauthier JA (eds,) Phylonyms: a Companion to the PhyloCode p:1225–1234. CRC Press
Felsenstein J (1985) Confidence limits on phylogenies: an approach using the bootstrap. Evolution 39:783–791
Ferreira-Cardoso S, Araújo R, Martins NE, Martins GG, Walsh S, Martins RMS, Kardjilov N, Manke I, Hilger A, Castanhinha R (2017) Floccular fossa size is not a reliable proxy of ecology and behaviour in vertebrates. Sci Rep 7(2005):s41598–017–01981–0
Fraser NC, Padian K, Walkden GM, Davis ALM (2002) Basal dinosauriform remains from Britain and the diagnosis of the Dinosauria. Palaeontology 45(1):79–95
Galton PM, Upchurch P (2004) Prosauropoda. In Weishampel DB, Dodson P, Osmolska H (eds) The Dinosauria (Second Edition) University of California Press Berkeley p232–258
Garcia MS, Müller RT, Pretto FA, Da-Rosa AAS, Dias-Da-Silva s (2021) Taxonomic and phylogenetic reassessment of a large-bodied dinosaur from the earliest dinosaur-bearing beds (Carnian, Upper Triassic) from southern Brazil. J Syst Palaeontol DOI: https://doi.org/10.1080/14772019.2021.1873433
Garcia MS, Pretto FA, Dias-Da-Silva S, Müller RT (2019) A dinosaur ilium from the Late Triassic of Brazil with comments on key–character supporting Saturnaliinae. An Acad Bras Ciênc 91:1–20

Gauthier JA, Langer MC, Novas FE, Bittencourt JS, Ezcurra MD (2020) Saurischia H. G. Seeley. In de Queiroz K, Cantino PD, Gauthier JA (eds,) Phylonyms: a Companion to the PhyloCode p:1219–1224. CRC Press

Godoy MM, Scherer OLB, Binotto RB, Kisvhlat EE, Dreher AM (2018) Geologia e Recursos Minerais da folha Santa Maria SH.22-V-C-IV: estado do Rio Grande do Sul. CPRM, Porto Alegre

Goloboff P, Catalano SA (2016) TNT version 1.5, including a full implementation of phylogenetic morphometrics. Cladistics 32(3):221–238

Goloboff PA, Farris JS, Nixon KC (2008) TNT a free program for phylogenetic analysis. Cladistics 24(5):774–786. https://doi.org/10.1111/j.1096-0031.2008.00217.x

Goloboff P, Farris J, Nixon K (2003a) T.N.T.: Tree Analysis Using New Technology. Program and documentation, available at http://www.zmuc.dk/public/phylogeny/tnt

Goloboff PA, Farris JS, Källersjö M, Oxelman B, Ramírez M, Szumik C (2003b) Improvements to resampling measures of group support. Cladistics 19:324–332

Griffin CT, Munyikwa D, Broderick TJ, Tolan S, Zondo M, Nesbitt SJ, Taruvinga H (2018) An exceptional new Late Triassic (Carnian) fossil assemblage from Zimbabwe and the biogeography of the earliest dinosaurs across Pangea. Society of Vertebrate Paleontology Meeting Program and Abstracts (Albuquerque, NM) p.137

Hennig EHW (1966) Phylogenetic Systematics. University of Illinois Press, Urbana

Horn BLD, Melo TM, Schultz CL, Philipp RP, Kloss HP, Goldberg K (2018) A new third-order sequence stratigraphic framework applied to the Triassic of the Parana Basin, Rio Grande do Sul, Brazil, based on structural, stratigraphic and paleontological data. J South Am Earth Sci 55:123–132

Irmis RB, Nesbitt SJ, Padian K, Smith ND, Turner AH, Woody D, Downs A (2007) A Late Triassic dinosauromorph assemblage from New Mexico and the rise of dinosaurs. Science 317(5836):358–361

Langer MC (2003) The pelvic and hindlimb anatomy of the stem-sauropodomorph *Saturnalia tupiniquim* (Late Triassic, Brazil). PaleoBios 23:1–40

Langer MC (2004) Basal Saurischia. In Weishampel DB, Dodson P, Osmolska H (eds) The Dinosauria (Second Edition) University of California Press Berkeley p25–46

Langer MC (2005) Studies on continental Late Triassic tetrapod biochronology. I. The type locality of *Saturnalia tupiniquim* and the faunal succession in south Brazil. J South Am Earth Sci 19:205–218

Langer MC, Benton MJ (2006) Early dinosaurs: a phylogenetic study. J Sys Palaeontol 4(4):309–358

Langer MC, Abdala F, Richter M, Benton MJ (1999) A sauropodomorph dinosaur from the Upper Triassic (Carman) of southern Brazil. CR Acad Sci IIA 329(7):511–517

Langer MC, França MAG, Gabriel S (2007) The pectoral girdle and forelimb anatomy of the stem-sauropodomorph *Saturnalia tuiniquim* (Upper Triassic, Brazil). Spec Pap Palaeontol 77:113–137

Langer MC, Ezcurra M, Bittencourt JS, Novas F (2009) The origin and early evolution of dinosaurs. Biol Rev 85:55–110

Langer MC, Bittencourt JS, Schultz CL (2011) A reassessment of the basal dinosaur *Guaibasaurus candelariensis* from the Late Triassic Caturrita Formation of south Brazil. Earth Environ Sci Trans R Soc Edinb 101(3–4):301–332

Langer MC, Ezcurra M, Rauhut OW, Benton MJ, Knoll F, McPhee BW, Brusatte SL (2017) Untangling the dinosaur family tree. Nature 551(7678):E1–E3

Langer MC, Ramezani J, Da Rosa AAS (2018) U-Pb age constraints on dinosaur rise from south Brazil. Gondwana Res 57:133–140

Langer MC, McPhee BW, Marsola JC, Roberto-da-Silva L, Cabreira SF (2019) Anatomy of the dinosaur *Pampadromaeus barberenai* (Saurischia—Sauropodomorpha) from the Late Triassic Santa Maria Formation of southern Brazil. PloS one 14(2):e0212543

Langer MC, Novas FE, Bittencourt JS, Ezcurra MD, Gauthier JA (2020) Dinosauria R. Owen. In de Queiroz K, Cantino PD, Gauthier JA (eds,) Phylonyms: a Companion to the PhyloCode p:1209–1217. CRC Press

Lehmann OE, Ezcurra MD, Butler RJ, Lloyd GT (2019) Biases with the generalized euclidean distance measure in disparity analyses with high levels of missing data. Palaeontol 62(5) 837–849. https://doi.org/10.1111/pala.12430

Lloyd GT (2016) Estimating morphological diversity and tempo with discrete character-taxon matrices: implementation, challenges, progress, and future directions. Biol J Linn Soc 118:131–151

Marsh AD, Rowe TB (2018) Anatomy and systematics of the sauropodomorph *Sarahsaurus aurifontanalis* from the Early Jurassic Kayenta Formation. PloS one 13(10):e0204007

Marsh AD, Parker WG, Langer MC, Nesbitt SJ (2019) Redescription of the holotype specimen of *Chindesaurus bryansmalli* Long and Murry 1995 (Dinosauria Theropoda) from Petrified Forest National Park Arizona. J Vert Paleontol 39(3):e1645682

Marsola JC, Bittencourt JS, Butler RJ, Da-Rosa ÁA, Sayão JM, Langer MC (2018) A new dinosaur with theropod affinities from the Late Triassic Santa Maria Formation South Brazil. J Vert Paleontol 38(5):e1531878

Martínez RN (2009) *Adeopapposaurus mognai* gen et sp nov (Dinosauria: Sauropodomorpha) with comments on adaptations of basal Sauropodomorpha. J Vert Paleontol 29(1):142–164

Martínez RN, Alcober OA (2009) A basal sauropodomorph (Dinosauria: Saurischia) from the Ischigualasto Formation (Triassic Carnian) and the early evolution of Sauropodomorpha. PLoS One 4(2):e4397

Martínez RN, Sereno PC, Alcober OA, Colombi CE, Renne PR, Montañez IP, Currie BS (2011) A basal dinosaur from the dawn of the dinosaur era in southwestern Pangaea. Science 331(6014):206–210

Martínez RN, Apaldetti C, Alcober OA, Colombi CE, Sereno PC, Fernandez E, Malnis PS, Correa GA, Abelin D (2012a) Vertebrate succession in the Ischigualasto Formation. J Vert Paleontol 32(sup1):10–30

Martínez RN, Apaldetti C, Abelin D (2012b) Basal sauropodomorphs from the Ischigualasto Formation. J Vert Paleontol 32(sup1):51–69

Martínez RN, Haro JA, Apaldetti C (2012c) Braincase of *Panphagia protos* (Dinosauria, Sauropodomorpha) J Vert Paleontol 32(sup1):70–82

McPhee BW, Choiniere JN (2018) The osteology of *Pulanesaura eocollum*: implications for the inclusivity of Sauropoda (Dinosauria). Zool J Linn Soc 182(4):830–861

McPhee BW, Yates AM, Choiniere JN, Abdala F (2014) The complete anatomy and phylogenetic relationships of *Antetonitrus ingenipes* (Sauropodiformes Dinosauria): implications for the origins of Sauropoda. Zool J Linn Soc 171(1):151–205

McPhee BW, Benson RB, Botha-Brink J, Bordy EM, Choiniere JN (2018) A giant dinosaur from the earliest Jurassic of South Africa and the transition to quadrupedality in early sauropodomorphs. Current Biol 28(19):3143–3151.e7. https://doi.org/10.1016/j.cub.2018.07.063

McPhee BW, Bonnan MF, Yates AM, Neveling J, Choiniere JN (2015) A new basal sauropod from the pre-Toarcian Jurassic of South Africa: evidence of niche–partitioning at the sauropodomorph-sauropod boundary? Sci Rep 5(1):1–12

McPhee BW, Bittencourt JS, Langer MC, Apaldetti C, Da-Rosa ÁAS (2020) Reassessment of *Unaysaurus tolentinoi* (Dinosauria: Sauropodomorpha) from the Late Triassic (early Norian) of Brazil with a consideration of the evidence for monophyly within non–sauropodan sauropodomorphs. J Sys Palaeontol 18(3):259–293

Müller RT (2020) Craniomandibular osteology of *Macrocollum itaquii* (Dinosauria: Sauropodomorpha) from the Late Triassic of southern Brazil. J Sys Palaeontol 18(10):805–841

Müller RT (2021) Olfactory acuity in early sauropodomorph dinosaurs. Hist Biol. https://doi.org/10.1080/08912963.2021.1914600

Müller RT, Garcia MS (2019) Rise of an empire: analysing the high diversity of the earliest sauropodomorph dinosaurs through distinct hypotheses. Hist Biol 32:1334–1339

Müller RT, Garcia MS (2020) A paraphyletic 'Silesauridae' as an alternative hypothesis for the initial radiation of ornithischian dinosaurs. Biol Lett 16:20200417

Müller RT, Langer MC, Cabreira SF, Sias-da-Silva S (2016a) The femoral anatomy of Pampadromaeus barberenai based on a new specimen from the Upper Triassic of Brazil. Hist Biol 28(5):656–665. https://doi.org/10.1080/08912963.2015.1004329

Müller RT, Langer MC, Dias-da-Silva S (2016b) Biostratigraphic significance of a new early sauropodomorph specimen from the Upper Triassic of southern Brazil. Hist Biol 29(2):187–202

Müller RT, Langer MC, Pacheco CP, Sias-da-Silva S (2017a) The role of ontogeny on character polarization in early dinosaurs: a new specimen from the Late Triassic of southern Brazil and its implications. Hist Biol 31(6):794–805

Müller RT, Pretto FA, Stefanello M, Silva-Neves E, Dias-da-Silva S (2017b) On a dinosaur axis from one of the oldest dinosaur–bearing sites worldwide. Acta Palaeontol Pol 62(3):543–548

Müller RT, Langer MC, Bronzati M, Pacheco CP, Cabreira SF, Dias-Da-Silva S (2018a) Early evolution of sauropodomorphs: anatomy and phylogenetic relationships of a remarkably well–preserved dinosaur from the Upper Triassic of southern Brazil. Zool J Linn Soc 184(4):1187–1248

Müller RT, Pretto F, Kerber L, Silva-Neves E, Dias-da-Silva S (2018b) Comment on 'A dinosaur missing–link? *Chilesaurus* and the early evolution of ornithischian dinosaurs'. Biol Lett 14(3):20170581

Müller RT, Langer MC, Dias-da-Silva S (2018c) An exceptionally preserved association of complete dinosaur skeletons reveals the oldest long–necked sauropodomorphs. Biol Lett 14(11):20180633

Müller RT, Ferreira JD, Pretto FA, Bronzati M, Kerber L (2020) The endocranial anatomy of *Buriolestes schultzi* (Dinosauria: Saurischia) and the early evolution of brain tissues in sauropodomorph dinosaurs. J Anat 238(4):809–827

Nesbitt SJ (2011) The early evolution of archosaurs: relationships and the origin of major clades. Bull Am Mus Nat Hist 2011(352):1–292

Nesbitt SJ, Chatterjee S (2008) Late Triassic dinosauriforms from the Post Quarry and surrounding areas, west Texas, USA. Neues Jahrb Geol Palaontol Abh 249:143–156

Nesbitt SJ, Ezcurra MD (2015) The early fossil record of dinosaurs in North America: a new neotheropod from the base of the Upper Triassic Dockum Group of Texas. Acta Palaeontol Pol 60(3):513–526

Nesbitt SJ, Barrett PM, Werning S, Sidor CA, Charig AJ (2013) The oldest dinosaur? A middle Triassic dinosauriform from Tanzania. Biol Lett 9(1):20120949. https://doi.org/10.1098/rsbl.2012.0949

Nesbitt SJ, Smith ND, Irmis RB, Turner AH, Downs A, Norell MA (2009) A complete skeleton of a Late Triassic saurischian and the early evolution of dinosaurs. Science 326(5959):1530–1533

Nesbitt SJ, Sidor CA, Irmis RB, Angielczyk KD, Smith RM, Tsuji LA (2010) Ecologically distinct dinosaurian sister group shows early diversification of Ornithodira. Nature 464(7285):95–98

Novas FE, Agnolin FL, Ezcurra MD, Müller RT, Martinelli AG, Langer MC (2021) Review of the fossil record of early dinosaurs from South America, and its phylogenetic implicances. J South Am Earth Sc 110:103341

Novas FE, Ezcurra MD, Chatterjee S, Kutty TS (2011) New dinosaur species from the Upper Triassic Upper Maleri and Lower Dharmaram formations of central India. Earth Environ Sci Trans R Soc Edinb 101(3–4):333–349

Novas FE (1996) Dinosaur Monophyly. J Vert Paleontol 16:723–741

Otero A, Pol D (2013) Postcranial anatomy and phylogenetic relationships of *Mussaurus patagonicus* (Dinosauria Sauropodomorpha). J Vert Paleontol 33(5):1138–1168

Otero A, Krupandan E, Pol D, Chinsamy A, Choiniere J (2015) A new basal sauropodiform from South Africa and the phylogenetic relationships of basal sauropodomorphs. Zool J Linn Soc 174(3):589–634

Pacheco C, Müller RT, Langer MC, Pretto FA, Kerber L, Silva SD (2019) *Gnathovorax cabreirai*: a new early dinosaur and the origin and initial radiation of predatory dinosaurs. PeerJ 7:e7963

Parry LA, Baron MG, Vinther J (2017) Multiple optimality criteria support Ornithoscelida. R Soc Open Sci 4(10) 170833

Pol D (2004) Phylogenetic relationshps of basal Sauropodomorpha. PhD thesis, Columbia University

Pol D, Escapa IH (2009) Unstable taxa in cladistic analysis: identification and the assessment of relevant characters. Cladistics 25:515–527

Pol D, Otero A, Apaldetti C, Martínez RN (2021) Triassic sauropodomorph dinosaurs from South America: The origin and diversification of dinosaur dominated herbivorous faunas. J South Am Earth Sci 107:103145

Pretto FA, Schultz CL, Langer MC (2015) New dinosaur remains from the Late Triassic of southern Brazil (Candelária Sequence *Hyperodapedon* Assemblage Zone). Alcheringa 39(2):264–273

Pretto FA, Veiga FH, Langer MC, Schultz CL (2017) A juvenile sauropodomorph tibia from the 'Botucaraí Hill' Late Triassic of Southern Brazil. Rev Bras Paleontol 19(3):407–414

Pretto FA, Langer MC, Schultz CL (2019) A new dinosaur (Saurischia: Sauropodomorpha) from the Late Triassic of Brazil provides insights on the evolution of sauropodomorph body plan. Zool J Linn Soc 185(2):388–416

Raath M (1996) Earliest evidence of dinosaurs from central Gondwana. Mem Queensl Mus 39:703–709

Remes K, Rauhut O (2005) The oldest Indian dinosaur *Alwalkeria maleriensis* Chaterjee revised: a chimera including remains of a basal saurischian. In Kellner AWA, Henriques DDR, Rodrigues T (eds) Boletim de Resumos do II Congresso Latino-americano de Paleontologia de Vertebrados, p218. Serie Livros 12, Museu Nacional, Rio de Janeiro

Rogers RR, Swisher CC III, Sereno PC, Forster CA, Monetta AM (1993) The Ischigualasto tetrapod assemblage (Late Triassic) and 40Ar/39Ar calibration of dinosaur origins. Science 260:794–797

Salgado L, Coria RA, Calvo JO (1997) Evolution of titanosaurid sauropods. I. Phylogenetic analysis based on the postcranial evidence. Ameghiniana 34:3–32

Schultz CL, Martinelli AG, Soares MB, Pinheiro FL, Kerber L, Horn BLD, Pretto FA, Müller RT, Melo TP (2020) Triassic faunal successions of the Paraná Basin, southern Brazil. J South Am Earth Sci 104:102846

Sereno PC (1998) A rationale for phylogenetic definitions, with application to the higher-level taxonomy of Dinosauria. Neues Jahrb Geol Palaontol Abh 210:41–83

Sereno PC (1999) The evolution of dinosaurs. Science 284(5423):2137–2147

Sereno PC (2007a) Basal Sauropodormorpha: historical and recent phylogenetic hypothesis, with comments on *Ammosaurus major* (Marsh, 1889). Spec Pap Palaeontol 77:261–289

Sereno PC (2007b) Logical basis for morphological characters in phylogenetics. Cladistics 23:565–587

Sereno PC, McAllister S, Brusatte SL (2005) TaxonSearch: a relational database for suprageneric taxa and phylogenetic definitions. PhyloInformatics 8:1–21

Sereno PC, Martínez RN, Alcober OA (2012). Osteology of *Eoraptor lunensis* (Dinosauria, Sauropodomorpha). Basal sauropodomorphs and the vertebrate fossil record of the Ischigualasto Formation (Late Triassic: Carnian-Norian) of Argentina. J Vert Paleontol 32(sup1):83–179.

Sereno PC, Forster CA, Rogers RR, Monetta AM (1993) Primitive dinosaur skeleton from Argentina and the early evolution of the Dinosauria. Nature 361:64–66

Smith ND, Makovicky PJ, Hammer WR, Currie PJ (2007) Osteology of *Cryolophosaurus ellioti* (Dinosauria: Theropoda) from the Early Jurassic of Antarctica and implications for early theropod evolution. Zool J Linn Soc 151(2):377–421

Sues HD, Nesbitt SJ, Berman DS, Henrici AC (2011) A late-surviving basal theropod dinosaur from the latest Triassic of North. America Proc R Soc Lond B Biol Sci 278(1723):3459–3464

Swofford DL, Begle DP (1993) User's manual for PAUP: phylogenetic analysis using parsimony, Version 3.1. Washington D.C.: Smithsonian Institution

Upchurch P (1997). Sauropodomorpha. In Currie PG, Padian K (eds) Encyclopedia of Dinosaurs Academic Press San Diego p658–660

Upchurch P, Barrett PM, Galton PM (2007) A phylogenetic analysis of basal sauropodomorph relationships: implications for the origin of sauropod dinosaurs. Spec Pap Palaeontol 77:57

Wang YM, You HL, Wang T (2017) A new basal sauropodiform dinosaur from the Lower Jurassic of Yunnan Province China. Sci Rep 7(1):1–11

Yates AM (2003) A definite prosauropod dinosaur from the lower Elliot Formation (Norian: Upper Triassic) of South Africa. Palaeontol Africana 39:63–68

Yates AM (2004) *Anchisaurus polyzelus* (Hitchcock): the smallest known sauropod dinosaur and the evolution of gigantism among sauropodomorph dinosaurs. Postilla 230:1–58

Yates AM (2007a) The first complete skull of the Triassic dinosaur *Melanorosaurus* Haughton (Sauropodomorpha: Anchisauria) Evolution and palaeobiology of early sauropodomorph dinosaurs. Spec Pap Palaeontol 77:9–55

Yates AM, (2007b) Solving a dinosaurian puzzle: the identity of *Aliwalia rex* Galton. Hist Biol 19(1):93–123

Yates AM, Kitching JW (2003) The earliest known sauropod dinosaur and the first steps towards sauropod locomotion. Proc R Soc Lond B: Biol Sci 270(1525):1753–1758

Zerfass H (2007) Geologia da Folha Agudo, SH.22-V-C-V, escala 1:100.000. Serviço Geológico do Brasil-CPRM

Zerfass H, Lavina EL, Schultz CL, Garcia AGV, Faccini UF, Chemale F Jr (2003) Sequence stratigraphy of continental Triassic strata of southernmost Brazil: a contribution to Southwestern Gondwana palaeogeography and palaeoclimate. Sediment Geol 161:85–180

Zhang QN, You HL, Wang T, Chatterjee S (2018) A new sauropodiform dinosaur with a 'sauropodan' skull from the Lower Jurassic Lufeng Formation of Yunnan Province China. Sci Rep 8(1):1–12

Non-sauropodiform Plateosaurians: Milestones Through the "Prosauropod" Bauplan

Alejandro Otero and Claire Peyre de Fabrègues

Abstract The early evolution of Sauropodomorpha is well recorded in Carnian beds of Argentina and Brazil. During the Norian and Rhaetian, sauropodomorphs notably diversified both taxonomically and ecologically, became abundant and ultimately dominated terrestrial ecosystems, adding to the information retrieved from the records from Europe, India, and Southern Africa. Despite the fact that the last decade witnessed an increase in taxonomic abundance of Carnian sauropodomorphs, their morphological disparity is low, characterized by small, gracile, and bipedal forms, with predatory/omnivorous feeding habits. By the Early Jurassic, this group had achieved their broadest geographical distribution and morphological disparity, ranging from small to medium-sized facultative bipedal basal sauropodomorphs to giant quadrupedal sauropods. The major changes in body plan after Carnian forms include the acquisition of features related to herbivory, large body size, and quadrupedality. This chapter is focused on the post-Carnian radiation of sauropodomorphs, for which the South American record accounts for about 25% of the world record. It has provided key information in understanding certain stages of this evolutionary radiation and has therefore highlighted the understanding of the evolution of this group.

Keywords Plateosauria · Body size · Herbivory · Locomotion · Evolution

A. Otero (✉)
Consejo Nacional de Investigaciones Científicas y Técnicas (CONICET), Buenos Aires, Argentina
e-mail: aleotero@fcnym.unlp.edu.ar

División Paleontología de Vertebrados, Facultad de Ciencias Naturales y Museo (Anexo Laboratorios), Calle 122 y 60, La Plata (B1900WA), Buenos Aires Province, Argentina

C. P. de Fabrègues
Centre for Vertebrate Evolutionary Biology, Yunnan University, Kunming, Yunnan Province, China

A. Otero et al. (eds.), *South American Sauropodomorph Dinosaurs*, Springer Earth System Sciences, https://doi.org/10.1007/978-3-030-95959-3_2

1 Introduction

Early-branching sauropodomorphs comprise the non-sauropod portion of Sauropodomorpha, which constitutes the first major dinosaurian group that radiated during the Triassic (e.g., Galton and Upchurch 2004; Langer et al. 2010). Traditionally, non-sauropod sauropodomorphs were thought to encompass an array of rather conservative omnivorous to herbivorous bipedal forms, characterized by small skulls with lanceolate teeth, long necks, and a robust hand bearing a well-developed digit one (Galton and Upchurch 2004; Martínez et al. 2013; Müller and Garcia 2020). This was certainly the big picture until the first decade of the 21st century, from when new discoveries and reinterpretations of previously known taxa brought the studies and knowledge of early sauropodomorphs to a state of flux in terms of their anatomy (e.g., Smith and Pol 2007; Martínez 2009; Bronzati and Rauhut 2018), phylogenetic relationships (Rowe et al. 2010; Yates et al. 2010; McPhee et al. 2015; Peyre de Fabrègues and Allain 2016, 2020; Zhang et al. 2018; Langer et al. 2019; Müller 2019; Peyre de Fabrègues et al. 2020; Otero and Pol in press) and paleobiology (Bronzati et al. 2017; Cerda et al. 2017; Otero et al. 2017, 2019). All this new information allowed having a wider picture of this group while recognizing previously unknown clades and the morphological disparity supporting them (e.g., Ezcurra 2010; Bronzati et al. 2019; Langer et al. 2019; Cabreira et al. 2016; Apaldetti et al. 2018, 2021). In this regard, on one hand, some of the basalmost branches of Sauropodomorpha are today known to putative be part of a monophyletic group (the Saturnaliidae/Saturnaliinae) recovered in several phylogenetic analyses (e.g., Ezcurra 2010; McPhee et al. 2014; Müller et al. 2018a; Garcia et al. 2019; Langer et al. 2019; Müller 2019; but see Chap. 1 for an alternative scenario). They are small, biped, gracile, predatory/faunivorous forms from the Carnian close to the primitive saurischian condition. On the other hand, the early-branching Sauropodiformes, constitute an array of post-Carnian forms, progressively gained sauropod features in a mosaic fashion, such as herbivorous diet, robust limbs, and shortening of manual phalanges. They include taxa like *Melanorosaurus*, *Aardonyx*, *Leonerasaurus*, and *Sefapanosaurus* (Bonnan and Yates 2007; Yates et al. 2010; Pol et al. 2011; Otero et al. 2015). The culmination of this trend is materialized by the Lessemsauridae: ~10 tons, quadruped, robust, herbivorous forms, representing the closest forms to sauropods (sensu Yates 2007b) in some studies (e.g., McPhee et al. 2018; Peyre de Fabrègues and Allain 2020) or the earliest branching sauropods (sensu Yates 2007b) in other studies (e.g., Pol and Powell 2007a; Apaldetti et al. 2018; see Chap. 3).

In between very early forms and Sauropodiformes are the so-called "core prosauropods" (Sereno 2007), or non-sauropodiform plateosaurians, a paraphyletic group that radiated during the Norian–Rhaetian and comprises forms with rather uniform proportions and body plan. They include the oldest early sauropodomorph families: Plateosauridae and Massospondylidae. Nowadays, this group incorporates most of the taxa that were traditionally included in the monophyletic "Prosauropoda" (Sereno 2007; Upchurch et al. 2007a; Apaldetti et al. 2014; Müller 2019; Peyre de Fabrègues et al. 2020; McPhee et al. 2020). Non-sauropodiform plateosaurians are found mostly in South America (6 genera), but also present in China (4 genera), followed by Southern Africa (3 genera), United States (3 genera), and Europe (1 genus). "Core prosauropods" represent an important evolutionary step because they started the early sauropodomorph diversification during the Late Triassic in terms of abundance, taxonomic and morphological diversity. In this regard, this group encompassed morphological changes that departed from the plesiomorphic saurischian body plan, mostly still retained by the very early forms such as *Eoraptor* and *Buriolestes*, from Carnian (see Chap. 1). After the Carnian, first steps toward herbivory, increasing body mass and the establishment of the typical "prosauropod" manus occurred during the ascendancy of "core prosauropods" (Pol et al. 2021).

The South American record of early sauropodomorphs well exemplifies the aforementioned transformations. That is even more true currently that they represent the majority (almost 30%) of taxa usually phylogenetically recovered as "core prosauropods". These taxa enable a better understanding of the morphological steps and evolutionary scenario that set the basis for the great diversification of this group that occurred after the Carnian. The occurrence of the South American taxa and the milestones of such transformations are presented in this chapter.

Institutional Abbreviations BPI: Evolutionary Studies Institute (ESI, formerly Bernard Price Institute), Johannesburg, South Africa; **CAPPA/UFSM:** Centro de Apoio à Pesquisa Paleontológica da Quarta Colônia da Universidade Federal de Santa Maria, São João do Polêsine, Brazil; **CPSGM**: Collections Paléontologiques du Service géologique du Maroc, direction de la Geologie, ministère de l'Énergie et des Mines, Rabat, Morocco; **FMNH**: Field Museum of Natural History, Chicago, USA; **GPIT**: Institut und Museum für Geologie und Paläontologie, Universität Tübingen, Tübingen, Germany; **IVPP**: Institute of Vertebrate Paleontology and Paleoanthropology, Beijing, People's Republic of China; **MB**: Institut für Paläontologie, Museum für Naturkunde, Humbolt-Universität, Berlin, Germany; **MCP**: Museu de Ciencias e Tecnologia PUCS, Porto Alegre, Brazil; **MLP**: Museo de La Plata, La Plata, Argentina; **NGMJ**: Nanjing Geological Museum, Nanjing, People's Republic of China; **NMQR**: National Museum, Bloemfontein, South Africa; **PVL**, Instituto "Miguel Lillo", San Miguel de Tucuman, Argentina; **PVSJ-UNSJ**: Paleontología de Vertebrados - Museo de Ciencias Naturales, Universidad Nacional de San Juan, San Juan, Argentina; **PULR**: Colección paleontológica del Museo de Ciencias Naturales de la Universidad Nacional de La Rioja, La Rioja, Argentina; **SAM:** Iziko South African Museum, Cape Town, South Africa; **UFSM**: Universidade Federal de Santa Maria, Santa Maria, Brazil.

2 Methods

2.1 Maturity Assessment

When possible, we provided an estimation of maturity (adult versus non-adult) for each specimen, based on the different criteria recognized thus far for assessing maturity in saurians in general and dinosaurs in particular (Hone et al. 2016; Griffin et al. 2020). Although there is not a single unambiguous method for assessing maturity in extant archosaurs, we followed the recommendations of Griffin et al. (2020) in using a set of methods that can be reasonably used in Sauropodomorpha, but with caveats, as follows: degree of cranial ossification, orbit size and proportions, snout elongation and cranial proportions, long bone histology, postcranial fusion, and co-ossification events, neurocentral suture fusion, vertebral morphology, and degree of postcranial ossification. Since assessing maturity is not a stated objective of this contribution, we only evaluate the ontogenetic stage in a qualitative fashion.

2.2 Body Mass Estimation

Analyses were carried out through the free software RStudio version 1.3.1093. Estimation of body mass was performed using the quadratic equation provided by Campione (2017), which utilizes the minimum shaft circumference of the femur (for bipedal taxa), whereas the limb comparisons through a three-variable graph were performed with Ggplot2 package (Wickham 2009).

2.3 Phylogenetic Relationships

The phylogenetic relationships scheme is taken from Pol et al. (2021), in which *Plateosaurus trossingensis* here replaced the no longer valid *P. engelhardti*, following the Case 3560 of the ICZN (2019; Galton 2012). In the same way, *P. "ingens"* is here denoted with single quotes denoting that this species is no longer valid (ICZN 2019).

3 Systematic Paleontology

Taxa below are given in chronological order, from Late Triassic to Early Jurassic. Table 1 summarizes taxon definitions used in this contribution. Table 2 depicts the record of non-sauropodomorph plateosaurians from South America, including relevant information on their occurrence.

Table 1 Phylogenetic definition of taxa included in this chapter

Taxon	Definition	Author
Sauropodomorpha Huene (1932)	The most inclusive clade containing *Saltasaurus* but not *Passer*, *Triceratops*, or *Herrerasaurus* (stem)	Baron et al. (2017)
Plateosauria Tornier (1913)	*Plateosaurus*, *Massospondylus*, their most recent common ancestor, and all descendants (node)	Sereno (1998)
Unaysauridae Müller et al. 2018a	The most inclusive clade including *Unaysaurus*, but neither *Plateosaurus* nor *Saltasaurus* (stem)	Müller et al. (2018a)
Riojasauridae Yates (2007a)	The most inclusive clade containing *Riojasaurus* but not *Plateosaurus*, *Massospondylus*, or *Anchisaurus* (stem)	Yates (2007a)
Massospondylidae Owen (1854)	All plateosaurians closer to *Massospondylus* than to *Plateosaurus* (stem)	Sereno (1998)
Sauropoda Marsh (1878)	The most inclusive clade containing *Saltasaurus* but not *Melanorosaurus* (stem)	Yates (2007a)

Dinosauria Owen 1842

Saurischia Seeley 1887

Sauropodomorpha Huene 1932

Plateosauria Tornier 1913

Definition "*Plateosaurus engelhardti*" von Meyer (1837) (*P. trossingensis*, see ICZN 2019, Nau et al. 2020; Lallensack et al. 2021), *Massospondylus carinatus* Owen 1854, their most recent common ancestor and all descendants (node-based taxon, Sereno 1998).

Comments The original node-based definition provided by Sereno (1998:63; see also Upchurch et al. 2007a:64) included *Plateosaurus* and *Massospondylus* in a phylogenetic context of a monophyletic Prosauropoda to encompass the two major and well-established families of prosauropod dinosaurs within a node-stem triplet (see Sereno 1998:52). In a similar phylogenetic context, Galton and Upchurch (2004:251) used *Jingshanosaurus* instead of *Massospondylus* for their definition, as the successive taxa were added to the stem supporting Plateosauria, whereas Sereno (2007:284) defined Plateosauria using the original internal specifiers (i.e., *Plateosaurus* and *Massospondylus*), adding an external specifier, *Saltasaurus*. All these definitions of Plateosauria have the same intention of defining a clade within a broad monophyletic Prosauropoda. To date, within the current paraphyletic scheme, such definitions are not applicable.

The german taxon *Ruehleia* could be used as a third specifier for Plateosauria since it is rather complete, include adult specimens (Galton and Upchurch 2004), and exhibits a fairly constant phylogenetic position (e.g., Müller 2019; McPhee et al. 2020; Peyre de Fabrègues et al. 2020; Pol et al. 2021). However, an exhaustive redescription of all the material would be first needed.

Unaysauridae Müller, Langer and Dias da Silva 2018a

Definition The most inclusive clade including *Unaysaurus tolentinoi* Leal et al. 2004, but neither "*Plateosaurus engelhardti*" von Meyer 1837 nor *Saltasaurus loricatus* Bonaparte and Powell 1980 (stem-based taxon, Müller et al. 2018a).

Diagnosis Unaysauridae was originally diagnosed as having a well-developed anterior expansion of the medial condyle of the astragalus and by the presence of a promaxillary fenestra (Müller et al. 2018a).

Comments In the original paper erecting Unaysauridae, *Jaklapallisaurus asymmetrica* was recovered as a member of this clade (see also Müller 2019). Hence, in the diagnosis of Unaysauridae, Müller et al. (2018a) included the presence of a promaxillary fenestra with the caveat that *J. asymmetrica* lacks cranial materials, thus remaining the anterior expansion of the medial condyle of the astragalus as the only putative unambiguous autapomorphy of Unaysauridae, as present in *Macrocollum*, *Unaysaurus*, and *Jaklapallisaurus*. *Unaysaurus* and *Macrocollum* were retrieved as sister taxa in two different datasets after the latter taxon was erected: the ones derived from Müller et al. 2018a (Müller 2019; Müller et al. 2018a, 2021) and that of Pol et al. (2021). *Jaklapallisaurus*, on the other hand, was only retrieved as a member of Unaysauridae in Müller's datasets. Since the Indian taxon seems to be problematic, with no clear phylogenetic position, future phylogenetic analyses confirming the inclusion or exclusion of *Jaklapallisaurus* from Unaysauridae will ultimately settle if the extreme anterior expansion of the astragalus constitutes a unique feature of Unaysauridae or a more widely spread character. If *Jaklapallisaurus* does not constitute a monophyletic group with *Macrocollum* and *Unaysaurus*, then the presence of a promaxillary fenestra will constitute the unique autapomorphy of Unaysauridae, plus other characters that optimize unambiguously for Unaysauridae: presence of prezygoparapophyseal laminae in trunk vertebrae and first phalanx of the pedal digit I 2.4 times longer (or more) than proximally high.

Unaysaurus Leal, Azevedo, Kellner, Da Rosa 2004

Unaysaurus tolentinoi Leal, Azevedo, Kellner, Da Rosa 2004

Holotype UFSM11069 (Fig. 2d–f), a single individual consisting of a partial skeleton (40% complete), including a skull and lower jaw, axis, dorsal vertebrae, ribs and gastralia, caudal vertebrae and some chevrons, scapulae, coracoid, humeri, ulnae, and radii, metacarpals, and near-complete right manus, incomplete tibiae, right astragalus, left metatarsals and some pedal elements (Leal et al. 2004:3; McPhee et al. 2020).

Locality, Horizon and Age Água Negra district, São Martinho da Serra, located 13 km north of Santa Maria, Rio Grande do Sul, Brazil (Fig. 1). The specimen was recovered from a reddish conglomerate, situated above a brownish to reddish shale with intercalated siltstone lenses, Caturrita Formation (Andreis et al. 1980). U–Pb radioisotopic age on a single zircon from this unit provided an early Norian maximum age of 225.42 ± 0.37 Ma (Langer et al. 2018).

Diagnosis The original diagnosis of *Unaysaurus* was modified by McPhee et al. (2020: 3) as follows (autapomorphies denoted with an asterisk): (1) a strongly developed eminence on the medial rim of the supratemporal fossa, at the frontal-parietal

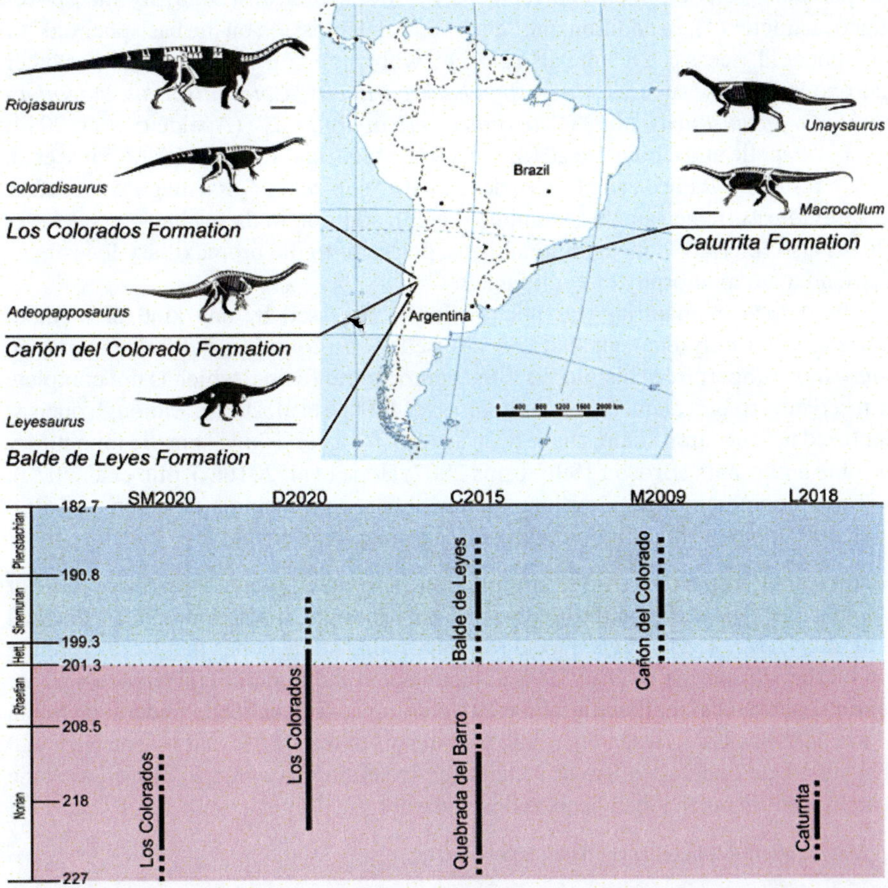

Fig. 1 Map of South America showing the geographic distribution of non-sauropodiform plateosaurians and temporal correlation of bearing formations. Scale bar equals 50 cm. Abbreviations: C2015, Colombi et al. (2015), D2020, Desojo et al. (2020), Hett., Hettangian, L2018, Langer et al. (2018), M2009, Martínez (2009), SM2020, Santi Malnis et al. (2020). Dashed lines indicate uncertain age limits (Silhouettes modified from; Paul 2010; Martínez 2009; Apaldetti et al. (2011; Müller and Garcia 2020; Pol et al. 2021)

juncture, that protrudes laterally over the fossa*; (2) a "sub-recess" at the juncture of the basisphenoid and basioccipital, between the basal tubera pedicles of the basisphenoid*; (3) fine, bifurcating ridges housed within the Meckelian groove of the dentary, delimiting two small fossae/foramina*.

Comments The original description of *Unaysaurus* provided by Leal et al. (2004) included a more expanded unique combination of characters than the one provided by McPhee et al. (2020). After such a revision, some of those original features, as stated by Leal et al. (2004), were recognized to be present in a wide array of taxa. In this regard, the presence of "well-developed laterodorsally oriented process formed by frontal and parietal and medial depression on mediodorsal surface of the parietals" was clarified and replaced by McPhee et al. (2020) by the above-stated character (1). In addition, the "deep ventral depression on the basisphenoid" or basisphenoid recess is a feature also present, with more or less development, in other sauropodomorphs, such as *Massospondylus carinatus*, *Coloradisaurus*, *Adeopapposaurus*, *Sarahsaurus,* and "*Plateosaurus erlenbergiensis*" (Apaldetti et al. 2014: Fig. 6; Chapelle and Choiniere 2018: Fig. 39A; Marsh and Rowe 2018; McPhee et al. 2020). Instead, McPhee et al. (2020) described a "sub-recess" as a diagnostic feature of *Unaysaurus* in the same basicranium area. In addition to these characters, Müller (2019) also referred to the additional fossa posterior to the promaxillary fenestra as a potential autapomorphy of *Unaysaurus.*

The degree of maturity was not previously assessed by any kind of method. Currently, the only element that can be used as a proxy for skeletal maturity in *Unaysaurus* (apart from histology) is the neurocentral fusion, which is not complete or fused in vertebral elements of *Unaysaurus* (McPhee et al. 2020). Although open or not fused neurocentral sutures have been a proxy for establishing somatic immaturity in some archosaurs (Brochu 1996; Irmis 2007; Hone et al. 2016; Griffin et al. 2020), it is not recommendable to evaluate maturity based on a single parameter (Griffin et al. 2020); hence, maturity in *Unaysaurus* remains not conclusive.

Historical Perspective Little information is known about the discovery of *Unaysaurus.* The skeleton of the type and only individual was found in 1998 (Leal et al. 2004). Mr. T. Marafiga, a local resident, found some bones that were cropping out on the side of the road in 1998 and contacted the Federal University of Santa Maria (Laboratório de Estratigrafia e Paleobiologia, Universidade Federal de Santa Maria, UFSM). Later, researchers and technicians of the UFSM and Museu Nacional (Universidade Federal do Rio de Janeiro) collected in situ remains, as well as some scattered elements that had been removed earlier.

Macrocollum Müller, Langer, Dias da Silva 2018a

Macrocollum itaquii Müller, Langer, Dias da Silva 2018a

Holotype CAPPA/UFSM 0001a (Fig. 2a), an almost complete and articulated skeleton (Müller et al. 2018a: 2).

Paratypes CAPPA/UFSM 0001b (Fig. 2b), an almost complete and partially articulated skeleton; CAPPA/UFSM 0001c (Fig. 2c), an articulated skeleton lacking skull and cervical series skeleton (Müller et al. 2018a: 2).

Referred Specimen CAPPA/UFSM 0001d, a partial skull found in close association with the holotype and paratypes (Müller 2019: 5).

Locality, Horizon and Age Wachholz site at Agudo, Rio Grande do Sul, Brazil; upper portion of the Candelária Sequence, Caturrita Formation, Paraná Basin (Fig. 1). Specimens come from massive fine sandstones at the base of the outcrop, above a carbonate concretion level. Architectural elements allow these lithologies to be attributed to crevasse splays, channel sandbar deposition, and further biogenic colonization, respectively (Müller et al. 2015). Stratigraphically correlated beds from a nearby site were dated as early Norian (ca 225.42 + 0.37), Late Triassic (Langer et al. 2018).

Diagnosis *Macrocollum* presents a unique combination of characters: antorbital fossa perforated by a promaxillary fenestra; medial margin of the supratemporal fossa with a simple smooth curve at the frontal/parietal suture; proximal width of metacarpal I is <65% of its length; iliac acetabulum not fully open, dorsally concave, but with a well-developed medial wall; ischiadic longitudinal groove not reaching the posterior half of the ischium; absence of a trochanteric shelf on the femur; medial condyle of the femur sub-rectangular in distal view; proximal end of metatarsal II with a straight medial margin (Müller et al. 2018a: 2).

Comments *Macrocollum* lacks unique diagnostic features (i.e., autapomorphies), but instead presents a unique combination of characters. It shares with *Unaysaurus* the presence of a promaxillary fenestra and, considering the phylogenetic closeness between both taxa, we believe it is important to pinpoint relevant morphological differences among them. One of these differences is the presence of the well-developed eminence on the medial rim of the supratemporal fossa, at the frontal-parietal juncture, present in *Unaysaurus* (McPhee et al. 2020), but that is absent in *Macrocollum* (Müller 2019). In addition, *Unaysaurus* presents a depression posterior to the promaxillary fenestra, which is also absent in *M. itaquii* (Müller 2019). Finally, despite the fact that both *Unaysaurus* and *Macrocollum* share the presence of a well-developed anterior expansion of the medial condyle of the astragalus (more than 30% of the anteroposterior width of the astragalar body measured at the lateral side), the expansion of *Macrocollum* represents 50% of the anteroposterior width of the astragalar body, whereas the expansion of *Unaysaurus* represents about 35% (Fig. 2a–f).

According to Müller (2019) the specimens CAPPA/UFSM 0001a/b/d are not fully-grown individuals, based on the presence of some visible neurocentral sutures in some of the trunk vertebrae (Brochu 1996; Irmis 2007), although this proxy has been demonstrated that cannot be used in isolation (Griffin et al. 2020). Orbit size, for example, is another method that can be used in combination with sauropodomorphs (Griffin et al. 2020). Though no allometry can be tested because of the lack of ontogenetic series for *Macrocollum*, if compared to the early mature material of

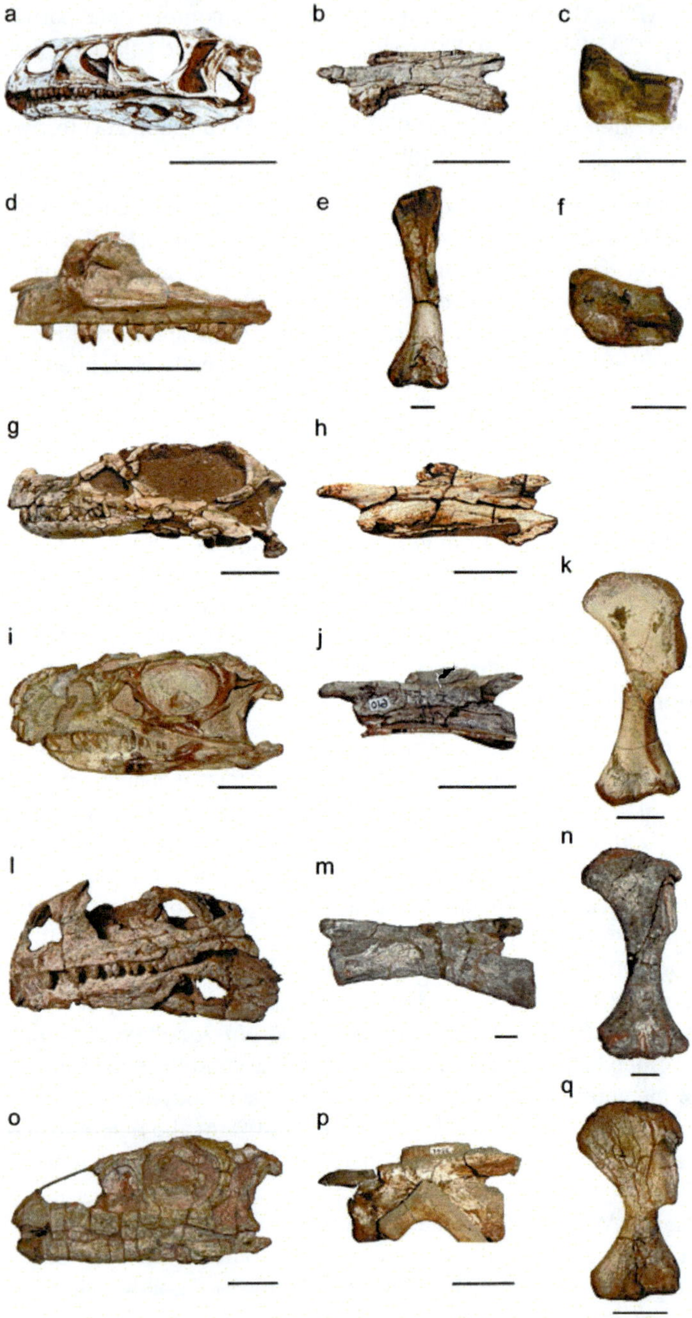
a
b
c
d
e
f
g
h
i
j
k
l
m
n
o
p
q

◂**Fig. 2** Selected bones of non-sauropodiform plateosaurians of South America. All skulls and cervical vertebrae are presented in left lateral view, whereas all humeri are presented in anterior view. **a–c**, *Macrocollum itaquii*. **a** Skull (CAPPA/UFSM 0001a). **b** Cervical vertebra 4 (CAPPA/UFSM 0001b). **c** Right astragalus (CAPPA/UFSM 0001c) in dorsal view. Scale bars equal 5 cm. **d–f**, *Unaysaurus tolentinoi* (UFSM 11,069). **d** Left maxilla. **e** Right humerus. **f** Right astragalus in dorsal view. Scale bars equal 3 cm (**d**) and 2 cm (**e, f**). **g–h** *Leyesaurus marayensis* (PVSJ 706). **g** Skull. **h** Cervical vertebra 6. Scale bars equal 3 cm. **i–k**, *Adeopapposaurus mognai* (PVSJ 610). **i** Skull. **j** Cervical vertebra 4. **k** Left humerus. Scale bars equal 3 cm. **l–n**, *Coloradisaurus brevis*. **l** Skull (PVL 3967). **m** Cervical vertebra 4? (PVL 5904). **n** Right humerus (PVL 5904). Scale bars equal 2 cm. **o–q** *Riojasaurus incertus*. **o** Referred skull (PULR 56). **p** Cervical vertebra 6 (PVL 3808). **q** Right humerus (PVL 3808). Scale bars equal 5 cm (o, q) and 10 cm (p). Elements in e, g, h, i, j, l, m, n, o, and q are mirrored

Mussaurus (Pol and Powell 2007b; Otero and Pol in press), it can be seen that available material of the former species does not correspond to a juvenile specimen, but only paleohistological studies will assess if somatic maturity in *Macrocollum* was reached at the time of death.

Historical Perspective The four specimens of *Macrocollum* were discovered and excavated by a Brazilian team prior to 2018 (Müller et al. 2018a).

Riojasauridae Yates 2007a

Definition The most inclusive clade containing *Riojasaurus incertus* Bonaparte 1969 but not "*Plateosaurus engelhardti*" von Meyer (1837), *Massospondylus carinatus* Owen (1854), or *Anchisaurus polyzelus* Hitchcock (1865) (stem-based taxon, Yates 2007a).

Riojasaurus Bonaparte 1967

Riojasaurus incertus Bonaparte 1967

Holotype PVL 3808 (Fig. 2p–q), incomplete skeleton which includes cervical and most dorsal vertebrae, sacrum, partial forelimb, partial pelvis, and hindlimb (Bonaparte 1971: 130).

Referred Specimens PULR 56 (Fig. 2o), almost complete skeleton, including skull and mandibles (Bonaparte and Pumares 1995: 342).

Comments Apart from the holotype, the original description of *Riojasaurus* provided by Bonaparte (1971) includes almost 20 specimens, most of them highly incomplete. Additional putative materials referred to this genus have been reported from the Quebrada del Barro Formation in San Juan Province (Bossi and Bonaparte 1978; Martínez et al. 2015). Although all these materials share similarities and a common heavy construction, some authors (McPhee and Choiniere 2017) questioned their affinities, particularly the juvenile specimen PULR 56 described by Bonaparte and Pumares (1995). Considering the plethora of specimens existing for this genus hitherto and the fact that no descriptions have been published since Bonaparte's contributions, a complete review of this taxon is needed and pending.

Locality, Horizon and Age The holotype comes from "Quebrada de los Jachaleros" locality, whereas the referred specimen comes from the "La Esquina" locality (Bonaparte 1971; Bonaparte and Pumares 1995), Talampaya National Park, near Villa Union town, La Rioja Province, Argentina (Fig. 1). The sequence is characterized by sandstone and mudstone, deposited by moderately sinuous fluvial systems. Crevasse splays in overbank settings characterize the top of the sequence, in which fossils were deposited (Caselli et al. 2001; Santi Malnis et al. 2020). Specimens come from the uppermost third of the Los Colorados Formation, traditionally regarded as Norian–Rhaetian, based on vertebrate biostratigraphy (Bonaparte 1982; Arcucci et al. 2004), although paleomagnetic data recently retrieved an age range no younger than 213 Ma (mid-late Norian, Kent et al. 2014).

Emended Diagnosis The following diagnosis is based on a unique combination of characters provided by Bonaparte and Pumares (1995: 342–343), emended from the original provided by Bonaparte (1971), although extremely vague characters (such as "pectoral girdle similar to that of *Plateosaurus* and *Lufengosaurus*") were omitted here. In addition, we included the autapomorphies listed by Galton and Upchurch (2004), marked with an asterisk: skull with broad antorbital fossa (anteroposterior length being 60% of the anteroposterior length of the orbit) and a sub-vertical lachrymal; rostral prominence on the premaxilla*; anteroposteriorly long maxilla (being 85% of the length of the dentary) including 24 sub-cylindrical teeth; alveolar row at the same level than the mandibular articulation; premaxilla and dentary bearing crests; a low, rounded lacrimal-prefrontal crest*; anteroventral surface of the dentary with a ventral projection (similar to that present in *Mussaurus*); an expanded distal end to the basipterygoid process*; robust cervical vertebrae lacking ventral keel; neural spines relatively short anteroposteriorly in cervicals 7–10; tall dorsal vertebrae bearing well-developed hyposphene-hypantrum complex and dorsoventrally low neural spines; three sacral vertebrae with the anteriormost two deeply fused; pelvic girdle with relatively anteroposteriorly short preacetabular process and dorsoventrally tall acetabulum; robust humerus (robustness index RI sensu Wilson and Upchurch 2003: 0.37) with transversely broad deltopectoral crest; humerus plus radius represent 67% of the length of the femur plus tibia; humerus is 80% of the length of the femur; tibia plus astragalus are 90% of the length of the femur.

Comments The antorbital fossa of PULR 56 (60% of the orbit) is proportionally larger than in other massopodans, such as *Leyesaurus* and *Adeopapposaurus* (PVSJ 610) but equal or smaller than in *Macrocollum* (Müller 2019), "*Plateosaurus erlenbergiensis*" (AMNH FARB 6810), *Massospondylus carinatus* (BP/1/5241) and *Lufengosaurus* (IVPP-V15). The anteroposterior length of the maxilla of PULR 56 is about 50% the length of the skull, which concurs with the values present in *M. carinatus* (BP/1/5241) and "*P. erlenbergiensis*", differing from the proportionally shorter maxilla present in *Adeopapposaurus, Leyesaurus,* and *Macrocollum.* The dentary alveolar row placed at the same level as the mandibular articulation is a shared feature with *Leyesaurus* and *Adeopapposaurus*, but different from the lower articulation reported in *Macrocollum, Plateosaurus* (AMNH FARB 6810, MB.R. 1937), and *Lufengosaurus*. Galton and Upchurch (2004:251) also listed "an expanded distal end to the basipterygoid process" as an autapomorphic feature of *Riojasaurus*. However, since they did not specify on which of the multiple specimens it was based on, we tentatively exclude this character from the emended diagnosis.

According to Bonaparte and Pumares (1995), PVL 3808 is about 35% larger than PULR 56, being the latter specimen conceived as a non-adult individual (McPhee and Choiniere 2017). However, neither of the specimens have been subjected to histological analyses; hence, information about sexual maturity is not accessible (but see Cerda et al. 2017 for analysis on other putative referred specimens).

Historical Perspective The first vertebrate remains from the Los Colorados Formation were uncovered by an Argentinean paleontological commission in 1960. Following these first discoveries, the Miguel Lillo Foundation, part of the National University of Tucuman (Argentina), carried out a series of fieldwork campaigns. The first rests attributed to *Riojasaurus* were discovered as early as 1962, but the bulk of the material was uncovered during the 1970 campaign and published the subsequent year (Bonaparte 1971). The following decade, a paleontological exploration project run by the Argentinean Museo Argentino de Ciencias Naturales together with the provincial University of La Rioja led to the discovery, in 1986, of an almost complete immature skeleton (Bonaparte and Pumares 1995, see comments above for referral issues).

Massospondylidae Huene 1914

Definition A stem-based taxon encompassing all plateosaurians closer to *Massospondylus carinatus* Owen 1854 than to *"Plateosaurus engelhardti"* von Meyer 1837 (Sereno 1998: 63).

Comments Although the definition of Massospondylidae remains unaltered since that provided by Sereno (1998), the taxonomic content of this clade may vary, according to different authors (see Discussion).

Coloradisaurus Galton 1990

Coloradisaurus brevis (Bonaparte 1978)

Holotype PVL 3967 (Fig. 2l), individual consisting of cranial and postcranial remains. It includes skull and mandible found in articulation, atlas-axis complex and third cervical vertebra, three other cervical vertebrae (possibly fourth to six), two anterior dorsal vertebrae, and three isolated caudal vertebrae; fragmentary remains of the scapula, coracoid, humerus, ulna, and radius, incomplete metacarpus and isolated manual non-terminal and ungual phalanges, right astragalus and calcaneum, distal tarsal and pedal unguals (Apaldetti et al. 2014: 1115).

Referred Specimen PVL 5904 (Fig. 2m, n), a partially articulated postcranial skeleton, which includes a single anterior cervical vertebra (third or fourth), an almost complete dorsal series (from second to thirteenth element), an isolated anterior caudal vertebra, and two fused middle caudal vertebrae. Appendicular material includes complete left scapula and coracoid, isolated right coracoid, right humerus, left metacarpal III, partial left ilium, articulated pubes, right ischium, both femora, both tibiae, proximal third of the left fibula, left astragalus, a medial left distal tarsal and complete left pes (Apaldetti et al. 2014:1115).

Locality, Horizon, and Age Both the holotype (PVL 3967) and the referred specimen (PVL 5904) were found by José Bonaparte at "La Esquina" locality, Ischigualasto–Villa Union basin, Talampaya National Park, La Rioja Province, Argentina (Bonaparte 1978) (Fig. 1). *Coloradisaurus*, as well as the majority of the vertebrate fauna recorded from there, were found toward the upper third of Los Colorados Formation, characterized by fossil-bearing sandstones and mudstones deposited in multistory and multilateral channelized bodies interspersed with thick floodplain deposits developed under seasonal semi-humid to semi-arid climate conditions (Caselli et al. 2001; Santi Malnis et al. 2020). The uppermost section of Los Colorados Formation was traditionally regarded as Norian based on vertebrate biostratigraphy (Bonaparte 1982, Arcucci et al. 2004; Martínez et al. 2015). In accordance with this, paleomagnetic data show that the age of the Los Colorados Formation ranges between 227 and 213 Ma, assigning an age no younger than 213 Ma to the La Esquina fauna (mid-late Norian, Kent et al. 2014). In addition, recent U–Pb dates of the uppermost Ischigualasto Formation (221.8 ± 0.1 Ma) indicate that the age of Los Colorados Formation could be several million years younger than previously thought (Desojo et al. 2020).

Diagnosis The diagnosis of *Coloradisaurus* was revised by Apaldetti et al. (2013a, 2014) as follows (autapomorphies denoted with an asterisk): a medium-sized basal sauropodomorph. Preorbital region shorter than in *Lufengosaurus* and *Plateosaurus*, with frontals longer than nasals, as reported in *Adeopapposaurus* (PVSJ 610); posteroventral corner of antorbital fossa projecting underneath the orbit and broad ascending process of the maxilla, both characters also present in *Ngwevu* (Chapelle et al. 2019: Fig. 1); circular supratemporal fenestrae*; tall and thin bony laminae forming the ventrolateral margins of a deep, ventrally opening fossa on the basisphenoid*; triangular bony laminae extend from the basipterygoid processes to the ventrolateral margins of the parasphenoid rostrum*; low mandibular articulation, reported also in *Sarahsaurus* (Rowe et al. 2010:Fig. 4A) and *Ngwevu* (Chapelle et al. 2019: Fig. 1); tab-like medial process on the medial border of the retroarticular process of the mandible*; low and elongated cervical vertebrae; deep depression with a sharply defined distal margin facing proximally on anterior surface of pubic apron*; semicircular cross-section of pubic apron, creating a deep trough along posterior surface of pubic apron*; posteromedial region of the distal surface of the tibia deflected and facing posterodistally*; and pyramidal dorsal process of the posteromedial corner of astragalus almost as high as ascending process* (Apaldetti et al. 2014: 1115).

Comments The authorship of the genus *Coloradisaurus* has traditionally been attributed to Lambert (1983). Greenfield et al. (2020) demonstrated that Lambert's work does not meet the requirements of Articles 13.1 and 13.3 of the International Code of Zoological Nomenclature, the name, therefore, being a *nomen nudum* in that publication. Thus, the valid authorship of the genus is *Coloradisaurus* (Galton 1990), which was made available a few years after the work of Lambert.

Despite the fact that the specimen PVL 5904 is more than twice as large as the holotype (PVL 3967) and also exhibits complete neurocentral suture closure in all the preserved vertebrae, the absence of a distinct outer circumferential layer (OCL) suggests that somatic maturity was not reached in PVL 5904 (Cerda et al. 2017). Thus, postcranial autapomorphies for *Coloradisaurus* should be regarded with caution.

Historical Perspective The first remains of *Coloradisaurus* were uncovered in October 1971 during a fieldwork campaign organized by the Miguel Lillo Foundation, National University of Tucuman (Argentina) and funded by the CONICET (Bonaparte 1978). Two years later, in 1973, Bonaparte discovered the referred postcranial skeleton (PVL 5904; Apaldetti et al. 2013a).

Adeopapposaurus Martínez 2009

Adeopapposaurus mognai Martínez 2009

Holotype PVSJ 610 (Fig. 2i–k), a fairly complete skeleton including complete skull with mandible in occlusion; complete cervical, dorsal, and sacral vertebral series; nine articulated caudal vertebrae with corresponding chevrons, pectoral girdle, forelimbs, incomplete pelvic girdle, and incomplete hindlimbs (Martínez 2009: 143).

Hypodigm PVSJ 568, individual including skull and postcranial remains. A partial skeleton consisting of partially articulated skull; atlas-axis complex; four articulated caudal vertebrae (presumably tenth to thirteenth), and fragments of indeterminate vertebral bodies; distal end of right femur; tibia, fibula, astragalus, calcaneum, distal tarsals III and IV, and articulated right foot. PVSJ 569, individual smaller than the holotype consisting of a partial skeleton. It includes second and third sacral vertebrae with no neural spines; six sacral ribs; sequence of first 46 caudal vertebrae, articulated; incomplete pelvis, lacking part of right ilium and pubis; hindlimbs lacking right tibia and the full complement of right phalanges. PVSJ 570, incomplete individual consisting of right and left humeri, left with only proximal posterior border, and centrum of second sacral vertebra (Martínez 2009: 143).

Locality, Horizon, and Age The holotype (PVSJ 610) and specimens PVSJ 568, PVSJ 569, and PVSJ 570 were found in the "Sierra de Mogna" locality, 70 km north of San Juan City, San Juan Province, Argentina (Martínez 2009) (Fig. 1). All specimens come from the Lower Jurassic Middle Member of the Cañón del Colorado Formation, which is composed of fine facies of sandstones and red siltstones with some conglomeratic intercalations (Martínez and Colombi 2011). The Upper Pelitic Member facies of this member yielded the *Adeopapposaurus* specimens. The holotype specimen, PVSJ 610, as well as PVSJ 570 and PVSJ 568 were found 148 m from the base of the formation, while PVSJ 569 was found 10 m below the first site (Martínez 2009: 143).

Comments The Lower Jurassic assignment for the Cañón del Colorado Formation is based on the phylogenetic proximity between *Adeopapposaurus* and the South African genus *Massospondylus*. The latter taxon is constrained to Lower Jurassic in all localities where it was found (e.g., Lesotho, Knoll 2005; South Africa, Owen

1854; Bordy et al. 2020; Viglietti et al. 2020; North America, Attridge et al. 1985; Zimbabwe, Cooper 1981).

Emended Diagnosis *Adeopapposaurus* was diagnosed by Martínez (2009) listing a unique combination of characters, but with no distinction of autapomorphies. Here, we denoted characters we consider autapomorphies with an asterisk: sauropodomorph with a series of large foramina in a sub-vertical row on the lateral surface of the premaxilla; strongly rugose depression bordered by a protuberance with a series of foramina in a sub-vertical row on the lateral surface of the anterior end of the dentary; 11 anteroposteriorly elongated cervical vertebrae and 13 dorsal vertebrae with neural arches taller than the respective centra* (Martínez 2009:143). Apart from these original features proposed by Martínez (2009), *Adeopapposaurus* presents a noteworthy development of the dorsoventral depth of the anterior ramus of the maxilla, which is 1.9 times the dorsoventral depth of the posterior ramus of the maxilla*. This value is considerably higher than those reported for other related taxa, such as *Leyesaurus* and *Coloradisaurus* (1), *Ngwevu* (1.25), *Lufengosaurus* (1.4), and *Macrocollum* (1.5). Only *M. carinatus* (BP/1/5241) presents a higher value than *Adeopapposaurus*, surpassing 2. In addition, the maxilla of *Adeopapposaurus* is notably short anteroposteriorly, being only 35% of the total length of the skull in PVSJ 610. Other early sauropodomorphs present values above 40% (*Leyesaurus, Macrocollum Massospondylus, Plateosaurus*).

Comments The presence of foramina in the anterior aspect of the premaxilla, arranged in a sub-vertical fashion, does not appear to be unique for *Adeopapposaurus*, since it is also reported in the massospondylid *Leyesaurus* (Apaldetti et al. 2011: Fig. 3); *M. carinatus* (BP/1/5241) displays three large foramina vertically arranged also in the anterior surface of the right side of the premaxilla, but scattered on the left side (Chapelle and Choiniere 2018: Fig. 5). Hence, it is also plausible that the arrangement of the foramina of the premaxilla does not follow uniform patterns on either side of the rostrum.

Historical Perspective The first unearthed specimen of *Adeopapposaurus* (PVSJ 568) was discovered by Ricardo Martínez in 1994 during an exploratory field trip to Sierra de Mogna locality. This specimen was the core subject of the docthoral thesis of R. Martínez. The specimen PVSJ 610 was discovered by Angel Praderio, former PhD student of R. Martínez, and designed posteriorly as the holotype of the species (Martínez pers. comm).

The specimen PVSJ 569, which is about 77% the size of the largest specimen (PVSJ 610), has been subjected to histological analyses, showing the absence of an OCL, but a distinct increase in the intrinsic fiber arrangement in the outer cortex, which suggests that sexual maturity may have already been attained (Cerda et al. 2017). In spite of this, specimen PVSJ 569 presents fully open neurocentral sutures of the sacral and anterior caudal vertebrae, and the transverse processes and sacral ribs are also unfused, which might be indicative, but not conclusive, of somatic immaturity (Hone et al. 2016; Griffin et al. 2020).

Leyesaurus Apaldetti, Martínez, Alcober, Pol 2011

Leyesaurus marayensis Apaldetti, Martínez, Alcober, Pol 2011

Holotype PVSJ 706 (Fig. 2g–h), individual consisting of a partial skeleton including partial skull with articulated mandible, but lacking both nasals, left prefrontal, middle section of the left maxilla, anterior half of the left lower jaw, supraoccipital, both exoccipitals, ophistotics, laterosphenoids, and vomers; atlas-axis complex articulated with anterior cervical vertebrae (third to seventh); an anterior and a middle caudal vertebra; proximal region of the left scapula, coracoid, and humerus; partial blade of the right pubis lacking distal and proximal ends; proximal region of both ischia; partial left pes that includes distal tarsals III and IV, metatarsal III lacking its distal end, complete metatarsals IV and V, first phalanx of digit I, second phalanx of digit II, and second phalanx of digit IV (Apaldetti et al. 2011: 3).

Referred Specimen PVSJ 1079, composed of several postcranial remains, including axial and appendicular ones (Cerda et al. 2017).

Comments PVSJ 1079 comes from the same monospecific level of PVSJ 706 (Balde de Leyes Formation, Marayes-El Carrizal Basin, Northwestern Argentina) and shares a clear phylogenetic signal with the holotype material albeit not sharing similar bone elements with autapomorphic features. Some of those features include presence of a projected posterodorsal corner of middle dorsal neural spines, concave lateral margins of the pubic apron, presence of a well-developed brevis fossa with sharp margins on the ventral surface of the postacetabular process of the ilium, pyramidal dorsal process on the posteromedial corner of the astragalus (Cerda et al. 2017).

Locality, Horizon, and Age The type specimen was found near the "Balde de Leyes" locality, in Balde de Leyes Formation, overlying the Quebrada del Barro Formation and 2 m below the contact with the Cretaceous unit, Marayes-El Carrizal Basin, Caucete Department of San Juan Province, Northwestern Argentina (Colombi et al. 2015; Martínez et al. 2015) (Fig. 1). The sequence is composed of reddish-brown coarse sandstone and conglomerate channels interlaid with fine clay-rich mudstones characterized by the development of calcisol, interpreted as the source area of a megafan (Colombi et al. 2015; Martínez et al. 2015: 382). The age of the unit is suggested to be Early Jurassic (Apaldetti et al. 2011; Martínez et al. 2015, but see Comments below).

Comments Originally, the age of the Quebrada del Barro Formation was regarded as Norian, upon the basis of the record of an incomplete sauropodomorph specimen (PVL 4087) with affinities with *Riojasaurus*, from the Norian Los Colorados Formation (Bossi and Bonaparte 1978). This material was considered as an indeterminate non-sauropod sauropodomorph (Apaldetti et al. 2011) and is currently lost. Consequently, the latter authors assigned a Lower Jurassic age to the Quebrada del Barro Formation (and as a result, to the overlying Balde de Leyes Formation as well), based on the similarities between *Leyesaurus* and the South African genus *Massospondylus*, recorded in the Lower Jurassic Upper Elliot Formation (Apaldetti et al. 2011). In this regard, the genus *Massospondylus* is recorded in Hettangian–Pliensbachian strata

(Viglietti et al. 2020), which tentatively can be ascribed for *Leyesaurus*, although more precise dating for the Quebrada del Barro Formation is pending.

Emended Diagnosis *Leyesaurus* is diagnosed by the following autapomorphies and combination of characters (asterisks denote autapomorphies): sharply acute angle (50°) formed by the ascending process of the maxilla with the alveolar margin* (see comments below); straight ascending process of the maxilla with a longitudinal ridge on its lateral surface (see comments below); noticeably bulging labial side of the maxillary teeth; neural arches of the cervical vertebrae with sinuous dorsal margin of the neural spines and short epipophyzes extending two-third of the length of the postzygapophyzes; and proximal articular surface of metatarsal III shelf-like and medially deflected* (Apaldetti et al. 2011: 5).

Comments Some characters originally described for *Leyesaurus* deserve discussion. For example, the "sharply acute angle (50°) formed by the ascending process of the maxilla with the alveolar margin" and "straight ascending process of the maxilla with a longitudinal ridge on its lateral surface" should be considered with caution since the whole skull of *Leyesaurus* is dorsoventrally crushed, so it is highly probable that the acute angle of the ascending process relative to the alveolar margin and the shape of the process would be a result of such a crushing. We removed from the original diagnosis the character "greatly elongated cervical vertebra (length/height ratio of sixth cervical centrum more than 5", which was originally stated as an autapomorphy by Apaldetti et al. (2011) since we now know that *Adeopapposaurus* (Martínez 2009) and *Massospondylus* (Barrett et al. 2019) have equivalent ratios.

Regarding maturity, the holotype of *Leyesaurus* (PVSJ 706) presents a skull somewhat distorted by taphonomic factors; hence, it is not possible to assess, for example, the orbit size in a confident way. Cervical vertebrae do not show evidence of open or not fully closed neurocentral sutures; however, no other proxies for assessing maturity are available for this specimen. The referred specimen PVSJ 1079, however, has been retrieved as sexually mature but not yet somatically matures, based on paleohistological studies (Cerda et al. 2017).

Historical Perspective In 2000, Luciano Leyes, a local resident from Balde de Leyes, reported the finding of dinosaur bones near the town. R. Martínez organized a short field trip to locate such bones. After a long search around the location, the reported bones could not be located. Nonetheless, Benito Leyes, Luciano's brother, found other fossil remains, including a skull and some postcranial remains, which turned out to be the specimen PVSJ 706.

4 The South American Record of Plateosauridae, Unaysauridae Riojasauridae and Massospondylidae

4.1 Plateosauridae

Plateosauridae was coined by Marsh (1895: 493) and first conceived as a European family under the "Order" Theropoda, long before the recognition of Prosauropoda and Sauropodomorpha by Huene (1932). After the stem-based phylogenetic definition given by Sereno (1998: 63: "all plateosaurians closer to *Plateosaurus* than to *Massospondylus*") and similar definitions provided by subsequent authors (e.g., Galton and Upchurch 2004: 252; Yates 2007a: 31), Plateosauridae encompassed forms from outside of Europe, including South American taxa, such as *Coloradisaurus* and *Riojasaurus*, from Los Colorados Formation (Upper Triassic), *Mussaurus*, from Laguna Colorada Formation (Lower Jurassic), all from Argentina, and *Unaysaurus*, from the Caturrita Formation (Upper Triassic) of Brazil. The three Argentinean genera were retrieved within this clade only under the monophyletic topology of Prosauropoda (Galton and Upchurch 2004; Upchurch et al. 2007a; see also Bonaparte 1978). *Unaysaurus*, although included inside Plateosauridae in several phylogenetic analyses (e.g., Leal et al. 2004; Yates 2007a; Ezcurra 2010; Langer et al. 2019; Peyre de Fabrègues et al. 2020; McPhee et al. 2020) does not share strong synapomorphic features with *Plateosaurus* (Leal et al. 2004; McPhee et al. 2020).

4.2 Unaysauridae

With the discovery of *Macrocollum*, from the Caturrita Formation, Müller et al. (2018a: 2) erected Unaysauridae as "the most inclusive clade including *Unaysaurus tolentinoi*, but not '*Plateosaurus engelhardti*' or *Saltasaurus loricatus*", clade also recovered by Pol et al. (2021). To date, it is not clear whether Unaysauridae optimizes better within or outside the Massopoda clade since its phylogenetic position in recent analyses has varied in different placements close to the base of Plateosauria (Müller et al. 2018a; Pol et al. 2021) or Massopoda (Müller 2019). Whether closer to basal plateosaurians or massopodans, Unaysauridae is characterized by unambiguous features that support it as a distinct clade relative to Plateosauridae (see Sect. 2.4.1). Moreover, members of this clade display a combination of plesiomorphic and derived features enshrining it as an "intermediate" taxon between the sauropodomorph origin and the first sauropodomorph radiation. Among the plesiomorphic features can be cited the presence of a relatively elongated antorbital fenestra, a craniomandibular joint that is relatively close to the tooth line, a partially closed acetabulum, and a gracile foot, characters present in *Macrocollum* (Fig. 2a–c). In addition, a maxilla bearing a promaxillary fenestra is a plesiomorphic feature also shared with *Unaysaurus*. On the other hand, other characters observed in *Unaysauridae* are closely associated with post-Carnian sauropodomorphs, such as wide external naris, absence of palatal teeth (feature reported in *Eoraptor*, *Buriolestes,* and *Pampadromaeus*), and presence of lanceolate teeth. Those morphological characters, coupled with the occurrence of the early Norian Caturrita Formation, point Unaysauridae as a key clade to understand a crucial step in the evolutionary scenario of early Sauropodomorpha.

4.3 Riojasauridae

Riojasaurus incertus was the first sauropodomorph to be described from the Late Triassic Los Colorados Formation, Argentina. The original description includes the holotype (a relatively complete postcranium; Bonaparte 1967, 1971) and multiple referred postcranial remains (Bonaparte 1971). A subadult specimen with a complete cranium was later described by Bonaparte and Pumares (1995) (Fig. 2o–q) and potential records of this genus have also been reported from the Quebrada del Barro Formation (Bossi and Bonaparte 1978; Martínez et al. 2015). To date, the only emended diagnosis of this taxon was provided by Bonaparte and Pumares (1995), with the addition of some autapomorphies by Galton and Upchurch (2004). An updated redescription of this genus that includes all known materials is necessary for the light of the recent new knowledge obtained for the group. Furthermore, the putative immature condition of PULR 56 (Bonaparte 1999) prompted some authors to cast doubts on the taxonomic identity of this specimen (McPhee and Choiniere 2017). As a matter of fact, the exclusion of the specimen PULR 56 in some phylogenetic analyses (McPhee et al. 2018; Müller 2019) has yielded a more derived position for this taxon, corroborating the plesiomorphic signal of the specimen. From a functional point of view, *Riojasaurus* appears separated from all other Norian sauropodomorphs in a biomechanical morphospace analysis on the basis of 29 craniodental features including mandible height over length ratios, symphysis angle, tooth slenderness index, and wear facets (Button et al. 2017). In terms of cranial and mandibular strength, *Riojasaurus* shows moderately robust and mechanically efficient jaws. This condition is common among early sauropodomorphs although, in *Riojasaurus,* jaws are more gracile and less efficient than in most others. *Riojasaurus* stands out strikingly against most Norian sauropodomorphs because of a narrower and more precisely occluding dentition. *Riojasaurus* was originally conceived as a member of Melanorosauridae on the basis of its overall robustness (Bonaparte 1971; Bonaparte and Pumares 1995). Later, under a phylogenetic definition of Melanorosauridae formally provided by Galton and Upchurch (2004), *Riojasaurus* was nested in this clade, as well as in Plateosauridae (Upchurch et al. 2007a), in the phylogenetic context of a monophyletic Prosauropoda. With the increased number of taxa in phylogenetic analyses and the general consensus of paraphyletic prosauropods, Melanorosauridae virtually became a synonym of Anchisauria, thus falling into disuse (Upchurch et al. 2007a).

The first comprehensive phylogenetic analysis retrieving early sauropodomorphs as paraphyletic successive groups leading to Sauropoda was provided by Yates (2003). Only a few years later, in Yates (2007a), Riojasauridae was erected as the most inclusive clade containing *Riojasaurus incertus* but not *Plateosaurus engelhardti*, *Massospondylus carinatus*, or *Anchisaurus polyzelus*. Riojasauridae, which also originally included the genus *Eucnemesaurus* from the lower Elliot Formation of South Africa (Yates 2007b; McPhee et al. 2015), is conceived as a low-diversity, pan-Gondwanan clade of large and robust sauropodomorphs at the base of Massopoda (see Ezcurra 2010; Otero and Pol 2013; McPhee et al. 2015; Apaldetti et al. 2018).

4.4 Massospondylidae

Massospondylidae includes all taxa closer to *Massospondylus* than to *Plateosaurus* (stem-based, Sereno 1998: 63). Phylogenetic analyses predating 2010 biased the taxonomic content of this clade to few taxa due to the scarce taxonomic and character sampling, *Coloradisaurus* being the only South American representative for this group (Yates and Kitching 2003; Yates 2007a). This taxon pushed the first record of Massospondylidae to the Norian of South America, as the rest of the clade had a widespread distribution in the Early Jurassic of Gondwana (Martínez 2009; Apaldetti et al. 2011; Chapelle and Choinière 2018; Chapelle et al. 2020), with the exception of the Laurasian *Lufengosaurus*, from China (Young 1941). After 2010, the taxon composition of Massospondylidae increased by virtue of data matrices enriched both in taxon and character sampling. The family then became a poorly resolved clade that has many different topologies depending on the characters and taxa included in different analyses (e.g., Ezcurra 2010; Novas et al. 2011; Pol et al. 2011; Apaldetti et al. 2011, 2013a, 2014; Müller 2019; Chapelle et al. 2020; Rauhut et al. 2020). In spite of this, Massospondylidae consistently included two additional South American forms: *Adeopapposaurus* and *Leyesaurus* (Apaldetti et al. 2011, 2014; Otero et al. 2015; McPhee et al. 2014, 2020; Müller et al. 2018a; Chapelle et al. 2020; Peyre de Fabrègues and Allain 2020). Features supporting this clade are highly variable depending on the phylogenetic analyses, and such variability relies upon the origin of the data matrix. In general terms, Massospondylidae includes the most gracile forms of Jurassic sauropodomorphs, characterized mainly by their relatively short skulls (in comparison to their body length) with a dorsoventrally high suborbital bar, great elongation of their cervical vertebrae (generally at least four times longer than high), and strong ventrolateral twisting of the manual phalanx I.1 (Fig. 2g–n). These features denote massospondylids occupied a differentiated region of the morphospace in comparison to other members of Sauropodomorpha (Apaldetti et al. 2013b).

5 Biogeographic Considerations

As seen in the previous chapter, the earliest unambiguous sauropodomorphs come from the Carnian strata of Argentina and Brazil, with the possible exception of *Nyasasaurus* from the Anisian(?) of Tanzania (Nesbitt et al. 2013; Baron et al. 2017). At the beginning of their evolution at the Carnian–Norian boundary, early-branching sauropodomorphs were not geographically dispersed, abundant, or morphologically disparate. Soon after, during the Norian, this group sauropodomorphs experienced an expansion both in geographical range and morphological repertoire. In this regard, members of Unaysauridae played a key role at this stage for two reasons: firstly, members of this clade come from the Caturrita Formation, dated as early Norian (225.42 ± 0.37 Ma; Langer et al. 2018); and secondly, *Unaysaurus* and *Macrocollum* show final steps in the assembly of the sauropodomorph body plan (e.g., body size increase, shifting to herbivory, neck elongation, and forelimb modifications, particularly the manus). Considering that the geographic expansion of early sauropodomorphs to Africa, India, and especially Europe was recorded in the late Norian, this implies that the establishment of the "prosauropod" body plan could have taken place before early sauropodomorph radiation outside South America, as depicted by Unaysauridae (Fig. 3).

Late Norian to early Rhaetian strata records the first link of South American-African sauropodomorph families, Massospondylidae and Riojasauridae, denoting some degree of faunal interchange between these two Gondwanan areas (Fig. 3). Considering the plethora of phylogenetic hypotheses published so far, the taxonomic content of Massospondylidae includes members from South America (*Coloradisaurus*, *Adeopapposaurus*, *Leyesaurus*), North America (*Sarahsaurus*), Africa (*Massospondylus*), Asia (*Lufengosaurus*, *Pradhania*), and Antarctica (*Glacialisaurus*), but only *Coloradisaurus*, from the Los Colorados Formation, is recorded in the Late Triassic, putatively pinpointing to the first step of the massospondylid radiation. By the Early Jurassic (Hettangian–Pliensbachian), Massospondylidae were widespread across Gondwana, with *Adeopapposaurus* and *Leyesaurus* as the South American representatives (Fig. 3).

Riojasauridae is another taxon that supposedly represents an example of faunal interchange between South American and African areas during the Late Triassic. Albeit the fact that *Riojasaurus* hypodigm may represent a chimera (McPhee et al. 2015; McPhee and Choiniere 2017), its holotype (PVL 3808) shows a clear distinct phylogenetic signal in respect to other coeval sauropodomorph taxa recorded in the Los Colorados Formation (Yates 2007a; Ezcurra 2010; Pol et al. 2011; Apaldetti et al. 2011; Otero et al. 2015; Langer et al. 2019; Müller 2019; McPhee et al. 2020). Also, when PVL 3808 is scored in isolation, it still nests together with the South African *Eucnemesaurus* (McPhee et al. 2020), thus confirming the existence of Riojasauridae as a taxonomic entity.

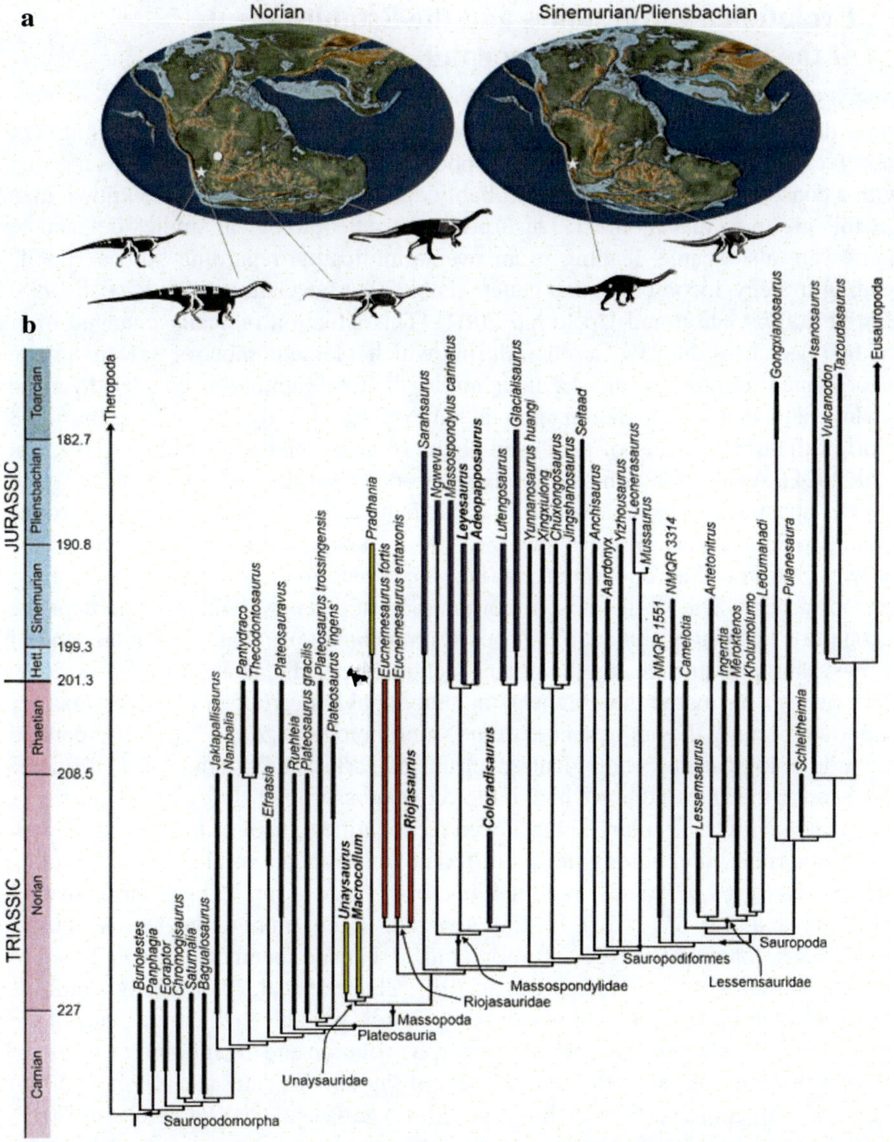

Fig. 3 Geographic and temporal distribution and phylogenetic relationships of non-sauropodiform plateosaurians from South America. **a** Geographic distribution during the Norian (left) and Sinemurian/Pliensbachian (right) ages. Stars denote Argentinean taxa; the circle denotes Brazilian taxa. Note that the temporal distribution of *Adeopapposaurus* and *Leyesaurus* is not conclusive, since no exact age is currently known. Not to scale. **b** Calibrated phylogeny highlighting (in bold) studied taxa in this contribution (modified from Pol et al. 2021). (silhouettes modified from Martínez 2009; Paul 2010; Apaldetti et al. 2011; Müller 2019; Pol et al. 2021; paleogeographic maps taken from www.gplates.org, Müller et al. 2018b)

6 Evolutionary Milestones and the Establishment of the Early Sauropodomorph Body Plan

Since the establishment of the genus *Plateosaurus* in 1837 and until the beginning of the 21st Century, non-sauropodan sauropodomorphs have been regarded as a group with a conservative bauplan, most probably due to the relatively scarce known taxa for this group to that moment. Unsurprisingly, paleobiological studies devoted to them were also scarce, leading to an oversimplification regarding "prosauropod" feeding strategy, locomotion, and general habits (but see Galton 1985; Parrish 1998; Barrett 2000; Galton and Upchurch 2004). This situation radically changed after the first decade of the 21st Century, during which a large number of new taxa were incorporated into this group. As an example, the total number of taxa used for the phylogenetic analysis in Galton and Upchurch (2004) is only 40% (representing 18 taxa) of the total number of taxa used about 16 years later (e.g., Müller 2019; Pol et al. 2021). As a result, non-sauropodan sauropodomorphs now encompass a much broader spectrum of taxa, arranged in a pectinate array of successive sister groups leading to Sauropoda. All these new forms have allowed us to better understand the early evolution of Sauropodomorpha (i.e., the evolution of non-eusauropod forms) in many aspects. One of the most important facts of this new dynamic is perhaps the recognition and establishment of very early-branching taxa, on the one hand, and of earliest sauropod lineages on the other (such as Saturnaliidae and Lessemsauridae, respectively). No matter their recovery as a monophyletic group or not, these taxa are important because they represent extreme "morphological grades", which expanded the morphological and ecological spectrum of early sauropodomorphs, formerly restricted to the knowledge we had of the "core-prosauropod" (e.g., *Plateosaurus* or *Massospondylus*) body plan. This allowed the identification of primitive forms, with a plesiomorphic morphology not so different from the ancestral dinosaurian plan, and derived forms, drastically modified in terms of size (from ~5 kg to more than 50 t, Campione and Evans 2020), feeding apparatus (from omnivorous/faunivorous to herbivorous, Sander et al. 2011; Button et al. 2017) and locomotion (from biped to quadruped, Bates et al. 2016; Otero et al. 2017; Chapelle et al. 2019), recognizable as the earliest sauropods. Within this scenario, it is now possible to illustrate many of the key anatomical transitions in sauropodomorph evolution and to pinpoint the origin of features that contributed to the later success of sauropods (Yates and Kitching 2003; Barrett and Upchurch 2007; Rauhut et al. 2011; Cerda et al. 2017; Otero et al. 2017, 2019; Langer et al. 2019; Apaldetti et al. 2018).

Sauropodomorpha was the first dinosaur clade to achieve a global distribution and largest sizes, being the early adoption of herbivory the driving factor behind this success (Sander et al. 2011, 2013). This section focuses on the Carnian–Rhaetian transition, which witnessed the anatomical transformations between the very early forms from the Carnian, such as *Saturnalia*, and the Norian–Rhaetian "core prosauropods". Such a transition is registered in the South American record and corresponds to the establishment of herbivory, a body mass increase, and the acquisition of the characteristic "prosauropod" manus.

6.1 First Steps Toward Herbivory

The conservative view of the "prosauropod" bauplan was traditionally also valid for the inferences about dietary habits. In this regard, non-sauropodan sauropodomorphs were assumed to be herbivorous/omnivorous with lanceolate, non-recurved, symmetrical, coarsely denticulate teeth (Barrett and Upchurch 2007) (Fig. 4). However, the knowledge about the feeding habits in dinosaurs in general and in non-sauropodan sauropodomorphs, in particular, is in a state of flux (Button and Zanno 2020). Until recently, the lack of information about Carnian–early Norian representatives precluded the study of the possible sequence of character transformations that occurred during early stages of sauropodomorph evolution. The discovery of new early forms and the reinterpretation of existing species under new methodological approaches allow us to develop a broad scenario of the origin and subsequent evolutionary trajectory of sauropodomorph dietary habits (Button et al. 2017; Button and Zanno 2020). Ultimately, understanding how they shifted to herbivory can potentially allow for the detection of the selection pressures involved in the evolution of

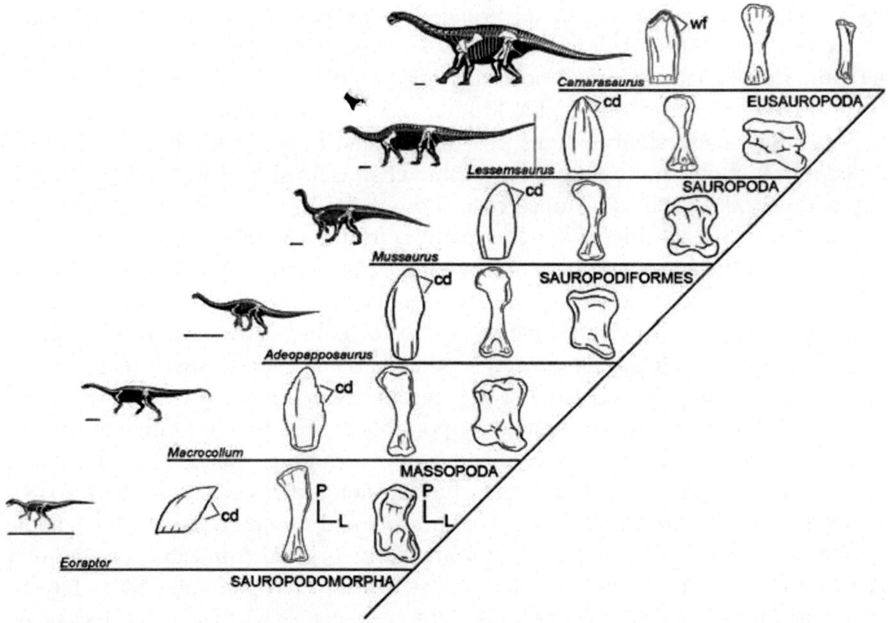

Fig. 4 Evolutionary trends in early-branching sauropodomorphs. Forms shift from small-bodied to heavily built ones. Teeth (labial view) are modified in terms of curvature, basal constriction, and denticles. Humeri (left side, anterior view) gain in robustness and the deltopectoral crest widens. Metacarpals I (right side, dorsal view, proximal towards top) present a drastic shortening. Abbreviations: cd, crown denticles, fd, L, lateral, P, proximal, wf, wear facet. Scale bars equal 50 cm except in *Eoraptor*, which is 20 cm. (Silhouettes modified from Wilson and Sereno 1998; Martínez 2009; Sereno et al. 2013; Müller 2019)

large body sizes (Stevens and Parrish 1999; Christian 2010; Hummel and Clauss 2011; Preuschoft et al. 2011; Sander et al. 2011, 2013).

It is not clear whether Carnian sauropodomorphs were either fully omnivorous or faunivorous, but certainly, the recurved and finely serrated teeth present in *Buriolestes*, *Panphagia,* and *Eoraptor* (Martínez and Alcober 2009; Sereno et al. 2013; Cabreira et al. 2016) resemble more with the ancestral dinosaurian condition, excluding them from an herbivorous morphospace (Sereno 1997; Barrett et al. 2011; Button et al. 2017; Müller and Garcia 2020) (Fig. 4). In addition, the palatal teeth present in *Eoraptor*, *Pampadromaeus,* and *Buriolestes* resemble the plesiomorphic condition of the early theropod *Eodromaeus murphy* (Martínez et al. 2011) and the early dinosauriform *Lewisuchus admixtus* (Romer 1972) and have been hypothesized to be a feature linked to retain the prey in the mouth. The loss of these kinds of teeth might be indicative of dietary shift.

Dental features related to herbivory among Sauropodomorpha first appear during the "upper" Carnian, within the "*Exaeretodon* Sub-AZ" of the Santa Maria Formation (Button et al. 2017; Pol et al. 2021). When comparing the tooth morphology of *Bagualosaurus* and *Pampadromaeus* with that of *Panphagia* or *Buriolestes,* the formers present teeth with basal constrictions, mostly lanceolate crowns, and slightly larger denticles most likely related to an herbivorous diet. In addition, the teeth of *Bagualosaurus* and *Pampadromaeus* share some large denticles on one carina and relatively small denticles on the opposite, which seems a unique condition among Sauropodomorpha (Müller, pers. comm.). Although *Bagualosaurus* exhibits additional cranial features possibly related to more efficient herbivorous habits, it still retains multiple plesiomorphic characters shared with other Carnian sauropodomorphs, such as a sloped rostral process of the maxilla, the presence of a marked longitudinal ridge extending parallel to the ventral margin of the maxillary portion of the antorbital fossa, a straight ventral margin of the iliac acetabulum and craniomedially set femoral head, among others (Pretto et al. 2018). While it is not clear whether body mass increases led to herbivory or vice versa in early sauropodomorph evolution, most traits related to large body mass (e.g., robust hindlimbs, especially the pes) are not yet present in *Bagualosaurus*, suggesting that some modifications in the skull anatomy possibly related to more efficient herbivorous habits, such as lanceolate teeth and coarse serrations, appeared earlier in the evolution of sauropodomorphs than their further increase in size (Pretto et al. 2018).

The anatomy of the early Norian *Unaysaurus* and *Macrocollum* presents additional features putatively related to increasing herbivory, either in a direct or indirect way. In this sense, these taxa resemble most Norian sauropodomorphs by having similar maxillary and dentary dentitions, with mesiodistally expanded tooth crowns, arranged in a more or less imbricated way, and bearing large, obliquely oriented marginal serrations, quite similar to those of extant iguanid lizards (Galton 1985, 1986; Crompton and Attridge 1986; Barrett 2000; Müller 2019; McPhee et al. 2020) (Fig. 4).

Further evidence of herbivorous diet is the presence of wear facets, which are fully present in Gravisaurian sauropods (Allain and Aquesbi 2008). Wear facets are rare among non-gravisaurian sauropodomorphs, although some isolated reports exist in Massospondylidae (Gow et al. 1990: Fig. 8; Martínez 2009: Fig. 12) and also in the putative *Yunnanosaurus* specimen from the Early to Middle Jurassic of Yunnan Province, China (Sekiya et al. 2014: Fig. 4). Nonetheless, such existence of scattered wear facets in some taxa is not indicative of precise dental occlusion or complex food processing. In a similar way, the inference of fleshy non-contractile cheeks, which are usually linked to a longitudinal ridge or carina on the posterolateral aspect of the dentary and/or large nutrient foramina on the lateral surfaces of the maxilla and dentary, have been cited for non-sauropodan sauropodomorphs (Galton 1973, 1985; Upchurch et al. 2007b), including *Riojasaurus*.

The presence of a large tongue, that may have been useful for manipulating food within the mouth, was suggested based on well-ossified hyoid bones in *Plateosaurus*, *Massospondylus*, and *Anchisaurus* (Barrett and Upchurch 2007) and later recognized in *Buriolestes* and *Macrocollum* (Cabreira et al. 2016; Müller 2019). Although ossified hyoid bones are well documented within Dinosauria (Holtz et al. 2004; Horner et al. 2004; Upchurch et al. 2004), their sole presence does not necessarily imply intraoral food processing like that of typical herbivorous tetrapods. In this regard, complex intraoral processing seems to be related to a more intricate hyoid apparatus ossification. Within archosaurs, only derived herbivorous ornithischian dinosaurs and taxa that evolved flight has a tongue structure that appears to be uniquely elaborated (Li et al. 2018).

Finally, evidence from paleoneurology postulated the relationship between the relative development of the floccular fossa lobes (FFL), the type of stance, and the acquisition of herbivory in the line to Sauropoda (Bronzati et al. 2017, see also Ballell et al. 2020; Müller et al. 2021). The FFL is part of the systems controlling the eyes, neck, and head movements (Witmer et al. 2003; Walsh et al. 2013; Ferreira-Cardoso et al. 2017). The volume variation detected for the FFL in the endocast of dinosaurs is related to locomotion and changes in feeding behavior, from active predators (large volume), such as *Herrerasaurus* and *Gnathovorax* (Stocker et al. 2016; Pacheco et al. 2019) to herbivores (reduced volume), such as *Brachiosaurus* and *Bonatitan* (Knoll and Schwarz-Wings 2009; Paulina Carabajal 2012). Hence, reduction of the FFL is potentially related to the adoption of a more restricted herbivorous diet. The presence of an enlarged protuberance in the region of the endocast corresponding to the FFL, such as that of *Saturnalia tupiniquim*, *Buriolestes schultzi*, and *Thecodontosaurus antiquus*, is indicative of a large FFL, which corresponds to the plesiomorphic condition (i.e., faunivorous behavior, Bronzati et al. 2017; Müller et al. 2021; Ballell et al. 2020). Furthermore, the small volume of the FFL in some taxa, such as *Plateosaurus trossingensis* (*P. engelhardti*, Galton 2012) and *Ngwevu intloko*, indicates that an initial volume reduction of these neural tissues occurred in the early evolutionary history of the group, before the origin of quadrupedalism and sauropods (Bronzati et al. 2017; Müller et al. 2021). In this context, the reduced volume of the FFL of sauropods might represent a vestigial trait. In this case, only a thorough investigation including a large sample of endocasts of non-sauropodan sauropodomorph taxa, alongside studies on the evolutionary drivers in extant taxa (see Walsh et al. 2013; Ferreira-Cardoso et al. 2017), will be able to clarify the evolution of the FFL in the lineage.

6.2 Body Mass Increase

The most evident morphological change that occurred between the lightly built, small-bodied bipedal sauropodomorphs from the Carnian and the heavily built, large quadrupedal sauropods that radiated in the Jurassic is undoubtedly a size increase (Fig. 4). In general, members of the Saturnaliidae or earliest sauropodomorphs (*Pampadromaeus*, *Saturnalia*, *Chromogisaurus* and their allies) did not surpass 1.5 m in length and weighed less than 20 kg based on femoral circumference (Martínez and Alcober 2009; Benson et al. 2014, 2018; Cabreira et al. 2016; Müller et al. 2021). These small bipedal forms from the Carnian can be regarded as a "first pulse" within the early sauropodomorphs' body mass evolution. *Bagualosaurus* represents an exception to the generalized pattern of Carnian small sauropodomorphs as its femur is 25% longer than other Carnian relatives and its body mass is here estimated to be about 23.5 *kg*, suggesting a "next step" toward body mass increase compared to "mid" Carnian forms. The latter, coupled with the fact that *Bagualosaurus* is consistently recovered in a more derived position than all other Carnian sauropodomorphs (Pretto et al. 2018; Langer et al. 2019; Müller 2019; Pol et al. 2021; see Chap. 1), suggests this genus may represent the earliest trend toward larger body sizes among Triassic sauropodomorphs.

The South American record provides the "second pulse" regarding body mass increase in Sauropodomorpha, which took place in the early Norian, recorded in the Caturrita Formation. The femur of *Macrocollum* is about 40% longer than that of *Bagualosaurus* and this genus has a body mass of 84 *kg*, estimated here through femoral circumference; *Unaysaurus*, on the other hand, does not preserve any femur, but has a scapula about the same length than *Macrocollum* (c. 200 *mm*), positioning both species as considerably larger than *Bagualosaurus*. In spite of this, information regarding the ontogenetic age of these taxa (i.e., via histological analysis) is needed to confirm the accuracy of size comparisons.

The next step of body mass increase occurred in the "mid-late" Norian and is recorded independently in different regions across the globe, including South Africa (e.g., *Eucnemesaurus*), Europe (*Plateosaurus*), and South America. In the latter continent, the Los Colorados Formation provides three taxa that successively display a body mass increase: *Coloradisaurus*, *Riojasaurus*, and *Lessemsaurus*. *Coloradisaurus* was estimated at 386 *kg* based on femoral circumference and was probably bipedal when compared to its close relative *Massospondylus* in terms of morphological similarity and phylogenetic proximity (Bonnan and Senter 2007; Chapelle et al. 2020). With an estimation of 2.2 t based on long bone circumferences, *Riojasaurus*, on the other hand, represents a robust quadruped (McPhee et al. 2018) approaching the "giant" sizes achieved by its coeval Lessemsauridae (see Chap. 3).

It is important to mention, however, that although a trend toward body size increased among Triassic early sauropodomorphs, some Early Jurassic forms, such as *Leyesaurus*, *Leonerasaurus*, and *Anchisaurus* seem to reverse this trend. These taxa exhibit quite small sizes, similar to late Carnian forms, demonstrating that size disparity was common among sauropodomorphs before the establishment of the sauropod body plan (see Chap. 12).

Increasing size has proved crucial for sauropodomorph evolutionary success for many reasons, such as predator evasion, but most notably because it led to a shift in browsing height (Bakker 1978; Parrish 2005; Tütken 2011; Sander 2013). The latter factor, coupled with additional traits, such as modification of the tooth battery (i.e., from fine serrated and posteriorly curved to coarse serrated and lanceolate teeth) and neck elongation, permitted this group to specialize in processing plants and to acquire an energetic advantage via high metabolic rates (Sander 2013).

6.3 The Establishment of the "Prosauropod" Forelimb

The evolution of Sauropodomorpha goes through multiple morphological transformations throughout about 160 myr. One of the most drastic changes during this process involved those occurring in the forelimb. In this regard, two milestones can clearly be pinpointed, each one characterizing the two main sauropodomorph groups. Firstly, all the transformations lead to the arch-like, non-supporting and robust hand typical of non-sauropodan sauropodomorphs (e.g., *Plateosaurus*) from the primitive, elongated and gracile saurischian hand (e.g., *Herrerasaurus*). Secondly, the successive shortening and reduction of manual elements until the acquisition of a graviportal (weight-bearing) hand, coupled with the long bone alignment, results in a columnar forelimb characteristic of the eusauropod bauplan (e.g., *Camarasaurus*) (Fig. 4). In this section we discuss the main anatomical changes that the forelimb of early sauropodomorph dinosaurs went through, highlighting the importance of the South American record on this matter (Table 2).

The most notable changes that occurred in the forelimb along the early evolution of sauropodomorphs are perhaps those related to a robustness increase. They are illustrated by the variations in the proportions of humeri. To quantify humerus robustness, we use the robustness index (RI) as defined by Wilson and Upchurch (2003: Table 3): average of the greatest widths of the proximal end, mid-shaft and distal end of the humerus divided by its length (Table 3).

Among early-branching forms, *Eoraptor* and *Saturnalia* are good examples for examining plesiomorphic patterns of the whole bauplan in general, and the forelimb in particular, since they slightly depart from the saurischian plesiomorphic condition. The humerus itself is characterized by a substantial increase of its robustness, with a distinct "jump" from earliest forms to plateosaurians. In this regard, Saturnaliidae has a RI under 0.30, with *Eoraptor* being the taxon with the lowest value (RI of 0.22). However, most taxa within Plateosauria exhibit robustness values around or above 0.30, albeit some extreme cases, such as *Riojasaurus*, reach a RI of 0.37 (Fig. 4, Table 3). Enlargement of the forelimb is most likely due to the shift from cursorial and gracile forms to bulkier ones that developed grasping and defense abilities. As a result of the enlargement of the humerus and deltopectoral crest, insertion areas for muscles increased, probably due to, or leading to, an augmentation in size of the

Table 2 Record of non-sauropodomorph plateosaurians from South America, in the order presented in Sect. 2.2 (from Late Triassic to Early Jurassic)

Taxon	Formation	Locality	Age	References
Unaysaurus tolentinoi	Caturrita	Água Negra district, São Martinho da Serra, located 13 km north of Santa Maria, Rio Grande do Sul state, Brazil	Norian (maximum age of 225.42 ± 0.37 Ma)	Leal et al. (2004), Langer et al. (2018)
Macrocollum itaquii	Caturrita	Wachholz site at Agudo, Rio Grande do Sul, Brazil	Norian (maximum age of 225.42 ± 0.37 Ma)	Müller et al. (2018a), Langer et al. (2018), Müller (2019)
Riojasaurus incertus	Los Colorados	Southern area of the Rajado hill; the holotype comes from "Quebrada de los Jachaleros" locality, whereas the specimen PULR 56 comes from the "La Esquina" locality, Talampaya National Park, nearby Villa Union town, La Rioja Province, Argentina	Norian (227–213 Ma)	Bonaparte (1971), Bonaparte and Pumares (1995)
Coloradisaurus brevis	Los Colorados	"La Esquina" locality, Ischigualasto–Villa Union basin, Talampaya National Park, La Rioja Province, Argentina	Norian (227–213 Ma)	Bonaparte (1978), Apaldetti et al. (2013a, 2014)
Adeopapposaurus mognai	Cañón del Colorado Formation, Middle Member	"Sierra de Mogna" locality, 70 km north of San Juan City, San Juan Province, Argentina	Lower Jurassic	Martínez (2009), Martínez and Colombi (2011)

(continued)

Table 2 (continued)

Taxon	Formation	Locality	Age	References
Leyesaurus marayensis	Balde de Leyes Formation	"Balde de Leyes" locality, San Juan Province, Argentina	Lower Jurassic	Apaldetti et al. (2011), Martínez et al. (2015), Colombi et al. (2015)

Table 3 Robustness index (sensu Wilson and Upchurch 2003) of the humerus of various sauropodomorphs (RI in ascending order)

Taxon	Humeral robustness index
Eoraptor (PVSJ 512)	0.22
Unaysaurus (UFSM11069)	0.27
Saturnalia (MCP 3844-PV)	0.28
Plateosaurus trossingensis (GPIT-PV-30784)	0.28
Mussaurus (MLP 68-II-27–1)	0.28
Antetonitrus (BP/1/4952)	0.29
Plateosaurus longiceps (MB.R. 4404)	0.30
Massospondylus carinatus (BP/1/4934)	0.30
Camarasaurus (FMNH 25,182)	0.30
Adeopapposaurus (PVSJ 610)	0.31
Ruehleia (MB.R.4718)	0.31
Lessemsaurus (PVL 4822)	0.31
Ingentia (PVSJ 1086)	0.33
Yunnanosaurus (NGMJ 004,546)	0.33
Coloradisaurus (PVL 5904)	0.34
Riojasaurus (PVL 3808)	0.37

corresponding muscles and, therefore, an additional opportunity to increase body size (Remes 2008; Otero et al. 2017; Otero 2018; see Chap. 12).

The radius/humerus ratio exhibits an interesting trend when plotted against humerus length. In non-eusauropod sauropodomorphs this ratio somewhat follows a negative correlation as the humerus is enlarged along the phylogeny (Fig. 5), and such a trend is well represented among South American forms. *Eoraptor* shows the highest radius/humerus ratio (above 0.7), some early-branching plateosaurians (e.g., *Unaysaurus*) and early massopodans (e.g., *Adeopapposaurus*) show ratios between 0.6 and 0.7, whereas early sauropodiforms (e.g., *Mussaurus*) start from values below 0.6 and some very early sauropods (e.g., *Ingentia*) fall to levels below 0.5. Interestingly, this trend is reversed in eusauropods, in which the radius is enlarged relative to the humerus length (see Wilson and Upchurch 2003: Table 2). Relative shortening

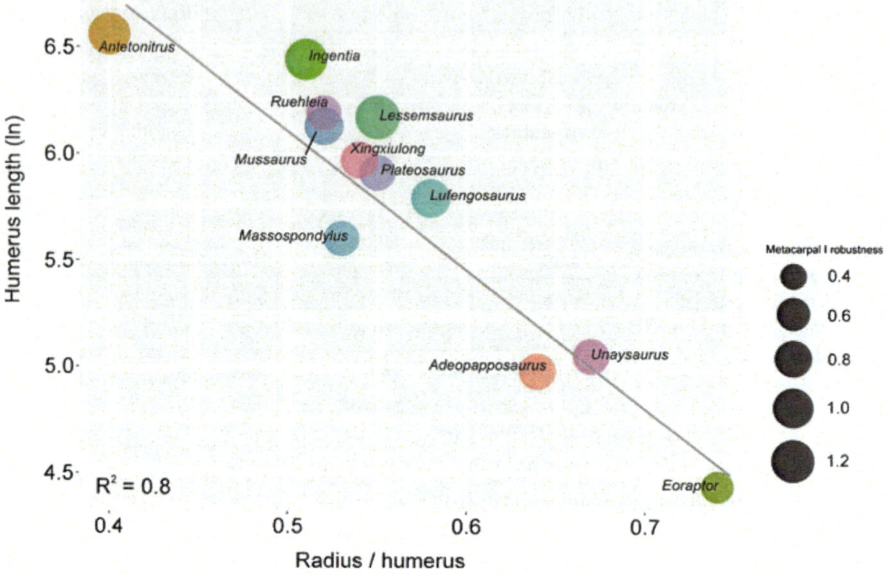

Fig. 5 Plot of humerus length against the radius/humerus length ratio in various non-eusauropodan sauropodomorphs. The robustness of the metacarpal I is included as a third variable (bubble size). Metacarpal robustness was measured as the proximal width/total length. The humeral length is used here as a proxy of size. Note that as the humeral length increases, the radius becomes proportionally short. Specimens used as follows: *Adeopapposaurus* (PVSJ 610), *Antetonitrus* (BP/1/4952), *Eoraptor* (PVSJ 512), *Ingentia* (PVSJ 1086), *Lessemsaurus* (PVL 4822), *Lufengosaurus* (IVPP V15), *Massospondylus carinatus* (BP/1/4934), *Mussaurus* (MLP 68-II-27–1), *Plateosaurus trossingensis* (GPIT-PV-30784), *Ruehleia* (MB.R.4718), *Unaysaurus* (UFSM11069), *Xingxiulong* (LFGT-D0003)

of the antebrachium is an adaptation going hand in hand with increasing body mass (Carrano 1999; Rauhut et al. 2011). This modification, coupled with changes within the radius-ulna articulation, allowed the antebrachium to act more as a stiffened unit during flexion/extension and thus reduced mechanical bone loading (Hutson and Hutson 2015; Otero et al. 2017; see Chap. 12).

Lessemsauridae depicts extreme shortening of the manus, showing extreme expression of metacarpal I robustness (Table 4; Figs. 4 and 5). It has been postulated that a switch occurred with the body mass increase: the non-essential functions (grasping, defense) of the manus were slowly supplanted by the locomotor function (Goussard 2009; Rauhut et al. 2011). Metacarpus shortening is a trend that, although accelerated in Sauropodiformes, seems to reach its climax in Lessemsauridae and does not last in Eusauropoda. The latter exhibit a reversal of the manus proportions in order to support the body mass and evolve toward graviportality (Bonnan 2003; Bonnan and Yates 2007).

Table 4 Metacarpal I width/metacarpal I length among various sauropodomorphs (ratio in ascending order)

Taxon	Metacarpal I width/metacarpal I length
Camarasaurus (FMNH 25,182)	0.37
Mussaurus (PVL 4587)	0.57
Tazoudasaurus (CPSGM To2-112)	0.6
Ruehleia (MB.R.4718)	0.68
Plateosaurus trossingensis (GPIT-PV-30784)	0.74
Unaysaurus (UFSM11069)	0.75
Massospondylus sp. (SAM-K5131)	0.75
Adeopapposaurus (PVSJ 610)	0.77
Mussaurus (MLP 68-II-27–1)	0.88
Lufengosaurus (IVPP V15)	0.93
Ingentia (PVSJ 1086)	1.09
Aardonyx (BP/1/5339)	1.12
Antetonitrus (BP/1/4952	1.17
Sefapanosaurus (BP/1/7438)	1.19
Lessemsaurus (PVL 4822)	1.3

While the proximal-most part of the hand was undergoing an extreme shortening of the metacarpus, the distalmost part was also subject to major changes. Most non-plateosaurian sauropodomorphs exhibit thin and elongate claws when non-sauropodiform plateosaurians possess a set of three high and developed digit claws. In earliest forms such as *Eoraptor* claws are gracile, almost raptorial, and practically subequal in length. In plateosaurians like *Adeopapposaurus* or *Riojasaurus*, the claw of digit I appear robust and dorsoventrally high, being almost twice the size of the claw of digit III. This key feature has been spotted and discussed early on (Charig et al. 1965; Cooper 1981) and quickly became the "trademark" of what we now call "core prosauropods" (Galton and Upchurch 2004). It has been one of the arguments that fed the bipedalism hypothesis for non-sauropodiform plateosaurians before proper biomechanical studies could be carried out, considering that the enlarged claw would have greatly hampered an obligatory quadrupedal stance (Cooper 1981; Upchurch 1997). The possible range of functions of this specialized manus was discussed by various authors, and tends to be consensually regarded as involved in defense, or participating in the feeding process (to grasp or hold onto branches, or to assist balancing) (Galton 1971; Upchurch 1997; Galton and Upchurch 2004; Bonnan and Senter 2007; Otero et al. 2017; Pol et al. 2021).

7 Summary

The record of non-sauropodiform plateosaurians is well documented in strata of Argentina and Brazil and is composed of post-Carnian forms, including the Unaysauridae, from the early Norian Caturrita Formation, Riojasauridae, from the Norian Los Colorados Formation, and Massospondylidae, which extend up to the Early Jurassic. The former clade is important because *Unaysaurus* and *Macrocollum* show the final steps in the assembly of the sauropodomorph body plan (i.e., body size increase, neck elongation, shifting to herbivory, forelimb modifications, particularly the manus). In the same way, the Norian age is a crucial time considering that the geographic distribution of early sauropodomorphs extends to Africa, India, and Europe during this period. This means that the establishment of the "prosauropod" body plan could have taken place before early sauropodomorph radiation outside South America and may have been driven by Unaysauridae. On the other hand, Massospondylidae and Riojasauridae, first recorded during Late Norian and Rhaetian, account for the first link of South American-African sauropodomorph families, implying some degree of faunal interchange within Gondwana.

The Carnian–Rhaetian transition witnessed the anatomical transformations between the gracile forms from the Carnian, such as *Eoraptor* and *Saturnalia* to the Norian–Rhaetian "core prosauropods", such as *Riojasaurus* and *Coloradisaurus*. Such transition is registered in the South American record and the age of this radiation is congruent with the only available radioisotopic date of the Caturrita Formation, highlighting the importance of this unit for understanding this key stage in sauropodomorph evolution. A suite of derived features that have long characterized later sauropodomorphs actually make their first appearance in these early Norian species from South America and are related to body mass increase, establishment of herbivory and acquisition of the typical "prosauropod" forelimb.

Acknowledgements We would like to thank Ricardo Martínez for providing historical information regarding *Leyesaurus* and *Adeopapposaurus*. R. T. Müller and K. Chapelle provided constructive suggestions that greatly improved this work.

References

Allain R, Aquesbi N (2008) Anatomy and phylogenetic relationships of *Tazoudasaurus naimi* (Dinosauria, Sauropoda) from the late Early Jurassic of Morocco. Geodiversitas 30:345–424

Andreis RR, Bossi GE, Montardo DK (1980) O Grupo Rosário do Sul (Triássico) no Rio Grande do Sul. In: Anais do XXXI Congresso Brasileiro de Geologia 2, Camboriú, Brazil, pp 659

Apaldetti C, Martínez RN, Alcober OA, Pol D (2011) A new Basal Sauropodomorph (Dinosauria: Saurischia) from Quebrada del Barro Formation (Marayes-El Carrizal Basin), Northwestern Argentina. PLoS ONE 6:e26964

Apaldetti C, Pol D, Yates A (2013a) The postcranial anatomy of *Coloradisaurus brevis* (Dinosauria: Sauropodomorpha) from the Late Triassic of Argentina and its phylogenetic implications. Palaeontology 56:277–301

Apaldetti C, Pol D, Martínez RN (2013b) Evolution patterns and morphological disparity on Sauropodomorpha during Triassic-Jurassic boundary. In: Abstracts of the 73rd annual meeting of the society of vertebrate paleontology. Los Angeles, USA, pp 79

Apaldetti C, Martinez RN, Pol D, Souter T (2014) Redescription of the skull of *Coloradisaurus brevis* (Dinosauria, Sauropodomorpha) from the Late Triassic Los Colorados Formation of the Ischigualasto-Villa Union Basin, northwestern Argentina. J Vertebr Paleontol 34:1113–1132

Apaldetti CG, Martínez RN, Cerda IA, Pol D, Alcober OA (2018) An early trend towards gigantism in Triassic sauropodomorph dinosaurs. Nature Ecology & Evolution 2:1227–1232

Apaldetti C, Pol D, Ezcurra MD, Martínez RN (2021) Sauropodomorph evolution across the Triassic–Jurassic boundary: body size, locomotion, and their influence on morphological disparity. Sci Rep 11:22534

Arcucci AB, Marsicano CA, Caselli AT (2004) Tetrapod association and palaeoenvironment of the Los Colorados Formation (Argentina): a significant sample from Western Gondwana at the end of the Triassic. Geobios 37(5):557–568. https://doi.org/10.1016/j.geobios.2003.04.008

Attridge J, Crompton AW, Jenkins FA Jr (1985) The Southern African Liassic prosauropod *Massospondylus* discovered in North America. J Vertebr Paleontol 5:128–132

Bakker RT (1978) Dinosaur feeding behavior and the origin of flowering plants. Nature 274:661–663

Ballell A, Rayfield EJ, Benton MJ (2020) J Vertebr Paleontol 40(2):e1770774. https://doi.org/10.1080/02724634.2020.1770774

Baron MG, Norman DB, Barrett PM (2017) A new hypothesis of dinosaur relationships and early dinosaur evolution. Nature 543:501–506

Barrett PM (2000) Prosauropod dinosaurs and iguanas: Speculations on the diets of extinct reptiles. In: Sues H-D (ed) Evolution of herbivory in terrestrial vertebrates: perspectives from the fossil record. Cambridge University Press, Cambridge, pp 42–78

Barrett PM, Upchurch P (2007) The evolution of herbivory in sauropodomorph dinosaurs. Spec Pap Palaeontol 77:91–112

Barrett PM, Butler RJ, Nesbitt SJ (2011) The roles of herbivory and omnivory in early dinosaur evolution. Earth Environ Sci Trans R Soc Edinb 101:383–396

Barrett PM, Chapelle KEJ, Staunton CK, Botha J, Choiniere JN (2019) Postcranial osteology of the neotype specimen of *Massospondylus carinatus* Owen, 1854 (Dinosauria:Sauropodomorpha) from the upper Elliot formation of South Africa. Palaeont. Afr. 53:114–178

Bates KT, Mannion PD, Falkingham PL, Brusatte SL, Hutchinson JR, Otero A, Sellers WI, Sullivan C, Stevens KA, Allen V (2016) Temporal and phylogenetic evolution of the sauropod dinosaur body plan. Royal Soc Open Sci 3(3):150636. https://doi.org/10.1098/rsos.150636

Benson RBJ, Campione NE, Carrano MT, Mannion PD, Sullivan C, Upchurch P, Evans DC (2014) Rates of dinosaur body mass evolution indicate 170 million years of sustained ecological innovation on the avian stem lineage. PLoS Biol 12:e1001853

Benson RBJ, Hunt G, Carrano MT, Campione NE (2018) Cope's rule and the adaptive landscape of dinosaur body size evolution. Palaeontology 61:13–48

Bonaparte JF (1967) Dos nuevas "faunas" de reptiles triásicos de Argentina. I International Symposium of Gondwana UNESCO (Mar del Plata) Actas:283–306

Bonaparte JF (1971) Los tetrápodos del sector superior de la Formación Los Colorados, La Rioja, Argentina (Triásico Superior). 1 Parte. Opera Lilloana 22:1–183

Bonaparte JF (1978) *Coloradia brevis* n. g. et n. sp. (Saurischia, Prosauropoda), dinosaurio Plateosauridae de la Formación Los Colorados, Triásico superior de La Rioja. Argentina. Ameghiniana 15:327–332

Bonaparte JF (1999) Evolución de las vértebras presacras en Sauropodomorpha. Ameghiniana 36:115–187

Bonaparte JF (1982) Faunal replacement in the Triassic of South America. J Vertebr Paleontol 2:362–371

Bonaparte JF, Powell JE (1980) A continental assemblage of tetrapods from the Upper Cretaceous beds of El Brete, northwestern Argentina (Sauropoda–Coelurosauria–Carnosauria–Aves). Mémoires De La Société Géologique De France 139:19–28

Bonaparte JF, Pumares JA (1995) Notas sobre el primer craneo de *Riojasaurus incertus* (Dinosauria, Prosauropoda, Melanorosauridae) del Triásico superior de La Rioja, Argentina. Ameghiniana 32:341–349

Bonnan MF (2003) The evolution of manus shape in sauropod dinosaurs: implications for fuctional morphology, forelimb orientation, and phylogeny. J Vertebr Paleontol 23:595–613

Bonnan MF, Senter P (2007) Were the basal sauropodomorph dinosaurs *Plateosaurus* and *Massospondylus* habitual quadrupeds? Spec Pap Palaeontol 77:139–155

Bonnan MF, Yates A (2007) A new description of the forelimb of the basal sauropodomorph *Melanorosaurus*: implications for the evolution of pronation, manus shape, and quadrupedalism in sauropod dinosaurs. Spec Pap Palaeontol 77:157–168

Bordy EM, Abrahams M, Sharman GR, Viglietti PA, Benson RJ, McPhee BW, Barrett PM, Sciscio L, Condon D, Mundil R, Rademan Z, Jinnah Z, Clark JM, Suarez CA, Chapelle KEJ, Choiniere JN (2020) A chronostratigraphic framework for the upper Stormberg Group: implications for the Triassic-Jurassic boundary in southern Africa. Earth Sci Rev 203:103120

Bossi GE, Bonaparte JF (1978) Sobre la presencia de un dinosaurio prosaurópodo en la formación Quebrada del Barro, en el borde austral de la cuenca de Marayes-El Carrizal (Triásico superior de San Juan). Acta Geológica Lilloana 15:41–47

Brochu CA (1996) Closure of neurocentral sutures during crocodilian ontogeny: implications for maturity assessment in fossil archosaurs. J Vertebr Paleontol 16:49–62

Bronzati M, Rauhut OWM, Bittencourt JS, Langer MC (2017) Endocast of the Late Triassic (Carnian) dinosaur *Saturnalia tupiniquim*: implications for the evolution of brain tissue in Sauropodomorpha. Sci Rep 7:11931

Bronzati M, Rauhut OWM (2018) Braincase redescription of *Efraasia minor* Huene, 1908 (Dinosauria: Sauropodomorpha) from the Late Triassic of Germany, with comments on the evolution of the sauropodomorph braincase. Zool J Linn Soc 182:173–224

Bronzati M, Müller RT, Langer MC (2019) Skull remains of the dinosaur *Saturnalia tupiniquim* (Late Triassic, Brazil): with comments on the early evolution of sauropodomorph feeding behaviour. PLoS ONE 14:e0221387

Button DJ, Barrett PM, Rayfield EJ (2017) Craniodental functional evolution in sauropodomorph dinosaurs. Paleobiology 43:435–462

Button DJ, Zanno LE (2020) Repeated evolution of divergent modes of herbivory in non-avian dinosaurs. Curr Biol 30:158–168

Cabreira SF, Kellner AWA, da-Silva SD, da Silva LR, Bronzati M, Marsola JCA, Müller RT, Bittencourt JS, Batista BJA, Raugust T, Carrilho R, Brodt A, Langer MC (2016) A unique late Triassic dinosauromorph assemblage reveals dinosaur ancestral anatomy and diet. Curr Biol 26:390–395

Campione NE (2017) Extrapolating body masses in large terrestrial vertebrates. Paleobiology 43:693–699

Campione NE, Evans DC (2020) The accuracy and precision of body mass estimation in non-avian dinosaurs. Biol Rev 95:1759–1797

Carrano MT (1999) What, if anything, is a cursor? Categories versus continua for determining locomotor habit in mammals and dinosaurs. J Zool Soc Lond 247:29–42

Caselli AT, Marsicano CA, Arcucci AB (2001) Sedimentología y paleontología de la Formación Los Colorados, Triásico Superior (provincias de La Rioja y San Juan, Argentina). Revista De La Asociación Geológica Argentina 56:173–188

Chapelle KEJ, Choiniere JN (2018) A revised cranial description of *Massospondylus carinatus* Owen (Dinosauria: Sauropodomorpha) based on computed tomographic scans and a review of cranial characters for basal Sauropodomorpha. PeerJ 6:e4224

Chapelle KEJ, Barrett PM, Botha J, Choiniere JN (2019) *Ngwevu intloko*: a new early sauropodomorph dinosaur from the Lower Jurassic Elliot Formation of South Africa and comments on cranial ontogeny in *Massospondylus carinatus*. PeerJ 7:e7240

Chapelle KEJ, Benson RBJ, Stiegler J, Otero A, Zhao Q, Choiniere JN (2020) A quantitative method for inferring locomotory shifts in amniotes during ontogeny, its application to dinosaurs and its bearing on the evolution of posture. Palaeontology 63:229–242

Charig AJ, Attridge J, Crompton AW (1965) On the origin of the sauropods and the classification of the Saurischia. Proc Linn Soc Lond 176:197–221

Cerda IA, Chinsamy A, Pol D, Apaldetti C, Otero A, Powell JE, Martínez RN (2017) Novel insight into the origin of the growth dynamics of sauropod dinosaurs. PLoS ONE 12:e0179707

Colombi CE, Santi-Malnis P, Correa GA, Martínez RN, Fernández E, Abelín D, Pradeiro, Apaldetti CG, Alcober O, Drovandi J (2015) La Formación Balde de Leyes, una nueva unidad estratigráfica de la cuenca Triásica de Marayes-El Carrizal, San Juan. Rev Asoc Geol Argentina 72(4):445–455

Cooper MR (1981) The prosauropod dinosaur *Massospondylus carinatus* Owen from Zimbabwe: Its biology, mode of life and phylogenetic significance. Occasional Papers of the Natural Museum of Rhodesia, B, Natural Sciences 6:689–840

Christian A (2010) Some sauropods raised their necks—evidence for high browsing in *Euhelopus zdanskyi*. Biol Let 6:823–825

Crompton AW, Attridge J (1986) Masticatory apparatus of the larger herbivores during Late Triassic and Early Jurassic time. In: Padian K (ed) The beginning of the age of the dinosaurs. Cambridge University Press, Cambridge, pp 223–236

Desojo JB, Fiorelli LE, Ezcurra MD, Martinelli AG, Ramezani J, Da Rosa AAS, von Baczko MB, Totteyn MJ, Montefeltro FC, Ezpeleta M, Langer MC (2020) The Late Triassic Ischigualasto Formation at Cerro Las Lajas (La Rioja, Argentina): fossil tetrapods, high-resolution chronostratigraphy, and faunal correlations. Sci Rep 10:10782

Ezcurra MD (2010) A new early dinosaur (Saurischia: Sauropodomorpha) from the Late Triassic of Argentina: a reassessment of dinosaur origin and phylogeny. J Syst Paleontol 8:371–425

Ferreira-Cardoso S, Araújo R, Martins NE, Martins GG, Walsh S, Martins RMS, Kardjilov N, Manke I, Hilger A, Castanhinha R (2017) Floccular fossa size is not a reliable proxy of ecology and behaviour in vertebrates. Sci Rep 7:2005

Galton PM (1971) The prosauropod dinosaur *Ammosaurus*, the crocodile *Protosuchus*, and their bearing on the age of the Navajo Sandstone of Northeastern Arizona. J Paleontol 45:781–795

Galton PM (1973) On the anatomy and relationships of *Efraasia diagnostica* (Huene) n. gen., a prosauropod dinosaur (Reptilia: Saurischia) from the upper triassic of Germany. Paläontol Z 47:229–255

Galton PM (1985) Diet of prosauropod dinosaurs from the late triassic and early jurassic. Lethaia 18:105–123

Galton PM (1986) Herbivorous adaptations of Late Triassic and Early Jurassic dinosaurs. In: Padian K (ed) The beginning of the age of the dinosaurs. Cambridge University Press, Cambridge, pp 203–221

Galton PM (1990) Basal Sauropodomorpha-Prosauropoda. In: Weishampel DB, Dodson P, Osmólska H (eds) The Dinosauria. University of California Press, Berkeley, pp 320–344

Galton PM (2012) *Plateosaurus engelhardti* Meyer, 1837 (Dinosauria, Sauropodomorpha): proposed replacement of unidentifiable name-bearing type by a neotype. Bull Zool NomClature 69:203–212

Galton PM, Upchurch P (2004) Prosauropoda. In: Weishampel DB, Dodson P, Osmólska H (eds) The Dinosauria, 2nd edn. University of California Press, Berkeley, pp 232–258

Garcia MS, Pretto FA, Dias-Da-Silva S, Müller RT (2019) A dinosaur ilium from the late triassic of Brazil with comments on key–character supporting Saturnaliinae. An Acad Bras Ciênc 91:1–20

Goussard F (2009) Étude morpho-fonctionnelle de la main des dinosaures sauropodomorphes: implications évolutives et apport de l'analyse en éléments finis. Thèse de doctorat, Muséum National d'Histoire Naturelle. Paris, France, pp 246

Gow CE, Kitching JW, Raath MA (1990) Skulls of the prosauropod dinosaur *Massospondylus carinatus* Owen in the collections of the bernard price institute for palaeontological research. Palaeontol Afr 27:45–58

Greenfield T, Bivens G, Fonseca A (2020) The correct authorship of *Coloradisaurus* (Dinosauria, Sauropodomorpha): Galton, 1990, not Lambert, 1983. Bull Zool Nomencl 77:153–155
Griffin CT, Stocker MR, Colleary C, Stefanic CM, Lessner EJ, Riegler M, Formoso K, Koeller K, Nesbitt SJ (2020) Assessing ontogenetic maturity in extinct saurian reptiles. Biol Rev 96:470–525
Hitchcock E Jr (1865) A supplement to the ichnology of New England. Wright and Potter, Boston
Holtz TR, Molnar RE, Currie PJ (2004) Basal Tetanurae. In: Weishampel DB, Dodson P, Osmólska H (eds) The Dinosauria, 2nd edn. University of California Press, Berkeley, pp 71–110
Hone DWE, Farke AA, Wedel MJ (2016) Ontogeny and the fossil record: what, if anything, is an adult dinosaur? Biol Lett 12:20150947
Horner JR, Weishampel DB, Forster CA (2004) Hadrosauridae. In: Weishampel DB, Dodson P, Osmólska H (eds) The Dinosauria, 2nd edn. University of California Press, Berkeley, pp 438–463
Huene F von (1914) Das natürliche System der Saurischia. Zentralblatt Mineralogie, Geologie, und Paläontologie B:154–158
von Huene F (1932) Die fossile Reptil-Ordnung Saurischia, ihre Entwicklung und Geschichte. Monographien Zur Geologie Und Paläontologie 4:1–361
Hummel J, Clauss M (2011) Sauropod feeding and digestive physiology. In: Klein N, Remes K, Gee CT, Sander PM (eds) Biology of the sauropod dinosaurs: understanding the life of giants. Indiana University Press, Bloomington, pp 11–33
Hutson JD, Hutson KN (2015) An examination of forearm bone mobility in *Alligator mississippiensis* (Daudin, 1802) and *Struthio camelus* Linnaeus, 1758 reveals that *Archaeopteryx* and dromaeosaurs shared an adaptation for gliding and/or flapping. Geodiversitas 37:325–344
ICZN (2019) Opinion 2435 (Case 3560)—*Plateosaurus* Meyer, 1837 (Dinosauria, Sauropodomorpha): new type species designated. Bull Zool Nomencl 76:144–145
Irmis RB (2007) Axial skeleton ontogeny in the Parasuchia (Archosauria: Pseudosuchia) and its implications for ontogenetic determination in archosaurs. J Vertebr Paleontol 27:350–361
Kent DV, Santi Malnis P, Colombi CE, Alcober OA, Martínez RN (2014) Age constraints on the dispersal of dinosaurs in the Late Triassic from magnetochronology of Los Colorados Formation (Argentina). Proc Natl Acad Sci 111:7958–7963
Knoll F (2005) The tetrapod fauna of the Upper Elliot and Clarens formations in the main Karoo Basin (South Africa and Lesotho). Bulletin De La Société Géologique De France 176:81–91
Knoll F, Schwarz-Wings D (2009) Palaeoneuroanatomy of *Brachiosaurus*. Annales De Paléontologie 95:165–175
Lallensack JN, Teschner EM, Pabst B, Sander PM (2021) New skulls of the basal sauropodomorph *Plateosaurus trossingensis* from Frick, Switzerland: Is there more than one species? Acta Palaeontol Pol 66:1–28
Lambert D (1983) A field guide to dinosaurs. Avon Books, New York, p 256
Langer MC, Ezcurra MD, Bittencourt JS, Novas FE (2010) The origin and early evolution of dinosaurs. Biol Rev 85:55–110
Langer MC, Ramezani J, Da Rosa AAS (2018) U-Pb age constraints on dinosaur rise from south Brazil. Gondwana Res 57:133–140
Langer MC, McPhee BW, Marsola JCA, Roberto-da-Silva L, Cabreira SF (2019) Anatomy of the dinosaur *Pampadromaeus barberenai* (Saurischia—Sauropodomorpha) from the Late Triassic Santa Maria Formation of southern Brazil. PLoS ONE 14:e0212543
Leal LA, Azevedo SAK, Kellner AWA, Da Rosa AAS (2004) A new early dinosaur (Sauropodomorpha) from the Caturrita Formation (Late Triassic), Paraná Basin, Brazil. Zootaxa 690:1–24
Li Z, Zhou Z, Clarke JA (2018) Convergent evolution of a mobile bony tongue in flighted dinosaurs and pterosaurs. PLoS ONE 13:e0198078
Marsh OC (1878) Principal characters of American Jurassic dinosaurs I. Am J Sci 16:411–416
Marsh OC (1895) On the Affinities and classification of the dinosaurian reptiles. Am J Sci 50:483–498
Marsh OC (1889) Notice of new American dinosaurs. Am J Sci, Series 3(37):331–336

Marsh AD, Rowe TB (2018) Anatomy and systematics of the sauropodomorph *Sarahsaurus aurifontanalis* from the Early Jurassic Kayenta Formation. PLoS ONE 13:e0204007

Martínez RN (2009) *Adeopapposaurus mognai*, gen. et sp. nov. (Dinosauria: Sauropodomorpha), with comments on adaptations of basal Sauropodomorpha. J Vertebr Paleontol 29:142–164

Martínez RN, Alcober OA (2009) A basal sauropodomorph (Dinosauria: Saurischia) from the Ischigualasto Formation (Triassic, Carnian) and the early evolution of Sauropodomorpha. PLoS ONE 4:e4397

Martínez RN, Colombi CE (2011) Evolución litofacial y edad de la Formación Cañón del Colorado (Jurásico inferior), Precordillera Oriental, San Juan. Revista De La Asociación Geológica Argentina 68:96–108

Martínez RN, Sereno PC, Alcober OA, Colombi CE, Renne PR, Montañez IP, Currie BS (2011) A basal dinosaur from the dawn of the dinosaur era in Southwestern Pangaea. Science 331:206–210

Martínez RN, Apaldetti C, Abelin D (2013) Basal sauropodomorphs from the Ischigualasto formation. J Vertebr Paleontolol 32(Supp. 1):51–69

Martínez RN, Apaldetti C, Correa G, Colombi CE, Fernández E, Santi Malnis P, Praderio A, Abelín D, Benegas LG, Aguilar-Cameo A, Alcober OA (2015) A new Late Triassic vertebrate assemblage from northwestern Argentina. Ameghiniana 52:379–390

von Meyer H (1837) Mittheilungen, an Professor Bronn gerichtet. Neues Jahrbuch Für Mineralogie, Geognosie, Geologie Und Petrefakten-Kunde 1837:314–317

McPhee BW, Yates AM, Choiniere JN, Abdala F (2014) The complete anatomy and phylogenetic relationships of *Antetonitrus ingenipes* (Sauropodiformes, Dinosauria): implications for the origins of Sauropoda. Zool J Linn Soc 171:151–205

McPhee BW, Choiniere JN, Yates AM, Viglietti PA (2015) A second species of *Eucnemesaurus* Van Hoepen, 1920 (Dinosauria, Sauropodomorpha): new information on the diversity and evolution of the sauropodomorph fauna of South Africa's lower Elliot Formation (latest Triassic). J Vertebr Paleontol 35:e980504

McPhee BW, Choiniere JN (2017) The osteology of *Pulanesaura eocollum*: implications for the inclusivity of Sauropoda (Dinosauria). Zool J Linn Soc 182:830–861

McPhee BW, Benson RBJ, Botha-Brink J, Bordy EM, Choiniere JN (2018) A giant dinosaur from the earliest Jurassic of South Africa and the transition to quadrupedality in early sauropodomorphs. Curr Biol 28:3143–3151

McPhee BW, Bittencourt JS, Langer MC, Apaldetti C, Da Rosa AAS (2020) Reassessment of *Unaysaurus tolentinoi* (Dinosauria: Sauropodomorpha) from the Late Triassic (early Norian) of Brazil, with a consideration of the evidence for monophyly within non-sauropodan sauropodomorphs. J Syst Paleontol 18:259–293

Müller RT, Da Rosa AAS, da Silva LR, Aires ASS, Pacheco CP, Pavanatto AEB, Dias da Silva S (2015) Wachholz, a new exquisite dinosaur bearing fossiliferous site from the Upper Triassic of southern Brazil. J S Am Earth Sci 61:120–128

Müller RT, Langer MC, Dias da Silva S (2018a) An exceptionally preserved association of complete dinosaur skeletons reveals the oldest long-necked sauropodomorphs. Biol Let 14:20180633

Müller RD, Cannon J, Qin X, Watson RJ, Gurnis M, Williams S, Pfaffelmoser T, Seton M, Russell SHJ, Zahirovic S (2018b) GPlates: Building a virtual Earth through deep time. Geochemistry. Geophys Geosyst 19:2243–2261

Müller RT (2019) Craniomandibular osteology of *Macrocollum itaquii* (Dinosauria: Sauropodomorpha) from the Late Triassic of southern Brazil. J Syst Paleontol 18:805–841

Müller RT, Garcia MS (2020) Rise of an empire: analyzing the high diversity of the earliest sauropodomorph dinosaurs through distinct hypotheses. Hist Biol 32:1334–1339

Müller RT, Ferreira JD, Pretto FA, Bronzati M, Kerber L (2021) The endocranial anatomy of *Buriolestes schultzi* (Dinosauria: Saurischia) and the early evolution of brain tissues in sauropodomorph dinosaurs. J Anat 238:809–827

Nau D, Lallensack JN, Bachmann U, Sander PM (2020) Postcranial osteology of the first early-stage juvenile skeleton of *Plateosaurus trossingensis* from the Norian of Frick, Switzerland. Acta Palaeontol Pol 65:679–708

Nesbitt SJ, Barrett PM, Werning S, Sidor CA, Charig AJ (2013) The oldest dinosaur? A middle triassic dinosauriform from Tanzania. Biol Let 9:20120949

Novas FE, Ezcurra MD, Chatterjee S, Kutty TS (2011) New dinosaur species from the Upper Triassic Upper Maleri and Lower Dharmaram formations of central India. Earth Environ Sci Trans Royal Soc Edinburgh 101(3–4):333–349

Otero A (2018) Forelimb musculature and osteological correlates in Sauropodomorpha (Dinosauria, Saurischia). PLoS ONE 13:e0198988

Otero A, Pol D (in press) Ontogenetic changes in the postcranial skeleton of Mussaurus patagonicus (Dinosauria, Sauropodomorpha) and their impact on the phylogenetic relationships of early sauropodomorphs. J Syst Palaeontol. https://doi.org/10.1080/14772019.2022.2039311

Otero A, Pol D (2013) J Vertebr Paleontol 33(5):1138–1168. https://doi.org/10.1080/02724634.2013.769444

Otero A, Krupandan E, Pol D, Chinsamy A, Choiniere J (2015) A new basal sauropodiform from South Africa and the phylogenetic relationships of basal sauropodomorphs. Zool J Linn Soc 174:589–634

Otero A, Allen V, Pol D, Hutchinson JR (2017) Forelimb muscle and joint actions in Archosauria: insights from *Crocodylus johnstoni* (Pseudosuchia) and *Mussaurus patagonicus* (Sauropodomorpha). PeerJ 5:e3976

Otero A, Cuff A, Allen V, Sumner-Rooney PD, Hutchinson JR (2019) Ontogenetic changes in the body plan of the sauropodomorph dinosaur *Mussaurus patagonicus* reveal shifts of locomotor stance during growth. Sci Rep 9:7614

Owen R (1842) Report on British fossil reptiles. Part II. Reports of the British association for the advancement of science 11:60–204

Owen R (1854) Descriptive catalogue of the fossil organic remains of reptilia and peces contained in the museum of the royal college of surgeons of England. Taylor and Francis, London, p 184

Pacheco C, Müller RT, Langer ML, Pretto FA, Kerber L, Dias da Silva S (2019) *Gnathovorax cabreirai*: a new early dinosaur and the origin and initial radiation of predatory dinosaurs. PeerJ 7:e7963

Parrish JM (1998) The origin of high browsing: The paleoecological significance of the sauropodomorph bauplan. J Vertebr Paleontol 18:69A

Parrish JM (2005) The origins of high browsing and the effects of phylogeny and scaling on neck length in sauropodomorphs. In: Carrano MT, Gaudin TJ, Blob RW, Wible JR (eds) Amniote paleobiology: perspectives on the evolution of mammals, birds, and reptiles. University of Chicago Press, Chicago, pp 201–223

Paul G (2010) The princeton field guide to dinosaurs. Princeton University Press, Princeton

Paulina Carabajal A (2012) Neuroanatomy of titanosaurid dinosaurs from the upper cretaceous of patagonia, with comments on endocranial variability within Sauropoda. Anat Rec 295:2141–2156

Peyre de Fabrègues C, Allain R (2016) New material and revision of *Melanorosaurus thabanensis*, a basal sauropodomorph from the Upper Triassic of Lesotho. PeerJ 4:e1639

Peyre de Fabrègues C, Allain R (2020) Kholumolumo ellenbergerorum, gen. et sp. nov., a new early sauropodomorph from the lower Elliot Formation (Upper Triassic) of Maphutseng, Lesotho. J Vertebr Paleontol 39(6):e1732996. https://doi.org/10.1080/02724634.2019.1732996

Peyre de Fabrègues C, Bi S, Li H, Li G, Yang L, Xu X (2020) A new species of early-diverging Sauropodiformes from the lower Jurassic Fengjiahe formation of Yunnan Province. China. Scientific Reports 10:10961

Pol D, Powell JE (2007a) New information on *Lessemsaurus sauropoides* (Dinosauria: Sauropodomorpha) from the Upper Triassic of Argentina. Spec Pap Palaeontol 77:223–243

Pol D, Powell JE (2007b) Skull anatomy of *Mussaurus patagonicus* (Dinosauria: Sauropodomorpha) from the Late Triassic of Patagonia. Hist Biol 19:125–144

Pol D, Garrido A, Cerda I (2011) A new sauropodomorph dinosaur from the Early Jurassic of Patagonia and the origin and evolution of the sauropod-type sacrum. PLoS ONE 6:e14572

Pol D, Otero A, Apaldetti C, Martínez (2021) Triassic sauropodomorph dinosaurs from South America: the origin and diversification of dinosaur dominated herbivorous faunas. J S Amn Earth Sci 107:103145

Pretto FA, Langer MC, Schultz CL (2018) A new dinosaur (Saurischia: Sauropodomorpha) from the Late Triassic of Brazil provides insights on the evolution of sauropodomorph body plan. Zool J Linn Soc 185:388–416

Preuschoft H, Hohn B, Stoinski S, Witzel U (2011) Why so huge? Biomechanical reasons for the acquisition of large size in sauropod and theropod dinosaurs. In: Klein N, Remes K, Gee CT, Sander PM (eds) Biology of the sauropod dinosaurs: understanding the life of giants. Indiana University Press, Bloomington, pp 195–216

Rauhut OWM, Fechner R, Remes K, Reis K (2011) How to get big in the Mesozoic: the evolution of the sauropodomorph body plan. In: Klein N, Remes K, Gee CT, Sander PM (eds) Biology of the sauropod dinosaurs: understanding the life of giants. Indiana University Press, Bloomington, pp 119–149

Rauhut OWM, Holwerda FM, Furrer H (2020) A derived sauropodiform dinosaur and other sauropodomorph material from the late triassic of canton Schaffhausen. Switzerland. Swiss Journal of Geosciences 113:8

Remes K (2008) Evolution of the pectoral girdle and forelimb in Sauropodomorpha (Dinosauria, Saurischia): osteology, myology and function. Dissertation zur Erlangung des Doktorgrades, Ludwig-Maximilians Universität, München, Deutschland, pp 355

Romer AS (1972) The Chañares (Argentina) Triassic reptile fauna; XIV, *Lewisuchus admixtus*, gen. et sp. nov., a further thecodont from the Chañares beds. Breviora 390:1–13

Rowe TB, Sues H-D, Reisz RR (2010) Dispersal and diversity in the earliest North American sauropodomorph dinosaurs, with a description of a new taxon. Proc R Soc Lond B 278:1044–1053

Sander PM (2013) An evolutionary cascade model for sauropod dinosaur gigantism—overview update and tests. PLoS One 8(10):e78573. https://doi.org/10.1371/journal.pone.0078573

Sander PM, Christian A, Clauss M, Fechner R, Gee CT, Griebeler E-M, Gunga H-C, Hummel J, Mallison H, Perry SF, Preuschoft H, Rauhut OWM, Remes K, Tütken T, Wings O, Witzel U (2011) Biology of the sauropod dinosaurs: the evolution of gigantism. Biol Rev 86:117–155

Santi Malnis P, Colombi C, Rothis LM, Alcober O (2020) Fluvial architecture and paleoenvironmental evolution of the los Colorados Formation (Norian): postrift stage of the Ischigualasto-Villa Unión Basin, NW Argentina. J Sediment Res 90:1436–1462

Seeley HG (1887) The classification of the Dinosauria. Geol Mag 3:562

Seeley HG (1888) On the classification of the fossil animals commonly named Dinosauria. Proc R Soc Lond 43:165–171

Sekiya T, Jin X, Zheng W, Shibata M, Azuma Y (2014) A new juvenile specimen of Yunnanosaurus robustus (Dinosauria: Sauropodomorpha) from Early to Middle Jurassic of Chuxiong Autonomous Prefecture, Yunnan Province, China. Hist Biol 26(2):252–277. https://doi.org/10.1080/08912963.2013.821702

Sereno PC (1997) The origin and evolution of dinosaurs. Annu Rev Earth Planet Sci 25:435–489

Sereno PC (1998) A rationale for phylogenetic definitions, with application to the higher-level taxonomy of Dinosauria. Neues Jahrbuch für Geologie und Paläontologie. Abhandlungen 210:41–83

Sereno PC (2007) Basal Sauropodomorpha: historical and recent phylogenetic hypotheses, with comments on Ammosaurus major (Marsh 1889). Spec Pap Palaeontol 77:261–289

Sereno PC, Martínez RN, Alcober OA (2013) Osteology of *Eoraptor lunensis* (Dinosauria, Sauropodomorpha). J Vertebr Paleontol 32(Supp. 1):83–179

Smith N, Pol D (2007) Anatomy of a basal sauropodomorph dinosaur from the early jurassic hanson formation of Antarctica. Acta Palaeontol Pol 52:657–674

Stevens KA, Parrish JM (1999) Neck posture and feeding habits of two jurassic sauropod dinosaurs. Science 284:798–800

Stocker MR, Nesbitt SJ, Criswell KE, Parker WG, Witmer LM, Rowe TB, Ridgely R, Brown MA (2016) A dome-headed stem archosaur exemplifies convergence among dinosaurs and their distant relatives. Curr Biol 26:2674–2680

Tornier G (1913) Reptilia. Paläontologie [Reptilia. Paleontology]. Handwörterbuch Der Naturwissenschaften 8:337–376

Tütken T (2011) The diet of sauropod dinosaurs: implications of carbon isotope analysis on teeth, bones, and plants. In: Klein N, Remes K, Gee CT, Sander PM (eds) Biology of the sauropod dinosaurs: understanding the life of giants. Indiana University Press, Bloomington, pp 57–82

Upchurch P (1997) Prosauropoda. In: Currie P, Padian K (eds) Encyclopedia of dinosaurs. Academic Press, London, pp 599–607

Upchurch P, Barrett P, Dodson P (2004) Sauropoda. In: Weishampel DB, Dodson P, Osmólska H (eds) The dinosauria, 2nd edn. University of California Press, Berkeley, California, pp 259–324

Upchurch P, Barrett P, Galton P (2007a) A phylogenetic analysis of basal sauropodomorph relationships: Implications for the origin of sauropod dinosaurs. Spec Pap Palaeontol 77:57–90

Upchurch P, Barrett P, Zhao X, Xu X (2007b) A re-evaluation of *Chinshakiangosaurus chunghoensis* Ye vide Dong 1992 (Dinosauria, Sauropodomorpha): implications for cranial evolution in basal sauropod dinosaurs. Geol Mag 144:247–262

Viglietti PA, McPhee BW, Bordy EM, Sciscio L, Barrett PM, Benson RBJ, Wills S, Chapelle KEJ, Dollman KN, Mdekazi C, Choiniere JN (2020) Biostratigraphy of the *Massospondylus* Assemblage Zone (Stormberg Group, Karoo Supergroup), South Africa. S Afr J Geol 123:249–262

Walsh SA, Iwaniuk AN, Knoll MA, Bourdon E, Barrett PM, Milner AC, Nudds RL, Abel RL, Sterpaio PD (2013) Avian cerebellar floccular fossa size is not a proxy for flying ability in birds. PLoS ONE 8:e67176

Wickham H (2009) Ggplot2: elegant graphics for data analysis. Springer-Verlag, New York, p 213

Wilson JA, Upchurch P (2003) A revision of *Titanosaurus* Lydekker (Dinosauria—Sauropoda), the first dinosaur genus with a 'Gondwanan' distribution. J Syst Palaentol 1:125–160

Wilson JA, Sereno PC (1998) Early evolution and higher-level phylogeny of sauropod dinosaurs. Soc Vertebr Paleontol Mem 5:1–79

Witmer LM, Chatterjee S, Franzosa J, Rowe T (2003) Neuroanatomy of flying reptiles and implications for flight, posture and behaviour. Nature 425:950–953

Yates AM (2003) A new species of the primitive dinosaur *Thecodontosaurus* (Saurischia: Sauropodomorpha) and its implications for the systematics of early dinosaurs. J Syst Paleontol 1:1–42

Yates AM (2007a) The first complete skull of the Triassic dinosaur *Melanorosaurus* Haughton (Sauropodomorpha: Anchisauria). Spec Pap Palaeontol 77:9–55

Yates AM (2007b) Solving a dinosaurian puzzle: the identity of *Aliwalia rex* Galton. Hist Biol 19:93–123

Yates AM, Kitching JW (2003) The earliest known sauropod dinosaur and the first steps towards sauropod locomotion. Proc R Soc Lond B 270:1753–1758

Yates AM, Bonnan MF, Neveling J, Chinsamy A, Blackbeard MG (2010) A new transitional sauropodomorph dinosaur from the Early Jurassic of South Africa and the evolution of sauropod feeding and quadrupedalism. Proc R Soc Lond B 277:787–794

Young CC (1941) A complete osteology of Lufengosaurus huenei Young (gen. et sp. nov.) from Lufeng, Yunnan, China. Palaeontol Sin Ser C 7:1–53

Zhang QN, You HL, Wang T, Chatterjee S (2018) A new sauropodiform dinosaur with a 'sauropodan' skull from the Lower Jurassic Lufeng Formation of Yunnan Province. China Scientific Reports 8:13464

South American Non-Gravisaurian Sauropodiformes and the Early Trend Towards Gigantism

Cecilia Apaldetti and Ricardo N. Martínez

Abstract Sauropodomorpha was an herbivorous clade of dinosaurs that lived through nearly the entire Mesozoic Era. Sauropodomorphs appeared at the end of the Triassic as small and gracile forms and evolved into the largest land animals that ever lived on Earth. This evolutionary path towards giant forms implied several skeletal modifications and anatomical innovations already recorded at the end of the Triassic that continued to develop throughout the rest of the Mesozoic. Recent discoveries of early-diverging Sauropodiformes from Gondwana shed light on this crucial stage in the history of the group, providing new information about the origin of gigantism among Sauropodomorpha. Non-gravisaurian sauropodiform is the grade of dinosaurs that—despite not being characterized by typical derived or 'sauropod-like' features, often regarded as a prerequisite for gigantism (long neck, columnar-limbed, obligated quadrupedal posture)—show the first clear steps towards gigantism. This group is known by moderately gracile to robust forms that attained large body size during the Late Triassic and Early Jurassic. The South American record is restricted to Argentina, including two early-branching forms from Patagonia (*Mussaurus patagonicus* and *Leonerasaurus taquetrensis*) and the lessemsaurids *Lessemsaurus sauropoides* and *Ingentia prima*, from the north-western region of that country. Here we provide a comprehensive review of these four taxa, highlighting the main plesiomorphic and derived traits that relate them to both early-diverging (Massopoda) and late-diverging (Gravisauria) sauropodomorphs, positioning them as the first steps towards gigantism on Earth.

Keyword Sauropodiformes · Gigantism · Quadrupedalism · Evolution

C. Apaldetti (✉) · R. N. Martínez
IMCN, Instituto y Museo de Ciencias Naturales, FCEFN, Universidad Nacional de San Juan, San Juan Province San Juan 5400, Argentina
e-mail: capaldetti@unsj.edu.ar

R. N. Martínez
e-mail: martinez@unsj.edu.ar

C. Apaldetti
Consejo Nacional de Investigaciones Científicas y Técnicas (CONICET) and Centro de Investigaciones de La Geósfera y Biósfera (CIGEOBIO), San Juan, Argentina

A. Otero et al. (eds.), *South American Sauropodomorph Dinosaurs*, Springer Earth System Sciences, https://doi.org/10.1007/978-3-030-95959-3_3

1 Introduction

Sauropodomorpha was the first successful group of herbivorous dinosaurs, dominating faunal assemblages and most terrestrial ecosystems for more than 140 myr, until their extinction at the end of the Cretaceous (Galton and Upchurch 2004; Upchurch et al. 2004; Mannion et al. 2011). From the Late Triassic (~230 myr), sauropodomorphs first appeared as small, gracile and bipedal forms recorded in South America (e.g. *Saturnalia*, *Eoraptor*, *Panphagia*) and quickly attained a global distribution. Moreover, within a few million years after their initial appearance, they had already evolved very large body sizes ($\geq$10 tonnes, Apaldetti et al. 2018; McPhee et al. 2018), eventually becoming the most massive land animals to ever live on Earth (Sander et al. 2011a, b; Benson et al. 2014, 2017; Carballido et al. 2017; Otero et al. 2021). The protagonists of this dramatic evolutionary transformation—one of the most extraordinary in the history of life—were laid during the Late Triassic and the Early Jurassic and are well-documented in taxa found mainly in South America and South Africa.

The evolutionary path towards gigantic body sizes entailed several key skeletal modifications, anatomical innovations and osteohistological novelties that allowed for almost continuous size increase throughout most of the Mesozoic Era (Rauhut et al. 2011; Sander et al. 2011a; Sander 2013a, b). The most visible changes related to their transformation, from the small-bodied and bipedal condition primitive to Dinosauria, to the gigantic, quadrupedal condition of Sauropoda, include decreased skull size, elongation of the neck and forelimbs, strengthening and columnization of the limbs, development of extensive postcranial pneumaticity, reduction of the phalanges, modification of the metacarpus into a U-shaped support structure and non-cyclical bone growth—among others (Wilson and Sereno 1998; Sander et al. 2004, 2011a, b; Klein et al. 2011; Rauhut et al. 2011; Sander 2013a, b; Cerda et al. 2017).

Non-gravisaurian sauropodiforms are a paraphyletic group (see Table 1) of medium-to-large-bodied taxa whose mosaic of plesiomorphic and derived traits (Table 2) relate them to both earlier-diverging (Massopoda) and later-diverging (Gravisaria) sauropodomorphs. It provides critical information on the acquisition of one of the key sauropodan adaptations: graviportal quadrupedalism—a unique locomotory strategy in which lateral shear forces are reduced via the parasagittal alignment of the limb bones in conjunction with reduced flexion at the knee/elbow (Wilson and Carrano 1999; Carrano 2001). Early-branching sauropodiforms are well-recorded within Gondwanan deposits (more than ten species known so far), mainly from the Late Triassic to Early Jurassic of South Africa and South America. The South American record includes the Triassic lessemsaurids *Lessemsaurus sauropoides* (Bonaparte 1999; Pol and Powell 2007b) and *Ingentia prima* (Apaldetti et al. 2018) from northwestern Argentina and the Jurassic Patagonian *Mussaurus patagonicus* (Bonaparte and Vince 1979; Pol and Powell 2007a; Otero and Pol 2013) and *Leonerasaurus taquetrensis* (Pol et al. 2011). The Patagonian Early Jurassic taxa represent some of the most basal forms among Sauropodiformes, whereas the Late Triassic

Table 1 List of the non-gravisaurian Sauropodiformes taxa based on the phylogenetic relationship of the latest phylogenetic analysis of Pol et al. (2021). South American taxa are highlighted in boldface

Taxon	Formation—Basin	Locality	Age	References
NMQR1551 (ex *Melanorosaurus*)	Lower Elliot Formation	Milner (now Swempoort 65) farm, South Africa	Late Triassic	Galton et al. (2005), Bonnan and Yates (2007), McPhee et al (2017)
Lessemsaurus sauropoides	**Los Colorados Formation, Ischigualasto-Villa Unión Basin**	**La Esquina, Villa Unión Town, La Rioja Province, Argentina**	**Late Triassic (Norian)**	Bonaparte (1999), Pol and Powell (2007b, Apaldetti et al. (2018)
Meroktenos thabanensis	Elliot Formation, Stormberg Group, Karoo Basin	Lesotho	Late Triassic	Peyre de Fabrègues and Allain (2016)
Kholumolumo ellenbergerorum	Lower Elliot Formation, Stormberg Group	Maphutsneg, Mohale's Hoek District, Lesotho	Late Triassic	Peyre de Fabrègues and Allain (2020)
Schleitheimia schutzi	Klettgau Formation, Gruhalde Member	Canton Schaffhausen, Switzerland	Early Jurassic	Rauhut et al. (2020)
Ingentia prima	**Quebrada del Barro Formation, Marayes-El Carrizal Basin**	**Quebrada del Puma, San Juan Province, Argentina**	**Late Triassic**	Apaldetti et al. (2018)
Camelotia borealis	Westbury Formation, Aust Cliff	England	Late Triassic	Galton (1985, 1998)
Blikanasaurus cromptoni	Lower Elliot Formation, Stormberg Group	Free State, South Africa	Late Triassic	Galton and Van Heerden (1985, 1998) Yates (2008)
Isanosaurus attavipachi	Nam Phong Formation,	Phu Nok Khian hill, Chaiyaphum Province, Thailand	Late Triassic-Lower Jurassic	Buffetaut et al. 2000
Sefapanosaurus zastronensis	Elliot Formation	Free State Province, South Africa	Late Triassic-Lower Jurassic	Otero et al. (2015)
NMQR3314 (ex *Melanorosaurus*)	Upper Elliot Formation	Damplaast 55 farm, South Africa	Early Jurassic	Galton et al. (2005), Bonnan and Yates (2007), McPhee et al (2017)

(continued)

Table 1 (continued)

Taxon	Formation—Basin	Locality	Age	References
Aardonyx celestae	Upper Elliot Formation	Spion Kop, Senekal District, Free State South Africa	Early Jurassic	Yates et al. (2010)
Antetonitrus ingenipes	Lower Elliot Formation, Stormberg Group	Ladybrand District, Free State, South Africa	Early Jurassic	Yates and Kitching (2003), McPhee et al. (2014)
Pulanesaura eocollum	Upper Elliot Formation	Spion Kop 932 farm, Senekal District, Free State, South Africa	Early Jurassic	McPhee et al. (2015), McPhee and Choiniere (2017)
Ledumahadi mafube	Upper Elliot Formation	Beginsel farm, Clarens, Free State Province, South Africa	Early Jurassic	McPhee et al. (2018)
Yizhousaurus sunae	Lufeng Formation, Zhangjiaao Member	Chuanjie Town, Lufeng County, Yunnan Province, China	Lower Jurassic	Zhang et al. (2018)
Mussaurus patagonicus	**Laguna Colorada Formation, El Tranquilo Group**	**Laguna Colorada locality, Santa Cruz Province, Patagonia Argentina**	**Early Jurassic**	Bonaparte and Vince (1979), Pol and Powell (2007a), Otero and Pol (2013)
Leonerasaurus traquetensis	**Las Leoneras Formation**	**Chubut Province, Patagonia Argentina**	**Early Jurassic**	Pol et al. (2011)
Gongxianosaurus shibeiensis	Ziliujing Formation, Dongyuemiao Member	Sichuan Province, China	Early Jurassic	He et al. (1998)
Irisosaurus yimenensis	Fengjiahe Formation	Jiaojiadian area, Shijie Township, Yimen County, Yunnan Province, China	Early Jurassic	Peyre de Fabrègues et al. (2020)

lessemsaurids are some of the more derived forms of the group, being potentially closely related to the first sauropods and demonstrating the first genuine experiments in gigantism (Apaldetti et al. 2018; McPhee et al. 2018). Here we provide a comprehensive review of these South American non-gravisaurian sauropodiforms exhibiting

Table 2 Main plesiomorphic and derived features among non-gravisaurian Sauropodiformes. Abbreviations: ch., morphological character (character list provided by Pol et al. 2021); PFB, parallel fibered bone; WFB, woven fibered bone

Elements/structure	Feature	Plesiomorphic condition	Derived condition
AXIAL SKELETON	Cervical vertebrae, morphology	Dorsoventrally low neural arches (ch. 133.0)	High neural arches, elongated neural spines (ch. 133.1)
	Cervical vertebrae, zygapophyseal level	Planar zygapophyseal facets (ch. 404.0)	Non-planar, postzygapophyses elevated relative to prezypapophyseal level (ch. 401.1–2)
	Cervico-dorsal hyposphenes	Low, < 70% the neural canal height (ch. 157.1)	Notably tall, ≥ than neural canal height (ch. 157.2)
	Cervico-dorsal pneumaticity	– (not ch.)	System of pneumatic structures
	Sacrum	3 elements (not ch.)	4 elements or more
SCAPULA	Scapular blade, robustness	Gracile, slender (not ch.)	Wide, extremely robust
HUMERUS	General morphology	Hourglasses, well-expanded ends	Not-hourglasses
	Deltopectoral crest, extension	Extended > 45% (ch. 207.0–1)	Extended < 45% (ch. 207.2)
	Deltopectoral crest, development	Well-developed (not ch.)	Low
ULNA	Radial fossa	Shallow (ch. 214.1)	Deep (ch. 214.2)
	Anterolateral process	Not well-developed (not ch.)	Well-developed, enlarged
	Olecranon	Well-developed (ch. 215.0)	Not well-developed (ch. 215.1)
MANUS	Metacarpal I, robustness	Elongated, proximal surface width ≤ 1 time the proximodistal length Broader and shorter than other metacarpals	Extremely robust, broad and short, mediolateral width of proximal surface > 1 times the total length
	Manual phalanx I.1	Twisted, ~ 60 degrees (ch. 234.2)	Twisted, < 60 degrees (ch. 234.0–1)
ILIUM	Preacetabular process	Short, low and triangular. Does not anteriorly enlarged (ch. 246.0; ch. 248.0)	Anteriorly projected and dorsoventrally expanded process (ch. 246.1; ch. 248.1)

(continued)

Table 2 (continued)

Elements/structure	Feature	Plesiomorphic condition	Derived condition
H/F (Humerus/Femur)	H/F length ratio	0.55–0.70 times (not ch.)	> 0.80 times
FL/HL (Forelimb/Hindlimb)	FL/HL length ratio	< 0.7 times (not ch.)	> 0.7 times
FEMUR	Shaft, curvature, lateral view	Sigmoid (ch. 280.0)	Weakly bent, straight (ch. 280.1–2)
	Mid-shaft cross-section	Circular (ch. 281.0)	Sub-ovoid or elliptical (ch. 281.1)
	Fourth trochanter, position	Proximal half of the shaft (ch. 293.0)	Mid-length of the shaft (ch. 293.1)
T/F (Tibia/Femur)	T/F length ratio	> 0.7 times	< 0.7 times
PES	Metatarsal I	Slender, twice as long as wide or greater (ch. 293.0)	Robust, length/width ratio < 1.5; or proximal surface between 0.4–0.7 times its proximodistal length (ch. 293.1)
	Metatarsal II, proximal articular surface	Hourglass shape (ch. 334.1, ch. 335.1)	Straight medial and lateral margins (ch. 334.0, ch. 335.0)
	Metatarsal V	Reduced, flat and triangular (not ch.)	Enlarged metatarsal
HISTOLOGY	Growth patterns	Cyclical, regularly spaced growth marks throughout entire cortex (ch. 418.0)	Acyclical, uninterrupted deposition of fibrolamellar bone tissue during early phase of development (ch. 418.1)
	Fibered bone disposition, abundance of WFB versus PFB	PFB > WFB (ch. 419.0)	WFB > WFB (ch. 419.1)

a mosaic of plesiomorphic and derived traits that capture the main evolutionary changes that developed in sauropodomorphs towards gigantism.

Institutional Abbreviations CRILAR: Centro Regional de Investigaciones Científicas y Tecnológicas, La Rioja, Argentina; **MACN**: Museo Argentino de Ciencias Naturales, Buenos Aires, Argentina; **MLP**: Museo de La Plata, La Plata, Argentina; **MPEF**: Museo Paleontológico 'Egidio Feruglio', Trelew, Argentina; **MPM**: Museo Regional Provincial 'Padre M. J. Molina', Rio Gallegos, Argentina;

NMQR: National Museum, Bloemfontein, South Africa; **PVL**: Instituto 'Miguel Lillo', Tucumán, Argentina; **PVSJ**: Paleovertebrate collection of the Instituto y Museo de Ciencias Naturales (IMCN), Universidad Nacional de San Juan, San Juan, Argentina.

2 Methods

2.1 Nomenclature and Terminology

We employed traditional anatomical and directional terms over veterinarian alternatives (Wilson 2006). For example, 'anterior' and 'posterior' are used as directional terms rather than the veterinarian alternatives 'rostral' or 'cranial' and 'caudal'.

Different sources for phylogenetic nomenclature are used to refer for taxa within Dinosauria (Table 3). The diagnosis of each taxon mentioned is based on the original descriptions and/or specific studies of each species.

2.2 Phylogenetic Relationships

The phylogenetic analysis was based on that recently published by Pol et al. (2021), focused on early sauropodomorphs including recently described species from the Late Triassic and Early Jurassic. A total of 79 taxa and 419 morphological characters were included in the data matrix (see below). The dataset was analysed with equally weighted parsimony in TNT 1.5 (Goloboff et al. 2008) using a heuristic search, and new technologies algorithms were applied until 100 hits to minimum length was reached. A subsequent search was conducted by performing a round of TBR branch swapping on the most parsimonious trees (MPTs).

3 Systematic Paleontology

Dinosauria Owen 1842

Saurischia Seeley 1888

Sauropodomorpha von Huene 1932

Anchisauria Galton and Upchurch 2004

Sauropodiformes Sereno 2007

Table 3 Phylogenetic definitions for Sauropodomorpha used in this study

Clade	Definition	Reference
Sauropodomorpha	The most inclusive clade containing *Saltasaurus loricatus* but not *Passer domesticus* or *Triceratops horridus*	Sereno (2007)
Massopoda	The most inclusive clade containing *Saltasaurus loricatus* but not *Plateosaurus engelhardti*	Yates (2007a, b)
Massospondylidae	The most inclusive clade containing *Massospondylus carinatus,* but not *Plateosaurus engelhardti* or *Saltasaurus loricatus*	Sereno (2007)
Anchisauria	*Anchisaurus polyzelus* and *Melanorosaurus readi*, their common ancestor, and all its descendants	Galton and Upchurch (2004)
Sauropodiformes	The least inclusive clade containing *Mussaurus patagonicus* and *Saltasaurus loricatus*	Sereno (2007)
Lessemsauridae	The clade defined as *Lessemsaurus sauropoides* and *Antetonitrus ingenipes*, and all the descendants from their most common ancestor	Apaldetti et al. (2018)
Sauropoda	The least inclusive clade containing *Vulcanodon karibaensis* and Eusauropoda	Salgado et al. (1997) Langer et al. (2010)
Gravisauria	A node-based taxon defined as the most recent common ancestor of *Tazoudasaurus naimi* and *Saltasaurus loricatus*, and all its descendants	Allain and Aquesbi (2008)
Eusauropoda	The least inclusive clade containing *Shunosaurus lii* and *Saltasaurus loricatus*	Upchurch et al. (2004)

Definition The least inclusive clade containing *Mussaurus patagonicus* (Bonaparte and Vince 1979) and *Saltasaurus loricatus* (Bonaparte and Powell 1980; Sereno 2007).

Diagnosis Based on the current analysis (Pol et al. 2021), Sauropodiformes is diagnosed by: absence of longitudinal ventral sulcus on proximal and middle caudal vertebrae; length of manual digit one greater than the length of manual digit two; longitudinal axis of the femur weakly bent with an offset of less than 10° in lateral view; femoral length between 60 and 79 cm; and the presence of the caudodistal tubercle of the radius. In some most parsimonious trees, the absence of a caudal end of dentary tooth row medially inset with a thick lateral ridge on the dentary forming a bucal emargination; and distal surface of tibiofibular crest wider mediolaterally than deep anteroposteriorly.

Mussaurus Bonaparte and Vince 1979

Mussaurus patagonicus Bonaparte and Vince 1979

Holotype PVL 4068, articulated skeleton of a post-hatchling individual.

Referred Specimens PVL 4208, partially articulated skeleton of a neonate individual. PVL 4209, partial skull and lower jaw of a neonate individual associated with partially articulated postcranial material. PVL 4210, almost complete skull associated with postcranial remains of a neonate individual. PVL 4211, incomplete skull and mandible and disarticulated postcranial elements of a neonate individual. PVL 4212, partially articulated elements of a neonate individual. PVL 4213, three closely associated skeletons of neonate individuals. PVL 5865, articulated skeleton and almost complete skull of a neonate individual. MACN-PV 4111, almost complete skeleton of a neonate individual. MPM-PV 1828, humerus and ulna of a juvenile individual. PVL 4587, partially articulated skeleton belonging to at least two different juvenile individuals. MPM-PV 1813/1, skull articulated with a cervicodorsal series of a juvenile individual. MPM-PV 1813/2, almost complete juvenile individual. MPM-PV 1813/3, skull, cervical and partial forelimb of juvenile individual. MPM-PV 1813/4, skull articulated with a cervicodorsal series, scapular girdle and forelimbs of a juvenile individual. MPM-PV 1813/5, dorsal vertebrae and partial hindlimb of a juvenile individual. MPM-PV 1813/6, dorsal and sacral vertebrae and partial hindlimb of a juvenile individual. MPM-PV 1813/7, dorsal and sacral vertebrae and partial hindlimb of a juvenile individual. MPM-PV 1813/8, precaudal vertebral series and partial hindlimb of a juvenile individual. MPM-PV 1813/9, dorsal vertebrae and partial forelimb of a juvenile individual. MPM-PV 1813/10, partial hindlimb of juvenile individual. MPM-PV 1813/11, dorsal vertebrae and partial forelimb of a juvenile individual. MPM-PV 1813/12, isolated femur of juvenile individual. MPM-PV 1813/13, isolated left pes of juvenile individual. MPM-PV 1836, incomplete and fragmentary juvenile individual, including axial and appendicular elements. MPM-PV 1838, fragmentary postcranial skeleton. MPM-PV 1822, incomplete right manus with articulated distal carpals of an adult individual. MPM-PV 1868, axial (including skull) and appendicular bones of an adult (?) individual. MPM-PV 1869, skull and fragmentary vertebral remains and appendicular bones of an adult (?) individual. MPM-PV 1901, dorsal and sacral vertebrae and appendicular bones of an adult (?) individual. MACN-SC 3379, partially articulated postcranial remains of an adult individual, consisting of several cervicodorsal vertebrae, forelimb and hind limb remains. MLP 61-III-20-22, postcranial skeleton, including vertebral remains and limb elements of an adult individual. MLP 61-III-20-23, postcranial skeleton, consisting of dorsal vertebrae, sacrum, ilia, ischia, pubis and fragmentary remains of the forelimb and hind limb of a putative subadult individual. MLP 68-II-27-1, postcranial skeleton consisting of articulated cervical (with fragmentary cranial bones associated), dorsal and caudal vertebrae, and girdle and limb elements, belonging to two incomplete adult individuals (specimens A and B).

Locality and Age The holotype and most of the referred material of *Mussaurus* come from 'Laguna Colorada' locality, Santa Cruz Province, Argentina, corresponding to the Laguna Colorada Formation, El Tranquilo Group (Late Sinemurian, Lower

Jurassic; Smith et al. 2014). The PVL, MACN and most MPM specimens were collected from the Colorada Lake, at 'Laguna Colorada' locality. Remaining adult specimens were collected from the 'Herbst' (MPM-PV 1901) and 'Di Persia' localities (MPM-PV 1868 and MPM-PV 1869), both putatively assigned to the Laguna Colorada Formation. Remains of *Mussaurus* were found in three successive loessite horizons, consisting on reddish to brownish siltstones and claystones, fine- to medium-grained sandstones and subordinate thin conglomerates deposited by a fluvial system with moderate sinuosity channels. The faunal remains are concentrated in a 3 m thick interval of structureless mottled light reddish-brown/olive-grey massive siltstone with scattered small mudrock pebbles which we interpret as floodplain loess. The vertebrate-bearing sediments are dated here as Early Jurassic (Sinemurian, Smith et al. 2014).

Diagnosis *Mussaurus patagonicus* is a non-gravisaurian sauropodiform with the following unique combination of characters (autapomorphies marked here with asterisks): anterior margin of the premaxilla directed posterodorsally, forming an angle of 45° with the alveolar margin; main body of premaxilla anteorposteriorly shorter than deep; ventral ramus of lacrimal straight and anteroposteriorly narrow, markedly constricted at midheight; the absence of lateral (superficial) lamina of lacrimal covering the posterodorsal region of antorbital sinus; presence of a thin ridge along the entire ventral ramus of the lacrimal; straight ventral end of descending process of postorbital; anterior extension of the infratemporal fenestra underneath the orbit; anterior and dorsal rami of quadratojugal sub-perpendicular to each other; presence of a dorsally projected peg-like process on the anterior tip of the dentary (immediately anterior to the first mandibular tooth)*; low and elongated posterior end of mandibular rami (posterior to the external mandibular fenestra) with only gently sigmoid dorsal margin*; dorsoventrally expanded anterior end of mandibular symphysis; premaxillary teeth and anterior maxillary teeth with sigmoid distal margin lacking serrations on their margins (or having weakly developed broad denticles at the apical region); posterior maxillary teeth lanceolate with serrations on the mesial and distal margins of the apical region; presence of a robust posterior infradiapophyseal lamina (subcircular in cross-section); presence of a well-developed prezygodiapophyseal lamina in middle dorsal vertebrae*; presence of a ventromedial robust ridge extending along the medial surface of the scapula reaching the distal third of the scapular blade*; distal carpal I smaller than distal carpal II*; presence of a well-developed convex bulge on the caudomedial margin of astragalus*; astragalus with concave distal lateromedial surface*.

Comments Bonaparte and Vince (1979) originally described *M. patagonicus* including remains of eight post-hatchling individuals. Some of these specimens are articulated, almost completely preserved, and found closely associated with each other. Additionally, two unhatched eggs and eggshell fragments were found in close association with these specimens (Bonaparte and Vince 1979). The individual age of these specimens has generally been interpreted as representing extremely young individuals rather than embryos (Bonaparte and Vince 1979; Weishampel and Horner 1994). Pol and Powell (2007a) expanded the original diagnosis, adding a suite of mainly craniodental characters through their work on the post-hatchling type material (Bonaparte and Vince 1979). Finally, Otero and Pol (2013) expanded the diagnosis of this taxon based on the specimens originally published by Casamiquela (1980) as well as new material from the type locality (MLP specimens), adding several autapomorphies and a unique combination of characters described from both adults and subadult individuals (Otero and Pol 2013).

Leonerasaurus Pol, Garrido and Cerda 2011

Leonerasaurus taquetrensis Pol, Garrido and Cerda 2011

Holotype MPEF-PV 1663, anterior region of right dentary and isolated teeth, articulated series of cervical and anterior dorsal vertebrae, partially articulated posterior dorsal vertebrae and articulated sacrum (preserved in natural contact with both ilia), right scapula and humerus, left and right ilia, right ischium, partially preserved femur, articulated metatarsal I and II, and pedal ungual. All vertebrae, as well as the scapula, humerus and pelvis, were found in natural position as a partially articulated specimen. The dentary, teeth, femur and pedal remains were found within one-metre radius of the centre of the articulated specimen. As no other remains were found at the site, these elements have been interpreted as belonging to a single individual (Pol et al. 2011).

Locality and Age Cañadón Las Leoneras, south of Cañadón del Zaino, southeast of Sierra de Taquetrén, Chubut Province, Central Patagonia, Argentina (Fig. 1c). The only known specimen was found approximately 42 m below the top of the Las Leoneras Formation, a unit considered to be Lower Jurassic in age (Nakayama 1973), and more specifically referred to the Pliensbachian–Toarcian (Figari and Courtade 1993) or upper Sinemurian–Toarcian (Page et al. 2000). However, no radiometric age for these sediments is currently available. The age of the Las Leoneras Formation is certainly constrained by the Middle Jurassic volcanic facies of the overlying Lonco Trapial Formation (Pol et al. 2011).

Diagnosis *Leonerasaurus taquetrensis* is a small non-gravisaurian sauropodiform diagnosed by a unique combination of characters including the following autapomorphy: anterior unserrated teeth with low, spoon-shaped crowns (slenderness index, SI = 1.3); dorsosacral rib attached to the preacetabular process of ilium (also present in *Lufengosaurus huenei*); neural arches of primordial sacrals positioned on the anterior half of the centrum; caudosacral rib directed anterolaterally; four sacral vertebrae (DS + S1 + S2 + CS)*; humeral deltopectoral crest low and medially deflected along its distal half; flattened ischial shafts (paralleled in *Anchisaurus polyzelus*).

Comments The single specimen of *Leonerasaurus* is partially articulated, mainly preserving postcranial elements. As in all non-gravisaurian sauropodiforms, *Leonerasaurus* is characterized by the presence of numerous plesiomorphic traits (typical of non-sauropod sauropodomorphs), as well as by more derived features, similar to the closest outgroups of Sauropoda. Derived features distinguishing *Leonerasaurus* from more basal sauropodomorph taxa include: a straight anterior region of the

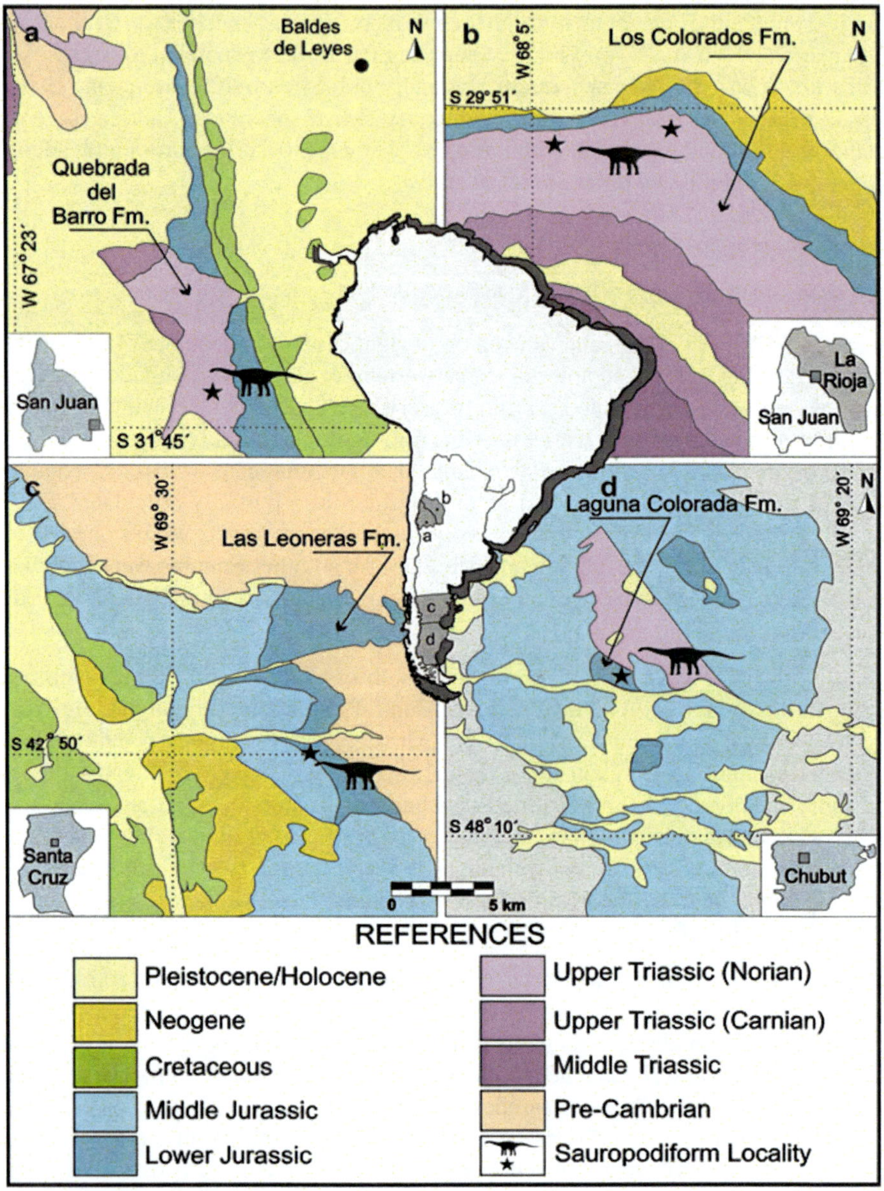

Fig. 1 Geological maps of the type localities of South American sauropodiform Sauropodomorpha. **a** Balde de Leyes (Quebrada del Barro Fm., Marayes-El Carrizal Basin): type locality of *Ingentia prima*. **b** La Esquina (Los Colorados Fm., Ischigualasto-Villa Unión Basin): type locality of *Lessemsaurus sauropoides*. **c** Las Leoneras (Las Leoneras Fm., Cañadón del Zaino): type locality of *Leonerasaurus taquetrensis*. **d** Laguna Colorada (Laguna Colorada Fm., El Tranquilo Group): type locality of *Mussaurus patagonicus*

dentary; slightly procumbent dentary teeth without marginal denticles and with convex labial surfaces and concave lingual surfaces, and a slightly developed wrinkling pattern; four sacral vertebrae with two primordial sacrals bounded by a dorsosacral and a caudosacral; preacetabular process of ilium dorsoventrally low and exceeding anterior margin of pubic peduncle (also present in *Anchisaurus* and *Mussaurus*). Finally, several plesiomorphic features distinguish *Leonerasaurus* from early-diverging sauropods: teeth lacking labial or lingual grooves; posterior teeth with large denticles oriented at 45° from the tooth margin; vertebral centra amphicoelous and acamerate; low and moderately elongated cervical vertebrae lacking postzygodiapophyseal lamina and with elongated prezygapophyses; dorsal vertebrae with low neural arches and neural spines elliptical in cross-section; absence of spinoprezygapophyseal laminae in all dorsal vertebrae and absence of prezygodiapophyseal lamina from in mid-dorsals onwards; posterior dorsal vertebrae with dorsoventrally low hyposphene-hypantrum complex; proximal metatarsal II hourglass shaped in proximal view.

Lessemsauridae Apaldetti, Martínez, Cerda, Pol and Alcober 2018

Definition The clade Lessemsauridae is defined as *Lessemsaurus sauropoides* (Bonaparte 1999) and *Antetonitrus ingenipes* (Yates and Kitching 2003), and all the descendants from their most common ancestor (Apaldetti et al. 2018).

Diagnosis Lessemsauridae differs from all other Sauropodomorpha in possessing the following unique character state combination (asterisks indicate apomorphies of the clade): robust scapulae with dorsal and ventral ends equally expanded*; bone growth characterized by the presence of thick zones of highly vascularized fibrolamellar bone, within a cyclical growth pattern*; slit-shaped neural canal of posterior dorsal vertebrae; anterior dorsal neural spines transversely expanded towards the dorsal end; and a minimum transverse shaft width of the first metacarpal greater than twice the minimum transverse shaft of the second metacarpal.

Lessemsaurus Bonaparte 1999

Lessemsaurus sauropoides Bonaparte 1999

Holotype PVL 4822-1, Bonaparte (1999) described and figured eight presacral neural arches. He mentioned additional presacral vertebrae and some appendicular elements as probably associated with this specimen. Owing to the lack of articulated remains, it cannot be determined which of the PVL 4822 elements belong to the same individual. Therefore, the holotype is now restricted to the eight presacral neural arches originally described and figured by Bonaparte (1999) and catalogued as PVL 4822-1. These eight neural arches are individually identified by the collection numbers PVL 4822-1/1, 4822-1/7 and PVL 4822-1/10 (Pol and Powell 2007b).

Referred Specimens Several elements belonging to the assemblage originally catalogued as PVL 4822 have been given additional numbers to allow identification of each individual element (PVL 4822/8, 4822/9 and 4822/11, 4822/79). These elements include dorsal and sacral vertebrae, scapulae, coracoid, humerus, ulna and radius, metacarpals I and II, manual digit I, ilium, ischium, pubes, femur, tibia, partial metatarsals I–V and pedal phalanges. Three other specimens were referred by Apaldetti et al. (2018): a distal third of a right femur (PVL 6580); an anterior caudal vertebra, complete left ilium and left ischium (CRILAR PV-302); and the ventral half of a right scapula along with the ventral half and partial dorsal blade of a left scapula (CRILAR PV-303).

Locality and Age Upper section of the Los Colorados Formation (Groeber and Stipanicic 1953), Ischigualasto-Villa Unión Basin. All the PVL specimens were found in 'La Esquina' locality, located south of Pagancillo Village, La Rioja Province, Argentina. The CRILAR-PV specimens were recovered from similar stratigraphic levels, albeit from an area located midway between the 'La Esquina' locality and Los Jachaleros Creek (Fig. 1b). The horizons where the remains were found are located approximately 145–160 m below the upper limit of the latter unit (Bonaparte 1999; Pol and Powell 2007b; Apaldetti et al. 2018). The age of the Los Colorados Formation has been considered as either Norian–Rhaetian (Bonaparte 1971) or the tetrapod-based biochron late Coloradian (Bonaparte 1973). Recently, the age of the unit has been magnetostratigraphically constrained to the lower–middle Norian in age (Kent et al. 2014; but see Desojo et al. 2020 for a proposed younger age for Los Colorados Formation).

Diagnosis *Lessemsaurus sauropoides* is a large non-gravisaurian sauropodiform with the following unique combination of characters (autapomorphies indicated with an asterisk): dorsal and middle-to-posterior cervicals with high neural arches; strong neural arch constriction below the postzygapophyses; deep postspinal fossa; dorsoventrally high infrapostzygapophyseal depression; middle and posterior dorsals with neural spines higher than long (with a height/length ratio of 1.5–2.0); robust scapula, with its blade markedly expanded; metacarpal I extremely short, with a proximal end lateromedially wider than metacarpal length; acute lateral process on proximolateral corner of metacarpal II*; medial flange in pubic peduncle of ilium, forming a narrow and marginal medial wall of the acetabulum*; brevis crest extending from the base of the ischial peduncle to the posterior tip of the postacetabular process*; and cross-section of the distal tibia subrectangular with its major axis orientated lateromedially and being twice as long as its anteroposterior extension (Pol and Powell 2007b). All the *Lessemsaurus* specimens share a unique histological feature, proposed as a synapomorphy of Lessemsauridae (i.e. *Ingentia*, *Antetonitrus*): bone growth characterized by the presence of thick zones of highly vascularized fibrolamellar bone, within a cyclical growth pattern (Apaldetti et al. 2018).

Comments All PVL specimens of *Lessemsaurus* were found and collected by José Bonaparte in 1970 during the Lillo expeditions to the Los Colorados Formation (Ischigualasto-Villa Unión Basin). The PVL specimens published by Bonaparte (1999) and later described in detail by Pol and Powell (2007b) were found in close association with other specimens collected during subsequent Bonaparte's expedition in 1971, with the entire assemblage interpreted as belonging to three different individuals (Bonaparte 1999; Pol and Powell 2007b). The partial femur PVL 6580, housed in PVL collection, was originally catalogued as 'prosauropod from Los Colorados' (textual PVL label written by Bonaparte). A recent description and detailed histology analysis of this specimen by Apaldetti et al. (2018) allowed for its referral to *Lessemsaurus* (see below). The CRILAR-PV specimens were collected by the IMCN-PVSJ team during the 2004–2005 field trips to the upper section of the Los Colorados Formation, in the area located between 'La Esquina' locality and Los Jachaleros Creek (Fig. 1b).

Ingentia Apaldetti, Martínez, Cerda, Pol and Alcober 2018

Ingentia prima Apaldetti, Martínez, Cerda, Pol and Alcober 2018

Holotype PVSJ 1086, six articulated posterior cervical vertebrae (C5–C10), glenoid region of right scapula and right humerus, radius and ulna, carpus and metacarpus lacking all phalanges (except phalanx IV.1 and V.1-2).

Referred Specimens PVSJ 1087, four anterior and a distal caudal vertebrae, proximal and distal ends of ulnae and radii, proximal third of left fibula, partial right foot including distal tarsal III and IV, proximal and distal ends of metatarsal I and II, two isolate non-terminal phalanges, and an ungual.

Locality and Age Southern of the Balde de Leyes village, San Juan Province, north-western Argentina. Southern outcrops of the Quebrada del Barro Formation (Marayes-El Carrizal Basin). Quebrada del Barro Formation has been proposed as late Norian–Rhaetian in age (Martínez et al. 2015). The horizon of the type and referred specimen is located 160 m below the top of the unit (Fig. 1a).

Diagnosis *Ingentia prima* is a large non-gravisaurian sauropodiform with the following unique combination of characters (autapomorphies indicated with an asterisk): mid-cervical neural arches almost twice as high as their respective centra; vertebrae C6–C10 with hyposphenes as dorsoventrally tall as the neural canal; pneumatic structures on posterior cervical neural arches, including deep fossae within the centrodiapophyseal fossa, with internal subfossae in C8–C9, and a complex of subfossae in the prezygapophyseal centrodiapophyseal fossa in C10*; expanded proximal end of the ulna with a posteromedial margin 1.5 times larger than the radial fossa margin*.

Comments Both specimens were discovered and collected by the IMCN-PVSJ team during the 2015 expedition to the Marayes-El Carrizal Basin, San Juan Province, north-western Argentina. The type material (PVSJ 1086) was found partially articulated. The referred specimen (PVSJ 1087) was found completely disarticulated several hundred metres away from the type specimen, but at a similar stratigraphic level.

4 Discussion

Diverse selection pressures and anatomical novelties that led Sauropodomorpha evolving from small bipedal forms (less than 10 kg weight) to giant quadrupedal (more than 50 tonnes) dinosaurs in the span of roughly 40–50 million years has been extensively researched in recent times. As a result, we can isolate key morphological changes essential to this transition, such as a decrease in skull size, lengthening of the neck, columnization of the limbs, lengthening of the forelimb relative to the hindlimbs, reduction of the distal limb segments, metacarpus into a U-shaped structure and a non-cyclical growth strategy (Wilson and Sereno 1998; Rauhut et al. 2011; Sander et al. 2004, 2011a, b; Sander 2013a; McPhee and Choiniere 2017). Early-diverging Sauropodiformes show the incipient stage of several of these changes and hence are of crucial importance to understanding the acquisition of obligate quadrupedalism and gigantism in Sauropodomorpha. In this context, the South American non-gravisaurian Sauropodiformes (*Mussaurus*, *Leonerasaurus*, *Lessemsaurus* and *Ingentia*) are of special importance as they reside at several key loci within Sauropodiformes, helping to unravel the polarity and/or stepwise adoption of traits most relevant to the origin of Sauropoda (Fig. 2 and 3).

4.1 Plesiomorphic Features Among Non-Gravisaurian Sauropodiformes

Given that only juvenile of *Mussaurus patagonicus* includes published complete cranial material (Pol and Powell 2007a), and only a few cranial fragments are preserved in *Leonerasaurus* (i.e. a fragment of the dentary and isolated teeth), the most relevant anatomical features observed among South American non-gravisaurian Sauropodiformes are related to the postcranial skeleton. Due to the immature nature of all the specimens of *Mussaurus* with cranial materials, evaluation of its characters as plesiomorphic or derived is controversial. Here we comment on some of the cranial plesiomorphic characters observed by Pol and Powell (2007a) in the cranium of *Mussaurus*.

4.1.1 Skull

The plesiomorphic cranial features observed by Pol and Powell (2007a) in *Mussaurus* include the presence of anteriorly located external nares, reduced antorbital fossa surrounding the antorbital fenestra, subcircular orbit with an extensive infraorbital region of the jugal separating the antorbital and infratemporal openings, frontal forming at least half of the orbital rim and extending posteriorly into the supratemporal fossa, dorsoventrally narrow anterior branch of quadratojugal that reaches the midpoint of the infratemporal opening (lacking contact with the maxilla), enlarged

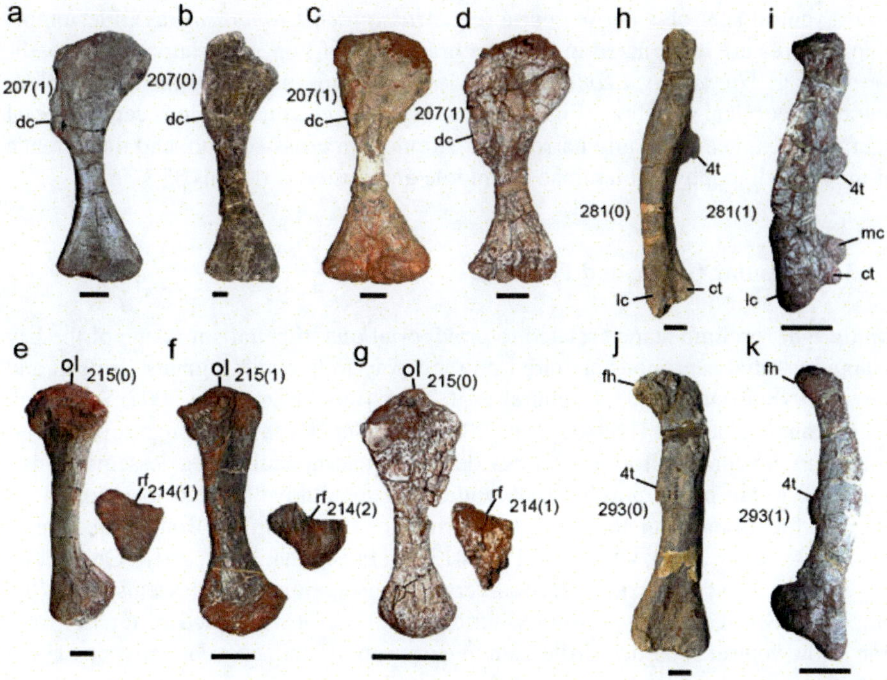

Fig. 2 Plesiomorphic versus derived features among South American sauropodiform Sauropodomorpha. **a–d** Humeri of *Mussaurus patagonicus* (**a**), *Leonerasaurus taquetrensis* (**b**), *Lessemsaurus sauropoides* (**c**), and *Ingentia prima* (**d**) in anterior view. **e–g** Ulnae of *Mussaurus patagonicus* (**e**), *Lessemsaurus sauropoides* (**f**), and *Ingentia prima* (**g**) in anterior (left) and proximal (right) view. **h**, **i** Femora of *Mussaurus patagonicus* (**h**), and *Lessemsaurus sauropoides* (**i**) in lateral view; **j**, **k** Femora of *Mussaurus patagonicus* (**j**), and *Lessemsaurus sauropoides* (**k**) in anterior view. Abbreviations: dc, deltopectoral crest; fh, femoral head; ol, olecranon; rf, radial fossa of the ulna; 4t, fourth trochanter. Characters and states (in parenthesis) numbers refer to morphological characters listed by Pol et al. (2021). Scale bars equal 10 **cm** (**d**, **i**, **k**), 5 **cm** (**a**, **c**, **e–h**, **j**), 1 **cm** (**b**)

supratemporal fenestra, semicircular facet that 'hooks' around the basipterygoid processes, mandibular rami meeting at an acute angle rather than forming a broad U-shaped symphysis, and dorsal margin of surangular poorly bowed dorsally (Pol and Powell 2007a).

4.1.2 Axial Skeleton

The progressive elongation of the neck is one of the most relevant transformations in sauropodomorph evolution, with the presence of elongated cervical vertebrae previously noted as a synapomorphic feature of the group (Gauthier 1986). Hence, several non-gravisaurian sauropodiforms show enlarged cervical vertebrae, and some even bear a ventral keel on anterior centra (e.g. *Mussaurus, Leonerasaurus*). Plesiomorphic features are recognized in the vertebral series of non-gravisaurian Sauropodiformes (e.g. *Mussaurus, Leonerasaurus*, lessemsaurids) including the lack of pleurocoels in cervical vertebrae (Fig. 3d–g), acamerate and amphicoelous centra, dorsal neural spines mediolaterally narrow and elliptical in cross-section, and the absence of the spinodiapophyseal laminae on middle and posterior dorsals.

4.1.3 Scapular Girdle and Forelimb

An upright forelimb stance related to graviportal quadrupedalism is one of the key adaptations that sauropods acquired on their way to becoming giants (Wilson and Sereno 1998; Bonnan 2003; Upchurch et al. 2004; Bonnan and Yates 2007; Yates et al. 2010; Rauhut et al. 2011; Sander et al. 2011b; Sander 2013a). Although several non-gravisaurian Sauropodiformes show evidence of quadrupedality (e.g. lessemsaurids), most still retain plesiomorphic traits indicating facultative bipedal locomotion or, at least, functions other than weight-bearing for the forelimb (Bonnan and Yates 2007; Yates et al. 2010; Otero and Pol 2013; McPhee et al. 2014, 2018; Otero et al. 2017, 2019). Hence, as in most early-diverging sauropodomorphs, lessemsaurids (the biggest body sizes among non-gravisaurian sauropodiforms) lacked an anterior rotation of the ventral shoulder girdle seen in eusauropods and, consequently, possessed a posteroventrally oriented scapular glenoid which precluded an erect posture of the humerus, and also lack a completely pruned hand (Yates and Kitching 2003; Bonnan and Yates 2007; Rauhut et al. 2011; see also Otero et al. 2017 for a similar pattern in *Mussaurus*).

It is, therefore, possible to observe some key plesiomorphic features in the forelimb of most non-gravisaurian sauropodiforms (Table 2), such as well-expanded humeral condyles and a well-developed deltopectoral crest (Fig. 2a–d), well-developed olecranon process (Fig. 2e, g), and a robust metacarpal I with a medially divergent pollex (Fig. 3l–n). The deltopectoral crest of non-gravisaurian sauropodiforms shows both the plesiomorphic (extended crest, > 45% total length of the humerus) and derived conditions (lesser extended, <45%) among Sauropodomorpha (Fig. 2a–d). Thus, the deltopectoral crest of *Mussaurus, Lessemsaurus, Ingentia* and *Antetonitrus* extends for approximately 35–40% of the total length of the humerus, resembling that of Gravisauria (e.g. *Vulcanodon*). In contrast, the deltopectoral crest in *Leonerasaurus* and most basal sauropodomorphs (e.g. massospondylids) extends approximately 50% of the humeral length (see above), whereas the deltopectoral crest is low and reduced among eusauropods.

The humeral/femoral length ratio among early-diverging sauropodomorphs varies between 0.55 (*Saturnalia*) and 0.70 (e.g. *Lufengosaurus huenei,* NMQR 3314, '*Melanorosaurus*'), whereas the ratio in eusauropods always exceeds 0.8 (Gauthier 1986). Similar to NMQR 3314 ('*Melanorosaurus*'), the humerus/femur ratio of *Mussaurus* is <0.8 (approximately 0.65, see Otero and Pol 2013); however, it is >0.8 in *Vulcanodon* and later-diverging sauropods.

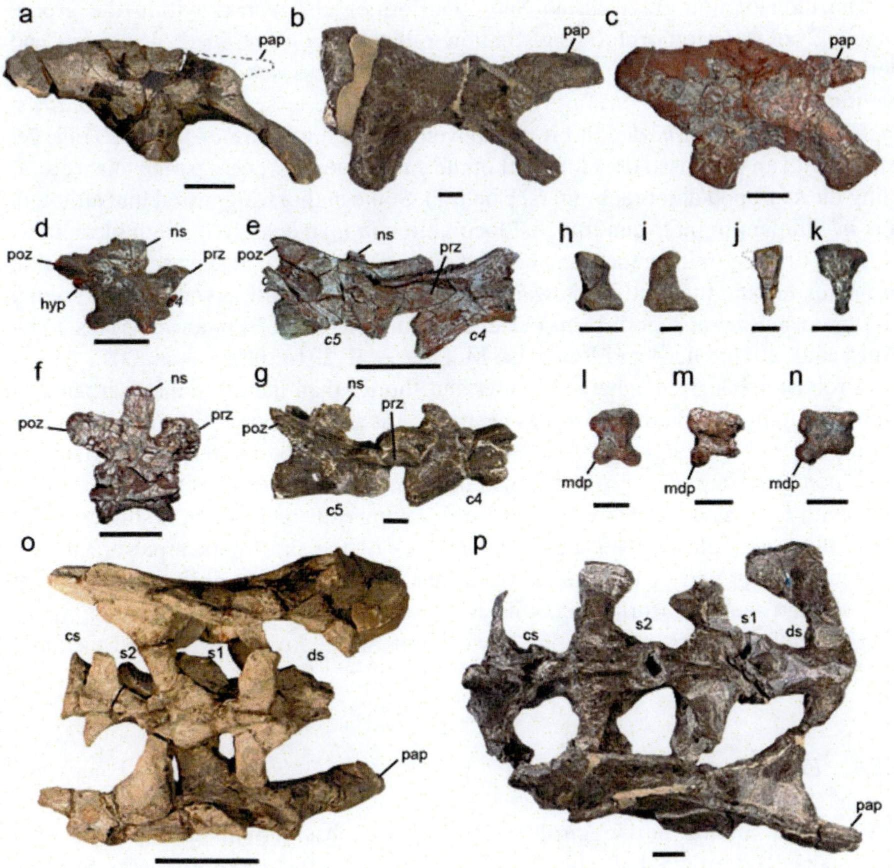

Fig. 3 Plesiomorphic and derived features among South American sauropodiform Sauropodomorpha. **a–c** Ilia of *Mussaurus patagonicus* (**a**), *Leonerasaurus taquetrensis* (**b**), *Lessemsaurus sauropoides* (**c**) in lateral view. **d–g** cervical vertebrae: 8 of *Lessemsaurus sauropoides* (**d**), 4, 5 of *Mussaurus patagonicus* (**e**), 8 of *Ingentia prima* (**f**), and 4, 5 of *Leonerasaurus taquetrensis* (**g**) in lateral view. **h–i** Metatarsal II of *Mussaurus patagonicus* (**h**), and *Leonerasaurus taquetrensis* (**i**) in proximal view. **j–k** metatarsal V of *Mussaurus patagonicus* (**j**), and *Lessemsaurus sauropoides* (**k**) in dorsal view (proximal points top). **l–n** metacarpal I of *Mussaurus patagonicus* (**l**), *Ingentia prima* (**m**), and *Lessemsaurus sauropoides* (**n**) in dorsal view (proximal points top). **o–p** sacrum of *Mussaurus patagonicus* (**o**), and *Leonerasaurus taquetrensis* (**p**) in dorsal view. Abbreviations: c4-5, cervical vertebra; cs, caudosacral vertebra; ds, dorsosacral vertebra; hyp, hyposphenes; mdp, medial divergent pollex; ns, neural spine; pap, preacetabular process; prz, prezigapophysis; poz, poszigapophysis; s1-2, primordial sacrals 1 and 2. Scale bars equal 10 **cm** (**a**, **c**, **e**, **f**, **o**), 5 **cm** (**d**, **h**, **j**, **k**, **l–n**), 1 **cm** (**b**, **I**, **p**)

The ulna of non-gravisaurian Sauropodiformes also varies within the group, showing both plesiomorphic (e.g. shallow radial fossa, developed olecranon) and derived features (e.g. well-developed craniolateral process; see Table 2). For example, the ulna of *Mussaurus*, *Aardonyx*, *Meroktenos* and the lessemsaurids *Antetonitrus*, *Ingentia* and *Ledumahadi* exhibit the abovementioned mixture of features (Fig. 2e, g), but with an increased development of the anterolateral process, somewhat resembling the sauropod antebrachium (see below). Some authors suggested that although it is insufficient to infer quadrupedal locomotion in taxa displaying both plesiomorphic and derived traits in the same bone, they pointed out that at least semi-pronation of the manus and the first step towards quadrupedal locomotion was present among non-gravisaurian sauropodiforms (Yates and Kitching 2003; Bonnan and Yates 2007; Yates et al. 2010; Otero and Pol 2013; McPhee et al. 2014; Otero et al. 2017).

A robust metacarpal I, being broader and shorter than the other metacarpals and with a medially divergent pollex, has been noted as a plesiomorphic condition among Sauropodomorpha (Sereno 1999; Bonnan and Yates 2007; Yates et al. 2010). In some non-gravisaurian sauropodiforms, this condition becomes even more exaggerated, with *Mussaurus* and lessemsaurids displaying especially squat first metacarpal (Fig. 3l–n; see below). The articular condyles of the first metacarpal are highly asymmetrical in early saurischians (Gauthier 1986), although in some early non-gravisaurian sauropodiforms taxa this asymmetry is extremely marked (Fig. 3l–n). Consequently, the first manual phalanx of digit I (phalanx I.1) is twisted medially concerning the rest of the manus.

4.1.4 Pelvic Girdle and Hindlimb

The changes in the pelvic girdle and hindlimb observed in eusauropods (e.g. anterodorsally developed iliac blade, reduction of brevis fossa, extremely short ischial process, among others) represent clear weight-bearing adaptations which evolved to resist their exceptionally high body masses, with up to 80% of the body mass supported by the pelvic girdle and hindlimbs in quadrupedal dinosaurs (Alexander 1985). In addition, the architecture associated with the pelvis and hindlimb underwent considerable alteration throughout the evolution of increasing body size (Wilson and Sereno 1998; Fechner 2009; Rauhut et al. 2011).

The ilium of non-gravisaurian Sauropodiformes is distinguishable from the morphology of that in Eusauropoda, which displays extensive dorsoventral development of the preacetabular process. However, as with most early-diverging sauropodomorphs, the ilia of basal Sauropodiformes retain a triangular and dorsoventrally low preacetabular process (e.g. *Leonerasaurus*, *Lessemsaurus*, *Antetonitrus*, NMQR 3314; Fig. 3a–c). The anterodorsal expansion of the ilium is related to the elongation and vertical orientation of the muscle *iliotibialis* that originates on the lateral surface of the ilium to attach to the cnemial crest of the tibia, which in turn is conditioned by the length and robustness of the hindlimb bones (e.g. robust, straight femur with ovoid cross-section; reduction of tibia/femur ratio; Fechner 2009). In the same way, the development of the expanded postacetabular process of the ilium among eusauropods is related to the lateral attaching of the muscle *caudofemoralis brevis*, which originates on this process to attach to the fourth trochanter of the femur (Fechner 2009). In non-gravisaurian sauropodomorphs, this muscle originates in the brevis fossa of the ilium (absent among eusauropods); therefore, the postacetabular process is not anteroposteriorly and dorsoventrally well-developed among basal forms (non-gravisaurian sauropodomorphs). The retention of primitive pelvic traits among non-gravisaurian sauropodiforms (especially among the large-bodied lessemsaurids) refutes suggestions that iliac modifications were strictly required for attaining gigantic body sizes (Wilson and Sereno 1998; Fechner 2009; Rauhut et al 2011).

The femoral shaft of *Mussaurus* and *Lessemsaurus* (and other basal Sauropodiformes, such as NMQR 1551 '*Melanorosaurus*') is curved in lateral view, showing the typical sigmoid femur of most non-eusauropod sauropodomorphs (Fig. 2h–i). Also consistent with the plesiomorphic condition, the femur of some non-gravisaurian sauropodomorphs (e.g. *Mussaurus*, *Leonerasaurus*, *Ledumahadi*) have a nearly circular mid-shaft cross-section. In contrast, *Lessemsaurus* presents an intermediate condition with a subovoid femoral shaft cross-section (Pol and Powell 2007b). This differs from the straight shaft with an elliptical cross-section of other forms (e.g. *Antetonitrus, Meroktenos, Schleitheimia*), which resemble the eusauropod condition (see below).

The first metatarsal of *Mussaurus* and *Leonerasaurus* is more than twice as long as wide, resembling the condition of most early-diverging non-sauropod sauropodomorphs. In sauropods and closely related taxa (e.g. *Aardonyx, Antetonitrus, Blikanasaurus, Vulcanodon*), metatarsal I is remarkably robust, being the widest element of the metatarsus and usually having a length/width ratio smaller than 1.5 (Upchurch et al. 2007; McPhee et al. 2015). The proximal surface of metatarsal I in *Mussaurus* and *Leonerasaurus* is ovoid, with the width of the first metatarsal shaft subequal to that of metatarsal II. This condition contrasts with the derived condition of eusauropods in which metatarsal I is much wider than the metatarsal II (Wilson and Sereno 1998). The metatarsal II of most non-gravisaurian sauropodomorphs (e.g. *Mussaurus, Leonerasaurus, Lessemsaurus* and NMQR 3314 '*Melanorosaurus*') has an hourglass-shaped proximal articular surface (Fig. 3h–i) with markedly concave lateral and medial margins for the articulation of metatarsals I and III, respectively (Sereno 1999). A dorsoventrally flat and triangular metatarsal V that is less than half of the length of all other metatarsals is a typical feature of all non-eusauropod sauropodomorphs, present in *Mussaurus* and *Lessemsaurus* (Fig. 3j–k).

4.2 Derived Features Among Non-Gravisaurian Sauropodiformes

Gravisaurian sauropods are unique among quadrupedal dinosaurs in having a columnar stance with erect, parasagittal limbs, providing graviportal support of body mass (Cooper 1984; Allain and Aquesbi 2008; McPhee et al. 2015, 2018). This involved a suite of morphological changes—mainly in the limbs—that characterize sauropods, including reduction of the deltopectoral crest, elongation and robustness of the humeral and femoral shafts, modification of the proximal end of the ulna, development of a U-shaped metacarpus structure, distal migration of the fourth trochanter to the midpoint of the femoral shaft, among others (Rauhut et al. 2011; Sander 2013a, b; McPhee and Choiniere 2017; McPhee et al. 2018). Some of the derived features acquired by Eusauropoda are, at least in incipient form, present among the non-gravisaurian sauropodiforms (Table 2).

4.2.1 Skull

The derived or sauropod-like features observed among the cranial elements of South American non-gravisaurian Sauropodiformes are restricted to a few traits observed on the dentition of *Leonerasaurus* (i.e. mesiodistally concave lingual surface of the crown; Pol et al. 2011). As was previously mentioned, the immature condition of all *Mussaurus* specimens precludes a critical evaluation of its cranial features.

4.2.2 Axial Skeleton

Elongation of the neck has been interpreted as providing a larger feeding envelope and increased browsing ranges for herbivores (e.g. Barrett 2014), and as such is one of the key traits that characterize the evolution of Sauropodomorpha (Wilson and Sereno 1998). However, in general, non-gravisaurian sauropodiforms lack an extreme elongated neck, with even lessemsaurids presenting proportionately short and high cervical vertebrae, indicating that the neck elongation was not a prerequisite for achieving body sizes comparable to those of early-diverging eusauropods or gravisaurians (Apaldetti et al. 2018; McPhee et al. 2018).

The cervicodorsal vertebral transition of *Pulanesaura* and lessemsaurids shows remarkably high neural arches with dorsoventrally elongated neural spines, which indicates an enlarged epaxial musculature relative to the plesiomorphic condition (Bonaparte 1999; Yates and Kitching 2003; McPhee et al. 2015, 2018; Apaldetti et al. 2018). Markedly expanded hyposphene-hypantrum connections are also present among the mid-cervical and anterior dorsal vertebrae of some lessemsaurids (the dorsoventral breadth is subequal or deeper than the height of the neural canal in *Lessemsaurus*, and *Ingentia,* Fig. 3d, f). Tall hyposphenes are also observed through the vertebral dorsal series of *Aardonyx,* NMQR 1551, *Antetonitrus*, *Meroktenos* and eusauropods (Yates 2007a; Yates et al. 2010). In contrast, most basal sauropodomorphs present less-developed hyposphenes ($<70\%$ the height of the neural canal in dorsal vertebrae). The presence of developed hyposphenes possibly indicates additional strengthening and reinforcement of the axial skeleton in large-bodied sauropodomorphs with short necks (Bonaparte 1999; McPhee et al. 2015; Stefanic and Nesbitt 2019). Thus, the non-planar cervical zygapophyseal facets observed in taxa like *Pulanesaura, Lessemsaurus* and *Ingentia* (Fig. 3d, f)—in which the postzygapophyses of the anterior to middle cervical vertebrae are elevated relative to the prezygapophyses (McPhee et al. 2015; McPhee and Choiniere 2017)—is possible related to the reorganizations of the axial musculature at nodes proximate to Sauropoda (McPhee et al. 2015). According to McPhee et al. (2015), this change suggests modifications in the kinematic potential of the neck, mainly associated with the degrees of flexion.

The cervico-dorsal vertebral transition of *Lessemsaurus* and *Ingentia* is also characterized by the presence of a novel system of pneumatic structures (i.e. a subfossae-within-fossae complex). Only cervical and abdominal air sac and lungs can produce pneumatic diverticula that invade the vertebrae (O'Connor and Claessens 2005; O'Connor 2006). The presence of a sophisticated pneumatic complex showing vertebral fossae that are in turn excavated by smaller, penetrative, subfossae (or internal chambers) is thought to be exclusively correlated with a well-developed cervical air sac system on these lessemsaurids. Yates et al. (2012) provided an in-depth examination of postcranial pneumaticity throughout the presacral vertebrae of Sauropodomorpha, demonstrating that only some dorsal vertebrae (posterior dorsal series) have pneumatic structures in some non-gravisaurian sauropodomorphs (i.e. *Eucnemesaurus, Aardonyx* and *Antetonitrus*). The presence of similar structures in the cervical series of primitive taxa like *Pantydraco* and the genus *Plateosaurus* has also been indicated; however, there are less certain (Galton and Kermack 2009; Yates et al. 2012). The pneumatic structures present in the cervical and dorsal vertebrae of lessemsaurids and other closely related taxa suggest the presence of an improved avian-like respiratory system early in the evolution of sauropodiform dinosaurs, well predating the rise of Eusauropoda (Yates et al. 2012; Apaldetti et al 2018).

4.2.3 Sacrum

Among the modifications related to the increase of body size during the early evolution of Sauropodomorpha, an increase in the number of sacral vertebrae (from three to four or more) represents a particularly notable trend. In this respect, the presence of four or more sacral vertebrae has traditionally been regarded as a diagnostic feature of quadrupedal sauropods (Upchurch 1998; Galton and Upchurch 2000; Sander et al. 2011a; Yates et al. 2010; Rauhut et al 2011). However, in recent years our understanding of the pattern of sauropodomorph sacral evolution has changed dramatically, with several discoveries showing that the four sacral arrangements of Sauropoda (or the clade of those large-bodied taxa such as NMQR 1551 ['*Melanorosaurus*'] + Sauropoda) appears independently on several occasions in the evolutionary history of the group and is thus diagnostic of more than one inclusive clade (Pol et al. 2011; Wang et al. 2017).

Pol et al. (2011) summarized this trend as three major stages in the early evolutionary history of the sauropodomorph sacrum: (1) the plesiomorphic condition for Sauropodomorpha in which the sacrum is almost exclusively composed of the two primordial sacrals (S1 + S2); (2) the condition present in most early-diverging sauropodomorphs, characterized by the incorporation of a third element identified as a dorsosacral (DS + S1 + S2; except for *Plateosaurus trossingensis* with S1 + S2 + CS); and (3) the condition typical of most early sauropod dinosaurs characterized by the presence of four sacrals (or more than four in derived taxa), in which one caudosacral element is incorporated (DS + S1 + S2 + CS). Although a specimen of '*Melanorosaurus*' (NMQR 1551) was regarded as exceptional for possessing upwards of five sacral vertebrae (DS1 + DS2 + S1 + S2(+CS); Yates 2007a, Yates et al. 2010), more recent discoveries evidence the presence of four sacral vertebrae among several other non-sauropod taxa. Specifically, the DS1 + S1 + S2 + CS configuration observed in the massospondylid *Adeopapposaurus* (Garcia et al. 2016, 2018), *Xingxiulong* (Wang et al. 2017, 2020), and the non-gravisaurian sauropodiforms *Leonerasaurus* (Pol et al. 2011) and *Mussaurus* (Otero and Pol 2013; Fig. 3o–p), reveals a more complex pattern of evolutionary convergence on the acquisition of the sauropod-like sacrum among Sauropodomorpha. Consequently, the four sacral conditions appeared relatively early in sauropodomorph evolution and, given the relatively small body size of some non-sauropod taxa (2–2.5 m long *Adeopapposaurus* and *Leonerasaurus*, and *4–5 m* long for *Xingxiulong*), demonstrates that the appearance of the sauropod-type sacrum predated the dramatic body size increase characteristic of Sauropoda.

4.2.4 Pectoral Girdle and Forelimb

The significant dorsal and ventral expanded ends of the scapulae of lessemsaurids (i.e. *Lessemsaurus* and *Antetonitrus*) is atypical among both early and later-diverging sauropodomorphs, which in general display a more gracile scapula. The extreme robustness of the lessemsaurid scapula has been related to a more horizontal neck posture instead of a trend towards a graviportal locomotory style (Remes 2008). McPhee et al. (2015) suggest that the markedly massive scapula of some non-sauropodan sauropodiform taxa (such as *Lessemsaurus sauropoides*, *Antetonitrus ingenipes* and possibly *Gongxianosaurus shibeiensis*) possibly represents a trade-off between the additional anchorage required to counteract the increased shear stresses experienced by a large-bodied quadruped with a less than erect forelimb on the one hand, and the necessary mobility of the forelimb for bipedal high-browsing on the other.

The well-developed anterolateral process of the ulna delimits a deep radial fossa that accommodates the radius in an anteromedial position concerning the ulna, allowing for a more pronated manus (Bonnan and Yates 2007). Among Gravisauria and more derived sauropods the radial fossa is notably deep (Wilson and Sereno 1998; Bonnan 2003; Yates and Kitching 2003; Bonnan and Yates 2007); however, some non-gravisaurian sauropodiforms also possess a moderately deep radial fossa bounded posteriorly by a prominent anterolateral process of the ulnar proximal end (e.g. *Lessemsaurus, Antetonitrus,* NMQR 3314 '*Melanorosaurus*'). The shallow radial fossa observed in *Antetonitrus* may have provided manual pronation positioning the radius anterior to the ulna, similar to the sauropodan-style forelimb (Yates and Kitching 2003; Bonnan and Senter 2007; Bonnan and Yates 2007; see McPhee et al. 2014 for further discussion).

The first metacarpal of *Mussaurus* and lessemsaurids is the stockiest in the metacarpus by a considerable margin, with a proximal end that is typically >1 times its proximodistal length (Fig. 3l–n), also seen in other non-sauropod Sauropodiformes such as *Aardonyx* and *Sefapanosaurus*. This differs from most early-diverging sauropodomorphs in which the width-length ratio of metacarpal I is approximately equal (e.g. massospondylids), or <1 among most non-eusauropod sauropodomorphs, and even it is markedly lengthened within Neosauropoda.

The mixture of plesiomorphic features and derived sauropod-like traits observed in the forelimb of basal Sauropodiformes is consistent with the mosaic evolution of the forelimb of 'transitional' sauropodomorphs in the acquisition of quadrupedal locomotion (Bonnan and Yates 2007; Yates et al. 2010; McPhee et al. 2018).

4.2.5 Hindlimb

The femur of non-gravisaurian Sauropodiformes exhibits a range of both plesiomorphic (e.g. sigmoid shape, circular cross-section) and derived (e.g. straightness, displacement of fourth trochanter, tibia/femur ratio) traits indicative of the transition or a progressive shift towards a more graviportal locomotory style and attendant shift in the centre of mass that this entailed (Wilson and Carrano 1999; Yates and Kitching 2003; Rauhut et al. 2011; see Table 2). The eccentric (=anteroposteriorly flattened cross-section) femoral shaft of many non-gravisaurian Sauropodiformes (e.g. it is subovoid in *Lessemsaurus* and notably ovoid in *Antetonitrus*, *Meroktenos*, *Camelotia*), as well as the reduction in the sigmoid curvature of the femur (especially in *Antetonitrus*), indicate a trend towards increased resistance to mediolateral bending at wider gaits (Wilson and Carrano 1999; Carrano 2001, 2005). The distal displacement of the fourth trochanter (from proximal to mid-way and posterior side of the femur) in some non-gravisaurian sauropodiforms (e.g. *Aardonyx*, *Lessemsaurus*, *Antetonitrus*, *Mussaurus*; Fig. 2k) and most of the derived taxa (e.g. *Vulcanodon*, *Barapasaurus*) reflect increasing leverage of the leg musculature, producing greater force albeit at a slower velocity (Gatesy 1990; Carrano 1999, 2005; Fechner 2009; Rauhut et al. 2011). The fourth trochanter of *Mussaurus*, instead, is located on the posteromedial margin of the proximal half of the femoral shaft, close to the mid-shaft (Fig. 2j), as in most non-sauropodiform sauropodomorphs. However, *Isanosaurus* exhibits the plesiomorphic condition in which the fourth trochanter is located at the proximal half of the femur (Buffetaut et al. 2000, see also McPhee and Choiniere 2017). The muscles *caudofemorales* are considered responsible for displacement movements (by retraction/adduction forces) in dinosaur locomotion (Gatesy 1990; Hutchinson and Gatesy 2000). A medial displacement of the fourth trochanter would have aligned the muscle *caudofemoralis brevis* with the iliac postacetabular process of the ilium, limiting the adductor forces in heavy-weighted sauropodomorphs (see McPhee et al. 2014). A noticeable rugose region of muscle attachment posterior and proximal to the fourth trochanter of *Antetonitrus* has been interpreted as the lateral incursion of the muscle *caudofemoralis brevis* on the posterior femoral shaft, that would have assisted with occasional bipedal locomotion of this taxon (McPhee et al. 2014, 2015; see also McPhee and Choiniere 2017, for further discussion about the femur attributes of *Antetonitrus*). The fourth trochanter bearing a medially deep depression for the insertion of the muscle *caudofemoralis longus* (Langer 2003) observed in *Lessemsaurus* and *Antetonitrus* (Yates and Kitching 2003; McPhee et al. 2014; Pol and Powell 2007b) indicates a more developed and distally shifted insertion site for this muscle related to the femoral retractor, as was observed among the most derived clades of Sauropodomorpha (Carrano 2005; Wilson 2005; Fechner 2009).

In addition, similar to later-diverging sauropods, the lesser trochanter of *Antetonitrus* (BP/1/4952) is laterally displaced, enlarging and shifting laterally the attachment area for the muscle *iliofemoralis* (see McPhee et al. 2014 for further discussion). This lateral location for the lesser trochanter (as well as reduced osteological correlates of muscular system) has been related to abductor-controlled postural support as an adaptation to a fully graviportal, columnar-limbed sauropod posture (Hutchinson and Gatesy 2000; Carrano 2005; Wilson 2005; Fechner 2009).

The reduction in tibial length relative to the femur (tibia (T)/femur (F) ratio; see Table 2) also represents a major contribution to the changes in hindlimb/muscle proportions, providing adaptations that work at high force but low velocity, consistent with increasing body mass (Alexander 1985; Fechner 2009). Most basal sauropodomorphs have a ratio ≥ 0.7; however, in progressively more proximate taxa to Sauropoda, this ratio reduces to around 0.6 (e.g. *Antetonitrus* (T/F = 0.64) and in *Lessemsaurus* which is estimated in 0.60 approximately [based on the largest elements of the PVL 4822 specimens, Pol and Powell 2007b), *Vulcanodon* (T/F = ~0.60). *Mussaurus* has a ratio of 0.72 (Otero and Pol 2013) resembling the general plesiomorphic state (Yates and Kitching 2003; Pol and Powell 2007b; McPhee et al. 2014).

4.3 Histology

Sauropod taxa exhibit uninterrupted deposition of bone tissue during the early phase of development, becoming periodically interrupted by growth marks (annuli and/or lines of arrested growth, LAGs) once sexual maturity is attained (Sander et al. 2004). The presence of growth marks throughout the entire cortex of both *Lessemsaurus* and *Ingentia* reveals that these taxa experienced periodically interrupted growth during most of their development, similar to the pattern inferred for most early-diverging sauropodomorphs (Sander et al. 2004; Cerda et al. 2017). Nevertheless, in lessemsaurids (i.e. *Lessemsaurus*, *Ingentia* and *Antetonitrus*), cortical bone exhibits histological features clearly different from early-diverging sauropodomorphs (Cerda et al. 2017; Krupandan et al. 2018). The high amount of woven fibered bone observed in these lessemsaurids appears similar to that seen in sauropod taxa, which is related—aside from their lack of growth marks—with fast and uninterrupted growth (Sander et al. 2004, 2011a). Therefore, together with a new trait (thick zones and an abundance of reticular bone), lessemsaurids combine plesiomorphic histological features typical of non-sauropod sauropodomorphs (i.e. well-defined growth marks throughout the cortex) with the derived condition (i.e. predominance of woven fibered bone in the primary matrix) characterized for sauropod taxa (Apaldetti et al. 2018; Krupandan et al. 2018).

4.4 Phylogenetic Relationships

We assessed the phylogenetic affinities of non-gravisaurian Sauropodiformes using the last dataset recently published by Pol et al. (2021). The analysis recovered 820 most parsimonious trees (MPTs) of 1693 steps of length (CI = 0.288, RI = 0.652; see Pol et al. 2021). The strict consensus resulted in a partially collapsed tree, including two main polytomies among early-diverging Sauropodiformes (see Pol et al. 2021). After the exclusion of six unstable taxa (including the sauropodiforms *Sefapanosaurus zastronensis*, *Irisosaurus yimenensis* and *Blikanasaurus cromptoni*), a reduced strict consensus recovered a less collapsed topology among non-gravisaurian sauropodiforms and closely related taxa (Fig. 4). Non-gravisaurian sauropodiformes were recovered as a paraphyletic group including two basal polytomies and several more resolved clades of apical nodes.

In this context, Sauropodiformes share: absence of a ventral sulcus on the proximal and middle caudal vertebrae (char. 190.1); length of the manual digit I greater than the length of manual digit II (char. 233.1); weakly bent longitudinal axis of the femur with an offset of less than 10° (char. 280.1); femoral length between 60 and 79 cm (char. 353.3); and the presence of the distal tubercle of the radius (char. 367.1). In some other MPTs, other synapomorphies support this clade: the presence of a posterior end of dentary tooth row medially inset with a thick lateral ridge on the dentary forming a buccal emargination (char. 97.0); and a mediolaterally wider than anteroposteriorly deep distal surface of the tibiofibular crest (char. 356.1).

The South American *Mussaurus patagonicus* and *Leonerasaurus taquetrensis* are recovered as some of the earliest-diverging members of Sauropodiformes (Fig. 4) in an unresolved clade together with *Aardonyx ingenipes* and *Yizhousaurus sunae*. The *Mussaurus* + *Leonerasaurus* clade is supported by a single unambiguous synapomorphy: parapophyses of the dorsal column completely shifted from the centrum to the neural arch, anterior to the thirteenth presacral vertebra (char. 154.0). Sauropoda (sensu Yates 2007a) is supported in this analysis by: the presence of coarsely wrinkled texture of the tooth enamel surface (char. 117.2); height of the mid-cervical neural arches greater than the height of the posterior centrum face (char. 133.1); shape of the anteromedial corner of the distal articular surface of the tibia forming a right angle (char. 310.0); lateral extent of the ventrolateral flange of the plantar surface of metatarsal II in proximal aspect well-developed, extending further laterally than ventromedial flange extends medially (char. 354.1). In some of the most parsimonious trees, two other synapomorphies support this clade: presence of lateral plates appressed to the labial side of the premaxillary, maxillary and dentary teeth

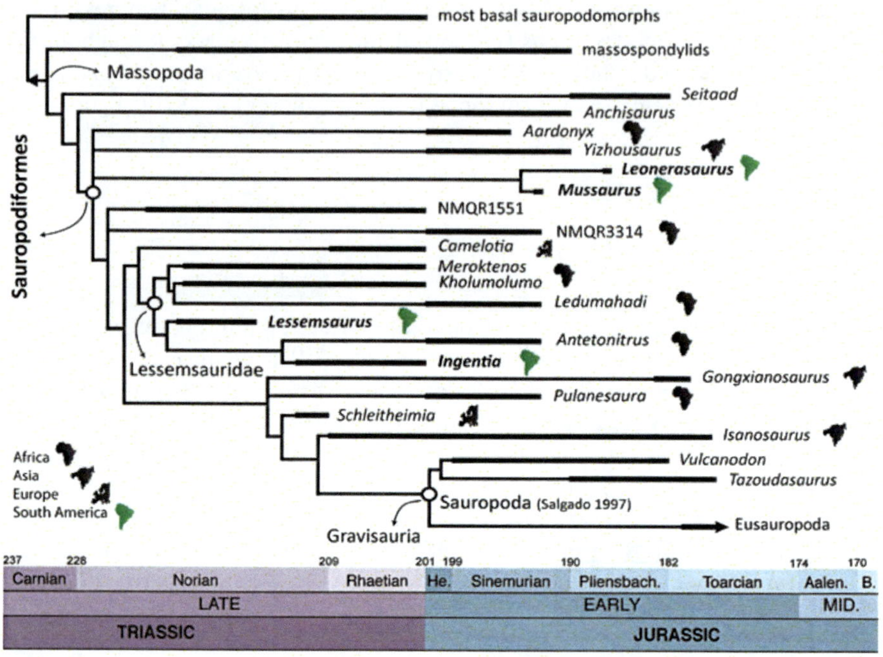

Fig. 4 Simplified reduced consensus tree of phylogenetic relationships of non-gravisaurian Sauropodiformes (based on the recent phylogenetic analysis of Pol et al. 2021) with their estimated chronostratigraphic ranges

(char. 2.1); and posterior extent of ascending process of the astragalus close to the posterior margin (char. 320.1).

Interestingly, some taxa that were originally proposed as non-sauropodiform sauropodomorphs, were here recovered as Sauropodiformes. Hence, *Kholumolumo ellenbergerorum* is depicted among Sauropodiformes, as the sister clade of *Ledumahadi mafube* (Fig. 4), this later clade is depicted as the sister-clade of *Meroktenos thabenensis*. This clade (*Meroktenos* + (*Kholumolumo* + *Ledumahadi*)) is united by the following synapomorphies: shape of femoral head roughly hemispherical with no sharp medial distal corner (char. 283.1); projection of the lesser trochanter just a scar upon the femoral surface (char. 287.0); position of the lesser trochanter near the centre of the anterior face of the femoral shaft in anterior view (char. 290.0); and not visible lesser trochanter in posterior view (char. 291.0). This new clade of South African taxa is reinterpreted as the sister-clade of *Lessemsaurus* and more derived sauropodiforms, forming the revised Lessemsauridae family (Fig. 4). Hence, Lessemsauridae is supported by dorsal neural arches in which the height from the neurocentral suture to the level of zygapophyseal facets is subequal to or greater than the height of the centrum (char. 164.1); and a slit-shaped posterior dorsal neural canal (char. 166.1). *Camelotia borealis* is recovered as the sister taxon to lessemsaurids, supported by two unambiguous synapomorphies: absence of ventral keels on anterior cervical centra (char. 132.1), and midpoint position of the fourth trochanter along the length of the femur (char. 293.1). The following synapomorphies are shared for non-gravisaurian Sauropodiformes: transversely compressed cross-sectional shape of the dorsal neural spines (char. 169.0); presence of spinodiapophyseal lamina on dorsal vertebrae separated from spinopostzygapophyseal lamina (char. 170.1); tall crest fourth trochanter (char. 292.1); and femoral length lesser than 100 cm (char. 353.4).

4.5 Biogeographical Distribution and Morphological Diversity During Triassic–Jurassic

During the Late Triassic and Early Jurassic, Sauropodomorpha was the most diverse and abundant dinosaur clade, being dominant and broadly distributed worldwide in Late Triassic continental ecosystems (Galton and Upchurch 2004; Upchurch et al. 2004; Pol et al. 2021; see Chaps. 1 and 2). Both massospondylids and sauropodiforms appeared during the Norian and were the only sauropodomorph clades that survived the end-Triassic extinction.

Non-gravisaurian sauropodiforms are mostly known from the Late Triassic and Early Jurassic of Gondwana (Fig. 4), in particular from South America and South Africa (Table 1). Moreover, even the basal gravisaurian taxa *Vulcanodon* and *Tazoudasaurus* come from the Early Jurassic of Zimbabwe and Morocco, respectively (Cooper 1984; Allain and Aquesbi 2008). Outside of Gondwana, *Camelotia* and *Schleitheimia* come from the Late Triassic of Europe (Galton 2005; Rauhut et al. 2020), and *Gongxianosaurus*, *Yizhousaurus* and *Irisosaurus* are recorded from the Early Jurassic of China (He et al. 1998; Zhang et al. 2018; Peyre de Fabrègues et al. 2020). *Isanosaurus* was named from the Late Triassic of Thailand (Buffetaut et al. 2000, 2002), although this age has been questioned, with a Late Jurassic age now suggested for the Nam Phong Formation (Racey and Goodall 2009).

Recent studies suggest that sauropodomorph morphological disparity shifted twice during its early evolution, first at the Carnian–Norian, through the rapid diversification of plateosaurians dinosaurs, and later at the end of Early Jurassic, when Gravisauria became the dominant herbivores in terrestrial ecosystems (Bronzati et al. 2019; McPhee et al. 2017; Apaldetti et al. 2021). Moreover, sauropodomorph abundance and diversity did not decrease globally across the Triassic–Jurassic boundary, suggesting that sauropodomorph evolution was not negatively affected by the end-Triassic extinction (McPhee et al. 2015, 2017, 2018; Rauhut et al. 2020; Apaldetti et al. 2021). In the case of non-gravisaurian Sauropodiformes, lessemsaurids demonstrates a continuous expansion of body size across the end-Triassic (*Lessemsaurus* ~ 10 tonnes) to Early Jurassic (*Ledumahadi* ~ 12 tonnes), indicating that the large-bodied lessemsaurids were unaffected by Triassic–Jurassic extinction. The abrupt increase of body mass and changes in locomotion styles were probably the most influential factors that led to the greatest disparity of Sauropodomorpha during the earliest Jurassic (McPhee et al. 2017, 2018; Apaldetti et al. 2021).

The mosaic of plesiomorphic and derived features observed among non-gravisaurian sauropodiforms seem to reveal that major changes observed along the sauropodomorph lineage through the Triassic–Jurassic boundary were not necessarily related to the emergence of anatomical novelties or new body plans, but related to change in the distribution of previously existing lineages (such as non-gravisaurian sauropodiforms; Apaldetti et al. 2021). Interestingly, the South American non-gravisaurian sauropodiforms notably show these patterns throughout the Late Triassic and Early Jurassic. The Early Jurassic Sauropodiformes *Mussaurus* and *Leonerasaurus*—the youngest taxa among Gondwanan non-sauropods—are more anatomically aligned with taxa-like massospondylids than to sauropods, exhibiting gracile bauplans and several features related to bipedal locomotion, at least in adulthood (Otero et al. 2019). On the other hand, lessemsaurids—the oldest branching clade of Sauropodiformes—are repeatedly resolved closer to the origin of Sauropoda, representing the first evidence of gigantism among Sauropodomorpha (Apaldetti et al. 2018).

Non-gravisaurian sauropodiforms coexisted for a short time with early Gravisauria during Toarcian, but they became extinct at the end of Early Jurassic for reason still unknown although some hypotheses aim to a competitive replacement, better use of the body's energy requirements and feeding apparatus (Barrett and Upchurch 2005; McPhee et al. 2015, 2018). This highlights the importance of the Late Triassic and Early Jurassic Gondwanan taxa—and among them the relevant South America record—for understanding the early evolutionary steps towards the origin and evolution of the eusauropod dinosaurs, the largest terrestrial animals that inhabited the Earth.

4.6 Lessemsaurids and the Origin of Gigantism

As was previously commented, gigantism in eusauropod dinosaurs has been proposed as the result of a suite of anatomical and physiological changes that led sauropodomorphs to progressively into massive bauplan, until they became the largest animals known in the history of life (Wilson and Sereno 1998; Klein et al. 2011; Rauhut et al. 2011; Sander et al. 2004, 2011b; Sander 2013a, b). From the Early Jurassic, the gravisaurian taxa (>10 tonnes) such as *Vulcanodon* and *Tazoudasaurus* exhibit the earliest unequivocal evidence for those series of novelties suggested necessary for achieving large body sizes (Allain and Aquesbi 2008). However, the mid-Norian lessemsaurids—the first sauropodomorphs to achieve ~10 tonnes—indicates the appearance of an early trend towards gigantism before the first large gravisaurians at the beginning of the Jurassic (Apaldetti et al. 2018). Even the lessemsaurid *Ledumahadi* exceeded 10 tonnes (it has been estimated in ~12 tonnes) at the earliest Jurassic, thus extending the total known size range for Early Jurassic

Sauropodomorpha (almost two orders of magnitude), and therefore being so far the largest animal to live on Earth up to that moment (McPhee et al. 2018).

Hence, lessemsaurids result in a key sauropodiform group for understanding novel pathways to achieve gigantism during the early evolution of Sauropodomorpha. They show that several anatomical modifications usually considered as a requirement to reach large body size (such as columnar limbs, the elongation of the neck and forelimb) were not strictly necessary, and new strategies and innovations probably contributed to attaining large body sizes without pre-adaptations towards gigantism (e.g. growth strategy and pneumatic diverticula invading cervical vertebrae; Apaldetti et al. 2018; McPhee et al. 2018). Thus, the pectoral girdle and forelimb of the lessemsaurids displays the posteroventral orientation of the scapulocoracoid glenoid that preludes an erect posture of the humerus and lack a completely pronated manus, with a relative flexed forelimb posture (shallow radial fossae and developed olecranon of the ulna, and twisted digit I; see above). The plesiomorphic ilium (small preacetabular process and deep brevis fossa, related to the magnitude of pelvic muscles; Fechner 2009; Rauhut et al. 2011) provides a primitive aspect to the pelvic girdle and hindlimbs of lessemsaurids. In this context, the presence of other derived features—such as the distinctive short but high cervical vertebrae bearing well-developed hyposphenes (see above)—possibly helped lesser shear stress in the neck when held at a low to horizontal angle (McPhee et al. 2015). A reinforced neck had probably something to do with supporting the additional weight (by providing additional muscle anchorage) of a large-bodied habitual quadruped that retained a laterally oriented forelimb for use in a non-locomotory context (i.e. when feeding; McPhee et al. 2015). These plesiomorphic features highlight that lessemsaurids were not fully columnar-limbed sauropodiforms with a short but robust neck, revealing that these features were gradually changing throughout the evolution to the gigantism among Sauropodomorpha (Apaldetti et al. 2018; McPhee et al 2018).

The presence of skeletal pneumaticity in cervico-dorsal vertebrae of lessemsaurids with cervical and abdominal diverticula invading the neural arches, potentially provides an improved avian-like respiratory system not developed by any other non-sauropod sauropodomorph. Because only cervical and abdominal air sacs and lungs can produce pneumatic diverticula that invade the vertebrae (O'connor and Claessens 2005), the presence of a sophisticated pneumatic complex along the neck of lessemsaurids (and in dorsal series of other non-gravisaurian sauropodiforms) suggests that a cervical air sac system was already well-developed among early-diverging Sauropodiformes during Norian times, 20 myr before the gravisaurians. This improved avian-style respiratory system observed among non-gravisaurian sauropodiforms (mainly in lessemsaurids) may also be related to the need to lose heat, especially if their extremely high bone growth rate necessitated a high metabolic rate (Sverdlova et al. 2012; Henderson 2013; Sander 2013b; Apaldetti et al. 2018). Cyclical and remarkably high bone growth rates represent a key lessemsaurids novelty which helped them attain large body sizes through a strategy of accelerated growth distinct from that associated with gigantism in eusauropods (Cerda et al. 2017; Apaldetti et al. 2018; Krupandan et al. 2018).

The derived traits among lessemsaurids (robust and pneumatic cervical vertebrae and ovoid cross-section of the femur) reveal that these derived femoral and vertebral traits maybe some of the key skeletal innovations that were acquired along lineage resulting in the largest terrestrial animals in the history of the Earth.

5 Conclusions

Non-gravisaurian sauropodiforms exhibit a mosaic of plesiomorphic and derived features, highlighting a path of progressive anatomical acquisitions towards a giant body-plan. In this context, the South American non-gravisaurian Sauropodiformes reveal key anatomical changes or, conversely, the retention of some primitive features that characterized this evolutionary path. The main suite of features that characterize non-gravisaurian Sauropodiformes are related to the plesiomorphic and/or derived nature of its traits, mainly observed on their limbs, girdles and neck. All the South American non-gravisaurian sauropodiforms come from the Late Triassic and Early Jurassic of Argentina. The earliest-diverging Sauropodiformes *Mussaurus patagonicus* and *Leonerasaurus taquetrensis* are the youngest records among Gondwanan non-sauropods, exhibiting a mix of features that suggest they are more closely related to massopondylids or early sauropodiform condition. The lessemsaurid *Lessemsaurus sauropoides* and *Ingentia prima*, instead, are the oldest South American Sauropodiformes, with some novel anatomical innovations that make them more closely related to early sauropods and to the first steps to gigantism on Earth.

Lessemsaurids show that quadrupedal sauropodomorphs lacking columnar limbs could attain sauropod-like body sizes (>10 tonnes), contrary to the generalized conception that columnar limb posture was a necessary attribute to reach giant body sizes (Apaldetti et al. 2018; McPhee et al. 2018). Beyond the plesiomorphic features that result in a non-sauropod appearance, lessemsaurids were able to achieve a body size as large as the first large gravisaurians from the middle of the Early Jurassic period (e.g. *Vulcanodon karibaensis*, *Tazoudasaurus naimi*). Even, lessemsaurids exceeded the size of their co-existing gravisaurian taxa at the earliest Jurassic (McPhee et al. 2018). This confirms that without pre-adaptation in the limbs (quadrupedal, columnar limbs), girdles (muscle-related modifications) or neck (elongation of vertebrae), a group of non-gravisaurian Sauropodiformes were able to attain gigantic body sizes in an evolutionary period of approximately 20 myr (Apaldetti et al. 2018; McPhee et al. 2018). The pneumaticity of the neck—related to an improved avian-style respiratory system—and a novel bone growth strategy were probably the key to attaining large sizes in the early evolutionary history of Sauropodomorpha to that point. Thus, the robust lessemsaurids from South America (*Lessemsaurus sauropoides* and *Ingentia prima*) also reveal that, although the eusauropod body-plan was fully established by the Early Jurassic, several features that have traditionally characterized the giant eusauropods, were already established by the Late Triassic (at least by the Norian).

The continuous increase in body size among lessemsaurids, from ~10 tonnes in the Late Triassic (*Lessemsaurus sauropoides*) to >10 tonnes in the earliest

Jurassic (*Ledumahadi mafube*), indicates that large-bodied Sauropodiformes were not affected by Triassic–Jurassic extinction. Even, sauropodomorph abundance and diversity did not decrease globally across the end-Triassic event, suggesting that sauropodomorph evolution was not negatively affected by the end-Triassic extinction, and even achieving their highest disparity during the earliest Jurassic (McPhee et al. 2015, 2017, 2018; Rauhut et al. 2020; Apaldetti et al. 2021). Thus, non-gravisaurian Sauropodiformes provide crucial information to understand the early evolutionary history of sauropodomorphs, with the South American taxa being some of the most relevant taxa related to the first steps towards the origin and evolution of gigantism in sauropod dinosaurs.

Acknowledgements We are grateful to the editors of this issue for the invitation to participate. We thank Adam Marsh and Blair McPhee for their valuable comments that improved this manuscript. We thank the support from the Agencia de Promoción Científica y Tecnológica—ANPCyT (PICT 2016-236), CICITCA-UNSJ grants (to CA and RNM), Sepkoski Grant (to CA), and Secretaría de Ciencia Gobierno de San Juan (SECITI to RNM).

References

Alexander RMN (1985) Mechanics of posture and gait of some large dinosaurs. Zool J Linn Soc 83:1–25

Allain R, Aquesbi N (2008) Anatomy and phylogenetic relationships of *Tazoudasaurus naimi* (Dinosauria, Sauropoda) from the late Early Jurassic of Morocco. Geodiversitas 30:345–424

Apaldetti C, Martínez RN, Cerda I, Pol D, Alcober OA (2018) An early trend towards gigantism in Triassic sauropodomorph dinosaurs. Nat Eco Evo 2:1227–1232. https://doi.org/10.1038/s41559-018-0599-y

Apaldetti C, Pol D, Ezcurra MD, Martínez RN (2021) Sauropodomorph evolution across the Triassic–Jurassic boundary: body size, locomotion, and their influence on morphological disparity. Sci Rep 11:22534

Barrett PM (2014) Paleobiology of herbivorous dinosaurs. Ann Rev Ear Plan Sci 42:207–230

Barrett PM, Upchurch P (2005) Sauropodomorph diversity through time: paleoecological and macroevolutionary implications. In: Wilson JA (ed) Curry Rogers KA. Evolution and Paleobiology. University of California Press, The Sauropods, pp 25–136

Benson RBJ, Campione NE, Carrano MT et al (2014) Rates of dinosaur body mass evolution indicate 170 million years of sustained ecological innovation on the avian stem linage. PLoS Biol 12. https://doi.org/10.1371/journal.pbio.1001853

Benson RBJ, Hunt G, Carrano MT, Campione NE (2017) Cope's rule and the adaptive landscape of dinosaur body size evolution. Palaeontology 61:13–48

Bonaparte JF (1971) Los tetrápodos del sector superior de la Formación Los Colorados, La Rioja, Argentina (Triásico superior). Opera Lilloana 22:1–183

Bonaparte JF (1973) Edades Reptil para el Triásico de Argentina y Brazil. Actas V Congreso Geológico Argentino 3:93–129

Bonaparte JF (1999) Evolución de las vértebras presacras en Sauropodomorpha. Ameghiniana 36:115–187

Bonaparte JF, Powell JE (1980) A continental assemblage of tetrapods from the Upper Cretaceous beds of El Brete, northwestern Argentina (Sauropoda – Coelurosauria – Carnosauria – Aves). Ecosysténes Continentaux du Mésozoique. Memoires de la Société Géologique de France (n.s.) 139: 19–28

Bonaparte JF, Vince M (1979) El hallazgo del primer nido de Dinosaurios Triásicos (Saurischia, Prosauropoda), Triásico Superior de Patagonia, Argentina. Ameghiniana 16:173–182
Bonnan MF (2003) The evolution of manus shape in sauropod dinosaurs: implications for functional morphology, forelimb orientationand sauropod phylogeny. J Vert Paleont 23:595–613
Bonnan MF, Senter P (2007) Were the basal sauropodomorph dinosaurs *Plateosaurus* and *Massospondylus* habitual quadrupeds? Spec Pap Palaeontol 77:139–155
Bonnan MF, Yates AM (2007) A new description of the forelimb of the basal sauropodomorph *Melanorosaurus*: implications for the evolution pronation, manus shape and quadrupedalism in sauropod dinosaurs. Spec Pap Palaeont 77:157–168
Bronzati M, Müller RT, Langer MC (2019) Skull remains of the dinosaur *Saturnalia tupiniquim* (Late Triassic, Brazil): With comments on the early evolution of sauropodomorph feeding behaviour. PLoS ONE 14(9). https://doi.org/10.1371/journal.pone.0221387
Buffetaut E, Suteethorn V, Cuny G, Tong H, Le Loeuff J, Khansubha S, Jongautchariyakul S (2000) The earliest known sauropod dinosaur. Nature 407:72–74
Buffetaut E, Suteethorn V, Le Loeuff J, Cuny G, Tong H, Khansubha S (2002) The first giant dinosaurs: a large sauropod from the Late Triassic of Thailand. C R Palevol 1:103–109
Carballido JL, Pol D, Otero A, Cerda IA, Salgado L, Garrido AC, Ramezzani J, Cúneo NR, Krause MJ (2017) A new giant titanosaur shed light on body mass evolution among sauropod dinosaurs. Proc R Soc B 20171219. https://doi.org/10.1098/rspb.2017.1219
Carrano MT (1999) What, if anything, is a cursor? Categories versus continua for determining locomotor habit in mammals and dinosaurs. J Zool 247:29–42
Carrano MT (2001) Implications of limb bone scaling, curvature and eccentricity in mammals and non-avian dinosaurs. J Zool 254:41–55
Carrano MT (2005) Body-size evolution in the Dinosauria. In: Carrano MT, Gauden TJ, Blob RW, Wible JR (eds) Amniote paleobiology: perspectives on the evolution of mammals, birds and reptiLessemsaurusUniversity of Chicago Press, pp 225–268
Casamiquela RM (1980) La presencia del género *Plateosaurus* (prosauropoda) en el Triásico Superior de la Formación El Tranquilo, Patagonia. In Actas II Congreso Argentino de Paleontología y Bioestratigrafía, I Congreso Latinoamericano de Paleontología, Vol 1, Buenos Aires, pp 143–158
Cerda IA, Chinsamy A, Pol D, Apaldetti C, Otero A, Powell JE, Martínez RN (2017) Novel insight into the origin of the growth dynamics of sauropod dinosaurs. PLoS ONE 12:e0179707
Cooper MRA (1984) A reassessment of *Vulcanodon karibaensis* Raath (Dinosauria: Saurischia) and the origin of the Sauropoda. Palaeontol Afr 25:203–231
Desojo JB, Fiorelli LE, Ezcurra MD, Martinelli AG, Ramezani J, Da Rosa AAS, von Baczko MB, Totteyn MJ, Montefeltro FC, Ezpeleta M, Langer MC (2020) The Late Triassic Ischigualasto Formation at Cerro Las Lajas (La Rioja, Argentina): fossil tetrapods, high-resolution chronostratigraphy, and faunal correlations. Sci Rep 10:10782
Figari E, Courtade SF (1993) Evolución tectosedimentaria de la Cuenca de Cañadón Asfalto, Chubut, Argentina. 13° Congreso Geológico Argentino y 2 Congreso de Exploración de Hidrocarburos. Actas 1:66–77
Fechner R (2009) Morphological evolution of the pelvic girdle and hindlimb of Dinosauromorpha on the lineage to Sauropoda. PhD thesis, Ludwigs Maximilians Universität, München
Galton PM (1985) Notes on the Melanorosauridae, a family of large Prosauropod Dinosaurs (Saurischia: Sauropodomorpha). Geobios 18:671–676
Galton PM (1998) Saurischian dinosaurs from the Upper Triassic of England: Camelotia (Prosauropoda, Melanorosauridae) and Avalonianus (Theropoda, ?Carnosauria). Palaeontogr Abt A 250:155–172
Galton PM (2005) Bones of large dinosaurs (Prosauropoda and Stegosauria) from the Rhaetic bone bed (Upper Triassic) of Aust Cliff, southwest England. Rev Paléobiol 24:51–74
Galton PM, Kermack D (2009) The anatomy of *Pantydraco caducus*, a very basal sauropodomorph dinosaur from the Rhaetian (Upper Triassic) of South Wales, UK. Rev Paléobiol 29:341–404
Galton PM, Upchurch P (2000) Prosauropod dinosaurs: homeotic transformations ("frame shifts") with third sacral as a caudosacral or a dorsosacral. J of Vert Paleont 20(3, Supplement):43A

Galton PM, Upchurch P (2004) Prosauropoda. In: Weishampel DW, Dodson P, Osmólska H (eds) The Dinosauria, 2nd edn. University of California Press, Berkeley, California, pp 232–258

Galton PM, Van Heerden J (1985) Partial hindlimb of *Blikanasaurus cromptoni* n. gen. and n. sp., representing a new family of prosauropod dinosaurs from the Upper Triassic of South Africa. Geobios 18:509–516

Galton PM, Van Heerden J (1998) Anatomy of the prosauropod dinosaur *Blikanasaurus cromptoni* (Upper Triassic, South Africa), with notes on the other tetrapods from the lower Elliot Formation. Paläontol Z 72:163–177

Galton PM, Van Heerden J, Yates AM (2005) Postcranial anatomy of referred specimens of the sauropodomorph dinosaur *Melanorosaurus* from the Upper Triassic of South Africa. In: Tidwell V, Carpenter K (eds) Thunder-lizards: the sauropodomorph dinosaurs. Indiana University Press, Bloomington, IN, pp 1–37

Garcia GA, Martínez RN, Apaldetti C (2016) Cuatro vértebras sacras en Sauropodomorpha: invento de Sauropodiformes? Ameghiniana 54(4):Suppl. R22

Garcia GA, Martínez RN, Apaldetti C (2018) Nuevos materiales del sauropodomorfo basal *Adeopapposaurus mognai* y sus implicancias en el origen de un sacro "tipo saurópodo". Publicaciones Electrónicas de la Asociación Paleontológica Argentina (PEAPA 19:1(R8). https://doi.org/10.5710/PEAPA.15.04.2019.296

Gatesy SM (1990) Caudofemoral musculature and the evolution of theropod locomotion. Paleobiology 16:170–186

Gauthier JA (1986) Saurischian monophyly and the origin of birds. In: Padian K (ed.) The origin of birds and the evolution of flight, vol 8. Memoirs of the California Academy of Sciences, pp 1–55

Goloboff PA, Farris JS, Nixon KC (2008) TNT, a free program for phylogenetic analysis. Cladistics 24:774–786

Groeber PFC, Stipanicic PN (1953) Triásico. In Groeber PFC (ed) Mesozoico Geografía de la Republica Argentina. Sociedad Argentina de Estudios Geográficos, Buenos Aires, pp 13–141

Hutchinson JR, Gatesy SM (2000) Adductors, abductors, and the evolution of archosaur locomotion. Paleobiology 26:734–751

He X, Wang C, Liu S, Zhou F, Liu T, Cai K, Dai B (1998) A new species of sauropod from the Early Jurassic of Gongxian Co. Sichuan Acta Geologica Sichuan 18:1–7

Henderson D (2013) Sauropod necks: are they really for heat loss? PLoS ONE 8:e77108

Kent DV, Santi Malnis P, Colombi CE, Alcober OA, Martínez RN (2014) Age constraints on the dispersal of dinosaurs in the late triassic from magnetochronology of Los Colorados Formation (Argentina). Proc Natl Acad Sci USA 111:7958–7963

Klein N, Remes K, Gee CT, Sander PM (2011) Biology of the Sauropod dinosaurs: understanding the life of giants. Indiana University Press, Bloomington

Krupandan E, Chinsamy-Turan A, Pol D (2018) The long bone histology of the Sauropodomorpha, *Antetonitrus ingenipes*. Anat Rec 301:1506–1518

Langer MC (2003) The pelvic and hind limb anatomy of the stem-sauropodomorph *Saturnalia tupiniquim* (Late Triassic, Brazil). Paleo Bios 23:1–30

Langer MC, Ezcurra MD, Bittencourt JS, Novas FE (2010) The origin and early evolution of dinosaurs. Biol Rev 84:1–56

Mannion PD, Upchurch P, Carrano MT, Barrett PM (2011) Testing the effect of the rock record on diversity: a multidisciplinary approach to elucidating the generic richness of sauropodomorph dinosaurs through time. Biol Rev 86:157–181

Martínez RN, Apaldetti C, Correa G, Colombi CE, Fernández E, Santi Malnis P, Praderio A, Abelín D, Benegas LG, Aguilar A, Alcober OA (2015) A new Late Triassic vertebrate assemblage from Northwestern Argentina. Ameghiniana 52:379–390

McPhee BW, Choiniere J (2017) The osteology of *Pulanesaura eocollum*: implications for the inclusivity of Sauropoda (Dinosauria). Zool J of the Linn Soc 182:830–861

McPhee BW, Benson RBJ, Botha-Brink J, Bordy EM, Choiniere JN (2018) A giant dinosaur from the earliest Jurassic of South Africa and the transition to quadrupedality in early sauropodomorphs. Curr Biol 8:15179. https://doi.org/10.1016/j.cub.2018.07.063

McPhee BW, Bonnan MF, Yates AM, Neveling J, Choiniere JN (2015) A new basal sauropod from the pre-Toarcian Jurassic of South Africa: evidence of niche-partitioning at the sauropodomorph–sauropod boundary? Sci Rep 5:13224. https://doi.org/10.1038/srep13224

McPhee BW, Bordy EM, Sciscio L, Choiniere JN (2017) The sauropodomorph biostratigraphy of the Elliot formation of southern Africa: tracking the evolution of Sauropodomorpha across the Triassic-Jurassic boundary. Acta Palaeontol Pol 62:441–465

McPhee BW, Yates AM, Choiniere JN, Abdala F (2014) The complete anatomy and phylogenetic relationships of *Antetonitrus ingenipes* (Sauropodiformes, Dinosauria): implications for the origins of Sauropoda. Zool J Lin Soc 171:151–205

Nakayama C (1973) Sedimentitas pre-bayocianas en el extremo austral de la Sierra de Taquetrén. Chubut (Argentina). 5to Congreso Geológico Argentino. Actas 3:269–277

O'Connor PM (2006) Postcranial Pneumaticity: an evaluation of soft-tissue influences on the postcranial skeleton and the reconstruction of pulmonary anatomy in Archosaurs. J Morphol 267:1199–1226

O'Connor PM, Claessens LPAM (2005) Basic avian pulmonary design and flow-through ventilation in non-avian theropod dinosaurs. Nature 436:253–256

Otero A, Pol D (2013) Postcranial anatomy and phylogenetic relationships of *Mussaurus patagonicus* (Dinosauria, Sauropodomorpha). J Vert Paleont 33:1138–1168

Otero A, Allen V, Pol D, Hutchinson JR (2017) Forelimb muscle and joint actions in Archosauria: insights from *Crocodylus johnstoni* (Pseudosuchia) and *Mussaurus patagonicus* (Sauropodomorpha). Peer J 5:e3976

Otero A, Cuff AR, Allen V, Summer-Rooney L, Pol D, Hutchinson JR (2019) ontogenetic changes in the body plan of the sauropodomorph dinosaur *Mussaurus patagonicus* reveal shifts of locomotor stance during growth. Sci Rep 9:7614. https://doi.org/10.1038/s41598-019-44037-1

Otero A, Krupandan E, Pol D, Chinsamy A, Choiniere JN (2015) A new basal sauropodiform from South Africa and the phylogenetic relationships of basal sauropodomorphs. Zool J Linn Soc 174:589–634

Otero A, Carballido JL, Salgado L, Canudo I, Garrido A (2021) Report of a giant titanosaur sauropod from the Upper Cretaceous of Neuquen Province, Argentina. Cretaceous Res 122:104754

Page R, Ardolino A, de Barrio RE, Franchi M, Lizuain A, Page S, Silva Nieto D (2000) Estratigrafía del Jurásico y Cretácico del Macizo de Somún Curá, provincias de Río Negro y Chubut. In: Caminos R (ed) Geología Argentina. Subsecretaría de Minería de la Nación, Buenos Aires, pp 460–488

Peyre de Fabrègues C, Allain R (2016) New material and revision of *Melanorosaurus thabanensis*, a basal sauropodomorph from the Upper Triassic of Lesotho. PeerJ 4. https://doi.org/10.7717/peerj.1639

Peyre de Fabrègues C, Bi S, Li H, Li G, Yang L, Xu X (2020) A new species of early-diverging Sauropodiformes from the Lower Jurassic Fengjiahe Formation of Yunnan Province. China Scient Rep 10:10961. https://doi.org/10.1038/s41598-020-67754-4

Peyre de Fabrègues C, Allain R (2020) *Kholumolumo ellenbergerorum*, gen. et sp. nov., a new early sauropodomorph from the lower Elliot Formation (Upper Triassic) of Maphutseng, Lesotho. J Vertebr Paleontol. https://doi.org/10.1080/02724634.2019.1732996

Pol D, Powell JE (2007a) Skull anatomy of *Mussaurus patagonicus* (Dinosauria: Sauropodomorpha). Late Triassic Patagonia Hist Biol 19:125–144

Pol D, Powell JE (2007b) New information on *Lessemsaurus sauropoides* (Dinosauria: Sauropodomorpha) from the Upper Triassic of Argentina. Spec Pap Palaeontol 77:223–243

Pol D, Garrido A, Cerda IA (2011) A new sauropodomorph dinosaur from the early Jurassic of Patagonia and the origin and evolution of the sauropod-type sacrum. PLoS ONE 6. https://doi.org/10.1371/journal.pone.0014572

Pol D, Otero A, Apaldetti C, Martínez RN (2021) Triassic sauropodomorph dinosaurs from South America and the origin and diversification of dinosaur dominated herbivorous faunas. J South Am Ear Sc 107. https://doi.org/10.1016/j.jsames.2020.103145

Racey A, Goodall JGS (2009) Palynology and stratigraphy of the Mesozoic Khorat Group red bed sequences from Thailand. Geol Soc Lond Spec Publ 315:69–83
Rauhut OWM, Fechner R, Remes K, Reis K (2011) How to get big in the Mesozoic: the evolution of the sauropodomorph body plan. In: Klein N, Remes K, Gee CT, Sander PM (eds) Biology of the Sauropod Dinosaurs: understanding the life of giants. Indiana University Press, Life of the Past, pp 119–149
Rauhut OWM, Holwerda FM, Furrer H (2020) A derived sauropodiform dinosaur and other sauropodomorph material from the Late Triassic of Canton Schaffhausen Switzerland. Swiss J Geosci 113:8. https://doi.org/10.1186/s00015-020-00360-8
Remes K (2008) Evolution of the pectoral girdle and forelimb in sauropodomorpha (Dinosauria, Saurischia): Osteology, Myology and Function. PhD thesis, Ludwig-Maximilians-Universität München
Salgado L, Calvo JO, Coria RA (1997) Evolution of the titanosaurid sauropods. 1. Phylogenetic analysis based on postcranial evidence. Ameghiniana 34:3–32
Sander PM (2013a) Sauropod gigantism: a cross-disciplinary approach. PLoS ONE Collections, Berkely
Sander PM (2013b) An evolutionary cascade model for sauropod dinosaur gigantism—overview, update and tests. PLoS ONE 8:e78573
Sander PM, Christian A, Clauss M, Fechner R, Gee CT, Griebeler E-M, Gunga H-C, Hummel J, Mallison H, Perry SF, Preuschoft H, Rauhut OWM, Remes K, Tütken T, Wings O, Witzel U (2011a) Biology of the sauropod dinosaurs: the evolution of gigantism. Biol Rev 86:117–155
Sander PM, Klein N, Buffetaut E, Cuny G, Suteethorn V, Le Loeuff J (2004) Adaptative radiation in sauropod dinosaurs: bone histology indicates rapid evolution of giant body size through acceleration. Org Div Evol 4:165–173
Sander PM, Klein N, Stein K, Wings O (2011a) Sauropod bone histology and its implicatios for sauropod biology. In Klein N, Remes K, Gee CT, Sander PM (eds) Biology of the Sauropod dinosaurs: understanding the life of giants. Life of the past. Indiana University Press
Sereno PC (1999) The evolution of dinosaurs. Science 284:2137–2147
Sereno PC (2007) Basal Sauropodomorpha: historical and recent phylogenetic hypotheses, with comments on *Ammosaurus major* (Marsh, 1989). Spec Pap Palaeontol 77:261–289
Smith RH, Marsicano CA, Pol D, Mancuso AC (2014) Ichnology of sauropodomorph nests from Patagonia indicates early Jurassic origin of herd-living and breeding site fidelity. In: Abstracts of the 4° International palaeontological congress, Mendoza, Argentina, pp 227
Stefanic CM, Nesbitt SJ (2019) The evolution and role of the hyposphene-hypantrum articulation in Archosauria: phylogeny, size and/or mechanics? R Soc Open Sci 6:190258
Sverdlova NS, Lambertz M, Witzel U, Perry SF (2012) Boundary conditions for heat transfer and evaporative cooling in the trachea and air sacs system of the domestic fowl: a two-dimensional CFD analysis. PLoS ONE 7:e45315
Upchurch P (1998) The phylogenetic relationships of sauropod dinosaurs. Zool J Linn Soc Lond 124:43–103
Upchurch P, Barrett PM, Dodson P (2004) Sauropoda. In: Dodson P, Osmólska H (eds) Weishampel DB. The dinosauria, University of California Press, pp 259–322
Upchurch P, Barrett PM, Galton PM (2007) A phylogenetic analysis of basal sauropodomorph relationships: implications for the origin of sauropod dinosaurs. Spec Pap Palaeontol 77:57–90
Wang Y-M, You H-L, Wang T (2017) A new basal sauropodiform dinosaur from the Lower Jurassic of Yunnan Province, China. Sci Rep 7:41881. https://doi.org/10.1038/srep41881
Wang Y-M, Wang T, Yang Z-W, You H-L (2020) Cranium and vertebral column of *Xingxiulong chengi* (Dinosauria: Sauropodomorpha) from the Early Jurassic of China. Anat Rec 303:772–789
Weishampel DB, Horner JR (1994) Life history syndromes, heterochrony, and the evolution of dinosauria. In: Carpenter K, Hirsch KF, Horner JR (eds) Dinosaur eggs and babies. Cambridge University Press, Cambridge, pp 229–243

Wilson JA (2005) Overview of sauropod phylogeny and evolution. In: Curry Rogers KA, Wilson JA (eds) The sauropods: evolution and paleobiology. University of California Press, Berkeley, CA, pp 15–49
Wilson JA (2006) Anatomical nomenclature of fossil vertebrates: standardized terms or lingua franca? J Vert Paleont 26:511–518
Wilson JA, Carrano M (1999) Titanosaurs and the origin of '"wide-gauge"' trackways: a biomechanical and systematic perspective on sauropod locomotion. Paleobiology 25:252–267
Wilson JA, Sereno PC (1998) Early evolution and higher-level phylogeny of sauropod dinosaurs. Mem Soc Vert Paleontol 5:1–68
Yates AM (2007a) The first complete skull of the Triassic dinosaur *Melanorosaurus* Haughton (Sauropodomorpha: Anchisauria). Spec Pap Palaeontol 77:9–55
Yates AM (2007b) Solving a dinosaurian puzzle: the identity of *Aliwalia rex* Galton. Hist Biol 19:93–123
Yates AM (2008) A second specimen of *Blikanasaurus* (Dinosauria: Sauropoda) and the biostratigraphy of the lower Elliot Formation. Palaeontol Afr 43:39–43
Yates AM, Kitching JW (2003) The earliest known sauropod dinosaur and the first steps towards sauropod locomotion. Proc R Soc Lond B 270:1753–1758
Yates AM, Bonnan MF, Neveling J, Chinsamy A, Blackbeard MG (2010) A new transitional sauropodomorph dinosaur from the early Jurassic of South Africa and the evolution of sauropod feeding and quadrupedalism. Proc R Soc Biol Sci B 277:787–794
Yates AM, Wedel MJ, Bonnan MF (2012) The early evolution of postcranial skeletal pneumaticity in sauropodomorph dinosaurs. Acta Palaeontol Polonica 57:85–100
Zhang Q-N, You H-L, Wang T, Chatterjee S (2018) A new sauropodiform dinosaur with a 'sauropodan' skull from the lower Jurassic Lufeng Formation of Yunnan Province China. Sci Rep 8:13464. https://doi.org/10.1038/s41598-018-31874-9

Sauropods from the Early Jurassic of South America and the Radiation of Eusauropoda

Diego Pol, Kevin Gomez, Femke M. Holwerda, Oliver W. M. Rauhut, and José L. Carballido

Abstract Eusauropods are large-bodied and long-necked dinosaurs that dominated the role of large herbivores in terrestrial ecosystems since at least the late Early Jurassic (Pliensbachian–Toarcian). Their early diversification is best recorded in South America where the best-preserved eusauropods and close relatives from this period of time have been found. The earliest sauropod from the Jurassic of South America is *Amygdalodon patagonicus* from the Cerro Carnerero Formation (Pliensbachian–early Toarcian), and its fragmentary remains suggest a position at the base of Gravisauria or as closely related to this clade. The Cañadón Asfalto Formation (middle–late Toarcian) has provided three named sauropods, although a higher diversity of sauropods may have existed. These are the basal eusauropod *Patagosaurus fariasi*, known from multiple specimens, the much more incompletely known early sauropod *Volkheimeria chubutensis*, and *Bagualia alba* that is known from multiple specimens and includes fairly complete craniomandibular remains. These taxa provide the earliest evidence of ecological predominance by large-bodied sauropods and are therefore significant for understanding the rise and success of this group in the Jurassic Period. The current knowledge of these sauropods from the late Early Jurassic of South America indicates that the evolutionary radiation of Eusauropoda occurred at least by the mid-Toarcian, subsequent to a large-scale

---

Electronic supplementary material The online version contains supplementary material available at (10.1007/978-3-030-95959-3_4).

D. Pol (✉) · K. Gomez · J. L. Carballido
Consejo Nacional de Investigaciones Científicas y Técnicas (CONICET), Buenos Aires, Argentina
e-mail: dpol@mef.org.ar

Museo Paleontológico Egidio Feruglio, 9100 Trelew, Chubut, Argentina

F. M. Holwerda
Royal Tyrrell Museum of Palaeontology, Drumheller, AB T0J 0Y0, Canada

O. W. M. Rauhut
SNSB—Bayerische Staatssammlung für Paläontologie und Geologie, 80333 Munich, Germany

Department for Earth and Environmental Sciences, Ludwig-Maximilians-Universität, 80333 Munich, Germany

GeoBioCenter, Ludwig-Maximilians-Universität Munich, 80333 Munich, Germany

A. Otero et al. (eds.), *South American Sauropodomorph Dinosaurs*, Springer Earth System Sciences, https://doi.org/10.1007/978-3-030-95959-3_4

volcanic event in the Southern Hemisphere that has been linked to global climatic change and the rise of conifers as the predominant components of Jurassic seasonal forests.

Keyword Eusauropoda · Patagonia · Cañadón Asfalto Formation · Toarcian

1 Introduction

The origin and diversification of sauropods is one of the most important events in the evolution of Sauropodomorpha because they became one of the most conspicuous groups of terrestrial vertebrates of the Mesozoic Era. Although a broad diversity of early sauropodomorph lineages existed during the Late Triassic and Early Jurassic, sauropods were the only surviving lineage during more than 100 million years in the Jurassic and Cretaceous periods. During this time, sauropods were abundant, present across the globe, and reached body sizes larger than any other land animal (Wilson and Sereno 1998; Upchurch et al. 2004; Barrett and Upchurch 2005; Barrett 2014; Bates et al. 2016).

The phylogenetic definition of Sauropoda has been debated in recent years, and the two most frequently used definitions result in placing the origin of this clade either in the Late Triassic or in the Early Jurassic. The definition proposed by Salgado et al. (1997) points to the most recent common ancestor of *Vulcanodon karibaensis* and Eusauropoda and all of its descendants, which implies the diversification of the group started in the Early Jurassic (based on the current fossil record). This has been adopted in some recent studies (McPhee et al. 2014; Peyre de Fabrègues et al. 2015; Lallensack et al. 2017; Bronzati et al. 2018; Rauhut et al. 2020). Other recent analyses followed the definition of Yates (2007) that points to the node of sauropodomorphs closer to *Saltasaurus loricatus* than to *Melanorosaurus readi*, which places the initial diversification of this clade in the latest Triassic (e.g. Pol and Powell 2007; Apaldetti et al. 2018; Pol et al. 2020, 2021). The latter is the definition accepted in this contribution it points to a clade that exclude all forms traditionally regarded as 'prosauropods' (e.g. *Melanorosaurus, Mussaurus, Anchisaurus*) and includes all species traditionally regarded as sauropods (e.g. *Amygdalodon, Gongxianosaurus, Volkheimeria*). Irrespective of the alternative phylogenetic definitions of Sauropoda, under both definitions it is clear that the earliest sauropods coexisted with non-sauropodan sauropodomorphs for at least 20 million years. Current knowledge of the anatomy and biology of these earliest sauropods is so far limited, not only for the latest Triassic taxa (e.g. *Ingentia, Lessemsaurus*; Apaldetti et al. 2018), but also for most of the earliest Jurassic species (e.g. *Gongxianosaurus, Pulanesaura, Sanpasaurus, Vulcanodon, Tazoudasaurus*; Luo and Wang 2000; McPhee et al. 2015; McPhee et al. 2016, Viglietti et al. 2018; Allain et al. 2004). It was only in the latest Early Jurassic when sauropods became predominant and completely replaced all other sauropodomorph lineages (i.e. 'prosauropods') that were the most abundant herbivores during the Late Triassic and Early Jurassic. The faunal replacement from 'prosauropod'-dominated to sauropod-dominated has been noted by many authors (e.g. Upchurch and Barrett 2005), but its precise timing and causes were poorly constrained until recently.

The sauropod group that radiated by the late Early Jurassic and occupied the niches of large herbivores during this time was Gravisauria, defined as the most recent common ancestor of *Tazoudasaurus naimi* and *Saltasaurus loricatus* and all of its descendants (Allain and Aquesbi 2008). This clade of large-bodied sauropods includes the highly successful group Eusauropoda and some other early sauropods from the Early Jurassic. The radiation of Gravisauria (and Eusauropoda) was the evolutionary event during which the sauropod body plan was finally assembled, characterized by a body size over 5 tons, skulls with deep and short rostrum, broad and robust U-shaped mandibles, large spoon-shaped teeth with thick enamel, long necks, increased pneumaticity in the axial skeleton, and robust columnar limbs (Wilson and Sereno 1998; Rauhut et al. 2011; Klein et al. 2011; Sander et al. 2011; Bates et al. 2016).

The events in the late Early Jurassic were therefore important in terms of the ecological dynamics of terrestrial ecosystems as well as for the important evolutionary modifications in sauropod anatomy and biology. Understanding these events, however, is hampered by the scarcity of sauropod fossils between the latest Early Jurassic and the earliest Middle Jurassic (180–170 Ma). Few sedimentary sequences in the world have preserved terrestrial ecosystems with dinosaur remains from this time. These include deposits in South America as well as other regions, such as India (e.g. Jain et al. 1975; Kutty et al. 1987; Yadagiri 2001; Bandyopadhyay et al. 2010), north Africa (Allain et al. 2004; Allain and Aquesbi 2008), and China (e.g. Dong et al. 1983; Wang and Sun 1983; Chen et al. 2006; Wang et al. 2018; Xu et al. 2018). Until recently, however, none of these basins had associated absolute dates to determine the precise timing of these events.

The South American record has provided critical information for understanding this event. In particular, the Cañadón Asfalto Basin in Central Patagonia (Chubut Province, Argentina) contains Jurassic sediments deposited in terrestrial ecosystems and with a diverse fossil record (Fig. 1). Dinosaurs have been known from this region since the mid-twentieth century (Cabrera 1947), and important sauropod remains were collected and published in the 1970s–1980s (Bonaparte 1979, 1986). More recent contributions have provided further information on the sauropods from these units (Rauhut 2003a, b; Carballido and Pol 2010; Pol et al. 2020; Holwerda et al. 2015, 2021; Becerra et al. 2017; Carballido et al. 2017). Furthermore, these sequences are now well constrained with precise absolute dates and a diverse fossil assemblage of plants and other vertebrate groups (Cúneo et al. 2013; Pol et al. 2020) that provide an ecological context for the rise and diversification of gravisaurian sauropods in the late Early Jurassic of Patagonia.

Here we present a summary of the currently available anatomical information of the Early Jurassic sauropods from Patagonia and discuss their phylogenetic position and synapomorphic features based on a recent analysis (Pol et al. 2020), as well as their importance for establishing the timing of the evolutionary radiation of sauropods.

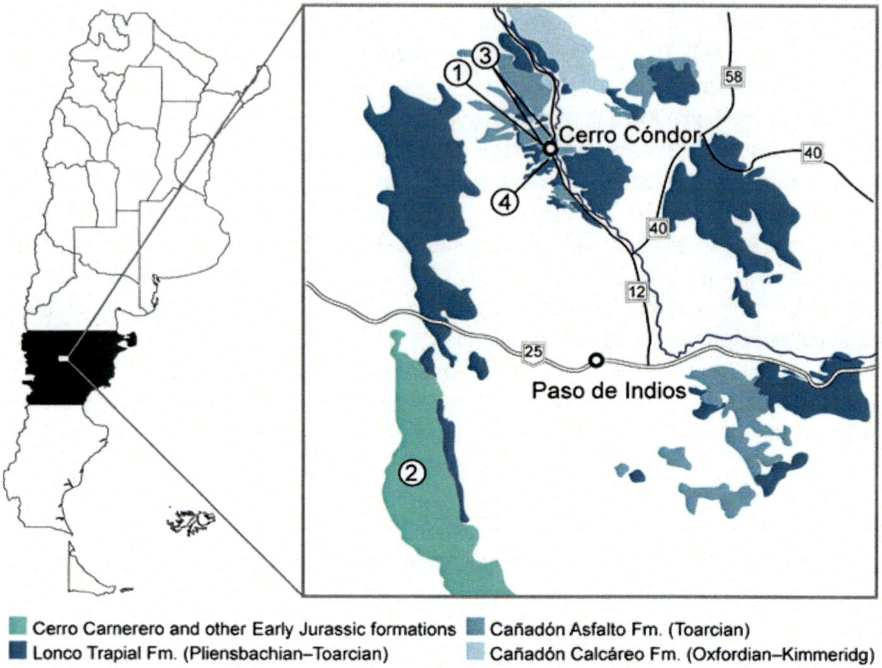

Fig. 1 Location map of the localities of the Early Jurassic sauropod from Patagonia (Chubut Province, Argentina). 1, *Volkheimeria chubutensis*; 2, *Amygdalodon patagonicus*; 3, *Patagosaurus fariasi*; 4, *Bagualia alba*

Institutional Abbreviations MACN: Museo Argentino de Ciencias Naturales 'Bernardino Rivadavia', Buenos Aires, Argentina; **MLP**: Museo de La Plata, La Plata, Argentina; **LEICT**: New Walk Museum and Art Gallery, Leicester Arts and Museum Service, Leicester, United Kingdom; **MNHN**: Musee National d'Histoire Naturelle, Paris, France; **MPEF-PV**: Museo Paleontológico Egidio Feruglio, Trelew, Argentina; **OUMNH**: Oxford University Museum of Natural History, Oxford, United Kingdom; **PVL**: Colección Paleontología de Vertebrados, Instituto Miguel Lillo, Tucuman, Argentina.

2 Systematic Palaeontology

Sauropodomorpha von Huene 1942

Sauropoda Marsh 1878

Definition Sauropodomorphs closer to *Saltasaurus loricatus* than to *Melanorosaurus readi* (Yates 2007).

Amygdalodon Cabrera 1947

Amygdalodon patagonicus Cabrera 1947

Lectotype MLP 46-VIII-21-1/2 (posterior dorsal vertebra).

Referred Specimens MLP 46-VIII-21-1/1 and MLP 46-VIII-21-1/3–11 (vertebral and rib remains), MLP 46-VIII-21-1/12, MLP 46-VIII-21-1/13, MLP 46-VIII-21-1/15, MLP 46-VIII-21-1/12 17 and MLP 46-VIII-21-1/12 18 (tooth crowns), MLP 46-VIII-21-1/12 14 and MLP 46-VIII-21-1/12 16 (tooth roots), MLP 46-VIII-21-1/19 (right pubis), MLP 36-XI-10-3/1 (posterior dorsal vertebrae with attached dorsal rib). The materials catalogued as MLP 46-VIII-21-1 were originally described by Cabrera (1947), and those of MLP 36-XI-10-3 were later added by Casamiquela (1963). Rauhut (2003a) noted that these remains may belong to more than one individual, due to the fact that the posterior dorsal vertebrae are about 70% the height of the sacral vertebra as well as the size difference of the preserved ribs. This observation led Rauhut (2003a) to designate the posterior dorsal vertebra as the lectotype, given the presence of diagnostic features and repeated elements that could be referred to this taxon based on apomorphic characters (see Diagnosis).

Locality and Age Cañadón Puelman, southwest of Cerro Carnerero, Sierra del Cerro Negro, Chubut Province, Argentina (Fig. 1). The remains of *Amygdalodon* were found in the Cerro Carnerero Formation, a unit that has been regarded as Early to Middle Jurassic by different authors. Although there are no radiometric dates for this unit, there are ammonites from underlying horizons biostratigraphically dated as Pliensbachian (Riccardi 1992). The Cerro Carnerero Formation is overlain by the Lonco Trapial Formation (radiometrically dated at 185 Ma; Pol et al. 2020), but this unit may not be synchronous in all areas in which it has been recognized. It is therefore likely that the Cerro Carnerero Formation was deposited at some point during the Pliensbachian or the early Toarcian.

Diagnosis *Amygdalodon* is a sauropod diagnosed by the following unique combination of characters (autapomorphy indicated with *): lateral walls of the neural canal and centropostzygapophyseal laminae flared laterally posteriorly; neural canal strongly flexed antero–posteriorly within the dorsal neural arches; spoon-shaped teeth with low slenderness index (SI = apicobasal length/mesiodistal width) values (1.34–1.49); enamel wrinkled, forming a pattern of pits and narrow apicobasal sulci*; total absence of denticles on both mesial and distal margins; wear facets extending mostly along one margin of the crowns.

Comments The description of *Amygdalodon patagonicus* (Cabrera 1947) marked the start of research and discovery of Jurassic dinosaurs from South America. This taxon therefore has a historic value for South American palaeontology. The anatomy of this taxon was reviewed by Casamiquela (1963) and more recently by Rauhut (2003a) and Carballido and Pol (2010). The teeth of *Amygdalodon* are mesiodistally broad and spatulate, slightly convex on the buccal surface and concave on the lingual surface (Fig. 2). It shares with other early sauropods with broad crowns (e.g. *Tazoudasaurus*, *Camarasaurus*; see Barrett and Upchurch 2005) the presence of an asymmetrical

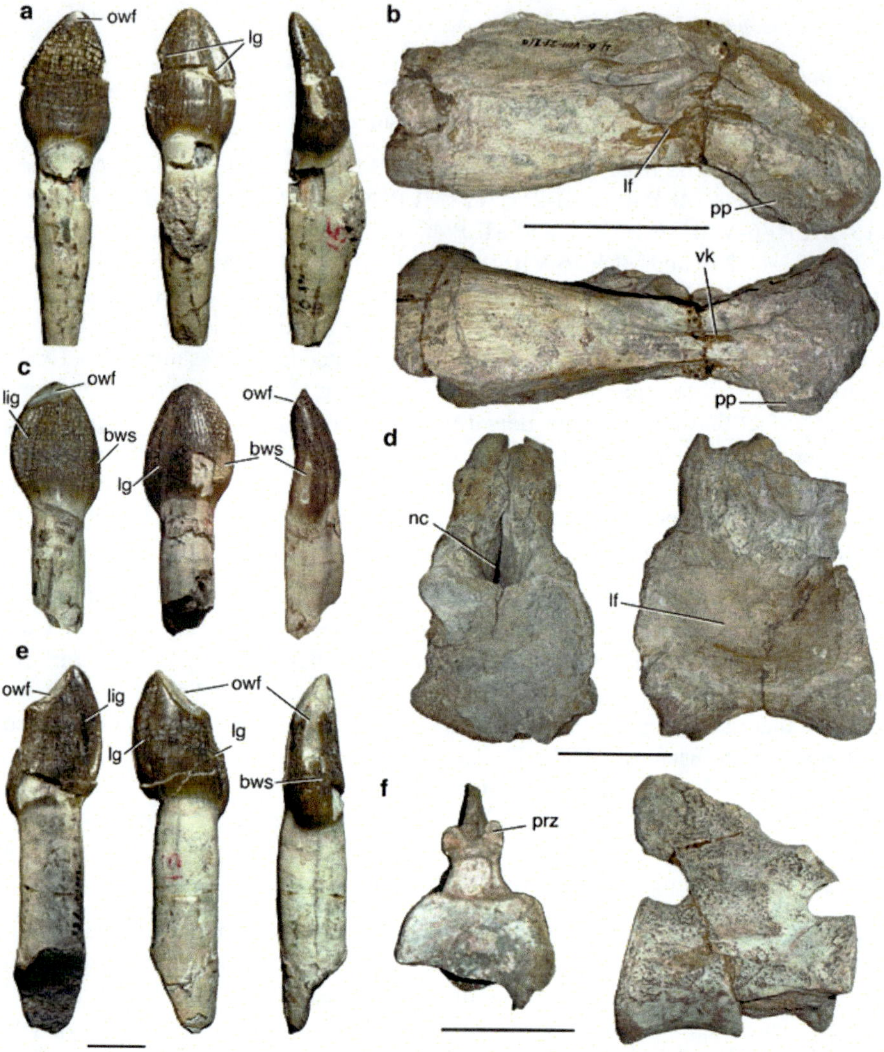

Fig. 2 Type material of *Amygdalodon patagonicus*. **a** Tooth (MLP 46-VIII-21-1/15) in lingual, labial and distal views. **b** Cervical vertebra (MLP 46-VIII-21-1/8) in right lateral and ventral views. **c** Tooth (MLP 46-VIII-21-1/12) in lingual, labial, and distal views. **d** Posterior dorsal vertebra (MLP 46-VIII-21-1/2), lectotype, in anterior and left lateral views. **e** Tooth (MLP 46-VIII-21-1/13) in lingual, labial, and distal views. **f** Caudal vertebra (MLP 46-VIII-21-1/3) in anterior and right lateral views. Scale bars, 10 cm for bones; 1 cm for teeth. Abbreviations: bws, basal wear facet; lf, lateral fossa; lg, labial groove; lig, lingual groove; nc, neural canal; owf, occlusal wear facet; pp, parapophysis; prz, prezygapophysis; vk, ventral keel

profile of the mesial and distal edges of the crown. The crown is broad relative to the narrow root and relative to its apicobasal height, resulting in SI (sensu Upchurch 1998) values below 1.5. The available teeth lack serrations, a rare feature for a non-neosauropodan sauropod, and have a unique pattern of wrinkling with aligned pits within a sulcus on the outer enamel surface (Carballido and Pol 2010). Extensive wear facets are present on several teeth (Fig. 2), indicative of tooth–tooth occlusion and resembling the V-shaped facets (sensu Wilson and Sereno 1998) of other sauropods.

The postcranial anatomy of *Amygdalodon* also contains relevant information (Rauhut 2003a). The single cervical centrum preserved is opisthocoelous as in early sauropods more derived than *Gongxianosaurus* (Wilson 2002) and is peculiar in being strongly flexed along its anterior part. The cervical centrum lacks a true pleurocoel and only has a shallow depression on its lateral surface (Fig. 2), unlike the condition of eusauropods (Upchurch 1998; Wilson 2002). Details of the cervical neural arch are limited, but the presence of anterior and posterior centrodiaopophyseal laminae, as well as prezygodiapophyseal and postzygodiapophyseal laminae can be confirmed, as in most sauropods. A posterior cervical prezygapophysis is present among the remains that preserves part of the prezygodiapophyseal, centroprezygapophyseal, and a thin intraprezygapophyseal laminae.

Remains of the dorsal vertebrae include centra and parts of the neural arches (Fig. 2). The neural spine of an anterior dorsal is mediolaterally compressed with a broader dorsal end and has remains of the spinoprezygapophyseal and spinopostzygapophyseal laminae, the latter of which is stouter. Posterior dorsal vertebrae have a massive centrum that is as high as long and lacks pleurocoels but has depressions on its lateral surfaces. The neural arch is as high as the centrum and lacks fossae on its anterior surface, contrasting with the derived condition of eusauropods (Upchurch 1998). The neural canal is high, slit-like and deeply incised into the centrum, as in several early sauropods (Rauhut et al. 2020). A fragment of a double headed rib was described by Rauhut (2003a) as having a bony web between capitulum and tuberculum.

Part of a sacral vertebra is preserved (Rauhut 2003a), although it had been previously identified as a caudal (Casamiquela 1963). The centrum has a wide anterior articular surface, it is strongly constricted posterior to it, and has a large transverse process arising from its anterior region. Two middle caudal vertebrae are preserved, with elongated centra that are lateromedially constricted at the midpoint and a neural arch placed slightly anteriorly from the middle of the centrum. The prezygapophyses arise from the anterior end of the neural arch, well separated from the posteriorly displaced and posterodorsally directed neural spine and the short postzygapophyses.

The shaft and distal end of a pubis is preserved that shows a transversely broad pubic apron that gradually expands towards the rounded distal end. A distal end of a right tibia is also present, which has a posterior expansion and a medially broad triangular outline in distal view.

Volkheimeria Bonaparte 1979

Volkheimeria chubutensis Bonaparte 1979

Holotype PVL 4077, incomplete cervical vertebra, two posterior dorsal vertebrae, an incomplete dorsal neural arch, an isolated dorsal vertebral centrum, two incomplete sacral vertebrae, two caudal neural arches and several caudal centra, two incomplete ilia, left ischium, pubis, femur, and tibia.

Locality and Age The only known specimen of *Volkheimeria chubutensis* was found in the Cerro Cóndor Sur locality (Fig. 1), approximately 1 km west from the Cerro Cóndor village (Chubut Province, Argentina). *Volkheimeria chubutensis* was found in the Cañadón Asfalto Formation, a continental unit consisting mainly of lacustrine deposits that crops out in west-central Chubut Province, Argentina (Cúneo et al. 2013; Figari et al. 2016). The age of the Cañadón Asfalto Formation was traditionally thought to be Callovian–Oxfordian, but recent radiometric dates bracketed the fossiliferous levels of this unit as Early Jurassic (Toarcian; Cúneo et al. 2013). The base of the Cañadón Asfalto Formation was dated at 179.17 ± 0.12 Ma, and the uppermost levels of this unit were dated at 178.07 ± 0.21 Ma (Pol et al. 2020). The locality where *Volkheimeria* and *Patagosaurus* were found is stratigraphically bracketed between the horizons of these two dates.

Emended Diagnosis Sauropod dinosaur diagnosed by the following unique combination of characters: large and posteriorly deep and sharply defined pleurocoel on cervical centrum; dorsal vertebrae with a neural arch dorsoventrally shorter than in *Patagosaurus*; neural spine with anteriorly flat and posteriorly thickened lateral surface; broad and rugose interspinal ligament attachments and thin spinoprezygapophyseal lamina; ilium with long pubic peduncle and broad preacetabular process; pubis very slender and with only slightly posteromedially canted pubic apron; ischium with a thin and subcylindrical shaft and a rounded expansion at its distal end; tibia only 58% of femoral length.

Comments Bonaparte (1979, 1986) described the remains of the holotype of *Volkheimeria* based on an incomplete specimen that preserved elements that could be distinguished from those of *Patagosaurus* (found in the same locality, see below), particularly in the mid-to-posterior dorsal vertebrae.

The sole cervical vertebra shows the presence of a deep but undivided pleurocoel on its lateral surface that is set ventral to a stout posterior centrodiapophyseal lamina (Fig. 3), resembling the condition of eusauropods. The diapophysis is short and placed just anterior to the mid-length of the centrum. The centrum is opisthocoelous and bears a subtle ventral ridge, as in other sauropods. In contrast to most eusauropods, the neural arch is considerably lower than the height of the centrum and the neural canal is large (>40% of centrum height).

The dorsal vertebrae have received much more attention in the descriptions of this taxon (Bonaparte 1986, 1999) because of their diagnostic and phylogenetically informative features. They have dorsoventrally elongated neural arches that form over 65% of the total vertebral length. The dorsal vertebrae have well-developed parapodiapophyseal and posterior centrodiapophyseal laminae as well as well-developed prezygodiapophyseal and postzygodiapophyseal laminae (Fig. 3), as in other sauropods. The fossae below the diapophyses are large and deep, although

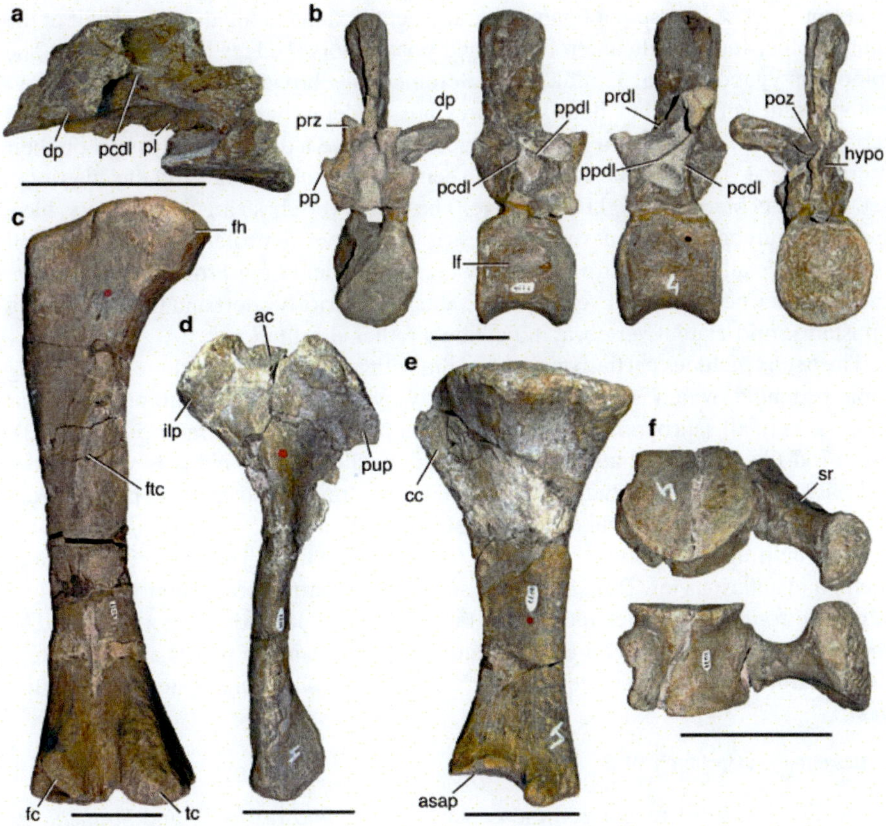

Fig. 3 Type material of *Volkheimeria chubutensis* (PVL 4077). **a** Cervical vertebra in left lateral view. **b** Dorsal vertebra in anterior, right lateral, left lateral, and posterior views. **c** Left femur in posterior view. **d** Right ischium in lateral view. **e** Left tibia in lateral view. **f** Sacral vertebra in anterior and ventral views. Scale bars 10 cm for bones. Abbreviations: ac, acetabulum; asap, articular surface for the ascending process of the astragalus; cc, cnemial crest; dp, diapophysis; fc, fibular condyle; fh, femoral head; ftc, fourth trochanter; hypo, hyposphene; ilp, iliac peduncle; lf, lateral fossa; pcdl, posterior centrodiapophyseal lamina; pl, pleurocoel; poz, postzygapophysis; pp, parapophysis; ppdl, paradiapophyseal lamina; prdl, prezygodiapophyseal lamina; prz, prezygapophysis; pup, pubic peduncle; sr, sacral rib; tc, tibial condyle

it is not clear if the centrodiapophyseal fossa ramifies within the neural arch, as in some eusauropods (e.g. *Patagosaurus*; Holwerda et al. 2021). The anterior surface of the neural arch is flat to slightly concave and has a broad centroprezygapophyseal lamina, and the posterior surface has a dorsoventrally long and lateromedially narrow hyposphene. The neural spine of the dorsal vertebrae is anteroposteriorly longer than lateromedially broad and has a mostly flat lateral surface (Fig. 3). This feature was highlighted by Bonaparte (1986) as denoting a more plesiomorphic configuration of the vertebral anatomy in comparison with *Patagosaurus*. Dorsally, the spine expands lateromedially so that it has a broad and rounded dorsal end. The broad and rugose interspinal ligament attachments extend along the dorsal half of the neural spine, forming its anterodorsal and posterodorsal edges in lateral view. The spinopostzygapophyseal laminae are lateromedially broader than the spinoprezygapophyseal laminae, but both structures are moderately developed in the dorsal vertebrae. The dorsal centra lack pleurocoels but have a distinct depression on their lateral surface. The articular facets of the centrum are subcircular, and the anterior is flat, but the posterior is slightly concave. The sacral vertebrae are fragmentary, have low neural arches, and transverse processes that curve ventrolaterally and are well expanded distally. The centra are poorly constricted and their articular surfaces are flat and subcircular. Caudal vertebrae are anteroposteriorly short and high, including a posterior mid-caudal vertebra, indicating a rather short tail.

The pelvis includes partially preserved ilia with a small ischial peduncle and a long pubic peduncle, which is surpassed anteriorly by a rounded and high preacetabular process, as in all sauropods. The ischium has a plate like an obturator plate (Fig. 3), a thin rod-like shaft, and a slightly expanded distal end. The pubis is very slender and only slightly shorter than the ischium, has a straight and laminar pubic apron, and a moderate anteroventral expansion at its distal end.

The hindlimb is only known from the femur and the tibia. The femur is straight in lateral and anterior view (Fig. 3), with a subcircular shaft in cross section, and a low ridge-like fourth trochanter located on the proximal half of the femoral shaft. The tibia is a robust element with a broad anterior end expansion that is approximately 30% the tibial proximodistal length and has only a small, laterally directed cnemial crest.

Eusauropoda Upchurch 1995

Definition The most recent common ancestor of *Shunosaurus lii* and *Saltasaurus loricatus* and all of its descendants (Upchurch et al. 2004).

Patagosaurus Bonaparte 1979

Patagosaurus fariasi Bonaparte 1979

Holotype PVL 4170

Referred Specimens PVL 4076; MACN-CH 932, 933, 935, 936 1299, 232; MPEF-PV 1670. This list of specimens is a subset of those previously referred to as *Patagosaurus fariasi*. After a re-study of the holotype (Holwerda et al. 2021), a revision of the specimens referred by Bonaparte (1986) to *Patagosaurus* were regarded as indeterminate and are no longer referred to this taxon (Holwerda 2019; see also Rauhut 2003b; Holwerda et al. 2015).

Locality and Age These remains of *Patagosaurus* have been found in three different localities near the Cerro Cóndor village (Chubut Province, Argentina). The type locality of the holotype of *Patagosaurus fariasi* is the locality Cerro Cóndor Norte, which lies approximately 2 km north from Cerro Cóndor village (Fig. 1). However, a large number of remains referred to *Patagosaurus* were found in the locality Cerro Cóndor South, west to the Cerro Cóndor village (Fig. 1). These two localities were mainly excavated by Bonaparte. The specimen MPEF-PV 1670 was found close to this second locality. *Patagosaurus fariasi* was found in the Cañadón Asfalto Formation (see above).

Diagnosis *Patagosaurus fariasi* is a non-neosauropodan eusauropod dinosaur that can be diagnosed based on the following morphological features, and the following combination of characters (autapomorphies indicated with *): cervical and anterior dorsal vertebrae with marked pleurocoel, which is deep in cervicals but shallower in dorsals; pleurocoel of cervical vertebrae deeper anteriorly with well-defined margins, but shallow posteriorly with well-defined dorsal and ventral margins; faint oblique accessory lamina in some cervicals, dividing the pleurocoel into an anterior deeper part and a shallower posterior part; cervicals with a relatively high neural spine; high dorsal placement of cervical postzygapophyses, with angle between the postzygodiapophyseal and posterior centrodiapophyseal laminae of about 55°; posterior dorsal neural arches with a centrodiapophyseal fossa that extends internally as a pneumatic structure, separated by the mirroring structure by a thin septum, and both of which connect into a ventral, oval-shaped internal pneumatic chamber located dorsally and separated from the neural canal; posterior dorsals with small round excavations on the posterior side of the distal extremity of the diapophyses*; posteriormost dorsals with a rudimentary aliform processes; dorsals without spinodiapophyseal lamina; ilium with round dorsal rim, hook-shaped anterior lobe and dorsoventrally elongated pubic peduncle; fused distal ischia with the paired distal shafts creating an angle of 110° to the horizontal; pubis with marked torsion and kidney-shaped pubic foramen.

Comments The premaxilla is a stout, square element, showing in lateral/labial view, a dorsoventrally high, oblique anterior rim of the snout, and a high, near vertical maxillary articular surface on the distal lateral side (Fig. 4). The dorsal side tapers to a point, giving the premaxilla a pentagonal appearance in lateral view. The dentaries are straight elements in lateral view and do not show the shallow concavity seen in *Shunosaurus, Camarasaurus, Giraffatitan or Europasaurus* (Janensch 1935; Madsen et al. 1995; Chatterjee and Zheng 2002; Marpmann et al. 2015). Both elements lack a prominent chin-like process on the anteroventral side, which is also absent in *Archaeodontosaurus* (though this element is incomplete) from the Middle Jurassic of Madagascar (Buffetaut 2005). Dentition in *Patagosaurus* is typically spoon-shaped, and D-shaped in cross section (Fig. 4). It shows rugose and striated wrinkles (Holwerda et al. 2015). Unerupted teeth show marginal denticles.

The axial skeleton is well represented in *Patagosaurus fariasi*. The cervical count is estimated to be *12–13* for *Patagosaurus* (based on comparisons with contemporary and anatomically similar sauropods such as *Cetiosaurus*, *Shunosaurus* and *Spinophorosaurus*). In comparison with other sauropods, cervicals are rather stout (Fig. 4), with an average elongation index (aEI; Chure et al. 2010) ranging from 1.9 to 2 in anterior to 1.2–1.4 in posterior cervicals and the 'traditional' elongation index (EI = vertebral length/posterior centrum width; Upchurch 1998) ranging from 2.1 in anterior to 1.2 in posterior cervicals. In juveniles, particularly in Morphological Ontogenetic Stage (MOS) 1, the elongation of cervicals is lower. In lateral

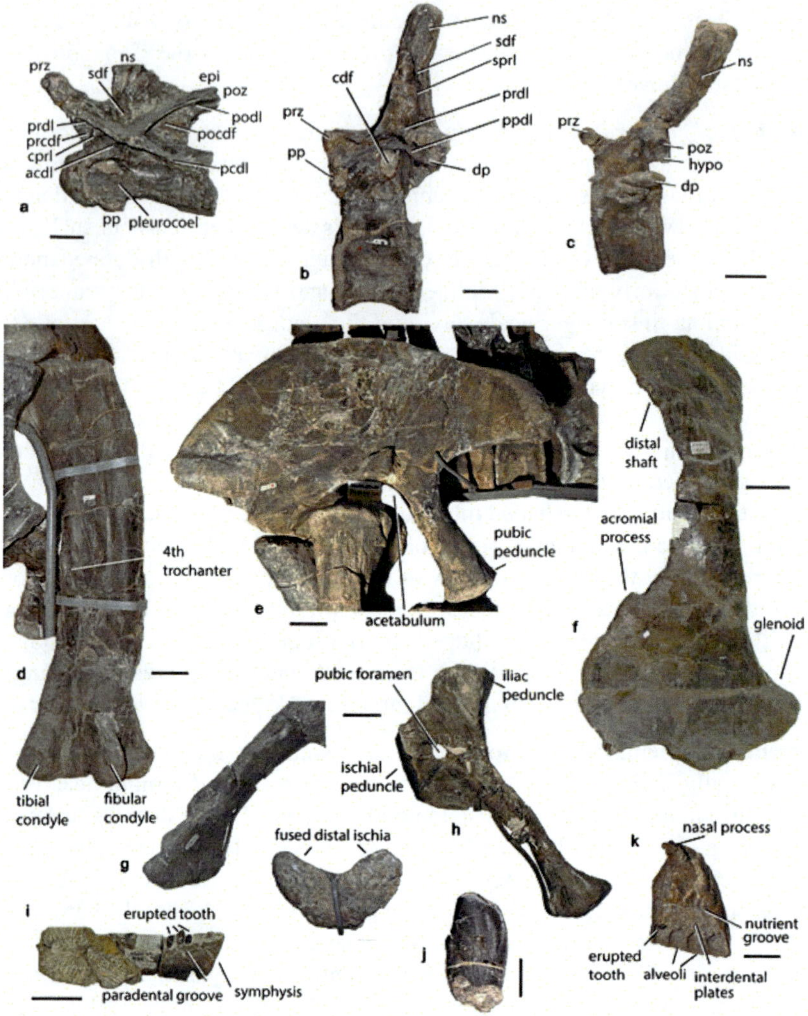

Fig. 4 *Patagosaurus fariasi* diagnostic elements. **a** Anterior cervical PVL 4170 (3) in left lateral view. **b** Posterior dorsal PVL 4170 (13) in right lateral view (image mirrored). **c** Anterior caudal PVL 4170 (19) in right lateral view (image mirrored). **d** Right femur PVL 4170 (37) in posterior view. **e** Right ilium PVL 4170 (34) in lateral view. **f** Right scapula MACN-CH 935 in lateral view. **g** Right pubis PVL 4170 (35) in lateral view. **h** Paired ischia PVL 4170 (36) in lateral and distal view. **i** Left premaxilla PVL 4076 in lingual view with dentition. **j** Isolated referred tooth MACN-CH 2008.3. **k** Left dentary MACN-CH 933 in medial view with dentition. Scale bars 10 cm for axial and appendicular, 5 cm for cranial elements, 1 cm for tooth. Abbreviations: acdl, anterior centrodiapophyseal lamina; df, centrodiapophyseal fossa; cprl, centroprezygapophyseal lamina; dp, diapophysis; epi, epipophysis; hypo, hyposphene; ns, neural spine; pcdl, posterior centrodiapophyseal lamina; pocdf, postzygapophyseal centrodiapophyseal fossa; podl, postzygapophyseal diapophyseal lamina; poz, postzygapophysis; pp, parapophysis; ppdl, paradiapophyseal lamina; prcdf, prezygapophyseal centrodiapophyseal fossa, prdl, prezygadiapophyseal lamina; prz, prezygapophysis; sdf, spinodiapophyseal fossa; sprl, spinoprezygapophyseal lamina

view, the centra are ventrally concave posterior to the parapophysis. The posteriormost 1/3rd. of the ventral side of the centra is convex, with the dorsoventral height of the centra increasing posteriorly. Cervicals are strongly opisthocoelous, with a prominent ventral keel accompanied by elliptical lateral fossae on the ventral side. Laterally, pleurocoels are developed as large, but only partially well-defined lateral depressions on the centra. In anterior cervicals, the pleurocoel is deeper than in posterior cervicals, and has a well-defined anterior, dorsal and ventral margin. In mid- and posterior cervicals the posterior margin of the pleurocoel is less clearly defined and the depression gradually fades into the lateral surface of the centrum. In some mid- to posterior cervicals, the left and right pleurocoels are only separated by thin septa (which are damaged or broken in some elements), but they do not invade the centrum and ramify within the bone, as is the case in neosauropods (Wedel 2005). The neural arches of *Patagosaurus* cervicals show steeply inclined postzygodiapophyseal laminae, a condition shared with mamenchisaurids (Mannion et al. 2019). Anteriorly, the intraprezygapophyseal laminae are separated medially, as in *Tazoudasaurus* (Allain and Aquesbi 2008) and the Rutland *Cetiosaurus* (LEICT 468.1968). Posteriorly, interpostzygapophyseal laminae (tpol) do meet at the midline. However, there are no centropostzygapophyseal laminae, as in *Tazoudasaurus* (Allain and Aquesbi 2008), but unlike the Rutland *Cetiosaurus* (LEICT 468.1968). Adult specimens show a single centropostzygapophyseal lamina (cpol) in posterior cervicals, whereas this lamina is absent in juveniles.

As there is no complete series of dorsals from any *Patagosaurus* individual, the total dorsal count can only be estimated as between 10 and 12. Anterior *Patagosaurus* dorsals are characterized by being opisthocoelous, having shallow pleurocoels, and displaying a vestigial ventral keel. Neural arches in anterior dorsals show ventrally projecting diapophyses, and oblique dorsally projecting prezygapophyses. All dorsal vertebrae are characterized by lacking a spinodiapophyseal lamina (Fig. 4), instead displaying a combination of a lateral spinopostzygapophyseal lamina and postzygapophyseal diapophyseal lamina (lspol + podl). Neural spines are transversely broad by ventrally flaring spinopostzygapophyseal laminae (spol).

Towards the posterior dorsals, the axial length decreases and especially the dorsoventral height of vertebrae increases drastically. Anteriorly, prezygapophyseal pedicels also increase in dorsoventral length, and develop deep, paired centroprezygapophyseal fossae, divided by a single interprezygapophyseal lamina (stprl), a condition shared with *Cetiosaurus oxoniensis* (Upchurch and Martin 2003, OUMNH J13644/2). With the elongation of the neural arch, lamination changes. At the transition from middle to posterior dorsals, anteriorly, centroprezygapophyseal laminae lengthen as the neural arch and the pedicels elongate. Posteriorly, first the intrapostzygapophyseal laminae meet, then the centropostzygapophyseal laminae disappear, and instead a single interpostzygapophyseal lamina (stpol) appears. Transverse processes project laterally, and not dorsally, in middle and posterior dorsals, distinguishing them from other contemporaneous sauropods. Hyposphenes are small and rhomboid in shape in adult individuals, but are much larger in juvenile specimens. The most notable feature of *Patagosaurus* dorsals is the presence of paired centrodiapophyseal fossae (cdf), or fenestrae, which appear in posteriormost dorsals. It was long thought that these were connected to the neural canal; however, CT data reveals that a thin septum separates the adjacent fenestrae from each other, and from the neural canal. Ventrally these fenestrae form a central chamber, still well above the neural canal. Interestingly, juvenile specimens also show the characteristic cdf.

Sacral vertebrae in *Patagosaurus* have elongated neural arches, resembling posterior dorsals in their lamination. The complete sacral count is 5. The first sacral vertebra is unfused, but the second and third are fused together, differing from neosauropods.

Anterior caudal vertebrae display elongated neural spines (Fig. 4). Together with the spines of sacral vertebrae, they show distinct 'saddle-shaped' neural spine summits, a condition shared with *Cetiosauriscus stewarti* and *Spinophorosaurus nigerensis*. Caudal vertebral centra show prominent chevron facets ventrally, as well as a faint ventral hollow.

Associated material of *Patagosaurus* has one scapula preserved. The scapula has a more prominent proximal expansion than in more basal sauropods; however, it is not as prominent as in neosauropods (Upchurch et al. 2004). It shows a rather small and flat acromion process (Fig. 4), as in *Cetiosaurus oxoniensis* and *Lapparentosaurus* MNHN-MAA 44, and unlike *Mamenchisaurus* (Ouyang and Ye 2002; Upchurch and Martin 2003). The shaft of the scapula is obliquely directed towards the posterior side, and slightly to the ventrolateral side, unlike in *Mamenchisaurus*, *Lapparentosaurus*, and *Cetiosaurus*.

Ilia in *Patagosaurus* show hook-shaped preacetabular processes, as in most sauropods. The acetabulum is relatively wide (Fig. 4), as in *Barapasaurus*, *Haplocanthosaurus*, and diplodocids (Hatcher 1903; Upchurch et al. 2004; Bandyopadhyay 2010), but differs in relative width from *Cetiosaurus*, *Tazoudasaurus* and titanosauriforms (Upchurch and Martin 2003; Allain and Aquesbi 2008; Díez Díaz et al. 2013; Poropat et al. 2015). Paired ischia are only found in the holotype, and show a distal fusion, forming a wide V-shape with an angle of 110° with the horizontal, representing an intermediate stage between the coplanar *Camarasaurus* ischial fusion state and that of diplodocoids, *Cetiosaurus*, *'Bothriospondylus madagascariensis'* and *Vulcanodon* (Janensch 1961; Cooper 1984; Upchurch and Martin 2003; Mannion 2010; Tschopp et al. 2015). There are no ontogenetic changes found in these elements between juvenile and adult individuals.

Appendicular: humeri and femora in *Patagosaurus* are robust and have low gracility indices, but in juveniles these elements are slender. There are more femora than humeri available from the holotype and specimens that can be confidently assigned to *Patagosaurus* (three confirmed femora and one or two humeri). The associated material only bears one confirmed humerus, belonging to a juvenile (MACN-CH 932). Another humerus, belonging to an adult individual, unfortunately cannot be confidently assigned to *Patagosaurus*. This latter element is robust, and both juvenile and adult humeri lack a prominent deltopectoral crest, although it is not clear whether this is a result of preservation. The juvenile individual MACN-CH 932 also has a small, kidney-shaped sternal plate, as well as a radius and ulna, which are slender elements. The anteromedial extremity of the ulna is hook-shaped as in *Lapparentosaurus*. The femora are robust, anteroposteriorly flattened, and transversely relatively wide (Fig. 4), as in other eusauropods (e.g. *Cetiosaurus oxoniensis*;

Upchurch and Martin 2003). The fourth trochanter is anteromedially placed, as in most eusauropods. There is a slight convex curvature in femora, which could suggest a slightly wider stance than strictly narrow-gauge, although more material is needed to test this. The tibiae, both juvenile and adult, lack a pronounced cnemial crest, as well as the 'step' for the medial malleolus. The juvenile MACN-CH 932 preserves a small incomplete pes, which resembles those in *Cetiosauriscus stewarti* and *Lapparentosaurus*. Unguals in *Patagosaurus* are massive and robust elements.

Bagualia Pol et al. 2020

Bagualia alba Pol et al. 2020

Holotype MPEF-PV 3301, posterior half of the skull articulated with seven cervical vertebrae, left lacrimal (MPEF-PV 3301/11), left prefrontal (MPEF-PV 3301/6), frontals (MPEF-PV 3301/3), right postorbital (MPEF-PV 3301/10), left parietal (MPEF-PV 3301/8), right squamosal (MPEF-PV 3301/9), left squamosal (MPEF-PV 3301/7), right quadrate (MPEF-PV 3301/4), braincase (MPEF-PV 3301/1), proatlas (MPEF-PV 3301/5), neurapophyses of atlas (MPEF-PV 3301/18), axis (MPEF-PV 3301/13), third cervical vertebra (MPEF-PV 3301/12), fourth cervical vertebra (MPEF-PV 3301/17), fifth cervical vertebra (MPEF-PV 3301/14); sixth cervical vertebra (MPEF-PV 3301/15), seventh cervical vertebra (MPEF-PV 3301/16).

Referred Specimens Several cranial and postcranial elements of at least three individuals (based on repeated elements) found at the same site. The specimens are similar sized (repeated humeri varying up to 15% in length). Since most of the elements are disarticulated, they are accessioned under different numbers: left premaxilla (MPEF-PV 3305), left maxilla (MPEF-PV 3304), left maxilla (MPEF-PV 3341a) right maxilla (MPEF-PV 3204), right nasal (MPEF-PV 3340), left quadrate (MPEF-PV3342), both dentaries and right surangular (MPEF-PV 3202), right surangular (MPEF-PV 3339), left pterygoid (MPEF-PV 11,017), several isolated teeth (MPEF-PV 3146/ 3174-3176/ 3203/ 3205/ 3207-3209/11,030-11,039/11,041-11,047/11,050), two middle cervical vertebrae (MPEF-PV 11,040), a mid-posterior cervical vertebra (MPEF-PV 3408), cervical centrum (MPEF-PV 3327), posterior cervical vertebrae (MPEF-PV 3349; MPEF-PV 3348), anteriormost dorsal vertebra (MPEF-PV 11,023), four dorsal centra (MPEF-PV 11,012/3343/3405/3403), dorsal neural arch (MPEF-PV 11,027), posterior dorsal vertebra (MPEF-PV 11,000), two posterior dorsal neural arches with sacrum, both ilia, right pubis and the first ten caudal vertebrae articulated (MPEF-PV 11,011), two segments of three articulated anterior caudal vertebrae (MPEF-PV 3316; MPEF-PV 11,044), twenty-nine isolated caudal vertebrae (MPEF-PV 3179/3300/3314/3315/3317-3326/3328-3331/3344-3346/3389/3401/3402/3404/3406/3407/3409/11,026), cervical rib (MPEF-PV 11,052), dorsal rib (MPEF-PV 11,058), five isolated transverse processes (MPEF-PV 11,001/11,002/11,003/11,004/11,005), thirteen haemal arches (MPEF-PV 3356/11,025/3351/3357/3353/3355/3352/3359/3358/11,009/3354/3390/11,010), two coracoids (MPEF-PV 11,015/3387), three scapulae (MPEF-PV 3382-384), two scapula-coracoids (MPEF-PV 3385/3386), five humeri (MPEF-PV 3311/3338/3380/3381/11,020), radius (MPEF-PV 3313), two ulnae

(MPEF-PV 3379/3312), ilium (MPEF-PV 3369), three pubes (MPEF-PV 11,011/11,019/11,051), two ischia (MPEF-PV 3337/11,016), five femora (MPEF-PV 3303/3371/11,024/11,021/11,022), tibia (MPEF-PV 3374), two fibulae (MPEF-PV 3376/3306), two astragali (MPEF-PV 3307/3308), calcaneum (MPEF-PV 11,018), four metapodials (MPEF-PV 3332/3333/3334/3309), four non-ungual phalanges (MPEF-PV 3335/11,028/11,029/11,049), two ungual phalanges (MPEF-PV 3310/3410).

Locality and Age Bagual Canyon, 5 km south of the village of Cerro Cóndor (Fig. 1). *Bagualia alba* was found in grey mudstone levels from the base of the Cañadón Asfalto Formation, which was dated at 179.17 ± 0.12 Ma (Early Jurassic, Toarcian). The holotype was found in close association with multiple postcranial remains of several individuals (minimum number of specimens based on repeated elements = 3), forming a bonebed of approximately 6 m wide, 3 m long, and 60 cm deep. The materials found in the bonebed include repeated elements that can be referred to *Bagualia alba* and show no differences between them, and thus support the interpretation of this assemblage as a monospecific accumulation of elements.

Diagnosis Basal eusauropod diagnosed by the following characters (autapomorphies indicated with *): pointed process on the anteroventral end of the premaxilla and anterodorsal end of the dentary*; anterior margin of the premaxilla without a marked step*; orbital margin of the frontal with a close V-shape pointed medially*, resulting in a short contribution to the orbit; supratemporal fenestra about as anteroposteriorly long as lateromedially wide*; strongly marked proatlantal facets on the laterodorsal margin of the foramen magnum; concave ventral margin of the distal portion of the cultriform process*; axis with an anterior process on the dorsal part of neural spine (convergent in *Jobaria* and *Europasaurus*); accessory lamina below the PCDL in middle cervical vertebrae*; EPRL present in middle cervical vertebrae (Pol et al. 2020).

Comments The rostrum of *Bagualia alba* is high and short and the anteroventral end of the premaxilla and anterodorsal end of the dentary bear a pointed ('beak-like') process (Fig. 5). The presence of a distinct process in the premaxilla has not been reported in other sauropods. Similarly, the presence of a moderate process in the anterior end of the dentary has been reported in some non-sauropod sauropodomorphs but it is not present in other sauropods and is therefore regarded as an autapomorphy of *Bagualia* given its phylogenetic context. *Bagualia* lacks a distinct step along the anterior margin of the premaxilla in lateral view, and its maxilla had a large participation in the margin of the external nares, contributing to more than one third of its border. The preserved margins of the antorbital fossa on the maxilla and on the lacrimal lack a well-developed antorbital fossa. Although the external nares are incomplete, we can infer they were posteriorly retracted due to the narial margin preserved in the nasal. The lacrimal has an extremely short anterior process located at the mid-dorsoventral length of the main body. The prefrontal of *Bagualia* is long and narrow and extends posterodorsally approaching the frontoparietal suture. Furthermore, the prefrontal lacks the anterior process. The postorbital descending process is long and transversely narrow. The supratemporal fenestra is exposed laterally. The orbital margin of the frontal has small indentations and a V-shaped profile in dorsal view (pointing medially), resulting in a short contribution to the orbit, which is unique among eusauropods. The frontal fails to enter the rim of the supratemporal fossa (Fig. 5). The configuration of the frontal is different from those of all other known sauropodomorphs and is regarded as an autapomorphy of *Bagualia alba*. The supratemporal fenestra is about as long as wide. The parietal occipital exposure is less than the height of the foramen magnum. The quadrate of *Bagualia* lacks a quadrate fossa on its lateral surface (Fig. 5). The braincase is very well preserved and shows many features that are phylogenetically relevant. The basal tubera are rounded and widely separated distally by a deep U-shaped incision visible in posterior view. They are placed only slightly ventral to the occipital condyle but dorsal to the basipterygoid processes and are narrower than the occipital condyle. In lateral view, the distal ends of the basipterygoid processes are located close to the level of the basal tubera (Fig. 5), rather than being located well anteriorly in respect to the tubera. The lateral surface of the basipterygoid processes bears a groove and a crest located distal to the preotic pendant, which indicates the course of the cerebral branch of the internal carotid artery. The preotic pendant covers the entrance of the internal carotid artery. The braincase of *Bagualia* shares with other eusauropods the absence of a well-developed basisphenoidal, subsellar, and lateral tympanic recess. The base of the cultriform process is anterodorsally oriented. Its ventral margin is straight towards its distal end, where it becomes concave in lateral view, which is not seen in any other known sauropodomorph. The distal end of the paroccipital process is not expanded and notably downturned, its mid-height being located close to the level of the dorsal margin of the occipital condyle. Well-developed proatlantal facets are present on the dorsolateral rim of the foramen magnum formed by the otoccipital. The supraoccipital is high, being twice as deep as the foramen magnum. The symphysis of the dentary is inclined with an angle of around 45 degrees to the horizontal and it is deep, the anterior ramus being around 1.5 times the height of the dentary at mid-length. In dorsal view, the dentaries of *Bagualia* form a U-shaped muzzle. The anteroventral margin of the dentary is gently rounded, without an expanded corner or 'chin'. The dentary bears 16 tooth positions bordered by a lateral wall and interdental plates (Fig. 5). The teeth of *Bagualia* are procumbent and spatulate, with their crown being broader than the root (the limit crown-root is marked by a step). They have a convex labial surface and a concave lingual one, resulting in a D-shaped cross section. The mesial and distal margins are asymmetrical and bear denticles, which are more numerous in the mesial margin than the distal one and distributed in the apical region, occupying close to the half dorsoventral height. Both lingual and labial surfaces have apicobasally oriented grooves. The SI values (sensu Upchurch 1998) of teeth from the premaxilla, maxillae and dentaries range between 1.1 and 1.6, with an average of 1.38. The enamel is wrinkled and is over 700 μm thick, which represents an increase in enamel thickness compared to non-sauropodan sauropodomorphs. CT scanning revealed up to three replacement teeth per position in the premaxilla, suggesting high dental replacement rates. Worn teeth show V-shaped wear facets.

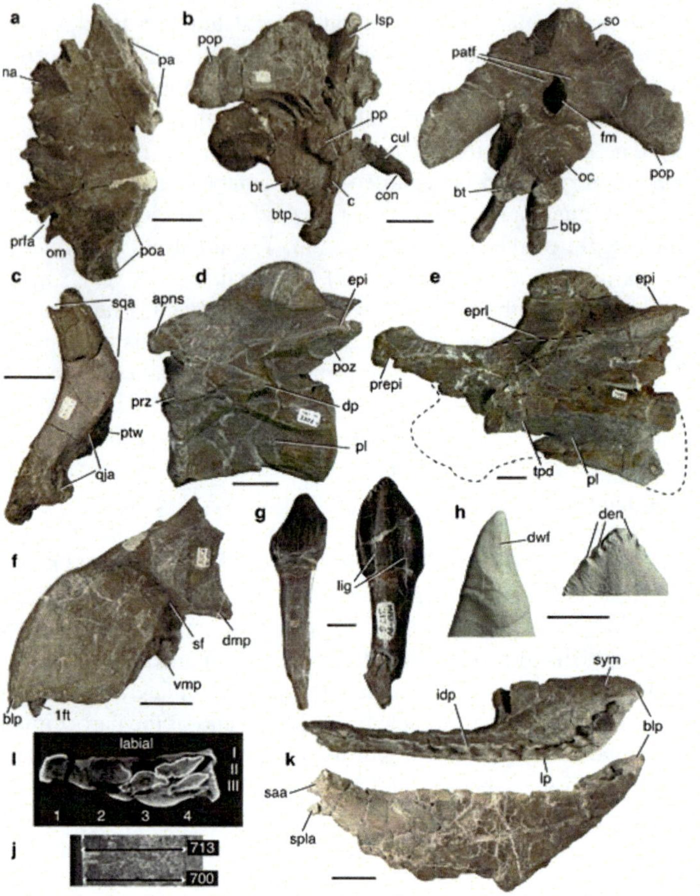

All cervical vertebrae are opisthocoelous with elongate centra (elongation index between 3 and 4.6 in the last preserved element of the holotype series). The centra present a well-developed ventral keel and marked pleurocoels (Fig. 5). The pleurocoels are anteriorly deep and become shallow posteriorly, and they are undivided. The anterior end of the neural spine of the axis is extended anteriorly as a marked, lobe-shaped process. In *Bagualia*, well-developed lateral vertebral laminae are present in the entire presacral series, including the axis. In the middle cervical vertebrae, there is an accessory lamina below the PCDL, which is parallel to the ACDL. The middle cervical vertebrae have an incipient development of the EPRL that extends longitudinally up to the anterior margin of the neural spine and the origin of the prezygapophyseal process. The diapophyses of cervical vertebrae have a triangular process on the posterior margin. The dorsal vertebrae lack pleurocoels. A PCPL and divided SPOL are present in middle-posterior dorsal vertebrae. The dorsal neural

◂**Fig. 5** Holotype and referred skull material of *Bagualia alba*. **a** Articulated frontals (MPEF-PV 3301-3) in dorsal view. **b** Braincase (MPEF-PV 3301-1) in right lateral and posterior views. **c** Right Quadrate (MPEF-PV 3301-4) in lateral view. **d** Axis (MPEF-PV 3301-13) in left lateral view. **e** Sixth cervical vertebra (MPEF-PV 3301-15) in left lateral view. **f** Premaxilla (MPEF-PV 3305) in lateral view. **g** Worn (MPEF-PV **11,036**) and non-functional (MPEF-PV 3176) teeth in lingual view. **h** Details of the wear facet of MPEF-PV **11,036** in distal view, and denticles of MPEF-PV 3176 in lingual view. **i** CT-generated transverse section of premaxilla MPEF-PV 3305 showing replacement teeth (I-III) in alveoli (1–4). **j** Section of enamel of a maxillary tooth (thickness shown in μm). **k** Right dentary (MPEF-PV 3302–3) in dorsal and lateral views. Scale bars 3 cm for bones; 1 cm for teeth. Dashed line for reconstructed parts. Abbreviations: apns, anterior process of neural spine; blp, beak-like process; bt, basal tubera; btp, basipterygoid process; c, crest; con, concavity; cul, cultriform process; den, denticles; dmp, dorsomedial process; dp, diapophysis; dwf, distal wear facet; epi, epipophysis; eprl, epipophyseal-prezygapophyseal lamina; fm, foramen magnum; idp, interdental plate; lig, lingual groove; lp, lateral plate; lsp, laterosphenoid; na, articular facet for nasal; oc, occipital condyle; om, orbital margin; pa, parietal articulation; patf, proatlantal facet; pl, pleurocoel; poa, postorbital articulation; pop, paroccipital process; poz, postzygapophysis; pp, preotic pendant; prepi, pre-epipophysis; prfa, articular facet for prefrontal; prz, prezygapophysis, ptw, pterygoid wing; qja, articular facet for quadratojugal; saa, surangular articulation; sf, subnarial foramen; so, supraoccipital; spla, splenial articulation; sqa, articular facet for squamosal; sym, symphysis; tpd, triangular process of diapophysis; vmp, ventromedial process; 1ft, first functional tooth

spines are wider lateromedially than anteroposteriorly. As in the dorsal vertebrae, the sacral vertebrae lack pleurocoels. In the anteriormost caudal vertebrae, there is a marked PRDL.

3 Discussion

3.1 Phylogenetic Relationships

The early sauropods recorded in the late Early Jurassic of South America are currently restricted to Central Patagonia. These forms belong to at least two separate lineages of large-bodied sauropods, rather than forming a clade of closely related taxa. *Bagualia* and *Patagosaurus* are known from multiple specimens and their phylogenetic affinities with eusauropods is well supported by multiple characters. *Amygdalodon* and *Volkheimeria*, however, are known from single and rather incomplete specimens, and therefore, their phylogenetic position is less certain. Nonetheless, the anatomy of these taxa is still phylogenetically informative at a certain level, as shown by their inclusion in recent phylogenetic analyses (Carballido and Pol 2010; Cerda et al. 2017; Pol et al. 2020; Rauhut et al. 2020). Here we discuss the phylogenetic affinities of these four taxa and their implications for understanding the early evolution of sauropods in the Early Jurassic, based on the dataset recently published by Pol et al. (2020). The reduced strict consensus (Fig. 6) represents the relevant part of the phylogeny recovered in that study. Here we also present jackknife support

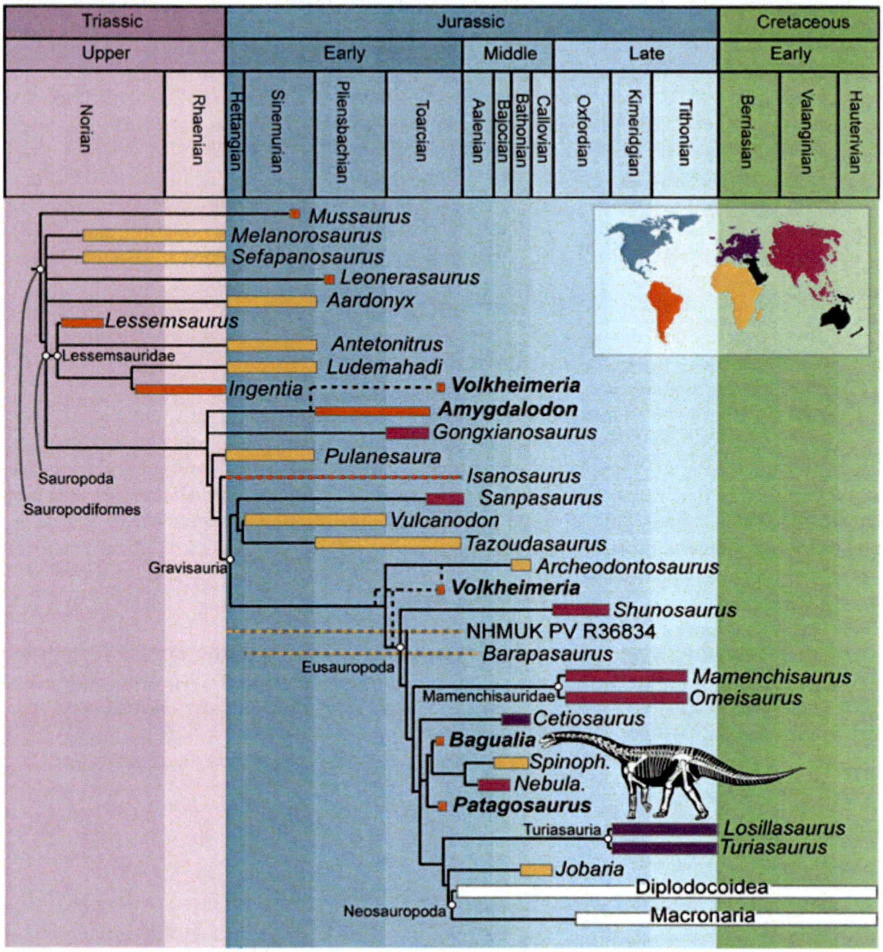

Fig. 6 Reduced strict consensus of the most parsimonious trees obtained for the dataset of Pol et al. (2020). The phylogenetic relationships of early sauropods are calibrated against geological time focusing on the Jurassic Period. Sauropod silhouette represents the anatomy of *Bagualia*

values for the complete tree (Fig. 7a) and a reduced majority rule tree that represents the jackknife support values ignoring the alternative position of unstable taxa (Fig. 7b; see below). The phylogenetic dataset and character list is also included in the Supplementary Materials.

In the reduced consensus, *Amygdalodon* is placed basally in the radiation of early sauropods, forming a polytomy with *Gongxianosaurus* from the Early Jurassic of China (Fig. 6). Its affinities with sauropods were evident since its description (Cabrera 1947), but synapomorphic features retrieved in the analysis that support its position are focused on the teeth: tooth–tooth occlusion (char. 101.1), extensive enamel wrinkling on the surface of the tooth crowns (char. 104.2), and presence of lingual

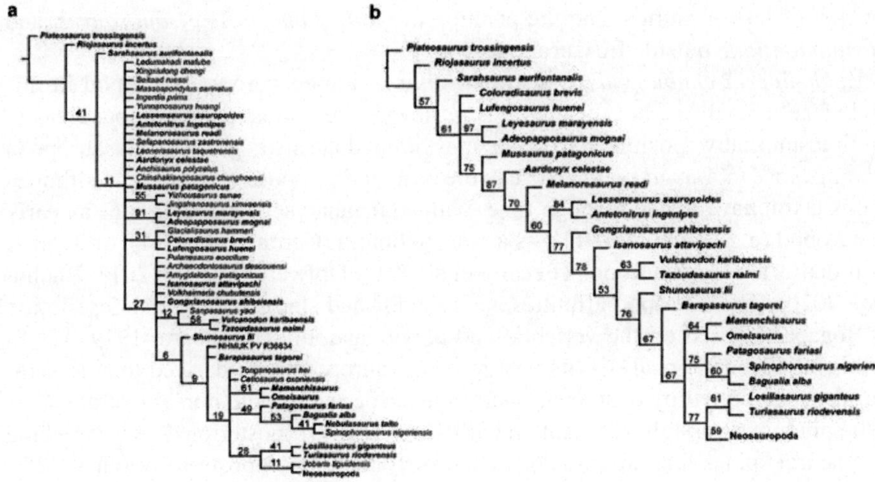

Fig. 7 Jackknife support values for early sauropodomorph nodes based on the dataset of Pol et al. (2020). **a** Standard result of jackknife parsimony with node values representing absolute frequencies in TNT (Goloboff and Catalano 2016). **b** Results of *pcrjak* procedure, reduced majority rule tree with values of absolute frequencies when the multiple positions of unstable taxa are ignored. Taxa in bold font are those detected as stable by *pcrjak* script (Pol and Goloboff 2020)

concavities on teeth (char. 383.1). These features are absent in more basal groups of sauropodomorphs and probably mark the presence of an early adaptation to bulk feeding. Placing *Amygdalodon* more basally within Sauropoda (e.g. allied to the Late Triassic–Early Jurassic lessemsaurids from Gondwana) implies three extra steps and more than five extra steps when placed more basally within Sauropodiformes (the least inclusive clade containing *Mussaurus* and *Saltasaurus*; Sereno 2007). *Amygdalodon* can take alternative positions within the basal nodes of Sauropoda with one or two extra steps. However, placing *Amygdalodon* within Eusauropoda, and closer to the Patagonian eusauropods *Patagosaurus* or *Bagualia*, implies over five extra steps. The characters involved in these extra steps not only include dental plesiomorphies of *Amygdalodon* (e.g. lack of labial grooves; char. 384) but also axial features such as the lack of a deep fossa on the anterior surface of the neural arch of mid-posterior dorsals (char. 155), neural spine lacking a median fossa (char 144), absence of cervical pleurocoels (char 114). The relationships of non-eusauropod sauropods are poorly supported, as revealed by the parsimony jackknife analysis in which all basal nodes of Sauropoda have frequency values below 50% (Fig. 7a). Many times, low values and lack of resolution in jackknife of bootstrap trees are due to the unstable placement of some taxa that take multiple alternative positions in the replicates of the support analysis. The *pcrjak* procedure (Pol and Goloboff 2020) aims to detect such cases and applying it to this dataset it identifies several early sauropods, including *Amygdalodon*, as unstable taxa that decrease support values in the jackknife analysis. Few nodes of early sauropods are well supported (e.g. >75%), such as the basal position of lessemsaurids, and the position of *Gongxianosaurus*, *Isanosaurus*, and *Vulcanodontidae* outside Eusauropoda (Fig. 7b).

Volkheimeria chubutensis is also placed as a non-eusauropod sauropod in this phylogenetic analysis, but this taxon is retrieved in the most parsimonious trees in multiple alternative positions, from the most immediate sister group of Eusauropoda to being close to *Amygdalodon* close to the base of Sauropoda (Fig. 6). The affinities of this taxon have varied through time, with some analyses retrieving it as an early eusauropod (e.g. Pol et al. 2011), as a vulcanodontid (Cerda et al. 2017), or in other positions as a non-eusauropod (Becerra et al. 2017; Holwerda and Pol 2018; Rauhut et al. 2020). The sauropod affinities were established since the original description by Bonaparte based on the vertebral and pelvic anatomy (Bonaparte 1979, 1999). In this phylogenetic analysis, its position as a sauropod more derived than lessemsaurids is supported by characters such as anterior and middle dorsal neural spines with spinoprezygapophyseal lamina (SPRL; char. 140), posterior dorsal vertebrae with neural spines that are broader transversely than anteroposteriorly (char. 175), ischial peduncle of ilium low and rounded (char. 280), and shape of the posterior margin of the postacetabular process of the ilium rounded (char. 488). Consequently, placing this taxon more basally than in the most parsimonious trees implies four extra steps. In the present analysis, *Volkheimeria* is excluded from Eusauropoda (Fig. 6) due to the presence of plesiomorphies such as middle and posterior dorsal neural arches lacking posterior centroparapophyseal lamina (PCPL; char. 161) and spinodiapophyseal lamina (SPDL; char. 164), and height of the dorsal neural arch less than 0.8 the height of the centrum (char. 425). Forcing *Volkheimeria* as a basal eusauropod, however, only requires two extra steps in this analysis. Further information on *Volkheimeria* is required to determine its phylogenetic position, not only because of its various placements in the most parsimonious topologies but also because it is identified by the pcrjak method as one of the taxa that becomes highly unstable during the nodal support analysis.

Patagosaurus fariasi is placed well nested within Eusauropoda and as the earliest branching member of a clade that includes *Bagualia*, the probably Middle Jurassic African taxon *Spinophorosaurus*, and the early Middle Jurassic Chinese *Nebulasaurus*. There are multiple eusauropod features in the anatomy of *Patagosaurus*, and its status as a eusauropod has never been questioned so far. Many of these characters were earlier noted by Bonaparte (1999) and include well-developed neural arch laminae in cervical vertebrae (char. 118), centroprezygapophyseal lamina dorsally divided in cervical middle and posterior cervical vertebrae (char. 127), middle and posterior dorsal vertebrae with the neural canal being enclosed in a deep fossa by pedicels in anterior view (char. 155), five sacral vertebrae (char. 181), scapular acromion process broad, width more than 150% minimum width of blade (char. 230), ulnar proximal condylar processes unequal, anterior arm longer (char. 262), and labial grooves on teeth (char. 384). As opposed to the previous two Patagonian sauropods, the phylogenetic placement of *Patagosaurus* is more robustly supported, placing it outside Eusauropoda requires over 13 extra steps. Similarly, the *pcrjak* does not indicate *Patagosaurus* is unstable and shows its placement in a node of Eusauropoda with high support value (75%; see Fig. 7b).

Patagosaurus is placed in a small clade of early eusauropods along with *Bagualia*, *Spinophorosaurus*, and *Nebulasaurus*. This position is supported by a combination of features, but most of the derived characters that influence the inclusion of *Patagosaurus* in this clade are similarities shared with *Spinophorosaurus*. These include the single lamina supporting the hyposphene and the absence of a spinodiapophyseal lamina in mid-to-posterior dorsals (chars. 154, 164), a relatively gracile humerus (RI < 0.27; char 256), slit-shaped posterior dorsal neural canal (char. 389; optimized as convergently acquired in other basal sauropods), and asymmetrical shape of the fourth trochanter of femur (char. 405). Placing *Patagosaurus* more basally within Eusauropoda implies at least four extra steps but placing it closer to Neosauropoda is highly suboptimal (over eleven extra steps).

Bagualia alba is also placed by the phylogenetic analysis within this clade of small eusauropods by the phylogenetic analysis. The position of *Bagualia* is closer to *Nebulasaurus* and *Spinophorosaurus* than *Patagosaurus*, due to the presence of single centropostzygapophyseal lamina in middle and posterior cervical vertebrae (char. 128) and laterally compressed anterior cervical vertebrae (centra approximately 1.25 times higher than wide; char. 386). Additionally, there are several cranial similarities shared with *Spinophorosaurus* (but currently unknown in *Patagosaurus*) such as the absence of a quadrate fossa (char. 52), basal tubera narrower than occipital condyle (char. 71), the floor of the braincase bent with the basipterygoid processes, and the parasphenoid rostrum below the level of the basioccipital condyle and the basal tuberae (char. 491). *Bagualia* is, however, excluded from the clade formed by *Nebulasaurus* and *Spinophorosaurus* because of the absence of features such as a postparietal foramen (char. 43), a wider than tall foramen magnum (char. 370), and a reduced crista interfenestralis (char. 423).

Bagualia is also robustly positioned within Eusauropoda and forcing this taxon outside this clade implies over 13 extra steps. Other positions within early eusauropods are plausible, as *Bagualia* can be placed slightly more basally or closer to neosauropods with five extra steps. The *pcrjak* analysis also endorses its relative stability as an early eusauropod, as it is left in the reduced jackknife tree as the sister group of *Spinophorosaurus* in 60% of the replicates (Fig. 7b).

3.1.1 Impact on Calibrated Phylogeny

The Early Jurassic sauropods from Patagonia are critical for calibrating the phylogeny of Sauropoda and for timing certain evolutionary events in the early evolution of the group. Their importance relative to other early sauropods lies not so much because of a unique phylogenetic position but mostly in the precise chronostratigraphic control for some of them, which contrasts with the less constrained ages of almost all Early Jurassic sauropods. High precision U-Pb dates from the Cañadón Asfalto Formation were published in recent years (Cúneo et al. 2013; Pol et al. 2020) that constrained the three named sauropods from this unit (*Patagosaurus*, *Bagualia*, *Volkheimeria*) into the middle Toarcian (ca 179–178 Ma).

The most influential of these in terms of calibrating the phylogenetic tree of early sauropods are *Bagualia* and *Patagosaurus* because they are the only eusauropods from the Early Jurassic that are dated with precision. Therefore, the mid-Toarcian age of these taxa provides a hard minimum radiation time for Eusauropoda (Pol et al. 2020). Furthermore, if the possible Toarcian age of the early neosauropod *Lingwulong* (Xu et al. 2018) is corroborated with radiometric dates, it would imply an even larger radiation of eusauropods took place at this time. These above-mentioned taxa are the most relevant eusauropods for dating the radiation of this node, because all other species are recorded either in the Middle or the Late Jurassic (e.g. *Shunosaurus*, *Omeisaurus*, *Cetiosaurus*, *Nebulasaurus*, *Spinophorosaurus*; Upchurch and Martin 2003; Rauhut and López Arbarello 2008; Xing et al. 2015; Wang et al. 2018). Two exceptions are present among the taxa retrieved as eusauropods in the phylogenetic analysis: *Barapasaurus* from India (Bandyopadhyay 2010) and the specimen NHMUK-PV R36834 from Morocco (Nicholl et al. 2018).

Barapasaurus tagorei from the Kota Formation (Bandyopadhyay et al. 2010) is known from a large number of remains and has long been regarded as a eusauropod. Although the Kota Formation has not been radiometrically dated, biostratigraphic data based on the vertebrate assemblage was used to assign the basal part of this unit to the Sinemurian–Pliensbachian (Bandyopadhyay and Roychowdhury 1996; Bandyopadhyay and Sengupta 2006; Bandyopadhyay and Ray 2020). An Early Jurassic age was also suggested based on data from palynomorphs (Prabhakar 1989) and charophytes (Bhattacharya et al. 1994). If this age assignment is correct, it would imply the initial radiation of Eusauropoda occurred earlier than the Toarcian. However, other authors have regarded the Kota Formation as Middle Jurassic or younger, based on data derived from ostracods (Govindan 1975; Misra and Satsangi 1979), palynomorphs (Vijaya and Prasad 2001), and mammals (Prasad and Manhas 2002, 2007). Therefore, there still is uncertainty regarding the age of the Kota Formation and consequently on the impact of *Barapasaurus* in timing the calibration of the eusauropodan radiation.

The much more fragmentary material from Morocco described by Nicholl et al. (2018) also shows vertebral features that are unique of eusauropods but unfortunately its precise geographic and stratigraphic provenance is uncertain. Nicholl et al. (2018) inferred that it must have been derived from Early Jurassic rocks of the Haute Moulouya Basin in Central Morocco. Despite the uncertainty associated to this material, it may provide evidence suggesting Eusauropoda was radiating and already broadly distributed at least by the late Early Jurassic. In sum, current data from the early eusauropods from Patagonia firmly establish that eusauropod radiation was ongoing by the mid-Toarcian and evidence from India and Morocco suggests this radiation may have started even earlier.

The non-eusauropod sauropods from the Early Jurassic of Patagonia (*Amygdalodon* and *Volkheimeria*) are less influential to the calibration of the sauropod phylogenetic tree. This is partly because there is more uncertainty regarding their position (see above), but mostly because the ages of most early sauropod nodes are determined by taxa that are older than *Volkheimeria* and *Amygdalodon*. Following the phylogenetic definition of Yates (2007), the origin of Sauropoda is placed in the Late

Triassic due to the age of *Lessemsaurus* and *Ingentia* (Apaldetti et al. 2018), as well as the Norian *Schleitheimia* (Rauhut et al. 2020). However, the radiation of Gravisauria likely represents an Early Jurassic event (Fig. 6). Arguably the most influential taxa for dating early gravisaurian diversification is *Vulcandodon* from the Early Jurassic Forest Sandstone Formation of southern Africa, which has been recently regarded as Sinemurian–Pliensbachian (Viglietti et al. 2018). Furthermore, another taxon potentially involved in dating the early sauropod radiation is *Isanosaurus* from the Nam Phong Formation of Thailand, originally regarded as Late Triassic (Buffetaut et al. 2000), but more recently suggested to be Early Jurassic in age (Racey and Goodall 2009; Laojumpon et al. 2017). Irrespective of the uncertainties in the age and position of these early sauropods, it seems certain that the two Patagonian sauropods *Volkheimeria* and *Amygdalodon* represent late Early Jurassic records of lineages that likely date back to the earliest Jurassic, if not to the latest Triassic.

Furthermore, apart from the named taxa, additional sauropods seem to be present in the Cañadón Asfalto Formation. A seemingly non-eusauropodan sauropodomorph tooth was described by Becerra et al. (2017), but it is unclear whether this element represents a so far unrecovered taxon or might be referable to the probable non-eusauropodan *Volkheimeria*. More importantly, Carballido et al. (2017) described another isolated tooth with neosauropod affinities, which differs from the teeth of *Patagosaurus* and *Bagualia* and thus certainly represents a so far unnamed further taxon of seemingly derived sauropod from the Cañadón Asfalto Formation. Furthermore, at least one specimen originally referred to *Patagosaurus* does not represent this taxon, nor does it seem to be referable to *Bagualia*, and might thus indicate a further eusauropod taxon. Thus, with at least three, and probably four taxa of eusauropods being recorded in the Cañadón Asfalto Formation, even including a possible neosauropod, this unit demonstrates a very rapid radiation of eusauropods towards the end of the Early Jurassic.

3.2 Biogeographical Considerations

According to current knowledge on the diversity and phylogeny of sauropods in the Early Jurassic, there does not seem to be a strong biogeographical signal. Remes et al. (2009) have suggested the Central Gondwanan Desert may have acted as a barrier to the dispersal of early sauropods based on putative affinities of *Patagosaurus* and *Barapasaurus*, both recorded in southern Gondwana. Although these taxa share some features, most parsimonious trees of most analyses have not supported a sister group relationship for them (Wilson 2002; Upchurch et al. 2004; Remes et al. 2009; Pol et al. 2011, 2020; Xing et al. 2015; Becerra et al. 2017; Holwerda and Pol 2018). The same pattern occurs for other Early Jurassic sauropods from Patagonia: they have never been retrieved allied to other taxa recorded in neighbouring regions. This lack of biogeographical pattern may be the result of poor sampling regimes (i.e. only a handful of well-known sauropods are recorded in the 25 million years of the Early

Jurassic). Alternatively, early sauropods may have been insensitive to the biogeographical barriers such as the Central Gondwanan Desert during the Early Jurassic (Hallam 1993; Remes et al. 2009) or radiated and attained a widespread distribution before the effective establishment of this barrier. More discoveries and early sauropod taxa from different regions of Pangea are needed to test these possible alternatives, which will require intensive sampling given sauropods seem to be numerically minor components of terrestrial ecosystems as preserved in currently known pre-Toarcian Jurassic rocks.

3.3 Palaeoecological Context in Patagonia

A drastic change in sauropodomorph faunal assemblages between the late Early and early Middle Jurassic has long been recognized as a major ecological evolutionary and faunal event in history of herbivorous dinosaurs (Barrett and Upchurch 2005; Barrett 2014). The discovery of large graviportal sauropods in the latest Early Jurassic led to the proposal of linking this faunal event to the large volcanic event of Karroo–Ferrar volcanism (Allain and Aquesbi 2008). Recent remains from Patagonia support this idea and identified a major change in sauropodomorph faunas before and after the extensive Lonco Trapial volcanic complex, which temporally coincided with the Karroo–Ferrar volcanism (Pol et al. 2020). Dating efforts in these deposits helped to constrain the timing of this volcanic event and faunal replacement to occur between ca. 189 Ma and 179 Ma (Cúneo et al. 2013; Pol et al. 2020). Studies on other taxonomic groups indicate this event also led to marked changes in theropod dinosaurs (Pol and Rauhut 2012; Rauhut and Pol 2019), indicating the restructuring of faunal components was not limited to large herbivorous dinosaurs. Furthermore, abundant plant remains found in deposits before and after the Pliensbachian–Toarcian volcanic event indicate major changes in the floral components of the Patagonian ecosystems (Escapa et al. 2008; Choo et al. 2016; Cúneo et al. 2013; Pol et al. 2020). Whereas humid conditions and a diverse flora composed of sphenophytes, ferns, seed ferns, cycads, bennetitaleans, and conifers existed in the pre-Pliensbachian of Patagonia, mid-late Toarcian sediments have preserved a flora indicative of a seasonally dry and warm climate dominated by conifers, such as Araucariaceae, Cheirolepidiaceae, and Cupressaceae. Sauropods predominated in the latter environments since the Toarcian but are unknown in Patagonia (and very rare at a global scale) prior to this time, when the niches of large herbivores were dominated by non-sauropodan sauropodomorphs. The post-Toarcian survival and success of eusauropods may have been influenced by their ability to feed on the conifer-dominated environments of Patagonia. In comparison with earlier sauropodomorphs, eusauropods had a longer neck that provided larger feeding envelopes and browsing heights, larger body sizes that may imply an expansion of gut capacity and fibre digestibility, deeper and more robust rostrum and mandibles indicative of higher bite force, and broader teeth with thicker enamel (>700 μm) and extensive shearing wear facets. These features have been interpreted as adaptations to obligate high-fibre herbivory and bulk feeding on tough, fibrous plant material (Sander et al. 2011; Barrett 2014; Button et al. 2017; Pol et al. 2020).

How far these changes in Patagonia reflect a local/regional phenomenon or a global change remains to be tested based on fossil and geochronological data from other geographical regions. Some information, however, argues for a global phenomenon during the Toarcian. Firstly, it is well known from marine sediments that the early Toarcian was a time during which important climate changes are recorded worldwide and temporally linked to the southern volcanic events (Hesselbo et al. 2000; Sell et al. 2014; Burgess et al. 2015). Secondly, recent evidence from terrestrial ecosystems in two disparate regions of the Northern Hemisphere (United Kingdom and China) shows a similar pattern as the one described for Patagonia, with a decrease in plant taxonomic diversity and increase in the dominance of conifers at the expense of ferns, seed ferns, and other plant groups (Slater et al. 2019; Deng et al. 2018). Thirdly, the currently available evidence on sauropodomorph faunal changes suggests that non-sauropodan sauropodomorphs became globally extinct (or extremely rare) before the Toarcian, whereas eusauropods became globally distributed and predominant after the Toarcian (Barrett and Upchurch 2005; Allain and Aquesbi 2008). More data are needed, however, to test if these changes actually represent a global trend for Toarcian terrestrial floras and if changes in sauropodomorph faunas during the Toarcian occurred following the pattern present in sequences from the Early Jurassic of South America (Table 1).

Table 1 Record of sauropod dinosaurs from the Early Jurassic South America, in the order presented in Sect. 2

Taxon	Formation	Age	Locality	References
Amygdalodon patagonicus	Cerro Carnerero	Pliensbachian–early Toarcian	Cañadón Puelman, Chubut Province, Argentina	Cabrera (1947), Casamiquela (1963), Rauhut (2003a), Carballido and Pol (2010)
Volkheimeria chubutensis	Cañadón Asfalto	Toarcian	Cerro Cóndor Sur, Chubut Province, Argentina	Bonaparte (1979, 1986)
Patagosaurus fariasi	Cañadón Asfalto	Toarcian	Cerro Cóndor Sur and Cerro Cóndor Norte, Chubut Province, Argentina	Bonaparte (1979, 1986), Holwerda et al. (2021)
Bagualia alba	Cañadón Asfalto	Toarcian	Bagual Canyon, Chubut Province, Argentina	Pol et al. (2020)

4 Conclusions

The sauropods recorded in the Early Jurassic of South America are restricted to the late Early Jurassic records from the Cañadón Asfalto Basin in Central Patagonia (Chubut Province, Argentina). The diversity of this group includes four named taxa that represent different evolutionary stages of the early history of Sauropoda. *Amygdalodon* and *Volkheimeria* show plesiomorphic features that placed them as non-eusauropods, and they likely represent members of long-lived lineages that originated close to the initial radiation of Sauropoda earlier during the Early Jurassic (or possibly in the latest Triassic). *Patagosaurus* and *Bagualia*, instead, have cranial and postcranial features that are exclusive of Eusauropoda and therefore are robustly placed within this clade. These taxa are important not only because they are known from multiple specimens (including well-preserved skull remains in the case of *Bagualia*) but also because they are the oldest definitive eusauropods for which we have precise geochronological information associated with the fossils. These dates place the eusauropod radiation at least in the mid-Toarcian after the massive volcanic event recorded in the Southern Hemisphere. This event, linked to climatic changes at a global scale, caused environmental changes in the Toarcian terrestrial ecosystems that led to conifer-dominated seasonal forests in which sauropods replaced non-sauropodan sauropodomorphs as the dominant large herbivores of terrestrial ecosystems.

Acknowledgements We would like to thank P. Mannion and M. Bronzati for their constructive reviews. We thank the Secretaría de Cultura (Provincia del Chubut) for granting collecting permits related to this project. The following grants contributed to the research and collection of materials included in this chapter: ANPCyT PICT 2006-1756, 2011-0808, 2014-1288, 2019-03834 (to DP), DFG RA 1012/9-1, 1012/13-1 (to OWMR).

Supplementary Information Available at: https://www.osf.io/h8qs3/?view_only=3100c6a2c8d54e9aa91fbf6655f9a2df.

References

Allain R, Aquesbi N (2008) Anatomy and phylogenetic relationships of *Tazoudasaurus naimi* (Dinosauria, Sauropoda) from the late Early Jurassic of Morocco. Geodiversitas 30:345–424

Allain R, Aquesbi N, Dejax J, Meyer C, Monbaron M, Montenat C, Richir P, Rochdy M, Russell D, Taquet P (2004) A basal sauropod dinosaur from the Early Jurassic of Morocco. CR Palevol 3:199–208

Apaldetti C, Martínez RN, Cerda IA, Pol D, Alcober O (2018) An early trend towards gigantism in Triassic sauropodomorph dinosaurs. Nat Ecol Evol 2:1227–1232

Bandyopadhyay S, Ray S (2020) Gondwana vertebrate faunas of India: their diversity and intercontinental relationships. Epis J Int Geosci 43:438–460

Bandyopadhyay S, Sengupta DP (2006) Vertebrate faunal turnover during the Triassic–Jurassic transition: an Indian scenario. In: Harris JD, Lucas SG, Kirkland JD, Milner ARC (eds) Terrestrial Triassic–Jurassic transition, New Mexico Museum of Natural History and Science Bulletin, New Mexico, pp 77–85

Bandyopadhyay S, Gillette DD, Ray S, Sengupta DP (2010) Osteology of *Barapasaurus tagorei* (Dinosauria: Sauropoda) from the early Jurassic of India. Palaeontology 53:533–569
Bandyopadhyay S, Roy Chowdhury T (1996) Beginning of the continental Jurassic in India—a palaeontological approach. Mus North Ariz Bull 60:371–378
Barrett PM (2014) Paleobiology of herbivorous dinosaurs. Annu Rev Earth Planet Sci 42:207–230
Barrett PM, Upchurch P (2005) Sauropodomorph diversity through time: paleoecological and macroevolutionary implications. In: Curry Rogers KA, Wilson JA (eds) The Sauropods. University of California Press, pp 125–156
Bates KT, Mannion PD, Falkingham PL, Brusatte SL, Hutchinson JR, Otero A, Sellers WI, Sullivan C, Stevens KA, Allen V (2016) Temporal and phylogenetic evolution of the sauropod dinosaur body plan. R Soc Open Sci 3:150636
Becerra MG, Gomez KL, Pol D (2017) A sauropodomorph tooth increases the diversity of dental morphotypes in the Cañadón Asfalto Formation (Early–Middle Jurassic) of Patagonia. CR Palevol 16:832–840
Bhattacharya A, Ray S, Datta PM, Maulick P (1994) Fossil Charophyta from the Kota Formation of the Pranhita-Godavari valley, Andhra Pradesh, India. Proc 9th Int Gond Symp 2:471–475
Bonaparte JF (1979) Dinosaurs: a Jurassic assemblage from Patagonia. Science 205:1377–1379
Bonaparte JF (1986) The dinosaurs (Carnosaurs, Allosaurids, Sauropods, Cetiosaurids) of the Middle Jurassic of Cerro Cóndor (Chubut, Argentina). Annales De Paléontologie 72:325–386
Bonaparte JF (1999) Evolución de las vértebras presacras en Sauropodomorpha. Ameghiniana 36:115–187
Bronzati M, Benson RBJ, Rauhut OWM (2018) Rapid transformation in the braincase of sauropod dinosaurs: integrated evolution of the braincase and neck in early sauropods? Palaeontology 61:289–302
Buffetaut E (2005) A new sauropod dinosaur with prosauropod-like teeth from the Middle Jurassic of Madagascar. Bulletin De La Societé Géologique De France 176:467–473
Buffetaut E, Suteethorn V, Cuny G, Tong H, Le Loeuff J, Khansubha S, Jongautchariyakul S (2000) The earliest known sauropod dinosaur. Nature 407:72–74
Burgess SD, Bowring SA, Fleming TH, Elliot DH (2015) High-precision geochronology links the Ferrar large igneous province with early-Jurassic ocean anoxia and biotic crisis. Earth Planet Sci Lett 415:90–99
Button DJ, Barrett PM, Rayfield EJ (2017) Craniodental functional evolution in sauropodomorph dinosaurs. Paleobiology 43:436–462
Cabrera A (1947) Un saurópodo nuevo del Jurásico de Patagonia. Notas Del Museo De La Plata 12:1–17
Carballido JL, Pol D (2010) The dentition of *Amygdalodon patagonicus* (Dinosauria: Sauropoda) and the dental evolution in basal sauropods. CR Palevol 9:83–93
Carballido JL, Holwerda FM, Pol D, Rauhut OWM (2017) An Early Jurassic sauropod tooth from Patagonia (Cañadón Asfalto Formation): implications for sauropod diversity. Publicación Electrónica De La Asociación Paleontológica Argentina 17:50–57
Casamiquela RM (1963) Consideraciones acerca de *Amygdalodon* Cabrera (Sauropoda, cetiosauridae) del Jurásico medio de la Patagonia. Ameghiniana 3:79–95
Cerda IA, Chinsamy A, Pol D, Apaldetti C, Otero A, Powell JE, Martínez RN (2017) Novel insight into the origin of the growth dynamics of sauropod dinosaurs. PLoS One 12:e0179707
Chatterjee S, Zheng Z (2002) Cranial anatomy of *Shunosaurus*, a basal sauropod dinosaur from the Middle Jurassic of China. Zool J Linn Soc 136:145–169
Chen PJ, Li J, Matsukawa M, Zhang H, Wang Q, Lockley MG (2006) Geological ages of dinosaur-track-bearing formations in China. Cretac Res 27:22–32
Choo T, Escapa I, Benjamin B (2016) Monotypic colonies of *Clathropteris meniscioides* (Dipteridaceae) from the Early Jurassic of central Patagonia, Argentina: implications for taxonomy and palaeoecology. Palaeontographica. Abteilung b, Palaophytologie 294:85–109

Chure D, Britt BB, Whitlock JA, Wilson JA (2010) First complete sauropod dinosaur skull from the Cretaceous of the Americas and the evolution of sauropod dentition. Naturwissenschaften 97:379–391

Cooper MR (1984) A reassessment of *Vulcanodon karibaensis* Raath (Dinosauria: Saurischia) and the origin of the Sauropoda. Palaeontol Afr 25:203–231

Cúneo R, Ramezani J, Scasso R, Pol D, Escapa I, Zavattieri AM, Bowring SA (2013) High-precision U-Pb geochronology and a new chronostratigraphy for the Cañadón Asfalto Basin, Chubut, central Patagonia: Implications for terrestrial faunal and floral evolution in Jurassic. Gondwana Res 24:1267–1275

Deng S, Lu Y, Zhao Y, Fan R, Ma X, Li X (2018) A warming and aridification event in terrestrial ecosystem of North China during the early Jurassic, possibly linked to the Toarcian Oceanic Anoxic Event. In: 5th International Paleontological Congress, París, pp 9–13

Díaz VD, Suberbiola XP, Sanz JL (2013) Appendicular skeleton and dermal armour of the Late Cretaceous titanosaur *Lirainosaurus astibiae* (Dinosauria: Sauropoda) from Spain. Palaeontol Electron 16:18

Dong Z, Zhou S, Zhang Y (1983) The dinosaurian remains from Sichuan basin, China. Palaeontologia Sinica 23:1–145 (in Chinese)

Escapa IH, Cúneo R, Cladera G (2008) New evidence for the age of the Jurassic Flora from Cañadón del Zaino, Sierra de Taquetrén, Chubut. Ameghiniana 45:633–637

Figari CE, Scasso RA, Cúneo NR, Escapa IH (2016) Estratigrafía y evolución geológica de la Cuenca de Cañadón Asfalto, Provincia del Chubut, Argentina. Lat Am J Sedimentol Basin Anal 22:135–169

Govindan A (1975) Jurassic freshwater ostracods from the Kota limestones of India. Palaeontology 18:207–216

Goloboff PA, Catalano SA (2016) TNT version 1.5, including a full implementation of phylogenetic morphometrics. Cladistics 32:221–238

Hallam A (1993) Jurassic climates as inferred from the sedimentary and fossil record. Philos Trans R Soc Lond Ser B Biol Sci 341:287–296

Hatcher JB (1903) Osteology of *Haplocanthosaurus*, with description of a new species and remarks on the probable habits of the Sauropoda and the age and origin of the *Atlantosaurus* beds: additional remarks on Diplodocus (Vol. 2, No. 1). Mem Carnegie Mus 2:1–75

Hesselbo SP, Gröcke DR, Jenkyns HC, Bjerrum CJ, Farrimond P, Bell HSM, Green OR (2000) Massive dissociation of gas hydrate during a Jurassic oceanic anoxic event. Nature 406:392–395

Holwerda FM (2019) 'Revision of basal sauropods from the Middle Jurassic of Patagonia and the early evolution of eusauropods'. PhD Thesis. Faculty of Geosciences, Ludwig-Maximilians University, Munich, Germany. 460, xlix pp

Holwerda FM, Pol D (2018) Phylogenetic analysis of Gondwanan basal eusauropods from the Early-Middle Jurasic of Patagonia, Argentina. Span J Palaeontol 33:298–298

Holwerda FM, Pol D, Rauhut OW (2015) Using dental enamel wrinkling to define sauropod tooth morphotypes from the Cañadón Asfalto Formation, Patagonia, Argentina. PLoS One 10:e0118100

Holwerda F, Rauhut OW, Pol D (2021) Osteological revision of the holotype of the Middle Jurassic sauropod dinosaur *Patagosaurus fariasi* Bonaparte, 1979 (Sauropoda: Cetiosauridae). Geodiversitas 43:575–643

Jain SL, Kutty TS, Roy-Chowdhury T, Chatterjee S (1975) The sauropod dinosaur from the Lower Jurassic Kota formation of India. Proc R Soc Lond Ser B Biol Sci 188:221–228

Janensch W (1935) Die Schädel der Sauropoden *Brachiosaurus*, *Barosaurus* und *Dicraeosaurus* aus den Tendaguruschichten Deutsch-Ostafrikas (Anfang). Palaeontographica-Supplementbände 7:145–248

Janensch W (1961) Die Gliedmaßen und Gliedmaßengürtel der Sauropoden der Tendaguru-Schichten. Palaeontographica-Supplementbände 7:177–235

Klein N, Remes K, Gee CT, Sander PM (2011) Biology of the sauropod dinosaurs: understanding the life of giants. Indiana University Press, Bloomington

Kutty TS, Jain SL, Chowdhury TR (1987) Gondwana sequence of the northern Pranhita-Godavari Valley: its stratigraphy and vertebrate faunas. The Palaeobotanist 36:214–229

Lallensack J, Klein H, Milan J, Wings O, Mateus O, Clemmensen L (2017) Sauropodomorph dinosaur trackways from the Fleming Fjord formation of East Greenland: evidence for Late Triassic sauropods. Acta Palaeontol Pol 62:833–843

Laojumpon C, Suteethorn V, Chanthasit P, Lauprasert K, Sutheethhorn S (2017) New evidence of sauropod dinosaurs from the Early Jurassic period of Thailand. Acta Geol Sin Engl Edition 91:1169–1178

Madsen JH, McIntosh JS, Berman DS (1995) Skull and atlas-axis complex of the Upper Jurassic sauropod *Camarasaurus* Cope (Reptilia: Saurischia). Bull Carnegie Mus Nat Hist 31:1–115

Mannion PD (2010) A revision of the sauropod dinosaur genus '*Bothriospondylus*' with a redescription of the type material of the Middle Jurassic form '*B. madagascarien*sis'. Palaeontology 53:277–296

Mannion PD, Upchurch P, Schwarz D, Wings O (2019) Taxonomic affinities of the putative titanosaurs from the Late Jurassic Tendaguru Formation of Tanzania: phylogenetic and biogeographic implications for eusauropod dinosaur evolution. Zool J Linnean Soc 185:784–909

Marpmann JS, Carballido JL, Sander PM, Knötschke N (2015) Cranial anatomy of the Late Jurassic dwarf sauropod *Europasaurus holgeri* (Dinosauria, Camarasauromorpha): ontogenetic changes and size dimorphism. J Syst Paleontol 13:221–263

McPhee BW, Bonnan MF, Yates AM, Neveling J, Choiniere JN (2015) A new basal sauropod from the pre-Toarcian Jurassic of South Africa: evidence of niche-partitioning at the sauropodomorph–sauropod boundary? Sci Rep 5:1–12

McPhee BW, Upchurch P, Mannion PD, Sullivan C, Butler RJ, Barrett PM (2016) A revision of *Sanpasaurus yaoi* Young, 1944 from the Early Jurassic of China, and its relevance to the early evolution of Sauropoda (Dinosauria). PeerJ 4:e2578

McPhee BW, Yates AM, Choiniere JN, Abdala F (2014) The complete anatomy and phylogenetic relationships of *Antetonitrus ingenipes* (Sauropodiformes, Dinosauria): implications for the origins of Sauropoda. Zool J Linn Soc 171:151–205

Misra RS, Satsangi PP (1979) Ostracods from Kota Formation. Geol Surv India Misc Publ 45:73–80

Nicholl CSC, Mannion PD, Barrett PM (2018) Sauropod dinosaur remains from a new Early Jurassic locality in the Central High Atlas of Morocco. Acta Palaeontol Pol 63:147–157

Ouyang H, Ye Y (2002) The first mamenchisaurian skeleton with complete skull: *Mamenchisaurus youngi*. Sichuan Science and Technology Press, Chengdu, China

Peyre de Fabrègues C, Allain R, Barriel V (2015) Root causes of phylogenetic incongruence observed within basal sauropodomorph interrelationships. Zool J Linn Soc 175:569–586

Pol D, Powell JE (2007) New information on *Lessemsaurus sauropoides* (Dinosauria: Sauropodomorpha) from the Upper Triassic of Argentina. Spec Pap Palaeontol 77:223–243

Pol D, Rauhut OW (2012) A Middle Jurassic abelisaurid from Patagonia and the early diversification of theropod dinosaurs. Proc R Soc B Biol Sci 279:3170–3175

Pol D, Goloboff PA (2020) The Impact of unstable taxa in Coelurosaurian phylogeny and resampling support measures for parsimony analyses. Bull Am Mus Nat Hist 440:98–115

Pol D, Garrido A, Cerda IA (2011) A new sauropodomorph dinosaur from the Early Jurassic of Patagonia and the origin and evolution of the sauropod-type sacrum. PLoS One 6:e14572

Pol D, Ramezani J, Gomez K, Carballido JL, Carabajal AP, Rauhut OWM, Cúneo NR (2020) Extinction of herbivorous dinosaurs linked to Early Jurassic global warming event. Proc R Soc B 287:20202310

Pol D, Otero A, Apaldetti C, Martínez RN (2021) Triassic sauropodomorph dinosaurs from South America: The origin and diversification of dinosaur dominated herbivorous faunas. J South Am Earth Sci 107:103145

Poropat SF, Upchurch P, Mannion PD, Hocknull SA, Kear BP, Sloan T, Sinapius GHK, Elliott DARevision of the sauropod dinosaur Diamantinasaurus matildae Hocknull, et al (2015) 2009 from the middle Cretaceous of Australia: implications for Gondwanan titanosauriform dispersal. Gondwana Res 27:995–1033

Prabhakar M (1989) Palynological evidence and its significance for the Kota Formation in the Pranhita-Godavari basin. In: Proceedings of 12th Ind Colloq Micropalaeontol Strat, pp 59–65
Prasad GVR, Manhas BK (2002) Triconodont mammals from the Jurassic Kota Formation of India. Geodiversitas 24:445–464
Prasad GVR, Manhas BK (2007) A new docodont mammal from the Jurassic Kota formation of India. Palaeontol Electron 10:1–11
Racey A, Goodall JGS (2009) Mesozoic red bed sequences from SE Asia and the significance of the Khorat Group of NE Thailand. Geol Soc Lond Spec Publ 315:41–67
Rauhut OW (2003a) Revision of *Amygdalodon patagonicus* Cabrera, 1947 (Dinosauria, Sauropoda). Fossil Rec 6:173–181
Rauhut OW (2003b) A dentary of *Patagosaurus* (Sauropoda) from the Middle Jurassic of Patagonia. Ameghiniana 40:425–432
Rauhut OWM, López-Arbarello A (2008) Archosaur evolution during the Jurassic: a southern perspective. Revista De La Asociación Geológica Argentina 63:557–585
Rauhut OW, Pol D (2019) Probable basal allosauroid from the early Middle Jurassic Cañadón Asfalto Formation of Argentina highlights phylogenetic uncertainty in tetanuran theropod dinosaurs. Sci Rep 9:1–9
Rauhut OW, Fechner R, Remes K, Reis K (2011) How to get big in the Mesozoic: the evolution of the sauropodomorph body plan. In: Klein N, Remes K, Gee CT, Sander PM (eds) Biology of the sauropod dinosaurs: understanding the life of giants. Indiana University Press, Bloomington and Indianapolis, pp 119–149
Rauhut OWM, Holwerda FM, Furrer H (2020) A derived sauropodiform dinosaur and other sauropodomorph material from the Late Triassic of Canton Schaffhausen, Switzerland. Swiss J Geosci 113:1–54
Remes K, Ortega F, Fierro I, Joger U, Kosma R, Ferrer JMM, Maga A (2009) A new basal sauropod dinosaur from the Middle Jurassic of Niger and the early evolution of Sauropoda. PLoS One 4:e6924
Riccardi AC, Gulisano CA, Mojica J, Palacios O, Schubert C, Thomson MRA (1992) Western South America and Antarctica. In: The jurassic of the circum-pacific, pp 122–161
Salgado L, Coria RA, Calvo JO (1997) Evolution of titanosaurid sauropods: Phytogenetic analysis based on the postcranial evidence. Ameghiniana 34:3–32
Sander PM, Christian A, Clauss M, Fechner R, Gee CT, Griebeler EM, Gunga H-C, Hummel J, Mallison H, Perry SF, Preuschoft H, Rauhut OWM, Remes K, Tütken T, Wings O, Witzel U (2011) Biology of the sauropod dinosaurs: the evolution of gigantism. Biol Rev 86:117–155
Sell B, Ovtcharova M, Guex J, Bartolini A, Jourdan F, Spangenberg JE, Schaltegger U (2014) Evaluating the temporal link between the Karoo LIP and climatic–biologic events of the Toarcian Stage with high-precision U-Pb geochronology. Earth Planet Sci Lett 408:48–56
Sereno PC (2007) The phylogenetic relationships of early dinosaurs: a comparative report. Hist Biol 19:145–155
Slater SM, Twitchett RJ, Danise S, Vajda V (2019) Substantial vegetation response to Early Jurassic global warming with impacts on oceanic anoxia. Nat Geosci 12:462–467
Tschopp E, Mateus O, Benson RB (2015) A specimen-level phylogenetic analysis and taxonomic revision of Diplodocidae (Dinosauria, Sauropoda). PeerJ 3:e857
Upchurch P (1995) The evolutionary history of sauropod dinosaurs. Philos Trans R Soc Lond Ser B Biol Sci 349:365–390
Upchurch P (1998) The phylogenetic relationships of sauropod dinosaurs. Zool J Linn Soc 124:43–103
Upchurch P, Barrett PM (2005) 3. Phylogenetic and Taxic Perspectives on Sauropod Diversity. In: Curry Rogers KA, Wilson JA (eds) The Sauropods: evolution and paleobiology. University of California, pp 104–124
Upchurch P, Martin J (2003) The anatomy and taxonomy of *Cetiosaurus* (Saurischia, Sauropoda) from the Middle Jurassic of England. J Vertebr Paleontol 23:208–231

Upchurch P, Barrett PM, Dodson P (2004) Sauropoda. In: Weishampel DB, Dodson P, Osmolska H (eds) The Dinosauria. University of California Press, Berkeley, pp 259–322

Viglietti PA, Barrett PM, Broderick TJ, Munyikwa D, MacNiven R, Broderick L, Choiniere JN (2018) Stratigraphy of the *Vulcanodon* type locality and its implications for regional correlations within the Karoo Supergroup. J Afr Earth Sc 137:149–156

Vijaya X, Prasad GVR (2001) Age of the Kota Formation, Pranhita Godavari Valley, India. A palynological approach. J Palaeontol Soc India 46:77–93

Wang J, Ye Y, Pei R, Tian Y, Feng C, Zheng D, Chang SC (2018) Age of Jurassic basal sauropods in Sichuan, China: a reappraisal of basal sauropod evolution. Bulletin 130:1493–1500

Wedel MJ (2005) Postcranial pneumaticity in sauropods and its implications for mass estimates. In: Curry Rogers K, Wilson JA (eds) The Sauropods: evolution and paleobiology. University of California Press, Berkeley, pp 201–228

Wilson J (2002) Sauropod dinosaur phylogeny: critique and cladistic analysis. Zool J Linn Soc 136:215–275

Wilson JA, Sereno PC (1998) Early evolution and higher-level phylogeny of sauropod dinosaurs. J Vertebr Paleontol 18:1–79

Xing L, Miyashita T, Currie PJ, You H, Zhang J, Dong Z (2015) A new basal eusauropod from the Middle Jurassic of Yunnan, China, and faunal compositions and transitions of Asian sauropodomorph dinosaurs. Acta Palaeontol Pol 60:145–154

Xu X, Upchurch P, Mannion PD, Barrett PM, Regalado-Fernandez OR, Mo J, Liu H (2018) A new Middle Jurassic diplodocoid suggests an earlier dispersal and diversification of sauropod dinosaurs. Nat Commun 9:1–9

Yadagiri P (2001) The osteology of *Kotasaurus yamanpalliensis*, a sauropod dinosaur from the Early Jurassic Kota Formation of India. J Vertebr Paleontol 21:242–252

Yaonan LUO, Changsheng WANG (2000) A new sauropod, *Gongxianosaurus*, from the Lower Jurassic of Sichuan, China. Acta Geol Sin Engl Edition 74:132–136

Yates AM (2007) The first complete skull of the Triassic dinosaur *Melanorosaurus* Haughton (Sauropodomorpha: Anchisauria). Evolution and palaeobiology of early sauropodomorph dinosaurs. Spec Pap Palaeontol 77:9–55

Yi-Gang W, Dong-Li S (1983) A survey of the Jurassic System of China. Can J Earth Sci 20:1646–1656

Highly Specialized Diplodocoids: The Rebbachisauridae

Leonardo Salgado, Pablo A. Gallina, Lucas Nicolás Lerzo, and José Ignacio Canudo

Abstract With 17 species formally identified throughout the world, Rebbachisauridae is, at present, the best-represented group of South American diplodocoids, and it has a temporal record ranging from the Barremian up to the Turonian. Defined as all diplodocoids more closely related to *Rebbachisaurus garasbae* than to *Diplodocus carnegii*, these sauropods are characterized by postcranial synapomorphies (e.g., absence of the hyposphenal ridge on anterior caudal vertebrae; presence of spinodiapophyseal lamina in caudal vertebrae). Although relatively complete skulls are known in only a few genera (*Limaysaurus, Lavocatisaurus,* and *Nigersaurus*), the whole cranial evidence indicates that they were highly specialized with respect to other diplodocoids (for instance Diplodocidae). South America counts ten genera of Rebbachisauridae, most of them from the Argentine Patagonia. They embrace a rather diverse group of basally branching forms (*Amazonsaurus*, *Zapalasaurus*, *Comahuesaurus,* and *Lavocatisaurus*), derived forms (as the limaysaurines *Limaysaurus* and *Cathartesaura* and the rebbachisaurines *Katepensaurus* and *Itapeuasaurus*), together with forms of uncertain phylogenetic filiation

L. Salgado (✉) · P. A. Gallina · L. N. Lerzo
Consejo Nacional de Investigaciones Científicas y Técnicas (CONICET), Buenos Ares, Argentina
e-mail: lsalgado@unrn.edu.ar

P. A. Gallina
e-mail: gallina.pablo@maimonides.edu

L. N. Lerzo
e-mail: lerzo.lucas@maimonides.edu

L. Salgado
Instituto de Investigación en Paleobiología y Geologia, Universidad Nacional de Rio Negro, Conicet. Av. General Julio A. Roca 1242, General Roca, Rio Negro 8332, Argentina

P. A. Gallina · L. N. Lerzo
Centro de Ciencias Naturales, Ambientales y Antropológicas, Fundación de Historia Natural Félix de Azara—Universidad Maimónides, Hidalgo 775, 7mo piso, Ciudad Autónoma de Buenos Aires C1405BCK, Argentina

J. I. Canudo
Grupo Aragosaurus-IUCA, Universidad de Zaragoza, Zaragoza 50009, Spain
e-mail: jicanudo@unizar.es

A. Otero et al. (eds.), *South American Sauropodomorph Dinosaurs*, Springer Earth System Sciences, https://doi.org/10.1007/978-3-030-95959-3_5

(*Rayososaurus*). Rebbachisaurids were important in South America toward the end of the Early Cretaceous, integrating, at that time, the sauropod faunas together with macronarians (Titanosauriformes) and other diplodocoids (Dicraeosauridae). They persisted up to at least the Turonian, being the last diplodocoids in becoming extinct globally.

Keywords Diplodocoidea · Rebbachisauridae · South America · Taxonomy and Systematics · Paleobiogeography

1 Introduction: A Brief Review on the Main Discoveries and Studies on South American Rebbachisaurid Sauropods

The Rebbachisauridae, a rather diverse group of diplodocoid sauropods, is one of the last groups of sauropods with familial rank in being recognized. Of the 17 species identified throughout the world, ten come from South America (Fig. 1), and eight from the Argentine Patagonia. All these taxa are included in Table 1, but those founded

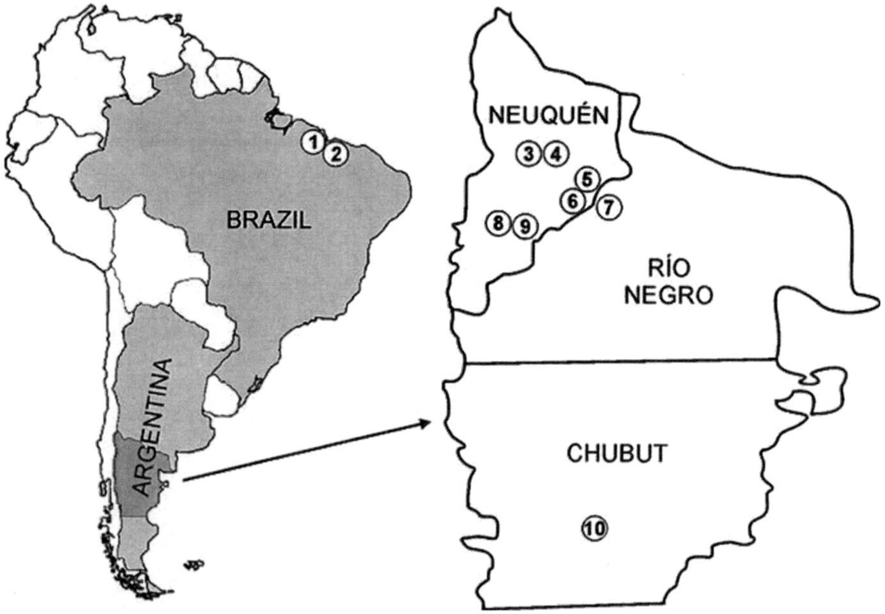

Fig. 1 Location map of the rebbachisaurid record in South America pointing to general areas of findings. 1, *Itapeuasaurus cajapioensis*; 2, *Amazonsaurus maranhensis*; 3, *Lavocatisaurus agrioensis*; 4, *Rayososaurus agrioensis*; 5, *Limaysaurus tessonei*; 6, *Nopcsaspondylus alarconensis*; 7, *Cathartesaura anaerobica*; 8, *Zapalasaurus bonapartei*; 9, *Comahuesaurus windhauseni* and 10, *Katepensaurus goicoecheai*

Table 1 Rebbachisaurid species listed according to the order in which they were published, with South American taxa in bold

Species name	Formation	Age and country
Rebbachisaurus garasbae	"Kem beds"	Cenomanian, Morocco
Limaysaurus tessonei	Candeleros and Huincul	Cenomanian–Turonian, Argentina
Rayososaurus agrioensis	Candeleros	Early Cenomanian, Argentina
Histriasaurus boscarollii	"Bale"	Hauterivian–Barremian, Croatia
Amazonsaurus maranhensis	Itapecuru	Aptian–Albian, Brazil
Nigersaurus taqueti	Elrhaz	Aptian–Albian, Niger
Demandasaurus darwini	Castrillo de la Reina	Late Barremian–early Aptian, Spain
Cathartesaura anaerobica	Huincul	Late Cenomanian–Turonian, Argentina
Zapalasaurus bonapartei	La Amarga	Late Barremian–early Aptian, Argentina
Nopcsaspondylus alarconensis	Candeleros and Huincul	Cenomanian–Turonian, Argentina
Comahuesaurus windhauseni	Lohan Cura	Aptian–Albian, Argentina
Katepensaurus goicoecheai	Bajo Barreal	Cenomanian–Turonian, Argentina
Tataouinea hannibalis	Ain el Guettar (Oum ed Diab M.)	Early Albian, Tunisia
Lavocatisaurus agrioensis	Rayoso	Aptian–Albian, Argentina
Maraapunisaurus fragillimus	Morrison	Late Jurassic, United States
Xenoposeidon proneneukos	Ashdown	Berriasian–Valanginian, England
Itapeuasaurus cajapioensis	Alcântara	Early Cenomanian, Brazil

on missing material, such as *Nopcsaspondylus alarconensis* or *Maraapunisaurus fragillimus*, are not considered in the body of the chapter. The type genus of the family is *Rebbachisaurus* Lavocat, 1954 originally described for the Early Cretaceous (Albian) of Morocco (Lavocat 1954), although these levels actually correspond to the Late Cretaceous (infra–Cenomanian, Table 1) or pre–late Cenomanian (Wilson and Allain 2015).

The oldest worldwide record of a rebbachisaurid is *Histriasaurus boscarollii* from the upper Hauterivian–lower Barremian of Croatia (Dalla Vecchia 1998) or, if it is finally confirmed as a Rebbachisauridae, *Xenoposeidon proneneukos* from the

Berriasian–Valanginian of England (Taylor and Naish 2007; Taylor 2018). On the other hand, the most recent records (Cenomanian–Turonian) are the South American *Katepensaurus goicoecheai* (Ibiricu et al. 2013), *Limaysaurus tessonei* (Calvo and Salgado 1995) and *Catharthesaura anaerobica* (Gallina and Apesteguía 2005). Rebbachisaurids were the last diplodocoids in becoming extinct, which would have occurred by the Turonian. After that, and until the end of the Cretaceous, sauropod faunas were only composed, at global level, by titanosauriform macronarians (see Chaps. 7–9, and 10).

Rebbachisaurids constitute a group of medium to small-sized sauropods, although some species would have reached relatively larger body sizes, such as *Rebbachisaurus* or *Limaysaurus*. Rebbachisaurids have cervical and dorsal vertebrae with single neural spines (unlike their sister group, the flagellicaudatans, see Chap 6) and are characterized by their highly modified skulls, which carried to the extreme many modifications of the diplodocoid skull-plan (Sereno et al. 2007).

Our ignorance of the existence of these sauropods at the beginning of the nineties of the 20th went hand in hand with our relative ignorance on the Early Cretaceous South American dinosaur faunas, since most of the knowledge about Cretaceous South American dinosaurs was, by that time, significantly skewed toward the latest Cretaceous. John McIntosh was who noticed for the first time that in South America there were sauropods related to the African species *Rebbachisaurus garasbae* (McIntosh 1990); indeed, he was who realized that a vertebra described by Franz Nopcsa in 1902 proceeding from the Upper Cretaceous of Neuquén, was anatomically similar to *Rebbachisaurus*, so that it "would appear to extend the range of this genus". Nopcsa (1902) had provisionally referred that vertebra to *Bothriospondylus* Owen (1875), whereas Hatcher (1903) later referred it to *Haplocanthosaurus*. More recently, that vertebra served for the foundation of *Nopcsaspondylus alarconensis* Apesteguía (2007), a taxon currently considered doubtful, since the material, formerly housed in Switzerland, is lost.

In the early nineties, the phylogenetic affinities of *Rebbachisaurus* were still uncertain. McIntosh (1990) was, again, the first to debate this issue, placing that genus among the Dicraeosaurinae within the family Diplodocidae. The diplodocid affinities of *Rebbachisaurus* remained unratified for years, more precisely until 1995, when J. Calvo and L. Salgado corroborated the existence of *Rebbachisaurus*-related sauropods in South America. In fact, their phylogenetic analysis, one of the first ones performed for the Sauropoda, recovered *Limaysaurus* ("*Rebbachisaurus*") *tessonei*, from the Cenomanian–Turonian of the Neuquén Province in the Argentine Patagonia (earlier considered as Albian–Cenomanian), and known on the basis of a relatively complete skeleton and other specimens, as the sister group of diplodocids + dicraeosaurines, a clade later named Flagellicaudata (Harris and Dodson 2004; Chap. 6). In turn, Calvo and Salgado (1995) included all those groups into a new one: the Diplodocimorpha (defined by them as "*Rebbachisaurus*" *tessonei*, Diplodocidae, and all descendants of their common ancestor), while they proposed the existence of a land bridge connection between Africa and South America in Albian–Cenomanian times.

Almost simultaneously, J. Bonaparte described *Rayososaurus agrioensis* from a well-preserved scapula and fragmentary appendicular bones and indeterminate fragments from Agrio del Medio, also in Neuquén Province, from sedimentary levels that he assumed belonged to the Lower Cretaceous Rayoso Formation (Bonaparte 1996), and which are currently considered to be the Cenomanian Candeleros Formation (Carballido et al. 2010), recognizing certain similarities with *Rebbachisaurus garasbae*. The following year, the same author proposed a new sauropod clade, naming it Rebbachisauridae, although without providing a phylogenetic definition or discussing its phylogenetic relationships within Sauropoda (Bonaparte 1997). Years later, Medeiros and Schultz (2004) recorded that genus in northeastern Brazil, but these materials are too fragmentary to be placed even in Rebbachisauridae (Wilson and Allain 2015).

The third rebbachisaurid described is the rather fragmentary *Amazonsaurus maranhensis* (Carvalho et al. 2003), the first member of the clade undoubtedly from the Lower Cretaceous, and the first one recorded in Brazil, although it was not originally considered as a rebbachisaurid but as a basal diplodocoid.

Amazonsaurus maranhensis was followed by *Cathartesaura anaerobica* from the Huincul Formation (upper Cenomanian–Turonian), one of the youngest rebbachisaurids, based on axial and appendicular elements (Gallina and Apesteguía 2005), and *Zapalasaurus bonapartei*, another Early Cretaceous (upper Barremian–lower Aptian) South American rebbachisaurid based on a single incomplete skeleton, and also recognized as a basal diplodocoid (Salgado et al. 2006). Like the aforementioned *Limaysaurus tessonei* and *Rayososaurus agrioensis*, both species come from the Neuquén Province.

The foundation of another Neuquenian species, *Comahuesaurus windhauseni* (Carballido et al. 2012), was based on materials belonging to many individuals from the Lower Cretaceous (Aptian–Albian) Lohan Cura Formation, previously described by Salgado et al. (2004) as *Limaysaurus* sp., a generic name proposed by them to be applied to *"Rebbachisaurus" tessonei*. These contributions provided evidence for the first time that rebbachisaurids, at least some species, had a gregarious behavior. In addition, Salgado et al. (2004) were the first to provide a phylogenetic definition for the clade, namely, Diplodocoidea more closely related to *Rebbachisaurus* than to *Diplodocus*.

The following year, L. Ibiricu and collaborators described and founded *Katepensaurus goicoecheai*, based on vertebral remains, being the first rebbachisaurid from the Golfo San Jorge Basin (which embraces the Argentine Patagonian provinces of Chubut and Santa Cruz), and the southernmost rebbachisaurid recorded up to the date (Ibiricu et al. 2013).

The last South American rebbachisaurids to be erected are *Lavocatisaurus agrioensis* Canudo et al. 2018, founded on at least three specimens (two juveniles and one adult) from the Lower Cretaceous of the Neuquén Province (Aptian–Albian), previously described by Salgado et al. (2012) as cf. *Zapalasaurus*, and *Itapeuasaurus cajapioensis* Matos Lindoso et al. (2019), based on several vertebral elements and a partial ischium from the Cajapió Municipality, northern Maranhão State, northern Brazil, Alcântara Formation, Sao Luís Basin, lower Upper Cretaceous (Cenomanian).

In this chapter, we offer a summary of the main anatomical characteristics of rebbachisaurids, a systematic list of the recognized taxa for South America, and a historical review of the main phylogenetic and paleobiogeographic analyzes performed on the group.

Institutional Abbreviations **MACN-N**: Museo Argentino de Ciencias Naturales, Buenos Aires, Argentina, colección Neuquén; **MOZ-Pv**: Museo Prof. Dr. Juan A. Olsacher, Zapala, Neuquén, Argentina; **MMCH-Pv**: Museo Municipal "Ernesto Bachmann" Villa El Chocón, Neuquén, Argentina; **MNHN-MRS**: Museum National d'Histoire Naturelle, Paris, France; **MPCA**: Museo Provincial "Carlos Ameghino", Cipolletti, Rio Negro, Argentina; **MUC**: Museo de Geologia y Paleontologia, Universidad Nacional del Comahue, Neuquén, Argentina; **UFMA**: fossil collection of the Universidade Federal do Maranhão; **UFRJ-DG**: Universidade Federal do Rio de Janeiro, Rio de Janeiro, Brazil, Department of Geology; **UFRJ-MN**: Universidade Federal do Rio de Janeiro, Museu Nacional, Departamento de Geologia e Paleontologia, Paleovertebrate collection, Rio de janeiro, Brazil; **UNPSJB-Pv**: Universidad Nacional de la Patagonia "San Juan Bosco", Comodoro Rivadavia, Chubut, Argentina, paleovertebrate collection; **WN-V**: Museum of Bale, Croatia.

2 Systematic Paleontology

The taxa listed in this section of the chapter were ordered following a phylogenetic criterion that takes into account the most recent phylogenies (Canudo et al. 2018; Matos Lindoso et al. 2019). They are, from the most basally branching taxa to the most derived: *Amazonsaurus maranhensis*; *Zapalasaurus bonapartei* (pruned in Canudo et al. 2018), *Comahuesaurus windhauseni*, *Lavocatisaurus agrioensis* (not considered in Matos Lindoso et al. 2019), *Limaysaurus tessonei*, *Cathartesaura anaerobica*, *Katepensaurus goicoecheai* (not considered in Matos Lindoso et al. 2019), *Itapeuasaurus cajapioensis* (not included in Canudo et al. 2018). Finally, it is listed *Rayososaurus agrioensis*, a taxon of uncertain relationships within the family.

Sauropoda Marsh 1878

Diplodocoidea Marsh 1878 (Upchurch 1995)

Definition *Diplodocus carnegii* Hatcher (1901) not *Saltasaurus loricatus* Bonaparte and Powell 1980 (Wilson and Sereno 1998).

Diplodocimorpha Calvo and Salgado 1995

Definition *Diplodocus carnegii* Hatcher 1901 + *Rebbachisaurus garasbae* Lavocat 1954 (Taylor and Naish 2005).

Rebbachisauridae Bonaparte 1997

Definition *Rebbachisaurus garasbae* Lavocat 1954, not *Diplodocus carnegii* Hatcher 1901 (Salgado et al. 2004).

Amazonsaurus Carvalho, Avilla and Salgado 2003

Amazonsaurus maranhensis Carvalho, Avilla and Salgado 2003

Holotype Two dorsal neural spines (MN 4558-V; UFRJ-DG 58-R/9); two dorsal centra (MN 4559-V; MN s/n(-V); neural spine of anterior caudal vertebra (UFRJ-DG 58-R/7) (Fig. 2); one mid-caudal vertebra (MN 4555-V); one mid-posterior caudal vertebra (MN 4560-V); one posterior caudal vertebra (MN 4556-V); one posterior caudal vertebra (UFRJ-DG 58- R/10); four chevrons (UFRJ-DG 58-R/2; 58-R/3; 58-R/4; 58-R/5); four chevrons (MN 4564-V); an ilium (UFRJ-DG 58-R/ 1); a partial pubis (MN s/n(-V); and three ribs (MN 4562-V).

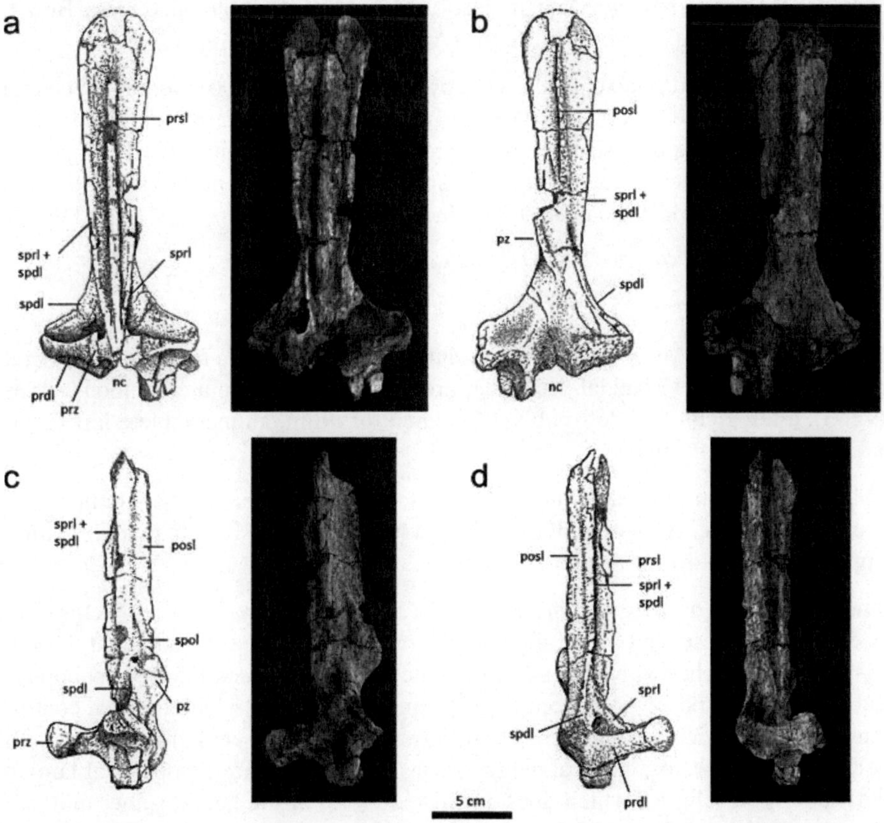

Fig. 2 *Amazonsaurus maranhensis* Part of the holotype, specimen UFRJ-DG 58-R/7. Anterior caudal neural spine, in anterior (**a**), posterior (**b**), left lateral (**c**) and right lateral (**d**) views. Abbreviations: nc, neural canal; prdl, prezygodiapophyseal lamina; prsl, prespinal lamina; prz, prezygapophysis; pz, postzygapophysis; spdl, spinodiapophyseal lamina; sprl, spinoprezygapophyseal lamina. Scale bar equals 5 cm (modified from Carvalho et al. 2003: Fig. 8)

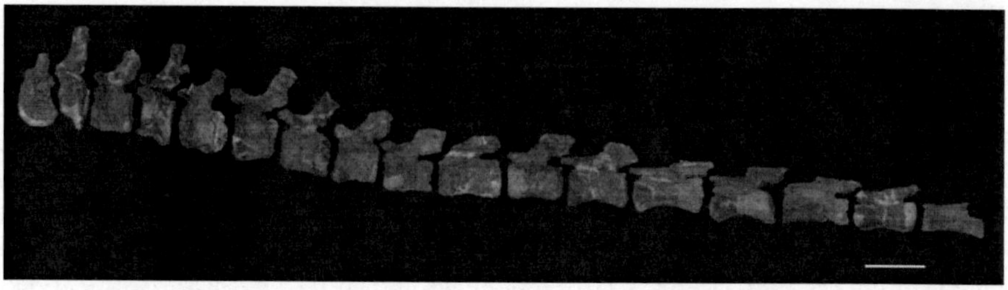

Fig. 3 *Zapalasaurus bonapartei* Part of the holotype Pv-6127-MOZ. Seventeen caudal vertebrae. Scale bar is 20 cm

Locality, Horizon and Age Mata, Itapecuru-Mirim County, Maranhão State, Brazil; Itapecuru Formation, Lower Cretaceous (Aptian–Albian).

Diagnosis Small sauropod characterized by Carvalho et al. (2003) as having caudal neural spines that are straight and posteriorly inclined, with "lateral" laminae formed by the spinoprezygapophyseal and postzygodiapophyseal laminae which, at least in the most anterior ones, bend anteriorly in such a way that the anterior surface of the lamina is concave while the posterior surface is convex.

Zapalasaurus Salgado, Carvalho and Garrido 2006

Zapalasaurus bonapartei Salgado, Carvalho and Garrido 2006

Holotype Pv-6127-MOZ; an anterior to mid-cervical vertebra; a fragment of a sacral transverse process, 17 caudal vertebrae, probably belonging to a continuous series (Fig. 3), a left ischium, a left pubis, a fragment of ilium, an incomplete left femur and a complete left tibia.

Locality, Horizon and Age Puesto Morales, La Picaza area, center–south of the Neuquén Province, Argentina; Puesto Parada Member of the La Amarga Formation (upper Barremian–lower Aptian).

Diagnosis *Zapalasaurus bonapartei* is distinguished from other rebbachisaurid species by the presence of the following characters listed by (Salgado et al. 2006): cervical neural arches with a lamina uniting the prezygapophysis and the zygapophyseal sector of the postzygodiapophyseal lamina (podl), with which forms a continuous lamina; cervical neural arches with postzygodiapophyseal lamina reduced on its diapophyseal sector; cervical neural arches with spinoprezygapophyseal lamina (sprl) poorly developed, which does not reach the tip of the neural spine; mid and posterior caudal vertebrae with anteroposteriorly elongated neural spine, whose anterior extreme is placed at a higher level than the posterior extreme; caudal vertebrae double their length in the first 20 vertebrae (convergent in diplodocines).

Comahuesaurus Carballido, Salgado, Pol, Canudo and Garrido 2012

Comahuesaurus windhauseni Carballido, Salgado, Pol, Canudo and Garrido 2012

Holotype MOZ-Pv 6722, posterior dorsal neural arch.

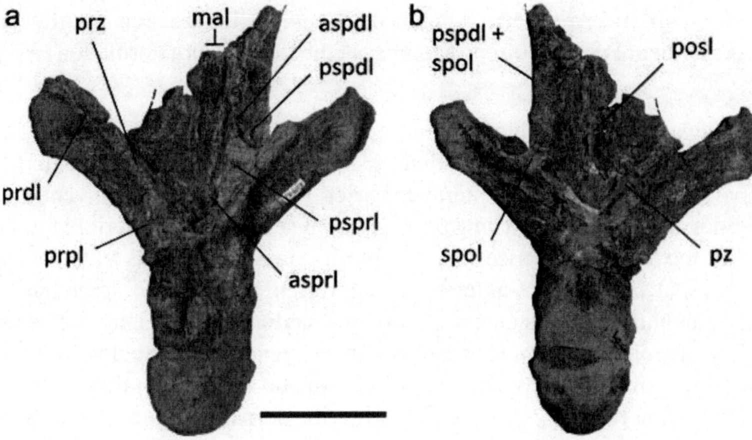

Fig. 4 *Comahuesaurus windhauseni*. Referred specimen MOZ 6650. Anterior dorsal vertebra in anterior (**a**) and posterior (**b**) views. Abbreviations: aspdl, anterior spinodiapophyseal lamina; asprl, anterior spinoprezygapophyseal lamina; mal, median anterior lamina; posl, postspinal lamina; prdl, prezygodiapophyseal lamina; prpl, prezygoparapophyseal lamina; prz, prezygapophysis; pspdl, posterior spinodiapophyseal lamina; psprl, posterior spinoprezygapophyseal lamina; pz, postzygapophysis; spol, spinopostzygapophyseal lamina. Scale bar equals 10 cm

Referred Specimens At least three individuals are represented in the holotype quarry (Salgado et al. 2004; Carballido et al. 2012). Both the holotype and referred specimens were excavated from a single bone bed that originated as a debris flow of an ephemeral river bed (Garrido and Salgado 2015). Anterior dorsal vertebra (MOZ-PV 6650) (Fig. 4), fragmentary dorsal centra (MOZ-PV 6645, MOZ-PV 6651, MOZ-PV 6653, MOZ-PV 6747, MOZ-PV 6751, MOZ-PV 6756), two neural arches (MOZ-PV 6652, MOZ-PV 6653), 35 caudal vertebrae (MOZ-PV 06,741, MOZ-PV 06,636, MOZ-PV 06,634, MOZ-PV 06,627, MOZ-PV 06,633, MOZ-PV 06,729, MOZ-PV 06,638, MOZ-PV 06,654, MOZ-PV 06,649, MOZ-PV 06,628, MOZ-PV 06,646, MOZ-PV 06,629, MOZ-PV 06,759, MOZ-PV 06,766, MOZ-PV 06,632, MOZ-PV 06,753, MOZ-PV 06,738, MOZ-PV 06,642, MOZ-PV 06,639, MOZ-PV 06,733, MOZ-PV 06,734, MOZ-PV 06,711, MOZ-PV 06,641, MOZ-PV 06,643, MOZ-PV 06,644, MOZ-PV 06,647), sternal plate (MOZ-PV 6717), one coracoid (MOZ-PV 6763), a complete right humerus (MOZ-PV 6762) and fragments of six other humeri (MOZ-PV 6664, MOZ-PV 6672, MOZ-PV 6673, MOZ-PV 6712, MOZ-PV 6714, MOZ-PV 6723), fragmentary ilium (MOZ-PV 6675), one complete pubis (MOZ-PV 6743) and seven fragments (MOZ-PV 6669a, MOZ-PV 6669b, MOZ-PV 6670, MOZ-PV 6659, MOZ-PV 6660, MOZ-PV 6667, MOZ-PV 6663), five ischia partially preserved (MOZ-PV 6676, MOZ-PV 6713, MOZ-PV 6719, MOZ-PV 6680, MOZ-PV 6658), two left femora (MOZ-PV 6728, MOZ-PV 6665), three right femora (MOZ-PV 6732, MOZ-PV 6761, MOZ-PV 6755), and four more fragmentary elements (MOZ-PV 6661, MOZ-PV 6666, MOZ-PV 6778, MOZ-PV 6721), proximal part of a tibia (MOZ-PV 6764), one left fibula partially preserved (MOZ-PV 6727).

Locality, Horizon and Age Cerro Aguada del León, La Picaza area, South-Central Neuquén, Argentina; Puesto Quiroga Member of the Lohan Cura Formation (Aptian–Albian).

Diagnosis *Comahuesaurus windhauseni* was characterized by Carballido et al. (2012) by the following characters (*indicates unique autapomorphic characters): 1*-anterior dorsal centra with strong lateral constriction, resulting in a thin ventral keel; 2*-anterior dorsal vertebrae with long prezygapophyses, which in anterior view cover around 3/4 of the transverse processes; 3*-anterior dorsal vertebrae with two spinoprezygapophyseal laminae; 4*-anterior dorsal vertebrae with two spinodiapophyseal laminae, an anterior and a posterior one; 5*- median anterior lamina formed by three different laminae, the anterior and posterior spinoprezygapophyseal laminae and the anterior spinodiapophyseal; 6*-posterior dorsal centra with the centroprezygapophyseal lamina medially divided; 7*-posterior dorsal neural arches with three spinopostzygapophyseal laminae; 8*-double contact between the posterior spinodiapophyseal lamina and the lateral spinopostzygapophyseal lamina; 9-anterior caudal vertebrae with well-developed prezygodiapophyseal fossa; 10-caudal vertebrae with short transverse process; 11-robust humerus, with a robustness index of 0.3 (sensu Wilson and Upchurch 2003); 12-ischium with straight shaft; 13-shaft of the ischium forming a right angle with the acetabulum; 14-iliac peduncle without a constriction or neck.

Comments The materials that served to erect *Comahuesaurus windhauseni* were originally assigned to *Limaysaurus* sp. by Salgado et al. (2004: 909) on the basis of three characters shared with *Limaysaurus tessonei*: (1) caudal vertebrae with the posterior articular surface more concave than the anterior; (2) a distally expanded pubis; (3) a pubic shaft that is oval in cross-section. A critique of this assignment can be found in Carballido et al. (2012: 646–647).

Lavocatisaurus Canudo, Carballido, Garrido, and Salgado, 2018

Lavocatisaurus agrioensis Canudo, Carballido, Garrido, and Salgado, 2018

Holotype MOZ-Pv 1232, partially articulated specimen, including cranial (Fig. 5a–g) and postcranial material: dentaries, left surangular, premaxillae and maxillae, left jugal, right squamosal, quadrates, 23 isolated teeth, and two series of 8 and 9 maxillary teeth, hyoid bone, 11 cervical vertebrae (including atlas and axis), 28 caudal vertebrae, cervical ribs, two dorsal ribs, humerus, fragment of radius.

Paratypes All from type locality, juvenile specimens (Salgado et al. 2012): MOZ-Pv 1248, posterior cervical centrum; MOZ-Pv 1249, cervical neural arch; MOZ-Pv 1251, dorsal neural arch; MOZ-Pv 1252, 1253, 1254, anterior caudal centra; MOZ-Pv 1255, scapula; MOZ-Pv 1267, left radius; MOZ-Pv 1256, left ulna; MOZ-Pv 1257, right metatarsal I; MOZ-Pv 1258, ?metatarsal V. An association of MOZ-Pv 1233, 1250, cervical centra; MOZ-Pv 1236, 1237, incomplete cervical neural arches; MOZ-Pv 1238, 1239, fragmentary neural arches; MOZ-Pv 1240, dorsal centrum; MOZ-Pv

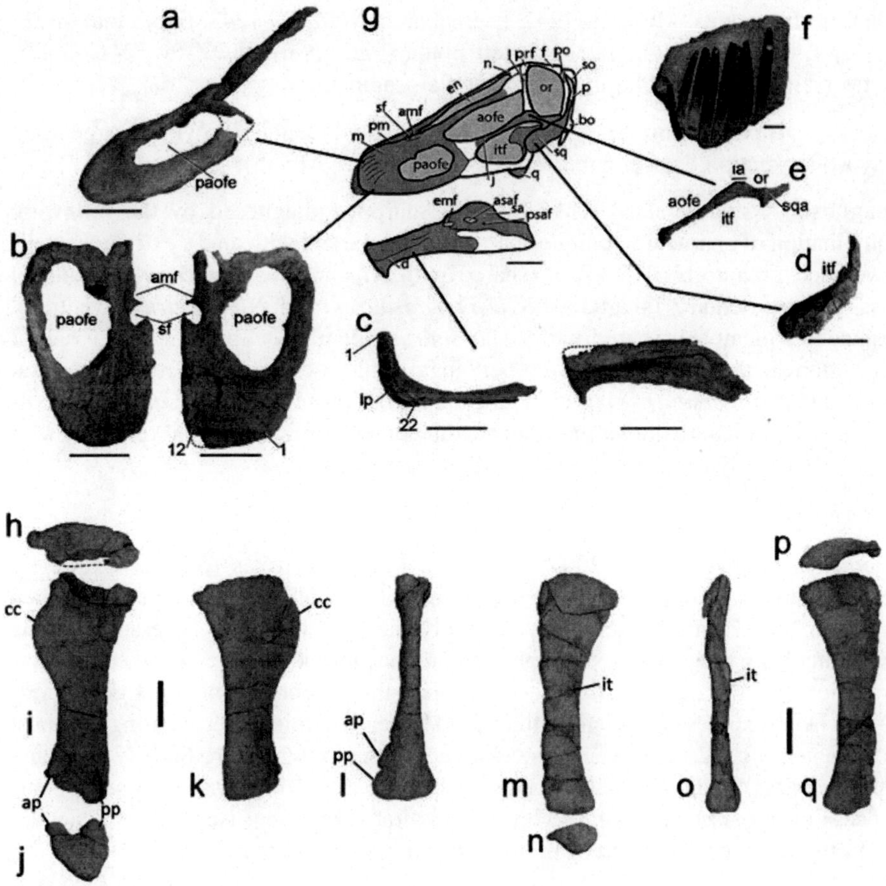

Fig. 5 *Lavocatisaurus agrioensis* (**a–g**) skull material, part of the holotype MOZ-Pv1232. **a** left maxilla in lateral view. **b** Right maxilla in anterodorsal (left) and posteroventral (right) views, numbers indicate tooth positions. **c** Left dentary in dorsal (left) and lateral (right) views, numbers indicate the tooth positions. **d** Right squamosal (inverted) in lateral view. **e** Right jugal (inverted) in lateral view. **f** Eight associated teeth in labial view. **g** Skeletal reconstruction based on the holotype MOZ-Pv1232. **h–k** Postcranial material, part of the paratypes MOZ-Pv 1244. Left tibia in proximal (**h**), lateral (**l**), distal (**j**), medial (**k**), and posterior (**l**) views. MOZ-Pv 1245 left fibula (**m–q**) in lateral (**m**), distal (**n**), anterior (**o**), proximal (**p**) and medial (**q**) views. Scale bars equal: **a–e**, 10 cm; F, 1 cm (modified from Canudo et al. 2018: Fig. 3); **h–q**, scale bar equals 7 cm (modified from Salgado et al. 2012: Fig. 9). Abbreviations: amf, anterior maxillary foramen; aofe, antorbital fenestra; ap, anterior process; asaf, anterior surangular foramen; bo, basioccipital; cc, cnemial crest; d, dentary; emf, external mandibular foramen; en, external naris; f, frontal; itf, infratemporal fenestra; j, jugal; l, lacrimal; la, lacrimal articulation; lp, lateral plate; lt, lateral tuberosity; m, maxilla; n, nasal; or, orbit; paofe, preantorbital fenestra; pm, premaxilla; po, postorbital; pp, posterior process; prf, prefrontal; psaf, posterior surangular foramen; q, quadrate; sa, surangular; sf, subnarial foramen; so, supraoccipital; sq, squamosal; sqa, squamosal articulation

1241, rib fragments; MOZ-Pv 1242, haemal arch; MOZ-Pv 1243, right ulna; MOZ-Pv 1244, left tibia; MOZ-Pv 1245, left fibula (Fig. 5h–q); MOZ-Pv 1246, end of metatarsal; MOZ-Pv 1247, indeterminate flat fragment.

Locality, Horizon, and Age Agrio del Medio site, Neuquén Province, Argentina; Rayoso Formation, Lower Cretaceous (Aptian–lower Albian).

Diagnosis A middle-sized rebbachisaurid sauropod diagnosed by the following combination of characters (unique characters are marked with an *): extremely well-developed preantorbital fenestra (shared with *Nigersaurus*); marked laterodorsal fossa in the dentary (shared with *Demandasaurus* and *Nigersaurus*); ventrally expanded squamosal (shared with *Limaysaurus*); dentary with pronounced ventral projection in the mesio-ventral corner; jugal long and contacting the squamosal (shared with *Nigersaurus*) but without foramina as present in *Nigersaurus*; *maxillary teeth significantly larger than the mandibular teeth; middle caudal vertebrae with anteriorly (nearly horizontally) projecting prezygapophysis.

Limaysaurus Salgado, Garrido, Cocca and Cocca 2004.

Diagnosis Salgado et al. (2004) diagnosed *Limaysaurus* as follows: Extremely reduced lateral temporal fenestra; supraoccipital height less than that of foramen magnum; basal tubera sheet-like; cervical neural arches with accessory lamina extending from the postzygodiapophyseal lamina anterodorsally; caudal centra with posterior articular surfaces more concave than their anterior counterparts; anterior caudal neural spines with distally thickened "lateral" laminae, terminating in robust bone; anterior caudal transverse processes composed of two laterodorsally projected osseous bars; distally expanded pubis; pubic shaft oval in cross-section; ambiens process of pubis placed distal to level of obturator foramen; ischium with slender shaft, twisted 90°; distal end of ischium virtually unexpanded.

Limaysaurus tessonei Calvo and Salgado 1995

Holotype MUCPv-205, articulated, well-preserved skeleton, including braincase, disarticulated cervical vertebrae, articulated vertebral column in posterior dorsals, and all caudals. Complete pelvic and pectoral girdles, nearly complete hind and forelimbs lacking a manus.

Referred Specimens MUCPv-206, a disarticulated skeleton composed of two posterior and two anterior cervical vertebrae, and one posterior dorsal vertebra, a sternal plate, four metacarpals, ribs. MUCPv-153, a partially articulated skeleton composed of two sacrals, the first six caudals, pubis, and ischium.

Locality, Horizon and Age All specimens come from the surroundings of Villa El Chocón (approximately 5 km south-west), Neuquén Province, Argentina; Candeleros (MUCPv 205, 206) and Huincul (MUCPv 153) formations, Cenomanian–Turonian (Garrido 2010).

Diagnosis Calvo and Salgado (1995) diagnosed *Limaysaurus* (*Rebbachisaurus*) *tessonei* as follows: rebbachisaurid possessing basipterygoid processes very thin and short (1); posterior process of the postorbital absent (2); anteroposteriorly elongate articular condyle of the quadrate (3); basal tubera very reduced (4); paroccipital processes not distally expanded (5); neural spine in posterior cervical and anterior dorsal vertebrae with an accessory lamina connecting the postzygodiapophyseal and spinoprezygapophyseal laminae (6); anterior dorsals with both spinoprezygapophyseal laminae contacting at the apex of the spine (7); transverse process in anterior caudal vertebrae formed by dorsally directed dorsal and ventral bars, differing from *R. garasbae* that has a true wing-like transverse process (8); shaft of the pubis oval in cross-section (9).

Comments Only characters 6 (reformulated as "cervical neural arches with accessory lamina extending from the postzygodiapophyseal lamina anterodorsally") and 8 were proposed by Wilson (2002: 274) as autapomorphies of *Rayososaurus* (considering *Limaysaurus* [="*Rebbachisaurus*"] *tessonei* as referable to that genus). In turn, only characters 8 and 9 were considered by Salgado et al. (2004) as autapomorphies of *Limaysaurus* (see above).

Cathartesaura Gallina and Apesteguía 2005

Cathartesaura anaerobica Gallina and Apesteguía 2005

Holotype MPCA-232, consisting of the following material from one quarry and presumably from a single specimen: a posterior cervical vertebra (Fig. 6a), a dorsal

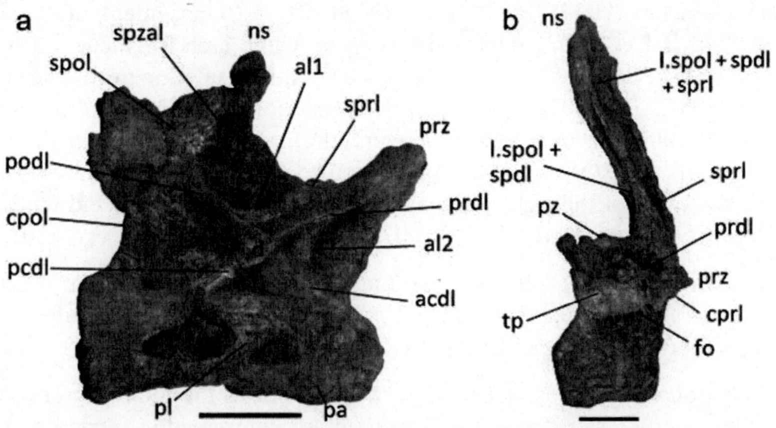

Fig. 6 *Cathartesaura anaerobica.* Part of the holotype MPCA 232. **a** Posterior cervical vertebra in right lateral view. **b** Anterior caudal vertebra in right lateral view. Abbreviations: acdl, anterior centrodiapophyseal lamina; al1, accessory lamina 1; al2, accessory lamina 2; cpol, centropostzygapophyseal lamina; cprl, centroprezygapophyseal lamina; d, diapophysis; fo, foramen; l.spol, lateral spinopostzygapophyseal lamina; ns, neural spine; pa, parapophysis; pcdl, posterior centrodiapophyseal lamina; pl, pleurocentral lamina; podl, postzygodiapophyseal lamina; prdl, prezygodiapophyseal lamina; prz, prezygapophysis; pz, postzygapophysis; spdl, spinodiapophyseal lamina; spol, spinopostzygapophyseal lamina; sprl, spinoprezygapophyseal lamina; spzal, suprapostzygapophyseal lamina; tp, transverse process. Scale bar equals 10 cm

vertebra, and an anterior caudal (Fig. 6b), a mid-caudal vertebra, left scapula, a left? ilium, and a right femur. A dorsal vertebra, a humerus, and one metatarsal were also collected, but they are poorly preserved.

Locality, Horizon and Age La Buitrera, 80 km SW of Cipolletti, Rio Negro, Argentina; Lower section of the Huincul Formation (upper Cenomanian–Turonian).

Diagnosis *Cathartesaura anaerobica* was diagnosed by Gallina and Apesteguía (2005) as a mid-sized sauropod dinosaur characterized by the following derived features: posterior cervical vertebra with an accessory lamina that arises from the middle of the prezygodiapophyseal lamina and reaches the centrum; thin wing-like transverse processes of anterior caudal mostly supported by the ventral bar forming a deep triangular fossa, also framed by the prezygodiapophyseal lamina and centroprezygapophyseal laminae; anterior caudal neural spine with the lateral lamina composed of the spinoprezygapophyseal lamina, the lateral spinopostzygapophyseal lamina, and the spinodiapophyseal lamina.

Katepensaurus Ibiricu, Casal, Martínez, Lamanna, Luna and Salgado 2013

Katepensaurus goicoecheai Ibiricu, Casal, Martínez, Lamanna, Luna and Salgado 2013

Holotype UNPSJB-PV 1007, an associated partial skeleton consisting of three anterior to middle cervical vertebrae (UNPSJB-PV 1007/1, 1007/2, and 1007/3), three middle to posterior dorsal vertebrae (UNPSJB-PV 1007/4, 1007/5, and 1007/6), two anterior caudal vertebrae (UNPSJB-PV 1007/7 and 1007/8), and two indeterminate elements (UNPSJB-PV 1007/9 and 1007/10), fragment of the right frontal (UNPSJB-PV 1007/29), an indeterminate neural arch fragment and partial cervical ribs (UNPSJB-PV 1007/35-36), an incomplete anterior to middle dorsal vertebra (UNPSJB-PV 1007/12), an incomplete anterior to middle dorsal neural arch (UNPSJB-PV 1007/31), three incomplete anterior caudal vertebrae (UNPSJB-PV 1007/9, 1007/10, 1007/11), possible metapodial fragment (UNPSJB-PV 1007/33), a possible fragment of the right astragalus (UNPSJB-PV 1007/32), and numerous indeterminate elements (UNPSJB-PV 1007/19-28,30,32,34) (Ibiricu et al. 2015).

Locality, Horizon and Age Estancia Laguna Palacios, south-central Chubut Province, central Patagonia, Argentina; Upper part of the Lower Member of the Bajo Barreal Formation, Upper Cretaceous (Cenomanian–Turonian).

Revised Diagnosis *Katepensaurus goicoecheai* possesses the following characters in the middle to posterior dorsal vertebrae that are interpreted as autapomorphies by Ibiricu et al. (2015) in their revised diagnosis of the species: (1) internal lamina divides lateral pneumatic fossa of centrum; (2) vertical ridges or crests present on lateral surface of vertebra, overlying neurocentral junction; (3) pair of laminae in parapophyseal centrodiapophyseal fossa; (4) well-defined, rounded fossae on lateral aspect of postzygapophyses; (5) posterior articular surface of centrum with ventral portion wider than dorsal portion, rendering it "teardrop shaped" in contour; (6) ovoid fossa on dorsal aspect of transverse processes in anterior to middle dorsal vertebrae and dorsal transverse processes perforated by elliptical fenestrae in middle to posterior dorsal vertebrae.

Comments The fragment of the right frontal (UNPSJB-PV 1007/29) has been regarded as an indeterminate element by Ibiricu et al. (2013). Many of the indeterminate elements of the holotype are probably fragmentary dorsal or anterior caudal neural spines. Ibiricu et al. (2013) postulated five autapomorphies of *Katepensaurus goicoecheai*, and Ibiricu et al. (2015) added two new autapomorphies corresponding to the anterior to mid-dorsal vertebrae.

Itapeuasaurus Matos Lindoso, Araújo Medeiros, Souza Carvalho, Araújo Pereira, Dienes Mendes, Vidoi Iori, Pinheiro Sousa, Souza Arcanjo, and Costa Madeira Silva 2019

Itapeuasaurus cajapioensis Matos Lindoso, Araújo Medeiros, Souza Carvalho, Araújo Pereira, Dienes Mendes, Vidoi Iori, Pinheiro Sousa, Souza Arcanjo, and Costa Madeira Silva 2019

Holotype An incomplete dorsal neural arch (UFMA. 1.10.1960e01); three anterior caudal vertebrae (UFMA. 1.10.1960e03, 1.10.1960e04, UFMA. 1.10.1960e05); and two middle caudal vertebrae (UFMA. 1.10.1960e07, 1.10.1960e08).

Referred Specimens A summit of dorsal neural spine (UFMA. 1.10.1960e02); an anterior caudal vertebra (UFMA. 1.10.1960e06); two chevrons (UFMA. 1.10.1960e10, 1.10.1960e11); an incomplete ischium (UFMA. 1.10.1960e09).

Locality, Horizon and Age Cajapió Municipality, northern Maranhão State; Alcântara Formation, Sao Luís Basin, Upper Cretaceous (Cenomanian).

Diagnosis Matos Lindoso et al. (2019) diagnosed *Itapeuasaurus cajapioensis* as follows: Rebbachisauridae distinguished by the following combination of characters on the dorsal and caudal vertebrae (autapomorphies are marked with an asterisk): presence of three shallow pneumatic fossae disposed vertically on the dorsal surface of the neural arch, which is separated by the spinoprezygapophyseal lamina and spinopostzygapophyseal accessory lamina; large and deep fossae on the ventrolateral aspect of the dorsal neural arch split by laminae obliquely oriented*; posterior centrodiapophyseal lamina forked ventrally forming the dorsal edge of the centrodiapophyseal fossa*; dorsal and ventral components of anterior caudal transverse process thinner than the usual bony bar associated to the presence of a prezygapophyseal centrodiapophyseal fossa lamina and prezygodiapophyseal centrodiapophyseal fossa accessory lamina*.

Rayososaurus Bonaparte 1997

Rayososaurus agrioensis Bonaparte 1997

Holotype Museo Nacional de Ciencias Naturales "Bernardino Rivadavia" (MACN) MACN-N 41, composed of a left scapula without the distal end of the scapular blade, an almost complete right scapular blade, the distal three-quarters of a left femur, and the proximal half of a left fibula.

Locality, Horizon and Age Agrio del Medio, Picunches Department, Neuquén Province, Argentina; Upper section of the Candeleros Formation (early Cenomanian).

Modified Diagnosis Sauropod characterized by the following autapomorphies: scapula that on its dorsal face presents a very well-developed acromion process directed in a markedly posterior direction; ventral margin of the scapula (evident in the right scapular blade) with a strong expansion directed ventrodistally (Carballido et al. 2010).

Comments Bonaparte's original diagnosis of *Rayososaurus agrioensis* included the following characters: (1) prominent spinous acromial process on scapula, internally flat, but externally with a rounded ridge running obliquely to scapular long axis, along the middle of the acromial process; (2) deep anterior (dorsal) border of scapular blade; (3) scapular blade well expanded distally; (4) acromial depression dorsoventrally elongated (Bonaparte 1996: 99).

Medeiros and Schultz (2004) assigned a series of materials from Ilha do Cajual (north of Maranhão, Albian–Cenomanian, Alcântara Formation) to *Rayososaurus*, but assuming that *Limaysaurus* was a junior synonym of *Rayososaurus*. To date, there are no elements that allow assigning the materials from Maranhão to a particular genus. Furthermore, *Rayososaurus agrioensis* material does not include caudal vertebrae, for which it is impossible to make comparisons between this species and the materials from Maranhão. Anyway, Mathos Lindoso et al. (2019) recognized that the sauropod materials from the Alcântara Formation are referable to *Limaysaurus* sp.

3 South American Rebbachisaurid: Main Anatomical Characteristics

In this section, we provide a summarized, comparative anatomical description of the Rebbachisauridae based on the South American taxa. Only when necessary, we will refer to non-South American species.

3.1 Cranial Skeleton

Apart from an isolated tooth collected by R. Lavocat, and included as part of the *Rebbachisaurus* material by Sereno et al. (2007) and Whitlock (2011) (see Wilson and Allain 2015), the first cranial elements of a rebbachisaurid to be described were those published by Calvo and Salgado (1995: Figs. 3–7). In fact, the cranial evidence was crucial for these authors to realize that *Limaysaurus tessonei* was related to the Diplodocidae. Beyond these remains (a tooth, a braincase, and a right quadrate), and until the recent description of *Lavocatisaurus agrioensis* (Canudo et al. 2018), most of the anatomical knowledge on rebbachisaurid skull was based on the African species *Nigersaurus taqueti* (Sereno et al. 2007). To date, *L. agrioensis* is the only South American rebbachisaurid (and the only one outside of Africa) whose rostral region has been preserved (Fig. 5a–g). Among South American rebbachisaurids, additional cranial materials, specifically, a partial right frontal, are known for *Katepensaurus goicoecheai* (Ibiricu et al. 2015). Both *L. agrioensis* and *N. taqueti* coincide in showing an extreme modification of the ground-level herbivory cranial pattern seen in diplodocoids in general (Sereno et al. 2007; Canudo et al. 2018) (Fig. 5).

Rebbachisaurids are characterized by their lightly constructed skulls, dental batteries placed at the ends of the dentaries, and maxillae (not observed in *Demandasaurus darwini*, Torcida Fernandez-Baldor et al. 2011), high rate of dental replacement, expanded muzzles, and skulls directed downwards. Like other diplodocoids, rebbachisaurids present an extreme reduction of the supratemporal fenestra, very elongated and only partially retracted external nostrils, and L-shaped jaws with very tight dental batteries (not observed in *Demandasaurus darwini*, Torcida Fernandez-Baldor et al. 2011). The snout of rebbachisaurids pointed down almost 70 degrees, which was determined mainly from the orientation of their semicircular channels (particularly, of their horizontal lateral semicircular channels). In turn, the nearly vertical orientation of the skull determines the posterior orientation of the occipital condyle.

The neurocranium of rebbachisaurids is known, albeit incompletely, only in *Limaysaurus tessonei* (Calvo and Salgado 1995: Figs. 3–7; Paulina Carabajal and Calvo 2021) and *Nigersaurus taqueti* (Sereno et al. 2007: Fig. 1). Paulina Carabajal et al. (2016) described additional materials (MMCH-PV 71) from the Candeleros Formation (lower Cenomanian) near Villa El Chocón (Neuquén Province). Recently, Paulina Carabajal and Calvo (2021) re-described the braincase of *Limaysaurus tessonei* and concluded, among other things, that South American rebbachisaurids are more similar to each other than to the African rebbachisaurid *Nigersaurus taqueti*.

Rebbachisaurids also show a characteristic tooth wear pattern and asymmetrical enamel (Sagado et al. 2004; Torcida Fernandez-Baldor et al. 2011), except in the basal form *Comahuesaurus windhauseni*, which have symmetrical enamel (Salgado et al. 2004; Carballido et al. 2012). Rebbachisaurids teeth exhibit wear facets at low angles on the lingual side, whereas on the labial side the wear facets are high angled (Sereno et al. 2007:Fig. 2A, D). These facets are thought to be caused by abrasion with soft plant matter.

As said, the description of *Lavocatisaurus agrioensis* resulted in a partial modification of the rebbachisaurid skull design proposed by Sereno et al. (2007). Also, the basal position of *L. agrioensis* with respect to *Nigersaurus taqueti* (these authors recovered it as the sister group of the Khebbashia, that is, Limaysaurinae+Rebbachisaurinae (see Sect. 4), allowed the authors who founded the species to explore the cranial evolution of the group (Canudo et al. 2018). In coincidence with its basal phylogenetic position, the skull of *Lavocatisaurus agrioensis* is more reminiscent of *Diplodocus* species than to *Nigersaurus taqueti* (Canudo et al. 2018). *Lavocatisaurus*'s cranial material has revealed that the previously presumed antorbital fenestra of *Nigersaurus* (Sereno et al. 2007: Fig 1A, B) is, instead, a highly developed preantorbital fenestra (Canudo et al. 2018: Fig. 3i) (Fig. 5g, paofe), a character that is present in other neosauropods. In *L. agrioensis* there are fewer maxillary teeth than in *N. taqueti* (12 vs 25). Also, the pattern of nutritious foramina piercing the maxilla and premaxilla is different in *L. agrioensis* and *N. taqueti*.

In *Lavocatisaurus agrioensis* it is possible to observe an articulation between the jugal and the squamosal, which excludes the postorbital (Fig. 5d, e, g) from the infratemporal fenestra, a character that has not been described in any other sauropod, although it can be inferred in *Limaysaurus agrioensis* and *Nigersaurus taqueti* (Canudo et al. 2018). The *Lavocatisaurus* skull also revealed that, at least in some rebbachisaurids, the squamosal is anteroventrally very expanded. This condition led Calvo and Salgado (1995) to misinterpret the squamosal of *L. tessonei* as the quadratojugal.

A large pre-antorbital fenestra is characteristic of *Lavocatisaurus agrioensis* but also, at least, of some Khebbashia, as it is present in *Nigersaurus taqueti*. Furthermore, *L. agrioensis* has revealed that the anterior displacement of the infratemporal fenestra, as well as the articulation between the squamosal and the jugal, occurred early in the evolution of the group.

The jaw of *Lavocatisaurus agrioensis* has 22 teeth, less than the 34 teeth observed in *Nigersaurus taqueti*, and surely more than in *Demandasaurus darwini*, whose exact number of mandibular teeth is uncertain (Torcida Fernandez-Baldor et al. 2011: Fig. 3). *Lavocatisaurus agrioensis, Limaysaurus tessonei, Nigersaurus taqueti,* and *Demandasaurus darwini* share tooth enamel asymmetry, thus differing from the symmetrical pattern observed in *Comahuesaurus windhauseni* (Salgado et al. 2004). This indicates that asymmetrical enamel evolved before the diversification of Khebbashia (Canudo et al. 2018). In sum, extremely modified architecture of the skull would have been achieved early in the evolution of the group.

3.2 Postcranial Skeleton

3.2.1 Cervical Vertebrae

The neck of Rebbachisauridae is apparently shorter than initially reconstructed by Calvo and Salgado (1995: Fig. 17) and Sereno et al. (2007). In this sense, the rebbachisaurid neck was apparently more similar to that of dicraeosaurids than that of diplodocids. This is coincident with the skull design inferred for these basal diplodocoids, specifically with the supposed impossibility of raising the head at high levels. Precisely, cranial and cervical anatomy were central in carrying to Sereno et al. (2007) to suggest that rebbachisaurids were the ecological equivalents of grazing mammals.

The first cervical vertebrae of a rebbachisaurid to be described were those of *Limaysaurus tessonei* (Calvo and Salgado 1995: Fig. 8A, B) (*Rebbachisaurus garasbae* did not preserve cervical vertebrae). Calvo and Salgado (1995) estimated the cervical vertebral formula of that species in 12 or 13 vertebrae. *Lavocatisaurus agrioensis* preserves 11 cervicals, but the sequence in the holotype is definitely not complete.

The axis is only known in *Nigersaurus taqueti*, *Demandasaurus darwini*, and among South American forms, in *Lavocatisaurus agrioensis*. Rebbachisaurid cervical centra have deep pleurocoels. *L. agrioensis* is the only rebbachisaurid that presents a single pleurocoel (Canudo et al. 2018: Fig. 2A3, A4). The undivided condition is also present in the anteriormost cervical vertebra (but not in the remaining cervical vertebrae, which have double pleurocoels) of the specimen MMCH-Pv 49, which comes from the surroundings of Villa El Chocón, in the Neuquén Province (Huincul Formation, upper Cenomanian–Turonian), although this is probably due to the obliteration of the anterior hemi-pleurocoel (Haluza et al. 2012: Fig. 2A, B). All other rebbachisaurids have double pleurocoels. In the case of *Nigersaurus*, it has two oval pleurocoels, the posterior being the largest (Sereno and Wilson 2005: Fig. 5.8; Sereno et al. 2007: Fig. 3). *Limaysaurus tessonei* and *Cathartesaura anaerobica* (at least in the posterior elements) the cervical vertebrae have double pleurocoels (Fig. 6a). The same occurs in *Katepensaurus goicoecheai*, where two oval pleurocoels divided by a pleurocentral lamina were described, which seems to be thicker than in *Limaysaurus tessonei* or *Cathartesaura anaerobica* (Ibiricu et al. 2013: Figs. 4, 5). *Zapalasaurus bonapartei* also shows the majority pattern reported in rebbachisaurids consisting of cervical vertebrae that have two pleurocoels separated by a thin bone septum (Salgado et al. 2006: Fig. 4).

Limaysaurus tessonei and *Cathartesaura anaerobica* have single, high cervical neural spines, similar in height to non-South American dicraeosaurids (Calvo and Salgado 1995: Fig.8A, B; Gallina and Apesteguía 2005: Fig. 2, Chap. 6). In total, the posterior cervical of *C. anaerobica* is twice as high as long. In this genus, the cervical neural spine is square in cross-section, and slightly inclined forward. The laminar pattern of the neural spines of *C. anaerobica* shows a great development of the epaxial musculature (Gallina and Apesteguía 2005) (Fig. 6a).

In other rebbachisaurids, the cervical vertebrae appear to be different. In *Zapalasaurus bonapartei* and *Lavocatisaurus agrioensis*, for instance, the cervical vertebrae have lower neural spines (Salgado et al. 2006: Fig. 4; Canudo et al. 2018: Fig. 2A). The only exception to this pattern is the anterior or middle cervical vertebra UNPSJB-PV 1005 from Estancia Ocho Hermanos, Chubut Province, Argentina, which has a low, dorsally bifid neural spine (Ibiricu et al. 2012).

The cervical vertebrae of the Rebbachisauridae exhibit the laminar basic configuration observed in other neosauropods. However, the cervical vertebrae of Rebbachisauridae have some characteristics that appear to be exclusive of this group. In *Limaysaurus tessonei*, for instance, Calvo and Salgado (1995) recognized, in the anterior and posterior cervical vertebrae, a lamina that runs from the middle of the spinoprezygapophyseal lamina (sprl) to the postzygodiapophyseal lamina (podl) named by them as accessory lamina (al) (Calvo and Salgado 1995: Fig. 8), proposing it as an autapomorphic character for *Limaysaurus tessonei*. Besides, Calvo and Salgado (1995), identified in the posterior cervical vertebrae of *L. tessonei* another lamina extending from the distal end of the sprl to the postzygapophysis (named suprapostzygapophyseal lamina, spzal), which persists in the anterior dorsal vertebrae (Calvo and Salgado 1995: Fig. 8). These accessory laminae, the spzal and the al, are also present in *Cathartesaura anaerobica.* In this species, these laminae are oriented in approximately the same way as in *L. tessonei* (Gallina and Apesteguía 2005: Fig. 2) (Fig. 6a).

In *Limaysaurus tessonei, Cathartesaura anaerobica,* as well as in specimen MMCH-Pv 49, where it receives the name of epipophyseal-prezygapophyseal lamina (eprl), the al unites the podl and the sprl. An identical connection is proposed for *Demandasaurus darwini,* where the so-called "accessory lamina" is described uniting the podl and the sprl (Torcida Fernandez-Baldor et al. 2011: Fig. 6). In turn, in *Zapalasaurus bonapartei,* this lamina, which did not receive a formal name but it was considered an autapomorphy of the species, is described as connecting the prezygapophysis with the podl (Salgado et al. 2006).

In *Nigersaurus taqueti* this lamina, which has received the name of prezygapophyseal-epipophyseal lamina (przepl), goes from the prezygapophysis to the epipophysis (which is not developed in other rebbachisaurids) (Sereno et al. 2007: Fig. 3B). In the African species, the przepl takes part of the podl, and is for this reason that it is described as extending far back, reaching the epipophysis (Sereno et al. 2007).

In *Lavocatisaurus agrioensis,* where the cervical epipophyses are absent, a lamina named epipophyseal-prezygapophyseal is described. Although the landmarks of this lamina were not specified by Canudo et al. (2018), there can be seen that it goes from a point equidistant from the sprl and the prezygodiapophyseal lamina (prdl) to the podl (Canudo et al. 2018: Fig. 2A4).

Evidently, the intersection of this accessory lamina produces changes in the trajectories of other laminae (sprl, prdl, podl), as occurs in *Nigersaurus taqueti* (Wilson et al. 2011: Fig. 6). For instance, in *Zapalasaurus bonapartei,* it intersects the podl about halfway through its trajectory, producing a break in its orientation so that the diapophyseal segment of the podl (podl (sd) in Salgado et al. 2006: Fig. 4C) becomes almost vertical. Seemingly, this segment of the podl is the only one that Sereno et al. (2007) would recognize as the true podl, whereas the zygapophyseal segment (podl (sz) in Salgado et al. 2006: Fig. 4C) would correspond, for Sereno et al. (2007), to the epipophyseal segment of the eprl. In *Z. bonapartei,* the new accessory lamina and the postzygapophyseal segment of the podl form an almost continuous lamina, which also occurs in *Lavocatisaurus agrioensis*; however, in this last species, the break in the trajectory of the podl is not as noticeable as in *Z. bonapartei* (Canudo et al. 2018: Fig. 2A4). In *Limaysaurus tessonei,* the prezygapophyseal segment of the accessory lamina is weakly developed, but it intersects the podl approximately at the same point as in *Z. bonapartei* or *L. agrioensis* (Calvo and Salgado 1995: Fig. 8B).

In *Cathartesaura anaerobica,* there is a lamina in the posterior cervical that Gallina and Apesteguía (2005) consider as an autapomorphy of the species: a second accessory lamina that joins the mid-length of the prdl with the centrum: this lamina is named accessory lamina 2 (al2) and would be also present in the MMCH-Pv 49 (Haluza et al. 2012: 220) (Fig. 6a).

In *Katepensaurus goicoecheai*, the ventral face of the centrum of the anterior cervical vertebra has a keel (Ibiricu et al. 2013: Fig. 3C, D), as in the anterior–mid-cervical of *Demandasaurus darwini* (Torcida Fernandez-Baldor et al. 2011), from the third cervical on, in *Lavocatisaurus agrioensis* (Canudo et al. 2018), and, at least in the posterior cervical vertebrae of specimen MMCH-Pv 49 (49/5, Haluza et al. 2012: 219).

3.2.2 Dorsal Vertebrae

Anterior Dorsal Vertebrae

The anterior dorsal vertebrae are known only in a few rebbachisaurids; among South American forms, in *Limaysaurus tessonei* (Calvo and Salgado 1995: Fig. 8C, D), *Comahuesaurus windhauseni* (MOZ-PV 6650, Carballido et al. 2012: Figs. 2, 3), *Katepensaurus goicoecheai* (Ibiricu et al. 2015: Figs. 5, 6), and, considering the rest of the world, in *Nigersaurus taqueti*. However, the cervico-dorsal transition is excellently documented in specimen MMCH-Pv 49, which was only preliminary described by Haluza et al. (2012: Fig. 3).

The anterior dorsal vertebrae are strongly opisthocoelous in *Limaysaurus tessonei* (Calvo and Salgado 1995: Fig. 8C, D), unlike the slightly opisthocoelous condition in the supposed fourth dorsal of *Comahuesaurus windhauseni* (Carballido et al. 2012: Fig. 2) and in the anterior dorsal vertebra of *Katepensaurus goicoecheai* (Ibiricu et al. 2015: Fig. 5). In *L. tessonei*, the vertebral centra are strongly compressed both laterally as dorsoventrally, with the ventral margin strongly concave, and one small, oval pleurocoel (Calvo and Salgado 1995: Fig. 8D). In *K. goicoecheai* the vertebral centrum is compressed anteroposteriorly with the ventral margin slightly concave and an ovate pleurocoel. This pleurocoel particularly resembles those described in *Amazonsaurus maranhensis* and MMCH- Pv 49 (Ibiricu et al. 2015). In *C. windhauseni*, on the other hand, the ventral margin of the centrum is straight. In this last species, in addition, and unlike other rebbachisaurids, a lateromedial compression of the ventral face of the centrum is observed in the anterior dorsal vertebrae, starting immediately ventral to the pleurocoels (Carballido et al. 2012).

In rebbachisaurids, the neural arch of the anterior dorsal vertebrae is high. Calvo and Salgado (1995) mentioned that, in the anterior dorsal of *Limaysaurus tessonei*, the postzygapophysis is elevated, being placed much higher than the prezygapophysis (which is possibly due to the inclination of the axis in this part of the column). In the anterior dorsal vertebra of *Comahuesaurus windhauseni*, the height of the neural arch is 3.5 times the height of the centrum (Carballido et al. 2012: Fig. 2).

In *Katepensaurus goicoecheai*, the anterior to mid-dorsal vertebrae (UNPSJB-PV 1007/12, 31) exhibit laterodiapophyseal fossae on the antero-dorsal surfaces of their transverse processes (Ibiricu et al. 2015, 2017: Fig. 6, 7; Fig. 3), which were considered an autapomorphy of the species (character 5, see above in the diagnosis of the species).

The neural spine of the anterior dorsal vertebrae of South American rebbachisaurids is single, high and straight. Carballido et al. (2012) argue that, in specimen MOZ-PV 6650 of *Comahuesaurus windhauseni*, the preserved part of the spine is six times the height of the centrum (measured from the posterior articulation) (Carballido et al. 2012: Figs. 2, 3) (Fig. 4). In *Limaysaurus tessonei*, in turn, the spine widens laterally toward the distal part (Calvo and Salgado 1995: Fig. 8C, D), which cannot be known in *C. windhauseni* because the distal portion of the spine has not been preserved.

In *Limaysaurus tessonei*, the neural spine is formed by two spol, two sprl, and a prespinal lamina that is placed between both sprl. In the anterior dorsal vertebra of this species, there is a lamina that is interpreted as the spzal observed in the posterior cervicals (Calvo and Salgado 1995: Fig. 8C, D.), similar to the condition present in the dorsal vertebra of *Itapeuasaurus cajapioensis* (Matos Lindoso et al. 2019: Fig. 4). However, at least in the dorsal vertebra, this supposed spzal is more probably the lateral spinopostzygapophyseal lamina (spol). In the anterior dorsal of *Comahuesaurus windhauseni* there are both anterior (aspdl) and posterior (pspdl) spinodiapophyseal laminae (Carballido et al. 2012: Fig. 2) (Fig. 4), which are not observed in the anterior dorsal of *L. tessonei* (Calvo and Salgado 1995: Fig. 8C, D). Carballido et al. (2012) argue that the simultaneous presence of the two spinodiapophyseal laminae (spdl) is rare in diplodocoids but common in titanosaurs. In specimen MMCH-Pv 49, a single spdl appears just in the third dorsal vertebrae (Haluza et al. 2012).

Also, the composition of the median laminae of the anterior dorsal vertebrae of rebbachisaurids is variable. In the basally branching taxon *Comahuesaurus windhauseni*, the anterior dorsal (MOZ-PV 6650, a probable fourth dorsal) presents a thick, "hybrid", median anterior lamina. The laminae that are involved in the integration of this median structure are the anterior (asprl) and posterior (psprl) spinoprezygapophyseal laminae, and, more distally, the aspdl (Carballido et al. 2012: Fig. 2A, 3). This conformation of the "hybrid" or complex median lamina is regarded as an autapomorphy of *C. windhauseni* by Carballido et al. (2012) and is very different from what is observed in *Limaysaurus tessonei*. In this last species, there is a prespinal lamina (prsl) (described but not figurate in Calvo and Salgado 1995: Fig. 8), which is separated from both sprl. In *Katepensaurus goicoecheai* the prsl is well-developed (Ibiricu et al. 2015: Fig. 6). According to Carballido et al. (2012: 642), the absence of a differentiated prsl is a character that *C. windhauseni* shares with other rebbachisaurids, such as *Demandasaurus darwini* (but here the vertebra is a mid-posterior dorsal, Torcida Fernandez-Baldor et al. 2011: Fig. 9) and *Nigersaurus taqueti*.

In the anterior dorsal of *Comahuesaurus windhauseni*, the median posterior lamina is integrated by the postspinal lamina (posl), which is well-developed (Carballido et al. 2012: Fig. 2B). On the contrary, this lamina is not reported in the anterior dorsal of *Limaysaurus tessonei* (Calvo and Salgado 1995).

Mid and Posterior Dorsals

Among South American rebbachisaurids, middle and posterior dorsal vertebrae are known only in five forms: *Limaysaurus tessonei* (Calvo and Salgado 1995: Fig. 9), *Comahuesaurus windhauseni* (Carballido et al. 2012: Figs. 4, 5), *Katepensaurus goicoecheai* (Ibiricu et al. 2013: Figs. 6–8), and in a partially preserved middle–posterior dorsal neural arch of *Itapeuasaurus cajapioensis* (Matos Lindoso et al. 2019: Figs. 3, 4), as well as in specimen MMCH-Pv 49 (Haluza et al. 2012: Fig. 2E–G). The morphology of the dorsal vertebrae of rebbachisaurids is distinctive. In fact, together with the morphology of the scapula, it was decisive, in the middle of the nineties, in linking *L. tessonei* with *Rebbachisaurus garasbae* (Calvo and Salgado 1995).

The posterior dorsal centrum of *Rebbachisaurus garasbae* is small and short, and the neural arch and neural spine are high, being the centrum less than 1/5 of the total vertebral height (Wilson and Allain 2015). Consequently, the postzygapophysis is placed in the neural arch at a height as that of the centrum itself. In both *R. garasbae* and *Limaysaurus tessonei*, as well as in the rest of the rebbachisaurids, the transverse processes are characteristically laterodorsally projected. In the middle to posterior dorsal vertebra of *Katepensaurus goicoecheai* (UNPSJB-PV 1007/4), the transverse processes exhibit a deep and well-defined ovoid fenestra. This structure is the laterodiapophyseal fenestra that Ibiricu et al. (2015) consider as an autapomorphy of the species (Ibiricu et al. 2015: Fig. 6–8).

The centrum of the posterior dorsal vertebrae of *Rebbachisaurus garasbae* is slightly opisthocoelous. In the holotype specimen (MNHN-MRS 1958), the anterior articular face is less convex than the posterior face is concave. The anterior articular face is convex especially on the dorsal part, being the ventral part almost flat, particularity that, according to Wilson and Allain (2015), is also present in other diplodocoids. In the African species, there is a subtriangular pleurocoel or pleurocentral foramen on each side of the centrum (Wilson and Allain 2015: Fig. 3). The centrum of the mid or posterior dorsal of *Katepensaurus goicoecheai* is similar to that of *R. garasbae*, although the pleurocoel is rather oval, less triangular than in *R. garasbae* (Ibiricu et al. 2013: Fig. 7). In *Katepensaurus goicoecheai*, the pleurocoel is internally divided by a vertical lamina (Ibiricu et al. 2013: Fig. 7B, specimen 1007/4).

In *Comahuesaurus windhauseni*, the centrum of the posterior dorsals is practically flat anteriorly, and strongly concave posteriorly (Carballido et al. 2012: Fig. 4). As in *Rebbachisaurus garasbae* and *Katepensaurus goicoecheai*, the articular face of the posterior dorsal vertebrae of *C. windhauseni* is oval, slightly higher than wide. As in the sauropod from Chubut, the pleurocoel of the posterior dorsal of *C. windhauseni* is oval with its ends acuminate.

The posterior dorsal vertebrae of Rebbachisauridae show hyposphene-hypantrum complexes of variable development. In *Limaysaurus tessonei* and *Rebbachisaurus garasbae* such an articular complex is definitely absent. This carried to Calvo and Salgado (1995) to postulate that this absence was a synapomorphy of "*Rebbachisaurus*" (which included *L. tessonei*) and, by extension, of all known rebbachisaurids at that time. This presumption began to change with the description of a hyposphene-hypantrum in the dorsal vertebrae WN-V6 of *Histriasaurus boscarollii* (Dalla Vecchia 1998), considered a "Rebbachisauroidea" by Apesteguía (2007) (that is, a form closely related to the Rebbachisauridae), and a basal rebbachisaurid by Sereno et al. (2007: Fig. 4). The description of *Demandasaurus darwini* and, overall, of *Comahuesaurus windhauseni*, clearly demonstrated that the absence of hyposphene-hypantrum was not an apomorphic character of rebbachisaurids more derived than *H. boscarollii* (Torcida Fernandez-Baldor et al. 2011; Carballido et al. 2012).

Carballido et al. (2012) claimed that the loss of hyposphene in the mid-dorsal vertebrae is convergent in Eutitanosauria and some Rebbachisauridae. Within the latter clade, that loss would have occurred only in limaysaurines, which included, according to them, *Limaysaurus tessonei*, *Rayososaurus agrioensis*, *Cathartesaura anaerobica,* and *Rebbachisaurus garasbae.* We must anticipate here that, in subsequent analyses (i.e., Fanti et al. 2015; Canudo et al. 2018), *R. garasbae* was recovered as more related to *Nigersaurus taqueti, Tataouinea hannibalis,* and *Demandasaurus darwini*; that is, as a Rebbachisaurinae (see below). The hyposphene described in the basal rebbachisaurid *Comahuesaurus windhauseni* (in the holotype specimen MOZ-PV 6722) by Carballido et al. (2012: Fig. 5) is, according to Wilson and Allain (2015), a structure definitely distinct from the delicate intrapostzygapophyseal lamina (tpol) reported in *R. garasbae* (Wilson and Allain 2015: Figs. 4, 5) and *Katepensaurus goicoecheai* (Ibiricu et al. 2013: Fig. 6), which, on the other hand, is absent in *L. tessonei.* The hyposphene of *C. windhauseni* is laminar, as in the nigersaurine *D. darwini,* and quite different from that of *Histriasaurus boscarollii,* where it is more developed. In this respect, Ibiricu et al. (2015) divided the Rebbachisauridae into three groups, according to the relative development of this accessory intervertebral articulation of the dorsal vertebrae (their categorization excludes those rebbachisaurids that do not possess hyposphene-hypantrum): (1) a group that comprises *H. boscarollii* and MACN-PV 35 (Apesteguía 2007: Fig. 2), with a well-developed subtriangular hyposphene; (2) a group with a "laminar hyposphene" or "hyposphenal ridge", which includes *C. windhauseni, D. darwini* and *N. taqueti*; and (3) a group with a well-defined infrapostzygapophyseal lamina (apparently, this lamina is the tpol of Wilson and Allain 2015) that comprises *K. goicoecheai* and MMCH-PV 49.

In *Rebbachisaurus garasbae* the prezygapophyses of the posterior dorsal vertebrae are fused to each other (Wilson and Allain 2015: Fig. 4A). The postzygapophyses are also fused to each other forming a simple, angled joint surface (Wilson and Allain 2015: Fig. 4C). Identical condition seems to be present in the mid or posterior dorsal of *Limaysaurus tessonei* (Calvo and Salgado 1995: Fig. 9), in *Katepensaurus goicoecheai* (Ibiricu et al. 2013: Fig. 6–8), and in specimen MMCH-Pv 49 (Haluza et al. 2012: Fig. 2). This particular anatomical condition led Apesteguía et al. (2010) to propose distinctive movement capabilities for the vertebral column of rebbachisaurids optimizing the torsion between successive vertebrae, allowing these sauropods a further mobility on their amphiplatyan centra. However, Wilson and Allain (2015) refused this idea, suggesting that the conjoined zygapophyseal articulations would serve to restrict, rather than facilitate, torsional motion, and coupled with other particular features of the dorsal vertebrae, would have been especially resistant to ventrally directed forces acting on this posterior region of the trunk of the rebbachisaurids.

The neural spine of the middle and posterior dorsal vertebrae of Rebbachisauridae is high, petal-shape, at least in *Rebbachisaurus garasbae* and *Limaysaurus tessonei*, and tetralaminated. This means that it is conformed by four laminae, anterior, posterior, and two lateral laminae, which can be either single or complex, and which normally vary in their conformation in the different taxa.

In *Katepensaurus goicoecheai* the median anterior lamina observed in the mid or posterior dorsal vertebra (UNPSJB-PV 1007/4) is interpreted as a prsl (Ibiricu et al. 2013: Fig. 6), as in the posterior dorsal of *Rebbachisaurus garasbae* (Wilson and Allain 2015: Fig. 5, 8, 9, 10). In *Comahuesaurus windhauseni* the asprl fuse to form an median anterior lamina (amedl in Carballido et al. 2012) (Fig. 4a). Supposedly, this structure is not homologous to the prsl.

In the mid or posterior dorsal of *Limaysaurus tessonei* (specimen MUCPv 206) the posterior median lamina, that is the postspinal lamina, has been interpreted as product of the fusion of the spol (the "suprapostzygapophseal" laminae of Calvo and Salgado 1995: Fig. 9). In the posterior dorsal of *Comahuesaurus windhauseni*, the medial spol fuse distally to form a single median hybrid lamina, the posterior median lamina, which probably, is not homologous to the posl (which, in fact, is present in the anterior dorsals) (Carballido et al. 2012: Fig. 5A, B). In the mid or posterior dorsal vertebrae of *Katepensaurus goicoecheai* there is a posterior median lamina formed by the posl and the m.spol (Ibiricu et al. 2013: Figs. 6, 8). Likeways, in the posterior dorsal vertebra of *Rebbachisaurus garasbae* (MNHN-MRS 1980), the posterior median lamina is formed by the fusion of the posl and the medial spol (Wilson and Allain 2015: Figs. 6, 8B).

In the posterior dorsal vertebra of *Rebbachisaurus garasbae* (MNHN-MRS 1958), the lateral lamina of the neural spine is a composed structure formed by the l.spol and the spdl (Wilson and Allain 2015: Fig. 5). In the mid or posterior dorsal of *Katepensaurus goicoecheai*, the l.spol appears to join the spdl to form the lateral lamina of the neural spine (UNPSJB-PV 1007/5) (Ibiricu et al. 2013: Fig. 6C). In the mid or posterior dorsal vertebra of *Limaysaurus tessonei* (MUCPv 206), the lateral lamina seems to be formed only by the spdl, since the existence of l.spol is not reported (Calvo and Salgado 1995: Fig. 9).

In the posterior dorsal of *Comahuesaurus windhauseni* (specimen MOZ-PV 6722, which is considered to be placed between the 10th and the 12th, Carballido et al. 2012: Fig. 5), the conformation of the lateral lamina of the neural spine is more complex than in other rebbachisaurids. In this species, there is a lamina named alspol (accessory lateral spinopostzygapophyseal lamina) that joins the single spdl to form another lamina, which, in turn, joins distally to the l.spol forming a complex structure that results from the fusion of these three laminae. This particular configuration of the lateral lamina of the neural spine in the posterior dorsal vertebrae of *C. windhauseni* was considered autapomorphic by Carballido et al. (2012).

Last, in some of the dorsal vertebrae of *Rebbachisaurus garasbae* there is a spinoparapophyseal lamina (sppl) that is short and runs parallel and ahead to the spdl (MNHN-MRS 1979 and 1980, Wilson and Allain 2015: Figs. 7B, 8A).

3.2.3 Sacral Vertebrae

The sacral vertebrae of Rebbachisauridae are poorly known (and to date unknown in South American forms), in part because only a few specimens preserve elements from this part of the skeleton. Without doubts, the best-preserved rebbachisaurid sacrum is that of *Tataouinea hannibalis*, which is composed of five vertebrae (Fanti et al. 2013, 2015: Fig. 3). The sacral centra show a variable degree of fusion (only vertebrae 1–2 and 2–3 are fully fused). These are highly pneumatized, and at least the sacral 5 has lateral pneumatopores. The neural arches of the sacral vertebrae of *T. hannibalis* are laminar, as result of the fusion of the prsl with the posl (a condition that is also reported in sacral 1–3 of *Rebbachisaurus garasbae* (Wilson and Allain 2015), and the fusion of the spdl with the sacral rib. The neural spines of the sacral vertebrae of the species from Tunisia have a semi camellate pneumatization structure. Even many of the neural spines have pneumatic foramen (Fanti et al. 2015: Fig. 5). The neural spines of the first three vertebrae are completely coalescing (as in *R. garasbae*): vertebrae 3 and 4 are partially coalescent, especially in their bases, whereas vertebra 5 is free (Fanti et al. 2015: Fig. 6).

3.2.4 Caudal Vertebrae

More or less complete caudal sequences are known in *Zapalasaurus bonapartei* (MOZ-Pv 6127, 17 caudal vertebrae), *Limaysaurus tessonei* (MUCPv-205, 40 caudal vertebrae), and *Lavocatisaurus agrioensis* (MOZ-Pv 1232, 28 caudal vertebrae). In these genera, the caudal centrum length increases significantly toward the end of the sequence. In *Zapalasaurus bonapartei*, for instance, the caudal centra doubled its length toward the caudal 15 of the preserved series with respect to the anterior ones (Salgado et al. 2006) (Fig. 3).

Calvo and Salgado (1995) mention that the caudal centra of *Limaysaurus tessonei* are platycoelous (Calvo and Salgado 1995), whereas for Gallina and Apesteguía (2005) the anterior and mid-caudal centra of *Cathartesaura anaerobica* are amphiplatyan. In turn, in *Zapalasaurus bonapartei* (Salgado et al. 2006), *Itapeuasaurus cajapioensis* (Matos Lindoso et al. 2019), and *Lavocatisaurus agrioensis* (Canudo et al. 2018), the caudal centra are described as amphicoelous.

Anterior Caudals

In *Lavocatisaurus agrioensis*, the articular faces of the anterior caudal vertebrae are concave, being the anterior faces less concave than the posterior ones, a condition that is present in the mid-caudals of this and other rebbachisaurids, although it is not reported in the mid (and posterior) caudals of *L. agrioensis* (in these vertebrae both articular faces are described as concave, Canudo et al. 2018: 684).

In *Limaysaurus tessonei* the lateral faces of the anterior caudal vertebrae are slightly concave, and their ventral faces are rather flat and are traversed by a longitudinal groove. In the anterior caudals of *Itapeuasaurus cajapioensis*, in turn, both the lateral and ventral faces are excavated (Matos Lindoso et al. 2019: Figs. 5–7), whereas in *Zapalasaurus bonapartei* and *Lavocatisaurus agrioensis*, the ventral face is reported to be transversely concave in those same vertebrae.

The anterior caudals of *Limaysaurus tessonei* have transverse processes composed of two laterodorsally projected osseous bars that leave an opening between them, which do not constitute true wing-like processes, as those that possess other diplodocoids (Calvo and Salgado 1995: Fig. 10). Calvo and Salgado (op. cit.) interpreted that the condition observed in *Limaysaurus tessonei* was an autapomorphy of the species, but this kind of transverse process also occurs in the anterior caudals of *Catharthesaura anaerobica*, where they are more laminar, at least more than in the species from Villa El Chocón (Gallina and Apesteguía 2005: Fig. 3A–C). Likeways, in *Itapeuasaurus cajapioensis* the dorsal and ventral components of the transverse processes of the anterior caudals are laminar instead of bar-shaped (Matos Lindoso et al. 2019: Figs. 5, 6). The anterior caudal transverse process were described in *Katepensaurus goicoecheai* and in the Rebbachisauridae indet. UNPSJB-PV 580 (Ibiricu et al. 2012: 224) as wing-like (Ibiricu et al. 2013:1359 and Figs. 9–11; Ibiricu et al. 2015:438 and Fig. 8), and their dorsal edges interpreted as the prdl. In turn, in *Comahuesaurus windhauseni*, the anterior caudal transverse processes were described by Carballido et al. (2012) as being dorsoventrally short, similar to those of *Zapalasaurus bonapartei* and *Limaysaurus tessonei*, and different from those of *Cathartesaura anaerobica*, *Katepensaurus goicoecheai,* and *Rebbachisaurus garasbae* (Carballido et al. 2012: 642).

In Rebbachisauridae, the neural spine of the anterior caudals is high, except in *Zapalasaurus bonapartei* and *Lavocatisaurus agrioensis,* where they are relatively low and not so vertical as in other diplodocoids (Salgado et al. 2006: Fig. 5; Canudo et al. 2018: Fig. A5).

As in the posterior dorsals, the neural spine of the anterior caudals of Rebbachisauridae follows a basic tetralaminated pattern, which in *Limaysaurus tessonei* is present in its simplest configuration: one anterior, one posterior, and two lateral laminae (Calvo and Salgado 1995: Fig. 10). In *Amazonsaurus maranhensis* (Carvalho et al. 2003: Fig. 8), *Limaysaurus tessonei* (Calvo and Salgado 1995: Fig. 10), *Katepensaurus goicoecheai* (Ibiricu et al. 2013: Figs. 9–11), *Cathartesaura anaerobica* (Gallina and Apesteguía 2005: Fig. 3), and *Itapeuasaurus cajapioensis* (Matos Lindoso et al. 2019: Figs. 5, 6), the neural spines of the anterior caudal neural vertebrae exhibit a petal-shape morphology in antero–posterior views, which results from the lateral expansion of the upper third of the lateral laminae (Fig. 2). This peculiar morphology can also be seen in *Rebbachisaurus garasbae* (Wilson and Allain 2015: Fig. 12) and in *Nigersaurus taqueti* (Sereno et al. 2007: Fig. 3). On the contrary, in *Zapalasaurus bonapartei* and *Lavocatisaurus agrioensis*, the neural spines of the anterior caudals are laterally unexpanded (personal observation, L.S.).

In most rebbachisaurids, the osseous laminae that conform the neural spine of the anterior caudals are complex, formed by the fusion of many components. With respect to the anterior and posterior axial laminae, there are differences from one species to another. In *Amazonsaurus maranhensis* (Carvalho et al. 2003: Fig.8A) (Fig. 2), *Limaysaurus tessonei* (Calvo and Salgado 1995: Fig. 10), *Cathartesaura anaerobica* (Gallina and Apesteguía 2005: Fig. 3A–C) (Fig. 6B) and *Katepensaurus goicoecheai* (Ibiricu et al. 2013: Figs. 9–11), the anterior lamina of the neural spine of the anterior caudals is claimed to be constituted only by the prsl. Identical situation is reported in *Rebbachisaurus garasbae* (Wilson and Allain 2015: Fig. 12). In *Itapeuasaurs cajapioensis*, the situation is ambiguous, since the prsl seems to bifurcate ventrally, being continuous with the medial sprl (Matos Lindoso et al. 2019: Figs. 5, 6).

With respect to the posterior lamina, in *Amazonsaurus maranhensis* (Carvalho et al. 2003: Fig. 8B) (Fig. 2b–d) and *Cathartesaura anaerobica* (Gallina and Apesteguía 2005: Fig. 3B) it is only conformed by the posl. In *Limaysaurus tessonei,* apparently, there is a posterior median lamina (called postspinal lamina by Calvo and Salgado 1995) that results from the fusion of both spol (Calvo and Salgado 1995: Fig. 10). In *Rebbachisaurus garasbae*, there is a posl apparently conjoined with both medial spol (Wilson and Allain 2015: Fig. 12), whereas in *Itapeuasaurus cajapioensis* the postspinal lamina is described as continuing down with the medial spol (Matos Lindoso et al. 2019: Figs. 5, 6).

Perhaps the greater variability and complexity are seen in the lateral laminae. For instance, Carvalho et al. (2003) report that, in *Amazonsaurus maranhensis*, the lateral laminae of the neural spine of the anterior caudals are formed, as in other diplodocoids, by the fusion of the sprl and the podl (which are actually the spdl), with the participation of a weakly developed spol (Carvalho et al. 2003: Fig. 8A) (Fig. 2). In turn, in *Katepensaurus goicoecheai,* Ibiricu et al. (2013) describe these laminae (named by them as lateral laminae ll) as being composed by the fusion of sprl and the spol (Ibiricu et al. 2013: Fig. 9B, 10B). The sprl are involved in the formation of the lateral laminae also in the anterior caudals of *Rebbachisaurus garasbae* (Wilson and Allain 2015) and *Cathartesaura anaerobica* (Gallina and Apesteguía 2005). In fact, in the first species, the lateral laminae are formed mainly by the sprl and the spdl, receiving distally the contribution of the weakly developed lateral spol (Wilson and Allain 2015: Fig. 12), and in the second one, the lateral laminae are integrated by the sprl, the spdl, and the l.spol, a conformation considered unique by Gallina and Apesteguía (2005: Fig. 3A) (Fig. 6b). On the contrary, in *Zapalasaurus bonapartei*, the sprl are described as independent structures that run up parallel to the tip of the spine (Salgado et al. 2006). Last, in the anterior caudals of *Itapeuasaurus cajapioensis*, the lateral laminae of the neural spines are interpreted as the spdl, without participation of any other laminae (Matos Lindoso et al. 2019: Figs. 5, 6).

Mid and Posterior Caudals

Mid and posterior caudal vertebrae are known in most South American rebbachisaurids. In general, these sauropods exhibit mid and posterior caudals with centra elongated with respect to the anterior caudal vertebrae. This can be clearly seen in *Zapalasaurus bonapartei* (Fig. 3), *Lavocatisaurus agrioensis,* and *Limaysaurus tessonei*; the three species where this section of the caudal sequence is better known. At least in *Limaysaurus tessonei* (Calvo and Salgado 1995: Fig. 11C, D) and *Lavocatisaurus agrioensis* (Canudo et al. 2018: Fig. 2A8) the distalmost caudal vertebrae are rod-like and with neural arches poorly developed.

In *Lavocatisaurus agrioensis*, *Limaysaurus tessonei* and *Zapalasaurus bonapartei*, the articular faces of the mid and posterior (but not the posteriormost) caudal centra are concave. In the mid-caudal vertebrae of *Comahuesaurus windhauseni*, the posterior articular faces are slightly more concave than the anterior ones (Carballido et al. 2012), and in *L. agrioensis* and *L. tessonei*, the caudal centra of the rod-like posteriormost caudal vertebrae are biconvex (Canudo et al. 2018: Fig. 2A8; Calvo and Salgado 1995: Fig. 11C, D).

The anterior–mid-caudals of *Comahuesaurus windhauseni* have a low longitudinal crest on the lateral faces (MOZ-PV 6634, Carballido et al. 2012: Fig. 8A), which is also observed in the mid-caudals of *Demandasaurus darwini* (Torcida Fernandez-Baldor et al. 2011: Fig. 12; in the Iberian species, in fact, there are two autapomorphic parallel longitudinal crests on the lateral face). In the mid-caudals of *Limaysaurus tessonei*, the lateral faces are slightly concave. In this last genus, the ventral faces of the mid-caudals are flat with a longitudinal groove (Calvo and Salgado 1995). Also, in *Itapeuasaurus cajapioensis* the ventral faces of the mid-caudal centra have a groove or deep notch (Matos Lindoso et al. 2019). On the contrary, the ventral faces of the mid-caudal centra of *Zapalasaurus bonapartei* are described as concave (Salgado et al. 2006), and in *Cathartesaura anaerobica* (Gallina and Apesteguía 2005) and *Lavocatisaurus agrioensis* (Canudo et al. 2018) as flat. In *Comahuesaurus windhauseni* the ventral surface of the mid-caudals is flat to slightly concave (Carballido et al. 2012: 644).

In *Comahuesaurus windhauseni*, the neural arch of the mid-caudal vertebrae is not positioned just on the middle of the centrum, but a little displaced anteriorly (Carballido et al. 2012: Fig. 8), as in *Limaysaurus tessonei* (Calvo and Salgado 1995: Fig. 11A) and *Lavocatisaurus agrioensis* (Canudo et al. 2018: Fig. 2), although not so much as in titanosaurs.

Prezygapophysis length is variable in mid and posterior caudals of Rebbachisauridae. For instance, in *Comahuesaurus windhauseni*, the prezygapophysis of these vertebrae are shorter than in *Zapalasaurus bonapartei* and *Limaysaurus tessonei*, and do not exceed the anterior margin of the anterior articular face (Carballido et al. 2012).

The neural spines of the mid-caudal vertebrae of *Limaysaurus tessonei* are wide and posteriorly projected (Calvo and Salgado 1995: Fig. 11A). In the posterior caudal vertebrae of this species (which was labeled as mid-posterior caudal by Calvo and Salgado 1995: Fig. 11B) the neural spine becomes less laminar and dorsoposteriorly projecting.

All caudal neural spines of *Zapalasaurus bonapartei* are relatively short and posteriorly inclined (Fig. 3). In this species, the neural spine becomes low and elongated anteroposteriorly toward caudal vertebrae 10–11. In turn, in *Amazonsaurus maranhensis,* the posterior caudal vertebrae have neural spines elongated and posteriorly inclined (Carvalho et al. 2003: Fig. 12). In *Itapeuasaurus cajapioensis*, the mid and posterior neural spines seem to be more vertical than in other rebbachisaurids (Matos Lindoso et al. 2019: Figs. 8, 9).

3.2.5 Appendicular Skeleton

Scapula

Complete or nearly complete scapulae are known in many South American rebbachisaurids: *Limaysaurus tessonei* (Calvo and Salgado 1995: Fig. 12A), *Rayososaurus agrioensis* (Carballido et al. 2010: Fig. 3), *Cathartesaura anaerobica* (Gallina and Apesteguía 2005: Fig. 4A), *Lavocatisaurus agrioensis* (the juvenile specimen MOZ-Pv 1255, Canudo et al. 2018: Fig. 2B) and specimen MMCH-Pv 49 (Haluza et al. 2012: Fig. 2H–J). The scapula of the South American rebbachisaurids is typically paddle or racquet-like, coincident with that of *Rebbachisaurus garasbae*, which has been recognized as a synapomorphy of Rebbachisauridae (Wilson 2002; Salgado et al. 2004; Sereno et al. 2007). The acromion is described as acute and directed upward in *Limaysaurus tessonei* (Calvo and Salgado 1995: Fig. 12A), posteriorly directed in *R. agrioensis* (Carballido et al. 2010: Fig. 3) and dorsally directed in *C. anaerobica* (Gallina and Apesteguía 2005: Fig. 4). In MMCH-Pv 49 the acromion process is described with a hook-like morphology (Haluza et al. 2012: Fig. 2I). The scapular blade is reported to be D shaped in cross-section in *R. agrioensis* (Carballido et al. 2010), *R. garasbae* (Wilson and Allain 2015), and *L. agrioensis* (Canudo et al. 2018).

A posterodorsally oriented acromion together with the expansion of the dorsal margin of the scapular blade (coincident with *Camarasaurus* and *Brachiosaurus*) produces the shortening of the distance between these two zones, which results in two morphologies: V-shaped (an ambiguous synapomorphy of Rebbachisauridae present in *Rebbachisaurus garasbae, Rayososaurus agrioensis, Cathartesaura anaerobica,* and *Lavocatisaurus agrioensis* (Carballido et al. 2010), and U-shaped, considered by Carballido et al. (2010) a derived condition present in *Limaysaurus tessonei* (in spite that Calvo and Salgado 1995, considered a V-shape angle between the acromion and the scapular blade as an autapomorphy of "*Rebbachisaurus*").

Coracoid

Among South American rebbachisaurids, coracoids are only known in *Limaysaurus tessonei* (Calvo and Salgado 1995: Fig. 12A), *Comahuesaurus windhauseni* (Salgado et al. 2004: Fig. 5A) and in MMCH-Pv 49, but in the last one the bone is only preserved in the area surrounding the glenoid fossa (Haluza et al. 2012: Fig. 2H–J). The glenoid cavity is thick and, at least in *Limaysaurus tessonei* (the condition is not observable due to bad preservation of specimen PV 6763 MOZ of *C. windhauseni*), the coracoid foramen is oval and closed.

Humerus

Among South American rebbachisaurids, the only species preserving femur and humerus from a single individual is *Limaysaurus tessonei* (holotype, MUCPv-205). Here, the humerus is relatively short, representing less than 70% of the femur length (Calvo and Salgado 1995: Fig. 12B). The holotype of *Rebbachisaurus garasbae* includes a right humerus (MNHN-MRS 1476, 1477, 2001, 2002), lacking its distal and proximal ends. Humeral robustness (shaft minimal perimeter-length ratio) is 0.43–0.45 in *R. garasbae*, based on the range of values estimated for the length of the humerus, a bit higher than that calculated for *L. tessonei* by Salgado et al. (2004), whose humeral robustness is of 0.42 (it must be underlined that the humeral robustness is not the robustness index considering in other studies, such as that of Carballido et al. 2012). In *Comahuesaurus windhauseni*, the humerus (MOZ-Pv 6762) is even more robust than in *R. garasbae* (Humeral robustness = 0.53, Salgado et al. 2004: Fig. 5 B, C). Among diplodocoids, only dicraeosaurids have relatively robust humeri (Carballido et al. 2012). In *C. windhauseni*, the humeral shaft is elliptical in cross-section, that is anteroposteriorly compressed, as in MMCH-Pv 49 (Haluza et al. 2012: Fig. 2K–M), instead of circular as in *L. tessonei* (Calvo and Salgado 1995; Wilson 2002; Salgado et al. 2004), or subcircular to circular as in *R. garasbae* (Wilson and Allain 2015). The holotype of *Cathartesaura anaerobica* (MPCA 232) is also integrated by a humerus, but it is badly preserved and unprepared (Gallina and Apesteguía 2005). In turn, in *Lavocatisaurus agrioensis,* the midshaft cross-section of the badly preserved humerus is elliptical (Canudo et al. 2018), although the proportions of the humeral midshaft can vary along the ontogeny, becoming more subcircular in largest specimens (Wilson and Allain 2015). In *L. agrioensis*, the humeral articular head is more developed than in *C. windhauseni* (Canudo et al. 2018).

Sternal Plate

Sternal plates were described in only one rebbachisaurid: *Limaysaurus tessonei* (referred specimen MUCPv 206). These are semilunar, with the lateral border concave and the medial one convex (Calvo and Salgado 1995). This shape, convergently acquired in some titanosaurs, was considered autapomorphic by Calvo and Salgado (1995). In *Comahuesaurus windhauseni*, there is a sternal plate among the referred specimens (MOZ-Pv 6717), but it was not described.

Radius and Ulna

The radius of *Limaysaurus tessonei* is elongated and oval in cross-section (Calvo and Salgado 1995: Fig. 12D). The ulna is proximally triradiate as in other sauropods. The ulna/tibia length ratio is 0.78 (Calvo and Salgado 1995). In *Lavocatisaurus agrioensis* there were preserved the left (MOZ-Pv 1256) and right (MOZ-Pv 1243) ulnae (Salgado et al. 2012: Fig. 8; Canudo et al. 2018). They are gracile and with their proximal end triradiate. Unlike titanosaurs, the olecranon process is low, which makes the proximal articulation nearly horizontal (Salgado et al. 2012).

Metacarpals

The only metacarpals described in a rebbachisaurid are those from the specimen MUCOv-206 assigned to *Limaysaurus tessonei* by Calvo and Salgado (1995), which were described but not illustrated. Evidence brought by these authors (op. cit.) points out that the Metacarpal II/III-Radius length ratio would have been lesser than 0.45 (Calvo and Salgado 1995: 24).

Ilium

Ilia are known in *Limaysaurus tessonei* (Calvo and Salgado 1995: Fig. 13), *Amazonsaurus maranhensis* (Carvalho et al. 2003: Fig. 17), and *Zapalasaurus bonapartei* (Salgado et al. 2006: Fig. 7). In the holotype of *L. tessonei* (MUCPv 205) the pubic peduncle of the ilium is laterotransversely wider than anteroposteriorly long, as is usual in sauropods (Calvo and Salgado 1995: Fig. 13). In *L. tessonei*, the higher point of the acetabulum is displaced anteriorly and the ischiatic peduncle is poorly developed, as in other sauropods. The left ilium of the holotype of *A. maranhensis* is also preserved (UFRJ-DG 58-R/1) (Carvalho et al. 2003: Fig. 17). Here, the anterior corner of the ilium is directed upward and forward, although this part of the bone cannot be seen in *L. tessonei*. Carvalho et al. (2003) reported that the internal structure of the ischiatic peduncle of *A. maranhensis* bears a pneumatic cavity, a feature which they consider unique among sauropods. These iliac chambers are also present in the Tunisian rebbachisaurid *Tataouinea hannibalis*, and could represent a synapomorphy of Rebbachisaridae (Fanti et al. 2015).

The holotype of *Zapalasaurus bonapartei* (Pv-6127-MOZ) has only preserved one fragment of ilium (Salgado et al. 2006: Fig. 7). Likewise, an acetabular portion of an ilium was reported in *Comahuesaurus windhauseni* (Pv-6765-MOZ) (Salgado et al. 2004, mentioned as Pv-6675-MOZ in Carballido et al. 2012), but this element was not described nor illustrated.

Ischium

Ischia are known in many rebbachisaurids. In the basally branching taxon *Zapalasaurus bonapartei* the proximal portion of the bone is laminar, and the distal one is thin and oval in cross-section (Salgado et al. 2006: Figs. 6, 7). In turn, in *Limaysaurus tessonei*, the ischiatic shaft is long, more flattened than in the rebbachisaurid from La Amarga Formation, and twisted at the level of the symphysis with its counterpart, which produces a nearly coplanar surface (Calvo and Salgado 1995: Figs. 13, 14). This condition is presumably the same in *Rebbachisaurus garasbae* (Wilson and Allain 2015: Fig. 15) and *Demandasaurus darwini* (Pereda Suberbiola et al. 2003: Fig. 3A–C). In *L. tessonei*, this can be seen in the referred specimen MUC-Pv 153 (Calvo and Salgado 1995: Fig. 14). In *Comahuesaurus windhauseni*, there is preserved a fragment of ischium (Pv-6713-MOZ) (Salgado et al. 2004: Fig. 5E) and several more fragments (Carballido et al. 2012), which are similar to that of *L. tessonei*. In *C. windhauseni* and more derived rebbachisaurids (Carballido et al. 2012) the central part of the acetabular portion of the ischium is narrow. In *L.*

tessonei as in *Demandasaurus darwini* and *Nigersaurus taqueti*, and unlike *C. windhauseni* and *Z. bonapartei*, the iliac peduncle of the ischium has a constriction of variable development.

Pubis

Amazonsaurus maranhensis has preserved two fragments of a right pubis with a paddle-like appearance (Carvalho et al. 2003: Fig. 16). The holotype of *Zapalasaurus bonapartei* has preserved a relatively complete left pubis, which is laminar, unlike other rebbachisaurids such as *Limaysaurus tessonei* and *Comahuesaurus windhauseni* (Salgado et al. 2006: Fig. 6). In fact, in *L. tessonei* (Calvo and Salgado 1995: Fig. 13) and *C. windhauseni* (Pv-6743-MOZ, right pubis) (Salgado et al. 2004; Fig. 5; Carballido et al. 2012; Fig. 10), the shaft is rather massive, oval in cross-section, with the origin scar for the *M. ambiens* weakly developed, unlike flagellicaudatans, and with its distal portion expanded. Both in *L. tessonei* and *C. windhauseni*, the *obturator* foramen is open, a condition considered by Salgado et al. (2004) as diagnostic only for *Limaysaurus* (genus to which the materials from Cerro Aguada del León were assigned by these authors), but later questioned by Carballido et al. (2012).

Femur

The holotype of *Limaysaurus tessonei* has preserved the complete left femur. It is robust and straight, with the lateral bulge not so prominent as in titanosauriforms (Calvo and Salgado 1995: Fig. 15A), which somehow coincides with the morphology observed in *Cathartesaura anaerobica* (Gallina and Apesteguía 2005) and *Demandasaurus darwini* (Pereda Suberbiola et al. 2003: Fig. 3). In *Comahuesaurus windhauseni*, the midshaft of the femur is elliptical in cross-section, and the lateral bulge is virtually nonexistent (Salgado et al. 2004; Fig. 5G; Carballido et al. 2012). As in *L. tessonei* (Calvo and Salgado 1995), *C. anaerobica* (Gallina and Apesteguía 2005) and *D. darwini* (Pereda Suberbiola et al. 2003), the fourth trochanter of *C. windhauseni* is reduced to a low ridge (Carballido et al. 2012). *Rayososaurus agrioensis* preserved the right femur, and only the proximal end and the fibular condyle are missing. In this species, the fourth trochanter is hardly developed (Carballido et al. 2010). So, the poorly developed fourth trochanter is a character shared within other rebbachisaurids.

Tibia and Fibula

The holotype of *Limaysaurus tessonei* (MUCPv-205) has preserved both tibiae and fibulae (Calvo and Salgado 1995: Fig. 15). The tibia/femur length ratio is nearly 0.55. Both tibiae are somewhat deformed, but it can be seen that the proximal articulation is ovoid, anteroposteriorly expanded, much more than its distal end, which is transversely expanded. *Comahuesaurus windhauseni* has preserved a proximal end of a right tibia (Pv-6764.MOZ) (Salgado et al. 2004: Fig. 5F). Unlike *L. tessonei*, the cnemial crest of *C. windhauseni* is low, but, similar to that species, the proximal articulation is ovoid. The holotype of *Zapalasaurus bonapartei* (Pv-6127-MOZ) has preserved the complete left tibia (Salgado et al. 2006: Fig. 8C). Unlike *L. tessonei*, the species from La Amarga Formation has both ends of the tibia equally expanded in anteroposterior sense. Although the proximal articulation is reported as ovoid, Salgado et al. (2006) consider the possibility that this is the result of deformation.

The paratype of *Lavocatisaurus agrioensis* includes a left tibia of a juvenile specimen (MOZ-Pv 1244) (Salgado et al. 2012: Fig. 9A–E; Canudo et al. 2018: Fig. 2A9) (Fig. 5h–l). The general proportions of this element resemble those of *Zapalasaurus bonapartei* (Salgado et al. 2006: Fig. 8C). As in other rebbachisaurids, the proximal articulation of the tibia of *L. agrioensis* is anteroposteriorly expanded, and oval instead of subcircular as in more derived diplodocoids, although Salgado et al. (2012) do not reject that this can be due to the early ontogenetic condition of the specimen. The anteroposterior expansion of the proximal articulation of the tibia reaches 30% of the total length of the bone in *L. agrioensis*. In this species, the cnemial crest of the tibia is lower and more angulated than in more derived diplodocoids (Salgado et al. 2012). In turn, the distal end of the tibia is triangular, so anteroposteriorly long as transversely wide (Salgado et al. 2012: Fig. 9B).

Fibula

The fibula of *Limaysaurus tessonei* (MUCPv-205) is sigmoid in lateral view, with both ends slightly expanded (Calvo and Salgado 1995: Fig. 15C). The left fibula that forms part of the paratype of *Lavocatisaurus agrioensis* (MOZ-Pv 1245) is expanded at both ends, a bit more than in *L. tessonei* (Salgado et al. 2012: Fig. 9F–J) (Fig. 5). *Rayososaurus agrioensis* has preserved a fibula but is badly damaged (Carballido et al. 2010).

Metatarsals and Phalanges The holotype of *Limaysaurus tessonei* (MUCPv-205) has preserved metatarsals II, IV, and V from both pes, and Metatarsal I only on the right foot (Calvo and Salgado 1995: Fig. 16). The paratypes of *Lavocatisaurus agrioensis* also comprise elements of the metatarsus: MOZ-Pv 1257, right Metatarsal I, and MOZ-Pv 1258,? Metatarsal IV (Salgado et al. 2012: Fig. 10; Canudo et al. 2018). As a noticeable character, the Metatarsal I in both species lack the typical laterodistal process observed in diplodocids (McIntosh 1990).

4 Rebbachisaurid Phylogenetic Relationships: A Historical Approach

Calvo and Salgado (1995) were the first to include a rebbachisaurid in a phylogenetic analysis, recovering *Limaysaurus tessonei* as the sister group of Diplodocidae (a group that, according to them, included Dicraeosaurinae plus Diplodocinae), recognizing 11 synapomorphies uniting that species with this group of familial rank. In turn, these authors erected Diplodocimorpha "to include *Rebbachisaurus tessonei* sp. nov. and all other Diplodocidae presenting characters 1–11." Later, Taylor and Naish (2005), although they did not provide a phylogenetic analysis of their own, modified this definition partially, providing the first formulation of the current definition: *Diplodocus* plus *Rebbachisaurus*.

Wilson (2002) was the first one in analyzing various rebbachisaurid taxa at the same time: *Rebbachisaurus*, *Nigersaurus*, and "*Rayososaurus*". On this last, it must be said that, in scoring "*Rayososaurus*", Wilson used *Limaysaurus tessonei* instead of *Rayosaurus agrioensis*, following a taxonomical interpretation based solely on scapular similarities (Wilson and Sereno 1998) that did not later prosper. Anyway, he could not progress in the resolution of the internal relationships of Rebbachisauridae, because he obtained a polytomy at the base of the clade (Wilson 2002: Fig. 13). He recognized 8 synapomorphic characters of the Rebbachisauridae, from which five were cranial: 1, orbital ventral margin rounded (reversal); 2, postorbital lacks posterior process; 3, frontal elongate anteroposteriorly, approximately twice transverse breadth (reversal); 4, supratemporal fenestra reduced or absent; 5, teeth with longitudinal grooves on lingual aspect; 6, "petal"-shaped posterior dorsal neural spines; 7, racquet-shaped scapular blade (Lavocat 1954; Calvo and Salgado 1995); 8, humerus with circular midshaft cross-section.

Rebbachisauridae was first defined by Salgado et al. (2004) as the group that contains diplodocoids more closely related to *Rebbachisaurus* than to *Diplodocus*. In the analysis of Salgado et al. (2006), *Zapalasaurus bonapartei* is the sister taxon to Diplodocimorpha (Salgado et al. 2006: Fig. 9), although *Amazonsaurus maranhensis* is recovered as a rebbachisaurid of uncertain affinities, forming a polytomy with the remaining members of the clade.

The analysis of Gallina and Apesteguía (2005) recovered *Limaysaurus tessonei* as the sister taxon of *Nigersaurus taqueti* (Gallina and Apesteguía 2005: Fig. 5), although with a low support, in a placement that was not maintained in posterior analyses.

Sereno et al. (2007) recovered *Zapalasaurus bonapartei* in a position more derived with respect to other analyses, as the sister group of *Limaysaurus tessonei* plus *Cathartesaura anaerobica* (Sereno et al. 2007: Fig. 4). Their analysis did not include *Amazonsaurus maranhensis*.

Whitlock (2011) was the first to recognize two subclades within Rebbachisauridae: the South American Limaysaurinae, integrated by *Limaysaurus* and *Cathartesaura* (and defined by him as *Limaysaurus,* not *Nigersaurus*), and the more cosmopolitan Nigersaurinae (defined as *Nigersaurus,* not *Limaysaurus*), integrated by *Nigersaurus*, the Spanish rebbachisaurid (*Demandasaurus*) and *Zapalasaurus.* Beyond the controversial position of *Zapalasaurus* assumed by this author, his analysis recovered *Histriasaurus* as the most basal Rebbachisauridae (a position that is maintained in current analyses), and *Rebbachisaurus* as the sister group of the later called Khebbaschia (Whitlock 2011: Fig. 7).

Torcida Fernandez-Baldor et al. (2011), employing the same matrix of Sereno et al. (2007), recovered *Zapalasaurus bonapartei* as the sister group of *Nigersaurus taqueti* plus *Demandasaurus darwini,* in a position derived with respect to other analyses (Torcida Fernandez-Baldor et al. 2011: Fig. 14). Also, they recovered *Rebbachisaurus garasbae* close to *Cathartesaura anaerobica,* instead of related to *Demandasaurus darwini* and *Nigersaurus taqueti* as in other analyses.

In Carballido et al. (2012), *Amazonsaurus maranhensis* is the most basal rebbachisaurid; *Comahuesaurus windhauseni* is the sister group of Limaysaurinae

(*Rayososaurus agrioensis*, *Rebbachisaurus garasbae*, *Cathartesaura anaerobica* and *Limaysaurus tessonei*) plus Nigersaurinae (*Demandasaurus darwini* and *Nigersaurus taqueti*), while *Zapalasaurus bonapartei* is placed together with *Histriasaurus boscarollii* forming a polytomy with *C. windhauseni* plus MDR (more derived rebbachisaurids) (Carballido et al. 2012: Fig. 12). Perhaps one of the greatest differences between the analysis of Carballido et al. (2012) and the previous ones is the falling down of *Zapalasaurus bonapartei* to more basal positions. The analyses of Wilson and Allain (2015), as well as that of Canudo et al. (2018), basically agree with that of Carballido et al. (2012), except that, in the first two analyses, *Rebbachisaurus garasbae* is placed within Rebbachisaurinae instead of Limaysaurinae (see below).

Ibiricu et al. (2013) did not perform a proper phylogenetic analysis but they followed Whitlock (2011) to evaluate the phylogenetic relationships of *Katepensaurus goicoecheai*, placing it in the Rebbachisauridae and, within them, in the Limaysaurinae subclade. This preliminary evaluation was confirmed by Ibiricu et al. (2015), whose phylogenetic analysis resulted in (1) *Amazonsaurus maranhensis* as the most basal rebbachisaurid; (2) a polytomy made up of *Histriasaurus boscarollii, Rebbachisaurus garasbae, Zapalasaurus bonapartei,* and Khebbashia, and (3) a polytomy at the base of Limaysaurinae integrated by *Katepensaurus goicoecheai, Cathartesaura anaerobica,* and *Limaysaurus tessonei* (strict consensus tree, Ibiricu et al. 2015: Fig. 10).

Rebbachisaurinae was defined by Wilson and Allain (2015) as all rebbachisaurids more closely related to *Rebbachisaurus garasbae* than to *Limaysaurus tessonei*. In this way, Nigersaurinae is a synonymous junior and Rebbachisaurinae has priority.

In Fanti et al. (2015), *Katepensaurus goicoecheai* is not a Limaysaurinae but the most basal Rebbachisaurinae (the new name that replaces Nigersaurinae: the sister group of limaysaurines). The strict consensus under implied weighting of the parsimony analysis of Fanti et al. (2015) also placed *Amazonsaurus maranhensis* at the base of Rebbachisauridae, and *Zapalasaurus bonapartei*, *Histriasaurus boscarollii*, *Comahuesaurus windhauseni,* and Khebbashia forming a polytomy (Fanti et al. 2015: Fig. 21). Khebbashia is the group that embraces Limaysaurinae and Rebbachisaurinae, defined as the least inclusive group containing *Limaysaurus tessonei*, *Nigersaurus taqueti* and *Rebbachisaurus garasbae.* According to Fanti et al. (2015), Limaysaurinae is exclusively South American and includes *L. tessonei* and *Cathartesaura anaerobica.* As said, *Katepensaurus goicoecheai* is the most basal Rebbachisaurinae, the sister group of *Nigersaurus taqueti* plus the trichotomy integrated by *Tataouinea hannibalis*, *Demandasaurus darwini* and *Rebbachisaurus garasbae.* As can be seen, in this analysis, all rebbachisaurines except *Katepensaurus goicoecheai* are Euro-African.

Regarding the Bayesian analysis of the same authors, it differs in that *Amazonsaurus maranhensis* and *Comahuesaurus windhauseni* are recovered as basal limaysaurines (Fanti et al. 2015: Figs. 22, 23). In contrast, in the parsimony analysis *A. maranhensis* is a basal rebbachisaurid and *C. windhauseni* is outside Khebbashia. The Bayesian analysis of Fanti et al. (2015) calculated simultaneously topology and the estimated time of cladogenesis. This analysis estimated the

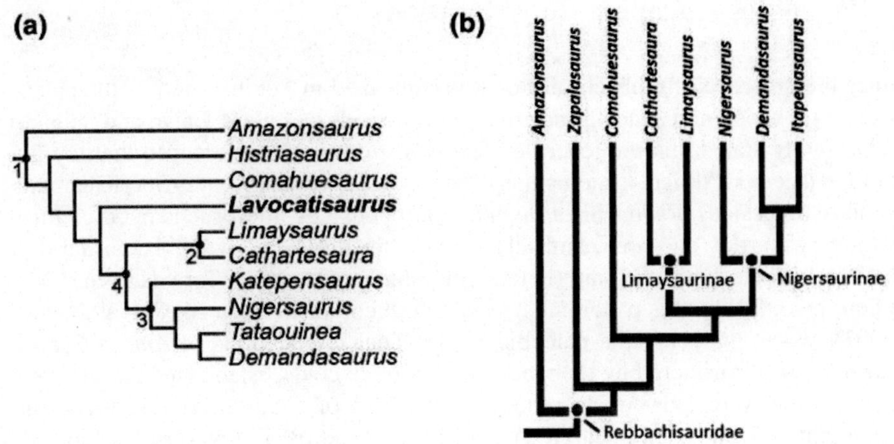

Fig. 7 Phylogenetic relationships of Rebbachisauridae. **a** Reduced consensus tree after pruning *Rayososaurus*, *Zapalasaurus*, and *Rebbachisaurus* after Canudo et al. (2018). Resolved clades are indicated with numbers: 1, Rebbachisauridae; 2, Limaysaurinae; 3, Rebbachisaurinae; 4, Khebbashia (modified from Canudo et al. 2018: Fig. 4). **b** Resulting topology of the agreement subtree analysis, excluding *Histriasaurus* and *Rebbachisaurus,* after Matos Lindoso et al. (2019) (modified from Matos Lindoso et al. op. cit.: Fig. 11)

origin of Rebbachisauridae by the 163 Ma, in the Oxfordian, and the divergence Limaysaurinae-Rebbachisaurinae by the 134 Ma, in the Hauterivian.

Canudo et al. (2018) placed *Lavocatisaurus agrioensis* between *Comahuesaurus windhauseni* and Khebbashia. As in practically all the analyses, *C. windhauseni* was recovered occupying an outer position with respect to Khebbashia (Fig. 7b). In this analysis, Rebbachisauridae was supported by the same four postcranial synapomorphies obtained by Wilson and Allain (2015): (1) the hyposphenal ridge on anterior caudal vertebrae (reversed in *Demandasaurus* and *Tataouinea*); (2) presence of spinodiapophyseal lamina (convergently present in lognkosaurian titanosaurs; (3) middle caudal vertebrae with a flat ventral surface; and 4) middle caudal vertebrae with anterodorsally oriented prezygapophyses (reverted in *Lavocatisaurus* and more derived rebbachisaurids (MDR)).

Finally, the analysis of Matos Lindoso et al. (2019) recovered *Nigersaurus taqueti* as the most basal Nigersaurinae (following to Mannion et al. 2019), as in Canudo et al. (2018) and *Comahuesaurus windhauseni*, *Zapalasaurus bonapartei,* and *Amazonsaurus maranhensis* as successive sister taxa of Khebbashia (Matos Lindoso et al. 2019: Fig. 11) (Fig. 7b).

5 Paleobiogeographic Considerations

Since the presence of rebbachisaurids was confirmed in South America, the paleobiogeographic significance of this group of sauropods was clear. Calvo and Salgado (1995) early stated that the record of *Limaysaurus tessonei* evidenced that, by the mid-Cretaceous (Albian–Cenomanian), there was an African-South American fauna composed, besides *Rebbachisaurus*-like sauropods, by mesosuchian crocodiles, araripemyd turtles, and coelacanths fishes. With the discovery of a rebbachisaurid in the Lower Cretaceous of Spain (Pereda Suberbiola et al. 2003; Torcida Fernandez-Baldor et al. 2011), this panorama became more complex. Even so, Carvalho et al. (2003), when analyzing the paleobiogeographical implications of *Amazonsaurus maranhensis* (considered by them as a basal diplodocoid), established that the new species they were introducing reinforced the idea of a South American-African community of dinosaurs integrated by titanosaurs and basal diplodocoids (not just rebbachisaurids) among sauropods, carcharodontosaurs, and spinosaurs among the theropods. Citing Canudo and Salgado (2003), they were prone to explain the virtual absence of rebbachisaurids in Laurasia as an effect of regional extinction. In turn, Gallina and Apesteguía (2005) put the focus more on the significance of rebbachisaurids as part of a global or Gondwanan fauna that became extinct by the "mid" Cretaceous. In this way, they rescued previous considerations that viewed the rebbachisaurids as the latest diplodocoids (which was unclear by that time, see Wilson 2002; Apesteguía et al. 2010), as part of a sauropod association that was replaced by other, integrated mainly by titanosaurs, which developed convergent characteristics with rebbachisaurids, mostly in the skull.

Just as Calvo and Salgado (1995) used the record of a South American rebbachisaurid to support the idea that South America and Africa were connected by the mid-Cretaceous by a land bridge connection, Canudo et al. (2009), based on the record in the Iberian Peninsula of an Early Cretaceous rebbachisaurid, promoted the existence of a land bridge connection between Africa and Europe at the late Barremian–early Aptian, although not ruling out an earlier dispersion from Africa.

The paleobiogeographic scenario of the rebbachisaurid diversification got complicated by the fact that the phylogenetic analyses recognized the existence of two well-defined clades within Rebbachisauridae: the South American Limaysaurinae, and the more cosmopolitan Rebbachisaurinae (formerly called Nigersaurinae) (which, in Whitlock's analysis of 2011, include *Zapalasaurus bonapartei*). That of Whitlock (op. cit) was the first attempt to get a global picture of the paleobiogeographic history of rebbachisaurids within the context of Diplodocoidea. Obviously, different phylogenies give rise to different paleobiogeographic interpretations. In this regard, the position assumed for *Amazonsaurus maranhensis* as a basal diplodocoid, made Whitlock propose that perhaps the Brazilian genus is the result of a late dispersal from North America (Whitlock 2011: 897). In the Cretaceous, the connection between Africa and Europe remained well after the breakup of Gondwana. In this case, for him, the estimated time of the split between, on one hand, the African-European rebbachisaurids, and, on the other hand, the South American rebbachisaurids, is much earlier (Hauterivian–Valanginian) than in other vicariant models based on different groups of dinosaurs, which would place the opening of the South Atlantic by the middle of the Cretaceous (Whitlock 2011: 897).

Mannion et al. (2012), based on the basal position of *Amazonsaurus maranhensis* and the instability of *Zapalasaurus bonapartei*, questioned the simplistic vicariant model of previous authors. In turn, Carballido et al. (2012) performed a complete biogeographic scenario of Rebbachisauridae using a Dispersal-Extinction-Cladogenesis analysis. This analysis recovered a long ghost lineage from the Late Jurassic, where these authors place the origin of the group (which, on the other hand, would have taken place in South America), up to the Hauterivian–Barremian, where is the oldest record (*Histriasaurus boscarollii*), and once originated, by the Early Cretaceous, basal forms of the group rapidly expanded into Africa and Europe. Carballido et al.'s analysis has not been able to establish the distribution of limaysaurines and rebbachisaurines (nigersaurines) due to the lack of resolution between these two clades.

In the scenario postulated by Carballido et al. (2012), South America would have witnessed early diversifications and local extinctions, including a local extinction in the early evolutionary stages of rebbachisaurines. South America would also be the place of origin of the clade integrated by *Comahuesaurus windhauseni* plus more derived rebbachisaurids. Based on the possible topologies that would result from the resolution of the existing trichotomy between *Rayososaurus agrioensis*, *Rebbachisaurus garasbae*, and Limaysaurine, Carballido et al. (2012: Fig. 14) contemplate a possible scenario in which South America experienced a local extinction prior to the origin of the Limaysaurinae plus Nigersaurine clade (Khebbashia), and a late expansion of Limaysaurinae by the late Early Cretaceous. The alternative scenario is a wider expansion and a local extinction in South America in the earliest stages of the evolution of the rebbachisaurines.

Ibiricu et al. (2015: Fig. 10) recovered *Katepensaurus goicoecheai* as a limaysaurine, postulating that limaysaurines were already diverse by the middle Cretaceous of southern South America. Ibiricu et al. (2015) supported that much of the early history of rebbachisaurids took place in Gondwana, more specifically in South America and Africa (considering *Histriasaurus boscarollii* as an immigrant from Africa).

Wilson and Allain (2015) agree with the vicariant scenario postulated by Sereno et al. (2007). The recovery of *Rebbachisaurus garasbae* within Rebbachisaurinae simplifies the scheme, in considering this subfamily as European and African, whereas Limaysaurinae is eminently South American. This model lacks local extinctions and dispersal events at the level of Khebbaschia, as does the model proposed by Carballido et al. (2012).

According to Fanti et al. (2015), *Katepensaurus* is the most basal Rebbachisaurinae, a clade that is widely distributed. In contrast, Limaysaurinae is exclusively from South America. For these authors, the rapid expansion and diversification of rebbachisaurids would have occurred between 135 and 130 Ma.

Fanti et al.'s BBM analysis supported South America as the most plausible ancestral area for Khebbashia, and the exclusive range of Limaysaurinae evolution and for the origin and early diversification of Rebbachisaurinae. The Diva tests coincided with BBM, with the difference that the ancestral area of early rebbachisaurine evolution is South America and Africa, rather than just South America.

With respect to the possible place of origin of rebbachisaurids, it must be said that, if it is confirmed that *Xenoposeidon proneneukos*, from the Ashdown Beds Formation (Berriasian–Valanginian), England (Taylor and Naish 2007), is a rebbachisaurid, as Taylor (2018) claims, it would be the oldest known member of Rebbachisauridae (not considering *Maraapunisaurus fragillimus*). Taylor (2018) assigned *X. proneneukos* to the family Rebbachisauridae based particularly in the "M"-shaped complex of laminae on the lateral faces of the neural arch, which is, on the other hand, diagnostic in *Rebbachisaurus garasbae*. In any case, future works focused on the relationships of Rebbachisauridae and inclusive of this taxon will likely influence the inferred center of origin for the clade (Whitlock and Wilson Mantilla 2020). At present, a South American origin for Rebbachisauridae seems to be the better-supported hypothesis as it was implied by the relationships of the group retrieved by several authors such as Carballido et al. (2010), Whitlock (2011), and Wilson and Allain (2015) mentioned above.

Recently, Carpenter (2018) kicked the chessboard of the paleobiogeography of the Rebbachisauridae adding the model to *Maraapunisaurus fragillimus,* a North American species from the Late Jurassic. This author supposes a North American origin for the clade, and a dispersal to Europe in the latest Jurassic and the earliest Cretaceous, and thereafter, to South America via Africa. The coeval presence of *Maraapunisaurus fragillimus* and diplodocids supports the hypothesis that the split of the Diplodocoidea into Diplodocidae and Rebbachisauridae occurred by the Middle Jurassic. However, as commented by Whitlock and Wilson Mantilla (2020), the hypothesis that this specimen is a rebbachisaurid merits close scrutiny, which represents a challenge because the specimen was lost and only a single figure, in posterior view, exists.

A new, more complex version of the vicariant, provincialistic model was supported by Matos Lindoso et al. (2019), based on the fact that they recovered *Itapeuasaurus cajapioensis* within the Rebbachisaurinae, unlike other South American rebbachisaurids which are or limaysaurines (*Limaysaurus* and *Cathartesaura*) or successive sister groups to Khebbashia (*Comahuesaurus, Zapalasaurus,* and *Amazonsaurus*). On these bases, they interpret this position as coherent with the hypothesis that North Brazilian dinosaur faunas recorded in the Alcântara Formation have more in common with Africa than with South America. According to these authors, citing work of Maisey (2011), the geographic barrier responsible for the first vicariance event between Limaysaurinae and Rebbachisaurinae was a hypothetical epicontinental seaway that separated the northeastern Brazil (still connected with Africa) from the rest of South America. It is important to highlight that Matos Lindoso et al. (2019) are not included in their phylogenetic analysis taxa such *Tataouinea hannibalis, Lavocatisaurus agrioensis,* and *Katepensaurus goicoecheai*. In this way, the information is not complete to conclude paleobiogeographic events.

6 Conclusions

South America is the continent where rebbachisaurid record is most abundant, with species ranging from the Lower Cretaceous to the Lower Upper Cretaceous. Recent phylogenetic analyses coincide in recognizing two groups of rebbachisaurids with different paleobiogeographic histories: rebbachisaurines and limaysaurines. In most recent analyses, *Amazonsaurus maranhensis* and *Zapalasaurus bonapartei* appear to persist in basal positions, as does *Comahuesaurus windhauseni.* Although the differentiation of the group occurred very early in the Late Jurassic, the evidence suggests that the main modifications in the body design of the group were established early on. Rebbachisaurid record has proven to be important to recognize dispersal events between different parts of Gondwana (Africa and South America) or between Laurasia and Gondwana.

Acknowledgements This contribution was financiated with funds from Project CGL2017-85038-P of the Spanish Ministerio de Economía y Competitividad-ERDF, as well as by the Aragón Regional Government ("grupo de Referencia Aragosaurus: reconstrucciones paleoambientales y recursos geológicos"; Plurianual Project CONICET No 11220130100683CO, "Diversidad y evolución de las asociaciones de dinosaurios del Cretácico Temprano de la Cuenca Neuquina. Faunas y floras asociadas" (Rodolfo Coria, director); and PICT ANPCyT 1925 "La fauna de saurópodos del Cretácico "medio" de Patagonia, evolución y diversificación de los Rebbachisauridae y Somphospondyli, aspectos evolutivos y ecológicos (José L. Carballido, director). Fidel Torcida Fernandez-Baldor, Lucio Ibiricu, and John Whitlock provided helpful comments that substantially improved the manuscript.

References

Apesteguía S (2007) The sauropod diversity of the La Amarga Formation (Barremian), Neuquén (Argentina). Gondwana Res 12:533–546

Apesteguía S, Gallina PA, Haluza A (2010) Not just a pretty face: anatomical peculiarities in the postcranium of Rebbachisaurids (Sauropoda: Diplodocoidea). Hist Biol 22(1–3):165–174

Bonaparte JF (1996) Cretaceous tetrapods of Argentina. Mün Geowiss Abhand A 30:73–130

Bonaparte JF (1997) *Rayososaurus agrioensis* Bonaparte, 1995. Ameghiniana 34:116

Calvo JO, Salgado L (1995) *Rebbachisaurus tessonei* sp. nov. a new Sauropoda from the Albian-Cenomanian of Argentina; new evidence on the origin of the Diplodocidae. Gaia 11:13–33

Canudo JI, Barco JL, Pereda-Suberbiola X, Ruiz-Omeñaca JI, Salgado L, Torcida Fernández-Baldor F, Gasulla JM (2009) What Iberian dinosaurs reveal about the bridge said to exist between Gondwana and Laurasia in the Early Cretaceous. B Soc Géol Fr 180(1):5–11

Canudo JI, Salgado L (2003) Los Dinosaurios del Neocomiense (Cretácico Inferior) de la Península Ibérica y Gondwana Occidental: Implicaciones biogeográficas. In: Pérez-Llorente F (coord) Dinosaurios y otros reptiles Mesozoicos en España. Instituto de Estudios Riojanos, Universidad de La Rioja, Logroño, España, pp 251–268

Canudo JI, Carballido JL, Garrido AC, Salgado L (2018) A new rebbachisaurid sauropod from the Lower Cretaceous (Aptian–Albian) of the Rayoso Formation (Neuquén, Argentina). Acta Palaeont Pol 63(4):679–691

Carballido JL, Garrido AC, Canudo JI, Salgado L (2010) Redescription of *Rayososaurus agrioensis* Bonaparte (Sauropoda, Diplodocoidea), a rebbachisaurid of the early Late Cretaceous of Neuquén. Geobios 43:493–502

Carballido JL, Salgado L, Pol D, Canudo JI, Garrido AC (2012) A new basal rebbachisaurid (Sauropoda, Diplodocoidea) from the Early Cretaceous of the Neuquén Group; evolution and biogeography of the group. Hist Biol 24(6):631–654

Carpenter K (2018) *Maraapunisaurus fragillimus*, N.G. (formerly *Amphicoelias fragillimus*), a basal Rebbachisaurid from the Morrison Formation (Upper Jurassic) of Colorado. Geol Inter W 5:227–244

Carvalho I, Avilla L, Salgado L (2003) *Amazonsaurus maranhensis* gen. et. sp. nov. (Sauropoda, Diplodocoidea) from the Lower Cretaceous (Aptian–Albian) of Brazil. Cretaceous Res 24:697–713

Dalla Vecchia FM (1998) Remains of Sauropoda (Reptilia, Saurischia) in the Lower Cretaceous (Upper Hauterivian/Lower Barremian) limestones of SW Istria (Croatia). Geol Croat 51(2):105–134

Fanti F, Cau A (2013) Contessi M (2013) A new sauropod dinosaur from the early Cretaceous from the Early Cretaceous of Tunisia with extreme avian-like pneumatization. Nat Commun 4:2080. https://doi.org/10.1038/ncomms3080

Fanti F, Cau A, Cantelli L, Hassine M, Auditore M (2015) New information on *Tataouinea hannibalis* from the Early Cretaceous of Tunisia and Implications for the tempo and Mode of Rebbachisaurid Sauropod Evolution. PLoS ONE. https://doi.org/10.1371/journal.pone.0123475

Gallina PA, Apesteguía S (2005) *Cathartesaura anaerobica* gen. et sp. nov., a new rebbachisaurid (Dinosauria, Sauropoda) from the Huincul Formation (Upper Cretaceous), Río Negro, Argentina. Rev Mus Argent Cienc Nat Ns 7:153–166

Garrido A (2010) Estratigrafía del Grupo Neuquén, Cretácico Superior de la Cuenca Neuquina (Argentina): nueva propuesta de ordenamiento litoestratigráfico. Revi Mus Argent Cienc Nat Ns 12:121–177

Garrido AC, Salgado L (2015) Taphonomy and depositional environment of a Lower Cretaceous monospecific sauropod bone assemblage (Puesto Quiroga Member, Lohan Cura Formation), Neuquén Province, Argentina. J S Am Earth Sci 61:53–61

Haluza A, Canale JI, Otero A, Pérez LM, Scanferla CA (2012) Changes in vertebral laminae across the cervicodorsal transition of a well-preserved rebbachisaurid (Dinosauria, Sauropoda) from the Cenomanian of Patagonia Argentina. J Vertebr Paleontol 32(1):219–224

Harris JH, Dodson P (2004) A new diplodocoid sauropod dinosaur from the Upper Jurassic Morrison Formation of Montana, USA. Acta Palaeontol Pol 49(2):197–210

Hatcher JB (1901) *Diplodocus* Marsh; its osteology, taxonomy, and probate habits, with a restoration of the skeleton. Mem Carnegie Mus 1:1–63

Hatcher JB (1903) Osteology of *Haplocanthosaurus*, with description of a new species, and remarks on the probable habits of the Sauropoda and the age and origin of the Atlantosaurus beds. Mem Carnegie Mus 2:1–75

Ibiricu LM, Casal GA, Lamanna MC, Martínez RD, Harris JD, Lacovara KJ (2012) The southernmost records of Rebbachisauridae (Sauropoda: Diplodocoidea), from early Late Cretaceous deposits in central Patagonia. Cretaceous Res 34:220–232

Ibiricu LM, Casal GA, Martínez RD, Lamanna MC, Luna M, Salgado L (2013) *Katepensaurus goicoecheai*, gen. et sp. nov., a Late Cretaceous rebbachisaurid (Sauropoda, Diplodocoidea) from central Patagonia, Argentina. J Vertebr Paleontol 33(6):1351–1366

Ibiricu LM, Casal GA, Martínez RD, Lamanna MC, Luna M, Salgado L (2015) New material of *Katepensaurus goicoecheai* (Sauropoda: Diplodocoidea) and its significance for the morphology and evolution of Rebbachisauridae. Ameghiniana 52(4):430–446

Ibiricu LM, Lamanna MC, Martínez RD, Casal GA, Cerda IA, Martínez G, Salgado L (2017) A novel form of postcranial skeletal pneumaticity in a sauropod dinosaur: implications for the paleobiology of Rebbachisauridae. Acta Palaeontol Pol 62(2):221–236

Lavocat R (1954) Sur les dinosauriens du continental intercalaire des Kem Kem de la Daoura. Comptes Rendus de la Dix-Neuvième Session, Con géol Inter, Alger 21(1952)3:65–68
Maisey JG (2011) Northeastern Brazil: out of Africa? In:Carvalho IS, Srivastava NK, Strohschoen Jr O, Lana CC (eds) Paleontologia: Cenários de Vida. Interciência, Rio de Janeiro, p 545,559
Mannion PD, Upchurch P, Mateus O, Barnes R, Jones MEH (2012) New information on the anatomy and systematic position of *Dinheirosaurus lourinhanensis* (Sauropoda: Diplodocoidea) from the Late Jurassic of Portugal, with a review of European diplodocoids. J Syst Palaeontol 10:521–551
Mannion PD, Upchurch P, Schwarz D, Wings O (2019) Taxonomic affinities of the putative titanosaurs from the Late Jurassic Tendaguru Formation of Tanzania: phylogenetic and biogeographic implications for eusauropod dinosaur evolution. Zool J Linn Soc 185(3):784–909
Marsh OC (1878) Principal characters of American Jurassic dinosaurs. Part I. Am J Sci (Series 3) 16:411–416
Matos Lindoso R, Araújo Medeiros MA, Souza Carvalho IS, Araújo Pereira A, Dienes Mendes I, Vidoi Iori F, Pinheiro Souza E, Souza Arcanjo SH, Costa Madeira Silva T (2019) A new rebbachisaurid (Sauropoda: Diplodocoidea) from the middle Cretaceous of northern Brazil. Cretaceous Res 104. https://doi.org/10.1016/j.cretres.2019.104191
McIntosh JS (1990) Sauropoda. In: Weishampel DB, Dodson P, Osmolska H (eds) The Dinosauria. University of California Press, Berkeley, pp 345–401
Medeiros MA, Schultz CL (2004) *Rayososaurus* (Sauropoda, Diplodocoidea) no meso-Cretáceo do norte-nordeste brasileiro. Revista Brasileira De Paleontologia 7(2):275–279
Nopcsa F (1902) Notizen über die Cretacischen Dinosaurier. Pt. 3. Wirbel eines südamerikanischen Sauropoden. Sitz Ber Aka Wissens 3:108–114
Owen R (1875) A monograph on the Fossil Reptilia of the Mesozoic Formations. Monograph on the Genus *Bothriospondylus*. Palaeontol Soc Monogr 29:15–26
Paulina-Carabajal AP, Canale JI, Haluza A (2016) New rebbachisaurid cranial remains (Sauropoda, Diplodocoidea) from the Cretaceous of Patagonia, Argentina, and the first endocranial description for a South American representative of the clade. J Vertebr Paleontol. https://doi.org/10.1080/02724634.2016.1167067
Paulina-Carabajal A, Calvo JO (2021) Re-description of the braincase of the rebbachisaurid sauropod *Limaysaurus tessonei* and novel endocranial information based on CT scans. An Acad Bras Cienc 93(Suppl. 2). https://doi.org/10.1590/0001-3765202120200762
Pereda Suberbiola X, Torcida F, Izquierdo LA, Huerta P, Montero D, Pérez G (2003) First rebbachisaurid dinosaur (Sauropoda, Diplodocoidea) from the early Cretaceous of Spain: palaeobiogeografical implications. B Soc Géol Fr 174(5):471–479
Salgado L, Canudo JI, Garrido AC, Carballido JL (2012) Evidence of gregariousness in rebbachisaurids (Dinosauria, Sauropoda, Diplodocoidea) from the Early Cretaceous of Neuquén (Rayoso Formation), Patagonia, Argentina. J Vertebr Paleont 32:603–613
Salgado L, Carvalho IS, Garrido AC (2006) *Zapalasaurus bonapartei*, un nuevo dinosaurio saurópodo de La Formación La Amarga (Cretácico Inferior), noroeste de Patagonia, Provincia de Neuquén, Argentina. Geobios 39:695–707
Salgado L, Garrido AC, Cocca S, Cocca R (2004) Lower Cretaceous Rebbachisaurid Sauropods from Cerro Aguada Del León (Lohan Cura formation), Neuquén Province, Northwestern Patagonia Argentina. J Vertebr Paleontol 24(4):903–912
Sereno PC, Wilson JA (2005) Structure and evolution of a sauropod tooth batery. In: Wilson JA (ed) Curry Rogers KA. University of California Press, Berkeley p, pp 157–177
Sereno PC, Wilson JA, Witmer LM, Whitlock JA, Maga A, Ide O, Rowe TA (2007) Structural extremes in a Cretaceous dinosaur. PLoS ONE 2. https://doi.org/10.1371/journal.pone.0001230
Taylor MP (2018) *Xenoposeidon* is the earliest known rebbachisaurid sauropod dinosaur. PeerJ 6:e5212
Taylor MP, Naish D (2005) The phylogenetic taxonomy of Diplodocoidea (Dinosauria: Sauropoda). PaleoBios 25:1–7
Taylor MP, Naish D (2007) An unusual new neosauropod dinosaur from the lower cretaceous hastings beds group of East Sussex England. Palaeontology 50(6):1547–1564

Torcida Fernández-Baldor F, Canudo JI, Huerta P, Montero D, Pereda Suberbiola X, Salgado L (2011) *Demandasaurus darwini*, a new rebbachisaurid sauropod from the Early Cretaceous of the Iberian Peninsula. Acta Palaeontol Pol 56(3):535–552

Upchurch P (1995) Evolutionary history of sauropod dinosaurs. Philos Trans R Soc Lond B 349:365–390

Whitlock JA (2011) A phylogenetic analysis of Diplodocoidea (Saurischia: Sauropoda). Zool J Linn Soc 161:872–915

Whitlock JA, Wilson Mantilla JA (2020) The Late Jurassic sauropod dinosaur 'Morosaurus' agilis Marsh, 1889 reexamined and reinterpreted as a dicraeosaurid. J Vertebr Paleontol. https://doi.org/10.1080/02724634.2020.1780600

Wilson JA (2002) Sauropod dinosaur phylogeny: critique and cladistic analysis. Zool J Linn Soc 136:217–276

Wilson JA, Allain R (2015) Osteology of *Rebbachisaurus garasbae* Lavocat, 1954, a diplodocoid (Dinosauria, Sauropoda) from the early Late Cretaceous-aged Kem Fem beds of southeastern Morocco. J Vertebr Paleontol. https://doi.org/10.1080/02724634.2014.1000701

Wilson JA, Sereno PC (1998) Early evolution and higher level phylogeny of sauropod dinosaurs. Soc Vertebr Paleontol Mem 5:1–68

Wilson JA, Upchurch P (2003) A revision of Titanosaurus Lydekker (Dinosauria-Sauropoda), the first dinosaur genus with 'gondwanan' distribution. J Syst Paleontol 1:125–160

Wilson JA, D'Emic MD, Ikejiri T, Moacdied EM, Whitlock JA (2011) A nomenclature for vertebral fossae in sauropods and other saurischian dinosaurs. PLoS ONE 6(2):e17114

Southernmost Spiny Backs and Whiplash Tails: Flagellicaudatans from South America

Pablo A. Gallina, Sebastián Apesteguía, José L. Carballido, and Juan P. Garderes

Abstract Flagellicaudatan diplodocoids include the two families Dicraeosauridae and Diplodocidae. Although different in sizes and relative proportions (e.g. neural arches height, neck length, tail length), they share several features, both cranial and postcranial, that recover them as a monophyletic group in updated phylogenies. The record of the group in South America was particularly scarce during the twentieth century, but their number and taxonomical diversity noticeably increased in the last decade. Up to now, five dicraeosaurid taxa (*Amargasaurus cazaui*, *Amargatitanis macni*, *Bajadasaurus pronuspinax*, *Brachytrachelopan mesai,* and *Pilmatueia faundezi*) and one diplodocid (*Leinkupal laticauda*) were recognized. Additionally, two presumably dicraeosaurid and three diplodocid records are known from fragmentary materials. Jurassic strata have provided both *Brachytrachelopan* and two of the indeterminate diplodocids, whereas the remaining five taxa, the third indeterminate diplodocid and the indeterminate dicraeosaurids come from the Early Cretaceous. Curiously, they are the only Cretaceous flagellicaudatan diplodocoids in the world, together with fragmentary records from South Africa, since the Jurassic–Cretaceous boundary marks a global extinction event for numerous species within the group. All these occurrences come from the only two countries of Patagonia: Argentina

P. A. Gallina (✉) · S. Apesteguía · J. L. Carballido · J. P. Garderes
Consejo Nacional de Investigaciones Científicas y Técnicas, Buenos Aires, Argentina
e-mail: gallina.pablo@maimonides.edu

S. Apesteguía
e-mail: sebastian.apesteguia@fundacionazara.org.ar

J. L. Carballido
e-mail: jcarballido@mef.org.ar

J. P. Garderes
e-mail: garderes.juanpablo@maimonides.edu

P. A. Gallina · S. Apesteguía · J. P. Garderes
Centro de Ciencias Naturales, Ambientales y Antropológicas, Fundación de Historia Natural Félix de Azara–Universidad Maimónides, Hidalgo 775, C1405BCK Buenos Aires, Argentina

J. L. Carballido
Museo Paleontológico Egidio Feruglio, Fontana 140, 9100 Trelew, Chubut, Argentina

A. Otero et al. (eds.), *South American Sauropodomorph Dinosaurs*, Springer Earth System Sciences, https://doi.org/10.1007/978-3-030-95959-3_6

and Chile. The currently rich record of South American flagellicaudatans demonstrates that they were a key component of the Late Jurassic to the earliest Cretaceous sauropod fauna, the Bajadan tetrapod assemblage, occupying the niches of narrow-crowned megaherbivores by a time when macronarian neosauropods only attained broad-crown forms.

Keywords Sauropods · Flagellicaudata · Dicraeosauridae · Diplodocidae · Patagonia

1 Introduction

Flagellicaudatan sauropods [defined by Harris and Dodson (2004) as the most recent common ancestor of *Dicraeosaurus* and *Diplodocus* and all of its descendants] are well known since the late 19th and the early twentieth centuries, mainly from the Upper Jurassic Morrison and Tendaguru formations in North America and Africa, respectively (e.g. Marsh 1878; Osborn 1899; Janensch 1914, 1929). For nearly a century, all the anatomical knowledge of the group, which includes the two families Dicraeosauridae and Diplodocidae, came from that Late Jurassic record. In fact, the rather complete skeletons of *Dicraeosaurus hansemanni* and *Dicraeosaurus sattleri* (Janensch 1914, 1929) were the only source of anatomical information for the family Dicraeosauridae, whereas diplodocid anatomy was based on the almost complete specimens of *Diplodocus* and those of *Apatosaurus, Barosaurus*, and *Tornieria* (see Tschopp et al. 2015 for precise specimens and taxonomy).

Although distinctive in size (dicraeosaurids are small to mid-size sauropods whereas most diplodocids are mainly large-bodied), relative proportions (dicraeosaurids generally have shorter necks and tails, and longer neural spines in their backs than diplodocids) and skeletal pneumatization (dicraeosaurids are much less pneumatized than diplodocids), they share more than a dozen of cranial and postcranial synapomorphies (see Whitlock 2011; Tschopp et al. 2015; Whitlock and Wilson Mantilla 2020) that recover them as a monophyletic group in successive phylogenies since the nineties of the twentieth century (e.g. McIntosh 1990; Calvo and Salgado 1995; Upchurch 1995).

In particular, the record of flagellicaudatans in South America (Fig. 1) remained elusive until 1984 when an expedition led by Dr. José Bonaparte in the Early Cretaceous La Amarga Formation, in the northern Argentinean Patagonia, found the first local evidence of a dicraeosaurid sauropod (i.e. *Amargasaurus cazaui* Salgado and Bonaparte 1991). That outstanding finding not only expanded the record both geographically (to the west of Gondwana and southernmost latitudes) and chronologically (to the Cretaceous) but also provided substantial insight into the anatomical knowledge of Dicraeosauridae and hence the flagellicaudatans. Actually, during the same field trip, a second dicraeosaurid was collected too, but remained unnamed for years on the shelves of the Museo Argentino de Ciencias Naturales collection (see *Amargatitanis* below).

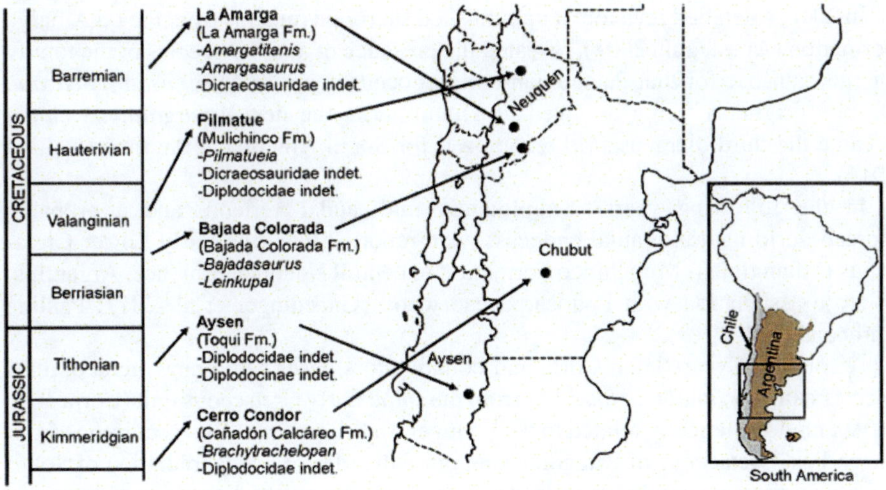

Fig. 1 Location map showing the provenance and chronological position of the record of South American flagellicaudatans

Almost fifteen years after the publication of *Amargasaurus*, another flagellicaudatan was recorded in earlier, Upper Jurassic strata of Chubut province, in the geographical heart of the Argentinean Patagonia (Rauhut et al. 2005). In this case, the new dicraeosaurid, named *Brachytrachelopan mesai*, notably contributed to our knowledge of the axial skeleton, in particular the cervical series, showing, for the first time, their peculiar shortening in the neck length, now considered as a derived feature within the group (Whitlock and Wilson Mantilla 2020).

Nine years later, the finding of a new flagellicaudatan revealed the presence of a different group in Patagonia: the Diplodocidae, being not only the first record of this group for South America but also the first convincing report of a diplodocid for the Cretaceous worldwide. The new *Leinkupal laticauda* from the Berrasian–Valanginian Bajada Colorada Formation (Gallina et al. 2014), in southern Neuquén Province (northern Patagonia, Argentina) also resulted in one of the smallest-bodied diplodocids in the world. This finding demonstrated that diplodocids survived in South America until, at least, the first half of the Early Cretaceous, exceeding their supposed global extinction as previously considered to have occurred at the Jurassic/Cretaceous boundary.

Almost simultaneously, in the following year, two Late Jurassic diplodocid vertebral remains were reported from the Chubut Province in Argentina and the Aysen region in Chile (Rauhut et al. 2015; Salgado et al. 2015). At the time, the latter was the only South American record of flagellicaudatans outside Argentina. These occurrences have shown that diplodocids were not only present but also more frequent and widely distributed throughout Gondwana during the Late Jurassic, only known until then from the East African *Tornieria africana*, from the Tendaguru Formation (Fraas 1908; Remes 2006).

In 2016, a detailed revision of unpublished sauropod remains from the La Amarga Formation (Apesteguía 2007), revealed the presence of another species of the family Dicraeosauridae for that Patagonian unit, co-occurring with *Amargasaurus cazaui*. Despite originally being assigned to Titanosauria, the new *Amargatitanis macni* became the third dicraeosaurid recognized for South America at the time (Gallina 2016).

In the following years, a diplodocid mid-caudal vertebra and a probable dicraeosaurid natural cranial endocast were reported from the Early Lower Cretaceous (Valanginian) Mulichinco Formation in central Neuquén province, Argentina, in close association with Podocarpaceae wood (Gnaedinger et al. 2017; Paulina Carabajal et al. 2018).

A couple of years later, Coria and collaborators described a new dicraeosaurid species named *Pilmatueia foundezi* from the same Early Cretaceous unit (Coria et al. 2019), phylogenetically related to the younger *Amargasaurus*. This occurrence from the earliest Cretaceous of Patagonia supported the idea of a local radiation of South American dicraeosaurids between the Late Jurassic and the late Early Cretaceous (Coria et al. 2019).

The last dicraeosaurid to be erected was *Bajadasaurus pronuspinax* (Gallina et al. 2019), from the earliest Lower Cretaceous terrestrial beds (Berriasian–Valanginian) of the Bajada Colorada Formation in south-eastern Neuquén Province. The recognition of this new species permitted not only to analyse novel features in the cranial and neck anatomy of the group, but also allowed to rekindle the discussion concerning the function, if any, of the elongated bifid neural spines that highlight the astonishing necks of dicraeosaurid sauropods.

Finally, two additional vertebral remains, presumably from dicraeosaurids (Windholz et al. 2021), were reported from the La Amarga Formation.

As evidenced, the number and taxonomic diversity of the flagellicaudatan record from South America noticeably increased in the last decade contributing in different ways (e.g. anatomy, phylogeny, biogeography, palaeobiology) to the knowledge of this distinctive group of sauropod dinosaurs (Table 1). A summary of the records, in the context of Gallina et al. (2019) phylogeny, as well as phylogenetic and palaeobiogeographical and chronological considerations are provided in this chapter.

Institutional Abbreviations MACN: Museo Argentino de Ciencias Naturales, Buenos Aires, Argentina; **MMCh**: Museo Municipal Ernesto Bachmann, Villa El Chocón, Neuquén, Argentina; **MPEF**: Museo Paleontológico Egidio Feruglio, Trelew, Chubut, Argentina; **MLL**: Museo Municipal de Las Lajas, Las Lajas, Neuquén, Argentina; **MOZ**: Museo Prof. Dr. Juan A. Olsacher, Zapala, Neuquén, Argentina; **SNGM**: Sernageomin, Santiago, Chile.

Table 1 Record of South America flagellicaudatans mentioned in this chapter, in the order present in Sect. 2

Taxon	Formation	Age	Locality	Reference
Amargatitanis macni	La Amarga	Barremian	La Amarga, Neuquén Province, Argentina	Apesteguía (2007), Gallina (2016)
Bajadasaurus pronuspinax	Bajada Colorada	Berriasian–Valanginian	Bajada Colorada, Neuquén Province, Argentina	Gallina et al. (2019)
Pilmatueia faundezi	Mulichinco	Valanginian	Pilmatue, Neuquén Province, Argentina	Coria et al. (2019)
Amargasaurus cazaui	La Amarga	Barremian	La Amarga, Neuquén Province, Argentina	Salgado and Bonaparte (1991)
Brachytrachelopan mesai	Cañadón Calcáreo	Oxfordian–Kimmeridgian	Cerro Cóndor, Chubut Province, Argentina	Rauhut et al. (2005)
Dicraeosauridae indet	La Amarga	Barremian	La Amarga, Neuquén Province, Argentina	Windholz et al. (2021)
Diplodocidae indet	Cañadón Calcáreo	Oxfordian–Kimmeridgian	Cerro Cóndor, Chubut Province, Argentina	Rauhut et al. (2015)
Diplodocidae indet	Toqui Formation	Tithonian	Aysén, Chile	Salgado et al. (2015)
Diplodocidae indet	Mulichinco	Valanginian	Pilmatue, Neuquén Province, Argentina	Gnaedinger et al. (2017)
Leinkupal laticauda	Bajada Colorada	Late Berriasian–Valanginian	Bajada Colorada, Neuquén Province, Argentina	Gallina et al. (2014)

(continued)

Table 1 (continued)

Taxon	Formation	Age	Locality	Reference
Diplodocinae indet	Toqui Formation	Tithonian	Aysén, Chile	Salgado et al. (2015)

2 Systematic Palaeontology

Sauropoda Marsh 1878

Diplodocoidea Marsh 1878 (Upchurch 1995)

Flagellicaudata Harris and Dodson 2004

Definition A node-based clade defined as the most recent common ancestor of *Dicraeosaurus* and *Diplodocus*, and all of its descendants (Harris and Dodson 2004)

Dicraeosauridae Janensch 1929

Definition A stem-based clade defined as all diplodocoid sauropods more closely related to *Dicraeosaurus* than to *Diplodocus* (Sereno 1998).

Amargatitanis Apesteguía 2007

Amargatitanis macni Apesteguía 2007

Holotype MACN PV N53; two caudal vertebrae, an incomplete right ischium, a right femur, an incomplete right tibia, an incomplete right fibula, a right astragalus, and an incomplete right metatarsal I.

Locality and Age The site is located 2.5 km SE from the bridge of National Route 40, crossing the La Amarga stream. The bridge is 80 km south of Zapala City, Neuquén Province, Argentina. The fossil remains come from fluvial channel sands deposited by braided rivers forming the rocks of the Puesto Antigual Member of the La Amarga Formation, Barremian, Lower Cretaceous (e.g. Leanza et al. 2004).

Revised Diagnosis Gallina (2016) rediagnosed *Amargatitanis macni* by the following autapomorphies (marked by an asterisk), as well as a unique combination of character states: anterior caudal prezygapophyseal centrodiapophyseal fossa deep and bearing an internal vertical lamina (*); proportionally wide proximal portion of the femur in relation to the distal end (*); fourth trochanter low, located at mid-length of the femur (shared with other flagellicaudatans); astragalus with a deep crescent-shaped posterior fossa, bearing two distant foramina (*).

Comments Apesteguía (2007) included a left scapula (MACN PV N34) and additional caudal vertebrae (MACN PV N51) as part of the holotype. However, Gallina (2016) excluded those specimens from the holotype, following article 73.1.5 of the International Code of Zoological Nomenclature, based on a detailed analysis of the field notes of José Bonaparte, plus the recognition of field numbers present in the collected material housed in the vertebrate collection of the MACN. Following the same criterion, this latter author also included a right ischium, a right tibia, a right fibula, and an incomplete right metatarsal I amongst the type material.

Bajadasaurus Gallina, Apesteguía, Canale, Haluza 2019

Bajadasaurus pronuspinax Gallina, Apesteguía, Canale, Haluza 2019

Holotype MMCh-PV 75; a nearly complete skull (including left maxilla, left lacrimal, both prefrontals, both frontals, both parietals, both postorbitals, both squamosals, left quadratojugal, both pterygoids, both quadrates, supraoccipital, exoccipital-opisthotic complex, basioccipital, basisphenoid, both prootics, both laterosphenoids, both orbitosphenoids, both dentaries, left surangular, both angulars, both splenials, left prearticular, left articular, isolated upper tooth row), both proatlases, atlantal neurapophyses, axis and the fifth cervical vertebra (Fig. 2).

Locality and Age Bajada Colorada locality, 40 km south of Picún Leufú town on the National Route 237, Neuquén Province, Argentina. The remains were found in outcrops of the Bajada Colorada Formation, Late Berriasian–Valanginian, Lower Cretaceous.

Diagnosis Gallina et al. (2019) diagnosed *Bajadasaurus pronuspinax* by the following autapomorphies (marked by an asterisk), as well as a unique combination of character states: post-temporal fenestra extended medially with a long parietal contribution (*); basipterygoid processes extremely slender and long, more than six times longer than lateromedially wide (*); elongate angular, longer than the anteroposterior length of the surangular; neural spine of the axis oriented vertically (*); paired, anteriorly curved, and extremely elongate bifid neural spines of anterior–mid-cervical vertebrae (*).

Comments *Bajadasaurus* represents a key-taxon within dicraeosaurid evolution by having the most complete cranial remains of the family, yielding data on the general proportion and aspect of the skull.

Pilmatueia Coria, Windholz, Ortega, Currie 2019

Pilmatueia faundezi Coria, Windholz, Ortega, Currie 2019

Holotype MLL-Pv-005, almost complete posterior dorsal vertebrae.

Paratype MLL-Pv-002, cervico-dorsal vertebra.

Referred Specimens MLL-Pv-004, mid-cervical vertebra; MLL-Pv-014, incomplete mid-dorsal neural arch; MLL-Pv-015, centrum of a mid-caudal vertebra; MLL-Pv-016, mid-caudal vertebra.

Locality and Age Pilmatue locality, 9 km north-east of Las Lajas, Neuquen Province, Argentina. The fossils are from the Mulichinco Formation, Valanginian, Lower Cretaceous.

Diagnosis Coria et al. (2019) diagnosed *Pilmatueia* by the following autapomorphies (marked by an asterisk), as well as a unique combination of character states:

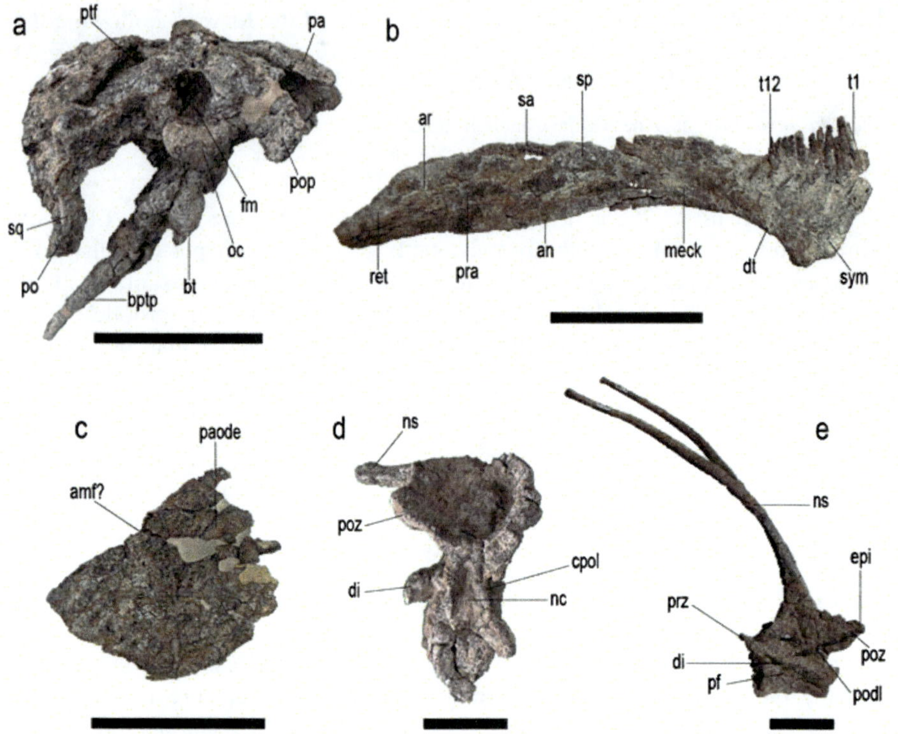

Fig. 2 Selected bones of *Bajadasaurus pronuspinax* Gallina et al. 2019. **a** Skull roof and braincase in posterior view. **b** Left lower jaw in medial view. **c** Left maxilar in lateral view. **d** Axis in posterior view. **e** Fifth cervical vertebra in left lateral view. Abbreviations: amf, anterior maxillary foramen; an, angular; ar, articular; bptp, basipterygoid process; bt, basal tubera; di, diapophysis; dt, dentary; epi, epipophysis; fm, foramen magnum; meck, meckelian canal; nc, neural canal; ns, neural spine; oc, occipital condyle; pa, parietal; paode, preantorbital depression; po, postorbital; podl, postzygodiapophyseal lamina; pf, pneumatic fossa; pop, paroccipital process; poz, postzygapophysis; pra, prearticular; prz, prezygapophysis; ptf, post-temporal fenestra; ret, retroarticular process; sa, surangular; sp, splenial; sq, squamosal; sym, mandibular symphysis; t, tooth. Scale (**a**, **b**, **c**, **e**) 10 cm, (**d**) 5 cm

cervico-dorsal vertebrae with dorsoventrally oriented ridges on the anterior surfaces of anterior centrodiapophyseal laminae(*); posterior dorsal vertebrae with deep fossae located posteriorly at the bases of the bifid neural spines separated by a thick, low, sagittal lamina (*); presence of mid- and posterior cervical vertebrae with bifid ventral keels that connect posterolaterally to the ventrolateral edges of the centra (also present in *Barosaurus*); the presence of vertical neural spines in mid-caudal vertebrae (also present in some derived diplodocids); mid-caudal neural arches positioned on the anterior half of the centrum; summit of mid- and posterior caudal neural spines more or less straight above postzygapophyses.

Comments Although retrieved as a derived form within Dicraeosauridae in Coria et al (2019), other analyses have recovered it in a more basal position; these are followed here. Further revision of the taxon would help in solving an accurate phylogenetic position (see Sect. 4).

Amargasaurus Salgado and Bonaparte 1991

Amargasaurus cazaui Salgado and Bonaparte 1991

Holotype MACN-N 15; partial skeleton comprised by the basicranium and temporal elements of the skull; 23 articulated presacral vertebrae in connection with the skull and the sacrum, almost completely preserved and in association with ribs; sacrum composed of five fused vertebrae; three mid-anterior caudal vertebrae; one posterior caudal vertebra; three haemal arches; one distal end of a cervical rib; incomplete dorsal ribs; many fragmentary ribs and caudal vertebral bodies; right scapula-coracoid; left humerus, radius and ulna; left ilium, femur, tibia and fibula; one astragalus and two metatarsal bones (Fig. 3).

Locality and Age The site is located 2.5 km SE from the bridge of National Route 40, crossing the La Amarga stream. The bridge is 80 km south of Zapala City, Neuquen Province, Argentina. The fossil remains come from fluvial channel sands of braided rivers from the Puesto Antigual Member of the La Amarga Formation, Barremian, Lower Cretaceous.

Diagnosis Salgado and Bonaparte (1991, p 335) diagnosed *Amargasaurus cazaui* as follows: Dicraeosauridae similar to *Dicraeosaurus hansemanni* in size. It presents taller presacral vertebrae than *Dicraeosaurus*, showing a bifurcation/partition of the neural spines more conspicuous, reaching the last presacral vertebra. Cervical neural spines are much longer than in *Dicraeosaurus*, subcylindrical in section. Pleurocoels are secondarily obliterated as in *Dicraeosaurus*. Five fused sacral vertebrae. Pelvic and hindlimb similar to *Dicraeosaurus*. Temporal region of the skull showing parietal and postparietal fenestrae as in *Dicraeosaurus*. Fused basal tubera. Shape and position of the external nostrils comparable to *D. hansemanni*. Stout orbitosphenoids with a wide exit for the CN I, surrounded by an osseous process.

Comments As the original diagnosis provided by Salgado and Bonaparte (1991) did not specify autapomorphic features but a combination of characters, further revision of both anatomy and diagnosis of this taxon will contribute in this aspect (Carballido et al., in prep.). In the context of that, 23 presacral vertebrae seem to be present in the axial skeleton of *Amargasaurus* (Carballido pers. obs.), as opposed to the 22 mentioned by Salgado and Bonaparte (1991).

Brachytrachelopan Rauhut, Remes, Flechner, Cladera, Puerta 2005

Brachytrachelopan mesai Rauhut, Remes, Flechner, Clader, Puerta 2005.

Holotype MPEF-PV 1716; an articulated partial skeleton, including 8 cervical, 12 dorsal, and 3 sacral vertebrae, the proximal parts of the posterior cervical and all

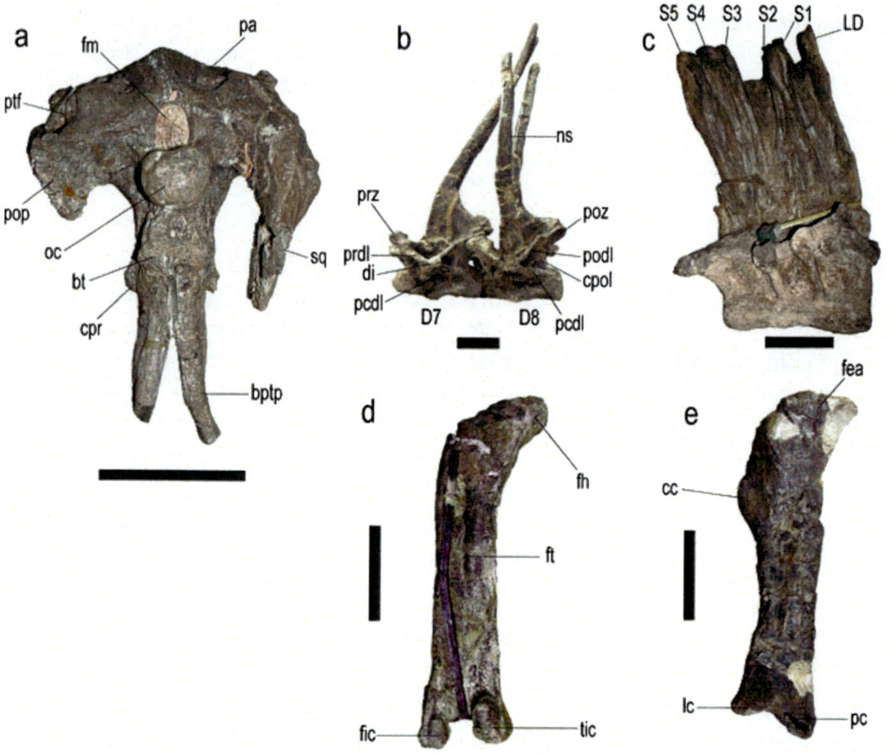

Fig. 3 Selected bones of *Amargasaurus cazaui* Salgado and Bonaparte 1991. **a** Skull roof and braincase in posterior view. **b** Dorsal vertebrae seven and eight in left lateral view, **c** Sacrum in right lateral view. **d** Left femur in posterior view. **e** Left tibia in lateral view. Abbreviations: bptp, basipterygoid process; bt, basal tubera; cc, cnemial crest; cpol, centropostzygapophyseal lamina; cpr, crista prootica; D, dorsal vertebra; di, diapophysis; fea, articulation with femur; fh, femoral head; fic, fibular condyle; fm, foramen magnum; ft, fourth trochanter; lc, lateral condyle; LD, last dorsal vertebra; ns, neural spine; oc, occipital condyle; pa, parietal; pc, posterior condyle; pcdl, posterior centrodiapophyseal lamina; podl, postzygodiapophyseal lamina; pop, paroccipital process; poz, postzygapophysis; prdl, prezygodiapophyseal lamina; prz, prezygapophysis; ptf, post-temporal fenestra; S, sacral vertebra; sq, squamosal; tic, tibial condyle. Scale (**a**, **b**, **c**) 10 cm, (**d**, **e**) 20 cm

dorsal ribs, the right ilium, distal part of the left femur, and proximal end of the left tibia (Fig. 4).

Locality and Age The fossil locality is on the summit of a hill about 25 km north-northeast of the village of Cerro Cóndor, Chubut Province, Argentina. The specimen comes from a fluvial sandstone within the Cañadón Calcáreo Formation, Oxfordian–Kimmeridgian, Upper Jurassic.

Diagnosis Rauhut et al. (2015) diagnosed *Brachytrachelopan mesai* as differing from all other sauropods in its very short neck, with individual cervical vertebrae

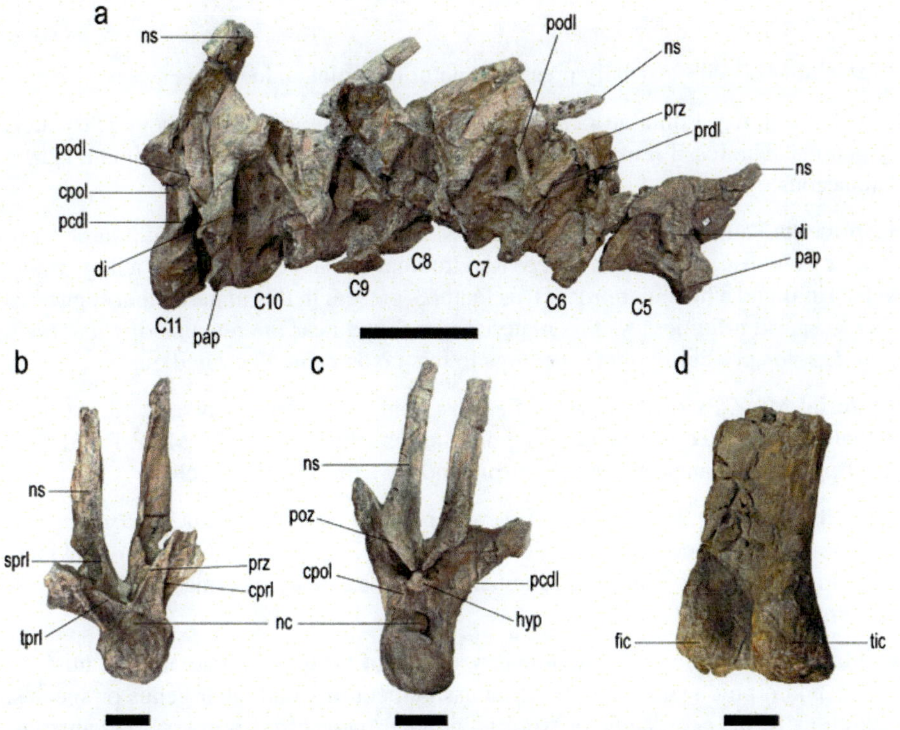

Fig. 4 Selected bones of *Brachytrachelopan mesai* Rauhut et al. (2005). **a** Articulated cervical series from the fifth to the eleventh in right lateral view. **b** Twelfth cervical vertebra in anterior view. **c** Second dorsal vertebra in posterior view. **d** Distal femur in posterior view. Abbreviations: C, cervical vertebra; cpol, centropostzygapophyseal lamina; di, diapophysis; fic, fibular condyle; hyp, hyposphene; nc, neural canal; ns, neural spine; pap, parapophysis; pcdl, posterior centrodiapophyseal lamina; podl, postzygodiapophyseal lamina; prdl, prezygodiapophyseal lamina; prz, prezygapophysis; sprl, spinoprezygapophyseal lamina; tic, tibial condyle; tprl, intraprezygapophyseal lamina. Scale (a) 20 cm, (b, c, d) 10 cm

being as long as, or shorter in anteroposterior length than, high posteriorly. Further apomorphies of the taxon include a pronounced, pillar-like centropostzygapophyseal lamina in the cervical vertebrae, a pronounced anterior inclination of the mid-cervical neural spines, with the tip of the spine extending beyond the anterior end of the centrum, and anterior dorsal neural spines one to six with vertical bases and anteriorly flexed tips.

Comments The holotype specimen of *Brachytrachelopan* was prepared in detail during 2020 and 2021 at the MEF labs in order to be fully described. During this procedure, several vertebrae (including all cervical and some dorsal elements) were disarticulated. A complete description of this specimen is still lacking, but would provide new anatomical information for a more accurate diagnosis and phylogenetic assessment of this taxon and the evolution of the family.

Dicraeosauridae? indet.

Material MLL 003; partially preserved natural cranial endocast.

Locality and Age Pilmatue locality, 9 km north-east of Las Lajas, Neuquen Province, Argentina. The fossil remains are from Mulichinco Formation, Valanginian, Lower Cretaceous.

Comments Paulina Carabajal et al. (2018) mentioned the presence of a large frontoparietal fenestra (related to a large longitudinal dorsal venous sinus), a large transverse sinus and a floccular process, as features present in this material that suggested dicraeosaurid affinities. As this material was found near the remains of *Pilmatueia foundezi*, the possibility that it belongs to this taxon cannot be ruled out.

Material MOZ-Pv 6126–1, was an almost complete anterior dorsal vertebra (now actually represented by isolated fragments after critical damage). MOZ-Pv 6126–2 is a fragmentary anterior dorsal centrum and base of the neural arch.

Locality and Age These remains were recovered near the *Amargasaurus cazaui* quarry (Windholz et al. 2021), 80 km south of Zapala City, Neuquen Province, Argentina, in Puesto Antigual Member of the La Amarga Formation, Barremian, Lower Cretaceous.

Comments The extremely fragmentary nature of these vertebrae and the absence of autapomorphic features preclude an assignment to a particular genus or species. Additionally, the geographic proximity with the *Amargasaurus* quarry, and anatomical similarities with both *Amargasaurus* and *Amargatitanis* suggests that the two vertebrae may represent additional and uninformative material of those taxa rather than a new species that would further increase dicraeosaurid diversity in the La Amarga Formation. However, as all comparative and taxonomical conclusions of Windholz et al. (2021) were based on a photograph and cannot be corroborated in the future, a precise taxonomic assignment becomes fairly speculative.

Diplodocidae Marsh 1884

Definition A stem-based clade defined as all diplodocoid sauropods more closely related to *Diplodocus* than to *Dicraeosaurus* (Sereno 1998).

Diplodocidae indet

Material Three dorsal vertebral centra, fragmentary scapula (not collected). The material is kept in the collections of the Museo Paleontologico Egidio Feruglio (MPEF) in Trelew, Argentina, under accession number MPEF-PV 1324 (Fig. 5a).

Locality and Age The locality lies 20 km north of the village of Cerro Condor on the eastern side of the Chubut river, Chubut Province, Argentina. The material comes from a coarse-grained sandstone within the upper part of the Cañadón Calcáreo Formation, Oxfordian–Kimmeridgian, Upper Jurassic.

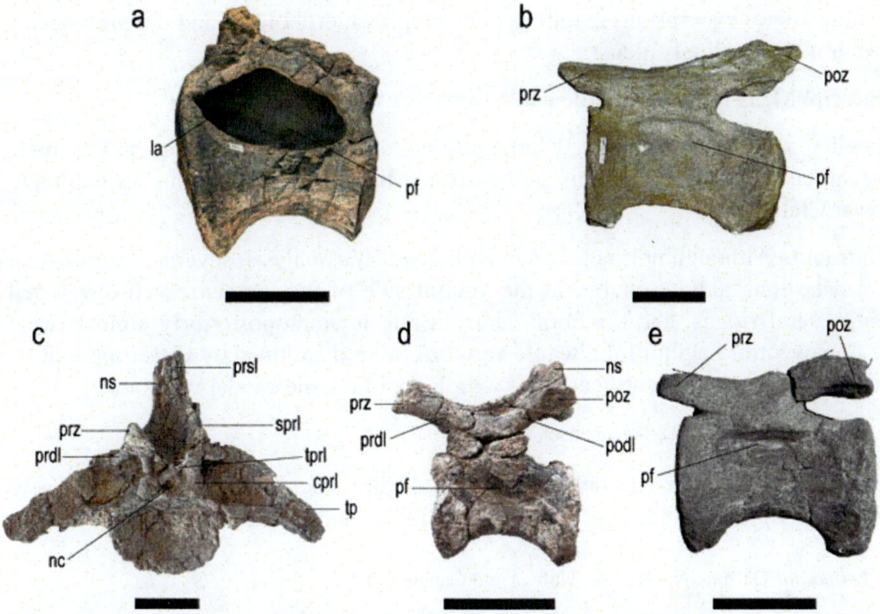

Fig. 5 Selected bones of *Leinkupal laticauda* and indeterminated diplodocids from South America. **a** Posterior dorsal centrum MPEF-PV 1324–1 in left lateral view modified from Rauhut et al. 2015). **b** Middle caudal vertebra MLL-PV-013 in left lateral view (modified from Gnaedinger et al. 2017). **c** Anterior caudal vertebra (Holotype) of *Leinkupal laticauda* in anterior view. **d** Middle caudal vertebra of *Leinkupal laticauda* in left lateral view. **e** Middle caudal vertebra SNGM-1979 in left lateral view (modified from Salgado et al. 2015). Abbreviations: cprl, centroprezygapophyseal lamina; la, lamina within the pneumatic fossa; nc, neural canal; ns, neural spine; pf, pneumatic fossa; podl, postzygodiapophyseal lamina; poz, postzygapophysis; prdl, prezygodiapophyseal lamina; prz, prezygapophysis; psrl, prespinal lamina; sprl, spinoprezygapophyseal lamina; tprl, intraprezygapophyseal lamina; tp, transverse process. Scale 10 cm

Comments These three dorsal vertebrae from Chubut Province represent the only diplodocid record from the Jurassic of South America together with those material from Chile (see below). Although recovered in three different equally parsimonious positions within Diplodocidae, forcing the specimen into different phylogenetic positions outside this clade required several extra steps. Therefore, although fragmentary, the phylogenetic position of the Cañadon Calcareo specimen amongst diplodocids is fairly well supported (Rauhut et al. 2015).

Material SNGM-1978, middle? cervical centrum.

Locality and Age Aysén, Patagonian Central Andes in Southern Chile. The bone-bearing unit is the Toqui Formation, dated as late Tithonian.

Comments Salgado et al. (2015) recovered SNGM 1978 as a diplodocid in their phylogenetic analysis based on the presence of cervical pleurocels divided in three or more lateral excavations resulting in a complex morphology, and the presence of a ventral longitudinal sulcus.

Material MLL-PV-013, middle caudal vertebra (Fig. 5b).

Locality and Age Pilmatue, 9 km north-east of Las Lajas, Neuquén Province, Argentina. This fossil remains come from Mulichinco Formation, Valanginian, Lower Cretaceous.

Comments Although not included in a phylogenetic analysis, several features such as a deep longitudinal cavity on the ventral side of the vertebra, well-developed ventrolateral ridges, and vertebral centrum that is anteroposteriorly almost twice the dorsoventral height of the whole vertebra, were mentioned by Gnaedinger et al. (2017) to support its identification as a diplodocid middle caudal vertebra.

Diplodocinae Marsh 1884; Janensch 1929

Definition A stem-based clade defined as all diplodocid sauropods more closely related to *Diplodocus* than to *Apatosaurus* (Taylor and Naish 2005).

Leinkupal Gallina, Apesteguía, Haluza and Canale 2014

Leinkupal laticauda Gallina, Apesteguía, Haluza and Canale 2014

Holotype MMCh-PV 63–1, includes one anterior caudal vertebra (Fig. 5c).

Paratype Two anterior cervical vertebrae (MMCh-PV 63- 2/3), one posterior cervical vertebra (MMCh-PV 63–4), one anterior dorsal vertebra (MMCh-PV 63–5), one anterior caudal vertebra (MMCh-PV 63–6), and two mid-caudal vertebrae (MMCh-PV 63–7/8) (Fig. 5d).

Locality and Age The remains were found in outcrops of the Bajada Colorada Formation (late Berriasian–Valanginian), at its type locality 40 km south of Picún Leufú town on the national route 237, in south-eastern Neuquén Province, Patagonia, Argentina.

Revised Diagnosis Tschopp et al. (2015) diagnosed *Leinkupal* by the following autapomorphies: (1) anterior caudal transverse processes have a single anterior centrodiapophyseal lamina; (2) anterior caudal transverse process extremely developed (about equal or wider to centrum width) with lateroventral expansions reinforced by robust dorsal and ventral bars; (3) very robust centroprezygapophyseal lamina in anterior caudal vertebra; (4) paired pneumatic fossae located on the base of the postzygapophysis, opposite to the articular side, in anterior-most caudal vertebra.

Comments Tschopp et al. (2015) mentioned that they included only the holotype specimen in their phylogenetic analysis without adding any autapomorphy proposed by Gallina et al. (2014) as a phylogenetic character. For that reason, autapomorphies (2) to (4) of the revised diagnosis are those proposed by Gallina et al. (2014). Regarding additional material referred to this taxon, a partial braincase and several postcranial elements were tentatively assigned to *Leinkupal*, but further studies are needed and in progress (Gallina et al. 2018; Garderes et al. 2022).

Diplodocinae indet

Material SNGM-1979, incomplete mid-posterior caudal vertebra (Fig. 5e).

Locality and Age Aysén, Patagonian Central Andes in Southern Chile. The bone-bearing unit is the Toqui Formation, dated as late Tithonian.

Comments The phylogenetic analysis performed by Salgado et al. (2015) unequivocally placed SNGM- 1979 within Diplodocinae by the presence of caudal pleurocoels, quadrangular middle caudal centra with flat ventral and lateral faces, and ventral longitudinal hollow in mid-caudal centra.

3 Anatomy of South American Flagellicaudatans: Brief Comments on the Main Contributions to the Cranial and Postcranial Knowledge of the Group

3.1 The Skull

Seven of the ten dicraeosaurid species worldwide registered (

Whitlock and Wilson Mantilla 2020) preserve cranial remains. All of these present the basicranium as the main source of skull information, but South American forms, and in particular *Bajadasaurus*, contribute strongly by adding novel information from other regions such as the snout and jaws. On the other hand, South American diplodocid cranial remains are restricted to the previously mentioned partial braincase tentatively assigned to *Leinkupal* (Garderes et al. 2022).

Amargasaurus shows a cranial morphology closely similar to *Dicraeosaurus*, which allowed, in 1992, the recognition of cranial synapomorphies of dicraeosaurids for the first time. In this sense, Salgado and Calvo (1992) recognized in the basicranium of this taxon an extreme length of the slightly diverging basipterygoid processes; an extreme reduction of the laterally facing supratemporal openings; fused frontals; a dorsolateral 'leaf-like' process of the crista prootica; deep dorsal sphenoidal pocket-like cavities; extensive frontoparietal contact; extreme anterior position of the lateral temporal fenestra; and presence of parietal and postparietal fontanelles (although a postparietal foramen occurs in a wide variety of sauropods, including some diplodocids; Tschopp et al. 2015). Besides, the ventrally oriented occipital condyle and the anteriorly projected, stout basipterygoid processes are also shared with other flagellicaudatans. The fused basal tubera and the paroccipital processes with marked ventral excavation seem to represent autapomorphies of this taxon (Salgado and Calvo 1992; Wilson 2002; Whitlock 2011).

In the case of *Bajadasaurus*, the cranial remains include the dermal bones (tooth-bearing, median roofing, circumorbital, and temporal bones) and palatal elements, braincase, and the nearly complete lower jaw. These allow to recognize an overall gracile skull of dicraeosaurid, with extremely elongated basipterygoid processes; elongate and slender anterior processes of the squamosal, probably in contact with the quadratojugal; a narrow, posteriorly elongated lateral temporal fenestra; dorsally exposed orbits; a medially extended post-temporal fenestra; anteroposteriorly extended and gracile lower jaw; and a reduced dentition in the maxilla and dentary (Gallina et al. 2019). Besides, a preantorbital fenestra seems to be absent or extremely reduced in the maxilla of *Bajadasaurus*, a similar condition occurs in *Dicraeosaurus*. This feature strongly contrasts with the observed in other groups of diplodocoids; the well-developed preantorbital fenestra present in diplodocids (Tschopp et al. 2015) and the extremely developed fenestra seen in rebbachisaurids (Canudo et al. 2018, see Chap. 5). Thus, its potential as a synapomorphy of the family cannot be discarded, as was recently corroborated by Whitlock and Wilson Mantilla (2020).

In addition, the skulls of *Bajadasaurus* and *Amargasaurus* show the cranial synapomorphies recovered for Dicraeosauridae in several phylogenetic analysis such as the crista prootica laterally expanded forming a dorsolateral process (Wilson 2002; Whitlock 2011; Tschopp et al. 2015; Whitlock and Wilson Mantilla 2020); the basal tubera narrower than the occipital condyle (Wilson 2002; Whitlock 2011; Tschopp et al. 2015); the basipterygoid processes narrowly diverging (Wilson 2002; Tschopp et al. 2015; Whitlock and Wilson Mantilla 2020); frontals fused in adults (Wilson 2002); postparietal foramen (Wilson 2002; Whitlock 2011; Whitlock and Wilson Mantilla 2020); supratemporal fenestra smaller than foramen magnum (Wilson 2002; Whitlock and Wilson Mantilla 2020); ventrally directed prong on squamosal (Whitlock 2011; Whitlock and Wilson Mantilla 2020); the area between the basipterygoid processes and parasphenoid rostrum forms a deep slot-like cavity that passes posteriorly between the bases of the basipterygoid processes (Tschopp et al. 2015). Besides that, the lower jaw of *Bajadasaurus* allows to corroborate several features proposed as synapomorphies from this region of the skull such as a subtriangular cross-sectional shape of the dentary symphysis, tapering sharply towards its ventral extreme (Whitlock 2011; Tschopp et al. 2015), and the presence of a tuberosity on the labial surface of the dentary, near the symphysis (Whitlock 2011; Tschopp et al. 2015).

3.2 *The Postcranium*

South American flagellicaudatan taxa added anatomical novelties to the postcrania of both dicraeosaurids and diplodocids.

On one hand, dicraeosaurids from the Lower Cretaceous such as *Bajadasaurus*, *Pilmatueia* and *Amargasaurus*, are characterized by the marked elongation of the bifid neural spines in their presacral vertebrae. Although fairly recognized as a dicraeosaurid synapomorphy restricted to dorsal elements (i.e. dorsal neural spines approximately four times centrum length; McIntosh 1990; Wilson 2002; Whitlock 2011), this feature seems to have been developed in an extreme fashion only in the presacral region of these South American forms, reaching the tallest elements in the cervical segment. In addition, the Late Jurassic *Brachytrachelopan* is the only dicraeosaurid, until now, with a particular shortening of cervical series that suggests niche partitioning within the group, ranging from low, ground level, to mid-height food-gathering strategies (Rauhut 2005; Paulina Carabajal et al. 2014).

On the other hand, *Leinkupal* shows a general small body size for a diplodocid, with an overall estimated length of 9 *m*. Diplodocids such as the Laurasian *Diplodocus* and *Apatosaurus*, and even the Gondwanan *Tornieria*, reached more than 20 m in length. In this context, *Leinkupal* could represent a lineage of small diplodocids that evolved restricted to western Gondwana, although more evidence is needed to support it.

Another unique feature to note from *Leinkupal* is the extreme lateral development of the anterior caudal transverse processes. The laterally extended and ventrally facing anterior caudal transverse processes in diplodocids have been interpreted as a good area for the attachment of the musculature related to the hindlimb and proximal region of the tail (Gallina and Otero, 2009). The condition seen in *Leinkupal* allows to infer that strong muscles were inserted in the base of the tail that probably allowed powerful lateral strokes of the (inferred) distal whiplash tail, in a way even more marked than that of other diplodocids.

4 Phylogenetic Considerations

Although recognized as distinct sauropod families during decades, since Upchurch (1995), both Dicraeosauridae and Diplodocidae were recovered as closely related monophyletic groups in successive phylogenies. In that pioneering work of sauropod phylogeny, Upchurch linked both Diplodocidae and Dicraeosauridae in a more inclusive group (the superfamily Diplodocoidea), based on several morphological features, mainly from the cranial anatomy. However, he also included nemegtosaurids sauropods (today widely accepted as part of Titanosauria) in that superfamily.

At the same time, Calvo and Salgado (1995) and subsequent phylogenies recognized and recovered Rebbachisauridae as another family within Diplodocoidea, on a more basal branch compared to Dicraeosauridae and Diplodocidae (e.g. Wilson 2002; Gallina and Apesteguía 2005; Salgado et al. 2006; Sereno et al. 2007; Whitlock 2011; Canudo et al. 2018). This allowed better resolution of the ingroup relationships of Diplodocoidea as well as the recognition of the Flagellicaudata (Dicraeosauridae + Diplodocidae) as a distinct clade supported by the common features in cranial and postcranial bones (Harris and Dodson 2004; Harris 2006; Whitlock 2011).

The general knowledge of dicraeosaurid diversity and evolution greatly increased in the last five years. Before 2015, only four genera were described and considered as part of this clade: *Dicraeosaurus*, *Amargasaurus*, *Brachytrachelopan*, and *Suuwassea*. This situation began to change with the dicraeosaurid affinities of *Dyslocosaurus* found by Tschopp et al. (2015) and the reinterpretation of *Amargatitanis* as a dicraeosaurid (Gallina 2016). After that, the description of three new taxa,

Bajadasaurus, *Pilmatueia*, *Lingwulong*, followed the same trend. Finally, the recent reinterpretation of '*Morosaurus*' as a basal dicraeosaurid under the new generic name of *Smitanosaurus* Whitlock and Wilson Mantilla, 2020 and the inclusion of *Kaatedocus* as a basal member of this clade, increase the taxonomic content of Dicraeosauridae more than twice than as previously recognized. Coupled with this increase in recognized diversity, new phylogenetic analyses are being carried out without obtaining a complete consensus, nor in the evolution of the group nor in the position of the South American taxa. Nevertheless, some general interrelationships seem to be better supported in the last and most complete analyses carried out (Gallina et al. 2019; Whitlock and Wilson Mantilla 2020; Fig. 6). Below we discuss, based on these two independent analyses, the positions of the South American dicraeosaurids, with some comments on previous analyses.

Bajadasaurus is recovered as the basalmost dicraeosaurid in both, Gallina et al. (2019) and Whitlock and Wilson Mantilla (2020), although with some differences.

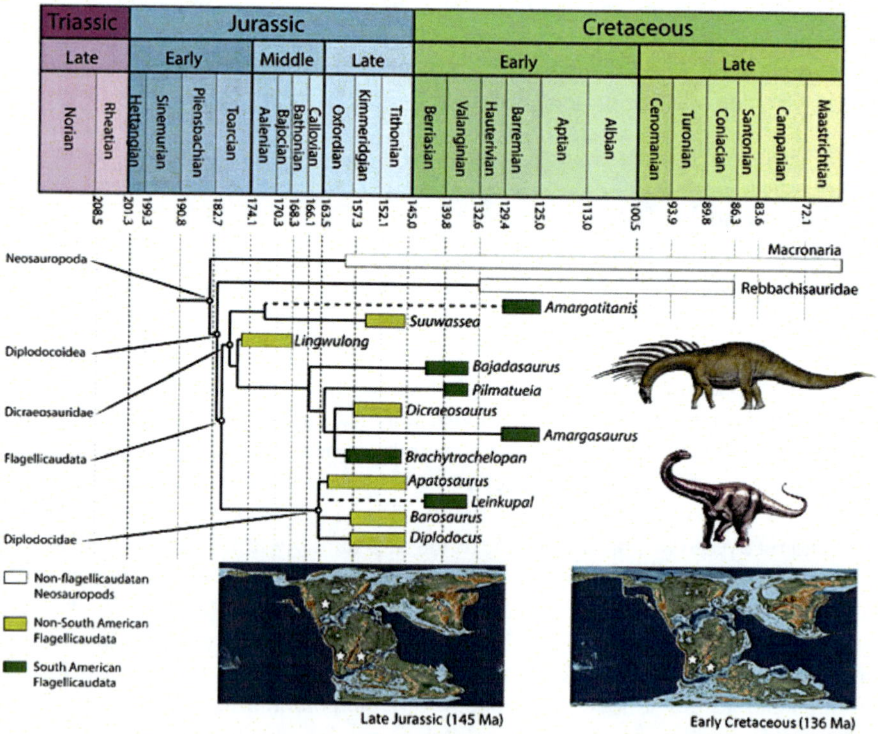

Fig. 6 Time calibrated tree showing the relationships of South American flagellicaudatans based on Gallina et al. (2019) phylogeny. Taxa in dashed lines show the tentative positions retrieved in other phylogenies (see Sect. 4). Geographic distribution of flagellicaudatans in Late Jurassic and Early Cretaceous is marked by white stars on the palaeomaps. Palaeogeographic reconstruction performed with GPlates software (Müller et al. 2018) using the PALEOMAP PaleoAtlas project (Scotese 2016). Life restoration of both *Bajadasaurus* and *Leinkupal* are courtesy of J.A. González

Contrary to the results of Gallina et al. (2019) and previous analyses (e.g. Coria et al. 2019; Xu et al. 2018), Whitlock and Wilson Mantilla (2020) found two main lineages of dicraeosaurids, with *Bajadasaurus*, *Lingwulong* and the specimen MOR 592 being recovered as part of one of these lineages. Irrespective of that, in both analyses *Bajadasaurus* and *Lingwulong* are closely related, as the Asian genus was recovered as the sister taxon of *Bajadasaurus* and more derived dicraeosaurids.

The reassignation of *Amargatitanis* as a dicraeosaurid was supported by the cladistic analysis of Gallina (2016). In that phylogeny, the taxon was recovered as the sister taxon of the North American *Suuwassea*, a form originally recovered as a basal flagellicaudatan, in an unresolved position amongst this clade (Harris and Dodson 2004; Harris 2006) but retrieved as a basal dicraeosaurid in numerous, independent phylogenies since then (Salgado et al. 2006; Whitlock 2011; Tschopp et al. 2015; Xu et al. 2018; Gallina et al. 2019). In the analysis of Gallina et al. (2019), *Amargatitanis* was found as an unstable taxon, with multiple positions amongst dicraeosaurids, or even as a basal diplodocid. Nevertheless, the close relationship of *Amargatitanis* and *Suuwassea* initially proposed by Gallina (2016) was also recovered by Whitlock and Wilson Mantilla (2020), with *Amargatitanis* being recovered as the sister taxon of *Suuwassea*, both as a basal clade of dicraeosaurids more closely related to *Dicraeosaurus* than to *Bajadasaurus*.

Amargasaurus cazaui was clearly related to *Dicraeosaurus* since its original description (Salgado and Bonaparte 1991), and a few years later recovered in the same way in the advent of sauropod phylogenies (e.g. Calvo and Salgado 1995; Upchurch 1995). Although the analysis of Calvo and Salgado (1995) followed the old conception of dicraeosaurids and diplodocids as subfamilies of Diplodocidae (Berman and McIntosh 1978), the close association of the Patagonian and the East African taxa was well supported. Successive phylogenetic analyses not only recovered *Amargasaurus* in close association with *Dicraeosaurus* but also showed high support values in the monophyly of the family Dicraeosauridae (e.g. Upchurch 1998; Wilson 2002). The abundant anatomical information that resulted from the nearly complete specimens of both taxa undoubtedly contributed to this consolidation of the family. For at least ten years, the family was represented only by these two genera.

The finding of *Brachytrachelopan* allowed the recognition of the first ingroup association, recovering this taxon more closely related to *Dicraeosaurus* than to *Amargasaurus* (Rauhut et al. 2005). Since then, the close association between these Jurassic dicraeosaurids was recovered in most phylogenies (e.g. Sereno et al. 2007; Whitlock 2011; Gallina 2016; Xu et al. 2018; Gallina et al. 2019). Whereas in the analysis of Gallina et al. (2019), *Amargasaurus*, *Brachytrachelopan*, and *Dicraeosaurus* were recovered in a trichotomy, the analyses of Whitlock and Wilson Mantilla (2020) recovered *Amargasaurus* as more related to *Brachytrachelopan* than to *Dicraeosaurus*.

Strikingly, Coria and collaborators included *Pilmatueia* in the analysis of Tschopp et al. (2015) whose phylogeny was focused in diplodocid ingroup relationships, but including several well-known dicraeosaurids (i.e. *Dicraeosaurus*, *Brachytrachelopan*, *Amargasaurus*, *Suuwassea*) and *Dyslocosaurus* (a fragmentary diplodocoid of uncertain position; Tschopp et al. 2015). In their analysis, Coria et al. (2019) recovered *Pilmatueia* in a derived position as the sister taxon of *Amargasaurus*. This close relationship between *Pilmatueia* and *Amargasaurus* was not confirmed by Gallina et al. (2019) or by Whitlock and Wilson Mantilla (2020). The ingroup topology recovered by Coria et al. (2019) also showed differences with the rest of phylogenies that recovered well-supported inner relationships (*Amargasaurus* (*Brachytrachelopan* + *Dicraeosaurus*) or even just that of *Brachytrachelopan* + *Dicraeosaurus*. It is interesting to note that in both analyses of Gallina et al. (2019) and Whitlock and Wilson Mantilla (2020), *Pilmatueia* was recovered as the sister lineage of the clade formed by *Dicraeosaurus*, *Amargasaurus,* and *Brachytrachelopan*. Future works, mainly focused on the complete description of some recently described dicraeosaurids (e.g. *Bajadasaurus, Lingwulong*) plus specimens and species pending complete studies (e.g. *Brachytrachelopan, Amargasaurus*), will certainly help in the knowledge of the evolution of dicraeosaurids and the search of a general consensus amongst phylogenies.

On the other hand, the only nominated diplodocid species, *Leinkupal laticauda*, was included in different phylogenies and recovered at a relatively stable position within Diplodocinae, in a basal (Gallina et al. 2014; Whitlock and Wilson Mantilla 2020) or more derived position (Tschopp et al. 2015; Tschopp and Mateus 2017). Meanwhile, the indeterminate records, mainly those of fragmentary specimens, from the Jurassic Cañadón Calcareo and Toqui Formations were retrieved as members of Diplodocidae or Diplodocinae in their original descriptions (Rauhut et al. 2015; Salgado et al. 2015).

5 Biogeographical and Chronological Considerations

From their early discoveries at the beginning of the twentieth century, both diplodocids and basal macronarians (mainly brachiosaurids) were recognized to have reached a wide distribution across the territories that constituted Pangea during Late Jurassic times, with well-known taxa such as *Diplodocus* Marsh (1878) and *Apatosaurus* Marsh, 1877 in Laurasia (Morrison Formation in USA), and *'Gigantosaurus'* Frass, 1908 (now *Tornieria* Sternfeld (1911) in Gondwana (Tendaguru Formation in Africa). Conversely, dicraeosaurids were considered as a fully Gondwanan lineage for many years after the recognition of *Dicraeosaurus* in the Late Jurassic of Africa (Janensch 1914) and *Amargasaurus* in the Early Cretaceous of Patagonia (Salgado and Bonaparte 1991). No clearly Laurasian records have been recovered until Salgado et al. (2006) considered *Suuwassea* as a basal dicraeosaurid. Such a position was later corroborated by several analyses (e.g. Whitlock 2011; Tschopp et al. 2015; Xu et al. 2018; Gallina et al. 2019), and the presence of this clade in the Morrison Formation was recently strengthened with the description of *Smitanosaurus* (Whitlock and Wilson Mantilla 2020). The recognition of *Lingwulong* in the Middle Jurassic of China not only reinforced the idea of a Pangean distribution of dicraeosaurids but also pulled the origin of Neosauropoda down to the Toarcian–Bajocian (Xu et al. 2018).

The idea of a Middle Jurassic origin for Neosauropoda and its major lineages, prior to the final fragmentation of Pangea, has been suggested by several authors (e.g. Upchurch et al. 2002; Carrano et al. 2012; Benson et al. 2014) in order to explain their widespread presence in Upper Jurassic outcrops of both Laurasia and Gondwana. In that scenario, the only putative neosauropod for the older, Middle–Late Jurassic strata (see Xu et al. 2018) is *Bellusaurus* (Upchurch et al. 2004; Carballido et al. 2015; Moore et al. 2018), which potentially indicated an earlier Jurassic origin for Neosauropoda and its two stem lineages, Macronaria and Diplodocoidea. Nevertheless, it was not until the recent discovery of the Asian dicraeosaurid that such hypothesis was proved with fossil evidence. In this context, the age of *Lingwulong* points to the origin of both diplodocoid lineages (Rebbachisauridae and Flagellicaudata), as well as the Macronaria during the Toarcian–Bajocian (Xu et al. 2018). Xu and collaborators noted that the diversification of Neosauropoda during the Early–Middle Jurassic was more notable than previously suspected, although its earliest evolution remains enigmatic because of the limited rock outcrops of this age. It must be noted that the Cañadón Asfalto Formation (traditionally interpreted as Middle Jurassic) which is rich in basal, non-neosauropod sauropods (see Chap. 4) is actually older than previously thought and must be considered as Toarcian in age (Pol et al. 2020).

Irrespective of the origin and speed of dispersion and diversification of the Neosauropoda, it is clear that by Late Jurassic times both dicraeosaurids and diplodocids were well-established and became the dominant sauropods across Laurasia and Gondwana (Fig. 6). In South America, these two lineages were recorded from the Upper Jurassic Cañadón Calcáreo Formation in Chubut Province, Argentina (Cúneo et al. 2013), with the recognition of *Brachytrachelopan* and an indeterminate diplodocid (see Sect. 2), whereas the Toqui Formation in southern Chile only produces remains of diplodocids so far (Salgado et al. 2015). These records, together with the presence of the basal camarasauromorph *Tehuelchesaurus* and basal titanosauriforms (Rauhut 2006; see Chap. 7), suggest that a similar, Pangean fauna inhabited Gondwana and Laurasia by Late Jurassic times (e.g. Apesteguía 2002; Whitlock 2011; D'Emic 2012; Xu et al. 2018; Mannion et al. 2019).

It was generally accepted that the Jurassic–Cretaceous boundary (145 Ma) produced a marked change in the dinosaur fauna, which is especially notable when comparing the peak in the diversity registered in the Late Jurassic and their decline by the Early Cretaceous (e.g. Upchurch and Barrett 2005; Barrett et al. 2009; Mannion et al. 2011). Nevertheless, as noted by Upchurch et al. (2015) the Early Cretaceous is characterized by a dearth of sauropod-bearing localities (McPhee et al. 2016), at least in a global record biased towards Laurasia. This conception was supported for many years by the total absence of flagellicaudatan sauropods in Cretaceous outcrops, a conception that started to change in the 90 s with the description of *Amargasaurus* and, more recently, with the newly recognized high diversity of dicraeosaurids in the Early Cretaceous of Patagonia and the presence of diplodocid sauropods, at least, during the beginning of the Cretaceous, which changed the panorama completely.

The records in Patagonia represent the most complete data of sauropod diversity throughout the earliest Early Cretaceous (Berriasian–Barremian), being of special importance to better understand the characteristics of this faunal transition.

The prospecting efforts carried out at the end of the twentieth century mainly by José Bonaparte parties and at the beginning of the twenty-first century by several researchers in the search of fossils in earliest Cretaceous outcrops of Patagonia resulted in the recognition of two units that yield the most exhaustive faunal evidence at present. The Bajada Colorada Formation (Berriasian–Valanginian; Foucault et al. 1987; Leanza and Hugo 1997) shows the oldest Cretaceous record for South America and has provided the highest diversity of sauropod dinosaurs described so far, with three genera including the diplodocid *Leinkupal*, the dicraeosaurid *Bajadasaurus*, and the titanosaur *Ninjatitan* (Gallina et al. 2021). This record indicates that both, dicraeosaurids and diplodocids, were most probably continuously present in South America since the Late Jurassic (taking into account the records of Cañadón Calcareo Formation, see above, which shows that the extinction of diplodocids is locally restricted to Laurasia, instead of being a global phenomenon as previously thought) (Gallina et al. 2014). When compared to the well-known Late Jurassic faunal components, the main difference with the fauna from the Bajada Colorada Formation is the presence of the earliest titanosaurian sauropods in the latter (Gallina et al. 2021). Therefore, in contrast to the presence of basal macronarians (e.g. *Tehuelchesaurus*, *Camarasaurus*) or basal titanosauriforms (such as the brachiosaurid *Giraffatitan* or the Titanosauriformes indet. from Cañadón Calcáreo; Rauhut, 2006), the presence of a derived somphospondylan in the Bajada Colorada Formation, together with the presence of both flagellicaudatan lineages, indicates a rather progressive replacement of the sauropod faunal associations between the Late Jurassic and the Early Cretaceous. In a similar way, the presence of a diplodocid caudal vertebra and the dicraeosaurid *Pilmatueia* in the Mulichinco Formation (Valanginian; Schwarz et al. 2011; Pino et al. 2017) support an increased flagellicaudatan diversity and abundance at the earliest Cretaceous. For the moment, and despite their reconstructed Jurassic origin (see above and Chap. 6), there are no rebbachisaurid records prior to the putative presence of this group in the La Amarga Formation (Barremian). A similar, but much more fragmentary fauna was recorded in the lowermost Cretaceous of Africa (Kirkwood Formation), with materials than can be clearly assigned to Dicraeosauridae, Diplodocidae, Camarasauromorpha, and Titanosauriformes (McPhee et al. 2016). Therefore, the record of Patagonia, plus that of Africa indicates a more widespread, Gondwanan, presence of flagellicaudatans at the beginning of the Cretaceous.

The La Amarga Formation (Barremian; Leanza and Hugo 1997; Apesteguía 2007) provided, up to this date, two dicraeosaurid taxa, *Amargasaurus* and *Amargatitanis*, plus additional fragmentary material that includes most probably basal titanosauriforms and rebbachisaurids (Salgado and Bonaparte 1991; Apesteguía 2007; Gallina 2016). Therefore, when compared to the earlier Cretaceous faunas, the main difference observed is the presence of rebbachisaurids (one of the oldest Cretaceous records of this clade) and the absence of diplodocids. Irrespective if the absence of diplodocids reflects a genuine final extinction of this lineage by Valanginian–Barremian times, it seems clear that none of the flagellicaudatans survived after the Barremian, as no signs of them were recovered in the relatively rich Aptian–Albian outcrops of Patagonia (Lohan Cura, Agrio, and Cerro Barcino formations; Leanza and Hugo 1995; 1997; Canudo et al. 2018; Krause et al. 2020). Indeed, the 'mid-Cretaceous' faunal assemblages of Patagonia are characterized in their sauropod components by the absence of flagellicaudatans, and a remarkable abundance of rebbachisaurids and titanosauriforms (Apesteguía 2002; Leanza et al. 2004; Krause et al. 2020). Therefore, and despite the previous hypothesis of a marked faunal change during the Jurassic/Cretaceous boundary, it is much more likely that, at least in Gondwana, a staggered faunal replacement occurred throughout the Early Cretaceous.

Finally, from the six tetrapod assemblages identified by Leanza et al. (2004) for Cretaceous tetrapods in the southern Neuquina Basin, the earliest was the Amargan assemblage (Barremian–Early Aptian), considered by the authors as a late Pangean fauna. At that time, no fossil vertebrates had been found from the Bajada Colorada Formation yet. Therefore, we decided to add here a brief characterization for a new, older tetrapod assemblage mainly represented by the Bajada Colorada and the Mulichinco formations: the Bajadan assemblage. The units that provide evidence to characterize this assemblage were deposited by Berriasian–Early Hauterivian times, all of them part of the Lower and Middle Mendoza Group, underneath the Coihuequican unconformity (Leanza 2009). This assemblage is characterized by the co-occurrence of mostly global clades originated in Pangean continents during the Late Jurassic (e.g. neosauropods, basal neotetanurans and neoceratosaurians theropods, eurypodans and basal iguanodontian ornithischians), and most of the taxa also occurring outside Gondwana, with no strong Gondwanan endemisms. The Bajadan assemblage thus includes flagellicaudatan (diplodocids and dicraeosaurids), and basal titanosaurian sauropods; abelisauroid ceratosaurian, megalosauroid, and carcharodontosaurid tetanuran theropods; and thyreophoran (either a eurypodan form or a new lineage), as well as basal iguanodontian ornithischians (Coria et al. 2013, 2019, 2020; Gallina et al. 2014, 2019, 2021; Apesteguía et al. 2015; Canale et al. 2017; Gnaedinger et al. 2017). The Bajadan fauna shows strong similarities with the Late Jurassic–earliest Cretaceous African fauna.

6 Conclusions

The number of specimens and taxonomical diversity of the flagellicaudatan record from South America noticeably increased in the last decade reaching five named dicraeosaurid taxa (*Amargasaurus cazaui*, *Amargatitanis macni*, *Bajadasaurus pronuspinax*, *Brachytrachelopan mesai*, and *Pilmatueia faundezi*), one diplodocid (*Leinkupal laticauda*), plus additional dicraeosaurid and diplodocid records known from fragmentary materials. All this record contributes in different ways to the knowledge of this distinctive and global group of sauropod dinosaurs.

Main anatomical contributions from the skull are provided from the basicrania of *Amargasaurus* and *Bajadasaurus* adding, in the latter, novel information from little explored regions such as the snout and jaws. While dicraeosaurids such as *Bajadasaurus*, *Pilmatueia,* and *Amargasaurus* are characterized by the marked elongation of their presacral bifid neural spines in an extreme fashion, and *Brachytrachelopan* shows a short neck by a particular shortening of the cervical series, the diplodocid *Leinkupal* presents a general small body size and an extreme lateral development of the anterior caudal transverse processes, adding novelties to the postcrania of flagellicaudatan diplodocoids.

The number of recognized species of Dicraeosauridae increased more than twice in the last five years, leading to new phylogenetic analyses, although we still lack a complete consensus for the evolutionary traits of the group and the relative position of some South American taxa. Nevertheless, the last and most complete analyses carried out seem to have achieved a good support in their general interrelationships.

By Late Jurassic times, flagellicaudatan diplodocoids became a well-established and quite dominant sauropod group across Laurasia and Gondwana. In South America, both dicraeosaurids and diplodocids were recorded from the Late Jurassic of Argentina and Chile. The conception of a marked change in the dinosaur fauna across Jurassic–Cretaceous boundary (145 Ma) shifted for southern continents with the unexpected high diversity of dicraeosaurids in the Early Cretaceous of Patagonia plus the presence of diplodocid sauropods at the beginning of the Cretaceous. These records from Patagonia actually represent the most complete record of sauropod diversity throughout the earliest Early Cretaceous (Berriasian–Barremian), being of special importance to better understand the characteristics of such faunal transition. Additionally, the similar African record indicates a more widespread, Gondwanan, presence of flagellicaudatans at the beginning of the Cretaceous, evidencing that a staggered faunal replacement occurred throughout the Early Cretaceous.

Finally, the new proposed Bajadan tetrapod assemblage (Berriasian–Early Hauterivian) is the oldest for the Neuquen basin tetrapod assemblages, showing strong similarities with the Late Jurassic–earliest Cretaceous African fauna.

Acknowledgements We thank the editors A. Otero, J. L. Carballido and D. Pol, for inviting us to participate in this book. The following institutions and projects are acknowledged for financial and/or logistic support for working with flagellicaudatans along the years: Fundación Azara, Universidad Maimónides, Museo Ernesto Bachmann Villa El Chocón (to PAG, SA and JPG), Agencia PICT 2013-0704, National Geographic/Waitt Grant W645-16 (to PAG), PIP-CONICET 114 201101 00314 (to SA), Agencia PICT 0668 and 1925 (to JLC). We are grateful to L. Salgado and E. Tschopp for their comments that greatly improved this chapter.

References

Apesteguía S (2002) Successional structure in continental tetrapod faunas from Argentina along the Cretaceous. In: Abstract of the 6º Simpósio sobre o Cretáceo do Brasil and 2º Simposio sobre el Cretácico de América del Sur, Sao Pedro, Brasil, 135–141 Nov 2002.

Apesteguía S (2007) The sauropod diversity of the La Amarga Formation (Barremian), Neuquén (Argentina). Gondwana Res 12:533–546

Apesteguía S, Gallina PA, Canale JI, Riguetti F, Garderes JP (2015) Nuevos restos de Stegosauria (Ornithischia, Thyreophora) del Cretácico Inferior temprano de Neuquén: el registro más antiguo de Sudamérica. Ameghiniana 52:R5

Barrett PM, McGowan AJ, Page V (2009) Dinosaur diversity and the rock record. P Roy Soc Lond B 276:2667–2674

Benson RBJ, Campione NE, Carrano MT, Mannion PD, Sullivan C, Upchurch P, Evans DC (2014) Rates of Dinosaur Body Mass Evolution Indicate 170 Million Years of Sustained Ecological Innovation on the Avian Stem Lineage. PLoS Biol 12(5):e1001853

Berman DS, McIntosh JS (1978) Skull and relationships of the Upper Jurassic Sauropod Apatosaurus (Reptilia, Saurischia). Carn Mus Bull 8:1–35

Calvo JO, Salgado L (1995) *Rebbachisaurus tessonei* sp. nov. a new Sauropoda from the Albian-Cenomanian of Argentina; new evidence on the origin of the Diplodocidae. Gaia 11:13–33

Canale JI, Apesteguía S, Gallina PA, Gianechini FA, Haluza A (2017) The oldest theropods from the Neuquén Basin: Predatory dinosaur diversity from the Bajada Colorada Formation (Lower Cretaceous: Berriasian–Valanginian), Neuquén, Argentina. Cret Res 71:63–78

Canudo JI, Carballido JL, Garrido AC, Salgado L (2018) A new rebbachisaurid sauropod from the Lower Cretaceous (Aptian-Albian) of the Rayoso Formation (Neuquén, Argentina). Acta Palaeontol Pol 63:679–691

Carballido JL, Pol D, Parra Ruge ML, Padilla Bernal S, Paramo-Fonseca ME, Etayo-Serna F (2015) A new Early Cretaceous brachiosaurid (Dinosauria, Neo- sauropoda) from northwestern Gondwana (Villa de Leiva, Colombia). J Vertebr Paleontol 35:e980505

Carrano MT, Benson RBJ, Sampson SD (2012) The phylogeny of Tetanurae (Dinosauria: Theropoda). J Syst Palaeontol 10:211–300

Coria RA, Ortega F, Currie P, Previtera E, Cardenas M (2013) An Early Cretaceous dinosaur assemblage from the Neuquen Basin, Patagonia. Argentina. Ameghiniana 50:R13

Coria RA, Windholz GJ, Ortega F, Currie PJ (2019) A new dicraeosaurid sauropod from the Lower Cretaceous (Mulichinco Formation, Valanginian, Neuquén Basin) of Argentina. Cret Res 93:33–48

Coria RA, Currie PJ, Ortega F, Baiano MA, 2020. An Early Cretaceous, medium-sized carcharodontosaurid theropod (Dinosauria, Saurischia) from the Mulichinco Formation (upper Valanginian), Neuquén Province, Patagonia, Argentina. Cret Res 111:104319

Cúneo R, Ramezani J, Scasso R, Pol D, Escapa I, Zavattieri AM, Bowring SA (2013) High-precision U-Pb geochronology and a new chronostratigraphy for the Cañadón Asfalto Basin, Chubut, central Patagonia: implications for terrestrial faunal and floral evolution in Jurassic. Gondwana Res 24:1267–1275

D'Emic (2012) The early evolution of titanosauriform sauropod dinosaurs. Zool J Linn Soc 166:624–671. https://doi.org/10.1111/j.1096-3642.2012.00853.x

Fraas E (1908) Ostafrikanische Dinosaurier. Palaeontographica 55:105–144

Foucault JE, Vaillard LC, Viñes RF (1987) Estratigrafía de las unidades aflorantes a lo largo del curso inferior del río Limay, provincias del Neuquén y Río Negro. 10º Congreso Geológico Argentino. Actas 1:139–142

Gallina PA (2016) Reappraisal of the Early Cretaceous sauropod dinosaur *Amargatitanis macni* (Apesteguía, 2007), from northwestern Patagonia, Argentina. Cret Res 64:79–87

Gallina PA, Apesteguía S (2005) *Cathartesaura anaerobica* gen. et sp. nov., a new rebbachisaurid (Dinosauria, Sauropoda) from the Huincul Formation (Upper Cretaceous), Río Negro, Argentina. Rev Mus Argentino Cienc Nat Ns 7:153–166

Gallina PA, Otero A (2009) Anterior caudal transverse processes in sauropod dinosaurs: morphological, phylogenetic and functional aspects. Ameghiniana 46:165–176

Gallina PA, Apesteguía S, Haluza A, Canale JI (2014) A diplodocid sauropod survivor from the Early Cretaceous of South America. PLoS ONE 9(5):e97128

Gallina PA, Canale JI, Apesteguía S (2018) New materials of the diplodocid sauropod Leinkupal laticauda from the Bajada Colorada Formation (Berriasian-Valanginian), Neuquén Province. Argentina. Peapa 18(2):R29

Gallina PA, Apesteguía S, Canale JI, Haluza A (2019) A new long-spined dinosaur from Patagonia sheds light on sauropod defense system. Sci Rep 9:1392

Gallina PA, Canale JI, Carballido JL (2021) The earliest known titanosaur sauropod dinosaur. Ameghiniana 58:35–51

Garderes JP, Gallina PA, Whitlock JA, Toledo N (2022) Neuroanatomy of a diplodocid sauropod dinosaur from the Lower Cretaceous of Patagonia, Argentina. Cret Res 129:e105024. https://doi.org/10.1016/j.cretres.2021.105024.

Gnaedinger S, Coria RA, Koppelhus E, Casadío C, Tunik M, Currie P (2017) First Lower Cretaceous record of Podocarpaceae wood associated with dinosaur remains from Patagonia, Neuquén Province, Argentina. Cret Res 78:228–239

Harris JD, Dodson PA (2004) New diplodocoid sauropod dinosaur from the Upper Jurassic Morrison Formation of Montana, USA. Acta Palaeontol Pol 49:197–210

Harris JD (2006) The significance of *Suuwassea emilieae* (Dinosauria: Sauropoda) for Flagellicaudatan intrarelationships and evolution. J Syst Palaeontol 4:185–198

Janensch W (1914) Übersicht über die Wirbeltierfauna der Tendaguru-Schichten, nebst einer kurzen Charakterisierung der neu aufgeführten Arten von Sauropoden. Archiv. Für Biontologie 3:81–110

Janensch W (1929) Die Wirbelsäule der Gattung Dicraeosaurus. Palaeontographica (Supplement 7):37–133

Krause JM, Ramezani J, Umazano AM, Pol D, Carballido JL, Sterli J, Puerta P, Cúneo NR, Bellosi ES (2020) High-resolution chronostratigraphy of the Cerro Barcino Formation (Patagonia): Paleobiologic implications for the mid-cretaceous dinosaur-rich fauna of South America. Gondwana Res 80:33–49

Leanza HA (2009) Las principales discordancias del Mesozoico de la Cuenca Neuquina según observaciones de superficie. Rev Mus Argentino Cienc Nat Ns 11:145–184

Leanza HA, Hugo CA (1995) Revisión estratigráfica del Cretácico inferior continental en el ámbito sudoriental de la Cuenca Neuquina. RAGA 50:30–32

Leanza HA, Hugo CA (1997) Hoja Geológica 3969-III - Picún Leufú, provincias del Neuquén y Río Negro. Instituto de Geología y Recursos Naturales. SEGEMAR. Boletín 218:1–135

Leanza HA, Apesteguía A, Novas FE, de la Fuente MS (2004) Cretaceous terrestrial beds from the Neuquén Basin (Argentina) and their tetrapod assemblages. Cret Res 25:61–87. https://doi.org/10.1016/j.cretres.2003.10.005

Mannion PD, Upchurch P, Carrano MT, Barrett PM (2011) Testing the effect of the rock record on diversity: a multidisciplinary approach to elucidating the generic richness of sauropodomorph dinosaurs through time. Biol Rev 86:157–181

Mannion PD, Upchurch P, Schwarz D, Wings O (2019) Taxonomic affinities of the putative titanosaurs from the Late Jurassic Tendaguru Formation of Tanzania: phylogenetic and biogeographic implications for eusauropod dinosaur evolution. Zool J Linn Soc 185:784–909

Marsh OC (1878) Principal characters of American Jurassic dinosaurs, Part I. Am J Sci (Series 3) 16:411–416

Marsh OC (1884) Principal characters of American Jurassic dinosaurs. Part VII. On the Diplodocidae, a new family of the Sauropoda. Am J Sci (Series 3) 27:160–168

McPhee BW, Mannion PD, de Klerk WJ, Cnoiniere JN (2016) High diversity in the sauropod dinosaur fauna of the Lower Cretaceous Kirkwood Formation of South Africa: Implications for the Jurassic-Cretaceous transition. Cret Res 59:228–248

McIntosh JS (1990) Sauropoda. In: Weishampel DB, Dodson P, Osmolska H (eds) The Dinosauria. University of California Press, Berkeley, pp 345–401

Moore AJ, Mo J, Clark J, Xu X (2018) Cranial anatomy of *Bellusaurus sui* (Dinosauria: Eusauropoda) from the Middle-Late Jurassic Shishugou Formation of northwest China and a review of sauropod cranial ontogeny. PeerJ 6:e4881

Müller RD, Cannon J, Qin X, Watson RJ, Gurnis M, Williams S, Pfaffelmoser T, Seton M, Russell S, Zahirovic S (2018) GPlates: building a virtual Earth through deep time. Geochem Geophy Geosy 19(7):2243–2261

Osborn HF (1899) A skeleton of *Diplodocus*. Mem Am Mus Nat Hist 5:191–214

Paulina Carabajal A, Carballido JL, Currie PJ (2014) Braincase, neuroanatomy and neck posture of *Amargasaurus cazaui* (Sauropoda: Dicraeosauridae) and its implications for understanding head posture in sauropods. J Vertebr Paleontol 34:870–882

Paulina Carabajal A, Coria RA, Currie PJ, Koppelhus EB (2018) A natural cranial endocast with possible dicraeosaurid (Sauropoda, Diplodocidae) affinities from the Lower Cretaceous of Patagonia. Cret Res 84:437–441

Pino DA, Tunik MA, Coria RA, Ison JI (2017) Análisis paleoambiental y de procedencia de los depósitos cretácicos aflorantes en la margen norte del Río Agrio en el área del Arroyo Pilmatue, Las Lajas, Cuenca Neuquina. In: Technical session 7 of the XX Congreso Geológico Argentino, Tucumán, Argentina, 102–109 August 2017

Pol D, Ramezani J, Gomez K, Carballido JL, Paulina Carabajal A, Rauhut OWM, Escapa I, Cúneo RN (2020) Extinction of herbivorous dinosaurs linked to Early Jurassic global warning event. P Roy Soc B 287:20202310

Rauhut OWM (2006) A brachiosaurid sauropod from the Late Jurassic Cañadón Calcáreo Formation of Chubut, Argentina. Foss Rec 2:226–237

Rauhut OWM, Remes K, Fechner R, Cladera G, Puerta P (2005) Discovery of a short-necked sauropod dinosaur from the Late Jurassic period of Patagonia. Nature 435:670–672

Rauhut OWM, Carballido JL, Pol D (2015) A diplodocid sauropod dinosaur from the Late Jurassic Canadón Calcáreo Formation of Chubut, Argentina. J Vertebr Paleontol 35:e982798

Remes K (2006) Revision of the Tendaguru sauropod dinosaur *Tornieria Africana* (Fraas) and its relevance for sauropod paleobiogeography. J Vertebr Paleontol 26:651–669

Salgado L, Bonaparte JF (1991) Un nuevo saurópodo Dicraeosauridae, *Amargasaurus cazaui* gen et sp. nov., de la Formación La Amarga, Neocomiano de la provincia del Neuquén. Argentina. Ameghiniana 28:333–346

Salgado L, Calvo JO (1992) Cranial osteology of *Amargasaurus cazaui* Salgado & Bonaparte (Sauropoda, Dicraeosuridae) from the Neocomian of Patagonia. Ameghiniana 29:337–346

Salgado L, De Souza CI, Garrido AC (2006) *Zapalasaurus bonapartei*, un nuevo dinosaurio saurópodo de La Formación La Amarga (Cretácico Inferior), noroeste de Patagonia, Provincia de Neuquén, Argentina. Geobios 39:695–707

Salgado L, Novas FE, Suarez M, De La Cruz R, Isasi M, Rubilar-Rogers D, Vargas A (2015) Late Jurassic sauropods in Chilean Patagonia. Ameghiniana 52:418–429

Scotese CR (2016) PALEOMAP PaleoAtlas for GPlates and the PaleoData plotter program. Geol Soc Am, Abstract with Programs 48(5)

Sereno PC, Wilson JA, Witmer LM, Whitlock JA, Maga A, Ide O, Rowe TA (2007) Structural extremes in a Cretaceous dinosaur. PLoS ONE 2:e1230

Sereno P (1998) A rationale for phylogenetic definitions, with application to the higher-level taxonomy of Dinosauria. Neues Jahrb Geol P-A 210:41–83

Schwarz E, Spalletti LA, Veiga GD (2011) La Formacion Mulichinco (Cretacico temprano) en la Cuenca Neuquina. Relatorio del XVIII Congreso Geologico Argentino 131–144

Sternfeld R (1911) Zur Nomenklatur der Gattung Gigantosaurus Fraas. Sitzungsberichte Der Gesellschaft Naturforschender Freunde Zu Berlin 1911:398

Taylor MP, Naish D (2005) The phylogenetic taxonomy of Diplodocoidea (Dinosauria: Sauropoda). PaleoBios 25:1–7

Tschopp E, Mateus O (2017) Osteology of *Galeamopus pabsti* sp. nov. (Sauropoda: Diplodocidae), with implications for neurocentral closure timing, and the cervico-dorsal transition in diplodocids. PeerJ 5:e3179

Tschopp E, Mateus O, Benson RBJ (2015) A specimen-level phylogenetic analysis and taxonomic revision of Diplodocidae (Dinosauria, Sauropoda). PeerJ 3:e857

Upchurch P (1995) Evolutionary history of sauropod dinosaurs. Philos T R Soc Lon B 349:365–390

Upchurch P (1998) The phylogenetic relationships of sauropod dinosaurs. Zool J Linn Soc 124:43–103

Upchurch P, Barrett PM (2005) A phylogenetic perspective on sauropod diversity. In: Curry Rogers KA, Wilson JA (eds) The sauropods: evolution and paleobiology. University of California Press, Berkeley and Los Angeles, pp 104–124

Upchurch P, Hunn CA, Norman DB (2002) An analysis of dinosaurian biogeography: evidence for the existence of vicariance and dispersal patterns caused by geological events. Proc Biol Sci 269:613–621

Upchurch P, Barrett PM, Dodson P (2004) Sauropoda. In: Weishampel DB, Dodson P, Osmólska H (eds) The Dinosauria, 2nd edn. University of California Press, Berkeley, pp 259–322

Upchurch P, Mannion PD, Taylor MT, Joger U (2015) The anatomy and phylogenetic relationships of "Pelorosaurus" becklesii (Neosauropoda Macronaria) from the early cretaceous of England. PLoS One 10(6):e0125819

Whitlock JA (2011) A phylogenetic analysis of Diplodocoidea (Saurischia: Sauropoda). Zool J Linn Soc 161:872–915

Whitlock JA, Wilson Mantilla JA (2020) The Late Jurassic sauropod dinosaur '*Morosaurus*' *agilis* Marsh, 1889 reexamined and reinterpreted as a dicraeosaurid. J Vertebr Paleontol e1780600 https://doi.org/10.1080/02724634.2020.1780600

Wilson JA (2002) Sauropod dinosaur phylogeny: critique and cladistics analysis. Zool J Linn Soc 136:217–276

Windholz GJ, Baiano MA, Bellardini F, Garrido A (2021) New Dicraeosauridae (Sauropoda, Diplodocoidea) remains from the La Amarga Formation (Barremian–Aptian, Lower Cretaceous), Neuquen Basin, Patagonia, Argentina. Cret Res 117:104629

Xu X, Upchurch P, Mannion PD, Barrett PM, Regalado-Fernandez OM, Mo J, Ma J, Liu H (2018) A new Middle Jurassic diplodocoid suggests an earlier dispersal and diversification of sauropod dinosaurs. Nat Comm 9:2700

The Rise of Non-Titanosaur Macronarians in South America

Jose L. Carballido, Flavio Bellardini, and Leonardo Salgado

Abstract Although the origin of Neosauropoda probably dates back to the Early–Middle Jurassic, it is not until the Late Jurassic that Macronaria become well represented in the fossil record. Unlike the great diversity of South American titanosaurs, basal macronarians are relatively scarce in the fossil record; even so, they provide valuable information for better understanding the first steps at the origin of this clade. The only non-titanosauriform macronarian sauropod from South America is *Tehuelchesaurus* from the Oxfordian-Tithonian of Argentina, while all other basal macronarians found up to date are titanosauriforms (either Brachiosauridae or Somphospondyli). Brachiosaurids were abundant in the Jurassic, but they apparently became extinct at the Jurassic/Cretaceous boundary all over the world except in North America. Isolated elements from the Late Jurassic of Argentina and *Padillasaurus* from the Early Cretaceous of Colombia were suggested as brachiosaurids, but these assignments are questionable. Until now, no clear somphospondylans have been recorded in Jurassic levels. In South America, basal, non-titanosaur somphospondylans are represented by three taxa registered in Argentina: *Chubutisaurus* and *Ligabuesaurus* from the Early Cretaceous (Aptian–Albian) and *Malarguesaurus* from the Late Cretaceous (Turonian–Coniacian). Here, we provide a complete revision on the fossil record of non-titanosaur macronarians from South America and the current phylogenetic status of them.

Keywords Macronaria · Late Jurassic · Early Cretaceous · Taxonomy · Evolution

J. L. Carballido (✉) · F. Bellardini · L. Salgado
Consejo Nacional de Investigaciones Científicas y Técnicas (CONICET), Buenos Aires, Argentina
e-mail: jcarballido@mef.org.ar

L. Salgado
e-mail: lsalgado@unrn.edu.ar

J. L. Carballido
Museo Paleontológico Egidio Feruglio, Fontana 140, Trelew (9100), Chubut, Argentina

F. Bellardini · L. Salgado
Instituto de Investigación en Paleobiología y Geología, Universidad Nacional de Río Negro, Av. Gral. J.A. Roca 1242, 8332 General Roca, Río Negro, Argentina

A. Otero et al. (eds.), *South American Sauropodomorph Dinosaurs*, Springer Earth System Sciences, https://doi.org/10.1007/978-3-030-95959-3_7

1 Introduction

Macronaria, defined as the clade formed by sauropods more closely related to *Saltasaurus* than to *Diplodocus* (Wilson and Sereno 1998), is the one of the two groups into which the Neosauropoda are divided. Taking into account the Middle Jurassic age of the dicraeosaurid neosauropod *Lingwulong*, the origin of the clade must be traced back to the Toarcian–Bajocian (Xu et al. 2018). Besides this taxon, undisputed macronarian sauropods are recorded from the Late Jurassic up to the latest Cretaceous. Camarasauromorpha, nested within Macronaria, is the basalmost macronarian, node-based, clade defined as: *Camarasaurus*, *Saltasaurus*, their most common ancestor and all its descendants (Salgado et al. 1997). Depending on the phylogenetic position of *Bellusaurus*, a putative macronarian from northwest China (Upchurch et al. 2004; Carballido and Sander 2014; Moore et al. 2018), their record could be traced back to the Callovian–Oxfordian. Nevertheless, in South America, the oldest records of camarasauromorph neosauropods are those from the Cañadón Calcáreo Formation (Oxfordian–Kimmeridgian; Cúneo et al. 2013), acquiring this clade its major diversity during the Late Cretaceous, when titanosaur flourished all over the world (Cerda et al. 2011). Camarasauromorpha was a diverse clade, both in their anatomy and in their body size, indicating it was an ecologically diverse clade. Towards the Late Jurassic, a great diversity of this lineage is observed, with a rich fossil record of basal camarasauromorphs (e.g. *Camarasaurus, Tehuelchesaurus, Europasaurus*) and the presence of Titanosauriformes, the lineage formed by *Saltasaurus*, *Giraffatitan*, their most recent common ancestor and all its descendants (Salgado et al. 1997). Brachiosaurids, one of the two stem lineages of Titanosauriformes (the other one is Somphospondyli), were unambiguously recovered from Upper Jurassic units of Africa (*Giraffatitan*), North America (*Brachiosaurus*) and Europe (*Vouivria, Lusotitan, Soriatitan*) (Mannion et al. 2017; Royo Torres et al. 2017). Its sister taxon, Somphospondyli, is the lineage that leads to Titanosauria, the most diverse clade of sauropods. Earliest titanosaur (*Ninjatitan*) comes from the Berriasian–Valanginian of Patagonia (Gallina et al. 2021). Additional but questionable records (see Mannion et al. 2019a) of titanosaurs from the lowest Cretaceous strata were described from Brazil (*Triunfosaurus*) and Russia (*Volgatitan* and *Tengrisaurus*) (Carvalho et al. 2017; Averianov and Efimov 2018; Averianov and Skutschas 2017). Nevertheless, it is not until the Late Cretaceous that this clade becomes dominant, being the Early Cretaceous dominated by basal forms of Somphospondyli.

Up to date, South American non-titanosaur macronarians are restricted to five genera: *Tehuelchesaurus* from the Late Jurassic, *Padillasaurus*, *Chubutisaurus*, *Ligabuesaurus*, from the Early Cretaceous, and *Malarguesaurus* from the earliest Late Cretaceous (Fig. 1; Table 1). Additionally, relevant non-diagnostic material at genus level was described from the Upper Jurassic outcrops of the Cañadón Asfalto Formation (Rauhut 2006) and from the Lower Cretaceous strata of the La Amarga Formation (Apesteguía 2007), both in Patagonia. Despite their current poor fossil record (when compared to that of Titanosauria), basal, non-titanosaur macronarians from South

Fig. 1 Political map of South America (black dots indicate capitals) showing the provenance of non-titanosaur macronarians from this continent. **1**, *Tehueclhesaurus*; **2**, *Ligabuesaurus*; **3**, *Padillasaurus*; **4**, *Chubutisaurus*; **5**, *Malarguesaurus*

Table 1 List of basal macronarian fossil records from South America and its most probable phylogenetic position, as is discussed here

Taxon/specimen	Taxonomical attribution	Geological unit	Locality	Age	Reference
Tehuelchesaurus benitezii	Camarasauromorpha	Cañadón Calcáreo Fm	Estancia Fernández, Chubut Province, Argentina	Oxfordian–Tithonian	Rich et al. (1999), Carballido et al. (2011a, b), Cúneo et al. (2013)
Padillasaurus leivaensis	Brachiosauridae	Paja Fm	Villa de Leyva, Ricaurte Province, Colombia	Barremian–Aptian	Carballido et al. (2015)
Chubutisaurus insignis	Somphospondyli	Cerro Barcino Fm	Estancia El dinosaurio, Chubut Province, Argentina	Cenomanian	del Corro (1975), Carballido et al. (2011a, b)
Ligabuesaurus leanzai	Somphospondyli	Lohan Cura Fm	Cerro de los Leones, Neuquén Province, Argentina	Albian	Bonaparte et al. (2006)
Malarguesasurus florenciae	Somphospondyli	Portezuelo Fm	Paso de las Bardas, Mendoza Province, Argentina	Upper Turonian–Lower Coniacian	González Riga et al. (2009)
MACN-PV-N97, 98, 68	Somphospondyli	La Amarga Fm	Estancia La Amarga, Neuquén Province	Late Barremian–Early Aptian	Apesteguía 2007
MACN-PV-N102	Titanosauria	La Amarga Fm	Estancia La Amarga, Neuquén Province	Late Barremian–Early Aptian	Apesteguía (2007)
MPEF-PV-3099	Titanosauriformes	Cañadón Calcáreo Fm	Cerro Condor, Chubut Province, Argentina	Tithonian	Rauhut (2006), Mannion et al. (2013)
MPEF-PV-3098	Brachiosauridae	Cañadón Calcáreo Fm	Cerro Condor, Chubut Province, Argentina	Tithonian	Rauhut (2006), Mannion et al. (2013)

America have a major impact, not only on the knowledge on the early evolution and diversification of the Camarasauromorpha as a whole, but also in relation to the previous steps to the origin of Titanosauria. Here, we provide a revision of the valid genera of the paraphyletic assemblage of non-titanosaur macronarians from South America and discuss their impact in the understanding of the evolution of Macronaria, the most diverse lineage of sauropod dinosaurs.

Institutional Abbreviations CHMO: Museo Provincial de Ciencias Naturales y Oceanografía, Puerto Madryn, Chubut, Argentina; **IANIGLA-PV**: Instituto Argentino de Nivología, Glaciología y Ciencias Ambientales, Colección Paleovertebrados, Mendoza, Argentina; **JACVM**, Junta de Acción Comunal Vereda Monquirá, Vereda Monquirá, Colombia; **MACN**: Museo Argentino de Ciencias Naturales Bernardino Rivadavia, Buenos Aires, Argentina; **MCF-PVPH**: Museo Carmen Funes, Paleontología de Vertebrados, Plaza Huincul, Neuquén, Argentina; **MPEF**: Museo Paleontológico Egidio Feruglio, Trelew, Chubut, Argentina.

2 Systematic Palaeontology

Neosauropoda Bonaparte 1986

Macronaria Wilson and Sereno 1998

Definition The most inclusive clade containing *Saltasaurus loricatus* Bonaparte and Powell (1980) but not *Diplodocus longus* Marsh (1878)

Comments The original, stem-based definition of Macronaria was provided by Wilson and Sereno (1998), which, together with Diplodocoidea (see Chaps. 5 and 6) and Neosauropoda, are forming the node-stem triplet (Sereno 1999). Given the stability of that node-stem triplet, Macronaria was preferred by many authors over Camarasauromorpha, a less inclusive, node-defined clade erected a year earlier (see below).

Camarasauromorpha Salgado, Coria and Calvo 1997

Definition *Camarasaurus supremus* Cope (1877), *Saltasaurus loricatus* Bonaparte and Powell (1980), their most common ancestor and all its descendants.

Common Synapomorphies Following Carballido et al. (2020), Macronaria is supported by five unambiguous synapomorphies: posterior articular surface of cervical vertebrae with ratio height/width between 0.9 and 0.7 (char. 132); transverse processes of dorsal vertebrae directed laterally or slightly upwards (char. 157); posterior dorsal centra with opisthocoelous articular surfaces (char. 207); length of puboischial contact one half total length of pubis (char. 337); ischia pubic articulation greater than the anteroposterior length of pubic pedicel (char. 342).

Comments The name Camarasauromorpha was coined by Salgado et al. (1997) and defined as the most recent common ancestor of Camarasauridae and Titanosauriformes and all its descendants. The name had limited use in the literature, as Macronaria (see above) was preferred because all (e.g. Wilson 2002) or most (e.g. Upchurch et al. 2004) of their taxonomic content was shared. In 2004, Upchurch et al. redefined this node-based clade as *Camarasaurus*, *Saltasaurus*, their most recent common ancestor and all its descendants (definition that is followed here but incorporating the species level names), recognizing the conceptual differences between Macronaria and Camarasauromorpha. Similarly, Taylor and Naish (2005) also considered both clades as non-equivalents. Indeed, these clades are clearly different entities, as such Macronaria had its origin in the Middle Jurassic (based on the age of the oldest neosauropod; Xu et al. 2018), whereas Camarasauromorpha would have originated in the Middle–Late Jurassic, as no older camarasauromorph was recovered, being the putative oldest taxon *Bellusaurus* (see below).

In view that Salgado et al. (1997) used Camarasauridae in their definition of Camarasauromorpha and given that Taylor and Naish (2005) used *Camarasaurus supremus* in their definition of Camarasauridae (sauropods phylogenetically closer to *Camarasaurus supremus* than to *Saltasaurus loricatus*), we slightly modified Upchurch's phylogenetic definition by replacing *Camarasaurus* with *Camarasaurus supremus*, since *Camarasaurus*, unlike *Saltasaurus*, is a non-monospecific genus.

Tehuelchesaurus Rich, Vickers-Rich, Giménez, Cúneo, Puerta, Vacca 1999

Tehuelchesaurus benitezii Rich, Vickers-Rich, Giménez, Cúneo, Puerta, Vacca 1999

Holotype MPEF-PV 1125 (Fig. 2), an incomplete postcranial skeleton partially articulated that includes 10 dorsal vertebrae, plus an eroded one, 4 sacral vertebrae, parts of the sacricostal yoke, several ribs, right scapulocoracoid, right humerus, left radius and ulna, fragment of the right ilium, right pubis and fragments of the left one, left ischium and the shaft of the right one, both femora and skin impressions.

Locality, Horizon and Age The holotype specimen was found at the base of the Cañadón Calcáreo Formation, Chubut, Argentina, in lacustrine silts and sandstones with tuffaceous content. For some years, the age of this unit (considered by some authors as an upper member of the Cañadón Asfalto Formation; Cabaleri et al. 2010) was debated. Nevertheless, based on new radiometric and stratigraphic analyses, the age of the formation was established as Oxfordian–Kimmeridgian by Cúneo et al. (2013), an age posteriorly ratified by Hauser et al. (2017).

Diagnosis The active diagnosis of *Tehuelchesaurus* was provided by Carballido et al. (2011a) and is mainly based on presacral vertebrae characters (autapomorphic characters are denoted with an *). *Middle dorsal vertebrae with two accessory laminae (al-1 and al-2) in the centrodiapophyseal fossa. The al-1 runs posterodorsally from the paradiapophyseal lamina and merges dorsally with the al-2, which in turn runs posteroventrally from this point to the posterior centrodiapophyseal lamina. *Dorsal vertebrae with an accessory and laterally oriented lamina on the lateral

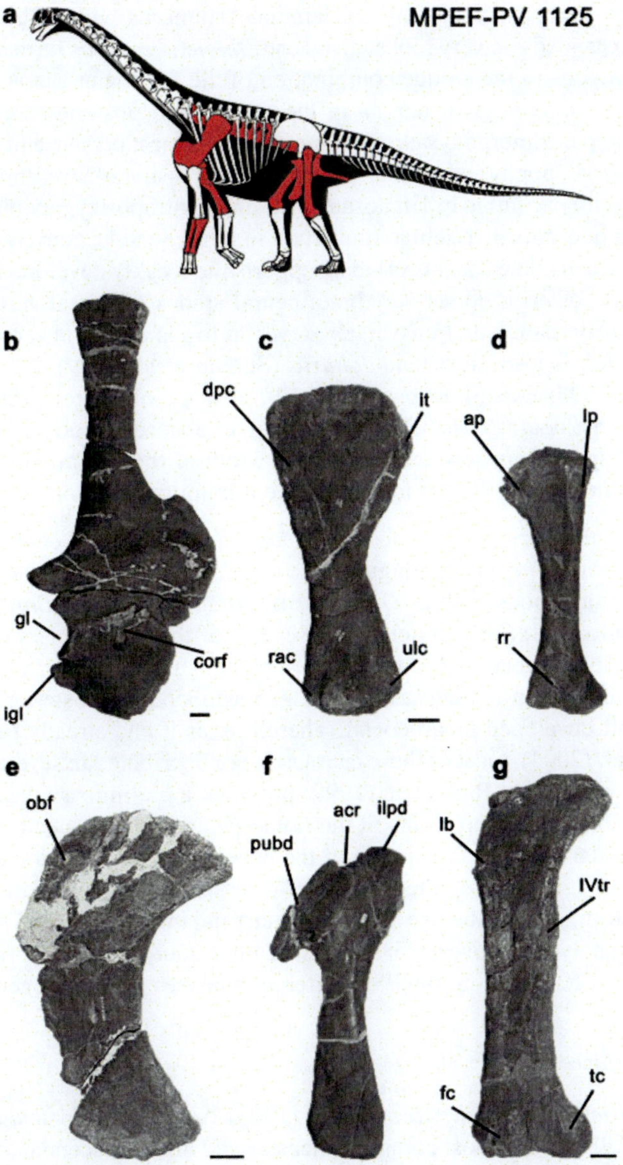

Fig. 2 *Tehuelchesaurus benitezii.* **a** Silhouette (modified from Taylor et al. 2011) showing the preserved elements. **b** Right scapulocoracoid in lateral view. **c** Right humerus in anterior view. **d** Left ulna in anterolateral view. **e** Right pubis in posterolateral view. **f** Left ischium in anterolateral view. **g** Left femur in posterior view. Abbreviations: **IVtr**, fourth trochanter; **acr**, acetabular rim; **ap**, anterior process; **corf**, coracoid foramen; **dpc**, deltopectoral crest; **fbc**, fibular condyle; **gl**, glenoid; **igl**, infraglenoid lip; **ilpd**, iliac peduncle; **it**, internal tuberosity; **lb**, lateral bulge; **lp**, lateral process; **obf**, obturator foramen; **pupd**, pubic peduncle; **rc**, radial condyle; **rr**, rugosity for contact with radius; **tc**, tibial condyle; **uc**, ulnar condyle. Scale bar equals 100 mm

surface of the intrapostzygapophyseal lamina. *Humerus with a strongly antero–posteriorly expanded and very robust distal end. *Tehuelchesaurus* further differs from all other sauropods in the unique combination of the following characters: absence of the prezygoparapophyseal lamina in the middle and posterior dorsal vertebrae (reverse of the plesiomorphic sauropodomorph condition); presence of an accessory posterior centrodiapophyseal lamina, giving the impression of a ventrally bifurcated pcdl (convergently acquired in Titanosauria); a single intrapostzygapophyseal lamina in at least the mid-dorsal vertebrae (uncertain in other dorsals; convergently present in diplodocids and some basal taxa) that supports the weakly developed hyposphene in middle and posterior dorsal vertebrae; neural spines of dorsal vertebrae longer anteroposteriorly than wide transversely, reversal to the ancestral sauropodomorph condition which is also present in *Jobaria* (Sereno et al. 1999) and *Galvesaurus* (Barco et al. 2006; Pérez-Pueyo et al. 2019); absence of lateral expansion in the dorsal end of the neural spine of dorsal vertebrae; absence of postspinal lamina in dorsal neural spines; greatest anteroposterior width of the acromion process of the scapula over the glenoid almost four times the minimum width of the shaft.

Comments In their original diagnosis of *Tehuelchesaurus*, Rich et al. (1999) listed a series of characters that distinguished *Tehuelchesaurus* from basal (i.e. non-neosauropod) sauropods such as *Omeisaurus*, *Patagosaurus* and *Barapasaurus*. In turn, Carballido et al. (2011a) noted that several of these characters corresponded to differences in the relative proportions of appendicular bones, or to the presence of derived characters in *Tehuelchesaurus* (e.g. opisthocoelous posterior dorsal vertebrae with well-developed pleurocoels). Therefore, as it had already been noted by Upchurch et al. (2004), most of these characters are distributed in several sauropods. In this first contribution, Rich et al. (1999) provided a preliminary description with an accurate table of measurements of the holotype and unique specimen known up to that date. After a major preparation of the bones (specially the dorsal sequence), Carballido et al. (2011a) presented a detailed description of *Tehuelchesaurus*, which includes a new diagnosis (the one here presented) and included *Tehuelchesaurus* in a phylogenetic analysis, recovering it as a basal camarasauromorph. This phylogenetic position was posteriorly confirmed by different analyses (see Phylogenetic Section below).

Titanosauriformes Salgado et al. 1997

Definition *Brachiosaurus altithorax* Riggs (1903), *Saltasaurus loricatus* Bonaparte and Powell (1980), their most common ancestor and all its descendants.

Common Synapomorphies Following Carballido et al. (2020), Titanosauriformes is supported by two unambiguous synapomorphies: advanced pneumatic structures in dorsal centra (char. 161); pubis larger (120% +) than ischium (char. 334).

Comments Salgado et al. (1997) coined the name Titanosauriformes and defined the clade to which they applied that name as the most recent common ancestor of '*Brachiosaurus*' *brancai* (actually *Giraffatitan brancai*), *Chubutisaurus insignis*, Titanosauria and all of its descendants. This node-based definition was slightly modified by Wilson and Sereno (1998), deleting *Chubutisaurus* and replacing Titanosauria with *Saltasaurus*. Posteriorly, Taylor (2009) proposed the generic separation of the North American brachiosaurid *Brachiosaurus altithorax* from the African taxon *Giraffatitan brancai* (formerly *Brachiosaurus brancai*) and replacing at the definition of Titanosauriformes, and its two stem clades (Somphospondyli and Brachiosauridae), *Giraffatitan brancai* by *Brachiosaurus altithorax*. As noted by Taylor (2009), this change was introduced in order to follow the Article 11.7 of the PhyloCode (Cantino and de Queiroz 2006) which requires that the species used as internal specifiers in a phylogenetic definition of a clade, whose name is based on a genus, must be the type species of that genus. Here and below, we follow this criterion, accepting and using the definitions of Titanosauriformes and its two stems (Brachiosauridae and Somphospondyli) as proposed by Taylor (2009).

Brachiosauridae Riggs 1903

Definition The most inclusive clade containing *Brachiosaurus altithorax* Riggs (1903) but not *Saltasaurus loricatus* Bonaparte and Powell (1980).

Common Synapomorphies Following Carballido et al. (2020), Brachiosauridae is supported by seventeen synapomorphies: premaxillary anterior margin with marked and long step (char. 2); premaxilla-maxilla suture twisted along its length, giving the contact a sinuous appearance in lateral view (char. 5); premaxilla, small finger-like, with vertically oriented premaxillary process near anteromedial corner of external naris (char. 6); dentary, with divided posteroventral process (char. 90); maxillary teeth with straight along axis (char. 101); dorsal vertebrae with single not bifid neural spines, and single, rough and wide prespinal lamina (PRSL) in the dorsal-most part of the neural spine (char. 165); anterior and middle dorsal vertebrae with horizontal or slightly posteroventrally oriented zygapophyseal articulation (char. 171); anterior dorsal vertebrae neural spine with triangular aliform processes projected far laterally (as far as caudal zygapophyses) (char. 173); anterior dorsal vertebrae with neural spine minimums width/length lower than 0.5 (thin and tall neural spines) (char. 174); middle and posterior dorsal vertebrae with transverse processes long (projecting along 1.5 the articular surface width) (char. 181); middle and posterior dorsal vertebrae neural spine with triangular aliform processes projected far laterally (as far as caudal zygapophyses) (char. 196); middle and posterior dorsal vertebrae with ventral spinodiapophyseal (SPDL) and spinopostzygapophyseal lamina (lSPOL) contact, well separated from the triangular aliform process (char. 203); posterior dorsal vertebrae with medial spinopostzygapophyseal lamina (mSPOL) that forms part of the median posterior lamina (char. 205); anterior and middle caudal vertebrae with a blind fossa in lateral centrum (char. 228); humerus-to-femur ratio greater than 0.90 (char 300); metatarsal IV with distal end bevelled upwards medially (char. 392); pubis with an ischiatic articular surface marked step formed by a proximal posterior directed surface and a more distal posterodorsal oriented surface (char. 416).

Comments Although the family name was first used by Riggs (1903), it was not until 1998 that Wilson and Sereno first proposed a formal stem definition for this clade. At this moment, Wilson and Sereno (1998) did not incorporate in their definition the species name. Here, we follow the definition of Brachiosauridae as proposed by Taylor (2009).

Padillasaurus Carballido, Pol, Parra Ruge, Padilla Bernal, Páramo-Fonseca and Etayo-Serna 2015

Padillasaurus leivaensis Carballido, Pol, Parra Ruge, Padilla Bernal, Páramo-Fonseca and Etayo-Serna 2015

Holotype JACVM 0001 (Fig. 3), a single specimen represented by an isolated posterior dorsal centrum, part of the last presacral dorsal vertebra articulated to the two anteriormost sacral vertebrae and the two posterior-most sacral vertebrae articulated with the first eight caudal vertebrae.

Locality, Horizon and Age The specimen was collected by a local farmer, and, although its exact position could not be known, it surely comes from the northeast of Villa de Leyva town, Department of Boyaca, Ricaurte Province, Colombia. The ammonoids preserved within the caudal vertebrae correspond to the *Gerhardtia galeatoides* subzone, which represents the lower part of the Barremian, which in the area is represented by the marine Paja Formation.

Diagnosis Medium-sized titanosauriform characterized by four autapomorphies: (1) first and second caudal vertebrae with high and dorsally directed prezygodiapophyseal laminae that converge with the centroprezygapophyseal laminae and form the lateroventral ventral margins of the prezygapophyseal processes; (2) anterior caudal vertebrae with weakly laterally expanded transverse processes; and (3) first caudal vertebrae with divided transverse process, the dorsal section of which is posterodorsally directed.

Comments Although *Padillasaurus* (Fig. 3) is known on the basis of fragmentary material, it is an important taxon for better understanding the evolution and geographic distribution of titanosauriform sauropods in South America. *Padillasaurus* not only represents the northernmost record of a titanosauriform in South America, but it also helps us to understand the sauropod diversity during the earliest Early Cretaceous. Originally considered as a brachiosaurid, given the presence of blind fossae in their caudal vertebrae (Carballido et al. 2015), it was later considered by Mannion et al. (2017) as a basal somphospondylan, as the presence of blind fossae has a broader distribution amongst Titanosauriformes, including Somphospondyli (e.g. *Savannasaurus*). Therefore, the phylogenetic position of *Padillasaurus* remains controversial, as recent analyses recovered it within Brachiosauridae (e.g. Carballido et al. 2019) or as a basal somphospondylan (e.g. Royo Torres et al. 2017; Mannion et al. 2019a). Based on the presence of blind fossae in caudal vertebrae and the 'polycamerate' internal pneumaticity in dorsal vertebrae, we provisionally considered *Padillasaurus* as a Brachiosauridae, although this assignment should be taken with caution (see Phylogenetic section below).

Somphospondyli Wilson and Sereno, 1998

Definition The most inclusive clade containing *Saltasaurus loricatus* Bonaparte and Powell (1980) but not *Brachiosaurus altithorax* Riggs (1903).

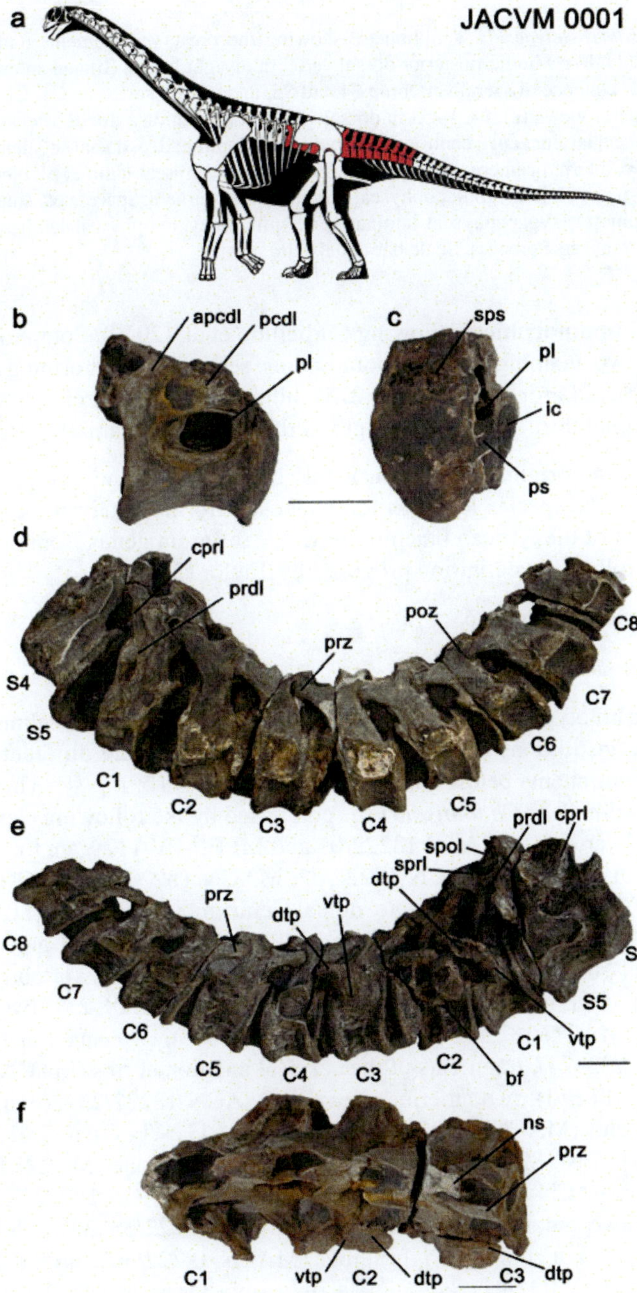
a
JACVM 0001
b
apcdl
pcdl
pl
c
sps
pl
ic
ps
d
cprl
prdl
poz
C8
prz
S4
C7
S5
C6
C1
C2
C5
C3
C4
e
prdl
cprl
spol
sprl
dtp
prz
dtp
vtp
C8
S4
C7
S5
C6
C1
vtp
C2
bf
C5
C4
C3
f
ns
prz
C1
vtp
C2
dtp
C3
dtp

◂**Fig. 3** *Padillasaurus leivaensis*. **a** Silhouette showing the preserved elements (modified from Taylor et al. 2011). **b–c** Middle-posterior dorsal vertebra in right lateral (**b**) and anterior views (**c**). **d–e** Articulated sequence of sacral vertebrae S4 and S5, and caudal vertebrae C1–C8 in left lateral (**d**) and right lateral views (**e**). **f** C1–C3 in dorsal view. Abbreviations: **apcdl**, accessory posterior centrodiapophyseal lamina; **cprl**, centroprezygapophyseal lamina; **dtp**, dorsal segment of the transverse process; **ic**, internal camera; **pcdl**, posterior centrodiapophyseal lamina; **pl**, pleurocoel; poz, postzygapophysis; **prdl**, prezygodiapophyseal lamina; **ps**, pneumatic space; **sps**, small pneumatic spaces; **spol**, spinopostzygapophyseal lamina; **sprl**, spinoprezygapophyseal lamina; **vtp**, ventral segment of the transverse process. Scale bar equals 100 mm

Common Synapomorphies Following Carballido et al. (2020), Somphospondyli is supported by two unambiguous synapomorphies: scapular blade forming a 45° angle respect to coracoid articulation (char. 275); humeral lateral margin almost straight until the proximal third of the total length of the humerus (char. 307).

Comments In the original definition of this clade Wilson and Sereno (1998) did not incorporate the species level (as for Titanosauriformes). Given the subsequent reconsideration of the African brachiosaurid as a different genus from that of North America, we follow the definition proposed by Taylor (2009).

Chubutisaurus del Corro 1975

Chubutisaurus insignis del Corro 1975

Holotype The holotype of *Chubutisaurus* is composed of a single specimen, actually housed in two institutions (MACN and MPEF) and with three different collection numbers (see comments below and Carballido et al. 2011b; Fig. 4). Thus, the type and unique specimen of *Chubutisaurus* is composed by the following elements: two anterior dorsal vertebrae (MACN 18222/01 and MPEF-PV 1129/A), four middle to posterior dorsal vertebrae (MACN 18222/02, MACN 18222/03, MACN 18222/05 and MPEF-PV 1129/B), a complete dorsal centrum (MPEF- PV 1129/B), two dorsal neural spines (MACN 18222/04 and MPEF-PV 1129/D), 11 anterior caudal centra (MACN 18222/06–13), fragments of four anterior caudal vertebrae (MACN 18222/14–17), four middle caudal centra (MACN 18222/18–21), two posterior caudal centra (MACN 18222/22 and MPEF-PV 1129/E), two caudal neural arches (MACN 18222/23–24), fragments of the cervical and dorsal ribs (MACN 18222/42 and MPEF-PV 1129/I), two anterior chevrons (MACN 18222/25–26) and isolated chevron fragments (MACN 18222/27 and MPEF-PV 1129/F–G), a nearly complete left scapula (MACN 18222/28) and fragments of the right one (MACN 18222/29), a left humerus (MACN 18222/30), left ulna and radius (MACN 18222/31–32), four complete and two incomplete metacarpals (MACN 18222/33–38), a left ischium (MACN 18222/39), a complete right femur (MACN 18222/40) and the dorsal half of the left one (CHMO-901), a complete right tibia (MACN 18222/41) and fragments of the left one (CHMO-565) and three histological samples of the right femur (MPEF-PV 1129/K1–K3).

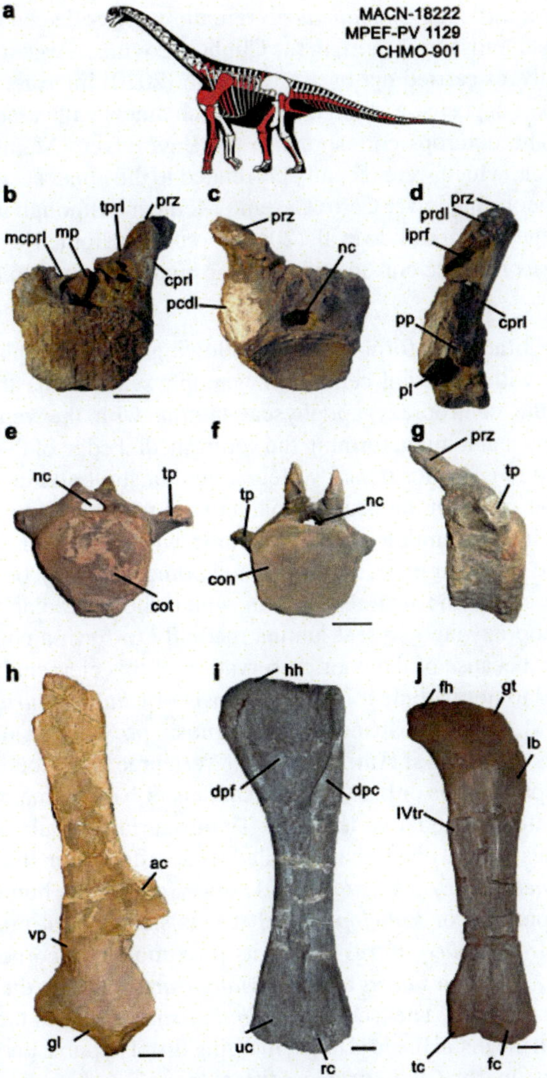

Fig. 4 *Chubutisaurus insignis*. **a** Silhouette showing the preserved elements (modified from Taylor et al. 2011). **b–d** Anterior dorsal vertebrae MACN 18,222/01 in anterior (**b**), posterior (**c**), and left lateral (**d**) views. **e–g** Anterior caudal vertebra MACN 18,222/07 in posterior (**e**), anterior (**f**) and left lateral (**g**) views. **h** Left scapula (MACN 18,222/28 in medial view. **i** Left humerus MACN 18,222/30 in anterior view. **j** Right femur MACN 18,222/40 in posterior view. Abbreviations: **IVtr**, fourth trochanter; **ac**, acromion; **cot**, cotyle; **con**, condyle; **cprl**, centroprezygapophyseal lamina; **dpc**, deltopectoral crest; **dpf**, deltopectoral fossa; **fc**, fibular condyle; **fh**, femur head; **gl**, glenoid; **gt**, great trochanter; **hh**, humeral head; **iprf**, infraprezygapophyseal fossa; **lb**, lateral bulge; **mcprl**, medial cprl; **mp**, medial pilar; **nc**, neural canal; **pcdl**, posterior centrodiapophyseal lamina; **pl**, pleurocoel; **pp**, parapophysis; **prdl**, prezygodiapophyseal lamina; **prz**, prezygapophysis; **rc**, radial condyle; **tc**, tibial condyle; **tp**, transverse process; **tprl**, intraprezygapophyseal lamina; **uc**; ulnar condyle; **vp**, ventral process. Scale bar equals 100 mm

Locality, Horizon and Age *Chubutisaurus* remains were collected in a single quarry at the 'Estancia el Dinosaurio' farm, in the Chubut Province, Argentina. The detailed stratigraphic analyses carried out by Krause et al. (2020) interpreted the section as fluvial channel belt deposits, with high sinuosity channels with attached margin bars in some cases. The outcrops correspond to the Bayo Overo Member of the Cerro Barcino Formation, which were recently correlated to the upper section of the Puesto la Paloma Member or the lower Cerro Castaño Member. Although it was considered Cenomanian in age by Carballido et al. (2011b), recent radiometric analyses indicate that *Chubutisaurus* must be considered as late Aptian to early Albian (Krause et al. 2020).

Diagnosis Large titanosauriform with the following autapomorphies: (1) anterior dorsal vertebrae with a medial centroprezygapophyseal lamina that connects the medial part of the centroprezygapophyseal lamina with the ventral half of the intraprezygapophyseal lamina, forming the ventromedial edge of the associated sub-rectangular fossa; (2) anterior dorsal vertebrae with a stout and internally pneumatized medial pillar between the neural canal and the ventral edge of the intraprezygapophyseal lamina; (3) middle dorsal vertebrae with large and deep pleurocoels that present three inner laminae. A further autapomorphy may be (4) the unusual arrangement present in the neural spine, in which the spinodiapophyseal lamina contacts the spinoprezygapophyseal lamina medially to form a composite anterior lamina. However, because of the poor preservation of this element, Carballido et al. (2011b) preferred to not include it in the diagnosis of *Chubutisaurus*.

Carballido et al. (2011b) expanded the diagnosis providing a unique combination of characters as follows. Anterior caudal vertebrae platycoelous/distoplatyan instead of the amphicoelous or platycoelous caudals of basal camarasauromorphs or the procoelous anterior caudal vertebrae of Titanosauria. Scapula with a D-shaped section and an unexpanded distal scapular blade that differs from the expanded distal end present in *Euhelopus*, *Brachiosaurus* and *Paluxysaurus*. The humerus of *Chubutisaurus* differs from that of *Euhelopus* in that it is slender (but not as much as in *Brachiosaurus*), from *Paluxysaurus* in that its proximolateral corner is square, and from *Wintonotitan*, *Ligabuesaurus* and *Phuwiangosaurus* in that the distal condyles are not anteriorly directed. The ischium of *Chubutisaurus* shows the plesiomorphic condition in which the distal blade is large, merging distal to pubic peduncle, differing from *Andesaurus*. Finally, *Chubutisaurus* differs from *Euhelopus* in that its femur is more anteroposteriorly compressed.

Comments As mentioned by Carballido et al. (2011b) *Chubutisaurus* (Fig. 4) was initially collected by G. del Corro in 1965, being the recovered bones cataloged as MACN 18222. Two elements, the dorsal half of the left femur plus fragments of the left tibia, were donated to the CHMO and are actually conserved at the MPEF collection (under the numbers CHMO-901 and CHMO-565), but still preserving their original 18222 number labelled. Additional materials were posteriorly collected from del Corro's excavation during MPEF field trips carried out in 1991 and 2007 (MPEF-PV 1129) (see Carballido et al. 2011b).

Ligabuesaurus Bonaparte, González Riga, Apesteguía 2006

Ligabuesaurus leanzaii Bonaparte, González Riga, Apesteguía 2006

Holotype MCF-PVPH-233 (Fig. 5), a single partially articulated individual represented by teeth and postcranial elements: ten maxillary teeth, a posterior cervical vertebra, an anterior dorsal vertebra, four posterior dorsal vertebrae, incomplete ribs, left and right scapulae, left humerus and proximal and distal thirds of the right one, right metacarpal II and III, distal epiphyses of left metacarpal II and IV, an incomplete right femur, a right tibia, a right fibula, a right astragalus and a nearly complete and articulated right foot with five metatarsals and three phalanges.

Referred Material Martinelli et al. (2007) described an isolated tooth MCF-PVPH-744 figured by Bonaparte et al. (2006: Fig. 2), that was found next to the *Ligabuesaurus* quarry, referring it to this genus, as they noted that several morphological features in the tooth row of the holotype (MCF-PVPH-233) were also present in the isolated tooth. This element is represented by a nearly complete and well-preserved functional tooth with a 'cone-chisel-like' crown and part of a cylindrical root. Following Martinelli et al. (2007), the crown is 'D-shaped' in cross section, with a labial convex surface, and has an apico–basally directed groove on the mesial margin of lingual surface and overall wrinkled enamel. Unlike what was reported by Bonaparte et al. (2006), no pseudo-denticles are regarded in the mesio-distal margins of MCF-PVPH-744 (nor of MCF-PVPH-233/01).

Locality, Horizon and Age The fossil remains come from the Cerro de los Leones locality, 10 kms to the south-west of Picún Leufú city, in the southern part of the Neuquén province, Patagonia, Argentina. This area is characterized by wide and thick fluvial outcrops referred to the lower section of the Cullín Grande Member, the upper unit of the Lohan Cura Formation (Bajada del Agrio Group), which is considered Albian (e.g. Leanza 2003; Leanza and Hugo 2001, 2011). The type material of *Ligabuesaurus leanzai* was found in the southern flank of the Cerro de los Leones, and the quarry was opened in coincidence with the fossiliferous level n°2 (sensu Martinelli et al. 2007), a layer at the base of the hill that is composed by reddish to purplish mudstones with intercalations of grey fine-grained sandstones. Following Martinelli et al. (2007), these outcrops are interpreted as floodplain deposits.

Diagnosis In the original diagnosis of *Ligabuesaurus*, Bonaparte et al. (2006) identified four autapomorphies, although in a recent revision Bellardini (2021) considered that only three of them are valid. Therefore, we follow this author and solely three characters are considered as autapomorphies of *Ligabuesaurus*: laminar and anteroposteriorly compressed neural spines on posterior cervical and anterior dorsal vertebrae that are rhomboid in shape and wider than the vertebral centra (convergently acquired in *Mendozasaurus*); spinoprezygapophyseal laminae in posterior cervical vertebrae forked to form two pairs of laminae: the medial pair unites them towards the top of the neural spine, and the lateral pair form the lateral border of the neural spine; dorsoventrally reduced neural arch pedicels in the posterior cervical and anterior dorsal vertebrae.

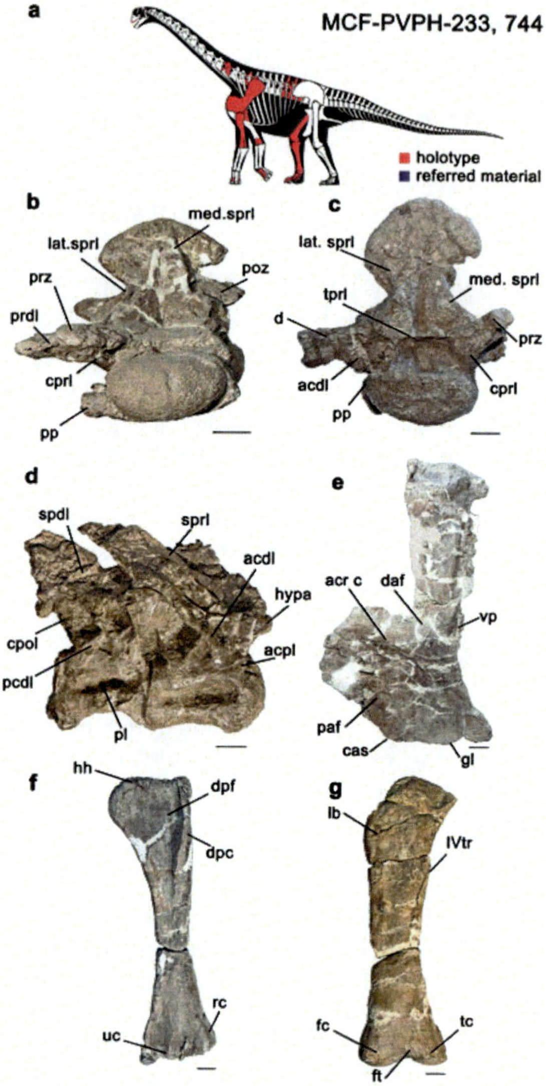

Fig. 5 *Ligabuesaurus leanzai.* **a** Silhouette showing the preserved elements (modified from Taylor et al. 2011). **b** Posterior cervical vertebra in anterior view. **c** Anterior dorsal vertebra in anterior view. **d** Posterior dorsal vertebrae in right lateral view. **e** Left scapula in lateral view. **f** Left humerus in anterior view. **g** Right femur in anterior view. Abbreviations: **acdl**, anterior centrodiapophysial lamina; **acpl**, anterior centroparapophysial lamina; **acr. c.**, acromial crest; **cas**, coracoid articular surface; **cpol**, centropostzygapophysial lamina; **cprl**, centroprezygapophyseal lamina; **d**, diapophysis; **daf**, distal acromial fossa; **dpc**, deltopectoral crest; **dpf**, deltopectoral fossa; **fc**, fibular condyle; **ft**, femoral trochlea; **gl**, glenoid; **hh**, humeral head; **lat. sprl**, lateral spinoprezygapophysial lamina; **lb**, lateral bulge; **med. sprl**, medial spinoprezygapophysial lamina; **paf**, proximal acromial fossa; **pcdl**, posterior centrodiapophysial lamina; **pl**, pleurocoel; **poz**, postzygapophysis; **pp**, parapophysis; **prdl**, prezygodiapophyseal lamina; **prz**, prezygapophysis; **rc**, radial condyle; **sprl**, spinoprezygapophysial lamina; **tc**, tibial condyle; **tprl**, intraprezygapophyseal lamina; **uc**; ulnar condyle; **vp**, ventral process. Scale bar equals 100 mm

Comments *Ligabuesaurus* (Fig. 5) was formalized by Bonaparte et al. (2006) on the basis of an incomplete, partially articulated skeleton collected from the southern flank of Cerro de los Leones, during fieldworks carried out between 1999 and 2004. However, at that time, different elements form the *'Ligabuesaurus'* quarry that, together with other sauropod bones from different sites of Cerro de los Leones, were sent to the laboratory of the Museo Argentino de Ciencias Naturales 'Bernardino Rivadavia' of Buenos Aires for preparation, not being included in the original description of the taxon. In 2014, these specimens were incorporated into the palaeontological collection of the Museo Municipal 'Carmen Funes' of Plaza Huincul, and the provenance of each bone in the quarry was established on the basis of field pictures, field-drawings and notes analyses (Bellardini 2021). Furthermore, during the fieldworks carried out by one of us (FB) in 2014–2016 at Cerro de los Leones, the original quarry was reopened, and new materials were collected. Therefore, several new bones of *Ligabuesaurus* are being prepared, which will be of great importance for enabling a complete and revised osteology, for the possibility that these bones offer to improve the diagnosis, and for the possibility of better establishing its phylogenetic relationships. For the moment, and awaiting a complete revision of the taxon, we exclude one character from the diagnosis: 'rudimentary prespinal lamina in posterior cervical and anterior dorsal vertebrae' (Bonaparte et al. 2006). In fact, in the posterior cervical vertebra MCF-PVPH-233/02 lacks the prespinal lamina, while in the anterior dorsal vertebra MCF-PVPH-233/03 there is a reduced lamina on the dorsal-most portion of the anterior face of the neural spine. Therefore, we consider, following Bellardini (2021), that the autapomorphic character described by Bonaparte et al. (2006) must be excluded from the diagnosis.

Malarguesaurus González Riga, Previtera, Pirrone 2009

Malarguesaurus florenciae González Riga, Previtera, Pirrone 2009

Holotype IANIGLA-PV 110 (Fig. 6), fragmentary specimen integrated on anterior, three middle and four posterior caudal vertebrae, chevrons, dorsal ribs, a fragmentary humerus, the proximal portion of the right femur and bone fragments.

Referred Material IANIGLA-PV 111, composed by two posterior caudal vertebrae and an incomplete fibula.

Locality, Horizon and Age The holotype specimen comes from the 'Quebrada Norte' quarry and was collected from massive mudstones with grey mottles and scarce pelitic clasts (González Riga et al. 2009) belonging to the Portezuelo Formation (Río Neuquén Subgroup, Neuquén Group, see Garrido 2010). The age of this unit is considered upper Turonian–lower Coniacian (see Garrido 2010). The referred specimen was found at 600 m from the 'Quebrada Norte' quarry at the 'Cerro la Torre' site, in the same stratigraphic level as the holotype specimen.

Diagnosis Based on the original diagnosis, *Malarguesaurus florenciae* is characterized by the following combination of characters (autapomorphic character denoted with an *): *anterior caudal neural spines vertically directed, with a concave caudal border and a caudodorsal corner forming a right angle; *procoelous/opisthoplatyan

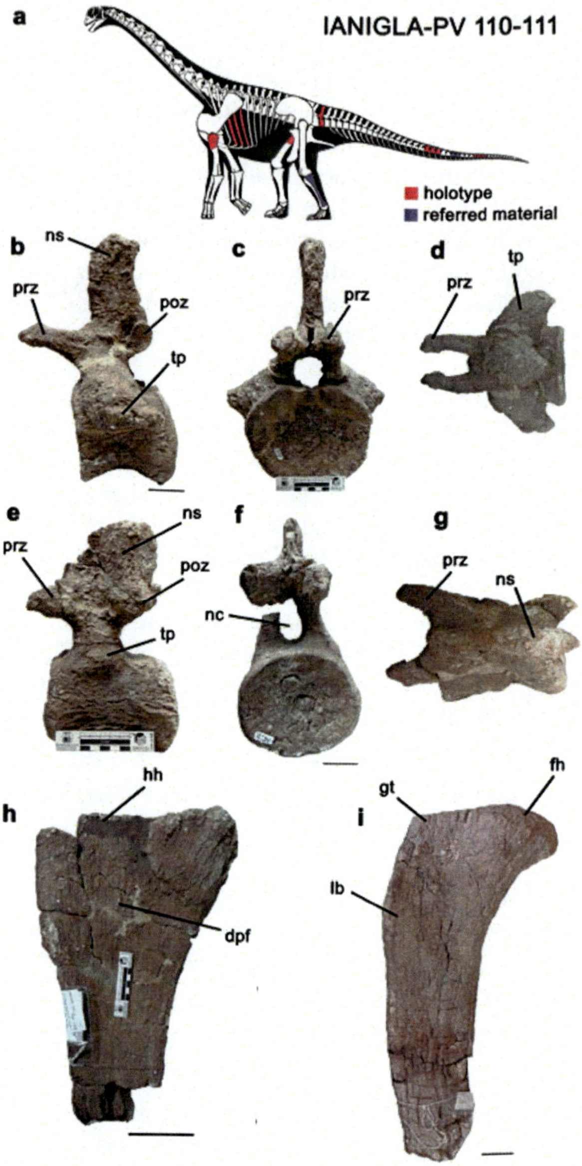

Fig. 6 *Malarguesaurus florenciae.* **a** Silhouette showing the preserved elements (modified from Taylor et al. 2011). **b–d** Anterior caudal vertebrae (IANIGLA-PV 110/1) in left lateral (**b**), anterior (**c**) and dorsal (**d**) views. **e–g** Middle caudal vertebra (IANIGLA-PV 110/3) in left lateral (**e**), anterior (**f**) and dorsal (**g**) views. **h** Proximal portion of a left humerus un anterior view (IANIGLA-PV 110). **i** Right femur in anterior view (IANIGLA-PV 110). Abbreviations: **dpf**, deltopectoral fossa; **fh**, femur head; **gl**, glenoid; **gt**, great trochanter; **hh**, humeral head; **lb**, lateral bulge; **nc**, neural canal; **ns**, neural spine; **poz**, postzygapophysis; **prz**, prezygapophysis; **tp**, transverse process. Scale bar equals 100 mm

anterior and middle caudal vertebrae associated with procoelous posterior caudal centra; *distal caudal neural spines having a concave and depressed dorsal border, combined with the presence of anteriorly placed neural arches on caudal vertebrae, relatively tall and anteroposteriorlly short anterior caudal neural arches, prominent postzygapophyseal processes in middle caudal vertebrae and lateral bulge on the femur below the major trochanter.

Comments Although included in this chapter of non-titanosaur macronarians, the phylogenetic position of *Malarguesaurus* is somewhat controversial. González Riga et al. (2009) recovered it as a basal somphospondylan, but recognized its weakly supported position. Most posteriorly published analyses confirmed this position (e.g. Mannion et al. 2013; Poropat et al. 2015; Gorscak et al. 2017), although with some differences and showing unstable positions amongst non-titanosaur somphospondylans or even as a basal titanosaur (e.g. Carballido et al. 2020; Gallina et al. 2021; Otero et al. 2021). Therefore, although for the moment we include *Malarguesaurus* as a non-titanosaur somphospondylan, its phylogenetic position is uncertain and its titanosaur affinities cannot be ruled out. It is interesting to note that, given its basal position within Somphospondyli and its stratigraphic provenance, *Malarguesaurus* represents the younger record of a non-titanosaur somphospondylan, being possibly the last survivor of this basal lineage of Titanosauriformes (see Phylogeny and Biogeographic sections).

3 Phylogeny

Our general, currently accepted, knowledge on Macronaria evolution starts with the pioneer work of Salgado et al. (1997) who recovered and erected Camarasauromorpha and Titanosauriformes for the first time, recognizing Titanosauria as a derived clade of titanosauriforms. Posterior analyses corroborated the phylogenetic hypothesis proposed by these authors (e.g. Wilson and Sereno 1998; Upchurch 1998; Wilson 2002; Upchurch et al. 2004). Based on the criterion of the node-stem triplet (Sereno 1998, 1999), Wilson and Sereno (1998) erected the stem lineage Macronaria together with Diplodocoidea and provided a node definition for Neosauropoda. Following the same criterion, these authors redefined the node-based lineage Titanosauriformes and its two sister lineages, Brachiosauridae and Somphospondyli, which were defined by their stem. Posterior phylogenetic analyses carried out increased the taxonomic content of these major and widely accepted clades (e.g. Carballido et al. 2011a, b; Mannion et al. 2013). Here, we provide a brief revision on the phylogenetic position of the basal macronarians from South America, as were previously recovered or suggested in different contributions. The simplified and calibrated phylogenetic tree shown in Fig. 7 shows the phylogenetic position of the taxa here mentioned, mainly following the results obtained by Carballido et al. (2020) but also considering alternative positions as is discussed below.

The Upper Jurassic (Oxfordian–Kimmeridgian; Cúneo et al. 2013) Cañadón Calcáreo Formation yielded the basalmost camarasauromoph recovered from South America, *Tehuelchesaurus benitezii* (Rich et al. 1999: Fig. 6), and more fragmentary materials of basal titanosauriforms (Rauhut 2006). *Tehuelchesaurus* were originally considered as a basal eusauropod closely related to mamenchisaurids, specially with *Omeisaurus* (Rich et al. 1999). Such hypothesis was later reinforced by the phylogenetic analyses of Alifanov and Averianov (2003) and Upchurch et al. (2004). Posteriorly, Rauhut et al. (2005) recovered it as a camarasauromorph, a position previously

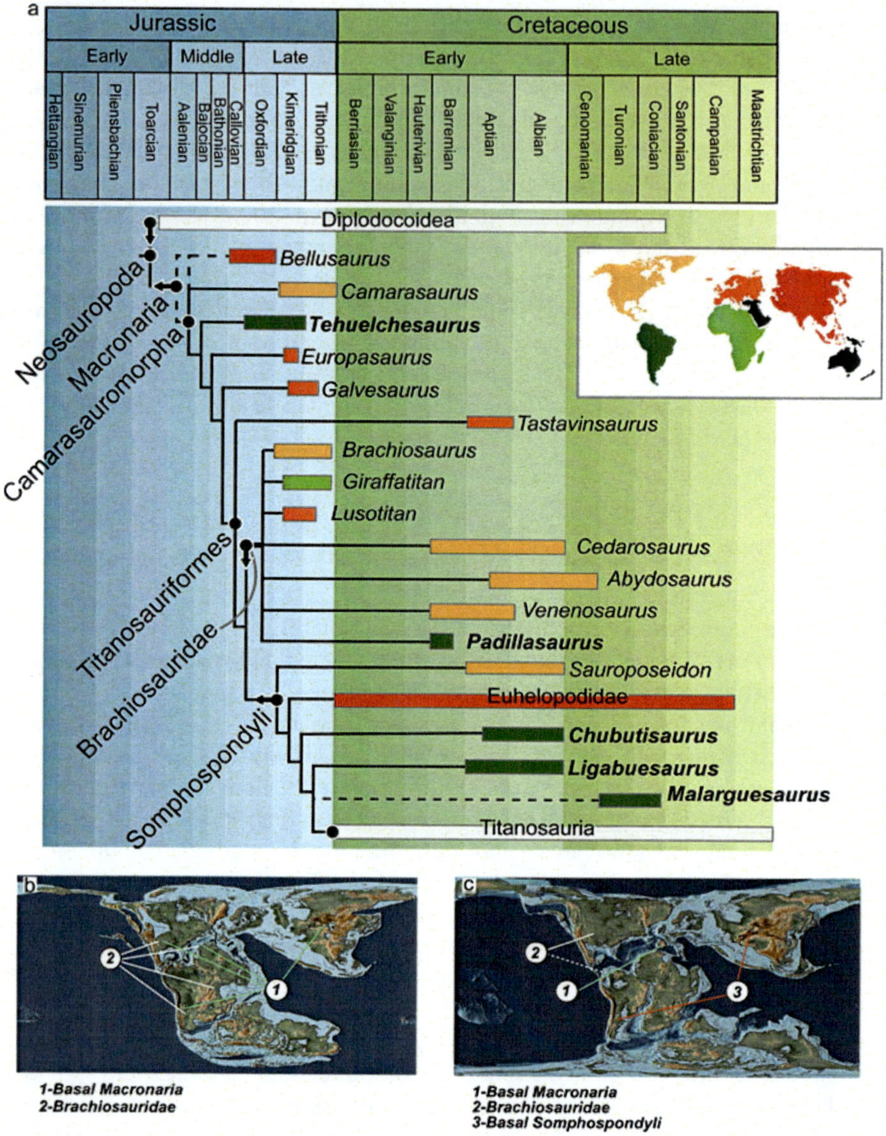

Fig. 7 **a** Simplified calibrated phylogenetic tree, mainly based on the results of Carballido et al. (2019), showing with dashed lines putative positions of unstable taxa (see Phylogenetic section for a more complete discussion on this) and their geographical location (colours indicate the continent of origin as shown in the map inlet). **b** Late Jurassic palaeogeographic map showing the precedence of basal (non-titanosauriforms) macronarians and brachiosaurids; **c** Cretaceous palaeogeographic map showing the precedence of basal macronarians, brachiosaurids (including with dashed lines to *Padillasaurus*), and basal somphospondylans (non-titanosaurs)

suggested by Rauhut (2002) and later followed by Salgado and Bonaparte (2007). The complete revision and description of the anatomy of *Tehuelchesaurus*, plus its inclusion in a phylogenetic analysis that used an array of taxa considering both, basal sauropods as well as Neosauropoda, recovered this Late Jurassic Patagonian taxon as a basal camarasauromorph (Carballido et al. 2011a), a position that was subsequently recovered and accepted by most authors (e.g. D'Emic 2012; Mannion et al. 2013). Recently, Mannion et al. (2019a), in their equally weighting analysis, recovered *Tehuelchesaurus* as a basal taxon of Turiasauria, a clade of basal eusauropods with a wide geographical distribution (North America, Africa and Europe) during the Late Jurassic–Early Cretaceous. However, in their implied weighting analysis it was recovered as a basal Macronaria (with *Camarasaurus* as a basal eusauropod). For the moment, and awaiting further clarification on the relationships of basal camarasauromophs and basal non-neosauropod eusauropods, we considered *Tehuelchesaurus* as a basal camarasauromorph, as this is the position widely accepted for this taxon. The results of Mannion et al. (2019a), as well as previous analyses that differ in the position of putative basal neosauropods such as *Haplocanthosaurus*, *Tehuelchesaurus*, *Bellusaurus* and *Camarasaurus* (Carballido et al. 2011a, b, 2015; Whitlock 2011a; D'Emic 2012; Mannion et al. 2013, 2019a), are alerting on the importance of having a wide taxonomic sampling that includes basal sauropods and neosauropods (including both Diplodocoidea and Macronaria lineages).

The oldest record of a South American titanosauriform also comes from the Upper Jurassic Cañadón Calcáreo Formation, exposed in the north–central part of Chubut Province. Rauhut (2006) described three different specimens assigned by him to the Brachiosauridae. One of these was described from materials observed in the field, so that no collection number was provided (Rauhut 2006: Fig. 2) Recent fieldwork carried out at this place resulted in the recognition of several elements at the place, which were fully excavated. Most of this material is still in plaster jackets awaiting for its preparation. One of the other two specimens described in this work (MPEF-PV 3099) is composed of caudal elements and incomplete appendicular bones (pubis and a humeral fragment). As noted by Mannion et al. (2013), there are no clearly derived brachiosaurid characters on the preserved elements of this specimen, although it can be clearly assigned to derived camarasauromorphs, based on the anteriorly placed neural arches (character that is present in *Europasaurus* and more derived camarasauromorphs; Carballido and Sander 2014; Carballido et al. 2020), and to Titanosauriformes, by the presence of lateromedially wide and anteroposteriorly short first haemal arch. The third specimen described by Rauhut is composed by a left humerus (MPEF-PV 3098), which was also considered as a brachiosaurid by D'Emic (2012), based on the rounded proximodistal corner of the element. Nevertheless, for Mannion et al. (2013), this element cannot be assigned beyond Titanosauriformes, due to the lack of diagnostic characters of Brachiosauridae. Rauhut (2006) justified his assignment to Brachiosauridae based on the slenderness of the humerus and the elongation of the deltopectoral crest, both supposed characters of that family. However, the robustness index (sensu Wilson and Upchurch 2003) of this element is of 0.23, being similar to that of *Europasaurus* and basal titanosauriformes (including both Brachiosauridae and Somphospondyli lineages; see Carballido et al. 2020), whereas the deltopectoral crest is similar in length to that of *Europasaurus* and several titanosauriformes, in which the crest extends up to almost the half of the humerus (see Carballido et al. 2020). Indeed, the proximolateral third of MPEF-PV 3098 is straight (Rauhut 2006), differing from basal, non-titanosauriform macronarians with have medially and laterally curved margins (e.g. *Europasaurus*, *Camarasaurus*), resembling the condition of several brachiosaurids (e.g. *Giraffatitan*, *Brachiosaurus*, *Cedarosaurus*) and differing from the more extended straight lateral margin of most somphospondylans (see Carballido et al. 2020 and references therein). It is for this reason that we accept provisionally the original brachiosaurid assignment of this element made by Rauhut (2006). Up to the moment, this is the only element that could indicate the presence of this clade in the Late Jurassic of South America.

Padillasaurus leivaensis from the Early Cretaceous of Colombia was described by Carballido et al. (2015) as the first brachiosaurid species nominated for South America, and the most recent record of this clade in Gondwana. This hypothesis was supported by the phylogenetic analysis carried out by these authors, although noting the low support of such a position. This position was sustained by the presence of a blind fossa on the lateral surface of the anterior caudal vertebra, a character recovered as a synapomorphy of Brachiosauridae (Carballido et al. 2015), or the clade formed by *Brachiosaurus* and more derived brachiosaurids (D'Emic 2012; Mannion et al. 2013). Nevertheless, Mannion et al. (2017) noted that the presence of a lateral fossa in anterior caudal vertebrae is more widespread amongst Titanosauriformes, being present in several somphospondylans (e.g. *Savannasaurus*; Poropat et al. 2016), and recovered it as a somphospondylan, a position posteriorly obtained in more recent augmented versions of this data set (e.g. Mannion et al. 2019a, b). Different data sets derived from Carballido et al. (2015) keep recovering *Padillasaurus* as a brachiosaurid, even when titanosaurs with a lateral fossa in anterior caudal vertebra are scored (e.g. *Patagotitan*, *Ninjatitan*, *Alamosaurus*; see Gallina et al. 2021). Besides the presence of a lateral fossa in the anterior caudal vertebrae, *Padillasaurus* shares with brachiosaurids (e.g. *Giraffatitan*) the presence of a 'polycamerate' internal pneumatic system in their dorsal vertebrae (Janensch 1947; Fig. 3; Carballido et al. 2015), differing thus from the more advanced and complex pneumatization pattern (camellate) that characterizes somphospondylans, or from larger camerae of basal camarasauromorphs (e.g. *Tehuelchesaurus*; Carballido et al. 2011a: Fig. 12). Although, for the moment we considered that *Padillasaurus* is most probably a brachiosaurid, such phylogenetic hypothesis is weakly supported and pendant of new analyses and additional materials (Fig. 7).

Earliest Cretaceous records of South American basal (non-titanosaurs) macronarians are restricted to the La Amarga Formation (Hauterivian–Barremian). Several isolated teeth from this lithostratigraphic unit were assigned to Titanosauria by Apesteguía (2007). These teeth were divided by this author into two morphotypes: 'broad-crowned' and 'narrow-crowned'. The only tooth of 'narrow-crowned' morphotype (MCN-PV N102) is circular in cross-section, having a high slenderness index (more than 4.5), two derived characters widespread present amongst titanosaurs (e.g. Salgado and Calvo 1997; Wilson 2002; D'Emic 2012). However, the teeth under the 'broad-crowned' morphotype (MACN-PV N97, 98, 68) are D-shaped and have low slender indexes, two plesiomorphic characters widely distributed amongst basal, non-titanosaur, titanosauriforms. Given that, and due to the fact that none of these last teeth show the twisted, typical condition of Brachiosauridae (D'Emic 2012), they more probably correspond to basal Somphospondyli.

Further basal somphospondylans from the Early Cretaceous come from the upper section of this epoch (Aptian–Albian). In his original description of *Chubutisaurus insignis*, del Corro (1975) erected the family 'Chubutisauridae', without discussing its affinities with other families of sauropods. Regardless of this, the term Chubutisauridae was hardly ever used again. The taxon was posteriorly redescribed by Salgado (1993) who noted several morphological similarities between *Chubutisaurus* and Titanosauriformes, at that time Brachiosauridae and Titanosauridae. *Chubutisaurus* was included for first time in a phylogenetic analysis by Salgado et al. (1997) who recovered it as a non-titanosaur titanosauriform (a basal somphospondylan), a position subsequently recovered in most of the phylogenetic analyses carried out since then (e.g. Bonaparte et al. 2006; Calvo et al. 2007; González Riga et al. 2009; Carballido et al. 2011a; Gallina and Apesteguía 2011; Mannion et al. 2013; González Riga and Ortiz David 2014). In most of the aforementioned analyses, *Chubutisaurus* is recovered as a taxon slightly more basally recovered than *Ligabuesaurus*, which is also included in this chapter (Fig. 7).

Ligabuesaurus leanzai was originally considered as a titanosaur following the stem definition proposed by Sereno (1998) and Salgado (2003), as all somphospondylans closer to *Saltasaurus* than to *Euhelopus* (but not including *Euhelopus* in their analysis). Nevertheless, based on currently accepted definition of Titanosauria (as *Andesaurus*, *Saltasaurus* their most recent common ancestor and all its descendants) and the phylogenetic results of Bonaparte et al. (2006) *Ligabuesaurus* was recovered as a non-titanosaur somphospondylan. In this analysis, *Ligabuesaurus* and *Phuwiangosaurus* were recovered within a polytomy together with the stem conducting to Titanosauria. This politomy is due to the unstable position of *Ligabuesaurus* as sister taxa of *Phuwiangosaurus* and, alternatively, in a more basal position as a somphospondylan less derived than *Phuwiangosaurus*. Posterior phylogenetic analyses reinforced the idea of *Ligabuesaurus* as a non-titanosaur somphospondylan, although with some differences. Thus, in the analyses of D 'Emic (2012), González Riga and Ortíz David (2014), Wick and Lehman (2014) and Poropat et al. (2020), the sauropod from Cerro de Los Leones occupies a more basal position than in those of Carballido et al. (2011a, 2017) and Mannion et al. (2019a). Up to the moment, only a few analyses have recovered *Ligabuesaurus* as a titanosaur. Mannion et al. (2013), for instance, found *Ligabuesaurus* within different positions inside Titanosauria (depending on different methodological criteria), although posterior versions of this data set recovered it as a basal somphospondylan (e.g. Mannion et al. 2017, 2019a, b). In a similar way, Gorscak and O'Connor (2019), recovered *Ligabuesaurus* as a lithostrotian titanosaur, although their results showed deep discrepancies with previous analyses on titanosaur evolution (see Chap. 8). Although a complete description of *Ligabuesaurus* is still pending, this taxon is here considered as a somphospondylan sauropod closely related to the origin of Titanosauria (Fig. 7).

Malarguesaurus is probably one of the youngest non-titanosaur somphospondylans, and, up to the moment, the unique member of this paraphyletic assemblage that survived up to the Late Cretaceous. It was described by González Riga et al. (2009) on the basis of, at least, two specimens from the Portezuelo Formation (upper Turonian–lower Coniacian). Originally recovered by these authors as a derived non-titanosaur somphospondylan, it is an unstable taxon mainly represented by amphiplatyan anterior caudal vertebrae and procoelous posterior caudal vertebrae. Carballido et al. (2011b) and Mannion et al. (2013) recovered *Malarguesaurus* as a titanosaur. Nevertheless, many other analyses (mainly derived from the data sets of Mannion et al. (2013) and Carballido and Sander (2014) show unstable relationships of *Malarguesaurus*, being generally pruned before the analysis (e.g. Mannion et al. 2017, 2019a; González Riga et al. 2019) or after it (e.g. Carballido et al. 2020; Otero et al. 2021; Gallina et al. 2021). In such analyses, *Malarguesaurus* is placed as both, a derived non-titanosaur somphospondylan or a basal titanosaur (e.g. Gallina et al. 2021). The instability of *Malarguesaurus* seems to be a product of its fragmentary record plus the unusual combination of articulations in its caudal vertebrae sequence. As noted by González Riga et al. (2009), while other sauropods with combined procoelous and amphicoelous caudal vertebrae have the ‘ball and socket’ articulation in their anterior caudal vertebrae (e.g. *Andesaurus*, *Phuwiangosaurus*, *Malawisaurus*), and the middle and or posterior caudal vertebrae with flat articular surfaces, *Malarguesaurus* shows the opposite pattern. Up to the moment, and pending new materials and/or phylogenetic analyses, *Malarguesaurus* is considered as the youngest and more derived non-titanosaur somphospondylan from South America, although it may have putative titanosaur affinities. Because of that, we placed *Malarguesaurus* in a polytomy at the base of Titanosauria in Fig. 7, although the non-titanosauria positions recovered by Gallina et al. (2021) indicated a more basal position amongst Titanosauriformes.

4 Biogeography and Paleobiological Considerations

It has been suggested that the origin of the Macronaria took place by the Middle Jurassic (e.g. Upchurch et al. 2002; Carballido et al. 2011b; Benson et al. 2014), although no indisputable neosauropod remains from this age have been documented until the recent description of the dicraeosaurid *Lingwulong* from the Toarcian–Bajocian of Asia (Xu et al. 2018). Based on this record, and if *Bellusaurus*, possibly from the Callovian–early Oxfordian (Xu et al. 2018), is effectively a macronarian (e.g. Carballido et al. 2015; Moore et al. 2018), the origin of this stem clade could be traced back to the upper Lower or the lower Middle Jurassic, with a ghost lineage of no less than 5 million years (Fig. 7). Nevertheless, the systematic position of *Bellusaurus* is far from a general consensus amongst the different phylogenetic analyses (e.g. Upchurch et al. 2004; Wilson and Upchurch 2009; Moore et al. 2018; Carballido et al. 2020; Mannion et al. 2019b), and we tentatively placed it as a basal macronarian in Fig. 7 (one of the two positions in which can be placed resolving the polytomy of *Bellusaurus* and *Camarasaurus* obtained by Carballido et al. 2020). For the Late Jurassic, macronarians were already distributed all around the world, being represented in South America by the basal camarasauromorph *Tehuelchesaurus*, from the Cañadón Calcáreo Formation, and by putative brachiosaurids and indeterminate titanosauriformes. Based on these records, the faunal assemblage from the Late Jurassic of South America resembles that of the Tendaguru and Morrison formations, in which both lineages of flagellicaudatan neosauropods are represented (see Chap. 6 and Rauhut et al. 2020) together with basal camarasauromorphs (e.g. *Camarasaurus*, *Tehuelchesaurus*), and basal titanosauriforms of brachiosaurid affinities (*Giraffatitan*, *Brachiosaurus*, MPEF-PV 3098). None of the Late Jurassic materials collected from the Cañadón Calcáreo Formation can be clearly assigned to Somphospondyli. Up to the moment, there is no unequivocal record of a somphospondylan titanosauriform in the Late Jurassic, despite that the inferred origin of this clade must temporally coincide with the origin of its sister lineage Brachiosauridae (Fig. 7).

Regarding its palaeoecology, the Late Jurassic faunal assemblages show forms adapted to browsing at ground level (diplodocids), at middle levels (dicraeosaurids) and at the high levels of the tree canopy (basal camarasauromorphs) (see Whitlock 2011b and references therein). Therefore, these sauropod faunal associations seem to feed on all plant strata.

The Jurassic/Cretaceous boundary (145 MA) is considered as a moment of a marked change in the faunal assemblages, characterized by a general decline in diversity (e.g. Upchurch and Barrett 2005; Barrett et al. 2009; Mannion et al. 2011) and by the extinction of typical Jurassic lineages (Flagellicaudata and Brachiosauridae). The discovery of the dicraeosaurid *Amargasaurus* in the lowest Lower Cretaceous and the rebbachisaurid *Limaysaurus* in the lower Upper Cretaceous, both in Patagonia, demonstrated that diplodocoids (both fragelicaudatans and basal forms) survived in South America beyond the Jurassic–Cretaceous boundary. This conception was recently strengthened with the recognition of an increase in diversity of dicraeosaurids and the existence of diplodocids in the Early Cretaceous of Patagonia and Africa (Gallina et al. 2014, 2019; McPhee et al. 2016; Coria et al. 2019; Chap. 6) Amongst Macronaria, Brachiosauridae seems to become extinct in all landmasses after the Jurassic, except in North America (e.g. Mannion et al. 2017). Nevertheless, the recent recognition of *Padillasaurus* as a putative brachiosaurid indicates that this clade could have survived into the earliest Cretaceous of South America (Carballido et al. 2015; Fig. 7). Up to the moment, there is no clear record of somphospondylans in the Late Jurassic, although this lineage shows an increase in diversity since the Early Cretaceous, reaching a major peak in diversity in the Late Cretaceous with the Titanosauria (Cerda et al. 2011; Chaps. 9 and 10).

Basal somphospondylans have a marked diversity in Asia and, in Gondwana, in South America and Australia (see Mannion et al. 2019b). During the Early Cretaceous of South America, somphospondylans coexisted with diplodocoids. The oldest record of such association containing both, the typical Jurassic fauna (mainly flagellicaudatans) and the typical Cretaceous fauna (mainly somphospondylans), comes from the Bajada Colorada Formation. Up to the moment, there is no record of rebbachisaurid diplodocoids in the earliest Cretaceous (pre-Barremian). This assemblage not solely shows that dicraeosaurids, diplodocids, and somphospondylans coexisted in the earliest Cretaceous, but also that titanosaurs were, by that time, part of the sauropod communities in Patagonia (Gallina et al. 2021). As said, despite this record, titanosaur somphospondylans did not become abundant in the fossil record until the Late Cretaceous (see Chap. 10). As mentioned for the Late Jurassic fauna, that from the earliest Cretaceous of Patagonia seems to feed on all vegetation strata.

The sauropod fauna from the La Amarga Formation (Hauterivian–Barremian) is, in general, similar to that of older Cretaceous units from Patagonia, like the Mulichinco and Bajada Colorada formations, except for the record of diplodocids (ground-height browsers), which are present only in the last unit. An additional difference is that the La Amarga Formation has the oldest fossil record of rebbachisaurids (see Chap. 5), which, based on several characters, are considered as ground-ground-height browser sauropods, and therefore surely, they exploited the resources previously used by diplodocids.

During the latest Early Cretaceous (Aptian–Albian), titanosaurs experienced a major diversification event (e.g. Zaher et al. 2011), although the records from Patagonia still show a dominance of basal (non-titanosaur) somphospondylans and rebbachisaurids. Apparently, somphospondylans were, by the late Early Cretaceous and the early Late Cretaceous, less abundant than rebbachisaurids, at least in semi-arid to arid environments, as recorded in some localities of the Neuquén and San Jorge basins. Indeed, although further studies are required, rebbachisaurids (e.g. *Comahuesaurus, Lavocatisaurus, Limaysaurus, Rayososaurus*) seem to prevail in semi-arid to arid environments. These differences can be seen even within the same formation. In fact, in the arid conditions of the Puesto Quiroga Member of the Lohan Cura Formation at Aguada del León (La Picaza area, see Chap 5) there is a prevalence of rebbachisaurids (*Comahuesaurus*), whereas at Cerro de los Leones (Neuquén Province), somphospondylans (*Ligabuesaurus*) seem to prevail in the humid environment of the Cullin Grande Member of the same unit. A similar scenario is recorded in the Cañadón Asfalto-Somún Curá basin, where solely somphospondylans have been recorded so far (*Chubutisaurus*, *Patagotitan*) in humid environments dominated by channels and floodplains (Krause et al. 2020). As recently suggested by Otero et al. (2021), the faunal assemblage from the Candeleros Formation has a marked vertical division of plant resources amongst different sauropod lineages. Nevertheless, the fact that different lineages have probably had different food preferences, it does not necessarily imply that those lineages are sympatric.

Non-titanosaur somphospondylans became extinct in the early Late Cretaceous as part of a larger, global extinction event at the Turonian–Coniacian boundary that involved the complete extinction of several clades (Coria and Salgado 2005; Novas et al. 2005, 2013). *Malarguesaurus*, from the Turonian–Coniacian, seems to represent the last survivor of this early Cretaceous lineage, although given its relatively unstable position this assumption must be taken with caution. Post Turonian sauropods are worldwide represented by titanosaur sauropods. This last record of a non-titanosaur somphospondylans is coincident with the last record of rebbachisaurids. The extinction of both, basal somphospondylans and rebbachisaurids, could have been caused by environmental changes, or be the product of a major titanosaur diversification. Whatever was the cause, titanosaurs, with marked differences in their general bauplan (e.g. neck elongations, robustness, skull differences) and body size, ended up occupying most or all of the available niches previously exploited by diplodocoids and basal macronarians.

5 Conclusions

South American basal non-titanosaur macronarians are known on the base of five species (*Tehuelchesaurus*, *Padillasaurus*, *Chubutisaurus*, *Ligabuesaurus* and *Malarguesaurus*) and a few additional elements from the Late Jurassic and the Early Cretaceous of Argentinian Patagonia that are not well enough preserved to be diagnostic at species level but bring information on the faunal diversity at these periods. While the basal position of *Tehuelchesaurus*, as a non-titanosauriform camarasauromorph, and *Chubutisaurus*, as a non-titanosaur somphospondylan, are well supported by most recent phylogenies, the positions of the other South American basal macronarians lack strong support. *Padillasaurus* is here considered as a brachiosaurid, but this is weakly supported. *Ligabuesaurus* is in need of a complete revision that includes new materials, which will provide, not only a better diagnosis but also the possibility of testing its phylogenetic position. Finally, *Malarguesaurus* is an unstable taxon, which is probably due to its unusual combination of anterior amphicoelous and posterior procoelous caudal vertebrae. Basal macronarians from South America are recorded from the Upper Jurassic up to the lowermost Upper Cretaceous deposits (if *Malarguesaurus* is a non-titanosaur macronarian) or up to the Lower/Upper Cretaceous boundary if it is considered a titanosaur. During the Late Jurassic and the early Late Cretaceous, there was an evident faunal succession of neosauropods, although they keep feeding on the same plant resources, since there were always forms of different height browsing. Finally, with the diversification of titanosaurs, all basal lineages of somphospondylans became extinct, which could have resulted from a better use of the plant resources by these derived macronarians, and leading to their greater size and morphological disparity.

Acknowledgements This contribution was financed with funds of the PICT ANPCyT 1925 'La fauna de saurópodos del Cretácico "medio" de Patagonia, evolución y diversificación de los Rebbachisauridae y Somphospondyli, aspectos evolutivos y ecológicos' (José L. Carballido, director) and the Plurianual Project CONICET N° 11220130100683CO, 'Diversidad y evolución de las asociaciones de dinosaurios del Cretácico Temprano de la Cuenca Neuquina. Faunas y floras asociadas' (Rodolfo Coria, director). We really appreciate the comments made by the reviewers J.I. Canudo and R. Santucci, whose suggestions improved the quality of this manuscript.

References

Apesteguía S (2007) The sauropod diversity of the La Amarga Formation (Barremian), Neuquén (Argentina). Gondwana Res 12:533–546

Alifanov VR, Averianov AO (2003) *Ferganasaurus verzilini gen. et sp. nov.*, a new neosauropod (Dinosauria, Saurischia, Sauropoda) from the Middle Jurassic of Fergana Valley Kirghizia. J Vertebr Paleontol 23:358–372

Averianov AO, Efimov V (2018) The oldest titanosaurian sauropod of the northern hemisphere. Biol Commun 63:145–162

Averianov AO, Skutschas P (2017) A new lithostrotian titanosaur (Dinosauria, Sauropoda) from the early Cretaceous of Transbaikalia, Russia. Biol Commun 62:6–18

Barco JL, Canudo JI, Cuenca-Bescós G (2006) Descripción de las vértebras cervicales de *Galvesaurus herreroi* Barco, Canudo, Cuenca-Bescós y Ruiz-Omeñaca, 2005 (Dinosauria, Sauropoda) del tránsito Jurásico-Cretácico en Galve (Teruel, Aragón, España). Revista Española De Paleontología 21(2):189–205

Barrett PM, McGowan AJ, Page V (2009) Dinosaur diversity and the rock record. Proc Royal Soc Lond B 276:2667–2674

Bellardini F (2021) Los saurópodos de Cerro de los Leones, Formación Lohan Cura (Cretácico Inferior), provincia del Neuquén: osteología y relaciones filogenéticas. Universidad Nacional del Comahue, Centro Regional Universitario, Bariloche, Tesis doctoral, p 574

Benson RBJ, Campione NE, Carrano MT, Mannion PD, Sullivan C, Upchurch P, Evans DC (2014) Rates of dinosaur body mass evolution indicate 170 million years of sustained ecological innovation on the avian stem lineage. PLoS Biol 12(5):e1001853

Bonaparte JF (1986) History of the terrestrial Cretaceous vertebrates of Gondwana. Actas IV Congreso Argentino de Paleontología y Bioestratigrafía 2:63–95

Bonaparte JF, Powell JE (1980) A continental assemblage of tetrapods from the Upper Cretaceous beds of El Brete, northwestern Argentina (Sauropoda-Coelurosauria- Carnosauria-Aves). Mémoires de la Société Géologique De France 139:19–28

Bonaparte JF, González Riga B, Apesteguía S (2006) *Ligabuesaurus leanzai* gen. et sp. nov. (Dinosauria, Sauropoda), a new titanosaur from the Lohan Cura Formation (Aptian, Lower Cretaceous) of Neuquén, Patagonia. Argentina. Cretaceous Res 27:364–376

Cabaleri N, Volkheimer W, Silva Nieto D, Armella C, Cagnoni M, Hauser N, Matteini M, Pimentel M (2010) U–Pb ages in zircons from Las Chacritas and Puesto Almada members of the Jurassic Cañadón Asfalto Formation, Chubut province, Argentina. In: VII South American symposium on isotope geology, Brasilia, pp 190–193

Calvo JO, Porfiri J, González Riga B, Kellner AWA (2007) A new Cretaceous terrestrial ecosystem from Gondwana with the description of a new sauropod dinosaur. An Acad Bras Cienc 79(3):529–541

Cantino PD, de Queiroz K (2006) International code of phylogenetic nomenclature (PhyloCode) Version 6. http://phylonames.org/code/

Carballido JL, Rauhut O, Pol D, Salgado L (2011a) Osteology and phylogenetic relationships of *Tehuelchesaurus benitezi* (Dinosauria, Sauropoda) from the Upper Jurassic of Patagonia. Zool J Linn Soc 163:605–662

Carballido JL, Pol D, Cerda IA, Salgado L (2011b) The osteology of *Chubutisaurus insignis* del Corro, 1975 (Dinosauria, Neosauropoda) from the "middle" Cretaceous of Central Patagonia Argentina. J Vertebr Paleontol 31(1):93–110

Carballido JL, Sander M (2014) Postcranial axial skeleton of *Europasaurus holgeri* (Dinosauria, Sauropoda) from the Upper Jurassic of Germany: implications for sauropod ontogeny and phylogenetic relationships of basal Macronaria. J Syst Palaeontol 12(3):335–387

Carballido JL, Pol P, Parra Ruge ML, Padilla Bernal S, Páramo-Fonseca ME, Etayo-Serna F (2015) A new Early Cretaceous brachiosaurid (Dinosauria, Neosauropoda) from northwestern Gondwana (Villa de Leiva, Colombia), J Vertebr Paleontol 35(5): e980505

Carballido JL, Pol D, Otero A, Cerda I, Salgado L, Garrido A, Ramezani J, Cúneo RN, Krause M (2017) A new giant titanosaur sheds light on body size evolution among sauropod dinosaurs. Proc Royal Soc B-Biol 284:20171219

Carballido JL, Scheli M, Knötschke N, Sander M (2020) The appendicular skeleton of the dwarf macronarian sauropod *Europasaurus holgeri* from the Late Jurassic of Germany and a re-evaluation of its systematic affinities. J Syst Paleontol 18(9):739–781

Carvalho IS, Salgado L, Lindoso RM, Araújo-Júnior HI, Costa Nogueira FC, Soares JA (2017) A new basal titanosaur (Dinosauria, Sauropoda) from the Lower Cretaceous of Brazil. J South Am Earth Sci 75:74–84

Cerda IA, Paulina Carabajal A, Salgado L, Coria RA, Reguero M, Tambussi CP, Moly JJ (2011) The first record of a sauropod dinosaur from Antarctica. Naturwissenschaften 99(1):83–87

Cope ED (1877) On a gigantic saurian from the Dakota epoc of Colorado. Pal Bull 25:5–10

Coria RA, Salgado L (2005) Mid-Cretaceous turnover of saurischian communities: evidence from the Neuquén Basin. In: Veiga G, Spalletti L, Howell JA, Schwartz E (eds) The Neuquén Basin: a case study in sequence stratigraphy and basin dynamics. Special Publications of the Geological Society 252, p 317–327

Coria RA, Windholz GJ, Ortega F, Currie PJ (2019) A new dicraeosaurid sauropod from the Lower Cretaceous (Mulichinco Formation, Valanginian, Neuquén Basin) of Argentina. Cretaceous Res 93:33–48

Cúneo R, Ramezani J, Scasso R, Pol D, Escapa I, Zavattieri AM, Bowring SA (2013) High-precision U-Pb geochronology and a new chronostratigraphy for the Cañadón Asfalto Basin, Chubut, central Patagonia: Implications for terrestrial faunal and floral evolution in Jurassic. Gondwana Res 24(3–4):1267–1275

del Corro G (1975) Un nuevo saurópodo del Cretácico Superior. *Chubutisaurus insignis* gen. et sp. nov. (Saurischia-Chubutisauridae nov.) del Cretácico Superior (Chubutiano), Chubut, Argentina. Actas I Congreso Argentino de Paleontología y Bioestratigrafia: 229–240

D'Emic MD (2012) The early evolution of titanosauriform sauropod dinosaurs. Zool J Linn Soc-Lond 166:624–671

Gallina PA, Apesteguía S, Haluza A, Canale JI (2014) A Diplodocid Sauropod Survivor from the Early Cretaceous of South America. PLoS ONE 9 (5):e97128

Gallina PA, Apesteguía S (2011) Cranial anatomy and phylogenetic position of the titanosaurian sauropod *Bonitasaura salgadoi*. A Pal Pol 56:45–60

Gallina PA, Apesteguía S, Canale JI, Haluza A (2019) A new long-spined dinosaur from Patagonia sheds light on sauropod defense system. Sci Rep 9:1392

Gallina PA, Canale JI, Carballido JL (2021) The earliest known titanosaur sauropod dinosaur. Ameghiniana 58(1):35–51

Garrido A (2010) Estratigrafía del Grupo Neuquén, Cretácico Superior de la Cuenca Neuquina (Argentina): nueva propuesta de ordenamiento litoestratigráfico. Rev Mus Arg C Nat, Nueva Serie 12:121–177

González Riga BJ, Ortíz David L (2014) A new titanosaur (Dinosauria, Sauropoda) from the Upper Cretaceous (Cerro Lisandro Formation) of Mendoza Province Argentina. Ameghiniana 51(1):3–25

González Riga BJ, Previtera E, Pirrone CA (2009) *Malarguesaurus florenciae* gen. et sp. nov., a new titanosauriform (Dinosauria, Sauropoda) from the Upper Cretaceous of Mendoza, Argentina. Cretaceous Res 30(1):135–148

González Riga BJ, Lamanna MC, Otero A, Ortiz David LD, Kellner AWA, Ibiricu LM (2019) An overview of the appendicular skeletal anatomy of South American titanosaurian sauropods, with definition of a newly recognized clade. An Acad Bras Cienc 91(2):1–42

Gorscak E, O'Connor PM (2019) A new African Titanosaurian Sauropod Dinosaur from the middle Cretaceous Galula Formation (Mtuka Member), Rukwa Rift Basin, Southwestern Tanzania. PLoS ONE 14(2):e0211412

Gorscak E, O'Connor PM, Roberts EM, Stevens NJ (2017) The second titanosaurian (Dinosauria: Sauropoda) from the middle Cretaceous Galula Formation, southwestern Tanzania, with remarks on African titanosaurian diversity. J Vertebr Paleontolo 37:e1343250

Hauser N, Cabaleri NG, Gallego OF, Monferran MD, Silva Nieto D, Armella C, Matteini M, Aparicio Gonzalez PA, Pimentel MM, Volkheimer W, Reimold WU (2017) U-Pb and Lu-Hf zircon geochronology of the Canadón Asfalto Basin, Chubut, Argentina: Implications for the magmatic evolution in central Patagonia. J South Am Earth Sci 78:190e212

Janensch W (1947) Pneumatizitat bei Wirbeln von Sauropoden und anderen Saurischien. Palaeontographica, Supplement 7(3):1–25

Krause JM, Ramezani J, Umazano AM, Pol D, Carballido JL, Sterli J, Puerta P, Cúneo NR, Bellosi E (2020) High-resolution chronostratigraphy of the Cerro Barcino Formation (Patagonia): Paleobiologic implications for the mid-cretaceous dinosaur-rich fauna of South America. Gondwana Res 80:33–49

Leanza HA (2003) Las sedimentitas huitrinianas y rayosianas (Cretácico inferior) en el ámbito central y meridional de la Cuenca Neuquina, Argentina. Servicio Geológico Minero Argentino, Serie De Contribuciones Técnicas-Geología 2:1–31

Leanza HA, Hugo CA (2001) Cretaceous red beds from southern Neuquén Basin (Argentina): age, distribution and stratigraphic discontinuities. Asociación Paleontológica Argentina Publicación Especial 7:117–122

Leanza HA, Hugo CA (2011) Las formaciones La Amarga y Lohan Cura (Cretácico Temprano) en el depocentro Picún Leufú. Relatorio del XVIII Congreso Geológico Argentino, Neuquén 223–230.

Mannion PD, Upchurch P, Carrano MT, Barrett PM (2011) Testing the effect of the rock record on diversity: a multidisciplinary approach to elucidating the generic richness of sauropodomorph dinosaurs through time. Biol Rev 86:157–181

Mannion PD, Upchurch P, Barnes RN, Mateus O (2013) Osteology of the Late Jurassic Portuguese sauropod dinosaur *Lusotitan atalaiensis* (Macronaria) and the evolutionary history of basal titanosauriforms. Zool J Linn Soc-Lond 168:98–20

Mannion PD, Allain R, Moine O (2017) The earliest known titanosauriform sauropod dinosaur and the evolution of Brachiosauridae. PeerJ 5:e3217

Mannion PD, Upchurch P, Jin X, Zheng W (2019a) New information on the Cretaceous sauropod dinosaurs of Zhejiang Province, China: impact on Laurasian titanosauriform phylogeny and biogeography. Royal Soc Open Sci 619105. https://doi.org/10.1098/rsos.191057

Mannion PD, Upchurch P, Schwarz D, Wings O (2019b) Taxonomic affinities of the putative titanosaurs from the Late Jurassic Tendaguru Formation of Tanzania: phylogenetic and biogeographic implications for eusauropod dinosaur evolution. Zool J Linn Soc-Lond 185(3):784–909

Martinelli AG, Garrido AC, Forasiepi AM, Paz R, Gurovich Y (2007) Notes on fossil remains from the Early Cretaceous Lohan Cura Formation, Neuquén Province, Argentina. Gondwana Res 11:537–552

Marsh OC (1878) Principal characters of American Jurassic dinosaurs, Part I. Am J Sci 27:160–168

McPhee BW, Mannion PD, de Klerk WJ, Choiniere JN (2016) High diversity in the sauropod dinosaur fauna of the Lower Cretaceous Kirkwood Formation of South Africa: Implications for the Jurassic-Cretaceous transition. Cretaceous Res 59:228–248

Moore A, Mo J, Clark JM, Xu X (2018) Cranial anatomy of *Bellusaurus sui* (Dinosauria: Eusauropoda) from the Middle-Late Jurassic Shishugou Formation of northwest China and a review of sauropod cranial ontogeny. PeerJ 6:e4881

Novas FE, de Valais S, Vickers-Rich P, Rich T (2005) A large Cretaceous theropod from Patagonia, Argentina, and the evolution of carcharodontosaurids. Naturwissenschaften 92:226–230

Novas FE, Agnolín FL, Ezcurra MD, Porfiri J, Canale JI (2013) Evolution of the carnivorous dinosaurs during the Cretaceous: the evidence from Patagonia. Cretac Res 45:174–215

Otero A, Carballido JL, Salgado L, Canudo JI, Garrido AC (2021) Report of a giant titanosaur sauropod from the Upper Cretaceous of Neuquén Province, Argentina. Cretaceous Res 122:104754

Pérez-Pueyo M, Moreno-Azanza BJL, Canudo JI (2019) New contributions to the phylogenetic position of the sauropod *Galvesaurus herreroi* from the late Kimmeridgian-early Tithonian (Jurassic) of Teruel (Spain). Bol Geol Min 130(3):375–392

Poropat SF, Upchurch P, Mannion PD, Hocknull SA, Kear BP, Sloan T, Sinapius GHK, Elliott DA (2015) Revision of the sauropod dinosaur *Diamantinasaurus matildae* Hocknull et al. 2009 from the mid-Cretaceous of Australia: implications for Gondwanan titanosauriform dispersal. Gondwana Res 27:995–1033

Poropat SF, Mannion PD, Upchurch P, Hocknull SA, Kear BP, Kundrát M, Tischler TT, Sloan T, Sinapius GHK, Elliott JA, Elliott DA (2016) New Australian sauropods shed light on Cretaceous dinosaur palaeobiogeography. Sci Rep-UK 6:34467

Poropat SF, Mannion PD, Upchurch P, Tischler TT, Sloan T, Sinapius GHK, Elliott JA, Elliott DA (2020) Osteology of the wide-hipped titanosaurian sauropod dinosaur *Savannasaurus elliottorum* from the Upper Cretaceous Winton Formation of Queensland, Australia. J Vertebr Paleontol 40: e1786836

Rauhut OWM (2002) Dinosaur evolution in the Jurassic: a South American perspective. J Vertebr Paleontol 22:89A

Rauhut OWM (2006) A brachiosaurid sauropod from the Late Jurassic Cañadón Calcáreo Formation of Chubut Argentina. Foss Rec 9(2):226–237

Rauhut OWM, Remes K, Fechner R, Cladera G, Puerta P (2005) Discovery of a short-necked sauropod dinosaur from the Late Jurassic period of Patagonia. Nature 435:670–672

Rauhut OWM, Carballido JL, Pol D (2020) First Osteological Record of a Stegosaur (Dinosauria, Ornithischia) from the Upper Jurassic of South America. J Vertebr Paleontol e1862133

Rich TH, Vickers-Rich P, Gimenez O, Cúneo R, Puerta P, Vacca R (1999) A new sauropod dinosaur from Chubut Province, Argentina. National Science Museum Monographs 15:61–84

Riggs ES (1903) *Brachiosaurus altithorax*, the largest known dinosaur. Am J Sci 15:299–306

Royo Torres R, Fuentes C, Meijide M, Meijide-Fuentes F, Meijide-Fuentes M (2017) A new Brachiosauridae sauropod dinosaur from the Lower Cretaceous of Europe (Soria Province, Spain). Cretaceous Res 80:38–55

Salgado L (1993) Comments on *Chubutisaurus insignis* del Corro (Saurischia, Sauropoda). Ameghiniana 30:265–270

Salgado L (2003) Should we abandon the name Titanosauridae? Some comments on the taxonomy of titanosaur sauropods (Dinosauria). Revista Española De Paleontología 18(1):13–20

Salgado L, Bonaparte JF (2007) Sauropodomorpha. In: Salgado L, Coria RA (eds) Gasparini Z. Patagonian Mesozoic Reptiles, Indiana University Press, pp 188–228

Salgado L, Coria RA, Calvo JO (1997) Evolution of titanosaurid sauropods. Part I. Phylogenetic Analysis based on the postcranial evidence. Ameghiniana 34(1):3−32

Salgado L, Calvo JO (1997) Evolution of titanosaurid sauropods. II: the cranial evidence. Ameghiniana 34:33–47

Sereno PC (1998) A rationale for phylogenetic definitions, with application to the higher-level taxonomy of Dinosauria. Neues Jahrb Geol P Abh 210:41−83

Sereno PC (1999) A rationale for dinosaurian taxonomy. J Verteb Paleontol 19(4):788–790

Sereno PC, Rose-Petruck C, Beck AL, Dutheil DB, Larsson HCE, Lyon GH, Moussa B, Sadleir RW, Sidor CA, Varricchio DG, Wilson GP, Wilson JA (1999) Cretaceous Sauropods from the Sahara and the Uneven rate of skeletal evolution among dinosaurs. Science 286:1342–1347

Taylor MP (2009) A re-evaluation of *Brachiosaurus altithorax* Riggs 1903 (Dinosauria, Sauropoda) and its generic separation from *Giraffatitan brancai* (Janensch 1914). J Vertebr Paleontol 29(3):787–806

Taylor MP, Naish D (2005) The phylogenetic taxonomy of Diplodocoidea (Dinosauria: Sauropoda). PaleoBios 25:1–7

Taylor MP, Wedel MJ, Cifelli RL (2011) A new sauropod dinosaur from the lower cretaceous cedar mountain formation, Utah, USA. Acta Pal Pol 56:75–98

Upchurch P (1998) The phylogenetic relationships of sauropod dinosaurs. Zool J Linn Soc-Lond 124:43–103

Upchurch P, Hunn CA, Norman DB (2002) An analysis of dinosaurian biogeography: evidence for the existence of vicariance and dispersal patterns caused by geological events. Proc Biol Sci 269:613–621

Upchurch P, Barrett PM (2005) Phylogenetic and taxic perspectives on sauropod diversity. In: Curry Rogers K, Wilson JA (eds) The Sauropods: Evolution and Paleobiology, pp 104–124

Upchurch P, Barnett PM, Dodson P (2004) Sauropoda. In: Weishampel DB, Dodson P, Osmólska H (eds) The Dinosauria (2nd edition), pp 252−322

Wick SL, Lehman TM (2014) A complete titanosaur femur from west Texas with comments regarding hindlimb posture. Cretaceous Res 49:39–44

Wilson JA (2002) Sauropod dinosaur phylogeny: critique and cladistic analysis. Zool J Linn Soc-Lond 136:217–276

Wilson JA, Sereno PC (1998) Early evolution and higher-level phylogeny of sauropod dinosaurs. Soc Vert Paleontol, Memoir 5:1–68

Wilson JA, Upchurch P (2003) A revision of *Titanosaurus* Lydekker (Dinosauria-Sauropoda), the first dinosaur genus with a Gondwanan distribution. J Syst Palaentol 1(3):125–160

Wilson JA, Upchurch P (2009) Redescription and reassessment of the phylogenetic affinities of *Euhelopus zdanskyi* (Dinosauria: Sauropoda) from the Early Cretaceous of China. J Syst Palaentol 7:199–239

Whitlock JA (2011a) A phylogenetic analysis of Diplodocoidea (Saurischia: Sauropoda). Zool J Linn Soc-Lond 161:872–915

Whitlock JA (2011b) Inferences of diplodocoid (Sauropoda: Dinosauria) feeding behavior from snout shape and microwear analyses. PLoS ONE 6(4):e 18304

Xu X, Upchurch P, Mannion PD, Barrett PM, Regalado-Fernandez OR, Mo J, Ma J, Liu H (2018) A new Middle Jurassic diplodocoid suggests an earlier dispersal and diversification of sauropod dinosaurs. Nat Commun 9:2700. https://doi.org/10.1038/s41467-018-05128-1

Zaher H, Pol D, Carvalho AB, Nascimento PM, Riccomini C, Larson P, Juárez-Valieri R, Pires-Domingues R, Silva NJD Jr, Campos DA (2011) A complete skull of an Early Cretaceous sauropod and the evolution of advanced titanosaurians. *PLOS One* 6: e16663

Titanosauria: A Critical Reappraisal of Its Systematics and the Relevance of the South American Record

José L. Carballido, Alejandro Otero, Philip D. Mannion, Leonardo Salgado, and Agustín Pérez Moreno

Abstract Our understanding of sauropodomorph evolution is continually improved with the recognition of new species and their inclusion in phylogenetic data sets, along with the incorporation of novel characters. Whereas some subclades remain stable in terms of their original taxonomic content, the definitions of others have proven to be less easily applicable as increased taxon and character sampling has changed sauropodomorph interrelationships. Relatively minor differences in the position of clade specifiers between alternative phylogenies result in substantial differences in clade composition in several instances. Titanosauria is the most successful and diverse clade of sauropodomorph dinosaurs, but its interrelationships are poorly understood. With an ever-growing number of recognized species and increased taxon sampling in phylogenies, resulting in the recovery of newly identified, diverse subclades,

---

Electronic supplementary material The online version contains supplementary material available at (10.1007/978-3-030-95959-3_8).

J. L. Carballido (✉) · A. Otero · L. Salgado · A. P. Moreno
Consejo Nacional de Investigaciones Científicas y Técnicas, Buenos Aires, Argentina
e-mail: jcarballido@mef.org.ar

L. Salgado
e-mail: lsalgado@unrn.edu.ar

A. P. Moreno
e-mail: aperezmoreno@mef.org.ar

J. L. Carballido
Museo Paleontológico Egidio Feruglio, Fontana 140, Trelew (9100), Chubut, Argentina

A. Otero · A. P. Moreno
División Paleontología de Vertebrados, Facultad de Ciencias Naturales y Museo (Anexo Laboratorios), Calle 122 y 60, La Plata (B1900WA), Buenos Aires Province, Argentina

P. D. Mannion
Department of Earth Sciences, University College London, Gower Street, London WC1E 6BT, UK

L. Salgado
Universidad Nacional de Río Negro. Instituto de Investigación en Paleobiología y Geología. UNRN-CONICET, Av. Julio A. Roca 1242, General Roca, Río Negro 8332, Argentina

A. Otero et al. (eds.), *South American Sauropodomorph Dinosaurs*, Springer Earth System Sciences, https://doi.org/10.1007/978-3-030-95959-3_8

the need to re-evaluate titanosaurian supra-specific taxonomy becomes increasingly important. Using fifteen phylogenetic data sets, we conducted a series of analyses to detect the most stable clades and taxa amongst Titanosauria. Based on this, we propose several changes to titanosaur systematics, including a newly defined node-stem triplet composed of Eutitanosauria and its constituent lineages, Colossosauria and Saltasauroidea, developing a new framework to improve stability and facilitate communication of phylogenetic results.

Keywords Titanosauria · Systematics · Eutitanosauria · Colossosauria · Saltasauroidea

1 Introduction

Our current understanding of the evolutionary relationships of sauropodomorph dinosaurs has experienced substantial changes since the earliest inclusion of the group in phylogenetic analyses (e.g. Gauthier 1986; Upchurch 1995, 1998; Salgado et al. 1997; Wilson and Sereno 1998). Such changes were accompanied by new systematic definitions of individual clades, but also by ongoing revision of previously proposed definitions. Notable examples include the alternative definitions proposed for Sauropodomorpha and Sauropoda, including stem- and node-based definitions and the use of different clade specifiers, resulting in contrasting compositions of these two clades (e.g. Salgado et al. 1997; Sereno 1998; Wilson and Sereno 1998; Galton and Upchurch 2004; Yates 2007; McPhee et al. 2018). By contrast, some clades were well defined on first usage and did not need to be redefined thereafter, such as the node-stem triplet formed by Neosauropoda and its two stems, Diplodocoidea and Macronaria (Wilson and Sereno 1998). Other robust clades required minor changes in order to accommodate species-level taxonomic revisions, such as the definitions of Titanosauriformes and its two stems (Brachiosauridae and Somphospondyli) (Taylor 2009). The definitions of some clades have proven to be less easily applicable as our understanding of relationships has developed, with relatively minor differences in the position of clade specifiers between alternative phylogenies resulting in substantial differences in clade composition. For example, minor differences in terms of node distances between alternative phylogenies result in a markedly different Lithostrotia lineage (see for example Mannion et al. 2019; Gallina et al. 2021). With an ever-growing number of recognized species and increased taxon sampling in phylogenies, resulting in the recovery of newly identified, diverse lineages, the need to re-evaluate the supra-specific taxonomy has become increasingly important. Within Neosauropoda, an important contribution to this issue was presented by Taylor and Naish (2005), who provided a revision of Diplodocoidea and the main lineages and definitions of its subclades. By contrast, no such systematic evaluation has occurred for the diplodocoid sister clade, Macronaria, with only the earliest diverging ('basal') lineages thus far considered (Taylor 2009; D'Emic 2012; Chap 7).

Despite being the most successful and diverse clade of sauropodomorph dinosaurs, with > 80 valid species currently known (e.g. Gorscak and O'Connor 2019) and a global fossil record (Upchurch et al. 2004; Cerda et al. 2011), the interrelationships of the macronarian clade Titanosauria are poorly understood (Curry Rogers 2005; Wilson 2006). The latter probably reflects, at least partly, the relatively low number of titanosaurian taxa incorporated into phylogenetic analyses. In the largest recent analyses, ~ 35% of putative titanosaur species were included by Carballido et al. (2020) and Poropat et al. (2021), with ~50% represented in that of Gorscak and O'Connor (2019). Until recently, the interrelationships of Titanosauria were also marked by an apparent absence of diverse lineages nested within this clade (e.g. Salgado et al. 1997; Upchurch et al. 2004; Wilson 2002; Wilson and Upchurch 2003; Curry Rogers 2005). This likely reflects the limited number of valid titanosaur species (<25) identified by the turn of the last century, with the last two decades marked by an exponential rise in the known diversity of Titanosauria, as well as the development of new and more inclusive phylogenetic analyses that have begun to shed light on the evolution of this diverse lineage of sauropods (see Chaps. 9 and 10). As a result, recent phylogenetic analyses have started to consistently recover a number of speciose titanosaurian subclades (e.g. Gallina and Apesteguía 2011; Santucci and Arruda Campos 2011; Carballido et al. 2017; Gorscak et al. 2017; Díez Díaz et al. 2018; González Riga et al. 2018; Mannion et al. 2019; Hechenleitner et al. 2020; Poropat et al. 2021). Whereas some of these clades are phylogenetically defined, others are not, and several existing definitions are problematic, leading to taxonomic instability. South American titanosaurs, in particular, play a central role in many of these clade definitions.

In this chapter, we discuss the systematic definitions for Titanosauria, their use and currently accepted definitions, and the problems that arise with their current usage. Finally, based on a series of analyses carried out to identify the most appropriate and stable clade specifiers, we propose the redefinition of several titanosaurian clades, providing a more applicable and stable phylogenetic framework for future studies.

2 Current Status of Titanosaur Systematics

Here we present an overview of the current status of the systematics of Titanosauria and its constituent clades, including potential problems. To facilitate discussion and help visualize definitions, we utilize an informal cladogram (Fig. 1), consisting solely of the specifier taxa of the titanosaurian subclades. This topology is mainly based on the results of the phylogenetic analyses presented by Carballido et al. (2017, 2020), Díez Díaz et al. (2018), Gorscak and O'Connor (2019), Mannion et al. (2019), Hechenleitner et al. (2020) and Poropat et al. (2021). As discussed below, some analyses vary in the placement of specifier taxa, which directly impacts upon the taxonomic content of the clades they define.

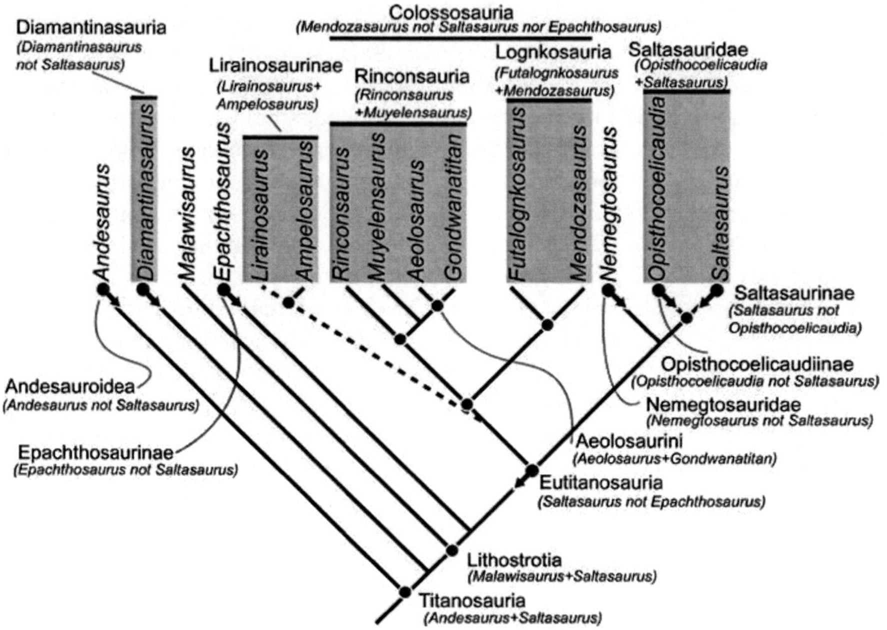

Fig. 1 Informal cladogram showing the hypothetical interrelationships of titanosaurs based on recent analyses, consisting solely of the specifier species of existing subclades. Node-based clades are marked with a circle and stem-based nodes with a circle and arrow

2.1 *Titanosauria*

Titanosauria was coined by Bonaparte and Coria (1993) to include the two families recognized by them: Andesauridae and Titanosauridae. Whereas Andesauridae was recognized by the presence of plesiomorphic characters and composed primarily of a paraphyletic assemblage of 'basal' species, Titanosauridae was based on the recognition of synapomorphies and consisted of more 'advanced' titanosaurs. The first formal definition of Titanosauria was provided by Salgado et al. (1997: 18) as: 'the most recent common ancestor of *Andesaurus delgadoi* and Titanosauridae and all of its descendants". The following year, Wilson and Sereno (1998: 22) presented a stem-based definition for Titanosauria: 'titanosauriforms more closely related to *Saltasaurus* than either *Brachiosaurus* or *Euhelopus*', which was partially followed by some authors (e.g. Wilson 2002). Subsequently, Wilson and Upchurch (2003) and Salgado (2003) presented a revision of the clade name, Titanosauria, in which they adapted these two phylogenetic definitions, presenting a node-based definition that replaced Titanosauridae with the species *Saltasaurus loricatus*. From 2003 onwards, nearly all authors have followed this node-based definition of Titanosauria (i.e. *Andesaurus*, *Saltasaurus*, their most recent common ancestor and all its descendants), with only a few contributions opting for the stem-based definition (e.g. Bonaparte et al. 2006). Based on this well-established node definition, the taxonomic content of Titanosauria is heavily dependent on the position of *Andesaurus*, known from the early Cenomanian of Argentina (Calvo and Bonaparte 1991). Although known only from a partial postcranial skeleton, *Andesaurus* has remained a relatively stable taxon in nearly all analyses of sauropod phylogenetic relationships over the last ~25 years. Thus, the original taxonomic content of Titanosauria proposed by Bonaparte and Coria (1993) and recovered by Salgado et al. (1997) has remained almost invariable despite the discovery of new taxa and greater taxon sampling in phylogenetic analyses (e.g. Gallina and Apesteguía 2011; D'Emic 2012; Carballido et al. 2013; Mannion et al. 2013; Gorscak et al. 2014; Gallina and Otero 2015; Tykoski and Fiorillo 2017; Carballido et al. 2017; González Riga et al. 2019; Diez Díaz et al. 2020; Poropat et al. 2021). Nevertheless, some recent analyses, especially those that applied Parsimony implied weighting or Bayesian approaches, rather than the 'standard' Parsimony with equal weighting approach, recovered *Andesaurus* in a more 'basal' position amongst Somphospondyli (e.g. Mannion et al. 2019; Gorscak and O'Connor 2019 [and previous iterations of these matrices]). Using the updated data set from Carballido et al. (2017), Hechenleitner et al. (2020) recovered *Andesaurus* in a similarly 'basal' position through an equal weighting parsimony phylogenetic analysis. These alternative positions for *Andesaurus* result in a much more inclusive Titanosauria that incorporates several taxa otherwise universally recovered as non-titanosaurian somphospondylans, such as *Chubutisaurus*, *Ligabuesaurus* and Euhelopodidae. Although we find no strong reason to use another definition for Titanosauria, and this discrepancy between equal and implied weighting has been reconciled in one recent analysis (Poropat et al. 2021), the apparent instability of *Andesaurus* requires further testing to determine whether a redefinition of Titanosauria should be considered. Several Linnaean ranks were also defined based on '*Titanosaurus*', the first titanosaur to be named (i.e. Titanosauroidea, Titanosauridae, Titanosaurinae; e.g. Salgado et al. 1997; Upchurch 1998; Sanz et al. 1999). However, their usage has been abandoned since the type genus *Titanosaurus* was considered invalid by Wilson and Upchurch (2003), which we follow here.

2.2 Andesauroidea and Related Subordinate Linnaean Ranks

Salgado (2003: 20) utilized Andesauroidea to apply to the clade defined as 'all titanosaurs closer to *Andesaurus* than to *Saltasaurus*'. Two other coordinated rank taxa of *Andesaurus* have been used in the literature, although neither has been phylogenetically defined. Andesaurinae was applied by Calvo and Bonaparte (1991) to distinguish *Andesaurus* from more derived titanosaurs, and Bonaparte and Coria (1993) used Andesauridae to group a subset of 'basal' titanosaurs (*Andesaurus*, *Argentinosaurus* and *Epachthosaurus*). Based on article 36.1 of the International Commission on Zoological Nomenclature (ICZN), the nominal authors of these three ranks are Calvo and Bonaparte (1991). Mannion and Calvo (2011) commented that no clear synapomorphies are present to consider any of these Linnaean taxa as applicable, with nearly all phylogenetic analyses recovering *Andesaurus*, *Argentinosaurus*, *Epachthosaurus* and other putative 'basal' titanosaurs, as a paraphyletic assemblage. However, a small number of phylogenetic analyses have recovered Andesauroidea as a 'basal' titanosaurian clade. Using implied weighting in a parsimony framework, Mannion et al. (2013) recovered *Andesaurus* in a 'basal' position amongst Somphospondyli (see above), forming a clade with several taxa traditionally considered as non-titanosaurs, including *Chubutisaurus*, *Ligabuesaurus* and *Sauroposeidon*. Similar results were later obtained with expanded versions of this data set using extended implied weighting (e.g. Mannion et al. 2019). Using a different data set, analysed within parsimony and Bayesian frameworks, a comparable andesauroid clade was also recovered by Gorscak and O'Connor (2016), as well as subsequent iterations of this data set (e.g. Gorscak and O'Connor 2019). However, the most recent analysis based on the data set of Mannion et al. (2013, 2019) produced congruent topologies under equal and extended implied weighting, with *Andesaurus* recovered in its more 'accepted' position, as a derived member of Somphospondyli that is the sister taxon to all other titanosaurs (Poropat et al. 2021). Therefore, although the definition of the clade is valid, Andesauroidea is currently restricted to *Andesaurus* in most phylogenetic analyses, with the notable exception of those based on the Gorscak and O'Connor (2016) data set. Although the latter includes the largest number of titanosaurs in any phylogenetic data set, it samples few taxa outside of the clade, limiting our understanding of the evolutionary transitions occurring close to this radiation. As such, we regard those data sets that sample somphospondylans more broadly to provide greater robusticity in reconstructions of the relationships of 'basal' titanosaurs. Following this, and given that there is no utility in a monospecific clade, the application of Andesauroidea currently remains unnecessary.

2.3 *Diamantinasauria*

Diamantinasauria is the most recently erected titanosaurian clade. It was coined and defined by Poropat et al. (2021) as 'the most inclusive clade that contains *Diamantinasaurus* but not *Saltasaurus*'. Based on the phylogenetic analyses of Poropat et al. (2021), Diamantinasauria is a 'basal' lineage of titanosaurs, as specifiers of other subordinate clades (e.g. *Epachthosaurus*, *Malawisaurus*) were recovered in a more nested position.

2.4 *Lithostrotia*

Upchurch et al. (2004: 311) erected Lithostrotia as a node-based clade, defined as '*Malawisaurus*, *Saltasaurus*, their most recent common ancestor and all its descendants' (see also Wilson and Upchurch 2003). In most phylogenetic analyses, *Malawisaurus* is recovered as one of the most 'basal' titanosaurs, such that Lithostrotia consists of nearly all titanosaurs (including Colossosauria and Saltasauridae), with the exception of *Andesaurus* (e.g. Upchurch et al. 2004; Calvo et al. 2007b; Salgado et al. 2015; Gorscak and O'Connor 2019; Poropat et al. 2021). However, in other analyses, Lithostrotia is far less inclusive, recovered as essentially the sister taxon to Colossosauria, with only Saltasauridae part of the lithostrotian lineage (e.g. Carballido et al. 2011a, b, 2017; Lacovara et al. 2014; Tykoski and Fiorillo 2017). Other analyses, such as that of Gallina and Apesteguía (2011), did not use Lithostrotia in their definitions, most probably because they retrieved it as a synonym of Titanosauria. It should be noted that a recent phylogenetic analysis published by Hechenleitner et al. (2020), which was based on the data set of Carballido et al. (2017), recovered *Malawisaurus* in its more 'accepted', 'basal' position, congruent with that obtained by most other phylogenies. As such, Lithostrotia is a widely used clade, but one that encloses a markedly different taxonomic content when alternative data sets are compared.

2.5 Eutitanosauria

Eutitanosauria was initially proposed by Sanz et al. (1999: 252) for a clade comprising 'the most recent common ancestor of *Saltasaurus*, *Argyrosaurus*, *Lirainosaurus*, plus the Peirópolis titanosaur and all its descendants'. This definition has mainly been ignored, in part because it was considered synonymous with Saltasauridae or Saltasaurinae (see below) by Wilson and Upchurch (2003). Eutitanosauria was redefined by Salgado (2003) as a stem group of sauropods more closely related to *Saltasaurus* than to *Epachthosaurus*. Although this definition is valid, a problem arises because of the different positions in which the Cenomanian–Turonian Argentinean titanosaur *Epachthosaurus* is recovered amongst various phylogenetic analyses. These differences drastically alter the taxa included within Eutitanosauria. The first phylogenetic analyses that included *Epachthosaurus* recovered it as a 'basal' titanosaur (e.g. Salgado et al. 1997) and as an early diverging member of Lithostrotia (e.g. Calvo et al. 2007a, b). Thus, Eutitanosauria was considered as an inclusive clade, broadly similar to Lithostrotia (Wilson and Upchurch 2003; Upchurch et al. 2004), that comprised two diverse stems: one leading to the clade including *Rinconsaurus* (i.e. Colossosauria) and the other to the clade including *Saltasaurus* (e.g. Calvo et al. 2007b; Hechenleitner et al. 2020). However, in analyses in which *Malawisaurus* is placed in a more nested position within Titanosauria, *Epachthosaurus* has been recovered outside of Lithostrotia (see above), such that Eutitanosauria is an even more inclusive clade that also incorporates Lithostrotia in its taxonomic content (e.g. Lacovara et al. 2014; Carballido et al. 2017; Tykoski and Fiorillo 2017). To further complicate matters, some phylogenetic analyses have recovered *Epachthosaurus* as a close relative or member of Colossosauria (e.g. Gallina and Apesteguía, 2011; Poropat et al. 2016; González Riga et al. 2018, 2019; Gorscak and O'Connor 2019), which implies a less inclusive Eutitanosauria. In general, many studies have not included Eutitanosauria amongst their phylogenetic definitions, nor labelled this clade on their topologies. In part, this likely reflects the labile position of the clade specifier, *Epachthosaurus*, even though it is known from most of the postcranial skeleton (Martínez et al. 2004). Although we consider Eutitanosauria as a valid clade, in the sense that it is correctly defined, it must be noted that its taxonomic content is highly labile amongst different phylogenetic analyses. As such, it is a problematic clade that is currently not very useful for communicating phylogenetic results.

2.6 Epachthosaurinae

Salgado (2003: 20) coined the subfamily Epachthosaurinae, defining it as 'all titanosaurs closer to *Epachthosaurus* than to *Saltasaurus*'. Despite being well defined, Epachthosaurinae has rarely been used since its original description. By definition, Epachthosaurinae (closer to *Epachthosaurus* than to *Saltasaurus*) is the sister taxon of Eutitanosauria (closer to *Saltasaurus* than to *Epachthosaurus*). However, as noted above, the labile phylogenetic position of *Epachthosaurus* is a problem in identifying the taxonomic content of both Eutitanosauria and Epachthosaurinae. For this reason, we do not recommend the use of Epachthosaurinae until the phylogenetic position of *Epachthosaurus* stabilizes.

2.7 Colossosauria

Colossosauria was recently erected by González Riga et al. (2019: 34) as 'the most inclusive clade containing *Mendozasaurus* but not *Saltasaurus* or *Epachthosaurus*'. The main intention of these authors was to name the lineage formed by Rinconsauria and Lognkosauria (see below), with several analyses having retrieved these two South American clades as closely related (see González Riga et al. 2019 and references therein). Given the unstable position of *Epachthosaurus* (see above), this clade could be understood as the complete stem leading to *Mendozasaurus* but not to *Saltasaurus* (e.g. González Riga and Ortiz 2014; Salgado et al. 2015; Carballido et al. 2017; Filippi et al. 2019), part of it (e.g. Gallina and Apesteguía 2011; Gallina and Otero 2015; Mannion et al. 2019), or it might only include *Mendozasaurus* (e.g. Gorscak and O'Connor 2019). These topological differences therefore result in a taxonomic composition that is often different to the original intention of the authors for Colossosauria, that is, naming a clade that unites Rinconsauria and Lognkosauria.

2.8 Lognkosauria

Lognkosauria was coined by Calvo et al. (2007b: 530) to be applied to a node-based clade formed by *Futalognkosaurus*, *Mendozasaurus*, their most recent common ancestor and all its descendants. For many years, Lognkosauria was composed solely of its two specifier taxa (e.g. Calvo et al. 2007b; Gallina and Apesteguía 2011; Salgado et al. 2015; Gallina and Otero 2015; González Riga et al. 2016; Tykoski and Fiorillo 2017). The first work to recover a major diversity of lognkosaur-related taxa was that of Gallina and Apesteguía (2011), showing for the first time a close relationship between Rinconsauria and Lognkosauria (formed by its two specifiers). Subsequently, Carballido et al. (2017) recognized a greater diversity within Lognkosauria, including the largest animals to have ever walked on land (e.g. *Argentinosaurus* and *Patagotitan*). A diverse Lognkosauria has subsequently been supported in other studies (e.g. González Riga et al. 2018, 2019; Mannion et al. 2019; Otero et al. 2021). As recently discussed by González Riga et al. (2019), Lognkosauria is recovered in most recent phylogenetic analyses as the sister lineage of Rinconsauria (Gallina and Apesteguía 2011; Salgado et al. 2015; Carballido et al. 2017; González Riga et al. 2018; Mannion et al. 2019).

2.9 *Rinconsauria*

Rinconsauria was defined by Calvo et al. (2007a: 487) as the clade formed by '*Muyelensaurus, Rinconsaurus*, their most recent common ancestor and all its descendants'. At the time the clade was defined, Rinconsauria was depicted as a lineage more derived than Lognkosauria and related to Aeolosaurini (see below). Most subsequent phylogenetic analyses that included both specifiers retrieved Rinconsauria with its original taxonomic content (e.g. Carballido et al. 2017; Tykoski and Fiorillo 2017; González Riga et al. 2018, 2019). However, in some analyses, Rinconsauria constitutes a more inclusive group, within which Lognkosauria is nested (e.g. Gallina and Apesteguía 2011; Gallina and Otero 2015; Díez Díaz et al. 2018). Whereas nearly all phylogenetic analyses have recovered the clade specifiers of Rinconsauria as sister taxa, or close relatives (e.g. Gallina and Apesteguía 2011; Gallina and Otero 2015; Carballido et al. 2017; Tykoski and Fiorillo 2017; González Riga et al. 2018, 2019; Poropat et al. 2021), a small number of analyses position *Rinconsaurus* and *Muyelensaurus* as distantly related, resulting in a markedly different taxonomic composition of Rinconsauria (e.g. Santucci and Arruda Campos 2011; Gorscak and O'Connor 2019 [Bayesian analysis]; Mannion et al. 2019 [analysis with extended implied weights]). Given that most analyses agree upon the taxonomic composition of Rinconsauria, as well as its relationships with other titanosaurs, we consider its definition generally non-problematic. However, it should be acknowledged that the anatomy of *Rinconsaurus* and *Muyelensaurus* has yet to be fully described (Calvo and González Riga 2003; Calvo et al. 2007a). An ongoing, comprehensive revision of these materials (Pérez-Moreno et al. 2018) should help clarify the phylogenetic content of Rinconsauria and its position amongst Titanosauria. Notably, the position of *Muyelensaurus* must be treated with caution given that a recent study of the material indicates the presence of more than one taxon (A.P.M. pers. obs.).

2.10 *Aeolosaurini*

Franco-Rosas et al. (2004: 332) coined Aeolosaurini for the stem-based clade formed by *Aeolosaurus rionegrinus* and *Gondwanatitan*, but not *Saltasaurus* or *Opisthocoelicaudia*. As noted by Carballido et al. (2017), the original definition of Franco-Rosas et al. (2004) was not strictly followed in many works, and in most phylogenetic analyses, the clade was incorrectly identified (e.g. Calvo et al. 2007a, b; González Riga et al. 2009; González Riga and Ortiz David 2014; Salgado et al. 2015; Gallina and Otero 2015). Most researchers seem to have implicitly followed a node-based definition for Aeolosaurini (with *Aeolosaurus* and *Gondwanatitan* as specifiers), although none of them presented a formal phylogenetic redefinition for this clade. Therefore, Carballido et al. (2017) proposed the node definition: *Aeolosaurus rionegrinus*, *Gondwanatitan faustoi*, their most recent common ancestor and all its descendants. The relationships of 'aeolosaurine' taxa are not clear, and different analyses show starkly contrasting positions for one or more of the species typically considered as aeolosaurines. Several analyses have recovered them as closely related to Rinconsauria (e.g. Santucci and Arruda Campos 2011; Carballido et al. 2017; Gorscak and O'Connor 2019), or Colossosauria more generally (e.g. Gallina and Apesteguía 2011; Coria et al. 2013; Salgado et al. 2015), whereas others have placed them closer to Saltasauridae (e.g. Poropat et al. 2016; González Riga et al. 2019). Nevertheless, most phylogenetic analyses that include a wide taxon sample of aeolosaurines recovered this clade as closely related to Rinconsauria. One possible issue is that the type species, *Aeolosaurus rionegrinus*, from the Campanian–Maastrichtian of Argentina, has never been described in detail (Powell 1987, 2003), with most authors scoring this species only on the available published literature. Further work is needed to understand the systematics of Aeolosaurini, including the identification of a stable definition that captures most authors' concept of the clade.

2.11 *Nemegtosauridae*

Nemegtosauridae was coined by Upchurch (1995) and subsequently defined by Upchurch et al. (2004: 303) as 'all diplodocoids more closely related to *Nemegtosaurus* than to *Diplodocus*'. However, robust support for a placement of *Nemegtosaurus* within Titanosauria (e.g. Curry Rogers and Forster 2001) resulted in the need for a new definition of this clade. The currently used definition was provided by Wilson (2005: 311) as: 'all titanosaurs more closely related to *Nemegtosaurus* than to *Saltasaurus*'. Based on this definition, Nemegtosauridae could include either a small number of taxa when recovered as closely related to Saltasauridae (e.g. Mannion et al. 2019; Hechenleitner et al. 2020; Poropat et al. 2021), or represent a diverse clade of titanosaurs when it is recovered as more closely related to Lognkosauria (e.g. Gallina and Apesteguía 2011; Díez Díaz et al. 2018, 2020; Sallam et al. 2018;

Gorscak and O'Connor 2019). The recent relocation of the quarry from which *Nemegtosaurus* was excavated resulted in the discovery of additional material of a taxon that had been represented solely by an isolated skull since its erection in 1971 (Currie et al. 2018). The incorporation of new postcranial elements in phylogenetic analyses will hopefully help to resolve the position of *Nemegtosaurus* and closely related taxa. Therefore, further analyses that incorporate these new data are needed to better understand the taxonomic content of this clade and to determine whether it requires redefinition, either using a node-based definition, or by choosing a more closely related specifier than *Saltasaurus*.

2.12 *Lirainosaurinae*

Lirainosaurinae was recently coined by Díez Díaz et al. (2018) as a node-based taxon defined as '*Lirainosaurus*, *Ampelosaurus*, their most recent common ancestor and all its descendants'. Only a small number of analyses have included both clade specifiers. Based on the phylogenetic results obtained by Díez Díaz et al. (2018, 2020), Lirainosaurinae could be considered as an early diverging clade within Colossosauria, recovered as more closely related to Lognkosauria than to Saltasaurinae. Although it would be a much more inclusive clade, topologies presented by Sallam et al. (2018) and Gorscak and O'Connor (2019) also support a position in which Lirainosaurinae is closer to Lognkosauria than to Saltasaurinae. However, these latter two analyses also recovered *Opisthocoelicaudia* within the *Lirainosaurus* + *Ampelosaurus* clade. This complicates the phylogenetic taxonomy given that Opisthocoelicaudiinae has formal priority over Lirainosaurinae (see below), and it contrasts with most other studies in which *Opisthocoelicaudia* is recovered as more closely related to Saltasaurinae than Lognkosauria. Given that *Nemegtosaurus* was also recovered as more closely related to Lirainosaurinae and Lognkosauria than to Saltasaurinae in the analyses of Díez Díaz et al. (2018, 2020), Sallam et al. (2018) and Gorscak and O'Connor (2019), the position and composition of Lirainosaurinae currently has to be considered as uncertain. Thus, this clade is marked with dashed lines in Fig. 1. Further analyses are needed to evaluate the robustness of this clade and its taxonomic diversity, as well as its possible synonymy with other clades.

2.13 *Saltasauroidea*

França et al. (2016: Fig. 4) used the name Saltasauroidea in their cladogram to refer to a polytomous clade comprising Saltasauridae, Aeolosaurini (as defined by Franco-Rosas et al. 2004), *Baurutitan*, *Diamantinasaurus* and *Isisaurus*. However, these authors did not present any discussion of their usage of Saltasauroidea, previously unpublished, nor did they provide a phylogenetic definition. As such, it is difficult to know the exact criterion followed by these authors in recognizing this clade.

Nevertheless, it seems logical to assume that they attempted to capture within Saltasauroidea a more inclusive lineage of titanosaurs related to Saltasauridae. To our knowledge, Saltasauroidea has not been used since.

2.14 *Saltasauridae*

Powell (1992) formally erected the subfamily Saltasaurinae (see below), and he is therefore the nominal author of Saltasauridae and Saltasauroidea as well, following ICZN article 36.1. Powell (1986) previously used Saltasaurinae in his unpublished thesis. Saltasauridae was first defined by Sereno (1998), establishing a node-stem triplet for derived titanosaurs, with Opisthocoelicaudiinae and Saltasaurinae as the two stem sister lineages. As such, Saltasauridae is defined as '*Opisthocoelicaudia, Saltasaurus,* their most recent common ancestor and all its descendants' (Sereno 1998: 63; Wilson and Upchurch 2003). In nearly every phylogenetic analysis that has included these two taxa, Saltasauridae has been recovered as a deeply nested clade of titanosaurs (e.g. Salgado et al. 1997; Upchurch 1998; Wilson and Sereno 1998; Wilson 2002; Upchurch et al. 2004). The exceptions to this are the analyses of Gorscak and O'Connor (2016 [and subsequent iterations of this data set]) and Díez Díaz et al. (2018, 2020). These studies recovered a much more diverse Saltasauridae, including specifiers of several other clades, e.g. Colossosauria, Lirainosaurinae, Lognkosauria and Nemegtosauridae.

2.15 *Saltasaurinae*

Saltasaurinae was first defined by Salgado et al. (1997: 24) as 'the clade formed by *Saltasaurus, Neuquensaurus*, their most recent common ancestor and all its descendants'. The following year, Sereno (1998) formed the node-stem triplet mentioned above and defined Saltasaurinae as all saltasaurids more closely related to *Saltasaurus* than to *Opisthocoelicaudia.* Since then, most authors have followed Sereno's stem definition of Saltasaurinae (e.g. Wilson and Upchurch 2003). Given its stability and facility for communication, especially in instances in which successive sister taxa to the clade formed by *Neuquensaurus* and *Saltasaurus* are recovered, the stem definition seems to be the better choice for defining Saltasaurinae. Saltasaurinae is almost always recovered as a low diversity clade of latest Cretaceous South American titanosaurs. The only exception to this comes from the analyses of Díez Díaz et al. (2018, 2020), in which Saltasaurinae is a diverse clade consisting of most saltasaurids (see above).

2.16 *Saltasaurini*

This clade was erected and defined by Salgado and Bonaparte (2007: 213) as 'the least inclusive clade containing *Saltasaurus loricatus* and *Neuquensaurus australis*', which essentially matches the original definition of Saltasaurinae given by Salgado et al. (1997). Salgado and Bonaparte (2007) erected Saltasaurini to reflect Powell's original concept of the composition of Saltasaurinae (i.e. a taxonomically limited family), contrasting with Sereno's (1998) potentially more inclusive definition, which incorporates taxa such as *Alamosaurus* and *Pellegrinisaurus* in some analyses. Although Saltasaurini is applicable in studies in which Saltasaurinae is diverse, and/or where a taxon other than *Neuquensaurus* is recovered as more closely related to *Saltasaurus* than *Opisthocoelicaudia*, in many analyses Saltasaurini is essentially synonymous with Saltasaurinae.

2.17 *Opisthocoelicaudiinae*

First coined by McIntosh (1990), Opisthocoelicaudiinae was applied by Sereno (1998) to the sister stem clade to Saltasaurinae, and it is currently defined as all saltasaurids closer to *Opisthocoelicaudia* than to *Saltasaurus. Alamosaurus* is one of the few additional taxa to have been recovered as a member of Opisthocoelicaudiinae (e.g. Wilson and Sereno 1998; Gallina and Apesteguía 2011; Salgado et al. 2015), although this placement has not been found consistently (see Mannion et al. 2019 for a discussion). By contrast, analyses based on iterations of the Gorscak and O'Connor (2016) data set recover a diverse Opisthocoelicaudiinae, including all saltasaurids except for Saltasaurini. Although the definition of this clade is valid and forms part of the node-stem triplet, the composition and phylogenetic position of Opisthocoelicaudiinae remain uncertain. Part of this might stem from the fact that *Opisthocoelicaudia* has never been included in a phylogenetic analysis based on first-hand study, with all scoring based on the information presented by Borsuk-Bialynicka (1977). An additional problem arises in that *Opisthocoelicaudia* and *Nemegtosaurus* might be closely related or even synonymous (Currie et al. 2018; though see Averianov and Lopatin 2019), in which case Nemegtosauridae would have priority over Opisthocoelicaudiinae.

3 Identifying More Stable Clades Within Titanosauria

In the previous section, we outlined the suprageneric clades previously erected within Titanosauria, highlighting the problems that arise for some of them when different topologies are compared. Regardless of the taxonomic content of Titanosauria and its interrelationships, the problem of unstable clades has a direct impact on scientific communication, which is the primary function of a clade name (de Queiroz 2007; Cantino and de Queiroz 2010). Additionally, the absence of stable names for diverse clades generates an additional communication problem (see Article 10.2, Cantino and de Queiroz 2010). The existence of diverse but unnamed clades precludes the possibility of easily referring to them and the general results of a phylogenetic analysis. For example, there is no defined name for referring to the diverse clade formed by the stems leading to Saltasauridae and Lognkosauria, nor a clear way for referring to these two stems. In this regard, the seemingly minor differences in the position of already defined specifiers (e.g. *Epachthosaurus*, *Malawisaurus*) result in markedly different clades uniting these 'derived' titanosaurs (in this case Eutitanosauria and Lithostrotia).

As pointed out by Sereno (1999: 341), 'stability of taxonomic content is enhanced by choosing: (1) effective reference taxa and (2) an effective configuration of phylogenetic definitions'. Support and stability are relevant, in particular when the definitions use different taxa as specifiers, which should ideally be well-known taxa with well-supported phylogenetic placements (Lee 2005a, b). Regarding the latter, it is preferable to choose a reference taxon that is as completely known as possible and that is well-nested within the clade to be defined (i.e. 'away' from the basal dichotomy), which should require more 'effort' to remove it from the clade in subsequent analyses (Sereno 1999). Although distance from the basal dichotomy is a good starting point, it still does not guarantee the stability of the selected taxon. The amount of available material preserved with derived characters (i.e. synapomorphies of successive basal nodes) is clearly also an important criterion. Therefore, in order to identify the main lineages consistently recovered in recent phylogenies, and to detect the most robustly placed specifiers, we ran a series of analyses to evaluate internal support and taxon stability within Titanosauria.

3.1 Identifying Relationships and Evaluating Robustness

We selected fifteen phylogenetic data sets that include a broad sample of titanosaurs (i.e. Gallina and Apesteguía 2011; Filippi et al. 2011, 2019; Salgado et al. 2015; Gallina and Otero 2015; Tykoski and Fiorillo 2017 [both versions presented therein]; González Riga et al. 2019; Mannion et al. 2019; Gorscak and O'Connor 2019; Carballido et al 2020; Hechenleitner et al. 2020; Díez Díaz et al. 2020; Gallina et al. 2021; Poropat et al. 2021). These data sets were each subjected to two analyses, using the PCR-Jackknife (Pol and Goloboff 2020) and a new script, the PCR-Bremer (available in supporting online material). These analyses aim to identify the main, underlying relationships amongst titanosaurs, as well as their robustness. The first PCR-Jackknife starts with an independent-removal jackknifing resampling of the data set, followed by IterPCR (Pol and Escapa 2009) and the subsequent removal of unstable taxa. The deletion of unstable taxa allows the recognition of the most robust, underlying topology, represented by a majority rule consensus (Pol and Goloboff 2020). The PCR-Bremer analysis implements a series of tree searches, saving suboptimal trees (here, up to four steps longer than the MPTs), followed by IterPCR and the removal of unstable taxa in order to obtain the most robust topology (i.e. the one that still persists in suboptimal trees four steps longer). In both sets of analyses, Bremer support is calculated once unstable taxa are removed. Both analyses have the same objective, namely to identify well-supported relationships amongst titanosaurs once unstable taxa are deleted. In addition, they therefore also provide a list of the most unstable taxa. For each taxon, a Recuperation Index (RcI) is calculated, which is the ratio of the number of analyses that conserve the taxon relative to the total number of analyses that include the taxon.

The complete results from these analyses are provided in the Supplementary Information, with selected results shown in Fig. 2. Overall, many titanosaurs are excluded, as a result of their relative instability amongst less parsimonious hypotheses (PCR-Bremer) or resampling of the data set (PCR-Jackknife). Under four extra steps, PCR-Bremer produces trees with fewer taxa than those from the PCR-Jackknife analyses (Fig. 2). Most of the main clade specifiers are recovered in < 50% of the analyses (RcI values; Table 1; Fig. 3). Although each data set differs in terms of taxon and character sampling, making comparison of the results difficult, several general observations can be made regarding the suprageneric taxonomy of Titanosauria.

Most analyses consistently recover the derived clade of titanosaurs, consisting of the two major stem lineages, i.e. those leading to Colossosauria and Saltasauridae. Support for these relationships is higher in data sets in which *Patagotitan* and other lognkosaurians are included, although such lineages were also recognized, and with relatively strong support, by Gallina and Apesteguía (2011). In the Saltasauridae stem, *Saltasaurus* is consistently recovered amongst the various topologies and is the taxon with the highest RcI (Table 1; Fig. 3). In the other stem, *Patagotitan* is recovered in 62% of the analyses, followed by *Futalognkosaurus* (50%) and *Mendozasaurus* (36%). In addition to their relative stability in the topology, both *Saltasaurus* and *Patagotitan* are well-known taxa in terms of anatomical regions preserved (e.g. Powell 2003; Carballido et al. 2017) and are always recovered as well-nested within their stems.

Besides these two stem lineages, most other previously defined clades are not recovered, or are recovered with an inconsistent taxonomic content. Here, we briefly provide comments on most of them, based on the presence or absence of their current clade specifiers:

(1) *Andesaurus*, one of the specifier taxa for Titanosauria, is recovered in only one-third of the analyses (RI: 0.33; Fig. 3; Table 1). This is particularly the case in more recent phylogenies, which implies instability of this taxon. Such a condition was previously noted (see Sect. 2), given the 'basal' position that this taxon can occupy under different analytical criteria.

(2) *Malawisaurus* is also retained in only 33% of the analyses (Fig. 3; Table 1), and there is no clear consensus on its phylogenetic position between different analyses.

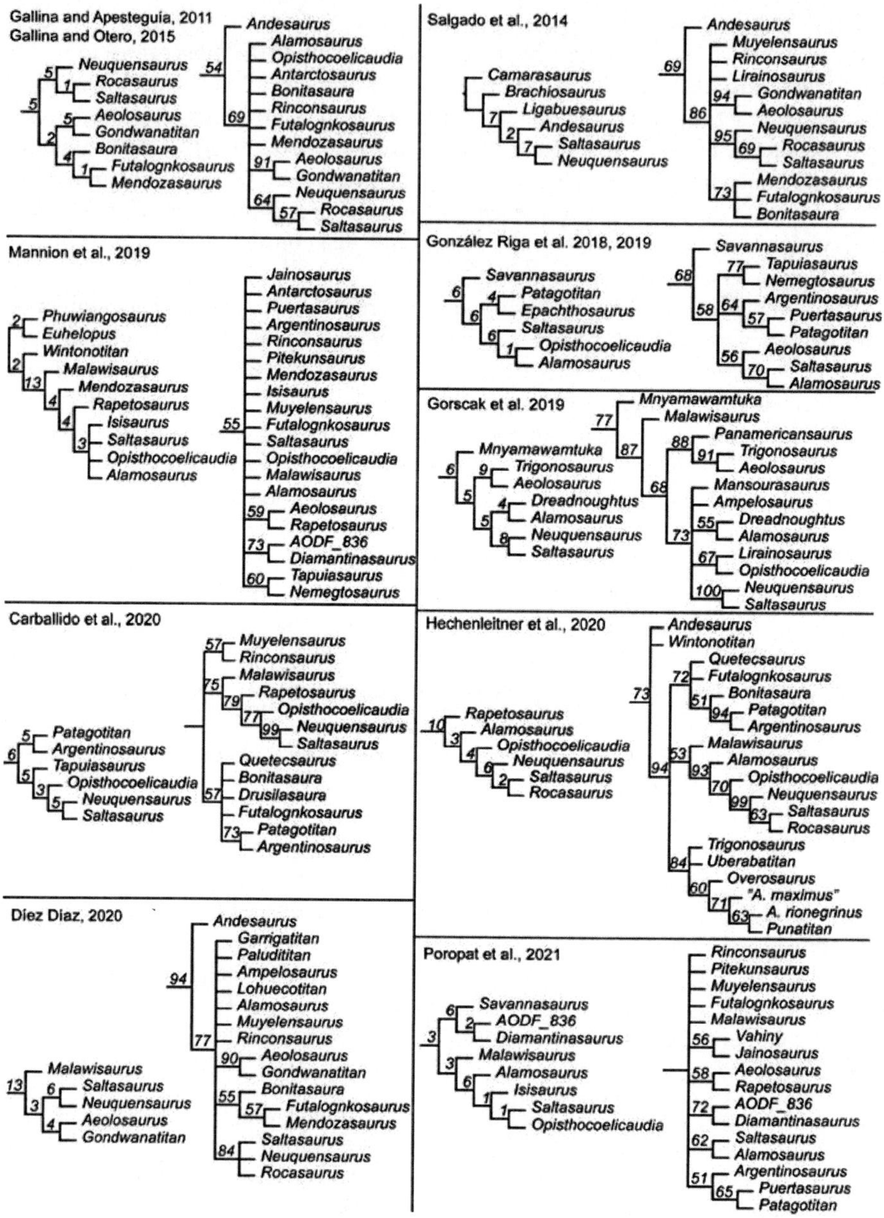

Fig. 2 Main results of the PCR-Bremer (left) and PCR-Jacknife (right) analyses of ten of the fifteen analysed data sets, showing solely the results recovered for titanosaur clades

Table 1 Recuperation Index (RcI) of the taxa used as specifiers in most of the clades amongst Titanosauria. RcI = number of analyses that conserve the taxon/total number of analyses to include the taxon

Specifier Taxon	RcI
Saltasaurus	0.97
Neuquensaurus	0.83
Patagotitan	0.62
Aeolosaurus	0.56
Opisthocoelicaudia	0.56
Gondwanatitan	0.55
Futalognkosaurus	0.50
Mendozasaurus	0.36
Andesaurus	0.33
Malawisaurus	0.32
Rinconsaurus	0.26
Muyelensaurus	0.23
Epachthosaurus	0.06

Fig. 3 Tag cloud resulting from the RcI values (see Table 1)

(3) *Epachthosaurus* has the lowest RcI (0.06; Fig. 3; Table 1) of all clade specifiers, indicating that this taxon is highly unstable and not currently suitable as a clade specifier (see also Sect. 2).

(4) *Rinconsaurus* and *Muyelensaurus* are recovered in approximately 25% of the analyses (Table 1), although they are not always found together as a clade. Thus, neither of them seems to be a good clade specifier, with *Muyelensaurus* additionally problematic given its possible chimeric nature.

(5) *Aeolosaurus* and *Gondwanatitan* are recovered in slightly more than half of the analyses in which both taxa were included (RI: 0.56 and 0.55; Fig. 3; Table 1). In many analyses, these two taxa are found as closely related, showing some support for the clade Aeolosaurini. Although this clade has not been widely sampled in published data sets, it seems to be a well-supported clade when its taxon sampling is expanded (e.g. Hechenleitner et al. 2020; Fig. 2).

(6) *Saltasaurus* and *Opisthocoelicaudia* are recovered in most of the analyses (RI: 0.97 and 0.56; Fig. 3; Table 1), and *Neuquensaurus* is recovered in 83% of the analyses (Table 1). Therefore, our analyses show high support for Saltasauridae and Saltasaurinae, although derived titanosaurs require more exhaustive sampling in phylogenetic data sets.

(7) Of the data sets analysed here, Diamantinasauria is only well sampled by González Riga et al. (2019), Mannion et al. (2019) and Poropat et al. (2021), but is here recovered as a well-supported clade based on the most recent of these iterations of the same underlying data set.

(8) Two of the Lirainosaurinae taxa (sensu Díez Díaz et al. 2020) are recovered in the PCR-Jackknife, but as part of a polytomy, and the clade is not recovered in trees one step longer than the MPTs (see Supplementary Material). Therefore, support for Lirainosaurinae is currently low.

4 Current Proposals and Recommendations

Based on conflicts amongst different phylogenies and the analyses presented above, here we present our recommendations for stabilizing Titanosauria systematics that should also improve the communication of phylogenetic results in future studies. Many of the clade definitions presented below follow previous work, but where a definition is modified we have attempted to minimize disruption of current and/or historical usage (see Article 10.1, Cantino and de Queiroz 2010). A short comment on each definition is also provided to highlight ongoing and/or potential future problems. A summary of the main clades is provided as an informal cladogram in Fig. 4 and Table 2.

4.1 Titanosauria Bonaparte and Coria (1993)

Definition: *Andesaurus delgadoi* Calvo and Bonaparte 1991, *Saltasaurus loricatus* Bonaparte and Powell 1980, their most recent common ancestor and all its descendants (see also Salgado et al. 1997; Wilson and Upchurch 2003).

Comments Although we have retained the most commonly used definition for Titanosauria, in which *Andesaurus delgadoi* is one of the clade specifiers, this species is not an optimal taxon to be used as a specifier, given that it was pruned from two-thirds of the PCR analyses. Further work is needed in order to clarify its relationship with other 'basal' somphospondylans. As noted earlier, there is no current applicable usage for any of the suprageneric rank taxa of *Andesaurus* (i.e. Andesaurinae, Andesauridae and Andesauroidea).

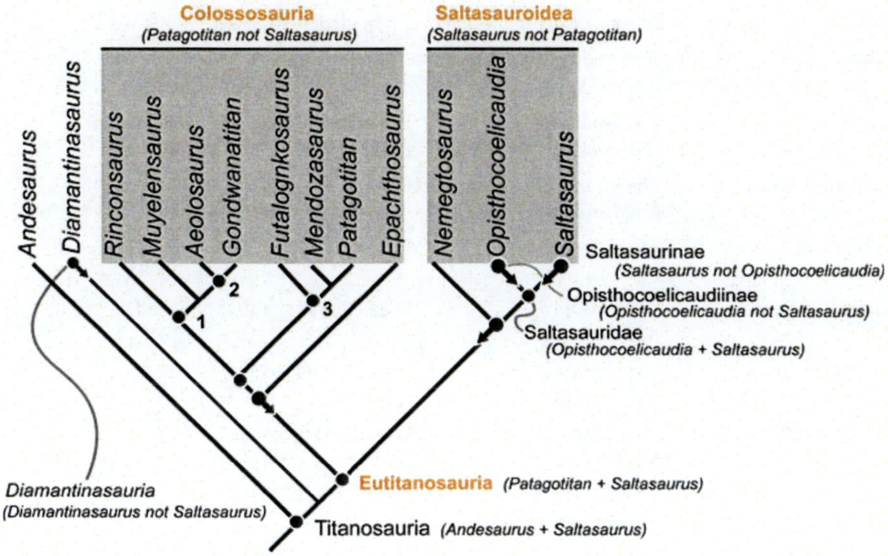

Fig. 4 Informal cladogram showing the hypothetical interrelationships of titanosaurs based on recent analyses, showing only the most stable clades and their definition, with orange font used for those that are here modified or newly proposed. Less well-supported or problematic clades are marked with numbers: 1, Rinconsauria; 2, Aeolosaurini; 3, Lognkosauria. Node-based clades are marked with a circle and stem-based nodes with a circle and arrow

4.2 *Diamantinasauria Poropat et al. (2021)*

Definition The most inclusive clade containing *Diamantinasaurus matildae* (Hocknull et al. 2009) but not *Saltasaurus loricatus* (Bonaparte and Powell 1980).

Comments Based on the data set of Poropat et al. (2021), Diamantinasauria appears to be a well-supported clade that represents an early diverging titanosaurian lineage. Although its content has been consistent in previous iterations of this phylogenetic data set, the position of Diamantinasauria within Titanosauria has fluctuated substantially (compare Poropat et al. (2015) with Poropat et al. (2016)). Further studies are needed to determine the stability of the clade, both in terms of definition and taxon content.

4.3 *Lithostrotia Upchurch et al. (2004)*

Definition *Malawisaurus dixeyi* (Haughton 1928), *Saltasaurus loricatus* (Bonaparte and Powell 1980), their most recent common ancestor and all its descendants.

Table 2 Phylogenetic definitions of the main titanosaurian clades discussed in this work

Clade name	Recommended definition	Nominal author	Current definition author
Titanosauria	*Andesaurus* + *Saltasaurus*	Bonaparte and Coria (1993)	Wilson and Upchurch (2003)
Diamantinasauria	*Diamantinasaurus* not *Saltasaurus*	Poropat et al. (2021)	Poropat et al. (2021)
Lithostrotia	*Malawisaurus* + *Saltasaurus*	Upchurch et al. (2004)	Upchurch et al. (2004)
Lirainosaurinae	*Lirainosaurus* + *Ampelosaurus*	Díez Díaz et al. (2018)	Díez Díaz et al. (2018)
Eutitanosauria	*Patagotitan* + *Saltasaurus*	Sanz et al. (1999)	New
Colossosauria	*Patagotitan* not *Saltasaurus*	González Riga et al. (2019)	New
Lognkosauria	*Mendozasaurus* + *Futalognkosaurus*	Calvo et al. (2007b)	Calvo et al. (2007b)
Rinconsauria	*Rinconsaurus* + *Muyelensaurus*	Calvo et al. (2007a)	Calvo et al. (2007a)
Aeolosaurini	*Aeolosaurus* + *Gondwanatitan*	Franco-Rosas et al. (2004)	Carballido et al. (2017)
Saltasauroidea	*Saltasaurus* not *Patagotitan*	Powell (1992)	New
Saltasauridae	*Opisthocoelicaudia* + *Saltasaurus*	Powell (1992)	Wilson and Upchurch (2003)
Saltasaurinae	*Saltasaurus* not *Opisthocoelicaudia*	Powell (1992)	Wilson and Upchurch (2003)
Opisthocoelicaudiinae	*Opisthocoelicaudia* not *Saltasaurus*	McIntosh (1990)	Wilson and Upchurch (2003)
Nemegtosauridae	*Nemegtosaurus* not *Saltasaurus*	Upchurch (1995)	Wilson (2005)

Comments The relative instability of *Malawisaurus*, which is evident through both comparison of phylogenies and the low RcI for this taxon (0.32; Table 1), indicates caution on the use of Lithostrotia and its current definition. Furthermore, it is possible that *Malawisaurus* represents a chimera (Gorscak 2016). A re-evaluation of the remains currently referred to *Malawisaurus* is needed to better clarify its status, as well as its relationship with other 'basal' titanosaurs, before any changes to the definition of Lithostrotia should be considered.

4.4 Lirainosaurinae Díez Díaz et al. (2018)

Definition *Lirainosaurus astibiae* (Sanz et al. 1999), *Ampelosaurus atacis* (Le Loeuff 1995), their most recent common ancestor and all its descendants.

Comments As noted above, Lirainosaurinae is a recently erected clade and has therefore not been evaluated in many analyses. Although we demonstrate that support for this clade is low, it requires further testing following the inclusion of putative lirainosaurine taxa in other phylogenetic data sets, as well as the development of novel characters to capture putative synapomorphies identified by Díez Díaz et al. (2018, 2020). In such a way, Lirainosaurinae might follow the path of Diamantinasauria, which has become better supported and its position more stable in the most recent iteration of the data set derived from Poropat et al (2016). If it does eventually prove to be a well-supported clade, this will also allow us to determine where Lirainosaurinae fits within Titanosauria, as well as whether it is synonymous with taxa such as Nemegtosauridae or Opisthocoelicaudiinae.

4.5 Eutitanosauria Sanz et al. (1999)

Definition *Patagotitan mayorum* (Carballido et al. 2017), *Saltasaurus loricatus* (Bonaparte and Powell 1980), their most recent common ancestor and all its descendants.

Comments Here we present a new definition for Eutitanosauria. Given that most analyses have recovered *Epachthosaurus* (the former specifier of this clade) close to the node formed by the common ancestor of *Patagotitan* and *Saltasaurus*, and that *Epachthosaurus* is a highly unstable taxon, we redefine this clade such that it is broadly in keeping with its original meaning. As such, Eutitanosauria is part of a newly defined node-stem clade, from which two stems (Colossosauria and Saltasauroidea) are defined below.

4.6 Colossosauria González Riga et al. (2019)

Definition The most inclusive clade containing *Patagotitan mayorum* (Carballido et al. 2017) 1980, but not *Saltasaurus loricatus* (Bonaparte and Powell 1980).

Comments Colossosauria was erected by González Riga et al. (2019: 34) as 'the most inclusive clade containing *Mendozasaurus* but not *Saltasaurus* or *Epachthosaurus*', with the clear intention of uniting Lognkosauria and Rinconsauria. As noted above, our analyses show that *Epachthosaurus* is an unsuitable clade specifier. Our analyses also demonstrate that *Mendozasaurus* is an unstable taxon to be used for defining the clade uniting Lognkosauria and Rinconsauria. Furthermore, both *Mendozasaurus* and *Rinconsaurus* (i.e. a clade specifier for Rinconsauria) are consistently recovered as 'basal' forms in their respective stem lineages, which implies that a minor change in their placement could drastically modify the taxonomic content of Colossosauria. Additionally, the validity and nature of Rinconsauria needs to be further evaluated, together with assessing the validity of *Muyelensaurus* (see above). Finally, the use of three reference taxa in a clade definition is unnecessary and is likely to result in further systematic problems in the future. As such, here we propose a new stem definition for Colossosauria that is largely in keeping with the intention of González Riga et al. (2019), using the well-known and deeply nested Argentinean titanosaurs *Patagotitan* and *Saltasaurus* as clade specifiers. The use of *Patagotitan*, rather than the unstable *Mendozasaurus*, as a clade specifier should mean that Colossosauria remains a readily applicable clade name regardless of its taxonomic content.

4.7 Lognkosauria Calvo et al. (2007a, b)

Definition *Mendozasaurus neuguyelap* (González Riga 2003), *Futalognkosaurus dukei* (Calvo et al. 2007a, b), their most recent common ancestor and all its descendants.

Comments Whereas *Futalognkosaurus* seems to be a relatively appropriate specifier, *Mendozasaurus* has a low RcI (0.36); in the future, this could result in a marked difference in terms of taxonomic content of Lognkosauria if *Mendozasaurus* is recovered in a position phylogenetically far from *Futalognkosaurus*. An additional 'problem' with Lognkosauria is that its definition means that several taxa are often recovered immediately outside of this clade (e.g. *Bonitasaura*, *Notocolossus* and the unnamed titanosaur from the Candeleros Formation in Neuquén; Otero et al. 2021), but are also not part of Rinconsauria. Therefore, a stem definition might be a better choice, in which Colossosauria could be split into two stems, creating a node-stem triplet. However, it is necessary to first identify a well-supported sister taxon to Lognkosauria, which is not currently possible, as commented below. As such, we do not make any proposed changes to the definition of Lognkosauria at this time.

4.8 Rinconsauria Calvo et al. (2007a, b)

Definition *Rinconsaurus caudamirus* (Calvo and González Riga 2003), *Muyelensaurus pecheni* (Calvo et al. 2007a, b), their most recent common ancestor and all its descendants.

Comments The two specifiers of Rinconsauria (*Rinconsaurus* and *Muyelensaurus*) were recovered in only ~25% of the analyses. Coupled with the possible chimeric nature of *Muyelensaurus* (see above), the status of Rinconsauria must be treated with caution. Furthermore, both species are typically recovered close to the basal colossosaurian node. Given these problems, neither species appears to be a good choice for defining a colossosaurian subclade. As with Lognkosauria, further evaluation of rinconsaurian taxa is needed before any revision to the clade definition.

4.9 Aeolosaurini Franco-Rosas et al. (2004)

Definition *Aeolosaurus rionegrinus* (Powell 1987), *Gondwanatitan faustoi* Kellner and Azevedo (1999), their most recent common ancestor and all its descendants.

Comments Although we follow the definition provided by Carballido et al. (2017), a revision of the clade specifier *Aeolosaurus rionegrinus* is needed, and further work is required to determine the most applicable definition of Aeolosaurini. When widely sampled, Aeolosaurini is usually recovered as a lineage closely related to, or within, Rinconsauria. Taking into account the possible problems with Rinconsauria, it is therefore advisable to revise the current definition of Aeolosaurini only once the taxonomic status of Rinconsauria is resolved. In this case, *Aeolosaurus rionegrinus* would comply with the recommendations of being far from the node, well-supported and based on fairly complete and well-preserved material. Furthermore, a stem definition could be considered if the stem lineage of *Aeolosaurus*-like forms can be well resolved.

4.10 Saltasauroidea Powell (1992)

Definition The most inclusive clade containing *Saltasaurus loricatus* (Bonaparte and Powell 1980), but not *Patagotitan mayorum* (Carballido et al. 2017).

Comments Although we prefer to use non-Linnaean rank clades for naming new lineages, here we opt for Saltasauroidea as it was recently used but never defined (França et al. 2016), and also because it already exists following the erection of the subfamily Saltasaurinae by Powell (1992). By definition, Saltasauroidea is the sister lineage of Colossosauria.

4.11 *Saltasauridae (Powell 1992)*

Definition *Opisthocoelicaudia skarzynskii* (Borsuk-Bialynicka 1977), *Saltasaurus loricatus* (Bonaparte and Powell 1980), their most recent common ancestor and all its descendants.

Comments Saltasauridae is a well-defined clade without nomenclatural problems. However, its content varies substantially between phylogenetic data sets. Future studies will need to sample a greater number of putative saltasauroid species in phylogenetic data sets to determine whether Saltasauridae is a depauperate or a diverse titanosaurian clade.

4.12 *Saltasaurinae*

Definition The most inclusive clade containing *Saltasaurus loricatus* (Bonaparte and Powell 1980), but not *Opisthocoelicaudia skarzynskii* (Borsuk-Bialynicka 1977).

Comments Saltasaurinae is also a well-defined clade. Currently, there is little applicability of Saltasaurini, which is synonomous with Saltasaurinae in most analyses. However, if future analyses provide support for a diverse Saltasauridae (see above), there might subsequently be use for Saltasaurini.

4.13 *Opisthocoelicaudiinae (McIntosh 1990)*

Definition The most inclusive clade containing *Opisthocoelicaudia skarzynskii* (Borsuk-Bialynicka 1977), but not *Saltasaurus loricatus* (Bonaparte and Powell 1980).

Comments By definition the sister taxon to Saltasaurinae, Opisthocoelicaudiinae is sometimes recovered as a non-monospecific lineage. A re-evaluation of *Opisthocoelicaudia skarzynskii*, the description of the postcranial skeleton of *Nemegtosaurus mongoliensis*, as well as greater incorporation of latest Cretaceous Eurasian titanosaurs, will be crucial in determining whether Opisthocoelicaudiinae is a useful and applicable clade name (see Mannion et al. 2019; Averianov and Lopatin 2020).

4.14 *Nemegtosauridae (Upchurch 1995)*

Definition The most inclusive clade containing *Nemegtosaurus mongoliensis* (Nowiński 1971), but not *Saltasaurus loricatus* (Bonaparte and Powell 1980).

Comments The phylogenetic position of *Nemegtosaurus* remains uncertain; as such, we do not propose any changes to the current definition of Nemegtosauridae. As noted above, future work needs to resolve the interrelationships of *Nemegtosaurus* and *Opisthocoelicaudia*, and thus whether Nemegtosauridae and Opisthocoelicaudiinae are synonymous (as well as possibly Lirainosaurinae), in which case Nemegtosauridae would have priority.

5 Conclusions

Here we present a critical reappraisal of titanosaurian systematics, quantitatively evaluating which clades are well supported and which clade specifiers are unstable. Based on these analyses, we provide a new framework that retains several clades and definitions, but that also proposes the use of new definitions, with the aim of stabilizing titanosaurian systematics, as well as aiding comparisons between different phylogenetic analyses. In this regard, we provide a new node-stem triplet composed of Eutitanosauria and its constituent lineages, Colossosauria and Saltasauroidea. Finally, we urge caution about the current usage of several less inclusive clade ranks (e.g. Rinconsauria, Aeolosaurini) based on the instability of their taxonomic content, outlining approaches to their future resolution.

Acknowledgements This contribution was supported by funds of the PICT ANPCyT 1925/2017 (to JLC and LS). PDM's contribution was supported by grants from the Royal Society (UF160216 and RGF\EA\201037). We thank Diego Pol for editing the manuscript and providing us with the PCR-Bremer script. We sincerely appreciate the comments made by the reviewers, S. Poropat and P. Gallina, whose contributions clearly improved the quality of this manuscript, as well as those provided by V. Díez Díaz on early versions of this work.

Supplementary Information Available at: https://osf.io/h8qs3/?view_only=3100c6a2c8d54e9aa91fbf6655f9a2df.

References

Averianov OA, Lopatin AV (2019) Sauropod diversity in the Upper Cretaceous Nemegt Formation of mongolia–a possible new specimen of *Nemegtosaurus*. Acta Pal Pol 64:313–321

Bonaparte JF, Coria RA (1993) Un nuevo y gigantesco saurópodo titanosaurio de la Formación Río Limay (Albiano-Cenomaniano) de la Provincia del Neuquén, Argentina. Ameghiniana 30:271–282

Bonaparte JF, Powell JE (1980) A continental assemblage of tetrapods from the Upper Cretaceous beds of El Brete, northwestern Argentina (Sauropoda-Coelurosauria- Carnosauria-Aves). Mémoires De La Société Géologique De France 139:19–28

Bonaparte JF, González Riga B, Apesteguía S (2006) *Ligabuesaurus leanzai* gen. et sp. nov. (Dinosauria, Sauropoda), a new titanosaur from the Lohan Cura Formation (Aptian, Lower Cretaceous) of Neuquén, Patagonia, Argentina. Cret Res 27:364–376

Borsuk-Bialynicka M (1977) A new camarasaurid sauropod *Opisthocoelicaudia skarzynskii* gen. n. sp. n. from the Upper Cretaceous of Mongolia. Pal Pol 37:5–64

Calvo JO, Bonaparte JF (1991) *Andesaurus delgadoi* gen. et sp. nov. (Saurischia-Sauropoda), dinosaurio Titanosauridae de la Formación Río Limay (Albiano-Cenomaniano), Neuquén, Argentina. Ameghiniana 28:303–310

Calvo JO, González Riga BJ (2003) *Rinconsaurus caudamirus* gen. et sp nov., a new titanosaurid (Dinosauria, Sauropoda) from the Late Cretaceous of Patagonia, Argentina. Rev Geol Chile 30(2):333–353

Calvo JO, González Riga BJ, Porfiri JD (2007a) A new titanosaur sauropod from the Late Cretaceous of Neuquén, Patagonia, Argentina. Arqu Mus Nac 65:485–504

Calvo JO, Porfiri JD, González Riga BJ, Kellner AWA (2007b) A new Cretaceous terrestrial ecosystem from Gondwana with the description of a new sauropod dinosaur. Acad Bras Cienc 79:529–541

Cantino PD, de Queiroz K (2010) International code of phylogenetic nomenclature. Version 4c [Internet]. http://www.ohio.edu/phylocode

Carballido JL, Rauhut O, Pol D, Salgado L (2011a) Osteology and phylogenetic relationships of *Tehuelchesaurus benitezi* (Dinosauria, Sauropoda) from the Upper Jurassic of Patagonia. Zool J Linn Soc 163:605–662

Carballido JL, Pol D, Cerda IA, Salgado L (2011b) The osteology of *Chubutisaurus insignis* del Corro, 1975 (Dinosauria, Neosauropoda) from the "middle" Cretaceous of Central Patagonia, Argentina. J Vertebr Paleontol 31(1):93–110

Carballido JL, Pol D, Otero A, Cerda IA, Salgado L, Garrido AC, Ramezani J, Cúneo NR, Krause JM (2017) A new giant titanosaur sheds light on body mass evolution among sauropod dinosaurs. P Roy Soc B 284:20171219

Carballido JL, Scheli M, Knötschke N, Sander M (2020) The appendicular skeleton of the dwarf macronarian sauropod *Europasaurus holgeri* from the Late Jurassic of Germany and a re-evaluation of its systematic affinities. J Syst Paleontol 18:739–781

Cerda IA, Paulina Carabajal A, Salgado L, Coria RA, Reguero M, Tambussi CP, Moly JJ (2011) The first record of a sauropod dinosaur from Antarctica. Naturwissenschaften 99(1):83–87

Coria RA, Filippi LS, Chiappe LM, García R, Arcucci AB (2013) *Overosaurus paradasorum* gen. et sp. nov., a new sauropod dinosaur (Titanosauria: Lithostrotia) from the Late Cretaceous of Neuquén, Patagonia, Argentina. Zootaxa 3683(4):357–376

Curry Rogers K, Forster CA (2001) The last of the dinosaur titans: a new sauropod from Madagascar. Letures Nat 412:530–534

Curry Rogers KA (2005) Titanosauria: a phylogenetic overview. In: Curry Rogers KA, Wilson JA (eds) The sauropods: evolution and paleobiology. University of California Press, Berkeley and Los Angeles, pp 50–103

Currie PJ, Wilson JA, Fanti F, Mainbayar B, Tsogtbaatar K (2018) Rediscovery of the type localities of the Late Cretaceous Mongolian sauropods *Nemegtosaurus mongoliensis* and *Opisthocoelicaudia skarzynskii*: Stratigraphic and taxonomic implications. Palaeogeogr Palaeoclimatol Palaeoecol 494:5–13

de Queiroz K (2007) Toward an integrated system of clade names. Syst Biol 56:956–974

D'Emic MD (2012) The early evolution of titanosauriform sauropod dinosaurs. Zool J Linn Soc-Lond 166:624–671

Díez Díaz V, Garcia G, Pereda Suberbiola X, Jentgen- Ceschino B, Stein K, Godefroit P, Valentin X (2018) The titanosaurian dinosaur *Atsinganosaurus velauciensis* (Sauropoda) from the Upper Cretaceous of southern France: new material, phylogenetic affinities, and palaeobiogeographical implications. Cret Res 91:429–456

Díez Díaz V, Garcia G, Pereda Suberbiola X, Jentgen-Ceschino B, Stein K, Godefroit P, Valentin X (2020) A new titanosaur (Dinosauria: Sauropoda) from the Upper Cretaceous of Velaux-La-Bastide Neuve (southern France). Hist Biol. https://doi.org/0.1080/08912963.2020.1841184

Filippi LS, Canudo JI, Salgado L, Garrido A, García R, Cerda IA, Otero A (2011) A new sauropod: *Petrobrasaurus puestohernandezi* gen. et sp. nov. from the Upper Cretaceous (Plottier Formation) of Rincón de los Sauces, Neuquén (Argentina). Geol Acta 9(1):1–12

Filippi LS, Salgado L, Garrido AC (2019) A new giant basal titanosaur sauropod in the Upper Cretaceous (Coniacian) of the Neuquén Basin, Argentina. Cret Res 100:61–81

Franco Rosas AC, Salgado L, Rosas CF, Carvalho IS (2004) Nuevos materiales de titanosaurios (Sauropoda) en el Cretácico Superior de Mato Grosso, Brasil. Rev Bras Paleontolog 7:329–336

França MAG, de A, Marsola JC, Riff D, Hsiou AS, Langer MC (2016). New lower jaw and teeth referred to *Maxakalisaurus topai* (Titanosauria: Aeolosaurini) and their implications for the phylogeny of titanosaurid sauropods. PeerJ 4:e2054

Gallina PA, Apesteguía S (2011) Cranial anatomy and phylogenetic position of the titanosaurian sauropod *Bonitasaura salgadoi*. Acta Pal Pol 56:45–60

Gallina PA, Otero A (2015) Reassessment of *Laplatasaurus araukanicus* (Sauropoda: Titanosauria) from the Upper Cretaceous of Patagonia, Argentina. Ameghiniana 52:487–501

Gallina PA, Canale JI, Carballido JL (2021) The earliest known titanosaur sauropod dinosaur. Ameghiniana 58(1):35–51

Gauthier JA (1986) Saurischian monophyly and the origin of birds. In: Padian K (ed), The origin of birds and the evolution of flight. Memoirs of the California academy of sciences, vol 8, pp 1–55

Galton PM, Upchurch P (2004) Prosauropoda. In: Weishampel DB, Dodson P, Osmólska H (eds) The Dinosauria, 2nd edn. University of California Press, Berkeley, pp 232–258

González Riga BJ (2003) A new titanosaur (Dinosauria, Sauropoda) from the Upper Cretaceous of Mendoza Province, Argentina. Ameghiniana 40:155–172

González Riga BJ, Ortiz David L (2014) A new titanosaur (Dinosauria, Sauropoda) from the Upper Cretaceous (Cerro Lisandro Formation) of Mendoza Province, Argentina. Ameghiniana 51:3–25

González Riga BJ, Lamanna M, Ortiz David L, Calvo J, Coria J (2016) A gigantic new dinosaur from Argentina and the evolution of the sauropod hind foot. Sci Rep 6:19165

González Riga BJ, Mannion P, Poropat S, Ortiz David L, Coria J (2018) Osteology of the Late Cretaceous Argentinean sauropod dinosaur *Mendozasaurus neguyelap*: implications for basal titanosaur relationships. Zool J Linn Soc 184:136–181

González Riga BJ, Lamanna MC, Otero A, Ortiz David LD, Kellner AWA, Ibiricu LM (2019) An overview of the appendicular skeletal anatomy of South American titanosaurian sauropods, with definition of a newly recognized clade. Acad Bras Cienc 91(2):1–42

Gorscak E (2016) Descriptive and comparative morphology of African Titanosaurian Sauropods: new information on the evolution of cretaceous African continental faunas. Unpublished PhD Dissertation, Ohio University, Biological Sciences (Arts and Sciences)

Gorscak E, O'Connor PM (2016) Time-calibrated models support congruency between Cretaceous continental rifting and titanosaurian evolutionary history. Biol Lett 12:20151047

Gorscak E, O'Connor PM, Stevens NJ, Roberts EM (2014) The basal titanosaurian *Rukwatitan bisepultus* (Dinosauria, Sauropoda) from the middle Cretaceous Galula Formation, Rukwa Rift Basin, southwestern Tanzania. J Vertebr Paleontol 34:1133–1154

Gorscak E, O'Connor PM, Roberts EM, Stevens NJ (2017) The second titanosaurian (Dinosauria: Sauropoda) from the middle Cretaceous Galula Formation, southwestern Tanzania, with remarks on African titanosaurian diversity. J Vertebr Paleontol 37:e1343250

Gorscak E, O'Connor PM (2019) A new African Titanosaurian Sauropod Dinosaur from the middle Cretaceous Galula Formation (Mtuka Member), Rukwa Rift Basin, Southwestern Tanzania. PLoS ONE 14:e0211412

Haughton SH (1928) On some remains from the Dinosaur Beds of Nyasaland. Trans Roy Soc S A 16:67–75

Hechenleitner EM, Leuzinger L, Martinelli AG, Rocher S, Fiorelli LE, Taborda JRA, Salgado L (2020) Two Late Cretaceous sauropods reveal titanosaurian dispersal across South America. Commun Biol 3:622

Hocknull SA, White MA, Tischler TR, Cook AG, Calleja ND, Sloan T, Elliott DA (2009) New mid-Cretaceous (latest Albian) dinosaurs from Winton, Queensland, Australia. PLoS ONE 4:e6190

Kellner AWA, Azevedo SAK (1999) A new sauropod dinosaur (Titanosauria) from the Late Cretaceous of Brazil. In: Tomida Y, Rich TH, Vickers-Rich P (eds), Proceedings of the second Gondwanan Dinosaur symposium, national science Museum monographs 15:111–142

Lacovara KJ, Lamanna MC, Ibiricu LM, Poole JC, Schroeter ER, Ullmann PV, Voegele KK, Boles ZM, Carter AM, Fowler EK, Egerton VM, Moyer AE, Coughenour CL, Schein JP, Harris JD, Martínez RD, Novas FE (2014) A gigantic, exceptionally complete titanosaurian sauropod dinosaur from southern Patagonia, Argentina. Sci Rep 4:6196

Le Loeuff J (1995) *Ampelosaurus atacis* (nov. gen., nov. sp.) un nouveau Titanosauridae (Dinosauria, Sauro- poda) du Crétacé supérieur de la Haute Vallée de l'Aude (France). Compt Rend Acad Sc Paris 32:693–699

Lee MSY (2005a) Choosing reference taxa in phylogenetic nomenclature. Zool Scripta 34:329–331

Lee MSY (2005b) Choosing reference taxa in phylogenetic nomenclature. Zool Scr 34:329–331

Mannion PD, Calvo JO (2011) Anatomy of the basal titanosaur (Dinosauria, Sauropoda) *Andesaurus delgadoi* from the mid-Cretaceous (Albian–early Cenomanian) Río Limay Formation, Neuquén Province, Argentina: implications for titanosaur systematics. Zool J Lin Soc 163:155–181

Mannion PD, Upchurch P, Barnes RN, Mateus O (2013) Osteology of the Late Jurassic Portuguese sauropod dinosaur *Lusotitan atalaiensis* (Macronaria) and the evolutionary history of basal titanosauriforms. Zool J Lin Soc 168:98–206

Mannion PD, Upchurch P, Jin X, Zheng W (2019) New information on the Cretaceous sauropod dinosaurs of Zhejiang Province, China: impact on Laurasian titanosauriform phylogeny and biogeography. R Soc Open Sci 6:191057

Martínez R, Giménez O, Rodríguez J, Luna M, Lamanna M (2004) An articulated specimen of the basal Titanosaurian (Dinosauria: Saurópoda) *Epachthosaurus sciuttoi* from the Early Late Cretaceous Bajo Barreal Formation of Chubut Province, Argentina. J Vertebr Paleontol 24(1):107–120

McIntosh JS (1990) Sauropoda. In: Weishampel DB, Dodson P, Osmólska H (eds) The Dinosauria, 2nd edn. University of California Press, Berkeley, pp 345–401

McPhee BW, Benson RBJ, Botha-Brink J, Bordy EM, Choiniere JN (2018) A giant dinosaur from the earliest Jurassic of South Africa and the transition to quadrupedality in early sauropodomorphs. Curr Biol 28:3143–3151

Otero A, Carballido JL, Salgado L, Canudo JI, Garrido AC (2021) Report of a giant titanosaur sauropod from the Upper Cretaceous of Neuquén Province, Argentina. Cret Res 2021:104754. https://doi.org/10.1016/j.cretres.2021.104754

Pérez-Moreno A, Carballido JL, Otero A, Salgado L, Calvo JO, Filippi LS (2018) Rediagnosis of *Rinconsaurus caudamirus* Calvo and Gonzáles Riga (Dinosauria, Sauropoda) from the Anacleto Formation (Upper Cretaceous: Campanian) of Rincon de los Sauces (Neuquén, Argentina). PEAPA 19:R12

Pol D, Goloboff PA (2020) The Impact of unstable taxa in Coelurosaurian phylogeny and resampling support measures for parsimony analyses. Bull Am Mus Nat Hist 440:98–115

Pol D, Escapa IH (2009) Unstable taxa in cladistic analysis: identification and the assessment of relevant characters. Cladistics 25:515–527

Poropat SF, Upchurch P, Mannion PD, Hocknull SA, Kear BP, Sloan T, Sinapius GHK, Elliott DA (2015) Revision of the sauropod dinosaur *Diamantinasaurus matildae* Hocknull et al. 2009 from the middle Cretaceous of Australia: implications for Gondwanan titanosauriform dispersal. *Gond Res* 27:995–1033

Poropat SF, Mannion PD, Upchurch P, Hocknull SA, Kear BP, Kundrát M, Tischler TR, Sloan T, Sinapius GH, Elliott JA, Elliott DA (2016) New Australian sauropods shed light on Cretaceous dinosaur palaeobiogeography. Sci Rep 6:34467

Poropat SF, Kundrát M, Mannion PD, Upchurch P, Tischler TR, Elliott DA (2021) Second specimen of the Late Cretaceous Australian sauropod dinosaur *Diamantinasaurus matildae* provides new anatomical information on the skull and neck of early titanosaurs. Zool J Linn Soc-Lond 192:610–674

Poropat SF, White MA, Vickers-Rich P, Rich TH (2019) Osteology of the wide-hipped titanosaurian sauropod dinosaur *Savannasaurus elliottorum* from the Upper Cretaceous Winton Formation of Queensland, Australia. J Vertebr Paleontol 40:e1786836

Powell JE (1986) Revisión de los titanosáuridos de América del Sur. Unpublished PhD Dissertation, Universidad Nacional de Tucumán, Argentina

Powell JE (1987) Morfología del esqueleto axial de los dinosaurios titanosauridos (Saurischia, Sauropoda) del Estado de Minas Gerais, Brasil. Anais Do X Congresso Brasileiro De Paleontologia, Rio De Janeiro: Sociedade Brasileira De Paleontologia 1:155–171

Powell JE (1992) Osteología de *Saltasaurus loricatus* (Sauropoda, Titanosauridae) del Cretácico Superior del Noroeste argentino. In: Sanz JL, Buscalioni AD (eds.) Los dinosaurios y su entorno biótico, Actas del Segundo Curso de Paleontología en Cuenca, Ayuntamiento de Cuenca, Instituto “Juan de Valdes”, pp 165–230

Powell JE (2003) Revision of South American titanosaurid dinosaurs: palaeobiological, palaeobiogeographical, and phylogenetic aspects. Rec Queen Vic Museum 111:1–173

Salgado L, Coria RA, Calvo JO (1997) Evolution of titanosaurid sauropods. I: phylogenetic analysis based on the postcranial evidence. Ameghiniana 34:3–32

Salgado L (2003) Should we abandon the name Titanosauridae? Some comments on the taxonomy of titanosaurian sauropods (Dinosauria). Rev Esp Paleontol 18:15–21

Salgado L, Bonaparte JF (2007) Sauropodomorpha. In: Salgado L, Coria RA (eds) Gasparini, Z. Patagonian Mesozoic Reptiles, Indiana University Press, pp 188–288

Salgado L, Gallina PA, Paulina-Carabajal A (2015) Redescription of *Bonatitan reigi* (Sauropoda: Titanosauria), from the Campanian-Maastrichtian of the Río Negro Province (Argentina). Hist Biol 27:525–548

Sallam HM, Gorscak E, O’Connor PM, El-Dawoudi IA, El-Sayed S, Saber S, Kora MA, Sertich JJW, Seiffert ER, Lamanna MC (2018) New Egyptian sauropod reveals Late Cretaceous dinosaur dispersal between Europe and Africa. Nat Ecol Evol 2:445–451

Santucci RM, Arruda Campos AC (2011) A new sauropod (Macronaria, Titanosauria) from the Adamantina Formation, Bauru Group, Upper Cretaceous of Brazil and the phylogenetic relationships of Aeolosaurini. Zootaxa 3085:1–33

Sanz JL, Powell JE, Le Loueff J, Martinez R, Pereda Suberbiola XP (1999) Sauropod remains from the Upper Cretaceous of Laño (northcentral Spain). Titanosaur phylogenetic relationships. Estud Mus Cienc Nat Alava 14:235–255

Sereno PC (1998) A rationale for phylogenetic definitions, with application to the higher-level taxonomy of Dinosauria. Neues Jahrb Geol P Abh 210:41–83

Sereno PC (1999) A rationale for dinosaurian taxonomy. J Verteb Paleontol 19(4):788–790

Taylor MP (2009) A re-evaluation of *Brachiosaurus altithorax* Riggs 1903 (Dinosauria, Sauropoda) and its generic separation from *Giraffatitan brancai* (Janensch 1914). J Vertebr Paleontol 29(3):787–806

Taylor MP, Naish D (2005) The phylogenetic taxonomy of Diplodocoidea (Dinosauria: Sauropoda). PaleoBios 25:1–7

Tykoski RS, Fiorillo AR (2017) An articulated cervical series of *Alamosaurus sanjuanensis* Gilmore, 1922 (Dinosauria, Sauropoda) from Texas: new perspective on the relationships of North America’s last giant sauropod. J Syst Palaeontol 15:339–364

Upchurch P (1995) The evolutionary history of sauropod dinosaurs. Philos T R Soc B 349:365–390

Upchurch P (1998) The phylogenetic relationships of sauropod dinosaurs. Zool J Linn Soc-Lond 124:43–103

Upchurch P, Barrett PM, Dodson P (2004) Sauropoda. In: Weishampel DB, Dodson P, Osmólska H (eds) The Dinosauria, 2nd edn. University of California Press, Berkeley, pp 259–322

Wilson JA (2002) Sauropod dinosaur phylogeny: critique and cladistic analysis. Zool J Linn Soc-Lond 136:217–276

Wilson JA (2005) Redescription of the Mongolian sauropod Nemegtosaurus mongoliensis Nowinski (Dinosauria: Saurischia) and comments on Late Cretaceous sauropod diversity. J Syst Paleo 3:283–318

Wilson JA (2006) An overview of titanosaur evolution and phylogeny. III Jornadas Internacionales sobre Paleontología de Dinosaurios y su Entorno. Salas de los Infantes, Burgos

Wilson JA, Sereno PC (1998) Early evolution and higher-level phylogeny of sauropod dinosaurs. Soc Vert Paleontol Memoir 5:1–68

Wilson JA, Upchurch P (2003) A revision of *Titanosaurus* Lydekker (Dinosauria-Sauropoda), the first dinosaur genus with a 'Gondwanan' distribution. J Syst Palaeontol 1:125–160

Yates AM (2007) The first complete skull of the Triassic dinosaur *Melanorosaurus* Haughton (Sauropodomorpha, Anchisauria). Spec Pap Palaeontol 77:9–55

Time for Giants: Titanosaurs from the Berriasian–Santonian Age

Pablo A. Gallina, Bernardo J. González Riga, and Leonardo D. Ortiz David

Abstract Titanosaurian sauropods were the most diverse and successful group of large-bodied terrestrial herbivores. Two aspects regarding their evolutionary history stand out, namely their great morphological diversity and their extensive record from various continental masses. In South America, and particularly in Argentina, the group has the richest record worldwide. This is mainly due to the conjunction of two factors: the extensive paleontological investigation carried out by South American researchers since the beginning of the twentieth century and the well-exposed outcrops of Cretaceous continental strata. With the exception of *Tapuiasaurus* from Brazil, the entire record of South American named titanosaurs discovered in the Berriasian–Santonian interval comes from Argentina, specifically from Patagonia, including the south of Mendoza, Neuquén, Río Negro, and Chubut Provinces. With a number of 22 valid taxa, the Early and 'Mid'-Cretaceous titanosaur record of South America includes basally branching ('basal') forms (e.g., *Andesaurus*, *Ninjatitan*), basal and derived lithostrotians (e.g., *Sarmientosaurus*, *Tapuiasaurus*, and *Epachthosaurus*), and most of the colossosaurs (mainly the giant lognkosaurs *Mendozasaurus*, *Futalognkosaurus*, *Patagotitan*, and *Notocolossus*). After their origin in the Early Cretaceous, titanosaurians experienced a rapid increase in taxonomic diversity, which is maintained toward the Late Cretaceous. However,

P. A. Gallina (✉) · B. J. González Riga · L. D. Ortiz David
Consejo Nacional de Investigaciones Científicas y Técnicas (CONICET), Buenos Aires, Argentina
e-mail: gallina.pablo@maimonides.edu

B. J. González Riga
e-mail: briga@mendoza-conicet.gob.ar

L. D. Ortiz David
e-mail: lortiz@mendoza-conicet.gob.ar

P. A. Gallina
Centro de Ciencias Naturales, Ambientales y Antropológicas, Fundación de Historia Natural Félix de Azara - Universidad Maimónides, Hidalgo 775, C1405BCK Buenos Aires, Argentina

B. J. González Riga · L. D. Ortiz David
Laboratorio y Museo de Dinosaurios, Facultad de Ciencias Exactas y Naturales, Universidad Nacional de Cuyo, Parque Gral. San Martín, Padre Contreras 1300, 5500 Mendoza, Argentina

Instituto Interdisciplinario de Ciencias Básicas, CONICET-UNCUYO, Mendoza, Argentina

A. Otero et al. (eds.), *South American Sauropodomorph Dinosaurs*, Springer Earth System Sciences, https://doi.org/10.1007/978-3-030-95959-3_9

during the Berriasian–Santonian interval, the size of titanosaurs reached its climax, representing the largest vertebrate animals ever to inhabit the earth. Some South American titanosaurs from this time have been widely used to define new clades (e.g., Colossosauria, Lognkosauria, Rinconsauria) that clarified the main phylogenetic relationships at lower level. Moreover, studies in some key paleobiological aspects related with the estimation of size contributed to better understanding the biology of some species in the context of the process of gigantism.

Keyword Titanosauria · Colossosauria · Titanosaur diversity · Cretaceous · South America

1 Introduction

The titanosaur sauropods are by far the most diverse and successful group of large-bodied terrestrial herbivores to walk the earth. Although recent evidence from the lowermost Cretaceous of Patagonia (i.e., Berriasian–Valanginian, Gallina et al. 2021) reinforces the South American origin for the group as previously inferred (D'Emic 2012; Gorsak and O´Connor 2016), it is not until the Aptian–Albian that the record of undisputed titanosaurs is evident in South America (Zaher et al. 2011; Carballido et al. 2017). Expeditions sent by the Museo de La Plata to Patagonia in the late 1800s and early 1900s resulted in the pioneering works of Lydekker (1893) and Huene (1929) who described forms from the uppermost part of the Cretaceous. Just around ninety years later, Berriasian–Santonian titanosaurs (hereafter BST) began to be described in the beginning of the nineties, in the twentieth century (except for the specimen later nominated as *Elaltitan*, see below), reaching a number of 22 valid taxa in the last three decades (Table 1). A brief summary of these findings, mentioning their geographic provenance (Fig. 1), is commented on below.

In the 90s, only three taxa were described from the Cenomanian–Turonian interval. In 1990, Powell nominated *Epachthosaurus sciuttoi* based on a posterior dorsal vertebra and a plaster cast of an articulated series of dorsal vertebrae and incomplete sacrum from 'Estancia Ocho Hermanos' in the Sierra de San Bernardo, Chubut Province, Central Patagonia (Powell 1990). In the following year, Calvo and Bonaparte (1991) described *Andesaurus delgadoi*, an incomplete but partially articulated specimen from Villa El Chocón, in Neuquén Province, Northern Patagonia. This taxon became extremely important for systematics and phylogeny of Titanosauria (see below). The third taxon was the iconic *Argentinosaurus huinculensis* (Bonaparte and Coria 1993), which was considered the largest terrestrial animal registered worldwide for many years.

During the first decade of the twenty-first century, five more titanosaurs increased the number of BST, from Coniacian–Santonian times. In 2003, two new species were nominated from the neighboring Mendoza and Neuquén Provinces, in Northwestern Patagonia. One was *Mendozasaurus neguyelap* (González Riga 2003, 2005), which includes several specimens ranging from mid- to large-sized bodies with a

Table 1 Record of Berriasian–Santonian titanosaurs from South America mentioned in this chapter, following the order present in Sect. 2

Taxon	Formation	Age	Locality	Reference
Andesaurus delgadoi	Candeleros	Cenomanian	El Chocón, Neuquén Province, Argentina	Calvo and Bonaparte (1991), Mannion and Calvo (2011)
Ninjatitan zapatai	Bajada Colorada	Berriasian–Valanginian	Bajada Colorada, Neuquén Province, Argentina	Gallina et al. (2021)
Choconsaurus baileywillisi	Huincul	Cenomanian–Turonian	El Chocón, Neuquén Province, Argentina	Simón et al. (2018)
Kaijutitan maui	Sierra Barrosa	Coniacian	Rincón de los Sauces, Neuquén Province, Argentina	Filippi et al. (2019)
Traukutitan eocaudata	Bajo de la Carpa	Santonian	Los Barreales, Neuquén Province, Argentina	Juárez Valieri and Calvo (2011)
Sarmientosaurus musacchioi	Bajo Barreal	Cenomanian–Turonian	Estancia Laguna Palacios, Chubut Province, Argentina	Martínez et al. (2016)
Tapuiasaurus macedoi	Quiricó	Aptian	Coração de Jesus City, Minas Gerais, Brazil	Zaher et al. (2011), Wilson et al. (2016)
Epachthosaurus sciuttoi	Bajo Barreal	Cenomanian–Turonian	Estancia Ocho hermanos, Chubut Province, Argentina	Powell (1990), Martínez et al. (2004)

(continued)

Table 1 (continued)

Taxon	Formation	Age	Locality	Reference
Petrobrasaurus puestohernandezi	Plottier	Coniacian–Santonian	Rincón de los Sauces, Neuquén Province, Argentina	Filippi et al. (2011)
Elaltitan lilloi	Lago Colhué Huapi	Coniacian–Santonian	Sierra de San Bernardo, Chubut Province, Argentina	Mannion and Otero (2012)
Bonitasaura salgadoi	Bajo de la Carpa	Santonian	Cerro Policia, Río Negro Province, Argentina	Apesteguía (2004), Gallina and Apesteguía (2011, 2015)
Baalsaurus mansillai	Portezuelo	Turonian–Coniacian	Los Barreales, Neuquén Province, Argentina	Calvo and González Riga (2019)
Quetecsaurus rusconii	Cerro Lisandro	Turonian	Cañada del Pichanal, Mendoza Province, Argentina	González Riga and Ortiz David (2014)
Mendozasaurus neguyelap	Sierra Barrosa	Coniacian	Arroyo Seco, Mendoza Province, Argentina	González Riga (2003, 2005), González Riga et al. (2018)
Futalognkosaurus dukei	Portezuelo	Turonian–Coniacian	Los Barreales, Neuquén Province, Argentina	Calvo et al. (2007a)
Drusilasaura deseadensis	Bajo Barreal	Cenomanian–Turonian	Estancia María Aike, Santa Cruz Province, Argentina	Navarrete et al. (2011)

(continued)

Table 1 (continued)

Taxon	Formation	Age	Locality	Reference
Argentinosaurus huinculensis	Huincul	Cenomanian–Turonian	Plaza Huincul, Neuquén Province, Argentina	Bonaparte and Coria (1993)
Patagotitan mayorum	Cerro Barcino	Albian	La Flecha, Chubut Province, Argentina	Carballido et al. (2017), Otero et al. (2020)
Notocolossus gonzalezparejasi	Plottier	Coniacian–Santonian	Cerro Guillermo, Mendoza Province, Argentina	Gonzalez Riga et al. (2016)
Muyelensaurus pecheni	Plottier	Coniacian–Santonian	Rincón de los Sauces, Neuquén Province, Argentina	Calvo et al. (2007b)
Rinconsaurus caudamirus	Bajo de la Carpa	Santonian	Rincón de los Sauces, Neuquén Province, Argentina	Calvo and Gonzalez Riga (2003), Pérez Moreno et al. (2022)
Overosaurus paradasorum	Bajo de la Carpa	Santonian	Rincón de los Sauces, Neuquén Province, Argentina	Coria et al. (2013)

particular broad and laterally expanded neural spines. The other was the slender small-sized *Rinconsaurus caudamirus* (Calvo and González Riga 2003) that shows an unusual morphology of posterior caudal vertebrae recognizing a short amphicoelous, biconvex, and opisthocoelous vertebrae in the usually procoelous series present in derived titanosaur taxa.

In the following year, Apesteguía (2004) named *Bonitasaura salgadoi* based on a mid-sized young adult specimen from Cerro Policía, in Río Negro Province. This taxon definitively demonstrated the presence of square jaws profile in titanosaurs sauropods and provided evidence for a unique keratinous beak to aid in cutting plant material. The colossal *Futalognkosaurus dukei* (Calvo et al. 2007a), represented by a partially articulated skeleton, was described two years later from Los Barreales lake, in Central Neuquén Province, recognizing for the first time a clade of giant titanosaurs named Lognkosauria. At the same time, another slender small-sized titanosaur was described as *Muyelensaurus pecheni* (Calvo et al. 2007b), from

Fig. 1 South American record of titanosaurs from the Berriasian–Santonian interval

Rincón de los Sauces, in the north of Neuquén Province, as well as the rise of a new clade called Rinconsauria within the group.

The last decade showed an exponential increase in the record of BST, with fourteen new taxa nominated. Four new taxa were described in 2011. The basal form *Traukutitan eocaudata* from Lake Los Barreales, Neuquén Province (Juárez Valieri and Calvo 2011), the large *Drusilasaura deseadensis* from Estancia María Aike in Santa Cruz Province, Southern Patagonia (Navarrete et al. 2011), another large titanosaur named *Petrobrasaurus puestohernandezi* from Rincón de los Sauces, in Neuquén Province (Filippi et al. 2011), and *Tapuiasaurus macedoi*, discovered in the northern part of Minas Gerais, Brazil. The latter is one of the two titanosaurs from South America for which the entire skull is preserved and the only BST from that country (Zaher et al. 2011). In the following year, Mannion and Otero (2012) erected *Elaltitan lilloi* based on a specimen collected by Bonaparte and collaborators more than three decades before from Sierra de San Bernardo, Chubut Province. The successive two years showed two new forms with anatomical and stratigraphical importance. *Overosaurus paradasorum* (Coria et al. 2013), from Rincón de los Sauces, Neuquén Province, is represented by nearly complete and articulated vertebral series where cervical, dorsal, sacral, and caudal segments are represented, meanwhile *Quetecsaurus rusconii* (González Riga and Ortiz David 2014) from Mendoza Province, close to the north of Patagonia, being the only sauropod recorded from the Cerro Lisandro Formation with peculiar characters of Lognkosauria. In 2016, two other important taxa were described contributing to the anatomy and evolution of titanosaurs. One of them, *Sarmientosaurus musacchioi* (Martínez et al. 2016) from Estancia Laguna Palacios, Chubut Province, is the second titanosaur from the continent that preserves a complete skull, which is associated with anterior cervical vertebrae, an exceptional taphonomic preservation. The other is the giant *Notocolossus gonzálezparejasi* (González Riga et al. 2016) from Cerro Guillermo in Southern Mendoza, one of the largest sauropods discovered which preserves the largest humerus known from any titanosaur and an articulated and complete pes which is distinct in its anatomy.

In 2017, another giant from Patagonia named *Patagotitan mayorum* (Carballido et al. 2017), from La Flecha ranch, in Chubut Province, was described based on several well-preserved specimens that also represent another of the largest sauropods recorded to date. Three new taxa were described from Neuquén Province in 2018 and 2019. From Villa El Chocón, *Choconsaurus baileywillisi* (Simón et al. 2018), adding new information on dental and appendicular anatomy, *Baalsaurus mansillai* (Calvo and González Riga 2019) from Los Barreal lake based on a L-shaped complete dentary, and the large form *Kaijutitan maui* (Filippi et al. 2019) from Rincón de los Sauces. Finally, in 2021, *Ninjatitan zapatai* was described from Bajada Colorada, in Neuquén Province, representing the earliest known titanosaur sauropod recorded to date.

In this chapter, we provide an updated systematic list of the South American BST in a general context of González Riga et al. (2019) phylogeny (or the original systematics when the taxon was not resolved in the mentioned analysis). Finally, a brief discussion regarding phylogenetic and paleobiogeographic considerations of these taxa within Titanosauria is performed.

Institutional Abbreviations IANIGLA-PV Instituto Argentino de Nivología, Glaciología y Ciencias Ambientales, Colección de Paleovertebrados, Mendoza, Argentina; **MACN-CH**: Museo Argentino de Ciencias Naturales, Colección Chubut, Buenos Aires, Argentina; **MAU-Pv**: Museo Municipal Argentino Urquiza, Paleontología de Vertebrados, Rincón de los Sauces, Neuquén, Argentina; **MCF-PVPH**: Museo Carmen Funes, Paleontología de Vertebrados, Plaza Huincul, Neuquén,

Argentina; **MDT-PV**:, Museo Desiderio Torres, Paleovertebrados, Sarmiento, Chubut, Argentina; **MLP**: Museo de La Plata, La Plata, Buenos Aires, Argentina; **MMCh-PV**: Museo Municipal Ernesto Bachmann, Villa El Chocón, Neuquén, Argentina; **MPCA**: Museo Provincial 'Carlos Ameghino', Cipolletti, Río Negro, Argentina; **MPEF-PV**: Museo Paleontológico Egidio Feruglio, Paleontología de Vertebrados, Trelew, Chubut, Argentina; **MPM-PV**: Museo Padre Molina, Paleontología de Vertebrados, Río Gallegos, Santa Cruz, Argentina; **MRS-Pv** and **MAU-PV**: Museo Paleontológico de Rincón de los Sauces (today Museo Municipal Argentino Urquiza), Rincón de los Sauces, Neuquén, Argentina; **MUCPv**: Museo de Geología y Paleontología Universidad Nacional del Comahue, Neuquén, Argentina; **MZSP-PV**: Museu de Zoologia da Universidade de São Paulo, São Paulo, Brazil; **PVL**: Colección de Paleontología de Vertebrados de la Fundación Instituto Miguel Lillo, Tucumán, Argentina; **UNCUYO-LD**: Universidad Nacional de Cuyo, Instituto de Ciencias Básicas, Laboratorio de Dinosaurios, Mendoza, Argentina; **UNPSJB-PV**: Universidad Nacional de la Patagonia San Juan Bosco, Comodoro Rivadavia, Chubut, Argentina, Colección de Paleovertebrados.

2 Systematic Paleontology

Sauropoda Marsh 1878

Titanosauriformes Salgado, Coria, and Calvo 1997

Titanosauria Bonaparte and Coria 1993

Definition A node-based clade defined as *Andesaurus delgadoi* (Calvo and Bonaparte 1991), *Saltasaurus loricatus* (Bonaparte and Powell 1980), their most recent common ancestor and all descendants (Wilson and Upchurch 2003) (Fig. 2).

Andesaurus Calvo and Bonaparte 1991

Andesaurus delgadoi Calvo and Bonaparte 1991 (Fig. 2a, b)

Holotype Designated as MUCPv 132, it is composed of a partially articulated skeleton which includes four posterior dorsal vertebrae, thoracic ribs, two sacral vertebrae, twenty-five caudal vertebrae, chevrons, right humerus, right metacarpal I, left metacarpal V, left pubis, ischia, left femur.

Horizon, Age, and Locality The specimen was found in the Candeleros Formation, Neuquén Basin, lower Cenomanian (Upper Cretaceous), near to Villa el Chocón, Neuquén Province, Argentina.

Revised diagnosis It is diagnosed by the following association of characters (autapomorphies are indicated by asterisks): (1)* posterior dorsal neural spine height greater than twice the centrum height (autapomorphic within Macronaria); (2)* anterior–middle caudal centra square shaped in lateral view; (3)* anteroposteriorly elongate,

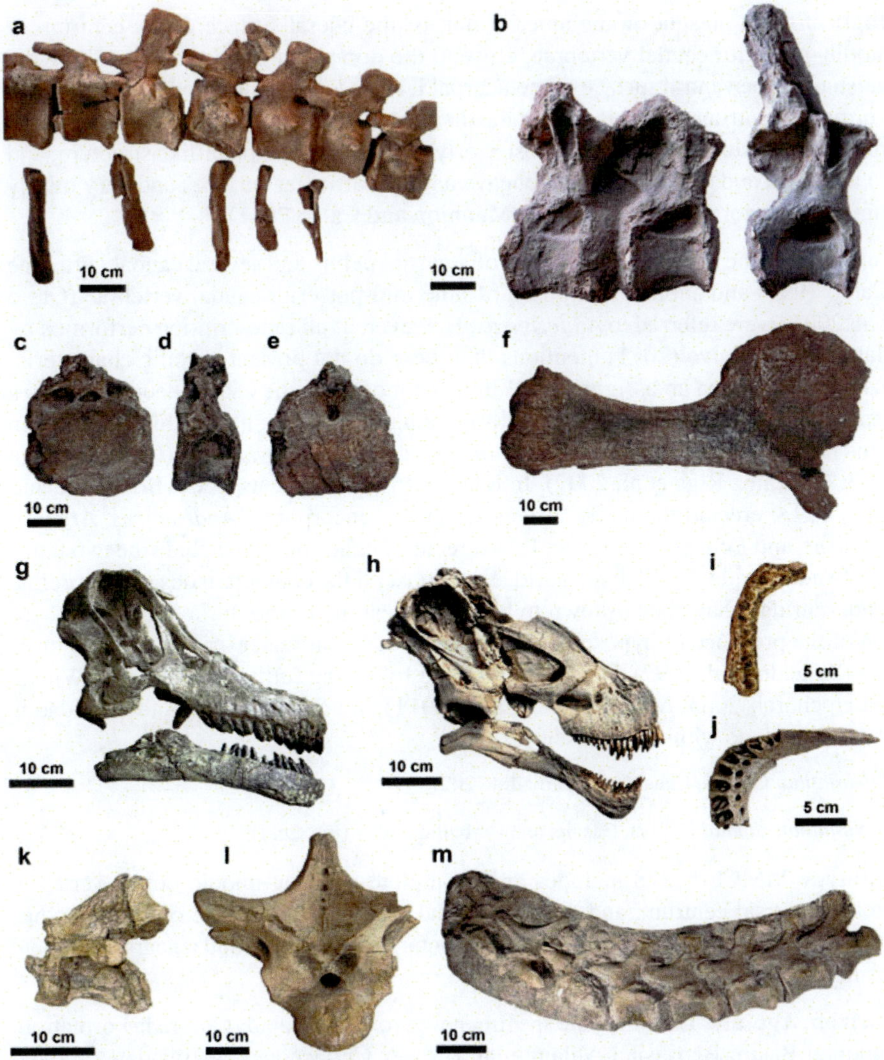

Fig. 2 Select bones of the titanosaurs *Andesaurus*, *Ninjatitan*, *Sarmientosaurus*, *Tapuiasaurus*, *Bonitasaura,* and *Baalsaurus*. **a** Articulated anterior to mid-caudal vertebrae of *Andesaurus* (MUCPv 132) in lateral view. **b** Dorsal vertebrae of *Andesaurus* (MUCPv 132) in lateral view. **c–e** Anterior caudal vertebra of *Ninjatitan* (MMCh-PV228) in anterior, right lateral, and posterior views, respectively. **f** Left scapula *Ninjatitan* (MMCh-PV228) in medial view. **g** Complete skull of *Sarmientosaurus* (MDT-PV 2) in right lateral view. **h** Complete skull of *Tapuiasaurus* (MZSP-PV 807) in right lateral view (photograph courtesy of H. Zaher), **i** Right dentary of *Baalsaurus* (MUCPv-1460) in dorsal view. **j** Right dentary of *Bonitasaura* (MPCA 460) in dorsal view. **k** Axis of *Bonitasaura* (MPCA 460) in lateral view. **l** Second dorsal vertebra of *Bonitasaura* (MPCA 460) in anterior view. **m** Articulated middle caudal vertebrae of *Bonitasaura* (MPCA 460) in lateral view. **i** Right dentary of *Baalsaurus* (MUCPv-1460) in dorsal view

elliptical fossa present on the anterior half of the lateral surface of the centrum of middle–posterior caudal vertebrae, close to the dorsal margin; (4)* ridge along the midshaft of the ventral surface of metacarpal I, close to the ventromedial margin; (5)* prominent ventromedial ridge along the distal half of metacarpal V; (6) caudal neural spine transversely narrow, anteroposteriorly elongate, with a plate-like structure; and (7) anterior caudal centra with a concave anterior articular surfaces and very mildly convex posterior surfaces (based on Mannion and Calvo 2011).

Comments One specimen composed of a partial pelvis and several caudal vertebrae (Calvo 1999) and another composed of nine mid-posterior caudal vertebrae (Otero et al. 2006) were referred to *Andesaurus*. However, a full redescription performed by Mannion and Calvo (2011) mentions that these do not present specific characteristics that allow such an assignment. Within Titanosauria, the caudal neural spine is a transversely narrow and anteroposteriorly elongate, forming a plate-like structure in some taxa like *Mendozasaurus, Epachthosaurus*, and *Malawisaurus* (González Riga 2003; González Riga et al. 2009). In relation to taxonomic aspects, (Bonaparte and Coria 1993) erected the name 'Andesauridae' to incorporate *Andesaurus, Argentinosaurus*, and *Epachthosaurus* and also erected Titanosauria to include the two families: 'Andesauridae' + 'Titanosauridae'. Almost all the characteristics used to define 'Andesauridae' have been shown to be widespread in Titanosauria, with the exception of the presence of hyposphene-hypantrum articulations restricted to these three taxa (Salgado et al. 1997). For this reason and following further studies (e.g., Wilson and Upchurch 2003; Mannion and Calvo 2011), we consider that Andesauridae is not a valid clade within Titanosauria.

Ninjatitan Gallina, Canale, and Carballido 2021

Ninjatitan zapatai Gallina, Canale, and Carballido 2021 (Fig. 2c, f)

Holotype MMCh-Pv228 includes an incomplete anterior–middle dorsal vertebra, a middle dorsal centrum, and anterior caudal centra with the base of neural arches preserved, a complete left scapula, a fragmentary distal femur, and a nearly complete left fibula of a single individual.

Horizon, Age, and Locality The specimen was found in Bajada Colorada Formation, Neuquén Basin, Berriasian–Valanginian (Lower Cretaceous), at Bajada Colorada, about 40 km south of Picún Leufú town on the National Route 237, Neuquén Province, Argentina.

Diagnosis It is diagnosed by the following association of characters (autapomorphies are indicated by asterisks): (1)* anterior–middle dorsal vertebra with well-marked lcpol and mcpol enclosing a marked and deep cpol-f; (2)* anterior–middle dorsal vertebra with mcpol dorsally and ventrally widened and extended to contact the dorsal rim of the cotyle; (3)* anterior– dorsal vertebra with mcpol delimiting a single fossa in which the neural canal is inset; (4) anterior–middle dorsal vertebra with dorsally acuminated neural canal fossa; and (5)* middle dorsal vertebra with three keels and two longitudinal shallow grooves in the ventral surface of the centrum (Gallina et al. 2021).

Comments As Gallina et al. (2021) recognized, the anterior caudal vertebra of *Ninjatitan* is slightly procoelous, differing from the strongly procoelous anterior caudal vertebra of most titanosaurs (except for *Andesaurus*). This condition fits better a basal position among titanosaurs (one of the possibilities recovered in their phylogenetic analysis). However, a more derived position related within Colossosauria cannot be discarded mainly supported by other features related to the presence of a large pneumatic foramen in the caudal centrum (a character shared with the first caudal centrum of *Patagotitan*). Taking account of the age of this taxon (the earliest known titanosaur to date), a better resolution of the phylogenetic position of *Ninjatitan* will impact other aspects of the group such as its paleobiogeography and evolution.

Choconsaurus Simón, Salgado, and Calvo 2018

Choconsaurus baileywillisi Simón, Salgado, and Calvo 2018

Holotype MMCh-PV 44 specimen is represented by an almost complete sequence of dorsal vertebrae, (1–4 and 6–11), a right scapula and coracoids, a right sternal plate, and five right metacarpals.

Referred Specimens Various skeletal elements were found at the La Antena site that were assigned as *Choconsaurus* paratype specimens because they present associations of characteristics described for the holotype. Paratype specimens of *Choconsaurus baileywillisi* include: MMCh-PV 80 (left premaxilla); MMCh-PV 81 (right dentary); MMCh-PV 82 (tooth); MMCh-PV 83 (maxillary tooth); MMCh-PV 84 (dentary tooth); MMCh-PV 85 (dentary tooth); MMCh-PV 86 (tooth); MMCh-PV 87 (maxillary tooth); MMCh-PV 88 (maxillary tooth); MMCh-PV 89, MMCh-PV90, MMCh-PV 91 (anterior/mid-cervical vertebrae); MMCh-PV 92 (posterior cervical vertebra); MMCh-PV 93–96 (dorsal rib fragment); MMCh-PV 97, MMCh-PV 98 (anterior caudal vertebrae); MUCPv-244/1–4 (mid-caudal vertebrae); MMCh-PV 99–111 (mid/posterior caudal vertebrae); MMCh-PV 112 (hemal arch).

Horizon, Age, and Locality All specimens come from the lower levels of the Huincul Formation, Neuquén Basin, Cenomanian–Turonian (Upper Cretaceous). The holotype and paratypes were found at the La Antena site, Villa el Chocón (Department of Confluencia), Neuquén Province, Argentina.

Diagnosis It is diagnosed by the following association of autapomorphies: (1) small notch in the dorsal margin of the posterior articular surface of the cervical centra; (2) dorsal neural spine with subparallel lateral margins in anterior and posterior views; accessory articulations lateral to the hyposphene, which are more developed in the anterior dorsal vertebrae; (3) dorsal vertebrae with a rough and wide prespinal lamina that extends for nearly the entire length of the neural spine; (4) mid- and posterior dorsal vertebrae (dorsal vertebrae ?sixth–?ninth) with accessory laminae placed between the posterior lamina and the accessory posterior centrodiapophyseal lamina; (5) absence of lateral spinopostzygapophyseal lamina in the mid- and posterior dorsal neural spines; (6) anterior caudal transverse processes shallow proximally, placed only on centrum; (7) anterior caudal vertebrae with hyposphene; (8) mid-caudal vertebrae with quadrangular, ventrally, and laterally flat centra; (9) scapular blade with rounded expansion on acromial side; (10) and acromial process located near to the midpoint of the scapular body (Simón et al. 2018).

Comments *Choconsaurus* is one of the best-preserved basal titanosaurs. It is also one of the youngest sauropods in showing tooth morphology intermediate between the broad-crown type of more basal titanosauriforms and the narrow-crown type of derived eutitanosaurs, a condition that is also found in *Abydosaurus mcintoshi*, a Lower Cretaceous basal titanosauriform from North America (Chure et al. 2010). This tooth morphology has relevance in the conceptual framework of faunal replacement in the Cenomanian, where the associations of basal diplodocoids and basal titanosaurs were progressively replaced by eutitanosaurs (Barrett and Upchurch 2005; Coria and Salgado 2005; Simón et al. 2018).

Kaijutitan Filippi, Salgado, and Garrido 2019

Kaijutitan maui Filippi, Salgado, and Garrido 2019

Holotype MAU-Pv-CM-522 is represented by an incomplete neurocranium (composed of supraoccipital, exoccipitals, left paraoccipital process, left exoccipital-opisthotic-prootic complex, left laterosphenoid, and orbitosphenoid, basioccipital/basisphenoid complex), anterior cervical vertebra, incomplete posterior cervical vertebra, proximal fragment of cervical rib, fragments from cervical ribs, left second dorsal rib, incomplete dorsal rib, fragment of dorsal rib, anterior caudal vertebra, left sternal plate, incomplete left coracoid, incomplete left scapula, incomplete left humerus?, right humerus, left ulna, incomplete right ulna, incomplete right radius?, right metacarpal II, right metacarpal III, a fragment of ilium?, incomplete right femur, right tibia, left astragalus, right distal epiphysis of metatarsal II, and indeterminate remains.

Horizon, Age, and Locality The specimen was found in Sierra Barrosa Formation, Neuquén Basin, upper Coniacian (Upper Cretaceous), at Cañadon Mistringa, about 9 km southwest of the city of Rincon de los Sauces, Neuquén Province, Argentina.

Diagnosis It is diagnosed by the following association of autapomorphies (those indicated with the asterisks were recovered in the phylogenetic analysis of Filippi et al. (2019)): (1)* width between the basal tuberosities almost four times the width of the foramen magnum; (2) foramen for the internal carotid artery located posteriorly on the basipterygoid processes, almost at the middle of the distance between these processes and the basal tuberosities; (3) bifid neural spine in anterior cervical vertebrae; (4) presence in the anterior cervical vertebrae of a medial tuberosity located posteriorly between both metapophyses; (5) spinopostzygapophyseal lamina (SPOL) in the anterior cervical, bifurcated in the spinal sector, generating anteroposteriorly elongated deep pneumatic cavities; (6) presence of a posteroventral keel in anterior cervical vertebrae, generated from the convergence of two ridges that originate from the posteroventral edge of the parapophysis; (7)* cervical vertebrae with an accessory lamina, which runs from the postzygodiapophyseal lamina (PODL) up to the spinoprezygapophyseal lamina (SPRL); (8)* absence of proximal pneumatopores in dorsal ribs; (9)* prespinal lamina triangular, product of a dorsal expansion in anterior caudal vertebrae; (10)* absence of a ventromedial process in the ventral margin of the scapula; (11)* tibial proximal condyle narrow, with its long axis anteroposteriorly oriented; (12)* tibial cnemial crest projecting anteriorly; and (13)* astragalus with foramina at the base of the ascending process (Filippi et al. 2019).

Comments *Kaijutitan maui* is a large titanosaur that represents a basal titanosaur. As Filippi et al. (2019) remark, it presents a singular combination of characters that includes the presence of bifid cervical neural spines, an unusual feature among titanosaurs. However, this feature does not show the degree of development seen in some diplodocoids.

Traukutitan Juárez Valieri and Calvo 2011

Traukutitan eocaudata Juárez Valieri and Calvo 2011

Holotype Designated as MUCPv 204, it presents a partially semi-articulated skeleton that includes both femora and thirteen anterior and mid-caudal centra.

Horizon, Age, and Locality The specimen was found in the lower section of the Bajo de la Carpa Formation, Neuquén Basin, Santonian (Upper Cretaceous), at the site called Trauku, north shore of Lake Los Barreales, Neuquén Province, Argentina.

Diagnosis It is diagnosed by the following association of characters: (1) proximal caudal vertebral centra higher than wider, with strongly procoelous center, with the ball placed in the middle dorsal part of the centrum; (2) prezygodiapophyseal lamina subvertical, with a ventrocaudal placement of the transverse process at the first caudal; (3) presence of a single deep foramen ventral to the base of the transverse process; (4) middle caudal vertebrae wider than elongated, with procoelous-opisthoplatyan centra, with the upper border of the posterior face slightly enlarged (Juárez Valieri and Calvo 2011).

Comments This species was represented by a very incomplete skeletal specimen which comprises both femora and an incomplete caudal series without neural spines. The original description has not phylogenetic analysis and neither autapomorphies in the diagnosis. This sauropod was originally studied by Salgado and Calvo (1993) and was assigned to the Río Colorado Formation. Later, Leanza et al. (2004) and Salgado and Bonaparte (2007) assign this specimento the Río Neuquén Subgroup (Coniacian–Santonian) and consider it as a representative of a 'faunal association' older than that proposed by the original authors. Juarez Valieri and Calvo (2011) carried out a review of the type locality and surrounding areas, which allowed them to confirm that the specimen comes from the base of the Bajo de la Carpa Formation (basal unit of the Río Colorado Subgroup, as originally proposed by Salgado and Calvo 1993). This contradicts the biostratigraphic interpretations made by Leanza et al. (2004) and by Salgado and Bonaparte (2007), among others, on the extinction period of the basal titanosaurs in the Neuquén Basin prior to the registration of the Bajo de la Carpa Formation deposits. *Traukutitan* represents the last definitive 'basal' titanosaurian that survived to the 'Senonian' times in Northern Patagonia.

Lithostrotia Upchurch, Barrett, and Dodson 2004

Definition A node-based clade defined as *Malawisaurus dixeyi* Haughton 1928, *Saltasaurus loricatus* Bonaparte and Powell 1980, their most recent common ancestor, and all its descendants (Upchurch et al. 2004).

Sarmientosaurus Martínez, Lamanna, Novas, Ridgely, Casal, Martínez, Vita, and Witmer 2016

Sarmientosaurus musacchioi Martínez, Lamanna, Novas, Ridgely, Casal, Martínez, Vita, and Witmer 2016 (Fig. 2g)

Holotype MDT-PV 2 is an articulated cranial and cervical skeleton consisting of the nearly complete skull, the partial axis associated with its rib from the right side and articulated with the cranial part of the third cervical vertebra, a fragment of the fifth cervical vertebra, the nearly complete sixth cervical vertebra and its right rib, the partial seventh cervical vertebra, and a section of ossified cervical tendon.

Horizon, Age, and Locality The specimen was found in the uppermost section of the lower Member of the Bajo Barreal Formation, Golfo de San Jorge Basin, Cenomanian–Turonian, (Upper Cretaceous), Estancia Laguna Palacios, Sierra Nevada Anticline, South-central Chubut Province, Central Patagonia, Argentina.

Diagnosis It is diagnosed by the following association of autapomorphies: (1) maximum (rostroventral–caudodorsal) diameter of orbit nearly 40% rostrocaudal length of cranium (as measured from tip of snout to occipital condyle); (2) complex maxilla-lacrimal articulation, with ascending ramus of maxilla embedded in and bordered laterally and medially by lacrimal dorsal process; (3) medial edge of caudal sector of maxillary ascending ramus bordering bony nasal aperture with low but well-defined ridge; (4) 'tongue-like' ventral process of quadratojugal that overlaps quadrate caudally; (5) separate foramina for all three branches of the trigeminal nerve; (6) absence of median venous canal connecting infundibular region to ventral part of brainstem; (7) premaxillary teeth subvertical, maxillary teeth procumbent, and dentary teeth recumbent; (8) middle cervical vertebrae with 'strut-like' (as opposed to 'sheet-like') centroprezygapophyseal laminae; (9) extremely elongate and slender ossified tendon extending along cervical series ventrolateral to vertebrae and ribs (Martínez et al. 2016).

Comments *Sarmientosaurus* is the first titanosaurian sauropod from the Upper Cretaceous of South America for which an articulated, virtually complete adult skull has been discovered. Phylogenetic analyses demonstrated that this taxon is a basal member of Titanosauria, being more derived than *Malawisaurus* but more basal than taxa frequently regarded as nemegtosaurids. *Sarmientosaurus* exhibits a previously undocumented cranial morphology that consists of a combination of plesiomorphic titanosauriform features. These characters offer novel cranial support for the phylogenetic hypothesis that titanosaurians are closely related to Brachiosauridae and other titanosauriforms (Martínez et al. 2016).

Tapuiasaurus Zaher, Pol, Carvalho, Nascimento, Riccomini, Larson, Juarez Valieri, Pires Domingues, da Silva, and Campos 2011

Tapuiasaurus macedoi Zaher, Pol, Carvalho, Nascimento, Riccomini, Larson, Juarez Valieri, Pires Domingues, da Silva, and Campos 2011 (Fig. 2h)

Holotype MZSP-PV 807 consists of an articulated partial skeleton composed of an almost complete skull and mandible, hyoid apparatus, atlas, axis, five cervical and five dorsal vertebrae and ribs, left sternal plate, right coracoid, right humerus, left radius, ulnae, metacarpals, femora, left fibula, and an almost complete left pes.

Horizon, Age, and Locality The skeleton was found in outcrops from the Quiricó Formation, Sanfranciscana Basin, lower Aptian (Lower Cretaceous), at Embira-Branca Hills near Coração de Jesus City, in Northern Minas Gerais, Brazil.

Revised Diagnosis It is diagnosed by the following association of autapomorphies: (1) hook-shaped posteroventral process of the quadratojugal; (2) maxillary process of the jugal tapering and forming most of the ventral margin of the antorbital fenestra; (3) anterolateral tip of the pterygoid contacting the medial surface of the ectopterygoid; (4) maxilla with a tapering post-dentigerous process of the maxilla that is elevated above the alveolar margin; (5) jugal with an elongate lacrimal process forming much of the posteroventral border of the antorbital fenestra; (6) lateral temporal fenestra divided by a second squamosal–postorbital contact, forming a small posterodorsal opening and an elongate anteroventral opening; (7) quadrate and quadratojugal with a narrow (c. 2 mm), end-on articulation; (8) maxillary teeth with labial wear; (9) absence of a posttemporal foramen; (10) flat overlapping articulation between squamosal and quadratojugal; and (11) basisphenoid with median ridge extending from contact with basioccipital onto ventral surface (Zaher et al. 2011; Wilson et al. 2016).

Comments *Tapuiasaurus* was originally recovered as an advanced titanosaurian in the Nemegtosauridae clade along with *Nemegtosaurus* and *Rapetosaurus* (Zaher et al. 2011). However, subsequent analyses recovered it as a basal titanosaur positioned adjacent to other Early Cretaceous forms, such as *Phuwiangosaurus* and *Malawisaurus* (Wilson et al. 2016). A key implication of the previous *Tapauisaurus*-as-nemegtosaurid hypothesis (Zaher et al. 2011) is that the 'classic' titanosaur skull morphology was viewed as restricted to a subgroup of titanosaurs that is geographically widespread (South America, Madagascar, Asia) and long lived (Aptian–Maastrichtian) but apparently not diverse. The revised hypothesis, which posits that *Tapuiasaurus* is a basal titanosaur, implies that the 'classic' titanosaurian skull is more widespread, features of which can be expected to be present in a broad array of titanosaurs, for which cranial remains are poorly known or completely unknown (Wilson et al. 2016).

Epachthosaurus Powell 1990

Epachthosaurus sciuttoi Powell 1990

Holotype MACN-CH 1317 is composed of an incomplete posterior dorsal vertebra (Powell 1990).

Paratype MACN-CH 18,689 consists of a cast, exposed in left lateral and ventral views, of six articulated mid- to posterior dorsal vertebrae, the partial sacrum, and a fragment of the pubic peduncle of the right ilium. Due to its occurrence in an extremely resistant matrix, the original fossil has not yet been recovered (Powell 1990).

Referred Specimens UNPSJB-PV 920 is composed of an exceptionally articulated skeleton composed by most of the axial and appendicular skeleton, including the complete pedes and manus in anatomical position. The elements not preserved are the skull, neck, four or five anterior dorsal, and distalmost caudal vertebrae (Martínez et al. 2004). Other specimens are UNPSJB-PV 1006, composed of an anterior dorsal vertebra (Casal and Ibiricu 2010), and UNPSJB-PV 956 that includes five partially preserved sacral vertebrae, an articulated incomplete right ilium, and a pubis fragment (Casal and Ibiricu 2010). Moreover, *Epachthosaurus* has a pedal phalangeal formula of 2–2-3–2-0 (Martínez et al. 2004), in contrast to a more reduced formula of 2–2-2–2-0 present presumably, in 'more derived' titanosaurs like in the Padrillo (UNCUYO-LD 313) and Invernada (MUCPv-1533) taxa, *Notocolossus* (González Riga et al. 2016), and probably in *Mendozasaurus* (González Riga et al. 2018).

Horizon, Age, and Locality Upper portion of the lower Member of the Bajo Barreal Formation, upper Cenomanian–lower Turonian (Upper Cretaceous). Holotype, paratype, and referred specimens were found in Estancia Ocho Hermanos, Sierra de San Bernardo, west of Lago Musters, South-central Chubut Province, Central Patagonia, Argentina.

Revised diagnosis It is diagnosed by the following association of autapomorphies: (1) mid- and posterior dorsal vertebrae possessing accessory articular processes extending ventrolaterally from the hyposphene; (2) a strongly developed intraprezygapophyseal lamina and aliform processes projecting laterally from the dorsal portion of the spinodiapophyseal lamina; (3) hyposphene-hypantrum articulations in caudal vertebrae 1–14; and (4) a pedal phalangeal formula of 2–2-3–2-0 (Martínez et al. 2004).

Comments *Epachthosaurus sciuttoi* is a taxon initially named by Powell (1990) based on the characters present in a single dorsal vertebra (MACN-CH 1317). Powell (1990) also assigned a series of dorsal vertebrae and an incomplete sacrum (a fragment of pubic peduncle and right ilium) to *Epachthosaurus*. Of this material, only one plastic cast taken in the field is available (MACN-CH 18,689). Salgado and Martínez (1993), Salgado (1996), and Sanz et al. (1999) considered that the characters initially proposed for *Epachthosaurus* were plesiomorphic. Later, Martínez and collaborators discovered an exceptionally articulated and mostly complete skeleton (UNPSJB-PV 920) which was assigned to *Epachthosaurus sciuttoi* (Martínez et al. 2004). Finally, Casal and Ibiricu (2010) studied two new specimens assigned to *Epachthosaurus* (UNPSJB-PV 1006 and UNPSJB-PV 956) and made several comments about the assignment of these new remains to the genus and its abundance in the site. These were collected at the same location and stratigraphic levels of other fossil materials. The study of the UNPSJB-PV 1006 dorsal vertebra (which is 1.5 times the size of the holotype vertebra) suggests that *Epachthosaurus* reached sizes much larger than originally considered.

Petrobrasaurus Filippi, Canudo, Salgado, Garrido, García, Cerda, and Otero 2011

Petrobrasaurus puestohernandezi Filippi, Canudo, Salgado, Garrido, García, Cerda, and Otero 2011

Holotype MAU-Pv-PH-449 is composed of two isolated teeth, posterior portion of possibly middle cervical vertebra, three incomplete posterior dorsal vertebrae, incomplete posterior dorsal arch, incomplete dorsal vertebral centrum, incomplete first caudal centrum, five anterior caudal vertebrae, right humerus, right and left sternal plates, four metacarpals, one of which is incomplete, right and left femur, left tibia lacking distal end, distal end of right tibia, distal fragment of iliac peduncle, left pubis, fragments of hemal arches, dorsal ribs, fragments of cervical ribs, and indeterminate remains.

Horizon, Age, and Locality *Petrobrasaurus* was found in the Plottier Formation, Neuquén Basin, Coniacian-Santonian (Upper Cretaceous), at Rincón de los Sauces, Neuquén Province, Argentina.

Diagnosis It is diagnosed by the following association of autapomorphies: (1) accessory posterior centrodiapophyseal lamina is more developed than the posterior centrodiapophyseal lamina; (2) posterior centroparapophyseal lamina extends from the point of contact with the accessory posterior centrodiapophyseal lamina; (3) presence of a broad concave surface delimited by the bony rim of the posterior centroparapophyseal lamina and a rim located on the dorsal edge of the pleurocoel; (4) presence of a deep fossa, sub-circular in outline, on a slightly concave surface delimited by the accessory posterior centrodiapophyseal lamina, the posterior centrodiapophyseal lamina, and a bony rim corresponding to the extension of the posterior centroparapophyseal lamina; (5) presence of a T-shaped anteroposterior expansion on the laminae of the neural arch (pcdl and apcdl); and the anterior caudal vertebrae characterized by: (6) the presence of a deep fossa between the spinoprezygapophyseal lamina and the postzygapophysis, bearing small fossae; (7) prespinal lamina distally thickened; (8) sub-circular small fossae on both sides of the base of the prespinal lamina; and (9) presence of a T-shaped anteroposterior expansion in the spinoprezygapophyseal lamina (Filippi et al. 2011).

Elaltitan Mannion and Otero 2012

Elaltitan lilloi Mannion and Otero 2012

Holotype PVL 4628 and MACN-CH 217 are associated partial skeleton comprising three dorsal vertebrae, two caudal vertebrae, left scapula, left humerus, left radius, both ulnae, right pubis, proximal half of right femur, distal part of left tibia, distal two-thirds of left fibula, right astragalus, and calcaneum.

Horizon, Age, and Locality The materials come from the bottom of the Lago Colhué Huapi Formation, Golfo de San Jorge Basin, Coniacian–Santonian (Upper Cretaceous), at south bank of the Río Senguerr, in the area between the bend of this river and the Pampa de María Santísima, southeast of the southernmost part of the Sierra de San Bernardo, Chubut Province, Argentina.

Diagnosis It is diagnosed by the following association of characteristics (autapomorphy is marked by an asterisk): (1) spinopostzygapophyseal laminae in middle–posterior dorsal vertebrae bifurcate into medial and lateral branches; (2)* dorsoventrally tall neural arch restricted to anterior half of centrum (excluding condylar ball) in anteriormost caudal vertebrae; (3) astragalar ascending process does not extend to the posterior margin of the astragalus; (4) presence of a calcaneum (Mannion and Otero 2012).

Comments Lydekker (1893) described the sauropod *Argyrosaurus superbus*, which consists of a partially articulated forelimb (MLP 77-V-29–1) discovered by Carlos Ameghino in 1888 in the province of Chubut, Southern Argentina. Subsequently, numerous specimens were referred to *Argyrosaurus* from elsewhere in Argentina and Uruguay (Lydekker 1893; Huene 1929, 1931; Powell 1986, 2003). These assignments were mainly based on the large size and robust proportions and of those isolated remains. The exception to this is a partial skeleton (PVL 4628/MACN-CH 217) recovered from the province of Chubut by J. F. Bonaparte. This skeleton was originally attributed to *Antarctosaurus* sp. (Bonaparte and Gasparini 1979), but then tentatively referred to *Argyrosaurus* by Powell (1986, 2003), who also provided a short description. Finally, Mannion and Otero (2012) validate *Argyrosaurus superbus* by expanding its diagnosis. On the other hand, these authors recognized that some fossil elements originally identified as *Argyrosaurus* (PVL 4628/MACN-CH 217) present different characteristics and were assigned to a new taxon that they named *Elaltitan lilloi.*

Colossosauria González Riga, Lamanna, Otero, Ortiz David, Kellner, and Ibiricu 2019

Definition A stem-based clade is defined as the most inclusive clade containing *Mendozasaurus neguyelap* Gonzalez Riga 2003 but not *Saltasaurus loricatus* Bonaparte and Powell 1980 or *Epachtosaurus sciuttoi* Powell 1990.

Bonitasaura Apesteguía 2004

Bonitasaura salgadoi Apesteguía 2004 (Fig. 2j–m)

Holotype Partially articulated young adult specimen (MPCA 460) including left frontal, left parietal, right lacrimal, left quadrate, right dentary with 15 teeth, isolated tooth, axis, three cervical, five dorsal, and 25 caudal vertebrae, several cervical and dorsal ribs, five chevrons, fragment of sternal plate, incomplete right humerus, two left metacarpals, left pubis, left ischium, left femur, right tibia, fragmentary left fibula, both astragali, both metatarsal Is, both metatarsal IIs, left metatarsal III, right metatarsal IV, both metatarsal Vs, one left and one right pedal phalanx, and single left ungual phalanx.

Referred Specimens MPCA 467 includes anterior caudal centrum, mid-caudal centrum, complete left ulna, distal fragment of left radius, proximal fragment of metacarpal III, distal fragment of metacarpal III?, right femoral head, distal fragment of left tibia, and complete left fibula. These correspond to a single individual that is different from the holotype, but equal in size to the holotype. A complete right tibia (MPCA 468) of a third, smaller specimen was also referred to *B. salgadoi* (Gallina and Apesteguía 2015).

Horizon, Age, and Locality The specimens were found at the top of Bajo de la Carpa Formation, Neuquén Basin, Santonian (Upper Cretaceous), at La Bonita Hill, near Cerro Policıa town, Río Negro Province, Argentina.

Revised Diagnosis It is diagnosed by the following association of characters (autapomorphies are indicated by asterisks): (1) frontal outer rim straight (not sigmoid); (2)* thin and enlarged maxillary process of lacrimal oriented downward and forward; (3)* posterior dentary edentulous, with sharp dorsal edge, and highly vascularized lateral surface; (4)* ventral longitudinal keel in posterior half of axial centrum; (5)* tongue-like process on spinoprezygapophyseal laminae of mid-cervical vertebrae; (6) very robust, diagonal neural arch pillars and bulging neural spine summits on anterior dorsal vertebrae; (7) longitudinal, paired fossae on sides of anterior median lamina in anterior dorsal vertebrae; (8)* circular, vertically oriented fossae aligned with anterior median lamina in mid-dorsal vertebrae; (9)* thin, longitudinal laminae diverging from anterior median lamina and postspinal laminae in anterior caudal vertebra; (10)* anterior, longitudinal ridge of the tibia with marked promontory just over anterior process of distal end (Gallina and Apesteguía 2015).

Comments When the holotype was designated, the material included by Apesteguía (2004) was a left frontal, left parietal, right dentary with 15 teeth, two cervical, six dorsal and 12 caudal vertebrae, radius, metacarpal, femur, tibia, two metatarsals, two chevrons, and several cervical and dorsal ribs. Further preparation, plus new collecting trips to the quarry, added a right lacrimal, a left quadrate, an isolated tooth, the axis, 13 additional caudal vertebrae, three chevron bones, an incomplete humerus, two metacarpals, the left pubis, the left ischium, a fragmentary left fibula, both astragali, four metatarsals, two pedal phalanges, and an ungual phalanx to the holotype specimen (Gallina and Apesteguía, 2011) and those material pertaining to the referred specimens (Gallina and Apesteguía 2015). The radius mentioned by Apesteguía (2004) was actually a metacarpal I. The anterior median lamina mentioned in the diagnosis refers to the commonly named prespinal lamina, not the anterior medial lamina applied by Gorscak and O'Connor (2019) to the median lamina behind the prezygapophyses of posterior dorsal vertebrae of *Mnyamawamtuka moyowamkia*.

Baalsaurus Calvo and González Riga 2019

Baalsaurus mansillai Calvo and González Riga 2019 (Fig. 2i)

Holotype The specimen MUCPv-1460 is represented by an almost complete right dentary with 13 alveoli.

Horizon, Age, and Locality Portezuelo Formation, Neuquén Basin, Turonian–Coniacian (Upper Cretaceous), North coast of Los Barreales lake, Neuquén Province, Argentina.

Diagnosis It is diagnosed by the following association of characters: (1) dentary alveoli with 10 alveoli on the anterior ramus; (2) dentary with ventrally and anteriorly inclined symphysis; and (3) wide ventral meckelian groove surrounded by a thin lamina that forms a keel in the ventral border of the dentary (Calvo and González Riga 2019).

Comments The discovery of *Baalsaurus* allows to recognize three different morphologies in sauropod dentaries in dorsal view: L-shaped, U-shaped, and smoothly curved-shaped. *Baalsaurus* is L-shaped, forming a strong 90 degrees angle between anterior and posterior ramus giving a square jaw morphology, different to *Nemegtosaurus* and *Rapetosaurus* but similar to *Antarctosaurus*. Based on similarities shared with *Bonitasaura*, this taxon is tentatively considered as a colossosaur.

Quetecsaurus González Riga and Ortiz David 2014

Quetecsaurus rusconii González Riga and Ortiz David 2014

Holotype UNCUYO-LD 300 is represented by associated bones belonging to a single individual: postorbital, two teeth, atlas, one posterior cervical vertebra, one incomplete anterior dorsal vertebra, one anterior caudal centrum, eight dorsal ribs, a coracoid, an incomplete humerus, distal fragments of a radius and an ulna, and five metacarpals.

Horizon, Age, and Locality Uppermost levels of the Cerro Lisandro Formation, Neuquén Basin, Middle–Upper Turonian (Upper Cretaceous), Cañada del Pichanal section, Malargüe Department, Mendoza Province, Argentina.

Diagnosis It is diagnosed by the following association of characters (autapomorphies are marked by an asterisk): (1)* intercentrum of atlas with prominent anteroventral border and expanded posteroventral processes; (2)* posterior cervical neural spines with incipient lateral laminae; (3)* humerus with marked sigmoid proximal border, rounded proximomedial corner, and angular proximolateral corner; (4) opisthocoelous cervical centra; (5) deep spinoprezygapophyseal and spinopostzygapophyseal fossae on posterior cervical vertebrae; (6) accessory posterior centrodiapophyseal lamina on posterior cervical vertebrae; (7) posterior cervical centra relatively elongate (total length/posterior cotyle height greater than 2.5); (8) cervical vertebra height/centrum length less than 1.5; (9) prespinal lamina extends to the base of the neural spine on anterior dorsal vertebrae; (10) opisthocoelous anterior dorsal centra; (11) procoelous anterior caudal centra; (12) neural arch located anteriorly on anterior caudal centra; (13) well-developed infraglenoid lip on the coracoid; (14) quadrangular coracoid with an anterodorsal angle around 90°; (15) hemal arches with open proximal articular facets; and (16) metacarpals without distal articular facets (González Riga and Ortiz David 2014).

Comments *Quetecsaurus* is the only sauropod recorded in the Cerro Lisandro Formation, covering a time range (Upper Turonian) little represented by titanosaurs in South America. *Quetecsaurus* is retrieved as a sister taxon of Lognkosauria (González Riga and Ortiz David 2014; Tykoski and Fiorillo 2017) or within that clade (Carballido et al. 2017; Hechetleiner et al. 2020). The recognition of posterior cervical neural spines with an incipient lateral spinodiapophyseal lamina represents a lower development of the typical wide and 'fan-shaped' cervical spines observed in lognkosaurs such as *Mendozasaurus* and *Futalognkosaurus*.

Lognkosauria Calvo, Porfiri, González Riga, and Kellner 2007a

Definition A node-based clade is defined as the most recent common ancestor of *Mendozasaurus neguyelap* and *Futalognkosaurus dukei* and all its descendants (Calvo et al. 2007a).

Mendozasaurus González Riga 2003

Mendozasaurus neguyelap González Riga 2003 (Fig. 3b–d)

Holotype IANIGLA-PV 065 consists of twenty-two mostly articulated caudal vertebrae and three anterior chevrons and fragments of posterior chevrons (González Riga 2003).

Referred specimens Five cervical vertebrae (IANIGLA-PV 076/1–3; 076/5, 084/1); two dorsal vertebrae (IANIGLA-PV 76/4, 066); one thoracic rib (IANIGLA-PV 084/2); a right scapula (IANIGLA-PV 068); a right sternal plate (IANIGLA-PV 067); a left and right humerus (IANIGLA-PV 069/1–2); a right radius (IANIGLA-PV 070/2); a right ulna (IANIGLA-PV 070/1); six metacarpals (IANIGLA-PV 071/1–5, 154); a fragment of pubis (IANIGLA-PV 072); the proximal half of a right femur (IANIGLA-PV 073/1) and a left femur (IANIGLA-PV 073/4); one left (IANIGLA-PV 074/2) and three right tibiae (IANIGLA-PV 073/2–3, 074/1); a left and right fibula (IANIGLA-PV 074/3 and 074/4, respectively); a right astragalus (IANIGLA-PV 155); 12 metatarsals (IANIGLA-PV 077/1–5, 100/1–6, 153); ten pedal phalanges (IANIGLA-PV 077/6–12, 078/1–2, 079); and four osteoderms (IANIGLA-PV 080/1–2, 81/1–2) (González Riga 2003, 2005; González Riga et al. 2018).

Horizon, Age, and Locality *Mendozasaurus* was found in the upper levels of the Sierra Barrosa Formation, Neuquina Basin; Middle–Upper Coniacian (Upper Cretaceous, Garrido 2010; González Riga and Ortiz David 2014). All the materials, both holotype and paratype, come from the same quarry and were found at the same stratigraphic level (González Riga and Astini 2007; González Riga et al. 2018) at the Arroyo Seco site, Malargüe Department, Mendoza Province, Argentina.

Revised Diagnosis It is diagnosed by the following association of characters (autapomorphies are marked by an asterisk): (1)* middle–posterior cervical vertebrae with tall and transversely expanded neural spines that are wider than the centra, with the lateral expansion formed by spinodiapophyseal laminae, without contribution of the pre- or postzygapophyses; (2)* anteriormost caudal neural spine dorsoventral height

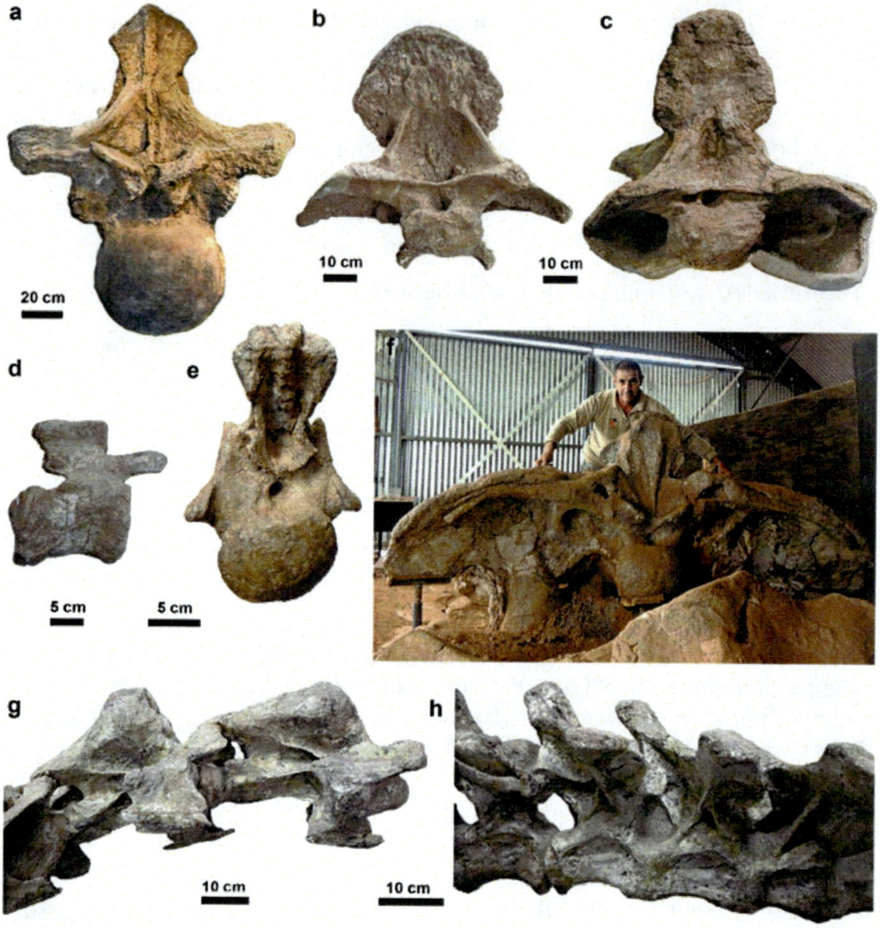

Fig. 3 Select bones of the titanosaurs *Argentinosaurus*, *Mendozasaurus*, *Futalognkosaurus,* and *Overosaurus*. **a** Anterior dorsal vertebra of *Argentinosaurus* (cast) in anterior view. **b** Cervical vertebra of *Mendozasaurus* (IANIGLA-PV 076/1) in anterior view. **c** Cervical vertebra of *Mendozasaurus* (IANIGLA-PV 084/1). **d** Middle caudal vertebra of *Mendozasarus* (IANIGLA-PV 065/15) in lateral view. **e** Anterior caudal vertebra of *Futalognkosaurus* (MUCPv-323) in posterior view. **f** Dr. Jorge Calvo next to the complete hip from *Futalognkosaurus* (MUCPv-323) in anterior view. **g** Articulated cervical vertebrae of *Overosaurus* (MAU-Pv-CO-439) in lateral view. **h** Articulated dorsal vertebrae of *Overosaurus* (MAU-Pv-CO-439) in lateral view.

divided by centrum height >1.2; (3)* anterior caudal vertebrae (excluding anteriormost) with ventrolateral thickening of prezygapophyses; (4)* middle caudal centra with greatly reduced posterior condyles displaced dorsally; (5)* laterally compressed and anteroposteriorly elongated middle caudal neural spines, with the horizontal dorsal margin forming a 90° angle with the dorsal portion of the anterior margin in lateral view; (6)* humerus with divided lateral distal condyle on anterior surface;

(7)* second ridge on posterior surface of distal third of radius, parallel to main interosseous ridge; (8) metacarpal I with ridge or tubercle on the dorsolateral margin at approximately two-thirds of length from proximal end; (9) large subconical to subspherical osteoderms, lacking a cingulum (González Riga et al. 2018).

Comments *Mendozasaurus* is a one best-studied South American titanosaurian, both taxonomically (González Riga 2003, 2005; González Riga et al. 2018) and taphonomically (González Riga and Astini 2007); this is used as specifier of Lognkosauria and Colossosauria clades. It is represented by five specimens of different sizes, four described in detailed in the osteology review (González Riga et al. 2008), and the last one, a large specimen of with a metatarsal III of 29.2 cm long (femoral length estimated at 1.6–2.0 m) is mentioned in González Riga et al. (2019: 26). A peculiar character of *Mendozasaurus*' cervicals is the transversely expanded neural spines that are wider than the centra, formed laterally by spinodiapophyseal laminae. This morphology is unique in this taxon but is also present with a less development in other taxa such as *Alamosaurus, Bonitasaura, Quetecsaurus*, and *Futalognkosaurus.*

Futalognkosaurus Calvo, Porfiri, González Riga, and Kellner 2007a

Futalognkosaurus dukei Calvo, Porfiri, González Riga, and Kellner 2007a (Fig. 3e, f)

Holotype MUCPv-323 comprises atlas, axis, twelve cervicals vertebrae (five anterior, four middle, and three posterior), ten dorsal vertebrae, ribs, complete sacrum, both ilia; right pubis, ischium; and one anterior caudal.

Horizon, Age, and Locality Portezuelo Formation, Neuquén Basin, Upper Turonian–Lower Coniacian (Upper Cretaceous). The holotype comes from the northern coast of the Los Barreales lake, 90 km northwest of Neuquén City, Neuquén Province, Patagonia, Argentina.

Diagnosis It is diagnosed by the following associations of autapomorphies: (1) neurapophysis of the atlas laminar and rectangular and posteriorly directed; (2) neural spine of the axis high and triangular; (3) posterior border of the neural spine on middle cervical elements concave; (4) ventral depression between parapophyses on middle cervical centra; (5) anterior dorsal vertebrae with horizontal and aliform diapophyses; (6) pre- and postzygapophyses of anterior dorsal vertebrae horizontal; (7) first caudal vertebra with prespinal lamina bifurcated on its base forming two small infraprespinal laminae; (8) supraspinal cavity in first caudal vertebra bordered by the prespinal and lateral laminae; (9) second and third sacral ribs fused; and (10) wide and well-developed iliac peduncle on ischia (Calvo et al. 2007a).

Comments *Futalognkosaurus* is one of a few exceptionally preserved giant titanosaurs characterized by an articulated axial sequence from the first cervical vertebra to the first caudal vertebra. In the taphonomic record, articulated cervical vertebrae are scarce, and this taxon fills a gap for the anatomical knowledge of lognkosaurians. This taxon was originally proposed as a specifier of Lognkosauria and is very important to better understand this clade. A further complete and detailed redescription of this taxon is in course by Calvo, González Riga, and Ortiz David.

Drusilasaura Navarrete, Casal, and Martínez 2011

Drusilasaura deseadensis Navarrete, Casal, and Martínez 2011

Holotype MPM-PV 2097 is composed of four dorsal vertebrae, one sacral vertebra, six caudal vertebrae, one left scapula, one incomplete dorsal rib, and indeterminate fragments.

Horizon, Age, and Locality The holotype comes from the upper part of the Upper Member of the Bajo Barreal Formation, Golfo de San Jorge Basin, Lower Cenomanian–Upper Turonian (Upper Cretaceous, Lamanna et al. 2002; Casal et al. 2015). The site is located on the southern bank of the Deseado river, Estancia María Aike, Lake Buenos Aires Department, Santa Cruz Province, Argentina.

Diagnosis It is diagnosed by the following association of autapomorphies: (1) anterior dorsal vertebrae with two robust spinodiapophyseal laminae, one anterior and other posterior, which delimit an elongate and deep supradiapophyseal cavity; (2) presence in the anterior dorsal vertebra of a small circumneural lamina surrounding the neural canal in posterior view; (3) the last sacral vertebra with postspinal lamina expanded toward the neural spine apex; (4) prespinal laminae, at least until 5°? caudal vertebra, expanded toward the neural spine apex; (5) anterior caudal vertebrae with tuberopostzygapophyseal laminae; (6) presence of ventral foramina, at least until the 4°? caudal vertebrae; and (7) prezygapophyseal tuberosity jointed the prespinal lamina until the 4°? caudal vertebrae (Navarrete et al. 2011).

Comments As occurred in other lognkosaurians, *Drusilasaura* seems to have dorsoventrally high and anteroposteriorly narrow transverse processes in the anterior caudal vertebrae. Thus, the mentioned tuberopostzygapophyseal lamina proposed by the authors should be recognized as a postzygodiapophyseal lamina. *Drusilasaura* is well-nested within Lognkosauria in the phylogenetic analysis of Carballido et al. (2017) (Fig. 3).

Argentinosaurus Bonaparte and Coria 1993

Argentinosaurus huinculensis Bonaparte and Coria 1993 (Fig. 3a)

Holotype MCF-PVPH-1 is composed of incomplete dorsal vertebrae, three anterior and three posterior; ventral portion of the sacrum including the vertebral centra 1 to 5 with most of the sacral ribs of the right side; one fragmented dorsal rib and the right tibia

Horizon, Age, and Locality The specimen was found in the Huincul Formation, Neuquén Basin, upper Cenomanian–lower Turonian (Upper Cretaceous), 8 km to the east of the city of Plaza Huincul, Neuquén Province, Argentina.

Diagnosis It is diagnosed by the following association of characters: (1) opisthocoelous dorsal vertebrae provided with well-developed hyposphene-hypantrum, with extra articulations between them; (2) anterior dorsal vertebrae with transversely wide and anteroposteriorly planar neural spines, with robust prespinal laminae; (3) middle and posterior dorsal vertebrae with low and wide vertebral bodies, with the ventral face almost planar, and with pleurocoels located in the anterior half of the body; (4) sacral vertebral bodies 2 to 5 very reduced; (5) dorsal ribs of tubular, cylindrical, and hollow structure; (6) bony macrocells in the sacrum and presacral vertebrae; and (7) slender tibia with short cnemial crest (Bonaparte and Coria 1993).

Comments There are two elements tentatively assigned to *Argentinosaurus* that have not been formally described, and their assignment is not clear (see Carballido et al. 2017 Supplementary Information). Both elements are femora, one 'complete' (partially reconstructed with plaster) with a length of 2.5 m and the other (MLP-DP 46-VIII-21–3) only composed by the diaphysis (Bonaparte 1996; Mazzetta et al. 2004). Estimates based on these elements place *Argentinosaurus* as the largest titanosaurian (Mazzetta et al. 2004; Benson et al. 2014; Paul 2019); however, none of the femora can be confidentially referred to this taxon (Carballido et al. 2017). Considering that the diagnosis of the taxon is that of 1993 and after almost three decades of superbly new information about titanosaurian sauropods, a complete revision of its anatomy and diagnosis is needed (Fig. 4).

Patagotitan Carballido, Pol, Otero, Cerda, Salgado, Garrido, Ramezani, Cúneo, and Krause 2017

Patagotitan mayorum Carballido, Pol, Otero, Cerda, Salgado, Garrido, Ramezani, Cúneo, and Krause 2017 (Fig. 4a–e)

Holotype MPEF-PV 3400 is an individual preserving an anterior and two middle cervical vertebrae, three anterior, two middle and two posterior dorsal vertebrae, six anterior caudal vertebrae, three chevrons, dorsal ribs, both sternal plates, right scapula and coracoid, both pubes, and both femora.

Referred specimens MPEF-PV 3399 is the most complete paratype and corresponds to a partially associated specimen composed of six posterior cervical vertebrae, one anterior, one middle and two posterior dorsal vertebrae, one anterior and sixteen posterior caudal vertebrae, ribs and chevrons, left ulna and radius, both ischia, left pubes, and one left femur. Additional paratypic specimens include: an isolated tooth (MPEF-PV 3372), an isolated posterior caudal vertebra (MPEF-PV 3393), two left humeri (MPEF-PV 3395, 3396) and one right humerus (MPEF-PV 3397), one left femur (MPEF-PV 3375) and one right femur (MPEF-PV 3394), and two partially preserved fibulae (MPEF-PV 3391, 3392).

Horizon, Age, and Locality Cerro Barcino Formation, Cañadón Asfalto Basin, Upper Albian (Lower Cretaceous; Carballido et al. 2017; Krause et al. 2020), 'La Flecha' ranch, Chubut Province, Argentina.

Revised diagnosis It is diagnosed by the following association of autapomorphies: (1) anterior dorsal vertebrae (D1–D3) with vertical prezygodiapophyseal lamina, due to the elevated position of the prezygapophysis with respect to the diapophysis; (2) anteriormost dorsal vertebrae (D1–D2) with ventral bulge in prespinal lamina; (3) hyposphene-hypantrum restricted to the articulation between D3 and D4; (4) middle and posterior dorsal vertebrae with vertical neural spine (reversal to the condition

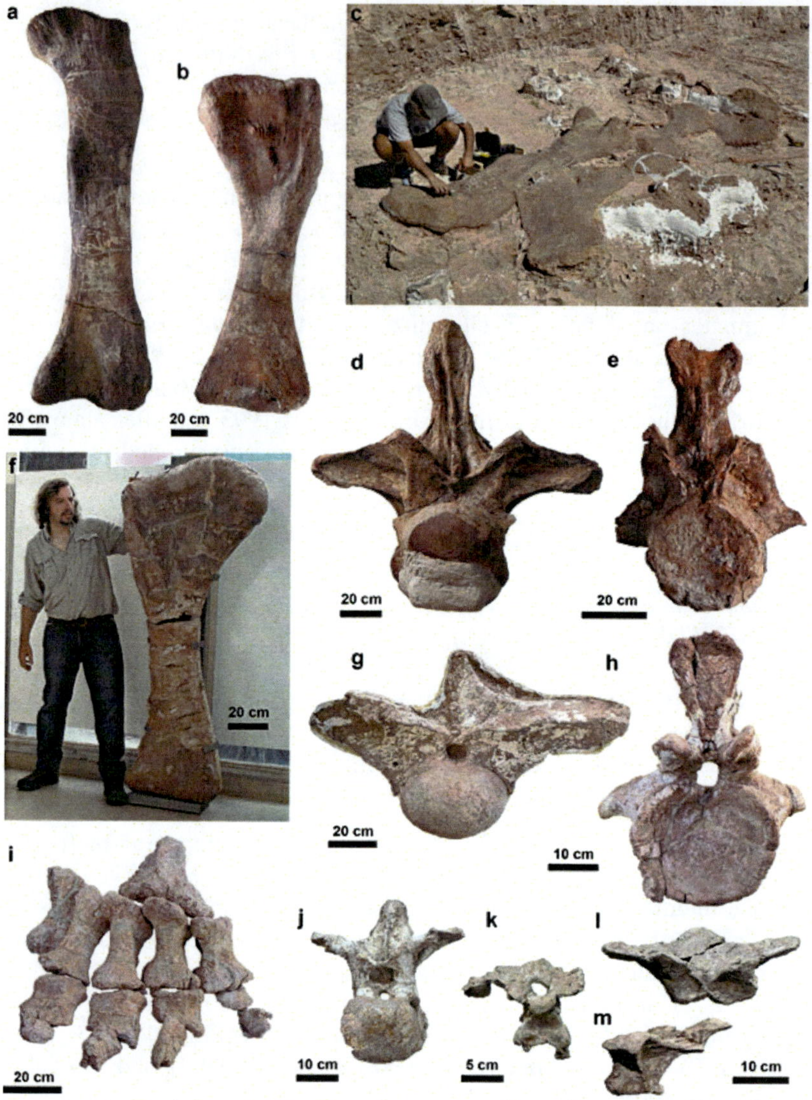

Fig. 4 Select bones of the titanosaurs *Patagotitan*, *Notocolossus*, *Muyelensaurus*, and *Rinconsaurus*. **a** Right femur of *Patagotitan* (MPEF-PV 4400/26) in posterior view. **b** Left humerus of *Patagotitan* (MPEF-PV 3397) in anterior view. **c** Dr. Leonardo Salgado at the *Patagotitan* site. **d** Anterior dorsal vertebra of *Patagotitan* (MPEF-PV 3400/5) in anterior view. **e** First caudal vertebra of *Patagotitan* (MPEF-PV 3400/11) in anterior view. **f** Dr. Bernardo González Riga next to the right humerus of *Notocolossus* (UNCUYO-LD 301). **g** Anterior dorsal vertebra of *Notocolossus* (UNCUYO-LD 301) in anterior view. **h** Anterior caudal vertebra of *Notocolossus* (UNCUYO-LD 301) in anterior view. **i** Articulated right astragalus and complete pes of *Notocolossus* (UNCUYO-LD 302) in anterior view. **j** Posterior dorsal vertebra of *Muyelensaurus* (MRS-Pv 68) in anterior view. **k** Braincase of *Muyelensaurus* (MRS-Pv 207) in occipital view. **l** Anterior–middle dorsal neural arch of *Rinconsaurus* (MAU-PV-CRS 05–3) in anterior view. **m** Anterior–middle dorsal neural arch of *Rinconsaurus* (MAU-PV-CRS 05–3) in lateral view

of non-titanosauriforms, convergently acquired in some derived titanosaurs); (5) first caudal vertebra with flat anterior and convex posterior articular surfaces; (6) anterior caudal vertebrae with neural spines four to six times transversally wider than anteroposteriorly long; (7) anterior caudal neural spines incipiently bifid with anteriorly directed tips; (8) lateral surface of the scapular blade with two divergent crests; (9) anterior surface of proximal humerus with a mediolaterally elongated and prominent muscle scar, paired with a smaller scar, placed below the former; (10) combined well-developed bulges for muscles supracoracoideus/deltoideus clavicularis and latissimus dorsi on the deltopectoral crest and posterolateral to the latter, respectively; (11) ischium with well-developed, sharp ridge projecting from the ischial tuberosity to the distal blade; and (12) coarsely rugose and straight lateral edge on distal femur (Otero et al. 2020).

Comments *Patagotitan* is the best-represented giant titanosaurian (by at least six specimens in a monospecific assemblage), and it is one of the largest dinosaurs in the word (Carballido et al. 2017; Paul 2019; Otero et al. 2020). In all phylogenetic analyses, *Patagotitan* is nested within Lognkosauria, a diverse clade that includes most of the largest titanosaurs known so far (Carballido et al. 2017; González Riga et al. 2018, 2019).

Notocolossus González Riga, Lamanna, Ortiz David, Calvo, and Coria 2016

Notocolossus gonzalezparejasi González Riga, Lamanna, Ortiz David, Calvo, and Coria 2016 (Fig. 4f–i)

Holotype UNCUYO-LD 301 is an associated partial skeleton of a very large individual consisting of an anterior dorsal vertebra, an anterior caudal vertebra, the right humerus, and the proximal end of the left pubis.

Referred Specimen UNCUYO-LD 302 is an associated partial skeleton of a second smaller-bodied individual. This includes an articulated anterior caudal series, consisting of seven incomplete vertebrae and hemal arches and the complete and articulated right astragalus and pes.

Horizon, Age, and Locality The specimens were found in the Plottier Formation, Neuquén Basin, upper Coniacian–lower Santonian (Late Cretaceous), in Cerro Guillermo, Malargüe Department, Mendoza Province, Argentina.

Diagnosis It is diagnosed by the following association of characters (autapomorphies are marked by an asterisk): (1)* anterior dorsal vertebra with parapophyseal centrodiapophyseal fossa subdivided by two 'accessory' laminae (one subvertical and visible in anterior and lateral views, the other anterodorsally oriented and visible only in lateral view); (2)* anterior caudal vertebrae with laminae that converge ventrally on the anterior surface of the neural spine, not reaching the prezygapophyses and forming a 'V-shaped' conformation in anterior view; humerus with: (3)* greatly expanded proximomedial process, the proximal apex of which lies well medial to the humeral midshaft; (4)* proportionally wide proximal end (proximal width: midshaft width ≈ 2.9); and (5)* proximolaterally–distomedially oriented ridge bounding the distal edge of 'coracobrachialis fossa'; and pes with: (6)* metatarsal I with proximal dorsoventral diameter greater than the proximodistal length of the bone; (7)* relatively short metatarsal III (only 1.2 times the length of metatarsal I); (8)* proximal phalanges more than half as wide as their corresponding metatarsals are long; and (9)* pedal unguals reduced, rugose, and distally truncated; anterior caudal vertebrae centra with (10) deeply concave anterior articular cotyles and strongly convex posterior articular condyles; (11) circular anterior articular surfaces and slightly quadrangular posterior articular surfaces; (12) anteroposteriorly concave lateral surfaces; (13) multiple vascular foramina on the lateral surfaces, ventral to the transverse processes; and (14) anteroposteriorly narrow, slightly concave ventral surfaces; transverse processes that are (15) robust, elongate, and posteroventrally directed, nearly reaching the anteroposterior level of the posterior condyle of the centrum; (16) wide and rounded at their lateral ends; and (17) ornamented by longitudinal ridges on their anteroventral margins at the approximate midlength of the process; and (18) neural arches that are anteriorly placed. *Notocolossus* also exhibits the following distinctive morphologies in the appendicular skeleton: (19) humerus with markedly asymmetrical proximal margin in anterior view (nearly straight laterally but strongly expanded and rounded proximomedially); metatarsal V (20) 90 percent the length of metatarsal IV; and (21) longer than metatarsal I; and (22) pedal phalangeal formula 2–2–2–2–0, with digits I–III bearing unguals (González Riga et al. 2016).

Comments *Notocolossus* is one of the largest land animals ever found, and it has the largest humerus discovered for any titanosaurians (González Riga et al. 2016; Paul 2019). Furthermore, this taxon is the largest titanosaurian that preserves a complete and articulated pes, which has a unique structure among titanosaurian sauropods. In contrast to other taxa, *Notocolossus* has a 'short-footed' pedal morphotype characterized by a massive structure with very short and robust metatarsals of almost equal length and small and blunt pedal unguals (González Riga et al. 2016). The discovery of *Notocolossus* has been important to analyze in detail the progressive reduction in the number of phalanges along the line to derived titanosaurs, resulting in the more reduced phalangeal formula of all sauropods.

Rinconsauria Calvo, González Riga, and Porfiri 2007b

Definition A node-based clade is defined as *Muyelensaurus pecheni* Calvo et al. 2007a, b, *Rinconsaurus caudamirus* Calvo and González Riga 2003, their most common ancestor, and all of its descendants (Calvo et al. 2007b).

Muyelensaurus Calvo, González Riga, and Porfiri 2007a, b

Muyelensaurus pecheni Calvo, González Riga, and Porfiri 2007a, b (Fig. 4j,k)

Holotype MRS-Pv 207 is represented by a braincase including partial frontal and parietal, basioccipital, incomplete basipterigoid process, supraoccipital, exoccipital, basisphenoid tubers, orbitosphenoids, and incomplete parasphenoids.

Paratype Several elements are associated with the holotype: a premaxilla (MRS-Pv 59, 60, 337), cervical vertebrae (MRS-Pv 65, 66,121, 122, 204, 230, 232, 229, 279,

391, 392, 420, 422, 428), dorsal vertebrae (MRS-Pv 67, 68, 224, 404, 412, 421), sacrum (MRS-Pv 355), caudal vertebrae (MRS-Pv 135, 137, 164,170, 171, 173, 174, 189, 190, 193, 200, 209, 214, 252, 377, 408), scapula (MRS-Pv 396, 397, 259), sternal plate (MRS-Pv 125), humerus (MRS-Pv 70, 132, 212, 352, 357, 387), ulnae (MRS-Pv 72, 243, 353, 182), radio (MRS-Pv 71, 139) metacarpals (MRS-Pv 127, 152, 157, 181, 198, 231, 235, 236), ischia (MRS-Pv 87, 199, 247, 251), ilia (MRS-Pv 131, 134, 202, 399), pubes (MRS-Pv 88, 154, 204, 371), femora (MRS-Pv 89, 91, 352, 356, 358, 389, 429), tibiae (MRS-Pv 161, 162, 257, 266), fibulae (MRS-Pv 90, 245, 246, 258, 271, 369, 375), astragalus (MRS-Pv 187), metatarsals (MRS-Pv 50–54, 128, 141, 142, 166, 168, 242, 273, 274, 378, 379), and phalanges (MRS-Pv 55, 56, 57, 58, 143, 144–147, 165, 237).

Referred specimens posterior dorsal vertebrae (MRS-Pv 123, 203, 419, and 431).

Horizon, Age and Locality The specimens come from the Plottier Formation (sensu Garrido 2010), Neuquén Basin, upper Coniacian–lower Santonian (Upper Cretaceous), 10 km west to Rincón de los Sauces, Neuquén Province, Argentina.

Diagnosis It is diagnosed by the following association of autapomorphies: (1) basal tubera diverge 70 degree from each other, (2) extensive, thin, and concave medial lamina that joins basal tubera ventrally, (3) basioccipital condyle wider than the proximal portion of the basal tubera, (4) posterior dorsal neural spines with large prespinal lamina reinforced by two small accessory laminae, and (5) distal end of pubic blade rectangular and medially thick (Calvo et al. 2007a).

Comments The holotype and paratypes of *Muyelensaurus* correspond to four adult and one juvenile individuals. The fossil materials were collected during 8 field trips in limestones and fine sandstones of overbank bone assemblage, where the elements were preserved completely disarticulated. In contrast to *Rinconsaurus*, *Muyelensaurus* has anterior dorsals with neural spines posteriorly directed less than 45 degrees with respect to the vertical, posterior dorsals with large and deep infradiapophyseal fossa and ventral face of posterior cervical centra narrow and strongly concave at level of the parapophysis (Calvo et al. 2007a). Preliminary studies (González Riga, pers. obs. 2018; Pérez Moreno et al. 2021) interpret that some caudal vertebrae referred to the paratype show some differences; this could support the hypothesis that some elements belong to other species different to *Muyelensaurus pecheni*.

Rinconsaurus Calvo and González Riga 2003

Rinconsaurus caudamirus Calvo and González Riga 2003 (Fig. 4l, m)

Holotype MAU-PV-CRS-26 is represented by eight articulated anterior and middle caudal vertebrae, six articulated posterior caudal vertebrae, and a right ilium.

Paratype The following bones associated with the holotype are included: MAU-PV-CRS-117, 263, two teeth; MAU-PV-CRS-21, one anterior cervical vertebra; MAU-PV-CRS 02, one middle cervical vertebra; MAU-PV-CRS-03, 04,08, three posterior cervical vertebrae; MAU-PV-CRS-05/3,06/1, 06/2, nine anterior dorsal vertebrae; MAU-PV-CRS-05/1, 05/2, middle dorsal vertebra; MAU-PV-CRS-17,

18,posterior dorsal vertebrae; MAU-PV-CRS-11, 13, 16, 19,dorsal vertebral centra; MAU-PV-CRS-112, dorsal rib; MAU-PV-CRS-41/1, 41/2, two sacral vertebrae; MAU-PV-CRS-22, one caudosacral vertebrae; MAU-PV-CRS-23,25/1, 25/2, 25/3, anterior caudal vertebrae; MAU-PV-CRS-27/1, 27/2, 27/3, 28/1, 28/2, 31, six middle caudal vertebrae; MAU-PV-CRS-29/1, 29/2, 29/3, 30/1, 30/2,30/3, 32, 34, 35, 36/1, 36/2, 38, 39, 40, fourteen posterior caudal vertebrae; MAU-PV-CRS-20, 42, 93, 99, 109, 110,113, chevrons; MAU-PV-CRS-43, right scapula and coracoid;MAU-PV-CRS-46, 103, 104, right and left sternal plates;MAU-PV-CRS-47, right humerus; MAU-PV-CRS-98/1, right metacarpal III; MAU-PV-CRS-98/2, right metacarpal IV;MAU-PV-CRS-98/3, right metacarpal V; MAU-PV-CRS-98/4, right metacarpal II; MAU-PV-CRS-98/5, right metacarpals I; MAU-PV-CRS-96/1, 96/2, 275/1, right and left ilia; MAU-PV-CRS-97/1, 97/2, 100/1, 100/2, right and left pubes; MAU-PV-CRS-94, 101, right ischia; MAU-PVCRS-49, 92, right and left femora; and MAU-PV-CRS-111,metatarsal.

Horizon, Age, and Locality The specimens were found in the Bajo de la Carpa Formation, Neuquén Basin, lower–middle Santonian (sensu Filippi 2015). The fossils come from Cañadón Río Seco locality, 2 km north of Rincón de los Sauces, Neuquén Province, Patagonia, Argentina.

Revised Diagnosis It can be diagnosed on the basis of the following autapomorphies (denoted with an asterisk) and a combination of character states not present in any other sauropod: (1)* Mid-anterior dorsal vertebrae with "stranded parapophyseal" lamina embedded between the paradiapophyseal lamina and the posterior centrodiapophyseal lamina; (2) neural spines in mid-anterior dorsal vertebrae inclined posteriorly more than 60° with respect to the vertical; (3) Posterior dorsal vertebrae without medial spinopostzygapophyseal lamina; (4)* Ratio between the height of the anterior articular surfaces and the posterior articular surfaces of the caudal vertebral centra <0.8 in anterior caudal vertebrae, between 0.8 and 0.9 in middle caudal vertebrae and >0.9 in posterior caudal vertebrae (convergent with, but more exaggerated than in *Muyelensaurus*); (5)* Ratio between the width of the ventral surface at the anterior border and its width at the midline in caudal vertebral centra: >4 in the anterior, between 4 and 2 in the middle, and <2 in the posterior caudal vertebrae; (6) Middle caudal vertebrae with bony processes that dorsally support the articular facets of postzygapophyses; (7) Procoelous posterior caudal vertebral centra with the eventual intercalation of a series of amphicoelous-biconvex or amphicoelous-opisthocoelous biconvex centra (Pérez Moreno et al. 2022).

Comments Calvo and González Riga (2003) suggested that the holotype and paratypes of *Rinconsaurus* correspond to three individuals (two adults and one juvenile). However, in a recent re-evaluation of the taxon Pérez Moreno et al. (2022) determined that all the material represents at least four individuals, but express that there are probably 41 or more. The minimum number of individuals has been calculated from the ilia, including left and right elements of different sizes. Except for the holotype and the series of articulated vertebrae, each preserved element is considered a different specimen, since neither obvious association nor articulation patterns can be established among them (Pérez Moreno et al. 2022). The paratypic specimen MAU-PV-CRS-102, originally identified as a prefrontal by Calvo and González Riga (2003), was excluded from the type series as it has been identified as an ilium of notosuchian crocodylomorph (Filippi et al. 2021). The paratype comprises disarticulated bones except for two unusual posterior caudal vertebrae that have procoelous, amphicoelous and biconvex centra, and the other one with opisthocoelous, biconvex, and procoelous centra. Similar morphological variation of the caudal centra has not been recorded at any other titanosaurian taxa.

Overosaurus Coria, Filippi, Chiappe, García, and Arcucci 2013

Overosaurus paradasorum Coria et al. 2013 (Fig. 3g, h)

Holotype MAU-Pv-CO-439 comprises forty articulated vertebrae, including the last four cervical vertebrae, ten dorsal vertebrae, six sacral vertebrae, and twenty caudal vertebrae, the ribs of the last three cervical vertebrae, six right dorsal ribs (dorsal vertebrae 2–5, 8, 9), five left dorsal ribs (dorsal vertebrae 2–5 and 8); proximal ends of dorsal ribs of both sides of Dorsal vertebrae 9 and 10, the complete right ilium, and fragments of the left ilium.

Horizon, Age, and Locality It was found in the Bajo de la Carpa Formation, Neuquén Basin, Santonian (Upper Cretaceous), Cerro Mesa de los Overo, 50 km southwest of Rincón de los Sauces, Neuquén Province, Argentina (Filippi 2015; Filippi et al. 2018).

Diagnosis It is diagnosed by the following association of characters: (1) posterior cervical vertebrae with long pre- and postzygapophyses that project beyond the anterior and posterior borders of the centrum, respectively; (2) postspinal lamina absent in all dorsal neural spines; (3) wide and massive 9th and 10th caudal centra that are slightly excavated laterally and have relatively flat ventral surfaces; (4) laminar projection on the posterior border of the second and third dorsal ribs; (5) ilium proportionally shorter anteroposteriorly and taller dorsoventrally than in other lithostrotians; (6) and the preacetabular process of the ilium strongly deflected laterally and with a ventrolaterally tapering end (Coria et al. 2013).

Comments *Overosaurus* is relevant from the anatomical point of view, since it is one of the few titanosaurian taxa that has preserved complete and articulated most of the vertebral sequence from the last cervicals to the posterior caudals. Originally, Coria et al. (2013) recovered *Overosaurus* as an Aeolosaurini; however, in recent analyses, it is considered as a sister taxon to Aeolosaurini and as a rinconsaur (Gallina and Otero 2015; Carballido et al. 2017; Hechenleitner et al. 2020).

3 Phylogenetic Considerations

Some titanosaurs mentioned in this chapter have been considered key species to understanding the phylogenetic relationships within the Titanosauria clade. In this upgrade, we follow the phylogenetic definitions published up to now.

The first is *Andesaurus delgadoi* (Calvo and Bonaparte 1991), a taxon from Lower Cenomanian recovered, by definition, as the most basal titanosaurian in phylogenetic analyses (e.g., Salgado et al. 1997; Upchurch 1998; González Riga 2003; Curry Rogers 2005; Gallina and Apesteguía 2011; González Riga and Ortiz David 2014; Tykoski and Fiorillo 2017; Carballido et al. 2017; González Riga et al. 2019). This is because it is used as specifier taxon to define the node-based Titanosauria as '*Andesaurus delgadoi, Saltasaurus loricatus*, their most recent common ancestor, and all its descendants' (Salgado et al. 1997; Wilson and Upchurch 2003) Thus, *Andesaurus* is an extremely important taxon for understanding the evolutionary trends of titanosaurs. Other studies use *Andesaurus* as specifier taxon of other clades like Andesauridae (rejected by Wilson and Upchurch 2003) and Andesauroidea (Salgado 2003), but today both clades are in disuse (Mannion and Calvo 2011; see Chap. 8).

Lognkosauria was erected by Calvo and colleagues as a node-based clade defined as 'the most recent common ancestor of *Mendozasaurus neguyelap* and *Futalognkosaurus dukei* and all descendants' (Calvo et al. 2007a). Originally, this clade was characterized by different characters, standing out the unusual morphology of the cervical vertebrae, which have amazing transversely expanded neural spines formed by the spinodiapophyseal lamina. This feature associated to a deep spinodoapophyseal fossa was first described in *Mendozasaurus* (González Riga 2005) and later in *Futalognkosaurus* (Calvo et al. 2007a) and in less degree of development, in *Quetectsaurus* (González Riga and Ortiz David 2004). Although the size of the neural spine that is wider than the centrum, is only present in *Mendozasaurus* and is considered an autapomorphy of this taxon (González Riga et al. 2018), this cervical morphology that suggests very wide and robust necks in the middle and posterior section is present at least in some lognkosaurians and related forms such as *Bonitasaura* (Gallina 2011) and is an interesting paleobiological aspect to study in detail. *Alamosaurus* is, in fact, a taxon that needs new comparative studies with *Mendozasaurus* and *Futalognkosaurus*. This taxon, from the Late Cretaceous of the USA, also shares with Lognkosauria the presence of laterally expanded posterior cervical neural spines (Tykoski and Fiorillo 2017). In one of the phylogenetic analyses carried out by these authors, which was based on the data matrix of González Riga and Ortiz David (2014), *Alamosaurus* is retrieved as the sister taxon of Lognkosauria.

The analysis of Carballido et al. (2017) recovered *Mendozasaurus, Futalognkosaurus, Quetecsaurus, Puertasaurus, Drusilasaura, Patagotitan*, and *Argentinosaurus,* as lognkosaurs, and the study of González Riga et al. (2019) includes *Mendozasaurus, Futalognkosaurus, Notocolossus, Argentinosaurus, Patagotitan*, and *Puertasaurus.* Although distinct analyses recovered Lognkosauria grouped by different sets of synapomorphies, the monophyletic nature of this clade is always supported.

Rinconsauria clade was erected by Calvo, González Riga, and Porfiri (2007b) when *Muyelensaurus* was described in 2007. This is a node-based clade defined as '*Muyelensaurus pecheni, Rinconsaurus caudamirus,* their most recent common ancestor, and all its descendants'. Rinconsauria comprises relatively small and slender species, and it was recovered close, as the sister taxa, or even including the Aeolosaurini clade (see Calvo et al. 2007b; González Riga et al. 2009; Carballido et al. 2017), although some studies recover both clades more distantly related (Poropat et al. 2016; González Riga et al. 2018; Mannion et al. 2019).

Aeolosaurini underwent some changes in the last few years. Originally, it was defined without any phylogenetic support as 'the most inclusive clade containing *Aeolosaurus rionegrinus* and *Gondwanatitan faustoi*, but not *Saltasaurus loricatus* and *Opisthocoelicaudia skarzynskii*' (Franco Rosas et al. 2004). Recently, Carballido et al. (2017) redefined the clade as '*Aeolosaurus rionegrinus*, *Gondwanatitan faustoi*, their most recent common ancestor, and all its descendants', improving the definition and capturing the traditional use of that clade, and recovered it well-nested within Rinconsauria in their phylogeny.

Within Titanosauria, González Riga et al. (2019) realized that independent analyses supported the existence of a derived titanosaurian lineage distinct from the 'Saltasaurinae line', which they named Colossosauria (Fig. 5). At present, this clade is composed of Lognkosauria and related taxa such as *Bonitasaura* and *Quetecsaurus* and Rinconsauria. Colossosauria is a stem-based clade originally defined by González Riga et al. (2019) as 'the most inclusive clade containing *Mendozasaurus neguyelap*, but not *Saltasaurus loricatus* or *Epachthosaurus sciuttoi*'. An alternative definition, using other specifiers, is proposed by Carballido et al. (see Chap. 8). This clade shows well-stability thorough recent different studies and phylogenetic analyses (Hechenleitner et al. 2020; Páramo et al. 2020; Averianov et al. 2021; Otero et al. 2021; Gallina et al. 2021) and is useful to better understand the lower-level relationships of titanosaurians.

Differences in the phylogenetic studies of González Riga et al. (2019) and Carballido et al. (2017) are related with the place of Lithostrotia clade, after the position of *Malawisaurus*, due to this taxon is included in the definition of the clade (Upchurch et al. 2004). However, this does not affect the existence of the Colossosauria clade, which is well supported in both analyses.

Titanosaurs show a truly extraordinary morphological diversity, and the recognition of the Lognkosauria, Rinconsauria, and Colossosauria clades, with most of their species belonging to the Aptian–Santonian interval, has been a relevant step to clarify the different lineages that evolved in this Cretaceous time interval. In particular, colossosaurs constitute one of the two great titanosaur lineages, and their recognition is essential to clarify phylogenetic relationships within Titanosauria. The other lineage comprises the Lithostrotia (González Riga et al. 2019) or Saltasauroidea (sensu Carballido et al., Chap 8), and its evolutionary history took place mainly at the end of the Cretaceous period (see Chap. 10).

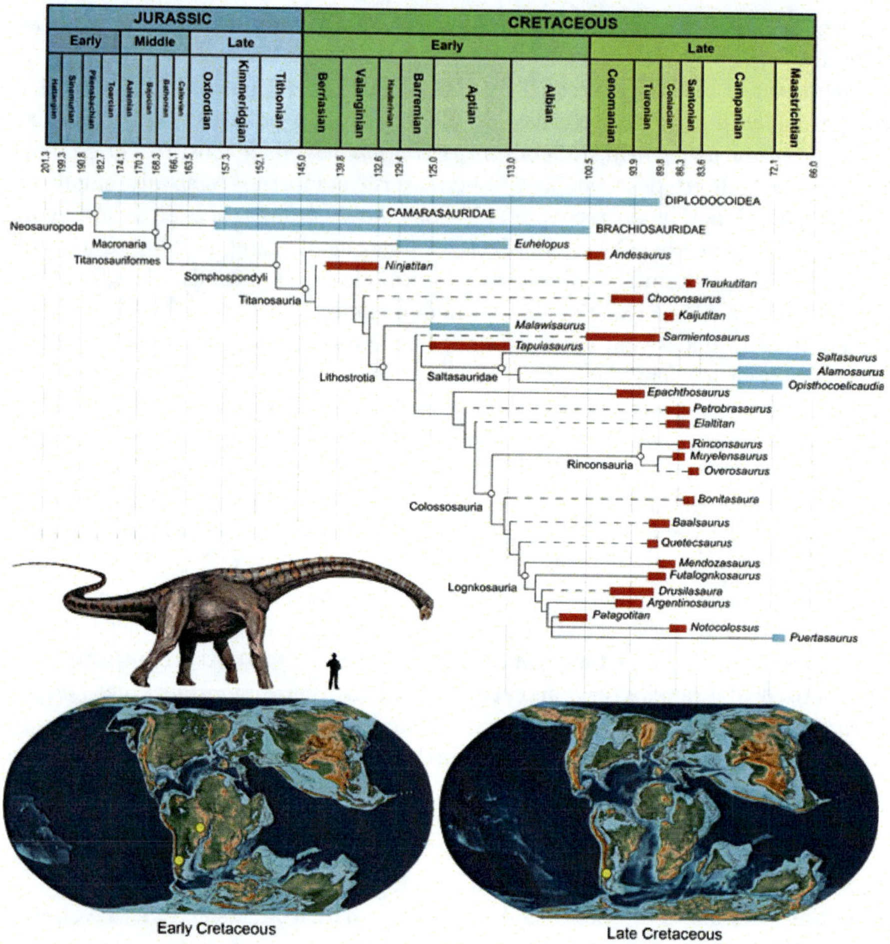

Fig. 5 Time-calibrated phylogenetic tree (based on González Riga et al. 2019) showing the relationships of South America titanosaurs from the Berriasian–Santonian interval. Taxa in dashed lines show the tentative positions retrieved in other phylogenies and/or the original systematic assignment. Geographic distribution of the titanosaurs mentioned in this chapter during the Early and Late Cretaceous, marked with a yellow circle on paleogeographic maps (reconstruction performed with GPlates software, Müller et al. 2018, using the PALEOMAP PaleoAtlas project, Scotese 2016). Life restoration of *Notocolossus* was made by B. González Riga

4 Biogeographical Considerations

The origin and early diversification of titanosaur sauropods or closely related titanosauriforms are still in debate, including the biogeographical distribution of the involved taxa. In this sense, the South American record has an important impact both in the origin and the biogeographical history of the group (Carballido et al. 2015; Carvalho et al. 2017; Gallina et al. 2021).

Some paleobiogeographic considerations regarding the diversification of some in-group clades, that involve several BST mentioned above, can be briefly analyzed here. Although predominantly registered in Southern South America (as portrayed in the systematic list above), the group seems to have a broad latitudinal geographic distribution on the continent prior and during the Aptian–Santonian interval, including fragmentary titanosaur records from Northern Brazil (Medeiros and Schultz 2001, 2002; Lindoso et al. 2013; Carvalho et al. 2017).

Taking into account that during Aptian–Santonian times, and in particular since Late Albian to Latest Santonian, South America tells us a history of progressive continental isolation to other landmasses until the latest Cretaceous continental re-connections (e.g., Hay et al. 1999; Wilf et al. 2013; Fanti 2012), the rise of clades such as Lognkosauria, Rinconsauria, and hence the recently recognized Colossosauria, seems to be good examples of South American endemism (Carballido et al. 2017; González Riga et al. 2018; González Riga et al. 2019; Hechenleitner et al. 2020; but, see Mannion et al. 2019).

Recent quantitative analyses focused on titanosaur sauropod paleobiogeography that shed light mainly in the origin and early geographic distribution of the group (Gorscak and O'Connor 2016; Poropat et al. 2016), including the hypothesis of the South America-Antarctica-Australia connection during the Albian–Turonian that shows that titanosaurs and other sauropods could have migrated through these passages (Ufnar et al. 2004; Hay 2008; Cantrill and Poole 2012; Hasegawa et al. 2012; Poropat et al. 2016). However, both phylogenies of Gorscak and O'Connor (2016) and Poropat et al. (2016) did recover different topologies of the in-group relationships, and adding that the phylogenetic position and constituent taxa of the Lognkosauria and Rinconsauria clades also showed differences in both studies, biogeographic considerations of them are difficult to ascertain.

The comparative time-calibrated approach from the González Riga et al. (2019) phylogeny (Fig. 5) would recognize the origin of Colossosauria at around Late Aptian times. In that context, the colossosaurian clade Lognkosauria seems to have originated in Patagonia no later than Early Albian, with *Patagotitan* as its oldest known member, evolving biogeographically restricted to the same region, in Southern South America (but, see Mannion et al. 2019, Lerzo et al. 2021). On the other hand, Rinconsauria seems to have been originated later in Patagonia, in Cenomanian–Turonian times, and the biogeographical distribution would extend to lower latitudes within South America, if aeolosaurines titanosaurs are included in that clade as recently suggested (Carballido et al. 2017; Hechenleitner et al. 2020; Gallina et al. 2021).

Even though the timing of origination of these titanosaur in-group clades agrees with recent hypotheses that most titanosaurian clades were already widespread by the Early–Middle Cretaceous (Gorscak and O'Connor 2016; Poropat et al. 2016; González Riga et al. 2019) and regardless the considered phylogeny, it is clear that colossosaurian sauropods were predominantly restricted to Southern South America.

At present, there is no strong evidence of colossosaurian titanosaur that remains outside South America, although this assumption would require further revision specially with the purported close affinities of other Gondwanan forms with rinconsaurian titanosaurs (González Riga et al. 2018; Mannion et al. 2019) and the recent interpretation of the Asian taxa *Tengrisaurus* as a Colossosauria (Averianov et al. 2021) and *Dzharatitanis* as a lognkosaur (Lerzo et al. 2021). However, as the origin of the clade Colossosauria seems to predate the final separation between both South America and Africa, at least in their northern region, the presence of representatives of this clade in other land masses cannot be ruled out.

A biogeographic analysis of the mentioned clades considering not only the phylogenetic approach but also the major geographic barriers recognized during Albian–Santonian times in South America (associated with the opening of the Atlantic Ocean such as mountain chains development, desert expanses, epicontinental seas, and marine ingressions; e.g., Barron 1987; Vergani 2005; Spalletti and Franzese 2007; Silvestro and Zubiri 2008) would help, in the future, to understand the role such barriers played in the distribution of colossosaurian titanosaurs.

5 Conclusions

So far, a total of twenty-two titanosaur valid taxa are recorded from the Berriasian–Santonian interval in South America, being one of the richest worldwide. Including basal forms, basal and derived lithostrotians, and the colossosaurs, almost the totality of this record comes from the Argentinean Patagonia, and only one taxon comes from Brazil. Main contribution of the BST to the titanosaur phylogeny falls in, firstly, *Andesaurus delgadoi* as a key taxon for understanding the evolutionary history of the clade and, secondly, the recognition of relatively stable in-group clades such as Colossosauria, Lognkosauria, and Rinconsauria. From a biogeographic point of view, colossosaurian sauropods, including the giant lognkosaurs and the slender rinconsaurs, seem to be good examples of endemism in the Cretaceous of South America, with a predominant distribution in the southernmost region of this Gondwanan land mass.

Acknowledgements We thank the editors A. Otero, J. L. Carballido and D. Pol for inviting us to participate in this book. J. L. Carballido kindly provided photographs of *Patagotitan*. We are grateful to R. Santucci, M. D'Emic, and Editor J.L. Carballido for their comments that greatly improved this chapter. The following institutions and projects are acknowledged for financial and/or logistic support for working with titanosaurs along the years: Fundación Azara and Universidad Maimónides (PAG), Facultad de Ciencias Exactas y Naturales, Universidad Nacional de Cuyo (BGR and LOD); Jurassic Foundation Grant 2008, Agencia PICT 2013-0704, National Geographic/Waitt Grant W645-16 (to PAG), CONICET PIP 0695 and Secretaría de Investigación, Internacionales y Posgrados (UNCuyo) SIIP-UNCUYO 2019-06/M112, SIIP-UNCUYO 2019-M085, SIIP-UNCUYO 2019-M069 (to BGR and LOD).

References

Apesteguía S (2004) *Bonitasaura salgadoi* gen. et sp. nov.: a beaked sauropod from the Late Cretaceous of Patagonia. Naturwissenschaften 91:493–497

Averianov AO, Sizov AV, Skutschas PP (2021) Gondwanan affinities of *Tengrisaurus*, Early Cretaceous titanosaur from Transbaikalia, Russia (Dinosauria, Sauropoda). Cret Res 2021:104731 https://doi.org/10.1016/j.cretres.2020.104731

Barron EJ (1987) Cretaceous plate tectonic reconstructions. Palaeogeogr Palaeocl Palaeoecol 59:3–29

Barrett PM, Upchurch P (2005) Sauropodomorph diversity through time. In: Curry Rogers KA, Wilson JA (eds) The sauropods: evolution and paleobiology. University of California Press, Berkeley and Los Angeles, pp 125–152

Benson RBJ, Campione NE, Carrano MT, Mannion PD, Sullivan C, Upchurch P, Evans DC (2014) Rates of dinosaur body mass evolution indicate 170 million years of sustained ecological innovation on the avian stem lineage. PLoS Biology 12(5): e1001853

Bonaparte JF (1996) Dinosaurios de América del Sur. Museo Argentino de Ciencias Naturales, Buenos Aires

Bonaparte JF, Coria RA (1993) Un nuevo y gigantesco saurópodo titanosaurio de la Formación Río Limay (Albiano-Cenomaniano) de la Provincia del Neuquén, Argentina. Ameghiniana 30:271–282

Bonaparte JF, Powell JE (1980) A continental assemblage of tetrapods from the Upper Cretaceous of Northwestern Argentina (Sauropoda-Coelurosauria-Carnosauria-Aves). Mémoires De La Société Géologique De France 139:19–28

Calvo JO (1999) Dinosaurs and other vertebrates of the Lake Ezequiel Ramos Mexía area, Neuquén-Patagonia, Argentina. Tokyo National Science Museum Monographs 15:13–45

Calvo JO (2014) New fossil remains of *Futalognkosaurus dukei* (Sauropoda, Titanosauria) from the Late Cretaceous of Neuquén, Argentina. In: Rubinstein CV, Marsicano CA, Waisfeld BG (eds) Fourth International Paleontological Congress Meeting, Mendoza, Argentina, 325 September 2014

Calvo JO, Bonaparte JF (1991) *Andesaurus delgadoi* gen. et sp. nov. (Saurischia-Sauropoda), dinosaurio Titanosauridae de la Formación Río Limay (Albiano-Cenomaniano), Neuquén. Argentina. Ameghiniana 28:303–310

Calvo JO, González Riga BJ (2003) *Rinconsaurus caudamirus* gen. et sp nov., a new titanosaurid (Dinosauria, Sauropoda) from the Late Cretaceous of Patagonia, Argentina. Rev Geol Chile 30(2):333–353

Calvo JO, González Riga BJ (2019) *Baalsaurus mansillai* gen. et sp. nov. a new titanosaurian sauropod (Late Cretaceous) from Neuquén, Patagonia, Argentina. An Acad Bras Cienc 91(2):1–11

Calvo JO, Porfiri JD, González Riga BJ, Kellner AWA (2007a) A new Cretaceous terrestrial ecosystem from Gondwana with the description of a new sauropod dinosaur. An Acad Bras Cienc 79:529–541

Calvo JO, González Riga BJ, Porfiri JD (2007b) A new titanosaur sauropod from the Late Cretaceous of Neuquén, Patagonia, Argentina. Arqu Mus Nac 65:485–504

Cantrill DJ, Poole I (2012) The Vegetation of Antarctica through Geological Time. Cambridge University Press

Carballido JL, Pol D, Parra Ruge ML, Padilla Bernal S, Páramo-Fonseca ME, Etayo-Serna F (2015) A new Early Cretaceous brachiosaurid (Dinosauria, Neosauropoda) from northwestern Gondwana (Villa de Leiva, Colombia). J Vertebr Paleontol 35(5): e980505

Carballido JL, Pol D, Otero A, Cerda IA, Salgado L, Garrido AC, Ramezani J, Cúneo NR, Krause JM (2017) A new giant titanosaur sheds light on body mass evolution among sauropod dinosaurs. P Roy Soc B 284:20171219

Carvalho I de S, Salgado L, Lindoso RM, Araújo-Júnior HI, Nogueira FCC, Soares JA (2017) A new basal titanosaur (Dinosauria, Sauropoda) from the Lower Cretaceous of Brazil. J S Am Earth Sci 75:74–84

Casal G, Ibiricu L (2010) Assignable material to *Epachthosaurus* Powell, 1990 (Sauropoda: Titanosauria), from the Bajo Barreal Formation, Upper Cretaceous, Chubut, Argentina. Rev Bras Paleontolog 13:247–256

Casal GA, Allard JO, Foix N (2015) Análisis estratigráfico y paleontológico de afloramientos del Cretácico Superior en la cuenca del Golfo San Jorge: propuesta de nueva unidad litoestratigráfica para el Grupo Chubut. Rev Asoc Geol Arg 72:81–99

Casal GA, Martínez RD, Luna M, Ibiricu L (2016) Ordenamiento y caracterización faunística del Cretacico Superior del Grupo Chubut, Cuenca del Golfo San Jorge. Argentina. Rev Bras Paleontolog 19(1):53–70

Chure D, Britt BB, Whitlock JA, Wilson JA (2010) First complete sauropod dinosaur skull from the Cretaceous of the Americas and the evolution of sauropod dentition. Naturwissenschaften 97:379–391

Coria RA, Salgado L (2005) Mid-Cretaceous turnover of saurischian communities: evidence from the Neuquén Basin. In: Veiga G, Spalletti L, Howell JA, Schwartz E (eds) The Neuquén Basin: A Case Study in Sequence Stratigraphy and Basin Dynamics. Special Publications of the Geological Society, vol 252. London, pp 317–327

Coria RA, Filippi LS, Chiappe LM, García R, Arcucci AB (2013) *Overosaurus paradasorum* gen. et sp. nov., a new sauropod dinosaur (Titanosauria: Lithostrotia) from the Late Cretaceous of Neuquén, Patagonia, Argentina. Zootaxa 3683(4):357–376

Curry Rogers KA (2005) Titanosauria: a phylogenetic overview. In: Curry Rogers KA, Wilson JA (eds) The sauropods: evolution and paleobiology. University of California Press, Berkeley and Los Angeles, pp 50–103

D'Emic (2012) The early evolution of titanosauriform sauropod dinosaurs. Zool J Linn Soc 166:624–671 https://doi.org/10.1111/j.1096-3642.2012.00853.x

Egerton VM (2011) The geology, paleontology and paleoecology of the Cerro Fortaleza Formation, Patagonia (Argentina). Drexel University, USA. Thesis unpublished 275pp

Fanti F (2012) Cretaceous continental bridges, insularity, and vicariance in the southern hemisphere: which route did dinosaurs take? In: Talent JA (ed) Earth and Life: global diversity, extinction intervals and biogeographic perturbations through time. Springer, New York, pp 883–911

Filippi LS (2015) Los dinosaurios Sauropoda del Cretácico Superior del Norte de la Cuenca Neuquina, Patagonia Argentina. Bol Inst Fisiogr Geol 85:19–28

Filippi LS, Canudo JI, Salgado L, Garrido A, García R, Cerda IA, Otero A (2011) A new sauropod: *Petrobrasaurus puestohernandezi* gen. et sp. nov. from the Upper Cretaceous (Plottier Formation) of Rincón de los Sauces, Neuquén (Argentina). Geol Acta 9(1):1–12

Filippi LS, Barrios F, Garrido AC (2018) A new peirosaurid from the Bajo de la Carpa Formation (Upper Cretaceous, Santonian) of Cerro Overo, Neuquén, Argentina. Cret Res 83:75–83

Filippi LS, Salgado L, Garrido AC (2019) A new giant basal titanosaur sauropod in the Upper Cretaceous (Coniacian) of the Neuquén Basin, Argentina. Cret Res 100:61–81

Filippi L, Juarez Valieri R, Barrios F (2021) The prefrontal of Rinconsaurus caudamirus (Sauropoda, Titanosauria) as a crocodyliform ilium. Cret Res 125:e104852

Franco Rosas AC, Salgado L, Rosas CF, Carvalho IS (2004) Nuevos materiales de titanosaurios (Sauropoda) en el Cretácico Superior de Mato Grosso, Brasil. Rev Bras Paleontolog 7:329–336

Gallina PA (2011) Notes on the axial skeleton of the titanosaur *Bonitasaura salgadoi* (Dinosauria-Sauropoda). An Acad Bras Cienc 83:235–245

Gallina PA, Apesteguía S (2011) Cranial anatomy and phylogenetic position of the titanosaurian sauropod *Bonitasaura salgadoi*. Acta Pal Pol 56:45–60

Gallina PA, Apesteguía S (2015) Postcranial anatomy of *Bonitasaura salgadoi* (Sauropoda: Titanosauria) from the Late Cretaceous of Patagonia. J Vertebr Paleontol 35(3):e924957

Gallina PA, Canale JI, Carballido JL (2021) The earliest known titanosaur sauropod dinosaur. Ameghiniana 58(1):35–51

Garrido AC (2010) Estratigrafía del Grupo Neuquén: Cretácico superior de Cuenca Neuquina: nueva propuesta de ordenamiento litoestratigráfico. Rev Mus Argentino Cienc Nat Ns 12(2):121–177

González Riga BJ (2003) A new titanosaur (Dinosauria, Sauropoda) from the Upper Cretaceous of Mendoza Province, Argentina. Ameghiniana 40:155–172

González Riga BJ (2005) Nuevos restos fósiles de Mendozasaurus neguyelap (Sauropoda, Titanosauria) del Cretácico tardío de Mendoza, Argentina. Ameghiniana 42:535–548

González Riga BJ, Astini RA (2007) Preservation of large titanosaur sauropods in overbank fluvial facies: A case study in the Cretaceous of Argentina. J South Am Earth Sci 23:290–303

González Riga BJ, Ortiz David L (2014) A new titanosaur (Dinosauria, Sauropoda) from the Upper Cretaceous (Cerro Lisandro Formation) of Mendoza Province, Argentina. Ameghiniana 51:3–25

González Riga BJ, Lamanna M (2014) Exceptionally-preserved titanosaurs from Mendoza Province, Argentina, provide new data on sauropod pedal evolution and diversity. In: Rubinstein CV, Marsicano CA, Waisfeld BG (eds) Fourth International Paleontological Congress Meeting, Mendoza, Argentina

González Riga BJ, Previtera E, Pirrone CA (2009) *Malarguesaurus florenciae* gen. et sp. nov., a new titanosauriform (Dinosauria, Sauropoda) from the Upper Cretaceous of Mendoza, Argentina. Cret Res 30(1):135–148

González Riga BJ, Lamanna M, Ortiz David L, Calvo J, Coria J (2016) A gigantic new dinosaur from Argentina and the evolution of the sauropod hind foot. Sci Rep 6:19165

González Riga BJ, Mannion P, Poropat S, Ortiz David L, Coria J (2018) Osteology of the Late Cretaceous Argentinean sauropod dinosaur *Mendozasaurus neguyelap*: implications for basal titanosaur relationships. Zool J Linn Soc 184:136–181

González Riga BJ, Lamanna MC, Otero A, Ortiz David LD, Kellner AWA, Ibiricu LM (2019) An overview of the appendicular skeletal anatomy of South American titanosaurian sauropods, with definition of a newly recognized clade. An Acad Bras Cienc 91(2):1–42

Gorscak E, O'Connor PM (2016) Time-calibrated models support congruency between Cretaceous continental rifting and titanosaurian evolutionary history. Biol Lett 12:20151047

Gorscak E, O'Connor PM (2019) A new African Titanosaurian Sauropod Dinosaur from the middle Cretaceous Galula Formation (Mtuka Member), Rukwa Rift Basin, Southwestern Tanzania. PLoS ONE 14(2): e0211412

Hasegawa H, Tada R, Jiang X, Suganuma Y, Imsamut S, Charusiri P, Ichinnorov N, Khand Y (2012) Drastic shrinking of the Hadley circulation during the mid-Cretaceous Supergreenhouse. Clim past 8:1323–1337

Haughton SH (1928) On some remains from the Dinosaur Beds of Nyasaland. Trans Roy Soc S A 16:67–75

Hay WW (2008) Evolving ideas about the Cretaceous climate and ocean circulation. Cret Res 29:725–753

Hay W, De Conto R, Wold C, Wilson K, Voigt S, Schulz M, Wold A, Dullo W-C, Ronov A, Balukhovsky A, Soding E (1999) Alternative global Cretaceous paleogeography. In: Barrera E, Johnson C (eds) Evolution of the Cretaceous ocean-climate system. Geological Society of America Special Paper, vol 332. Geological Society of America, Boulder, pp 1–47

Hechenleitner EM, Leuzinger L, Martinelli AG, Rocher S, Fiorelli LE, Taborda JRA, Salgado L (2020) Two Late Cretaceous sauropods reveal titanosaurian dispersal across South America. Commun Biol 3:622

Huene F von (1929) Los Saurísquios y Ornitísquios del Cretácico Argentino. An Mus La Plata, 2nd serie 3:1–196.

Huene F von (1931) Verschiedene Mesozoische wirbeltierreste aus Südamerika. Neues Jb Miner Geol P A 66:181–198

Juárez Valieri RD, Calvo JO (2011) Revision of MUCPv 204, a Senonian Basal Titanosaur from Northern Patagonia. In: Calvo JO, Porfiri JD, González Riga BJ, Dos Santos D (eds) Paleontología y dinosaurios desde América Latina. EDIUNC, Neuquén, p 143–152

Krause JM, Ramezani J, Umazano AM, Pol D, Carballido JL, Sterli J, Puerta P, Cúneo RN, Bellosi ES (2020) High-resolution chronostratigraphy of the Cerro Barcino Formation (Patagonia): paleobiologic implications for the mid-Cretaceous dinosaur-rich fauna of South America. Gondwana Res 80:33–49

Lacovara KJ, Lamanna MC, Ibiricu LM, Poole JC, Schroeter ER, Ullmann PV, Voegele KK, Boles ZM, Carter AM, Fowler EK, Egerton VM, Moyer AE, Coughenour CL, Schein JP, Harris JD, Martínez RD, Novas FE (2014) A gigantic, exceptionally complete titanosaurian sauropod dinosaur from southern Patagonia. Argentina. Sci Rep 4:6196

Lamanna M, Martínez R, Smith J (2002) A definitive abelisaurid theropod dinosaur from the Early Late Cretaceous of Patagonia. J Vertebr Paleontol 22(1):58–69

Leanza HA, Apesteguía S, Novas FE, De la Fuente MS (2004) Cretaceous terrestrial beds from the Neuquén basin (Argentina) and their tetrapod assemblages. Cret Res 25:1–96

Lerzo LN, Carballido JL, Gallina PA (2021) Rebbachisaurid sauropods in Asia? A re-evaluation of the phylogenetic position of Dzharatitanis kingi from the Late Cretaceous of Uzbekistan. PeAPA 21(1):18–27

Lindoso RM, Marinho TS, Santucci RM, Medeiros MA, Carvalho IS (2013) A titanosaur (Dinosauria: Sauropoda) osteoderm from the Alcântara Formation (Cenomanian), São Luís Basin, Northeastern Brazil. Cret Res 45:43–48

Lydekker R (1893) The dinosaurs of Patagonia. An Mus La Plata 2:1–14

Mannion PD, Calvo JO (2011) Anatomy of the basal titanosaur (Dinosauria, Sauropoda) *Andesaurus delgadoi* from the mid-Cretaceous (Albian–early Cenomanian) Río Limay Formation, Neuquén Province, Argentina: implications for titanosaur systematics. Zool J Lin Soc 163:155–181

Mannion PD, Otero A (2012) A reappraisal of the Late Cretaceous Argentinean sauropod dinosaur *Argyrosaurus superbus*, with a description of a new titanosaur genus. J Vertebr Paleontol 32(3):614–638

Mannion PD, Upchurch P, Barnes RN, Mateus O (2013) Osteology of the Late Jurassic Portuguese sauropod dinosaur *Lusotitan atalaiensis* (Macronaria) and the evolutionary history of basal titanosauriforms. Zool J Lin Soc 168:98–206

Mannion PD, Upchurch P, Jin X, Zheng W (2019) New information on the Cretaceous sauropod dinosaurs of Zhejiang Province, China: impact on Laurasian titanosauriform phylogeny and biogeography. R Soc Open Sci 6:191057

Marsh OC (1878) Principal characters of American Jurassic dinosaurs, Part I. Am J Sci (Series 3) 16:411–416

Martínez R, Giménez O, Rodríguez J, Luna M, Lamanna M (2004) An articulated specimen of the basal Titanosaurian (Dinosauria: Saurópoda) *Epachthosaurus sciuttoi* from the Early Late Cretaceous Bajo Barreal Formation of Chubut Province. Argentina. J Vertebr Paleontol 24(1):107–120

Martínez R, Lamanna M, Novas FE, Ridgely RC, Casal GA, Martínez JE, Vita JR, Witmer LM (2016) A Basal Lithostrotian Titanosaur (Dinosauria: Sauropoda) with a Complete Skull: Implications for the Evolution and Paleobiology of Titanosauria. PLoS ONE 11(4):e0151661

Mazzetta GV, Christiansen P, Fariña RA (2004) Giants and Bizarres: Body Size of Some Southern South American Cretaceous Dinosaurs. Hist Biol 16(2–4):71–83

Medeiros MA, Schultz CL (2001) Uma paleocomunidade de vertebrados do Cretáceo Médio, Bacia de São Luís. In: Rossetti DF, Góes AM, Truckenbrodt W (eds) O Cretáceo na Bacia de São Luís-Grajaú, Coleção Friedrich Katzer. Museu Paraense Emílio Goeldi, Belém, p 209–221

Medeiros MA, Schultz CL (2002) A fauna dinossauriana da Laje do Coringa, Cretáceo médio, Bacia de São Luís. Arqu Mus Nac 60:155–162

Müller RD, Cannon J, Qin X, Watson RJ, Gurnis M, Williams S, Pfaffelmoser T, Seton M, Russell S, Zahirovic S (2018) GPlates: building a virtual Earth through deep time. Geochem Geophy Geosy 19(7):2243–2261

Navarrete C, Casal GA, Martínez RD (2011) *Drusilasaura deseadensis* gen. et sp. nov., un nuevo titanosaurio (Dinosauria-Sauropoda) de la Formación Bajo Barreal, Cretácico Superior del Norte de Santa Cruz. Argentina. Rev Bras Paleontolog 14:1–14

Novas FE, Salgado L, Calvo JO, Agnolin F (2005) Giant titanosaur (Dinosauria, Sauropoda) from the Late Cretaceous of Patagonia. Rev Mus Argentino Cienc Nat Ns 7:37–41

Otero A, Canale JI, Haluza A, Calvo JO, Pérez LM, Ochoa J (2006) New remains of *Andesaurus* (Sauropoda-Titanosauria): contribution to the knowledge of its caudal anatomy. Ameghiniana 43(4):50R

Otero A, Carballido JL, Pérez Moreno A (2020) The appendicular osteology of Patagotitan mayorum (Dinosauria, Sauropoda). J Vertebr Paleontol 40(4):e1793158

Otero A, Carballido JL, Salgado L, Canudo JI, Garrido AC (2021) Report of a giant titanosaur sauropod from the Upper Cretaceous of Neuquén Province. Argentina Cret Res 2021:104754. https://doi.org/10.1016/j.cretres.2021.104754

Paul G (2019) Determining the largest known land animal: a critical comparison of differing methods for restoring the volume and mass of extinct animals. Ann Carnegie Mus 85(4):335–358

Páramo A, Mocho P, Ortega F (2020) Three-dimensional analysis of the titanosaurian limb skeleton: implications for systematic analysis. J Iber Geol 46:369–402

Pérez Moreno A, Carballido JL, Otero A, Salgado L, Calvo JO, Filippi LS (2021) Estatus taxonómico de Muyelensaurus pecheni (Dinosauria, Sauropoda) de la Formación Plottier (Cretácico Superior: Coniaciano) Rincón de los Sauces. Jornadas Argentinas de Paleontología de Vertebrados (San Rafael, Argentina), Resúmenes

Pérez Moreno A, Carballido JL, Otero A, Salgado L, Calvo JO (2022) The axial skeleton of Rinconsaurus caudamirus (Sauropoda: Titanosauria from the Late Cretaceous of Patagonia, Argentina. Ameghiniana 59:1–46

Poropat SF, Mannion PD, Upchurch P, Hocknull SA, Kear BP, Kundrát M, Tischler TR, Sloan T, Sinapius GH, Elliott JA, Elliott DA (2016) New Australian sauropods shed light on Cretaceous dinosaur palaeobiogeography. Sci Rep 6:34467

Powell JE (1990) *Epachthosaurus sciuttoi* (gen. et sp. nov.) un dinosaurio saurópodo del Cretácico de Patagonia (Provincia de Chubut, Argentina). In: Actas V Congreso Argentino de Paleontología y Bioestratigrafía, vol 1. Tucumán, pp 123–128

Salgado L (1996) *Pellegrinisaurus powelli* nov. gen. et sp. (Sauropoda, Titanosauridae) from the Upper Cretaceous of Lago Pellegrini, northwestern Patagonia. Argentina Ameghiniana 33:355–365

Salgado L (2003) Should we abandon the name Titanosauridae? Some comments on the taxonomy of titanosaurian sauropods (Dinosauria). Rev Esp Paleontol 18:15–21

Salgado L, Calvo JO (1993) Report of a sauropod with amphiplatyan mid-caudal vertebae from the Late Cretaceous of Neuquén province (Argentina). Ameghiniana 30:215–218

Salgado L, Martínez RD (1993) Relaciones filogeneticas de los titanosauridos basales *Andesaurus delgadoi* y *Epachthosaurus* sp. Ameghiniana 30(3):339

Salgado L, Coria RA, Calvo JO (1997) Evolution of titanosaurid sauropods. I: phylogenetic analysis based on the postcranial evidence. Ameghiniana 34:3–32

Salgado L, Bonaparte JF (2007) Sauropodomorpha. In: Salgado L, Coria RA (eds) Gasparini, Z. Patagonian Mesozoic Reptiles, Indiana University Press, pp 188–288

Sanz JL, Powell JE, Le Loueff J, Martinez R, Pereda Suberbiola XP (1999) Sauropod remains from the Upper Cretaceous of Laño (northcentral Spain). Titanosaur phylogenetic relationships. Estud Mus Cienc Nat Alava 14:235–255

Schroeter ER, Boles ZM, Lacovara KJ (2011) The histology of a massive titanosaur from Argentina and implications for maximum size. J Vertebr Paleontol 31(2):189A

Scotese CR (2016) PALEOMAP PaleoAtlas for GPlates and the PaleoData plotter program. Geol Soc Am, Abstract with Programs 48(5)

Silvestro J, Zubiri M (2008) Convergencia Oblicua: Modelo estructural alternativo para la Dorsal Neuquina (39°S) – Neuquén. Rev Asoc Geol Arg 63(1):49–64

Simón E, Salgado L, Calvo JO (2018) A new titanosaur sauropod from the upper Cretaceous of Patagonia, Neuquén Province. Argentina Ameghiniana 55(1):1–29

Spalletti LA, Franzese J (2007) Mesozoic palaeogeography and palaeoenvironmental evolution of Patagonia (southern South America). In: Gasparini Z, Coria RA, Salgado L (eds) Patagonian Mesozoic Reptiles. Indiana University Press, pp 29–49

Tykoski RS, Fiorillo AR (2017) An articulated cervical series of *Alamosaurus sanjuanensis* Gilmore, 1922 (Dinosauria, Sauropoda) from Texas: new perspective on the relationships of North America's last giant sauropod. J Syst Palaeontol 15:339–364

Ufnar DF, González LA, Ludvigson GA, Brenner RL, Witzke BJ (2004) Evidence for increased latent heat transport during the Cretaceous (Albian) greenhouse warming. Geology 32:1049–1052

Ullmann PV, Lacovara KJ (2016) Appendicular osteology of *Dreadnoughtus schrani*, a giant titanosaurian (Sauropoda, Titanosauria) from the Late Cretaceous of Patagonia, Argentina. J Vertebr Paleontol 36:e1225303

Upchurch P (1998) The phylogenetic relationships of sauropod dinosaurs. Zool J Linn Soc 124:43–103

Upchurch P, Barrett PM, Dodson P (2004) Sauropoda. In: Weishampel DB, Dodson P, Osmólska H (eds) The Dinosauria, 2nd edn. University of California Press, Berkeley, pp 259–322

Vergani GD (2005) Control estructural de la sedimentación jurásica (Grupo Cuyo) en la Dorsal de Huincul, Cuenca Neuquina, Argentina, Modelo de falla lístrica rampa-plano, invertida. Bol Inf Petrol 1:32–42

Voegele KK, Lamanna MC, Lacovara KJ (2017) Osteology of the dorsal vertebrae of the giant titanosaurian sauropod dinosaur *Dreadnoughtus schrani* from the Late Cretaceous of Argentina. Acta Palaeontol Pol 62(4):667–681

Wilf P, Cúneo NR, Escapa IH, Pol D, Woodburne MO (2013) Splendid and Seldom Isolated: the paleobiogeography of patagonia. Ann Rev Earth Pl Sc 41(1):561–603

Wilson JA, Upchurch P (2003) A revision of *Titanosaurus* Lydekker (Dinosauria-Sauropoda), the first dinosaur genus with a 'Gondwanan' distribution. J Syst Palaeontol 1:125–160

Wilson JA, Pol D, Carvalho AB, Zaher H (2016) The skull of the titanosaur *Tapuiasaurus macedoi* (Dinosauria: Sauropoda), a basal titanosaur from the Lower Cretaceous of Brazil. Zool J Linn Soc 178:611–662

Zaher H, Pol D, Carvalho AB, Nascimento PM, Riccomini C, Larson P, Juarez-Valieri R, Pires-Domingues R, da Silva Jr NJ, de Almeida Campos D (2011) A complete skull of an early cretaceous sauropod and the evolution of advanced titanosaurians. PLoS ONE 6(2):e16663

Last Titans: Titanosaurs From the Campanian–Maastrichtian Age

Rodrigo M. Santucci and Leonardo S. Filippi

Abstract After the extinction of rebbachisaurids during the Cenomanian–Turonian interval, titanosaurs were the only group of sauropods to face the K–Pg event. This same global pattern also holds for the end-Cretaceous (Campanian–Maastrichtian) titanosaur record in South America, where their remains can be found from southern Argentina to Ecuador, with more frequent findings in Argentina and Brazil. In this chapter, we review these fossil findings and the main aspects of the taxonomy, systematics, and paleogeographic implications of this record and briefly discuss the importance of these occurrences for the understanding of titanosaur evolution. The diversity and abundance of end-Cretaceous titanosaur taxa in South America represent about 25% of the known Titanosauria species in the world, which makes them the most common group of large terrestrial herbivores of that time. Cretaceous titanosaurs from South America also vary highly in morphology and size, comprising small to large-sized taxa, for example. Their record mainly consists of appendicular and axial remains, including rare skull material, but also comprises eggs, nests, footprints, and coprolites. In South America, by the end of the Late Cretaceous, titanosaurs were generally represented by more derived titanosaurians that are mainly taxonomically assigned to more derived species within Aeolosaurini and Saltasaurinae.

Keyword Titanosauria · Colossosauria · Aeolosaurini · Saltasaurinae · Late Cretaceous

R. M. Santucci (✉)
University of Brasília, Faculdade UnB Planaltina, Brasília, DF 73300-000, Brazil
e-mail: rodrigoms@unb.br

L. S. Filippi
Museo Municipal Argentino Urquiza, Rincón de los Sauces, Chos Malal 1277, 8319 Neuquén, Argentina

A. Otero et al. (eds.), *South American Sauropodomorph Dinosaurs*, Springer Earth System Sciences, https://doi.org/10.1007/978-3-030-95959-3_10

1 Introduction

The global diversity of sauropod dinosaurs abruptly decreased at the end of the Jurassic and a gradual recovery is witnessed throughout the Cretaceous period (Hunt et al. 1994; Mannion et al. 2011). This increase in sauropod diversity during Cretaceous times is complex, because it encompasses the final demise of rebbachisaurids at the Cenomanian–Turonian and the striking diversification of titanosaurs. Additionally, this global trend is different for some Lower Cretaceous deposits in South America, such as the Berriasian–Valanginian Bajada Colorada Formation from Patagonia that provided evidence that at least some diplodocids survived into the Early Cretaceous and were coeval to the earliest titanosaurs (Gallina et al. 2021). In this scenario, titanosaurs evolved and became the last group of sauropods to walk on the Earth with a number of known species that rival the diversity of non-titanosaur sauropods before the end-Jurassic extinction (Hunt et al. 1994; Mannion et al. 2011).

Titanosaurs also had a wide geographic distribution, since their remains have been found in all continents from sedimentary deposits formed in different environments, especially in inland fluvial systems during the end of the Late Cretaceous (Mannion and Upchurch 2010; Powell 2003). In South America, titanosaurs comprise common dinosaur findings in most of Cretaceous strata (see also Chap. 9). This is particularly true when one considers the Campanian–Maastrichtian titanosaur record that encompasses formally named species mainly from Argentina and Brazil, but also with records in Chile and Ecuador (Fig. 1 and Table 1). Taken solely, the diversity of end-Cretaceous titanosaurs in South America (nearly 30 named species) represents about 25% of the known global titanosaur species.

In this chapter, the main aspects of the taxonomy and phylogenetic implications of Campanian–Maastrichtian titanosaurs from South America are reviewed. We also briefly discuss the paleogeographic aspects of this record and its importance for the understanding of the titanosaur evolution.

Institutional Abbreviations: CPPLIP: Centro de Pesquisas Paleontológicas L. Ivor Price, Minas Gerais, Brazil; **CRILAR**: Paleovertebrate Collection of the Centro Regional de Investigaciones Científicas y Transferencia Tecnológica de La Rioja, Argentina; **MACN**: Museo de Ciencias Naturales Bernardino Rivadavia, Buenos Aires, Argentina; **MAU**: Museo Municipal Argentino Urquiza, Rincón de los Sauces, Neuquén, Argentina; **MBC**: Universidade Federal de Uberlândia, Minas Gerais, Brazil; **MCF**: Museo Carmen Funes, Plaza Huincul, Neuquén, Argentina; **MCS**: Museo Cinco Saltos, Río Negro, Argentina. **MCT**: Museu de Ciências da Terra, Rio de Janeiro, Brazil; **MJG**: Museo Jorge Gerold, Ingeniero Jacobacci, Río Negro, Argentina; **MLP**: Museo de Ciencias Naturales de La Plata, La Plata, Argentina; **MN**: Museu Nacional, Rio de Janeiro, Brazil; **MPM**: Museu de Paleontologia de Marília, São Paulo, Brazil; **MPCA**: Museo Provincial Carlos Ameghino, Cipolletti, Río Negro, Argentina; **MPM-PV**: Museo Padre Molina, Río Gallegos, Santa Cruz, Argentina; **MPMA**: Museu de Paleontologia de Monte Alto, São Paulo, Brazil; **MUCPv**: Museo Universidad Nacional del Comahue, Neuquén, Argentina;

Fig. 1 Map showing the geographical occurrences of formally named Campanian–Maastrichtian titanosaurs from Argentina, Brazil, Chile, and Ecuador

MUGEO: Museu Geológico Valdemar Lefèvre, São Paulo, Brazil; **PVL**: Collection of Vertebrate Paleontology of the Instituto Miguel Lillo, Tucumán, Argentina; **SGO**: Museo Nacional de Historia Natural, Santiago, Chile; **SNGM**: Sección de Paleontología y Estratigrafía, Servicio Nacional de Geología y Minería, Santiago, Chile; **UNPSJB**: Universidad Nacional de la Patagonia, San Juan Bosco, Comodoro Rivadavia, Chubut, Argentina; **YM**: Yamana locality collection, Ecuador.

Table 1 Campanian–Maastrichtian titanosaurs from South America

Taxon	Occurrence	Age	References
TITANOSAURIA Bonaparte and Coria, 1990 A node-based clade defined as *Andesaurus delgadoi* (Calvo and Bonaparte 1991), *Saltasaurus loricatus* (Bonaparte and Powell 1980), their most recent common ancestor and all descendants (Salgado et al. 1997; Wilson and Upchurch 2003)			
Dreadnoughtus (Lacovara et al. 2014)			
D. schrani (Lacovara et al. 2014)	Cerro Fortaleza Fm., Argentina	Campanian–Maastrichtian (LK)	Lacovara et al. (2014), Ullmann and Lacovara (2016), Voegele et al. (2017)
Narambuenatitan (Filippi et al. 2011)			
N. palomoi (Filippi et al. 2011)	Anacleto Fm., Argentina	early–middle Campanian (LK)	Filippi et al. (2011)
TITANOSAURIA *incertae sedis*			
Adamantisaurus (Santucci and Bertini 2006a)			
A. mezzalirai (Santucci and Bertini 2006a)	Adamantina Fm., Brazil	Campanian–early Maastrichtian (LK)	Santucci and Bertini (2006a)
Arackar (Rubilar-Rogers et al. 2021)			
A. ***licanantay*** (Rubilar-Rogers et al. 2021)	Hornitos Fm., Chile	Campanian–Maastrichtian (LK)	Rubilar-Rogers et al. (2021)
Argyrosaurus (Lydekker 1893)			
A. superbus (Lydekker 1893)	Lago Colhué Huapi Fm., Argentina	Campanian–Maastrichtian (LK)	Lydekker (1893), Huene (1929), Powell (2003), Mannion and Otero (2012)
Atacamatitan (Kellner et al. 2011)			
A. chilensis (Kellner et al. 2011)	Tolar Fm., Chile	Cenomanian–Maastrichtian (LK)	Kellner et al. (2011), Rubilar-Rogers et al. (2012)
Austroposeidon (Bandeira et al. 2016)			
A. magnificus (Bandeira et al. 2016)	Adamantina Fm., Brazil	Campanian–early Maastrichtian (LK)	Bandeira et al. (2016)

(continued)

Table 1 (continued)

Taxon	Occurrence	Age	References
Barrosasaurus (Salgado and Coria 2009)			
B. casamiquelai (Salgado and Coria 2009)	Anacleto Fm., Argentina	early–middle Campanian (LK)	Salgado and Coria (2009)
Baurutitan (Kellner et al. 2005)			
B. britoi (Kellner et al. 2005)	Marília Fm., Brazil	Maastrichtian (LK)	Powell (1986, 1987), Kellner et al. (2005)
Bonatitan (Martinelli and Forasiepi 2004)			
B. reigi (Martinelli and Forasiepi 2004)	Allen Fm., Argentina	late Campanian–early Maastrichtian (LK)	Martinelli and Forasiepi (2004), Salgado et al. (2015)
Brasilotitan (Machado et al. 2013)			
B. nemophagus (Machado et al. 2013)	Adamantina Fm., Brazil	Campanian–early Maastrichtian (LK)	Machado et al. (2013)
Laplatasaurus (Huene 1929)			
L. araukanicus (Huene 1929) (=*Titanosaurus araukanicus* (Huene 1929)	Anacleto Fm., Argentina	early–middle Campanian (LK)	Huene (1929), Bonaparte and Gasparini (1979), Powell (2003), Gallina and Otero (2015)
Nullotitan (Novas et al. 2019)			
N. glaciaris (Novas et al. 2019)	Chorrilo Fm., Argentina	late Campanian–early Maastrichtian (LK)	Novas et al. (2019)
Pellegrinisaurus (Salgado 1996)			
P. powelli (Salgado 1996)	Anacleto Fm., Argentina	early–middle Campanian (LK)	Powell (1986), Salgado (1996), Cerda et al. (2021)
Pitekunsaurus (Filippi and Garrido 2008)			
P. macayai (Filippi and Garrido 2008)	Anacleto Fm., Argentina	early–middle Campanian (LK)	Filippi and Garrido (2008)
Trigonosaurus (Campos et al. 2005)			

(continued)

Table 1 (continued)

Taxon	Occurrence	Age	References
T. pricei (Campos et al. 2005)	Marília Fm., Brazil	Maastrichtian (LK)	Powell (1986, 1987), Campos et al. (2005)
Uberabatitan (Salgado and Carvalho 2008)			
U. ribeiroi (Salgado and Carvalho 2008)	Marília Fm., Brazil	Maastrichtian (LK)	Salgado and Carvalho (2008), Silva Jr. et al. (2019)
SALTASAURIDAE (Bonaparte and Powell 1980) A node-based clade defined as *Opisthocoelicaudia skarzynskii* (Borsuk-Białynicka 1977), *Saltasaurus loricatus* (Bonaparte and Powell 1980), their most recent common ancestor, and all its descendants			
SALTASAURINAE (Bonaparte and Powell 1980) A stem clade defined as the most inclusive clade containing *Saltasaurus loricatus* (Bonaparte and Powell 1980), but not *Opisthocoelicaudia skarzynskii* Borsuk-Białynicka 1977			
Yamanasaurus (Apesteguía et al. 2020)			
Y. lojaensis (Apesteguía et al. 2020)	Río Playas Fm., Ecuador	late Maastrichtian (LK)	Apesteguía et al. (2020)
Neuquensaurus (Powell 1992) (=*Microcoelus* Lydekker 1893; *Loricosaurus* Huene 1927)			
N. australis (Lydekker 1893) (=*Titanosaurus australis* Lydekker 1893; *Microcoelus patagonicus* Lydekker 1893; *Titanosaurus robustus* Huene 1927; *Loricosaurus scutatus* Huene 1927)	Anacleto Fm, Argentina	early–middle Campanian (LK)	Lydekker (1893), Bonaparte and Gasparini (1979), Powell (1986), Powell (1992), Powell (2003). Salgado et al. (2005), Otero (2010), D'Emic and Wilson (2011)
Saltasaurus (Bonaparte and Powell 1980)			
S. loricatus (Bonaparte and Powell 1980)	Lecho Fm., Argentina	late Campanian?–early Maastrichtian (LK)	Bonaparte and Powell (1980), Powell (1992), Powell (2003)
Rocasaurus (Salgado and Azpilicueta 2000)			

(continued)

Table 1 (continued)

Taxon	Occurrence	Age	References
R. muniozi (Salgado and Azpilicueta 2000)	Allen Fm., Argentina	Campanian–Maastrichtian (LK)	Salgado and Azpilicueta (2000)
COLOSSOSAURIA González-Riga et al. 2019 The most inclusive clade containing *Patagotitan mayorum* (Carballido et al. 2017), but not *Saltasaurus loricatus* (Bonaparte and Powell 1980)			
Antarctosaurus (Huene 1927)			
A. wichmannianus (Huene 1927)	Anacleto Fm., Argentina	early–middle Campanian (LK)	Huene (1927), Powell (2003)
LOGNKOSAURIA (Calvo et al. 2007a) The most recent common ancestor of *Mendozasaurus neguyelap* González Riga 2005 and *Futalognkosaurus dukei* (Calvo, Porfiri, González Riga and Kellner 2007a) and all its descendants			
Puertasaurus (Novas et al. 2005)			
P. reuili (Novas et al. 2005)	Cerro Fortaleza Fm., Argentina	Campanian (LK)	Novas et al. (2005)
RINCONSAURIA Calvo, González Riga and Porfiri 2007b *Rinconsaurus caudamirus* (Calvo and González Riga 2003), *Muyelensaurus pecheni* (Calvo et al. 2007b), their most recent common ancestor and all its descendants			
Maxakalisaurus (Kellner et al. 2006)			
M. topai (Kellner et al. 2006)	Adamantina Fm., Brazil	Campanian–early Maastrichtian (LK)	Kellner et al. (2006)
Panamericansaurus (Calvo and Porfiri 2010)			
P. schroederi (Calvo and Porfiri 2010)	Allen Fm., Argentina	late Campanian–early Maastrichtian (LK)	Calvo and Porfiri (2010)
AEOLOSAURINI Costa Franco-Rosas et al. 2004; Carballido et al. 2017 *Aeolosaurus rionegrinus* (Powell 1987), *Gondwanatitan faustoi* (Kellner and Azevedo 1999), their most recent common ancestor and all its descendants			
Aeolosaurus (Powell 1987)			
A. rionegrinus (Powell 1987)	Angostura Colorada Fm., Argentina	late Campanian–early Maastrichtian (LK)	Powell (1987, 2003)
A. colhuehuapensis (Casal et al. 2007)	Lago Colhué Huapi Fm., Argentina	Campanian–Maastrichtian (LK)	Casal et al. (2007)

(continued)

Table 1 (continued)

Taxon	Occurrence	Age	References
A. maximus (Santucci and Arruda Campos 2011)	Adamantina Fm., Brazil	Campanian–early Maastrichtian (LK)	Santucci and Arruda Campos (2011)
Bravasaurus (Hechenleitner et al. 2020)			
B. arrierosorum (Hechenleitner et al. 2020)	Ciénaga del Río Huaco Fm., Argentina	Campanian–Maastrichtian (LK)	Hechenleitner et al. (2020)
Gondwanatitan (Kellner and Azevedo 1999)			
G. faustoi (Kellner and Azevedo 1999)	Adamantina Fm., Brazil	Campanian–early Maastrichtian (LK)	Kellner and Azevedo (1999)
Punatitan (Hechenleitner et al. 2020)			
P. coughlini (Hechenleitner et al. 2020)	Ciénaga del Río Huaco Fm., Argentina	Campanian–Maastrichtian (LK)	Hechenleitner et al. (2020)

2 Systematic Paleontology

In this section, the updated diagnoses from literature for each taxon mentioned in this chapter were used. The phylogenetic tree of Carballido et al. (2020) was considered as the main reference for the relationships of the main groups within Titanosauria. The works of Coria et al. (2013), Santucci and Arruda Campos (2011), Hechenleitner et al. (2020) were used for the relationships of Rinconsauria and Aeolosaurini, and the works of Calvo et al. (2007a), González Riga and David (2014), and Gallina and Apesteguía (2011) were used for the relationships within Saltasauridae and Saltasaurinae. The phylogenetic definitions of Carballido et al. (see Chap. 8) were also adopted in our discussions.

Sauropoda Marsh 1878

Titanosauriformes Salgado, Coria and Calvo 1997

Titanosauria Bonaparte and Coria 1993

Definition A node-based clade defined as *Andesaurus delgadoi* Calvo and Bonaparte 1991, *Saltasaurus loricatus* Bonaparte and Powell 1980, their most recent common ancestor and all descendants (Salgado et al. 1997; Wilson and Upchurch 2003).

Dreadnoughtus Lacovara, Lamanna, Ibiricu, Poole, Schroeter, Ullman, Voegele, Boles, Carter, Fowler, Egerton, Moyer, Coughenour, Schein, Harris, Martínez and Novas 2014

Dreadnoughtus schrani Lacovara, Lamanna, Ibiricu, Poole, Schroeter, Ullman, Voegele, Boles, Carter, Fowler, Egerton, Moyer, Coughenour, Schein, Harris, Martínez and Novas 2014 (Figs. 3a, 4b, m, w).

Holotype MPM-PV 1156, a partially articulated skeleton that comprises a fragment of maxilla, a tooth, a posterior cervical vertebra (~ninth), up to eight dorsal vertebrae, a partial sacrum, thirty-two caudal vertebrae, cervical and dorsal ribs, eighteen haemal arches, both sternal plates, left scapula, coracoid, humerus, ulna, radius, a complete pelvis, left femur, both tibiae, left fibula and astragalus, right metatarsals I and II, and the ungual of right pedal digit I.

Referred Specimens MPM-PV 3546, a partially articulated postcranial skeleton of a slightly smaller individual that includes a partial anterior cervical vertebra (~fourth), several dorsal vertebrae and ribs, sacrum, seven caudal vertebrae and five haemal arches, both ilia, pubes, ischia, and the left femur.

Locality and Age Outcrop in the northern Río La Leona Valley between Lago Argentino and Lago Viedma, Santa Cruz Province, southern Argentina (Fig. 1). Holotype and paratype were found in the Cerro Fortaleza Formation, Magallanes-Austral Basin, Campanian–Maastrichtian (Upper Cretaceous, sensu Sickmann et al. 2018) (Fig. 2).

Diagnosis A gigantic titanosaurian sauropod diagnosed by eight autapomorphies (* observable in both known specimens): (1) first caudal centrum with axial ventral keel; (2) anterior caudal neural spines with extensively subdivided pneumatocoel between spinoprezygapophyseal and spinopostzygapophyseal laminae*; (3) anterior caudal neural spines with hypertrophied prespinal and postspinal laminae*; (4) mid-caudal neural spines with triangular anterodorsal process that extends well beyond anterior margin of centrum*; (5) anterior haemal arches with paddle-shaped distal expansion*; (6) posterodorsally anteroventrally oriented ridge on medial surface of anterior end of scapular blade; (7) posteromedial surface of radius concave in proximal view; and (8) distal end of radius nearly square, with subequal anteroposterior and mediolateral dimensions (Lacovara et al. 2014).

Comments Ullmann and Lacovara (2016) and Voegele et al. (2017) mentioned that the paratype (MPM-PV 3546) includes a partial anterior cervical vertebra (~4th), six anterior caudal vertebrae, both ilia, pubes, and ischia, and the deformed left femur. Therefore, it does not match Lacovara et al. (2014), since both works do not include multiple dorsal vertebrae or ribs. They also mentioned that there are six caudal vertebrae (not seven) and did not include the presence of haemal arches.

Together with *Puertasaurus* (Novas et al. 2005), *Dreadnoughtus* is the only giant titanosaur found in the upper part of the Upper Cretaceous deposits of Gondwana. Both taxa were found in the same formation, and although the remains of *Puertasaurus* are very fragmentary, *Dreadnoughtus* represents one of the most complete large titanosaur known to date. The holotype comprises several axial and appendicular elements, including an articulated caudal sequence of seventeen vertebrae with their respective haemal arches. Furthermore, its remains are in better conditions

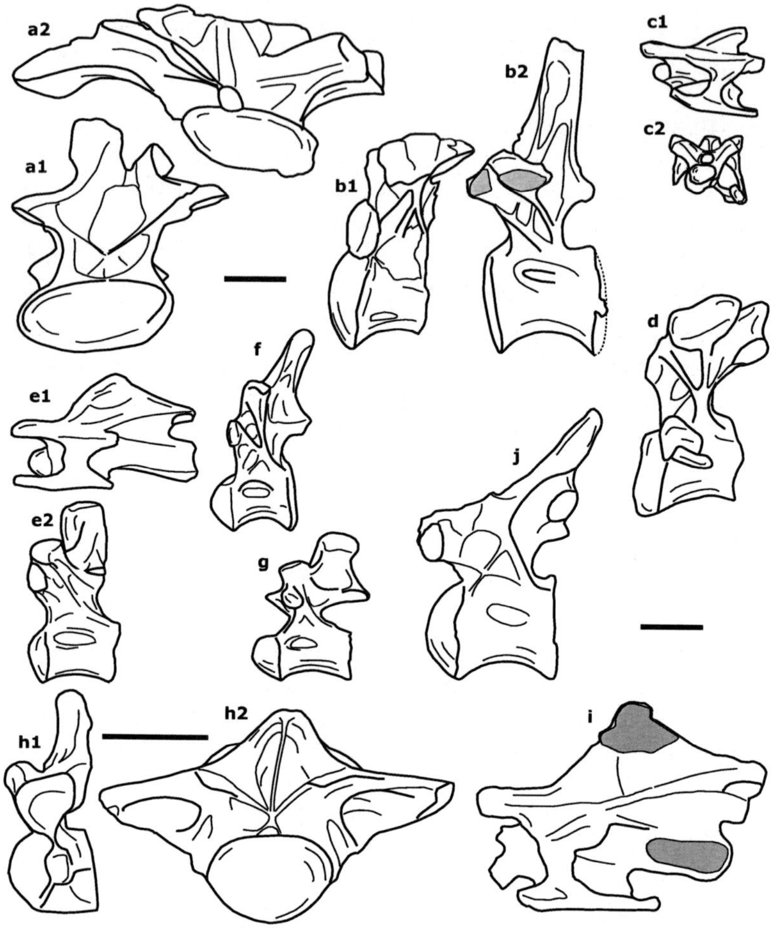

Fig. 2 Cervical and dorsal vertebrae of Campanian–Maastrichtian titanosaurs from South America in left lateral view (unless stated otherwise). **a** *Austroposeidon*, posterior cervical in posterior view (**a1**) and anterior cervical in anterior view (**a2**) (MCT 1628-R, Bandeira et al. 2016). **b** *Barrosasaurus*, anterior dorsal (**b1**) and posterior (**b2**) dorsal in reversed right lateral view (MCF-PVPH-447/3 and MCF-PVPH-447/1, Salgado and Coria 2009). **c** *Brasilotitan*, posterior cervical in left lateral (**c1**) and anterior (**c2**) views (MPM 125R, Machado et al. 2013). **d** *Pitekunsaurus*, anterior dorsal (MAU-Pv-AG-446/9, Filippi and Garrido 2008). **e** *Trigonosaurus*, posterior cervical (**e1**) and posterior dorsal (**e2**) in reversed right lateral view (MCT 1719-R, Campos et al. 2005). **f** *Neuquensaurus*, posterior dorsal in reversed right lateral view (MCS-5/20–23, Salgado et al. 2005). **g** *Saltasaurus*, posterior dorsal in reversed right lateral view (PVL 4017–135, Powell 2003). **h** *Puertasaurus*, anterior dorsal vertebra in posterior (**h1**) and left (**h2**) lateral views (MPM-1002, Novas et al. 2005). **i** *Maxakalisaurus*, posterior cervical in reversed right lateral view (MN 5013-V, Kellner et al. 2006); **j** *Punatitan*, mid-dorsal vertebrae in reversed right lateral view (CRILAR-Pv 614, Hechenleitner et al. 2020). Scale bar equals 50 cm in H and 10 cm in other images

than the exceptionally preserved titanosaur *Futalognkosaurus* (Calvo et al. 2007a, c, see also Chap. 9). Therefore, *Dreadnoughtus* is a particularly important taxon for understanding the anatomy of giant titanosaurs. Histological analyses of the humerus and femur of the holotype (Schroeter et al. 2011; Lacovara et al. 2014) showed that the holotype was not a fully grown individual, at most having achieved histological ontogenetic stage 9 of the 14-stage sequence proposed by Klein and Sander (2008) at its time of death (Ullmann and Lacovara 2016).

Narambuenatitan Filippi, García, and Garrido 2011

Narambuenatitan palomoi Filippi, García, and Garrido 2011 (Figs. 3b, 4a, l)

Holotype MAU-Pv-N-425, is represented by one single incomplete skeleton comprising left premaxilla and maxilla, braincase, left and right quadrates, anterior cervical vertebra, cervical rib fragments, posterior dorsal vertebra, three dorsal ribs, eleven anterior caudal vertebrae, six middle caudal vertebrae, two caudal transverse processes, two fragments of haemal arches, left sternal plate, right coracoid, left humerus, left ulna, left and right pubes, pubic peduncle of the left ilium, proximal fragment of the right ischium, and incomplete left femur.

Locality and Age Narambuena oilfield, Rincón de los Sauces, Northern Neuquén, Argentina (Fig. 1). Neuquén Group (Upper Cretaceous), Subgroup Río Colorado, Anacleto Formation (lower-middle Campanian) (Filippi et al. 2011).

Diagnosis A medium titanosaur characterized by the following autapomorphies: (1) large participation of the frontal to the rostral margin of the supratemporal fossa; (2) absence of ornamentation on the supraorbital or lateral margin of the frontal; (3) presence of a lamina joining each parietal crest with the rostral surface of the supratemporal; (4) pleurocoel dorsal margin defined by a prominent convex bony edge in posterior dorsal vertebrae; (5) distal enlargement of the prespinal lamina in posterior dorsal vertebrae, matching with a lateral enlarged of the neural spine; (6) kidney-shaped prezygapophysis in posterior dorsal vertebrae; (7) presence of numerous pneumatic cavities located posterior to the centroparapophyseal lamina and centrodiapophyseal lamina in posterior dorsal vertebrae; (8) mid-anterior caudal vertebrae with the anterodorsal margin of the neural spine anteriorly inclined; (9) mid-caudal vertebra with a slender prezygapophysis and neural arch (Filippi et al. 2011).

Comments The phylogenetic relationships of *Narambuenatitan palomoi* were analyzed using the data matrix Calvo et al. (2007a). The analysis recovered *Narambuenatitan* as a titanosaur nested in an unresolved trichotomy with *Epachthosaurus* and Eutitanosauria, supported by the presence of a laminar and posterior elongated neural spine in mid-caudal vertebrae. In a recent work about its neuroanatomy (Paulina-Carabajal et al. 2020), *Narambuenatitan* exhibits a remarkable similarity in endocast morphology with an approximately contemporaneous putative saltasaurid from Fox-Amphoux-Métisson (France) (Knoll et al. 2019). New analyses that include a greater number of taxa and characters can clarify the phylogenetic relationships of *Narambuenatitan* in the future.

Titanosauria incertae sedis

Adamantisaurus Santucci and Bertini 2006a

Adamantisaurus mezzalirai Santucci and Bertini 2006a (Fig. 3c1–2)

Holotype MUGEO 1282 and 1295 are represented by one single incomplete skeleton comprising six articulated anterior caudal vertebrae (MUGEO 1282), the first of which probably corresponds to the second caudal vertebra, and two haemal arches (MUGEO 1289 and 1295).

Locality and Age Flórida Paulista, São Paulo State, Brazil (Fig. 1). Bauru Group (Upper Cretaceous), Adamantina Formation, Campanian–early Maastrichtian (Batezelli 2017; Dias et al. 2020). While some proposals based on fossils provided controversial age estimates, recent works based on the development of regional discordances and depositional sequences (Batezelli 2017) and fission track dating on zircon grains (Dias et al. 2020) indicate that the deposition of Bauru Group units is related to the development of the extensive alkaline magmatism in the north/northeast area of the basin.

Diagnosis Titanosaur characterized by the following combination of characteristics: (1) anterior caudal vertebrae with straight or slightly backward-projecting neural spines, with distal end strongly expanded laterally; (2) stout prespinal lamina; (3) pre- and postzygapophyses with wide articular facets; (4) concave postzygapophyseal articular facets on anterior caudal vertebrae (Santucci and Bertini 2006a).

Comments The material was collected by Sérgio Mezzalira from the fluvial deposits of the Adamantina Formation and was originally illustrated and regarded by Mezzalira (1959) as Titanosauridae indet. At the same site, a nearly complete turtle (S. Mezzalira *pers. information*), some isolated titanosaur teeth, and an isolated titanosaur femur were also found. Considering its relatively small size when compared to the anterior caudal series of *Adamantisaurus*, the femur clearly belongs to another individual of smaller size and, because of that, was not assigned to this taxon.

In its original description, *Adamantisaurus* was scored in the data matrix provided by Salgado et al. (1997). However, this analysis returned a large polytomy within Titanosauriformes, except for a monophyletic group composed of *Alamosaurus*, *Neuquensaurus*, and *Saltasaurus* (Santucci and Bertini, 2006a). So far, *Adamantisaurus* were not considered in any other phylogenetic analysis and its phylogenetic relationships within Titanosauria remains uncertain.

Arackar Rubilar-Rogers, Vargas, González Riga, Soto-Acuña, Alarcón-Muñoz, Iriarte-Díaz, Arévalo and Gutstein 2021

Arackar licanantay Rubilar-Rogers, Vargas, González Riga, Soto-Acuña, Alarcón-Muñoz, Iriarte-Díaz, Arévalo and Gutstein 2021 (Fig. 4c, n)

Holotype SNGM-1/1–23, a well-preserved partial postcranial skeleton comprising two cervical centra, two anterior dorsal neural arches, one posterior dorsal neural arch, three dorsal centra, right humerus, left ischium, left femur, and various unidentified fragments.

Locality and Age La Higuera ravine, nearly 75 km south Copiapó, Atacama Region, northern Chile. Hornitos Formation, Campanian–Maastrichtian, Upper Cretaceous (Rubilar-Rogers et al. 2021).

Diagnosis A small sub-adult titanosaur diagnosed by the following association of characteristics (* indicates autapomorphies): anterior dorsal vertebrae with neural arches with (1) spinopostzygapophyseal lamina inclined laterally less than 20° to the horizontal, with an anterodorsal concave (saddle-shaped) face; (2) absence of intra-postzygapophyseal lamina; (3) prezygapophyseal facets well separated from each other and located at half the distance between the diapophysis and the sagittal plane; (4) lateromedially narrow pedicles of the neural arch, as wide as the neural canal. The mid-dorsal vertebrae with neural arches showing (5) anteroposterior distance of the parapophysis-diapophysis longer than the articular surfaces of the pedicle in lateral view; (6) wide and tall centroprezygapophyseal fossa + parapophyseal centroprezygapophyseal fossa extending on the entire anterior faces of the pedicles ∗ ; (7) reduced spinopostzygapophyseal laminae, shorter than the postzygapophyseal facet length ∗ ; (8) absence of postzygodiapophyseal laminae; (9) presence of anterior spinodiapophyseal laminae; (10) neural spines strongly directed posteriorly more than 40° with respect to the vertical; (11) absence of triangular aliform processes in neural spines. The ischium presents (12) a relatively long pubic articulation, greater than the anteroposterior length of the proximal plate peduncle; and (13) a relatively short and narrow posterior process (Rubilar-Rogers et al. 2021).

Comments *Arackar* is a small-sized titanosaur known from a few remains from different portions of the skeleton including axial and appendicular elements from a single individual. So far, together with *Atacamatitan*, it is the only titanosaur formally described from Chile. *Arackar* is considered as a sub-adult individual due to incomplete neurocentral sutures of the axial elements. Its phylogenetic relationships were evaluated using the dataset of Carballido et al. (2017) that showed *Arackar* is well nested within Titanosauria and more related to other derived titanosaurs such as *Isisaurus*, its sister taxon, and *Rapetosaurus* (Rubilar-Rogers et al. 2021).

Argyrosaurus Lydekker 1893

Argyrosaurus superbus Lydekker 1893 (Fig. 4d)

Holotype MLP 77-V-29–1, is composed of a partly articulated left forelimb consisting of the humerus, ulna, radius, probably two carpals (now lost), and all five metacarpals.

Locality and Age Left bank (northwest) of the Río Chico, near Pampa Pelada, northeast of Lago Colhué Huapi, Chubut Province, Argentina (Fig. 1). Chubut Group (Upper Cretaceous), Lago Colhué Huapi Formation (Campanian–Maastrichtian) (Ibiricu et al. 2020).

Emended Diagnosis Mannion and Otero (2012) diagnosed *Argyrosaurus*, assigning it five autapomorphies (marked by an asterisk), as well as a unique combination of character states: (1) medial margin of humerus forms a transversely wide ridge that projects prominently anteriorly*; (2) midshaft of humerus extremely compressed anteroposteriorly (transverse to anteroposterior width ratio = 2.6)*; (3) transverse width of distal end of radius only slightly greater than midshaft width (ratio = 1.3)*; (4) radius is subtriangular in distal view*; 5) presence of ossified carpals; (6) extreme elongation of metacarpals (longest metacarpal to radius length ratio = 0.6)*; (7) metacarpals II and III are the longest elements in metacarpus.

Comments Lydekker (1893) did not provide a diagnosis for *Argyrosaurus,* merely noting several differences with *Neuquensaurus* ('*Titanosaurus*') *australis* that he deemed worthy of generic distinction. Other authors defined the diagnosis based on the robustness of the limbs (Bonaparte and Gasparini 1979). Powell (2003) relied on a series of characters none of which being unique to *Argyrosaurus* because they are present in other titanosaurs. On the other hand, authors such as Upchurch et al. (2004) and Novas (2009) have proposed additional autapomorphies but based on the referenced materials and not on the holotype (Mannion and Otero 2012).

The material assigned to *Argyrosaurus* was illustrated and reviewed by Lydekker (1893), Huene (1929), and Powell (2003) and comprises the holotype and other related remains (PVL 4628/MACN-CH 2017). However, Mannion and Otero (2012) distinguished these remains from *Argyrosaurus* based on an unusual character combination plus one autapomorphy, erecting a new taxon, *Elaltitan lilloi.*

At the moment, the phylogenetic relationships of *A. superbus* are still uncertain, and even though *Argyrosaurus* shows some titanosaur-derived features, most of its characteristics are present in basal titanosaurs. Also, *Argyrosaurus* has two characters that suggest a relatively basal position within Titanosauria (Mannion and Otero 2012), presence of carpal bones (Huene 1929), which are unknown in all other titanosaurs, and metacarpal II and III as the longest metacarpals. Alternatively, Gorscak and O'Connor (2019) recovered it as the sister taxa to *Trigonosaurus* within Aeolosaurini in their Bayesian phylogenetic analysis.

Atacamatitan Kellner, Rubilar-Rogers, Vargas and Suárez 2011

Atacamatitan chilensis Kellner, Rubilar-Rogers, Vargas and Suárez 2011 (Figs. 3d, 4o)

Holotype SGO-PV-961, the specimen includes a right femur, the proximal end of a humerus, two dorsal vertebrae, posterior caudal vertebrae, dorsal ribs and a fragmentary element regarded as part of the sternum, besides other incomplete bones.

Locality and Age Tolar Formation, Upper Cretaceous (Kellner et al. 2011; Rubilar-Rogers and Gutstein 2012; Rubilar-Rogers et al. 2012), near Conchi Viejo City, Atacama Desert, Antofagasta, Chile (Kellner et al. 2011) (Fig. 1).

Diagnosis Titanosaur with the following characteristics (* indicates autapomorphies): (1) centrum of dorsal vertebrae pierced by pleurocoels, which are rounded or elliptical but do not acuminate posteriorly; (2) ventral surface of the dorsal vertebra strongly concave; (3) posterior caudals with a laterally compressed neural spine that bears a blade-like anterior margin*; and (4) femur with the shaft getting gradually narrower until two thirds of its length.

Comments Although *Atacamatitan* was not included in a phylogenetic analysis at the time of its description, it was considered by Kellner et al. (2011) as more basal than Saltasauridae, but more derived than *Malawisaurus*, and not a member of Aeolosaurini. *Atacamatitan chilensis* comprises fragmentary material and was included in only one phylogenetic analysis so far (Rubilar-Rogers and Gutstein 2012). The results are inconclusive but suggest that *Atacamatitan* is a titanosaur more derived than *Malawisaurus* and possibly more related to Saltasauridae. The fragmentary nature of this species precludes further evaluation of its phylogenetic relationships.

Until the discovery of *Atacamatitan*, the inferred age for the Tolar Formation had not been precisely defined. This dinosaur specimen implies a Cretaceous age at least for the outcrops of this formation near the city of Conchi Viejo.

Austroposeidon Bandeira, Simbras, Machado, Campos, Oliveira and Kellner 2016

Austroposeidon magnificus Bandeira, Simbras, Machado, Campos, Oliveira and Kellner 2016 (Fig. 2a1–2)

Holotype MCT 1628-R, consisting of two incomplete posterior cervical vertebrae, one fragment of cervical rib, seven fragments of dorsal vertebrae, and one fragment of sacral rib.

Locality and Age Presidente Prudente City, southeastern São Paulo State, Brazil (Campos and Castro 1978; Bandeira et al. 2016) (Fig. 1). Adamantina Formation (=Presidente Prudente Formation), Campanian–early Maastrichtian (Batezelli 2017; Dias et al. 2020).

Diagnosis *Austroposeidon magnificus* is characterized by the following combination of characteristics (*indicates autapomorphies): (1) *columnar-like centropostzygapophyseal laminae in the 13th cervical vertebrae; (2) *13th cervical vertebra bearing a bifurcated posterior centrodiapophyseal lamina; (3) *first dorsal vertebra with anterior and posterior centrodiapophyseal laminae curved ventrolaterally, with the diapophysis reaching the dorsal margin of the centrum; (4) *posterior dorsal vertebra with the anterior most portion of the spinoprezygapophyseal laminae forked; (5) presence of medial crest placed on the ventral surface of the 13th cervical centrum; (6) presence of a suprapleurocoel lamina limiting the pleurocoel from the centrodiapophyseal fossae in the 13th cervical vertebrae; (7) presence of developed centrodiapophyseal fossa in posterior cervical vertebra; (8) posterior cervical vertebrae with tall neural spines; (9) presence of triangular centropostzygapophyseal fossae around the neural channel in the posterior cervical vertebra; (10) robust spinoprezygapophyseal laminae in the anterior dorsal vertebrae; (11) developed spinodiapophyseal laminae in the anterior dorsal vertebrae; (12) strongly developed postzygaphophysis in the first dorsal vertebra; (13) neural spine of the first dorsal vertebrae in vertical position and anteriorly located; (14) well-developed prespinal lamina in the anterior dorsal vertebrae until the base of the neural spine; (15) diapophyses in the anterior dorsal vertebrae expanded anteroposteriorly and strongly inclined ventrolaterally; (16) presence of well-developed pneumatized camellae tissue in the cervical and dorsal vertebrae; (17) absence of hyposphene-hypantrum articulation in the dorsal vertebra; (18) short and robust cervical ribs (Bandeira et al. 2016).

Comments *Austroposeidon* is known from fragmentary axial remains, mainly from posterior cervical and anterior dorsal vertebrae. Considering the best preserved elements, *Austroposeidon* can be regarded as a large-sized titanosaur. CT scan images revealed its cervical vertebrae have concentrical internal structures which were considered as putative growth marks (Bandeira et al. 2016). Despite its fragmentary nature, *Austroposeidon* was considered to be close to Lognkosauria sharing with this clade the presence of a ventral keel on the posterior cervical centra separating two concavities or fossae (Bandeira et al. 2016). Such character only concerning posterior cervical vertebrae was not considered in subsequent and more comprehensive phylogenetic analyses and, because of that, further evaluation of *Austroposeidon* phylogenetic relationships cannot be made.

Barrosasaurus Salgado and Coria 2009

Barrosasaurus casamiquelai Salgado and Coria 2009 (Fig. 2b1–2)

Holotype Is composed of one anterior dorsal (MCF-PVPH-447/3) and two posterior dorsal vertebrae (MCF-PVPH-447/1 and 2).

Locality and Age Sierra Barrosa, Neuquén Province, Argentina (Fig. 1). Neuquén Group (Upper Cretaceous), Río Colorado Subgroup, Anacleto Formation (lower-middle Campanian) (Salgado and Coria 2009).

Diagnosis Large titanosaur characterized by the following autapomorphies: (1) posterior dorsal vertebrae bearing paired and short spinal laminae, nearly parallel to the prespinal lamina; (2) well-developed anterior spinodiapophyseal laminae in posterior dorsal vertebrae; (3) anterior spinodiapophyseal laminae divided into at least two, and up to three, branches in posteriormost dorsal vertebrae; (4) aliform processes with broad, planar and rugose laterodorsal surfaces; (5) portion of neural spine distal to the aliform process is longer than the portion of the spine that is proximal to the aliform process, at least in the posterior dorsals (Salgado and Coria 2009).

Comments The Sierra Barrosa locality includes extensive outcrops of the Portezuelo, Plottier, Bajo de la Carpa, and Anacleto formations. Abundant terrestrial vertebrates, including saurischian dinosaurs, crocodiles, and turtles have been recovered from multiple stratigraphic levels (Coria et al. 2000, 2001, 2004; Coria and Currie 2002). The materials of the titanosaur sauropod *B. casamiquelai* were collected from a single horizon and were deeply embedded in reddish fluvial sandstones, positioned close to one another. Therefore, they were interpreted as belonging to a single individual.

Although the exact phylogenetic relationships of *B. casamiquelai* have not yet been established, Salgado and Coria (2009) considered this species as a Titanosauria based on opisthocoelous dorsal centra, middle and posterior dorsal neural arches that lack fossa between the neural canal and prezygapophyses, eye-shaped pleurocoels in dorsal vertebrae, presence of an accessory posterior centrodiapophyseal lamina in dorsal vertebrae, and absence of hyposphene-hypantrum articulations in dorsal vertebrae.

Baurutitan Kellner, Campos and Trotta 2005

Baurutitan britoi Kellner, Campos and Trotta 2005 (Fig. 3e1–3)

Holotype MCT 1490-R, a nearly complete anterior and mid-caudal sequence comprising the last sacral, a sequence of 18 caudals and 15 haemal arches.

Locality and Age Peirópolis ('Caieira' quarry), Uberaba, Minas Gerais State, Brazil (Powell 1987, 2003; Campos et al. 2005) (Fig. 1). Serra da Galga Member, Marília Formation, Bauru Group, Maastrichtian (Dias-Brito et al. 2001; Dias et al. 2020).

Diagnosis *Baurutitan britoi* is characterized by the following characteristics (* indicates autapomorphies): (1) *strongly pointed laterally directed process intercepting the spinoprezygapophyseal lamina on the caudal 1; (2) *anterolaterally directed spinoprezygapophyseal lamina; (3) first caudal biconvex; (4) developed dorsal tuberosity on the neural arch that turns into a lateral ridge in the mid-caudals; (5) neural spines in the most anterior caudals (1–4) inclined posteriorly, changing to a more vertical to subvertical inclination in caudal 5 to 10; (6) prezygapophyseal tuberosity on the lateral margin of the prezygapophysis of caudals 2–4, which becomes less developed in the posterior elements, smoothing out in the mid-caudals; (7) sub-rectangular outline of anterior articulation of the caudal centrum, becoming more squared toward the end of the tail (Kellner et al. 2005).

Comments The material comprising the holotype of *Baurutitan* was informally referred in the literature as 'Series C from Peirópolis' (e.g., Powell 1987, 2003) and was found only a few meters away from the materials of *Trigonosaurus* (see Kellner et al. 2005, Fig. 3). However, the caudals of *Baurutitan* are quite distinct from the caudals of the paratype of *Trigonosaurus*. For example, the neural spine is shorter and directed backwards in *Baurutitan*.

Kellner et al. (2005) did not provide a phylogenetic analysis about relationships of *Baurutitan*, but this species was included in some recent phylogenetic analyses on titanosaurs as in Carballido et al. (2017), Hechenleitner et al. (2020), and Gorscak and O'Connor (2019), being considered as a basal Titanosauria, a basal Colossosauria or as a member of Saltasauridae, respectively.

Bonatitan Martinelli and Forasiepi 2004

Bonatitan reigi Martinelli and Forasiepi 2004 (Figs. 3f, 4e, p, x)

Holotype MACN-PV- RN 821 (sensu Salgado et al. 2015) comprises a complete neurocranium (Individual C, Salgado et al. 2015).

Referred Material (sensu Salgado et al. 2015; Individual A, B, D, and E) (MACN-PV-RN 821): centrum of a dorsal vertebra, a neural arch of a mid-dorsal vertebra, a sacral rib, a neural arch of a mid-caudal vertebra, humerus, both femora, left tibia,

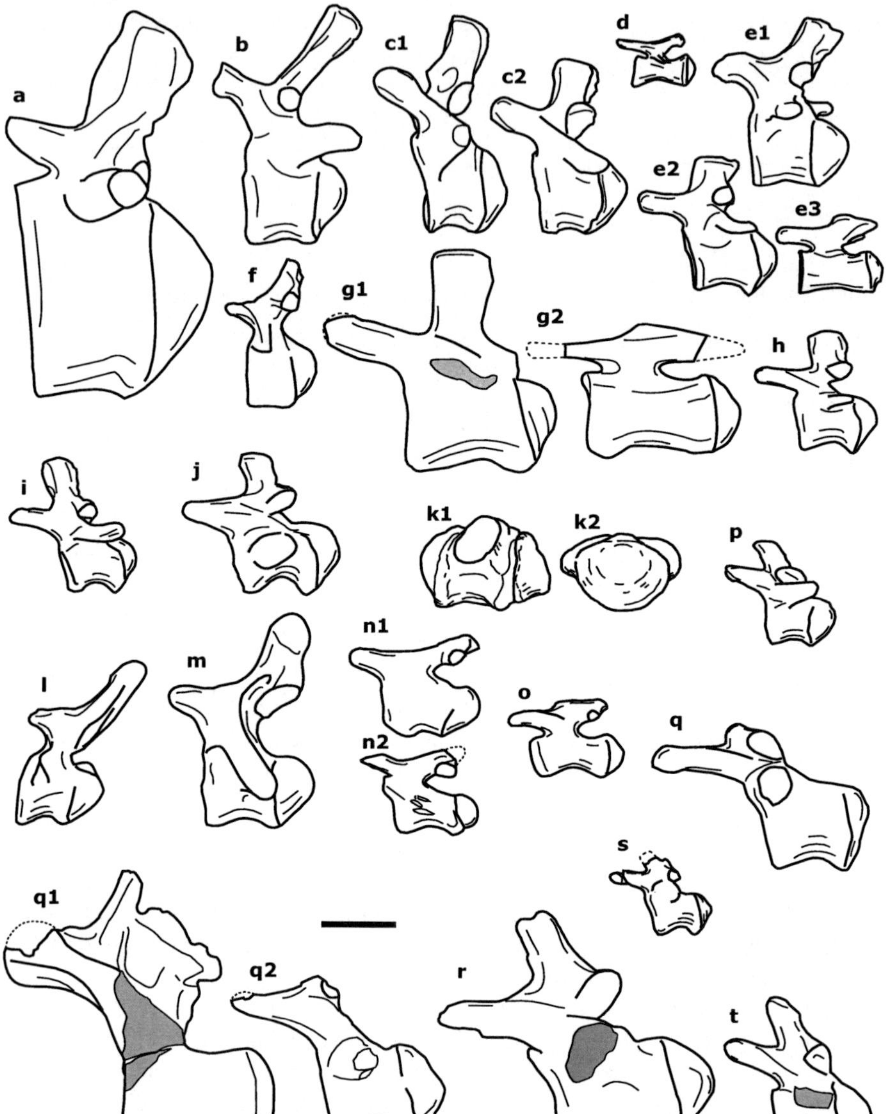

incomplete right tibia, left fibula, astragalus (referred as a calcaneus by Martinelli and Forasiepi 2004), metacarpal I, metacarpal IV (described as an ulna by Martinelli and Forasiepi 2004). (MACN-PV-RN 1061): incomplete neurocranium, a centrum of a cervical vertebra, six fragments of chevrons, metacarpal I, metacarpal V, femur, left

◀**Fig. 3** Caudal vertebrae of Campanian–Maastrichtian titanosaurs from South America in left lateral view (unless stated otherwise). **a** *Dreadnoughtus*, anterior caudal (MPM-PV 1156, Lacovara et al. 2014). **b** *Narambuenatitan*, anterior caudal (MAU-Pv-N-425, Filippi et al. 2011). **c** *Adamantisaurus*, anterior caudals (MUGEO 1289, Santucci and Bertini 2006a). **d** *Atacamatitan*, posterior caudal (SGO.PV.961c, Kellner et al. 2011). **e** *Baurutitan*, anterior (**e1**), mid-anterior (**e2**), and posterior (**e3**) caudals in reversed right lateral view (MCT 1490-R, Kellner et al. 2005). **f** *Bonatitan*, anterior caudal in reversed right lateral view (MACN-PV RN 821, Salgado et al. 2015). **g** *Pellegrinisaurus*, anterior (**g1**) and posterior (**g2**) caudal vertebrae (MPCA 1500, Salgado 1996). **h** *Pitekunsaurus*, anterior caudal (MAU-Pv-AG-446/25, Filippi and Garrido 2008); **i** *Trigonosaurus*, mid-anterior caudal reversed right lateral view (MCT 1719-R, Campos et al. 2005). **j** *Uberabatitan*, mid-anterior caudal (CPPLIP-1019, Silva Jr. et al. 2019); **k** *Yamanasaurus*, caudal centra in right (**k1**) and posterior (**k2**) views (Apesteguía et al. 2020). **l** *Neuquensaurus*, anterior caudal (MCS-5/2, Salgado et al. 2005). **m** *Saltasaurus*, anterior caudal (PVL 4017–23, Powell 2003). **n** *Rocasaurus*, mid-caudal in reversed right lateral view (MP-CA-Pv 46 and MPCV-Pv 58, Salgado and Azpilicueta 2000). **o** *Maxakalisaurus*, posterior caudal in reversed right lateral view (MN 5013-V, Kellner et al. 2006). **p** *Panamericansaurus*, mid-anterior caudal (MUCPv-417, Calvo and Porfiri 2010). **q** *Aeolosaurus rionegrinus*, mid-anterior caudal (MJG-R1, Powell 2003). **q1–2** *Aeolosaurus colhuehuapensis*, anterior and mid-anterior caudals (UNPSJB-PV 959/1 and UNPSJB-PV 959/11, Casal et al. 2007). **r** *Aeolosaurus maximus*, mid-anterior caudal (MPMA 12–0001-97, Santucci and Arruda Campos 2011). **s** *Gondwanatitan*, anterior caudal (MN 4111-V, Kellner and Azevedo 1999). **t** *Punatitan*, mid-caudal in reversed right lateral view (CRILAR-Pv 614, Hechenleitner et al. 2020). Scale bar equals 10 cm

tibia, astragalus, a proximal fragment of a metatarsal I, distal fragment of a metatarsal III, a complete metatarsal III, a proximal fragment of a metatarsal V.

Locality and Age Bajo de Santa Rosa, Río Negro, Argentina (Fig. 1). Malargüe Group (Upper Cretaceous), Allen Formation (late Campanian–early Maastrichtian) (Martinelli and Forasiepi 2004).

Emended Diagnosis Small-size titanosaur characterized by the following association of characters: (1) small paired pits on the frontals; (2) basisphenoid tubera log and narrow (more than twice long as wide); (3) dorsal to mid-caudal vertebrae with deep oval to circular pits on both side of the prespinal lamina; (4) anterior caudal vertebra with spinopostzygapophysial and spinoprezygapophysial laminae; (5) neural arch of anterior caudals with deep interzygapophysial fossae with numerous pits; (6) anterior caudal vertebra with an accessory sub-horizontal lamina extending from the anteroventral portion of the postzygapophysis to the mid-portion of the spinoprezygapophysial lamina; (7) anterior caudal vertebra with a prominent axial crest on the ventral surface of the centrum; (8) posterior ridge of the metacarpal IV (Salgado et al. 2015).

Comments Martinelli and Forasiepi (2004) identified two specimens (MACN-PV RN 821 and MACN-PV RN 1061) in their original description. Posteriorly, a redescription made by Salgado et al. (2015) distinguished five individuals (individuals A-E) and re-defined the neurocranium, MACN-PV RN 821 (Individual C), as the holotype. The rest of the material labeled as specimen (MACN-PV RN 821) is considered as part of the referred material. Several undescribed materials from

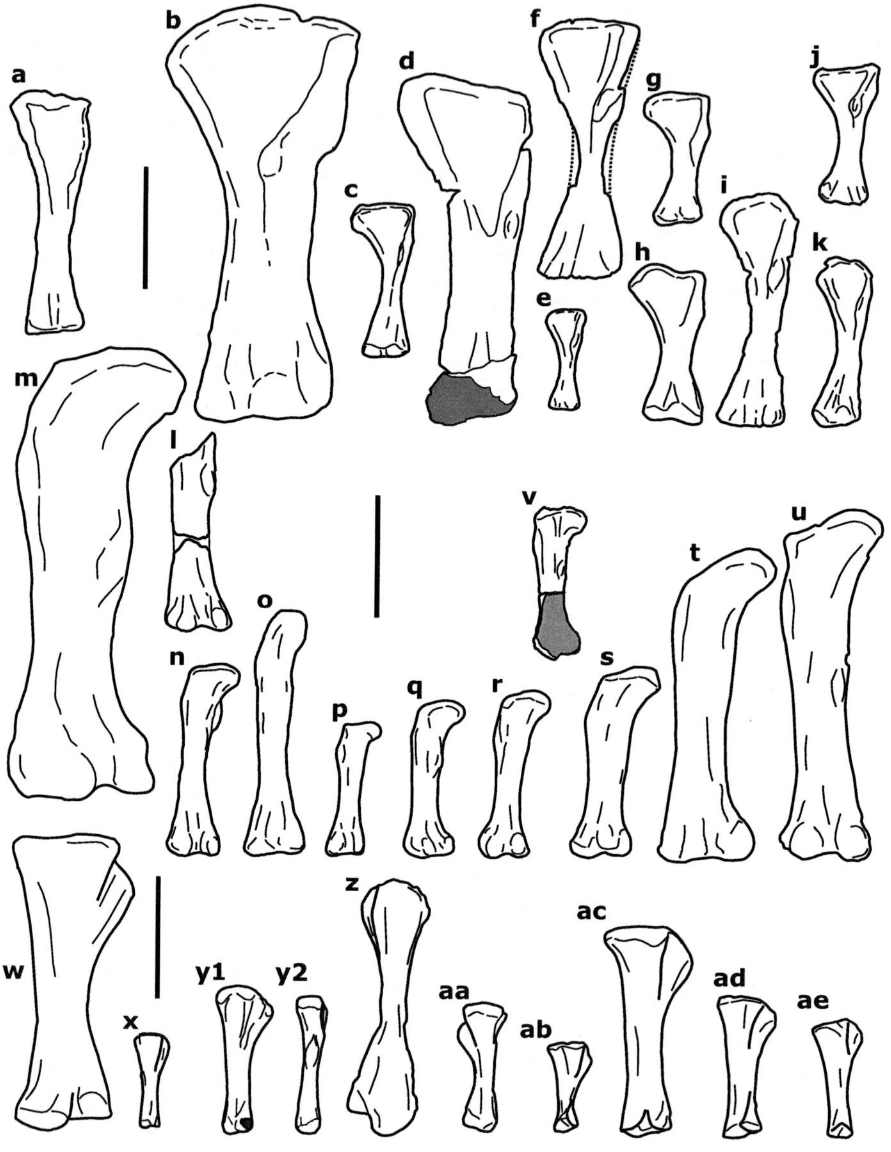

the appendicular and axial skeleton were included and described such as sacral ribs, haemal arches, metatarsals, metacarpals, and astragali.

In the original description, Martinelli and Forasiepi (2004) considered *B. reigi* as a Saltasaurinae based on the following features: (1) anterodorsal edge of the neural spine at the posterior level of the postzygapophyses in the mid-caudal neural arch; (2) distal femoral condyles anteriorly exposed; (3) cancellous osseous tissue in the

◀**Fig. 4** Appendicular elements of Campanian–Maastrichtian titanosaurs from South America. **a–k** left humerus in anterior view. **a** *Narambuenatitan* (MAU-Pv-N-425, Filippi et al. 2011). **b** *Dreadnoughtus* (MPM-PV 1156, Lacovara et al. 2014). **c** *Arackar* (right reversed, SNGM-1/1, Rubilar-Rogers et al. 2021). **d** *Argyrosaurus* (MLP 77-V-29–1, Mannion and Otero 2012). **e** *Bonatitan* (MACN-PV RN 821, Salgado et al. 2015). **f** *Nullotitan* (right reversed, MPM 21,546, Novas et al. 2019). **g** *Neuquensaurus* (MLP – CS 1050, Otero 2010). **h** *Saltasaurus* (PVL 4017–67, Powell 2003). **i** *Panamericansaurus* (MUCPv-417, Calvo and Porfiri 2010). **j** *Bravasaurus* (CRILAR-Pv 613, Hechenleitner et al. 2020). **k** *Gondwanatitan* (right reversed, MN 4111-V, Kellner and Azevedo 1999). **l-s** left femur in posterior view. **l** *Narambuenatitan* (MAU-Pv-N-425, Filippi et al. 2011). **m** *Dreadnoughtus* (MPM-PV 1156, Lacovara et al. 2014). **n** *Arackar* (SNGM-1/2, Rubilar-Rogers et al. 2021). **o** *Atacamatitan* (SGO-PV-961, Kellner et al. 2011). **p** *Bonatitan* (MACN-PV RN 821, Salgado et al. 2015). **q** *Uberabatitan* (CPPLIP-1238, Silva Jr. et al. 2019). **r** *Neuquensaurus* (MLP-CS 1118, Otero 2010). **s** *Saltasaurus* (PVL 4017–79, Powell 2003). **t** *Antarctosaurus* (MACN 6904, Powell 2003). **u** *A. maximus* (MPMA 12–0001-97, Santucci and Arruda Campos 2011). **v** *Bravasaurus* (CRILAR-Pv 613, Hechenleitner et al. 2020). **w–ae** right tibia in lateral view. **w** *Dreadnoughtus* (MPM-PV 1156, Ullmann and Lacovara 2016). **x** *Bonatitan* (MACN-PV RN 821, left tibia reversed, Salgado et al. 2015). **y1** and **y2** *Laplatasaurus* (MLP-CS 1128, right tibia, and MLP-CS 1127, right fibula in lateral view, Gallina and Otero 2015). **z** *Nullotitan* (left reversed, MPM 21,548, Novas et al. 2019). **aa** *Uberabatitan* (CPPLIP-917, left tibia reversed, Silva Jr. et al. 2019). **ab** *Saltasaurus* (PVL 4017–84, Powell 2003). **ac** *Antarctosaurus* (MACN 6904, left tibia reversed, Huene 1929). **ad** *A. rionegrinus* (MJG-R 1, Powell 2003). **ae** *Gondwanatitan* (MN 4111-V, left tibia reversed, Kellner and Azevedo 1999). Scale bars equals 50 cm

presacral and caudal vertebrae, mainly observed in the neural arch. In a phylogenetic analysis made by Salgado et al. (2015), *B. reigi* is recovered within Titanosauria as a basal member of a wide clade recovered as the sister group of Saltasaurinae, rejecting its inclusion within Saltasaurinae as postulated by Martinelli and Forasiepi (2004).

Brasilotitan Machado, Avilla, Nava, Campos and Kellner 2013

Brasilotitan nemophagus Machado, Avilla, Nava, Campos and Kellner 2013 (Fig. 2c1–2)

Holotype MPM 125R, right dentary, two cervical vertebrae, three incomplete sacral vertebrae, fragmentary ilium, fragments of an ischium, one incomplete ungual, and fragmentary elements. Casts of the dentary are housed at the Museu Nacional/UFRJ (MN 7371-V).

Referred Specimens MPM 126R, an isolated tooth.

Locality and Age Presidente Prudente, southeastern São Paulo State, Brazil (Machado et al. 2013) (Fig. 1). Adamantina Formation, (Campanian–early Maastrichtian (Batezelli 2017; Dias et al. 2020).

Diagnosis *Brasilotitan nemophagus* is characterized by the following combination of characteristics (*indicates autapomorphies): (1) *dorsal symphyseal region of the dentary twisted medially; (2) *dorsal portion of the dentary symphyseal contact thicker anteroposteriorly than the ventral part; (3) *presence of anteriorly directed accessory prezygapophyseal articulation surface on cervical vertebrae; (4) *intraprezygapophyseal laminae of cervical vertebrae V-shaped in dorsal view; (5) mandible with an 'L'-shaped morphology in dorsal view; (6) sharp anteroposteriorly oriented crest on the dorsal margin at the posterior region of the mandibular ramus; (7) cervical vertebrae with dorsoventrally low neural spines (approximately 36% and 38% of the total size of the vertebrae); (8) ventral portion of the cervical centrum arched dorsally; (9) prezygapophyses of cervical vertebrae placed anteriorly relative the anterior articulation of the centrum (Machado et al. 2013).

Comments *Brasilotitan nemophagus* was collected near the site where *Austroposeidon* was found, and although being also known from fragmentary remains, it comprises some cranial and appendicular elements. The phylogenetic affinities of *Brasilotitan* were not evaluated in its original description, but the authors argued that it could be more related to *Antarctosaurus* and/or *Bonitasaura* due to the L-shaped dentary morphology (Machado et al. 2013). *Brasilotitan* was depicted as closely related to *Uberabatitan* in a phylogenetic analysis provided by Bandeira et al. (2016). A CT scan of the preserved anterior portion of the right dentary showed *Brasilotitan* has at least two pencil-like reposition teeth per alveolus where the basal most replacement tooth is displaced lingually and the more apical replacement tooth is set at the pulp cavity of the functional tooth like in *Nemegtosaurus* (Wilson 2005).

Laplatasaurus Huene 1929

Laplatasaurus araukanicus Huene 1929 (=*Titanosaurus araukanicus* Huene 1929) (Fig. 4y1, 2)

Lectotype MLP-CS 1128, right tibia and MLP-CS 1127, right fibula.

Locality and Age Cinco Saltos, Río Negro Province, Argentina (Fig. 1). Neuquen Group (Upper Cretaceous), Anacleto Formation (early–middle Campanian).

Emended Diagnosis *L. araukanicus* is characterized by the following combination of characters (* indicates autapomorphies): (1) slender tibia with deep cnemial fossa, extending more than one third the length of the shaft*; (2) fibula with extremely developed anterior fossa on the proximal end, laterally delimited by a thick margin* (a slightly less developed fossa is present in *Uberabatitan* and *Jainosaurus*); (3) fibula with a well-developed lateral tuberosity, similar to that one present in *Neuquensaurus* (Gallina and Otero 2015).

Comments The material referred to *L. araukanicus* (Huene 1929) comes from different Patagonian localities (e.g., close to Neuquén city in Neuquén Province; General Roca, Rancho de Ávila, and Cinco Saltos in Río Negro Province; Sierra de San Bernardo in Chubut Province). Taking into account that Gallina and Otero (2015) considered as the materials referred to *L. araukanikus* the lectotype designated by Bonaparte and Gasparini (1979), the only valid geographic and stratigraphic occurrence for this taxon is Cinco Saltos.

Due to the uncertainty surrounding the material referred to the genus, *L. araukanicus* was not included in previous phylogenetic analyses. Gallina and Otero (2015) considered that the only assurance of assignment to the species corresponds to the lectotype and included *Laplatasaurus* in a phylogenetic analysis using a modified version of Salgado et al. (2015) data matrix. In this analysis, *Laplatasaurus* is recovered together with *Uberabatitan* within Titanosauria. The clade is supported by the presence of a longitudinal fossa on the posterior face of the fibular midshaft.

Nullotitan Novas, Agnolin, Rozadilla, Aranciaga-Rolando, Brisson-Egli, Motta, Cerroni, Ezcurra, Martinelli, D´Angelo, Alvarez-Herrera, Gentil, Bogan, Chimento, García-Marsà, Lo Coco, Miquel, Brito, Vera, Perez Loinaze, Fernández, and Salgado 2019

Nullotitan glaciaris Novas, Agnolin, Rozadilla, Aranciaga-Rolando, Brisson-Egli, Motta, Cerroni, Ezcurra, Martinelli, D´Angelo, Alvarez-Herrera, Gentil, Bogan, Chimento, García-Marsà, Lo Coco, Miquel, Brito, Vera, Perez Loinaze, Fernández, and Salgado 2019 (Fig. 4f, z)

Holotype MACN-PV 18,644 and MPM 21,542, an isolated cervical centrum (MACN-PV 18,644, presumably the 3rd cervical vertebra), fragmentary cervical rib shaft, dorsal rib shaft fragments, several caudal vertebrae, fragmentary left scapula, proximal and distal ends of right femur, almost complete right tibia, fibula, and astragalus (MPM 21,542).

Referred Specimens MPM 21,545, complete humerus, lacking cortex of the midshaft, partial rib, and vertebra (found about 100 m away from the holotype). MPM 21,546, isolated and partially preserved distal caudal centra. MPM 21,547, sequence of five articulated mid-caudal vertebrae with their respective haemal arches (material remains in the field). MPM 21,548, isolated complete left tibia. MPM 21,549, anterior to mid-caudal centrum. These specimens were collected from nearby localities, from levels above and below the level the holotype was found.

Locality and Age Approximately, 25 km southwestern El Calafate, Santa Cruz, Argentina (Fig. 1). Austral-Magallanes Basin, Chorrillo Formation (upper Campanian–lower Maastrichtian) (Novas et al. 2019).

Diagnosis A large titanosaur characterized by the following combination of characters (* indicates autapomorphies): (1) anterior caudal centra notably anteroposteriorly short, its transverse diameter duplicating its anteroposterior length; (2) proximal and mid-caudal centra with lateral and ventral surfaces profusely excavated by large blind depressions*; (3) mid-caudal vertebrae with lateral surface with a tabicated large fossa below the transverse process; (4) centra of all available caudal vertebrae lacking signs of pneumatization; (5) mid-caudals with a deep ventral longitudinal furrow surrounded by two longitudinal thick ridges; (6) fibula with a pronounced sigmoid curvature when viewed anteriorly or posteriorly*; and (7) distal end of tibia strongly anteroposteriorly compressed and more transversely expanded than in other titanosaurs (Novas et al. 2019).

Comments *Nullotitan glaciaris* is a large titanosaur known from fragmentary remains of a couple of different individuals. Novas et al. (2019) did not evaluate the phylogenetic relationships of this titanosaur in their original description. However, they assigned it to Colossosauria due to the absence of Saltasaurinae and Aeolosaurini synapomorphies and because this species has a humerus with minimum mediolateral width divided by proximodistal length less than 0.15 and deltopectoral crest with mediolateral thickness of anterior attachment surface with distal half mediolaterally expanded relative to proximal half. Cerda et al. (2021) included *Nullotitan* in a large data matrix and recovered this taxon as more closely related to *Pellegrinisaurus*, forming with *Alamosaurus* and *Baurutitan*, a clade that is the sister group of Saltasauridae (equally weighted analysis) or in a more basal position, being the sister species of the monophyletic group composed of Saltasauridae and the group comprising *Baurutitan*, *Pellegrinisaurus*, *Alamosaurus*, and *Trigonosaurus* (implied weighted analysis). In both cases, *Nullotitan* is not recovered as a Colossosauria as originally proposed by Novas et al. (2019). *Nullotitan* is also one of the southernmost Upper Cretaceous titanosaur records known so far.

Pellegrinisaurus Salgado 1996

Pellegrinisaurus powelli Salgado 1996 (Fig. 3g1, 2)

Holotype MPCA1500/1–31, is represented by four dorsal vertebra centra, 26 incomplete caudals, and an incomplete right femur.

Locality and Age Lago Pellegrini, northwestern of Río Negro Province, Argentina (Fig. 1). Neuquén Group (Upper Cretaceous), Anacleto Formation (lower-middle Campanian) (Salgado 1996; Heredia and Salgado 1999).

Emended Diagnosis Large titanosaur characterized by the following combination of characteristics (* indicates autapomorphies): presence of a prominent blind lateral fossa ('pleurocoel') with well-developed dorsal rim in the anterior caudal centra*; femur with longitudinal ridge located on the posterior surface of the shaft and laterally displaced, originated from the trochanteric shelf and extended to the distal third of the shaft (Cerda et al. 2021).

Comments This titanosaur was discovered and collected in 1975 by Mr. Roberto Abel, at the time Director of the Museo Provincial Carlos Ameghino (Cipolleti, Río Negro, Argentina) and assistants. Powell (1986) assigned it as cf. *Epachthosaurus* sp. based on certain characters of the dorsal vertebrae.

The original diagnosis, proposed by Salgado (1996), was mostly based on the morphology of caudal vertebrae, and regarded the presence of the following autapomorphies: mid-posterior and posterior caudal vertebra with anteroposteriorly elongated and dorsoventrally depressed neural spines, the anterior ends of which are at a higher position than the posterior ones; dorsal vertebra centra strongly depressed; and transverse width of the centrum of the posterior dorsal vertebrae approximately twice the maximum dorsoventral depth. However, in a more recent evaluation of the holotype, Cerda et al. (2021) noted that the characteristics of the original diagnosis were influenced by taphonomic processes or are present in other titanosaur species.

Although *Pellegrinisaurus* was not originally included in a phylogenetic analysis, it was considered by Salgado (1996) as the putative sister taxon of Saltasaurinae, as they share the presence of low mid- and posterior caudal centra with dorsoventrally convex lateral surface. It was included in few phylogenetic studies, being considered the sister taxon of *Alamosaurus* (Upchurch et al. 2004) and as more related to the clade composed of *Alamosaurus* and *Baurutitan* (Gorscak and O'Connor 2019, Bayesian analysis). In this latter study, *Saltasaurus* and *Neuquensaurus* form a more basal group that is not closely related to *Pellegrinisaurus*. Cerda et al. (2021), using a larger data matrix, recovered *Pellegrinisaurus* in a clade that also included *Alamosaurus* and *Baurutitan*, being this clade the sister group of Saltasauridae.

Pitekunsaurus Filippi and Garrido 2008

Pitekunsaurus macayai Filippi and Garrido 2008 (Figs. 2d, 3h)

Holotype MAU-Pv-AG-446, is represented by one single incomplete skeleton comprising braincase; left frontal; one tooth axis; two anterior cervical vertebrae; one posterior cervical; two anterior dorsal vertebrae; one posterior dorsal vertebra; one anterior caudal vertebra; three posterior caudal vertebrae; right scapula; right ulna and an incomplete left femur.

Locality and Age Aguada Grande, Rincón de los Sauces, Northern Neuquén, Argentina. (Fig. 1). Neuquén Group (Upper Cretaceous), Subgroup Río Colorado, Anacleto Formation (lower-middle Campanian) (Filippi and Garrido 2008).

Diagnosis Small titanosaur characterized by four autapomorphies: (1) basipterygoid processes broadly separated and parallelly projected; (2) presence of small depressions or longitudinal grooves in the spinal sector of spinopostzygapophyseal lamina in anterior cervical vertebrae; (3) centropostzygapophyseal lamina forked proximally in anterior dorsal vertebrae, and (4) posterior centrodiapophyseal lamina with accessory lamina in anterior dorsal vertebrae (Filippi and Garrido 2008).

Comments Originally, *Pitekunsaurus* was not included in a phylogenetic analysis, but the morphological variation observed in the caudal vertebrae, especially in the posterior caudal vertebrae, made Filippi and Garrido (2008) suspect the existence of a probable affinity with *Rinconsaurus caudamirus* (Calvo and González Riga 2003), a titanosaur characterized by procoelous posterior caudal centra with intercalation of a series of amphicoelous-biconvex or amphicoelous-opisthocoelous-biconvex centra.

Pitekunsaurus was included by Coria et al. (2013) in a phylogenetic analysis for the first time, and this species was recovered in a polytomy with *Gondwanatitan, Aeolosaurus rionegrinus*, and *A. maximus,* as a possible Aeolosaurini.

In the last years, González Riga et al. (2018) recovered *Pitekunsaurus* as the sister taxon of the clade comprising *Epachthosaurus* + (Rinconsauria + Lognkosauria) placed near the base of Titanosauria. This clade, Rinconsauria + Lognkosauria, was defined through a new analysis by González Riga et al. (2019) and named as Colossosauria (see Chap. 9). Moreover, Mannion et al. (2019), using an updated version of the dataset used in the analyses mentioned above, also recovered this species as a Titanosauria, providing additional evidence that *Pitekunsaurus* is closely related to Colossosauria.

Trigonosaurus Campos, Kellner, Bertini and Santucci 2005

Trigonosaurus pricei Campos, Kellner, Bertini and Santucci 2005 (Figs. 2e, 3i)

Holotype MCT 1488-R, comprises an incomplete vertebral column formed by five cervical vertebrae, ten dorsals (last cervical and all dorsals articulated), sacrum, and left ilium.

Paratype MCT 1719-R, ten caudal vertebrae corresponding to elements of the anterior and mid-portion of the tail.

Locality and Age Peirópolis ('Caieira' quarry), Uberaba, Minas Gerais State, Brazil (Powell 1987, 2003; Campos et al. 2005) (Fig. 1). Serra da Galga Member, Marília Formation, Bauru Group, Maastrichtian (Dias-Brito et al. 2001).

Diagnosis *Trigonosaurus pricei* is characterized by the following combination of characteristics: (1) elongated mid-cervical vertebrae, with low neural spine and concave ventral margin; (2) elongated mid-dorsal vertebrae with neural spine strongly inclined posteriorly; (3) 9th and 10th dorsal vertebrae with incipient postzygodiapophyseal lamina; (4) anterior caudal vertebrae with thin base broadening toward the top; (5) anteriormost caudals (2nd-5th) with two, and mid-caudals with one, pronounced dorsal depression on the transverse process; (6) prezygapophyses on the caudal vertebrae extended forward (but not to the same degree as *Aeolosaurus rionegrinus* and *Gondwanatitan faustoi*) with wide articular faces; (7) haemal arches with articular surfaces strongly developed starting on caudal 3 until the last preserved element (caudal 20); (8) well-developed transverse processes throughout the sequence formed by anterior and medial caudals up to, at least, caudal 20 (Campos et al. 2005).

Comments The material comprising the holotype and the paratype of *Trigonosaurus* was informally referred to in the literature as 'Series B from Peirópolis' (e.g., Powell 1987, 2003; Salgado et al. 1997; Campos and Kellner 1999). Considering that neither the caudal vertebrae were found articulated to each other nor to the sacrum, they were regarded as paratypes by Campos et al. (2005) because of their proximity with the articulated presacral elements, compatible size (mainly when compared to the sacral vertebrae), and similar morphology among the caudal elements. These authors, however, did not rule out the possibility that both the presacral and caudal series belong to the same individual.

One large dorsal vertebra (CPPLIP 494) from the same outcrop described by Santucci and Bertini (2006b) were assigned to *Trigonosaurus pricei* because of the similarities shared by CPPLIP 494 with the posteriormost dorsal of the holotype (Juárez Valieri and Díaz 2013). However, as stated originally (Santucci and Bertini 2006b), CPPLIP 494 corresponds to a mid-posterior dorsal vertebra, since its parapophyses are clearly located in front of the diapophyses (and not displaced in a more ventral position as in the 9th and 10th dorsals of *Trigonosaurus*). Even considering CPPLIP 494 as a posterior dorsal, the spinodiapophyseal lamina is much more robust than in *Trigonosaurus,* and the diapopostzygopophyseal lamina is well developed and thin, which differs from the condition seen in *Trigonosaurus* that has a more robust, but incipiently developed, diapopostzygopophyseal lamina.

The phylogenetic relationships of *Trigonosaurus* have been evaluated by several different authors and even so far it is not well established. Salgado et al.

(1997) suggested that *Trigonosaurus* (Series B, at that time) is closely related to *Opisthocoelicaudia* due to the absence of postzygodiapophyseal laminae. Although not performing a phylogenetic analysis, Campos et al. (2005) suggested that *Trigonosaurus* could be related to *Aeolosaurus* and *Gondwanatitan* because both the neural arch and spine of the anterior caudals are inclined forward. More recently, *Trigonosaurus* was included in more complete data matrices being considered either as more related to Saltasauridae (e.g., Carballido and Sander 2013) or to Aeolosaurini (e.g., Gorscak and O'Connor 2019; Hechenleitner et al. 2020).

Uberabatitan Salgado and Carvalho 2008

Uberabatitan ribeiroi Salgado and Carvalho 2008 (Figs. 3j, 4q, aa)

Holotype CPPLIP-912, CPPLIP-1107, and CPPLIP-1082 consist of a left tibia, fibula, and astragalus. This is a new definition of the materials attributed to the holotype made by Silva Jr. et al. (2019) considering the original assignment of Salgado and Carvalho (2008). According to Silva Jr. et al. (2019), these three elements represent the most complete likely articulated set of elements of the original holotypic series and additionally have two of the originally proposed autapomorphies of the taxon.

Referred Specimens CPPLIP-1058, CPPLIP-1057, CPPLIP-914, CPPLIP-919 (anterior cervical vertebrae); CPPLIP-1091, CPPLIP-1104 (anterior cervical neural arches); CPPLIP-992, CPPLIP-1023 (mid-cervical vertebrae); CPPLIP-993, CPPLIP-915 (posterior cervical centra); CPPLIP-922, CPPLIP-917, CPPLIP-1081, CPPLIP-921, CPPLIP-929, CPPLIP-1105 (cervical ribs); CPPLIP-1077 (anterior dorsal vertebra); CPPLIP-1068 (mid-dorsal neural arch); CPPLIP-923 (dorsal rib); CPPLIP-1099 (sacral centrum); CPPLIP-1079 (anterior caudal vertebra); CPPLIP-1017 (mid-caudal vertebra); CPPLIP-1009, CPPLIP-1010, CPPLIP-1011, CPPLIP-1012 (posterior caudal vertebrae); CPPLIP-1056 (anterior haemal arch); CPPLIP-1006 (posterior haemal arch); CPPLIP-1027 (sternal plate); CPPLIP-1109 (right coracoid); CPPLIP-1030 (left humerus); CPPLIP-1032 (left radius); CPPLIP-911 (right radius); CPPLIP-1080 (right metacarpal); CPPLIP-1029, CPPLIP-1103 (left and right pubes). All these elements were originally part of the holotype as proposed by Salgado and Carvalho (2008) and were considered as referred material by Silva Jr. et al. (2019). The following isolated remains were originally considered as referred specimens by Salgado and Carvalho (2008): CPPLIP-1075, CPPLIP-1022 (anterior cervical vertebrae); CPPLIP-1085 (anterior/mid-cervical vertebra); CPPLIP-994 (mid-cervical vertebra); CPPLIP-1070 (mid-cervical centrum); CPPLIP-1024, CPPLIP-1108 (posterior cervical vertebrae); CPPLIP-918 (cervical vertebra); CPPLIP-991 (posterior cervical neural arch); CPPLIP-1014 (posterior caudal vertebra); CPPLIP-1078 (fragment of vertebra); CPPLIP-1065 (dorsal rib); CPPLIP-1018 (mid-caudal vertebra); CPPLIP-1019 (mid-caudal vertebra); CPPLIP1020 (two fused mid-caudal vertebrae); CPPLIP-1008 (posterior caudal centrum); CPPLIP-1005, CPPLIP-1003, CPPLIP-1004 (haemal arches); CPPLIP-1120 (left coracoid); CPPLIP-913 (fragment of right pubis); CPPLIP1026 (fragment of ischium), CPPLIP-898 (distal end of a left femur); CPPLIP-1106 (left fibula), CPPLIP-1116 (mid-dorsal centrum); CPPLIP-894 (partial right femur). These new materials, collected from the same site, were referred to *U. ribeiroi* by Silva Jr. et al. (2019): CPPLIP-1238 (left femur), CPPLIP-1690 (mid-cervical vertebra), CPPLIP-1189 (left femur), CPPLIP-1043 (metatarsal II), CPPLIP-971 (ungual phalanx), and CPPLIP-1691 (proximal chevron).

Locality and Age Uberaba, Minas Gerais State, Brazil (Salgado and Carvalho 2008; Silva Jr. et al. 2019) (Fig. 1). Serra da Galga Member, Marília Formation, Bauru Group, Maastrichtian (Dias-Brito et al. 2001).

Emended Diagnosis *Uberabatitan ribeiroi* is characterized by the following autapomorphies: (1) anterior and mid-cervical vertebrae with epipophyseal-prezygapophyseal lamina segmented in two unconnected segments, zygapophyseal and diapophyseal, of which the former extends anterodorsally over the latter; (2) mid-dorsal vertebrae with a robust composite lateral lamina formed mainly by a diapophyseal lamina, probably homologous to the postzygodiapophyseal lamina and, to a lesser extent, by a relic of the spinodiapophyseal lamina; (3) mid-(and possibly posterior) dorsal vertebrae with neural accessory laminae parallel to the prespinal lamina, which are probably the spinoprezygapophyseal laminae; (4) mid-caudal centra with deeply excavated lateral surfaces; (5) pubis with a very stout longitudinal crest on its external (ventral) surface; (6) proximal end of the tibia with a very robust lateral protuberance that articulates with an equally robust medial knob in the fibula; (7) astragalus with a well-developed anteroposterior crest that mediodistally delimits the tibial articulation. All autapomorphies were originally proposed by Salgado and Carvalho (2008) except for the last one which was included by Silva Jr. et al. (2019).

Comments *Uberabatitan ribeiroi* is composed mainly of disarticulated remains of at least five different individuals since three left femora (CPPLIP-1238, CPPLIP-1189, and CPPLIP-898), one large cervical vertebra (CPPLIP-1690) and remains of at least one juvenile specimen were recovered from the same site (Silva Jr. et al. 2019). Although collected from the same Serra da Galga Member (Marília Formation), *Uberabatitan* was found nearly 35 km away from the site where *Trigonosaurus* and *Baurutitan* were excavated.

Originally, Salgado and Carvalho (2008) considered the materials of *Uberabatitan* represented three different individuals, but Silva Jr. et al. (2019) suggested, based on morphometric measurements of axial elements, that more individuals were present. Because of that, they excluded several elements from the holotypic series.

Uberabatitan has been considered in some few phylogenetic analyses, but only the last one (provided by Silva Jr. et al. 2019) included all elements known for this species. The analyses show *Uberabatitan* is a non-Saltasauridae titanosaur (Silva Jr. et al. 2019).

Saltasauridae Sereno 1998

Definition A node-based clade defined as *Opisthocoelicaudia skarzynskii* Borsuk-Białynicka 1977, *Saltasaurus loricatus* Bonaparte and Powell 1980, their most recent common ancestor, and all its descendants.

Saltasaurinae Sereno 1998

Definition A stem clade defined as the most inclusive clade containing *Saltasaurus loricatus* Bonaparte and Powell 1980 but not *Opisthocoelicaudia skarzynskii* Borsuk-Białynicka 1977.

Yamanasaurus Apesteguía et al. 2020

Yamanasaurus lojaensis Apesteguía et al. 2020 (Fig. 3k1, 2)

Holotype YM-UTPL-002, two fused sacral vertebrae; YM-INPC-014, a fragment comprising the posterior half of a procoelous caudal vertebra; YM-INPC-016, the proximal half of a left humerus; YM-INPC-015, the proximal half of a radius; and YM-INPC-017, a badly damaged fragment of a putative proximal part of a tibia.

Locality and Age Yamana region, Paltas Cantón, Loja Province, southwestern Ecuador (Apesteguía et al. 2020) (Fig. 1). Río Playas Formation, late Maastrichtian (Upper Cretaceous).

Diagnosis *Yamanasaurus lojaensis* is a small-sized titanosaurian characterized by the following association of characteristics (*indicates autapomorphies): (1) anterior to mid-caudal vertebrae with a dorsoventrally compressed condyle, with the posterior tip elevated respect to the midline, with no longitudinal ventral ridge, and camellate spongy bone (putatively shared with *Neuquensaurus*); (2) last sacral centrum as long as tall, with small ovoid, shallow blind fossa in the lateral side*; (3) radius robust with flattened diaphysis and a marked neck or cingulum right under the epiphysis, with an heptagonal proximal articular surface* (Apesteguía et al. 2020).

Comments *Yamanasaurus* is known by only one individual composed of fragmentary remains. This taxon represents the northernmost latest Cretaceous titanosaur formally named in South America. Due to its fragmentary state, *Yamanasaurus* was not included in any cladistic dataset available so far. However, the presence of proportionally short limb elements suggests *Yamanasaurus* could be a Saltasaurinae and the presence of a narrow last sacral vertebra, together with marked dorsoventral angulation between the last two sacral elements, dorsoventrally compressed condyle in the anterior to mid-caudal vertebra with posterior tip elevated respect to the midline, and absence of a longitudinal ventral ridge and spongy inner structure (camellate bone), suggest that *Yamanasaurus* could be more related the Argentinian saltasaurine *Neuquensaurus* (Apesteguía et al. 2020).

Neuquensaurus Powell 1992

Neuquensaurus australis (Lydekker 1893) (Figs. 2f, 3l, 4g, r)

N. australis Lydekker 1893 (=*Titanosaurus australis* Lydekker 1893; *Microcoelus patagonicus* Lydekker 1893; *Titanosaurus robustus* Huene 1927; *Loricosaurus scutatus* Huene 1927).

Holotype MLP Ly 1–7, an incomplete sacrum, consisting of six co-ossified centra and one unfused biconvex centrum, and six partial caudal vertebrae.

Neuquensaurus robustus (Huene 1929) *nomen dubium.*

Lectotype Left ulna (MLP–CS 1094), right ulna (MLP–CS 1095), left radius (MLP–CS 1171), left femur (MLP–CS 1480) (Bonaparte and Gasparini 1979).

Locality and Age The holotype and limb elements of *N. australis* (Lydekker 1893) came from Neuquén Province, Argentina. The materials of *N. australis* and *N. robustus* (Huene 1929) came from General Roca and Cinco Saltos, Río Negro Province, Argentina. The specimen of *N. australis* and associated elements (Salgado et al. 2005) came from Cinco Saltos, Río Negro Province, Argentina (Fig. 1). Neuquén Group (Upper Cretaceous), Subgroup Río Colorado, Anacleto Formation (early–middle Campanian) (D'Emic and Wilson 2011).

Referred material Referred materials of *N. australis* are composed by cervical, dorsal, and caudal vertebrae (Series Ly 1–3 and Series of Cinco Saltos 1–3) and appendicular elements (Series of Cinco Saltos 4) (Powell 2003; Otero 2010). On the other hand, there is also a partially articulated skeleton (MCS-5) integrated by axial and appendicular elements and two osteoderms, and other isolated elements: a right tibia (MCS-6), a left scapula (MCS-7), a left humerus (MCS-8), a right femur (MCS-9), and a metatarsal (MCS-10) (Salgado et al. 2005).

Emended diagnosis Powell (1986, 2003) proposed a diagnosis which includes fifteen diagnostic characters of *N. australis.* Salgado et al. (2005) reviewed and modified the diagnosis of *N. australis* as follows: (1) fibula with strong lateral tuberosity and bent shaft (Powell 2003); (2) anteriormost dorsal vertebrae lacking centroprezygapophyseal and centropostzygapophyseal laminae (Wilson 2002: 275); (3) seven sacral vertebrae; (4) sacral centra 3 to 5 narrowed (modified from Powell 2003); (5) mid- and posterior caudal vertebrae bearing transversely wide, non-keeled ventral depression, limited by lateral rounded ridges culminating in articular facets of hemapophyses (Powell 2003); and (6) lateral walls of caudal vertebral centra little exposed in ventral view (Powell 2003).

Recently, D'Emic and Wilson (2011) reviewed the diagnosis of *N. australis,* indicating the following unique features of the holotypic material: (1) length of sacral vertebral series about 1.5 times the width between the sacricostal yokes (Powell 1986; Salgado et al. 2005). Additionally, the assignment of other remains to this species such as (MCS–5) allowed the emendation of the diagnosis to include features present in other parts of the skeleton that are not preserved in the holotype: (2) mid-caudal vertebrae prezygapophyses with a non-articulating anterior process, (3) longitudinal ridge below transverse processes, lamina postzygodiapophiseal present and elongate, (4) fibula rotated antero-medially and slightly distorted posteriorly so that the lateral trochanter is visible in anterior view (i.e., when the distal portion is oriented transversely).

Some of the features listed as diagnostic of *N. australis* by Powell (1986) and Salgado et al. (2005) present a wider distribution within Titanosauria.

Comments The specimen of *N. robustus* (Huene 1929) is considered as a *nomen dubium* because differences between the latter and *N. australis* can be considered as individual variation or owing to sexual dimorphism (Otero and Vizcaíno 2008).

The holotype and limb elements of *N. australis* ('*Titanosaurus*' *australis* Lydekker 1893) came from Neuquén, but unfortunately, Lydekker did not provide a precise location of the bones nor a specific stratigraphic position. However, Huene (1929: (4) provided more detailed locality information: '... in the elevated right bank of the Río Neuquén, some kilometers (2–4) from the railway bridge and in the confluence before Neuquén...'. Thus, the holotypic locality is near 38°58' S, 68°00' W (D'Emic and Wilson 2011). Other specimens of *N. australis* and *N. robustus* ('*T*'. *australis* and '*T*'. *robustus* Huene 1929) came from General Roca and Cinco Saltos (Río Negro, Province), from strata belonging to the 'Dinosaurier schichten' (Keidel 1917). Those 'Dinosaur beds,' where remains of *Neuquensaurus* were found, correspond to the Anacleto Formation ('Senonense inferior,' Huene 1929: 11) (Otero 2010). The specimen of Cinco Saltos (MCS-5) came from a heterolithic succession composed of mudstones and siltstones alternating with sandstone beds, in a channel–floodplain deposit of red siltstones and red and green mudstones alternating with light yellow sandstones displayed in thin beds. This evidence provides additional support for the previously proposed depositional system in which *Neuquensaurus* was found.

Neuquensaurus was included, as is the case of *Saltasaurus,* in several phylogenetic analyses (e.g., Wilson 2002; Calvo and González Riga 2003; Santucci and Arruda Campos 2011; Lacovara et al. 2014; Carballido et al. 2017; González Riga et al. 2018; Filippi et al. 2019) with strong support indicating it is a Saltasaurinae, sharing with *S. loricatus* a series of features like well-developed cervical neural arch lamination, spongy caudal bone texture, posterior caudal centra dorsoventrally flattened, and breadth of posterior centrum at least twice height and femoral distal condyles exposed on anterior portion of femoral shaft.

Saltasaurus Bonaparte and Powell 1980

Saltasaurus loricatus Bonaparte and Powell 1980 (Figs. 2g, 3m, 4h, s, ab)

Holotype PVL. 4017–92, is represented by a complete sacrum fused to both ilia.

Referred Material Postcranial elements and osteoderms identified as PVL. 4017 and also the specimens CNS-V 10.023 and 10.024 corresponding to at least 5 adult and sub-adult individuals, which include 3 distinct cranial fragments, an axis, 14 cervical vertebrae, 20 dorsal vertebrae, 3 sacral vertebrae, 26 caudal vertebrae, 4 scapulae, 3 coracoids, 4 sternal plates, 10 humeri, 4 radii, 5 ulnae, 5 metacarpals, 5 ilia, 4 pubes, 2 ischia, 5 femora, 5 tibiae, 4 fibulae, 7 metatarsals, 6 dermal plates, and 4 associated small dermal ossicles in natural position.

Locality and Age El Brete, southern Salta province, Argentina (Fig. 1). Balbuena Subgroup, Salta Group (Upper Cretaceous), Lecho Formation, Senonian (upper Campanian?–lower Maastrichtian) (Bonaparte and Powell 1980).

Emended diagnosis Originally, Bonaparte and Powell (1980) defined *Saltasaurus* as a titanosaurid closely related to *Titanosaurus* (*T. australis* and *T. robustus*), with the first caudal not biconvex but procoelous, the ilia with a strongly developed antero-lateral projection; with a few dermal scutes, and thousands of small dermal ossicles distributed in the skin. Powell (1992) modified the diagnosis as: (1) skull with the upper temporal aperture far more reduced than in *Antarctosaurus*, dorsally closed by the coalescence of the parietal and the frontal; (2) complete basisphenoid with a triangular shape in the ventral view, the base is transversally narrow with a pronounced sagittal crest, (3) the fenestra ovalis opens for the passage of the IX–XI cranial nerves; (4) the basipterygoid processes are united at their proximal ends; (5) cervical vertebrae relatively shorter and broader than in the known titanosaurs; (6) dorsal vertebrae centra without 'keels'; (7) neural spine extremely stout and rounded distally; (8) diapophyses with a flat dorsal surface at their distal end; (9) sacrum with six fused vertebrae, with convex anterior and posterior articulations; (10) first caudal vertebrae procoelous; (11) neural spine broad and anteroposteriorly compressed, inclined backwards in the first caudal vertebrae; (12) scapula with a medial prominence close to the anterodorsal edge; (13) long bones relatively short and robust; (14) metacarpals much shorter than in *Antarctosaurus* and *Argyrosaurus*; (15) ilium with long preacetabular lamina and strongly curved outward; (16) intradermic ossicles and dermal scutes.

Comments *Saltasaurus* is part of the El Brete assemblage that also includes the remains of a small abelisaurid (*Noasaurus leali* Bonaparte and Powell 1980), isolated teeth of a supposed carnosaur, and numerous remains of three different types of birds (e.g., *Yungavolucris brevipedalis* Chiappe 1993; *Elbretornis bonapartei* Walker and Dyke 2010). The sediments observed were interpreted as a coastal fluvial-lake system, with frequent swampland and abundant vegetation (Bonaparte et al. 1977).

Saltasaurus is recovered in various phylogenetic analyses (e.g., Wilson 2002; Calvo and González Riga 2003; Santucci and Arruda Campos 2011; Lacovara et al., 2014; Carballido et al. 2017; González Riga et al. 2018; Filippi et al. 2019) firmly related with *Neuquensaurus* by a series of characters (see comments on *Neuquensaurus australis* above).

Rocasaurus Salgado and Azpilicueta 2000

Rocasaurus muniozi Salgado and Azpilicueta 2000 (Fig. 3n1, 2)

Holotype MPCA-Pv 46, is represented by a cervical centra, cervical neural arch, two dorsal centra, three dorsals neural arch, two sacral neural arches, a middle caudal vertebra, a mid-posterior caudal vertebra, two ischia, a left pubis, a left ilium, a fragment of right ilium, and a left femur.

Refereed Material Three anterior caudal centra (MPCA-Pv 47, 48, 60), a middle caudal centra (MPCA-Pv 49), a posterior caudal centra (MPCA-Pv 50), and six distal caudal centra (MPCA-Pv 51–56). Cf. *Rocasaurus* sp.: an anterior caudal vertebra (MPCA-Pv 57), a middle caudal vertebra (MPCA-Pv 58), and posterior caudal vertebra (MPCA-Pv 59).

Locality and Age Salitral Moreno, 25 km from General Roca, Río Negro Province, Argentina (Fig. 1). Malargüe Group (Upper Cretaceous), Allen Formation (Campanian–Maastrichtian) (Salgado and Azpilicueta 2000).

Diagnosis Saltasaurinae characterized by the following autapomorphies: (1) presence of a distal expansion on the lateral pubic margin, (2) ischium with a broad lamina, (3) caudal vertebrae with a deep ventral cavity divided by a longitudinal septum and with the posterior articular surface markedly depressed and extended ventrally forward (Salgado and Azpilicueta 2000).

Comments Although *Rocasaurus* was not originally included in a phylogenetic analysis, but was considered by Salgado and Azpilicueta (2000) as closely related to *Saltasaurus*. This would be supported by the presence of a ventral cavity divided by a longitudinal septum in the anterior and middle caudal vertebrae, and a poorly developed ischial pedicel of the ilium. In a recent phylogenetic analysis carried out by Hechenleitner et al. (2020), *Rocasaurus* is recovered as a Saltasauridae in a polytomy with South American and Laurasian taxa. The phylogenetic analysis presented by González Riga (2003) depicted *Rocasaurus* as more related to *Saltasaurus* than to *Neuquensaurus* within Saltasaurinae.

Colossosauria González-Riga, Lamanna, Otero, David, Kellner and Ibiricu 2019.

Definition The most inclusive clade containing *Patagotitan mayorum* Carballido et al. 2017, but not *Saltasaurus loricatus* Bonaparte and Powell 1980.

Colossosauria incertae sedis.

Antarctosaurus Huene 1929

Antarctosaurus wichmannianus Huene 1929 (Fig. 4t, ac)

Holotype MACN 6904, is composed of an incomplete cranium and mandible.

Referred Material Rib fragments, left scapula, incomplete right humerus, two fragments of the distal radius, a fragment of the proximal end of an ulna, seven metacarpals (two of these incomplete), a phalanx, a fragment of an ilium, an almost complete right ischium and part of the left, a fragment of the pubis, left femur, left tibia, left fibula, five left metatarsals and some phalanges from the foot.

Locality and Age Fifteen kilometers to the southwest of the city of General Roca, Río Negro, Argentina (Fig. 1). Neuquén Group (Upper Cretaceous), Subgroup Río Colorado, Anacleto Formation (early–middle Campanian) (Powell 2003).

Emended Diagnosis A large titanosaur characterized by the following combination of characters: (1) high skull at the occipital region, large orbits, wide muzzle, straight and broad symphyseal region perpendicular to the mandibular rami; (2) broad and short parietal and frontal; (3) upper temporal fenestra reduced but partially open toward the dorsal side; (4) triangular basipresphenoid complex wide and short; (4) foramen for nerves IX–XI separated from the fenestra ovalis; basipterygoid process long, forming a very divergent bar; (5) basioccipital tuberosities separated; (6) paraoccipital process long and recurved as in *Saltasaurus*; (7) narrow scapular blade, with wide supraglenoid region; axis of the scapular lamina almost perpendicular to the major axis of the supraglenoid part of the scapula; (8) ischiatic lamina narrow and distally expanded; (9) well-developed pubic peduncle; (10) tibia with stout articular ends; (11) fibula with lateral tuberosity formed by two parallel rugosities that are parallel to the long axis of the bone (Powell 2003).

Comments As it happens with the descriptions of the first records of sauropods from Patagonia (e.g., '*Titanosaurus*' (=*Neuquensaurus*) *australis* and *Laplatasaurus araukanicus*), *A. wichmannianus* was collected during several field trips and from more than one locality, with no quarry maps depicting their skeletal associations, designation of type material, and/or supported diagnoses based on autapomorphic traits (Gallina and Otero 2015). Additionally, considering the absence of such field information, it is uncertain whether the skull elements are associated or not to the postcranial remains (Wilson 2005) and, depending on the situation, only the cranial remains could be regarded to this species. This made the taxonomic position of the specimens problematic. The finding of new materials or future taxonomic revisions may solve many of these taxonomic and phylogenetic problems presented by these taxa which were based on abundant material collected during several field campaigns.

At the moment, the phylogenetic relationships of *A. wichmannianus* have not been clarified, but it has been included in at least two analyzes (Gallina and Apesteguía 2011 and Mannion et al. 2019). Both analyzes coincide in recovering *Antarctosaurus* as a member of Colossosauria.

Lognkosauria Calvo, Porfiri, González Riga and Kellner 2007a

Definition The most recent common ancestor of *Mendozasaurus neguyelap* González Riga 2005 and *Futalognkosaurus dukei* Calvo, Porfiri, González Riga and Kellner 2007a and all its descendants.

Puertasaurus Novas, Salgado, Calvo and Agnolin 2005

Puertasaurus reuili Novas, Salgado, Calvo and Agnolin 2005 (Fig. 2h1, 2)

Holotype MPM 10,002, consisting of four disarticulated vertebrae, including most part of a cervical vertebra, a complete anterior dorsal (second dorsal), and the centra of two caudal vertebrae.

Locality and Age Cerro Los Hornos, La Leona, Santa Cruz Province, Argentina (Leanza 1972; Arbe 2002; Novas et al. 2005; Egerton 2011; Varela et al. 2012; Sickmann et al. 2018) (Fig. 1). Cerro Fortaleza Formation (=Pari Aike Formation), Magallanes-Austral Basin, Campanian (Upper Cretaceous).

Diagnosis *Puertasaurus reuili* is diagnosed based on the following combination of characters: (1) gigantic size; (2) cervical neural spine considerably inflated, being transversally wider than the vertebral centrum and bearing strong dorsolateral ridges; (3) caudal cervicals with spinoprezygapophyseal laminae transversely thick and dorsoventrally deep; (4) cranial dorsal vertebrae extremely short (shorter than in other sauropods) (Novas et al. 2005).

Comments Although *Puertasaurus* was not included originally in a phylogenetic analysis, in recent works, it was recovered as an unstable member of Lognkosauria (Carballido et al. 2017; González Riga et al. 2018; Hechenleitner et al. 2020).

Rinconsauria Calvo, González Riga and Porfiri 2007b

Definition *Rinconsaurus caudamirus* Calvo and González Riga, 2003, *Muyelensaurus pecheni* Calvo, González Riga and Porfiri 2007b, their most recent common ancestor and all its descendants.

Maxakalisaurus Kellner, Campos, Azevedo, Trotta, Henriques, Craik and Paula Silva 2006

Maxakalisaurus topai Kellner, de Almeida Campos, de Azevedo, Trotta, Henriques, Craik and Paula Silva 2006 (Figs. 2i, 3o)

Holotype MN 5013-V, an incomplete skeleton comprising right maxilla and teeth, 12 fragmentary cervical vertebrae, including portions of cervical ribs, seven fragmentary dorsal vertebrae, several rib fragments, one sacral neural spine, one sacral centrum, six caudal vertebrae, haemal arches, part of both scapulae, both sternal plates, distal end of the left ischium, both humeri, 2nd and 4th right metacarpals, fragmentary fibula, and one osteoderm (Kellner et al. 2006).

Referred Specimens The distal end of a right scapula (MN 7048-V) and two sternal plates (MN 7049-V and MN 7050-V, respectively) regarded as a second specimen (Kellner et al. 2006). An incomplete right dentary (MBC-42-PV) and isolated teeth collected at the type locality (França et al. 2016).

Locality and Age Prata, Minas Gerais State, Brazil (Kellner et al. 2006; França et al. 2016) (Fig. 1). Adamantina Formation, Campanian–early Maastrichtian (Batezelli 2017; Dias et al. 2020).

Emended Diagnosis *Maxakalisaurus topai* is characterized by the following association of characteristics: (1) Meckelian channel does not enter the symphyseal area of the dentary, (2) U-shaped tooth row, (3) teeth with high-angled planar facets and suboval in cross section, (4) two replacement teeth per alveolus, (5) anterior and mid-posterior caudal vertebrae dorsoventrally compressed, including both articular ends, (6) mid-posterior caudal vertebrae with lateral surface of the centrum strongly concave (spool-shaped), (7) mid-posterior caudal vertebrae with neural spine directed anteriorly, (8) presence of at least one mid-posterior caudal with biconvex centrum, (9) metacarpal IV about 12% shorter than metacarpal II, and (10) sacral centrum with keel-shaped ventral surface (Kellner et al. 2006; emended by França et al. 2016).

Comments *Maxakalisaurus topai* is represented by an incomplete skeleton mostly composed of fragmentary axial remains. Some caudal centra are unfused to the neural arch, which could indicate that the specimen was not fully grown. *Maxakalisaurus* was included in few phylogenetic analyses so far, but in these studies, it is regarded as a species more related to Aeolosaurini (Santucci and Arruda Campos 2011; França et al. 2016). In such analyses, Rinconsauria is not retrieved as a monophyletic group and *Maxakalisaurus* is nested in a polytomy with *Panamericansaurus*, *Rinconsaurus*, and the clade composed of *Gondwanatitan* and *Aeolosaurus*. In Gorscak and O'Connor (2019), *Maxakalisaurus* was considered either as the sister species of *Paralititan*, forming a monophyletic group with Saltasaurinae (parsimony analysis) or as the sister species of the clade composed of Saltasaurinae plus *Paralititan* (Bayesian analysis). However, as more recent analyses based on different datasets recovered *Rinconsaurus* and *Muyelensaurus* as successive basal taxa more related to Aeolosaurini (e.g., Carballido et al. 2017; Gorscak and O'Connor 2019; Rubilar-Rogers et al. 2021) and considering the phylogenetic relationships of *Maxakalisaurus* it can be considered as a member of Rinconsauria.

Panamericansaurus Porfiri and Calvo 2010

Panamericansaurus schroederi Porfiri and Calvo 2010 (Figs. 3p, 4i)

Holotype MUCPv-417, is represented by five caudal vertebrae, one sacral vertebra, left humerus, two chevrons, and rib fragments.

Locality and Age San Patricio del Chañar, Neuquén Province, Argentina (Fig. 1). Malargüe Group (Upper Cretaceous), Allen Formation (upper Campanian–lower Maastrichtian) (Porfiri and Calvo 2010).

Diagnosis Slender titanosaur characterized by the following autapomorphies: (1) mid-posterior caudal vertebrae with prezygapophyses anterodorsally oriented and equal or shorter than the length of the base of the vertebral centra; (2) postzygapophysis located medial to the vertebral centra; (3) spinoprezygapophyseal lamina dorsoventrally expanded in middle caudals; (4) humerus with a robustness index less than 0.40 (Porfiri and Calvo 2010).

Comments The first remains of this specimen were originally assigned to the genus *Aeolosaurus* (Calvo and Porfiri 2003). Subsequent modifications made in the diagnosis of this genus (Casal et al. 2007) allowed a reevaluation of this material and the establishment of a new species.

Panamericansaurus is rarely included in phylogenetic analyses; it is considered by Porfiri and Calvo (2010) as belonging to Aeolosaurini due to the presence of the characters proposed by Costa Franco-Rosas et al. (2004) for this clade. Phylogenetic analysis carried out by Santucci and Arruda Campos (2011) recovered *Aeolosaurus, Gondwanatitan, Panamericansaurus, Rinconsaurus,* and *Maxakalisaurus* as aeolosaurines (sensu Costa Franco-Rosas et al. 2004), a clade supported by one synapomorphy: posterior caudal centra dorsoventrally flattened. A more recent evaluation using Bayesian phylogenetic analysis and a different dataset recovered *Panamericansaurus* as the sister species of the monophyletic group composed of *Gondwanatitan* plus *Aeolosaurus* and *Trigonosaurus* plus *Argyrosaurus* (Gorscak and O'Connor 2019, Bayesian analysis). In such analysis, Rinconsauria, that is generally recovered as closely related to Aeolosaurini, would also include *Bonitasaura* and is depicted in a more basal position in the tree. Finally, considering the phylogenetic definition of Aeolosaurini used here (see Carballido et al. this book, Chap. 8), this clade is less inclusive than its original definition and, according to the phylogenetic proposals made so far, *Panamericansaurus* cannot be considered as a member of Aeolosaurini.

Aeolosaurini Costa Franco-Rosas, Salgado, Rosas and Carvalho 2004

Definition *Aeolosaurus rionegrinus* Powell 1987, *Gondwanatitan faustoi* Kellner and Azevedo 1999, their most recent common ancestor and all its descendants.

Aeolosaurus Powell 1987

Aeolosaurus rionegrinus Powell 1987 (Figs. 3q, 4ad)

Holotype MJG-R 1, seven anterior caudal vertebrae, right and left incomplete scapulae, right and left humeri, right ulna and radius, right and left ischia, five metacarpals, right tibia and fibula, an astragalus, and indeterminate fragments.

Locality and Age Casa de Piedra, Estancia Maquinchao, Río Negro Province, Argentina (Fig. 1). Angostura Colorada Formation (Upper Cretaceous, late Campanian–early Maastrichtian) (Volkheimer 1973).

Emended Diagnosis *A. rionegrinus* is defined by the following characteristics: (1) prezygapophyses directed mainly upward on anteriormost caudal vertebrae and slightly curved downward on the remaining anterior caudal vertebrae; (2) prezygapophyseal articular facets widened by the presence of both a dorsal and a ventral protuberance on the prezygapophyses of the anterior caudal vertebrae; and (3) apex of the convexity of the posterior articulation strongly displaced upward, so that the apex is flushed to the level of the dorsal margin of the centrum on anterior and middle caudal vertebrae (Santucci and Arruda Campos 2011).

Comments *Aeolosaurus rionegrinus* was the first species of the genus to be described and, at the time of its description, part of its diagnosis was based on the conspicuous morphology of its anterior caudal vertebrae. With the assignment of other materials to the genus *Aeolosaurus* (or to other Aeolosaurini species), it has been noticed that some of the features of the caudal vertebrae are not unique to *A. rionegrinus*. *Aeolosaurus rionegrinus* is generally depicted as more related to other Aeolosaurini like *Gondwanatitan* (e. g. Calvo et al. 2007b; Gallina and Apesteguía 2011) or to *A. colhuehuapensis* when this second Argentinian species is considered in the analysis (e.g., Santucci and Arruda Campos 2011; Coria et al. 2013; Hechenleitner et al. 2020).

Aeolosaurus colhuehuapensis Casal, Martínez, Luna, Sciutto and Lamanna 2007 (Fig. 3q1, 2).

Holotype UNPSJB-PV 959/1–27 is composed of 21 articulated vertebrae caudal and 7 articulated hemapophyses.

Locality and Age Island emerged on the south-east of Colhué Huapi Lake, Sarmiento, Chubut Province, Argentina (Fig. 1). Colhue Huapi Formation, Campanian–Maastrichtian (Casal et al. 2007; Ibiricu et al. 2020).

Diagnosis *A. colhuehuapensis* is distinguished from *A. rionegrinus* by the following characters: (1) anterior caudal vertebrae possessing deep fossae between the transverse processes and the base of the neural spine; (2) anterior caudal vertebrae with a slender lamina joining the base of the prespinal lamina with the lower edge of the articular facet in the medial part of the prezygapophysis, producing a fossa on both sides of the prespinal lamina; (3) caudal vertebrae with the posterior articular cone centrally placed (Casal et al. 2007).

Comments *A. colhuehuapensis* is a mid-sized titanosaur known from an articulated caudal series (more posterior caudals are not preserved) and respective haemal arches (Casal et al., 2007). The few phylogenetic analyses that included this species showed it is an Aeolosaurini more closely related to *A. rionegrinus* by sharing the presence of pneumatopores in anterior caudal centra and the presence of prezygapophyses with wide articular facets in anterior caudal vertebrae (Santucci and Arruda Campos 2011).

Aeolosaurus maximus Santucci and Arruda Campos 2011 (Figs. 3r, 4u)

Holotype MPMA 12–0001-97, two incomplete posterior cervical vertebrae; seven incomplete cervical ribs; a fragmentary anterior dorsal centrum; a fragment of a middle dorsal vertebra; a fragmentary posterior dorsal vertebrae; several incomplete diapophysis of dorsal vertebrae; 12 incomplete dorsal ribs; six articulated anterior caudal vertebrae; a mid-caudal centrum; two posterior caudal vertebrae; six anterior, one mid, and one posterior haemal arches; a probable fragmentary scapula; an incomplete right humerus; a probably fragmentary left humerus; a probably incomplete radius; incomplete right femur; left femur; and a partial left ischium.

Locality and Age Monte Alto, São Paulo State, Brazil (Santucci and Arruda Campos 2011) (Fig. 1). Adamantina Formation, Campanian–early Maastrichtian (Batezelli 2017; Dias et al. 2020).

Diagnosis *Aeolosaurus maximus* is characterized by the following association of characteristics: (1) anterior and middle haemal arches with a well-developed posterior protuberance below the articular area; (2) mid-posterior haemal arches with a lateral bulge on the distal portion of the articular process; (3) presence of posterior centrodiapophyseal lamina at least 50% thicker than the postzygodiapophyseal lamina in posterior cervical vertebrae; (4) posterior dorsal vertebrae with intrapostzygapophyseal lamina and an oblique anterior centropostzygapophyseal lamina which bifurcates from the proximal portion of the centropostzygapophyseal lamina; (5) mid-thoracic ribs bearing well-developed anterior and posterior crests with a D-shaped cross section (Santucci and Arruda Campos 2011).

Comments *Aeolosaurus maximus* is a relatively large titanosaur from Bauru Group deposits. Besides the analysis presented in the original description, some other phylogenies depict this species as the close relative of the clade formed by the Argentinian species *Aeolosaurus rionegrinus* and *Aeolosaurus colhuehuapensis* (e.g., Coria et al. 2013).

The original diagnosis of *Aeolosaurus* was modified by Casal et al. (2007) who considered the presence of postzygapophyses in mid-caudal vertebrae located in anterior position with respect to the vertebral body, reaching the centrum articular margin, and open hemapophyses with articular facets arranged into two planes as synapomorphies of the genus. Subsequently, Santucci and Arruda Campos (2011) proposed a new diagnosis based on a phylogenetic evaluation of these species that considered the presence of prezygapophyses curved downward on anterior caudal vertebrae and haemal arches with double articular facets set in a concave posterodorsal surface on anterior and mid-caudal vertebrae. With that said, *A. maximus* does not show prezygapophyses as displaced anteriorly as in *A. rionegrinus*, *Aeolosaurus* sp. (Salgado and Coria 1993), and *A. colhuehuapensis* as proposed by Casal et al. (2007), but shares with *A. rionegrinus* and *A. colhuehuapensis* the synapomorphic characteristics proposed by Santucci and Arruda Campos (2011).

Some other phylogenies placed *A. maximus* as more related to other Aeolosaurini species, these analyses however neither included *A. colhuehuapensis* nor considered all synapomorphies that were proposed for the genus (e.g., Bandeira et al. 2016; Hechenleitner et al. 2020). Considering the new findings of Aeolosaurini species and the availability of larger data matrices, a new evaluation of this clade is necessary, including the status of *A. maximus.*

Bravasaurus Hechenleitner, Leuzinger, Martinelli, Rocher, Fiorelli, Taborda and Salgado 2020

Bravasaurus arrierosorum Hechenleitner, Leuzinger, Martinelli, Rocher, Fiorelli, Taborda and Salgado 2020 (Fig. 4i, v)

Holotype CRILAR-Pv 612, is composed of a right quadrate and quadratojugal, four cervical, five dorsal, and three caudal vertebrae, few dorsal ribs, three haemal arches, left humerus, fragmentary ulna, metacarpal IV, partial left ilium with sacral ribs, right pubis, partial ischium, left femur, and both fibulae.

Locality and Age La Rioja Province, Argentina (Fig. 1). Ciénaga del Río Huaco Formation (Campanian–Maastrichtian) (Hechenleitner et al. 2020).

Diagnosis A small-sized titanosaur characterized by the following association of features (* indicates autapomorphies): (1) quadrate with articular surface entirely divided by a medial sulcus*; (2) spinoprezygapophyseal lamina forming a conspicuous step between neural spine and prezygapophyses in middle cervical vertebrae*; (3) strongly depressed centra (up to twice as wide as tall) in posterior dorsal vertebrae; (4) robust dorsal edge of pneumatic foramen in dorsal centra, forming prominent shelf that extends laterally beyond the level of the ventral margin of the centum*; (5) posterior dorsal vertebrae with a rough postspinal lamina, ventrally interrupted by middle spinopostzygapophyseal laminae that contact the postzygapophyses; (6) posterior dorsal vertebrae with small ventral spinopostzygapophyseal fossa delimited dorsally by the middle spinopostzygapophyseal laminae and ventrally by the interpostzygapophyseal lamina; (7) humerus with narrow midshaft, with midshaft/proximal width ratio of 0.36; (8) deltopectoral crest of the humerus expanded distally; (9) slender fibula (Robustness Index [RI] 49 = 0.15); (10) distal condyle of the fibula transversely expanded, more than twice the midshaft breadth (Hechenleitner et al. 2020).

Comments In Hechenleitner et al. (2020), the phylogenetic analysis placed *Bravasaurus* within Colossosauria, and this species additionally shows three of the five ambiguous synapomorphies of the clade Rinconsauria. *Bravasaurus* is included in a collapsed clade comprising the Brazilian taxa *Trigonosaurus, Uberabatitan,* and *Gondwanatitan,* which is supported by a single synapomorphy: height/width ratio of the posterior articular surface of cervical centra smaller than 0.7. In such analysis, and also considering the definition adopted here, *Bravasaurus* is nested within Aeolosaurini being more related to *Gondwanatitan* than to *Aeolosaurus.*

Gondwanatitan Kellner and Azevedo 1999

Gondwanatitan faustoi Kellner and Azevedo 1999 (Figs. 3s, 4k, ae)

Holotype MN 4111-V, an incomplete skeleton consisting of two cervical, seven dorsal, six sacral, and 24 caudal vertebrae partially preserved, proximal portion of a left scapula, incomplete left ilium, shaft portion of both pubis, incomplete ischia, both humeri and tibiae, and several rib fragments.

Locality and Age Álvares Machado, southeastern São Paulo State, Brazil (Cunha et al. 1987; Kellner and Azevedo 1999) (Fig. 1). Adamantina Formation, Campanian–early Maastrichtian (Batezelli 2017; Dias et al. 2020).

Diagnosis *Gondwanatitan faustoi* is characterized by the following association of characteristics: (1) proximal and mid-caudal vertebrae with 'heart-shaped' distal articulation; (2) humerus with well-developed and medially curved deltopectoral crest; (3) tibia with anterior portion of the proximal articulation projected dorsally, and (4) cnemial crest only slightly curved laterally (Kellner and Azevedo 1999).

Comments *Gondwanatitan faustoi* is a small-sized sauropod and consists of one of the most complete titanosaur from Brazil, with several axial and appendicular remains preserved. Originally, it was not included in any cladistic analysis but was considered to be closely related to the genus *Aeolosaurus* because they share the presence of anterior and mid-caudal vertebrae with the neural spine strongly projected forward (Kellner and Azevedo 1999). Subsequent works that included both taxa in phylogenetic analyses generally depict these genera as sister taxa (e.g., Calvo et al. 2007a,c; Gallina and Apesteguía 2011; Santucci and Arruda Campos 2011; González Riga and David 2014).

Punatitan Hechenleitner, Leuzinger, Martinelli, Rocher, Fiorelli, Taborda and Salgado 2020

Punatitan coughlini Hechenleitner, Leuzinger, Martinelli, Rocher, Fiorelli, Taborda and Salgado 2020 (Figs. 2j, 3t)

Holotype CRILAR-Pv 614, is composed of a partial skeleton composed of the anterior portion of posterior cervical vertebra (C12?), two middle dorsal vertebrae (D6–D7?), partial sacrum, 13 articulated caudal vertebrae with some articulated haemal arches, right pubis, left ischium, and several dorsal ribs.

Locality and Age La Rioja Province, Argentina (Fig. 1). Ciénaga del Río Huaco Formation (Campanian–Maastrichtian) (Hechenleitner et al. 2020).

Diagnosis *P. coughlini* is characterized by the following combination of features (* indicates autapomorphies): (1) middle dorsal vertebrae (likely D6–D7) with anterior and posterior spinodiapophyseal laminae forming a wide and flat surface between aliform and transverse processes*; (2) accessory posterior centrodiapophyseal lamina crossed over by the posterior centroparapophyseal lamina, forming a X-shaped intersection in D6–D7; (3) posterior centroparapophyseal lamina reaches the bottom of posterior centrodiapophyseal lamina in dorsals 6 and 7*; (4) extra-depression placed ventrally to the intersection of posterior centroparapophyseal lamina and accessory posterior centrodiapophyseal lamina in dorsals 6 and 7*; (5) deep postzygodiapophyseal centrodiapophyseal fossa in D6–D7; (6) neural spine of D6 tapering dorsally, forming an inverted-'V' profile in anterior/posterior view; (7) caudal transverse processes persist beyond the caudal 15; (8) slightly anteriorly inclined neural spines in mid-anterior caudal vertebrae (Ca5–6 to Ca10); and (9) distally expanded prezygapophyses in mid-anterior caudal vertebrae (Hechenleitner et al. 2020).

Comments *Punatitan* is one of the few (together with *Bravasaurus* and *Saltasaurus*) formally described titanosaur species from NW Argentina. In Hechenleitner et al. (2020), the phylogenetic analysis recovered *Punatitan* within Colossosauria, showing three of the seven ambiguous synapomorphies of the clade. Moreover, it is nested within Aeolosaurini with other Argentinian and Brazilian Upper Cretaceous species and is closely related to *Aeolosaurus rionegrinus* by sharing the presence of anterior and mid-caudal vertebrae with distally expanded prezygapophyses, being *Aeolosaurus maximus* the sister taxon of these two species (Hechenleitner et al. 2020). Its phylogenetic relationships with *Aeolosaurus colhuehuapensis* are unknown so far.

3 Discussion

3.1 Phylogenetic Relationships

By the end of the Late Cretaceous (Campanian–Maastrichtian), titanosaurs achieved their greatest diversity in South America with about 30 species formally described so far. Unfortunately, the phylogenetic relationships of most of these species are not well established yet, which also precludes a more detailed evaluation of their biogeographic history. The reasons for this lack of resolution regarding several taxa in titanosaur phylogeny are not completely clear, but considering that several species are known from fragmentary and incomplete material, this lack of information can, at least in part, explain this trend. Moreover, the use of too distinct data matrices, with highly different numbers of both characters and taxa, can also be responsible for topological differences in these analyses. However, in general, the record of Campanian–Maastrichtian titanosaurs comprises species from different well-supported clades, such as Saltasaurinae, Rinconsauria (and its less inclusive group Aeolosaurini), lognkosaurians as *Puertasaurus* (or species closely related to that clade as *Austroposeidon*), and few more basal Titanosauria such as *Narambuenatitan* and *Dreadnoughtus*.

To date, Saltasaurinae were only found in Argentina, being represented by three species: *Neuquensaurus australis*, *Saltasaurus loricatus*, and *Rocasaurus muniozi* (Powell 2003; Salgado and Azpilicueta 2000; Salgado et al. 2005). The only exception are the fragmentary remains of *Yamanasaurus*, from Ecuador (Apesteguía et al. 2020). Another Argentinian species, *Bonatitan reigi,* was originally regarded as more related to Saltasaurinae (Martinelli and Forasiepi 2004), but a recent evaluation of its morphology and phylogenetic relationships considered it as a more basal species within Titanosauria (Salgado et al. 2015). However, it is noteworthy the striking similarity among the anterior caudal vertebrae of *Bonatitan* and other Saltasaurinae.

In Brazil and Chile, some species were regarded as more related to Saltasaurinae and potentially could represent species more related to Saltasauridae (e.g., *Atacamatitan*, Kellner et al. 2011; *Trigonosaurus*, Carballido and Sander 2013). A fragmentary material from Ecuador, *Yamanasaurus lojaensis,* was also considered to be more related to Saltasaurinae (Apesteguía et al. 2020). These findings show that while the less inclusive clade Saltasaurinae is so far known from Patagonia and NW Argentina, Saltasauridae titanosaurs could have had a more widespread geographic distribution in South America.

The record of both Rinconsauria and Aeolosaurini (a less inclusive clade within Rinconsauria, see Fig. 5) comprises species found in Argentina and Brazil. While some more basal Rinconsauria (e.g., *Rinconsaurus* and *Muyelensaurus* Calvo and González Riga 2003; Calvo et al. 2007b) were found in Turonian–Santonian strata, the other species assigned to this group are Campanian–Maastrichtian in age. Although the phylogenetic relationships within Rinconsauria and Aeolosaurini remain disputed, the known record of these Campanian–Maastrichtian species is restricted to South America. Rinconsauria is also considered as the sister group of another clade, Lognkosauria, that only comprises South American taxa. This contrasts with the phylogenetic relationships of Saltasaurinae that, within Saltasauridae, are closely related to Laurasian titanosaurs (e.g., *Alamosaurus* and *Opisthocoelicaudia* Calvo et al. 2007a; Filippi et al. 2011; Lacovara et al. 2014; Carballido et al. 2017).

3.2 Biogeographical Considerations and Macroevolutionary Patterns

The record of Campanian–Maastrichtian titanosaurs was also widespread in terms of geographic distribution in South America, with species formally described in Argentina, Brazil, Chile, and Ecuador, besides other unnamed records encompassing fragmentary bones, eggs, and footprints suggesting they were also present at least in Bolivia (e.g., Meyer et al. 2021), Peru (e.g., Mourier et al. 1986), and Uruguay (e.g., Soto et al. 2012). They were also found in a variety of sedimentary rocks representing

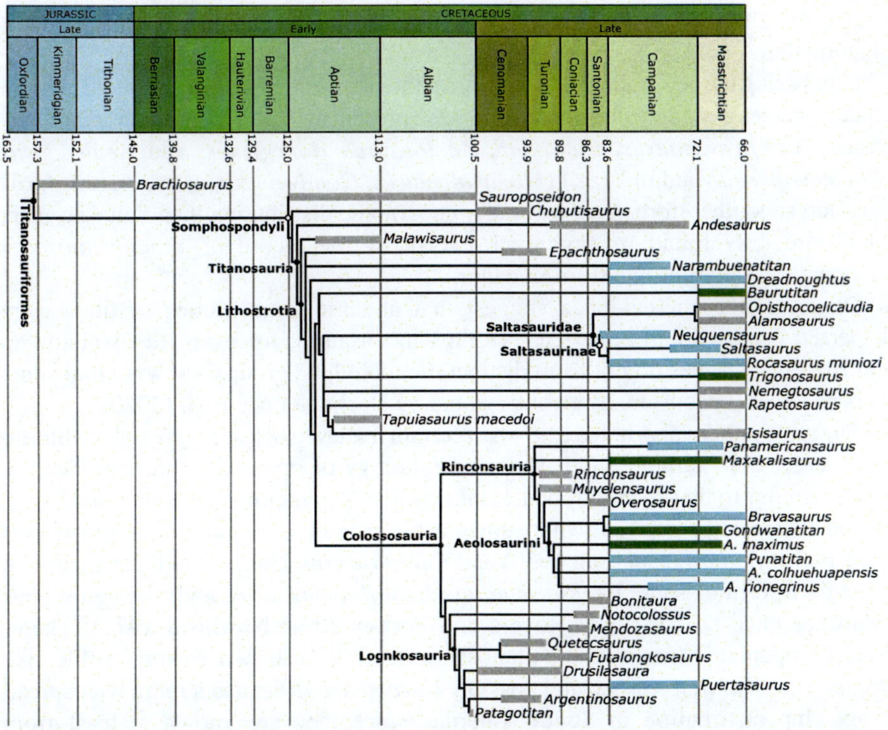

Fig. 5 Hypothetical phylogenetic relationships of Campanian–Maastrichtian titanosaurs from South America. Relationships of the main clades (e.g., Titanosauria, Saltasauridae, Rinconsauria, and Lognkosauria) are based in Carballido et al. (2020). Relationships within Rinconsauria and Aeolosaurini are based in Coria et al. (2013) and Santucci and Arruda Campos (2011). Relationships within Saltasauridae and Saltasaurinae are based in Calvo et al. (2007a), González Riga and David (2014), Gallina and Apesteguía (2011). Argentinian and Brazilian species highlighted in blue and green, respectively. Some species with more controversial phylogenetic relationships are not depicted in this figure. Black circles and open semi-circle indicate node-based and stem-based clades, respectively

different depositional systems such as lacustrine, fluvial, alluvial, and deltaic (e.g., Campos et al. 2005; Filippi et al. 2011; González Riga and Tomaselli 2019; Kellner and Azevedo 1999; Meyer et al. 2021; Powell 2003; Salgado and Carvalho 2008), which suggest titanosaurs were not only the most diverse group of dinosaurs in South America at the end of Cretaceous but also were able to inhabit different environments.

Only a few species of more basal Titanosauria are present in Campanian–Maastrichtian deposits, *Dreadnoughtus* and *Narambuenatitan* (Lacovara et al. 2014; Filippi et al. 2011) and, considering this record, more basal titanosaurs were restricted to Argentina by the end of the Cretaceous.

The Campanian–Maastrichtian record of Lognkosauria comprises only one species, *Puertasaurus*, known by fragmentary axial remains from Argentina (Novas et al. 2005). Therefore, *Puertasaurus* seems to be the last member of this clade of gigantic titanosaurs.

Comparing the geographic distribution of the two South American titanosaur main clades, species assigned to Rinconsauria were present at least in Argentina (*Panamericansaurus*, *Bravasaurus*, *Punatitan*, *Aeolosaurus rionegrinus*, and *Aeolosaurus colhuehuapensis*) and in Brazil (*Maxakalisaurus*, *Gondwanatitan*, and *Aeolosaurus maximus*) and, therefore, have a broader geographic distribution than Saltasaurinae, which are only found in Argentina, except by the record of *Yamanasaurus*, a fragmentary Saltasaurinae from Ecuador.

Different lines of evidence suggest that at least some groups of titanosaurs dispersed across South America. Within Rinconsauria, for example, Argentinian and Brazilian species do not form distinct monophyletic groups, showing that some dispersal events occurred, as recently noted by Hechenleitner et al. (2020).

The same pattern also holds true when certain features such as the rostral profile are considered. For example, the L-shape morphology of premaxillar-maxilar/dentary bones, similar to that seen in rebbachisaurids like *Nigersaurus* (Sereno et al. 2007), can be considered a derived feature since is not seen in more basal titanosauriforms such as *Giraffatitan* (Janensch 1950) and *Europasaurus* (Sander et al. 2006), and in more derived titanosaurs like *Sarmientosaurus*, *Rapetosaurus*, and *Nemegtosaurus* (Martínez et al. 2016; Curry Rogers and Forster 2001; Nowinski 1971). Therefore, Campanian–Maastrichtian titanosaurs bearing L-shaped rostral profile like *Brasilotitan*, *Antarctosaurus*, and possibly *Narambuenatitan* had a more widespread geographic distribution in South America, since they are known at least from Argentina and Brazil (Powell 2003; Machado et al. 2013; Filippi et al. 2016).

Body size also greatly varied among titanosaurs at the end of the Cretaceous in South America. Saltasaurinae species like *Saltasaurus* and *Neuquensaurus* are considered as small-sized sauropods and the same holds true for most species assigned to Rinconsauria or Aeolosaurini. On the other hand, species like *Antarctosaurus*, *Dreadnoughtus* and the lognkosaurian *Puertasaurus* are regarded as large-sized forms. A trend of increase in body size variation within Titanosauria during the Cretaceous period was previously reported by Coria and Salgado (2005) and Souza and Santucci (2014) who suggested that body size variation, mainly toward smaller size, took place after the extinction of rebbachisaurids. The titanosaur fossil record in South America is in line with this global trend.

Little is known about the titanosaur diet. Carbon stable isotope analyses of sauropod tooth enamel (including European titanosauriforms) show they mostly feed on C_3 plants (Tütken 2011). The same holds for the analyses of stable isotopes from several titanosaur eggshells from La Rioja Province, Argentina (Leuzinger et al. 2021). Additionally, the study of some titanosaur coprolites from the Upper Cretaceous Lameta Formation, India, showed the presence of Poaceae, conifer, and palm remains, which indicates they feed on a variety of plants (Prasad et al. 2005). In South America, some coprolites from the Marília Formation (Maastrichtian, Brazil) were considered as produced by titanosaurs (Oliveira and Santucci 2020) and showed the presence of Gnetophyta, arboreal conifers, and cycads pollen grains.

Tooth morphology alone could not furnish more specific evidence on Campanian–Maastrichtian titanosaur feeding habits. However, recovered teeth can be assigned to two different morphotypes: the pencil-like morphology, like that observed in diplodocids, although clearly distinct given the inclination of their wear facets (Calvo and González Riga, 2018) and a kind of spoon-shaped (or flame-like) tooth (e.g., Santucci and Bertini 2001). These two tooth morphologies show the last titanosaurs from South America could have had different feeding strategies.

Acknowledgements We thank the editors for inviting us to contribute to this book. We also thank Jose Carballido, Pablo Gallina, and Philip Mannion for their comments that greatly improved an earlier version of this chapter. RMS also thanks to the University of Brasília, Brazilian National Council for Scientific and Technological Development (CNPq), and Coordination for the Improvement of Higher Education Personnel (CAPES) for supporting part of this work.

References

Apesteguía S, Soto Luzuriaga JE, Gallina PA, Granda JT, Jaramillo, GAG (2020) The first dinosaur remains from the Cretaceous of Ecuador. Cretaceous Res 104345. https://doi.org/10.1016/j.cretres.2019.104345

Arbe HA (2002) Análisis estratigráfico del Cretácico de la Cuenca Austral. In: Haller MJ (ed) Geología y Recursos Naturales de Santa Cruz, vol 8. Relatorio del XV Congreso Geológico Argentino, El Calafate, pp 103–128

Bandeira KLN, Simbras FM, Machado EB, Campos DA, Oliveira GR, Kellner AWA (2016) A new giant Titanosauria (Dinosauria: Sauropoda) from the Late Cretaceous Bauru Group. Brazil Plos ONE 11(10):e0163373. https://doi.org/10.1371/journal.pone.0163373

Batezelli A (2017) Continental systems tracts of the Brazilian Cretaceous Bauru Basin and their relationship with the tectonic and climatic evolution of South America. Basin Res 29:1–25

Bonaparte JF, Coria RA (1993) Un nuevo y gigantesco sauropodo titanosaurio de la Formacion Rio Limay (Albiano-Cenomaniano) de la Provincia del Neuquen, Argentina. Ameghiniana 30:271–282

Bonaparte JF, Gasparini Z (1979) Los saurópodos de los grupos Neuquén y Chubut, y sus relaciones cronológicas. Actas Del VII Congreso Geologico Argentino, Neuquén 2:393–406

Bonaparte JF, Powell JE (1980) A continental assemblage of tetrapods from the Upper Cretaceous beds of El Brete, northwestern Argentina (Sauropoda-Coelurosauria-Carnosauria-Aves). Mem S Geo F 139:19–28

Bonaparte JF, Salfity JA, Bossi G, Powell JE (1977) Hallazgo de dinosaurios y aves cretácicas en la Formación Lecho de El Brete (Salta), próximo al límite con Tucumán. Acta Geologica Lilloana 14:5–17

Calvo JO, González Riga B (2003) *Rinconsaurus caudamirus* gen. et sp. nov., a new titanosaurid (Dinosauria, Sauropoda) from the late Cretaceous of Patagonia, Argentina. Revista Geológica de Chile 30(2):333–353

Calvo JO, González Riga B (2018) *Baalsaurus mansillai* gen. et sp. nov. a new titanosaurian sauropod (Late Cretaceous) from Neuquén, Patagonia, Argentina. Anais da Academia Brasileira de Ciências. https://doi.org/10.1590/0001-3765201820180661

Calvo JO, Porfiri JD (2003) Primer registro de *Aeolosaurus* en la provincia de Neuquén. Ameghiniana 4(40):34–60

Calvo JO, Porfiri JD, González Riga BJ, Kellner AWA (2007a) A new Cretaceous terrestrial ecosystem from Gondwana with the description of a new sauropod dinosaur. An Acad Bras Cienc 79:529–541
Calvo JO, González Riga BJ, Porfiri JA (2007b) A new titanosaur sauropod from the Late Cretaceous of Neuquén, Patagonia, Argentina. Arq do Museu Nacional 65:485–504
Calvo JO, Porfiri JD, González Riga BJ, Kellner AWA (2007c) Anatomy of *Futalognkosaurus dukei* Calvo, Porfiri, González Riga & Kellner, 2007c (Dinosauria, Titanosauridae) from the Neuquén Group (Late Cretaceous), Patagonia, Argentina. Arq do Museu Nacional 65:511–526
Campos DA, Castro HEF (1978) Localidades fossilíferas da Folha Paranapanema (SF—22). In: Schobbenhaus C, Coordinator. Carta geológica do Brasil ao Milionésimo, Folha Paranapanema (SF. 22). Departamento Nacional da Produção Mineral, Divisão de Geologia e Mineralogia, Brasília, pp 46–77
Campos DA, Kellner AWA (1999) On some sauropod (Titanosauridae) pelves from the continental Cretaceous of Brazil. In: Tomida Y, Rich TH, Vickers-Rich P (eds) Proceedings of the second gondwanan dinosaur symposium, vol 15. National Sciences Museum Monographs, Tokyo, pp 143–166
Campos DA, Kellner AWA, Bertini RJ, Santucci RM (2005) On a titanosaurid (Dinosauria, Sauropoda) vertebral column from the Bauru Group, Late Cretaceous of Brazil. Arq Do Museu Nacional 63(3):565–596
Carballido JL, Sander PM (2013) Postcranial axial skeleton of *Europasaurus holgeri* (Dinosauria, Sauropoda) from the Upper Jurassic of Germany: implications for sauropod ontogeny and phylogenetic relationships of basal Macronaria. J Syst Paleontol 12(3):335–387
Carballido JL, Pol D, Otero A, Cerda IA, Salgado L, Garrido AC, Ramezani J, Cúneo NR, Krause JM (2017) A new giant titanosaur sheds light on body mass evolution among sauropod dinosaurs. Proc R Soc B 284:20171219. https://doi.org/10.1098/rspb.2017.1219
Carballido JL, Scheil M, Knötschke N, Sander M (2020) The appendicular skeleton of the dwarf macronarian sauropod *Europasaurus holgeri* from the Late Jurassic of Germany and a re-evaluation of its systematic affinities. J Syst Paleontol 18:739–781
Carballido JL, Otero A, Mannion PD, Salgado L, Moreno AP. (in press, this book) Titanosauria: a critical reappraisal of its systematics and the relevance of the South American record. In: Otero A, Carballido JL, Pol D (ed) South American Sauropodomorph dinosaurs: record, diversity and evolution, 1st edn. Springer, Heidelberg
Casal G, Martínez R, Luna M, Sciutto JC, Lamanna M (2007) *Aeolosaurus colhuehuapensis* sp. nov. (Sauropoda, Titanosauria) de la Formación Bajo Barreal, Cretácico Superior de Argentina. Revista Brasileira De Paleontologia 10:53–62
Cerda I, Zurriaguza VL, Carballido JL, González R, Salgado L (2021) Osteology, paleohistology and phylogenetic relationships of *Pellegrinisaurus powelli* (Dinosauria: Sauropoda) from the Upper Cretaceous of Argentinean Patagonia. Cret Res 128:104957
Chiappe L (1993) Enantiornithine (Aves) Tarsometatarsi from the Cretaceous Lecho Formation of Northwestern Argentina. Am Mus Novit 3083:27
Coria RA, Currie PJ (2002) Un gran terópodo celurosaurio en el Cretácico Superior de Neuquén. Ameghiniana 39:9R
Coria RA, Currie PJ, Eberth D, Garrido A (2000) Bird footprints from the Anacleto Formation (Late Cretaceous) in Neuquén Province, Argentina. Ameghiniana 39:453–463
Coria RA, Currie PJ, Eberth D, Garrido A, Koppelhus E (2001) Nuevos vertebrados fósiles del Cretácico Superior de Neuquén. Resúmenes XVII Jornadas Argentinas de Paleontología de Vertebrados. Ameghiniana 38(Suppl):6R
Coria RA, Currie PJ, Paulina-Carabajal A, Garrido A, Koppelhus E (2004) Un nuevo linaje de terópodos en el Cretácico Tardío de Argentina. Ameghiniana 41:42R
Coria RA, Filippi LS, Chiappe LM, García R, Arcucci AB (2013) *Overosaurus paradasorum* gen. et sp. nov., a new sauropod dinosaur (Titanosauria: Lithostrotia) from the Late Cretaceous of Neuquén, Patagonia, Argentina. Zootaxa 3683(4):357. https://doi.org/10.11646/zootaxa.3683.4.2

Coria RA, Salgado L (2005) Mid-Cretaceous turnover of saurischian dinosaur communities: evidence from the Neuquen Basin. In: Veiga GD, Spalletti LA, Howell JA, Schwarz E (eds) The Neuquen Basin, Argentina: a case study in sequence stratigraphy and basins dynamics. Special Publication, London, Geological Society, pp 317–327

Costa Franco-Rosas A, Salgado L, Rosas C, Carvalho IS (2004) Nuevos materiales de titanosaurios (Sauropoda) en el Cretácico Superior de Mato Grosso. Brasil Revista Brasileira De Paleontologia 3(7):329–336

Cunha FS, Rêgo DD, Capilla R (1987) Nova ocorrência de répteis cretácicos no 'Sítio Myzobushi', da Formação Bauru, em Álvares Machado, SP. *Anais do 10º Congresso Brasileiro de Paleontologia.* Rio de Janeiro 155–171

Curry Rogers K, Forster CA (2001) The last of the dinosaur titans: a new sauropod from Madagascar. Nature 412(6846):530–534

D'Emic MD, Wilson JA (2011) New remains attributable to the holotype of the sauropod dinosaur *Neuquensaurus australis*, with implications for saltasaurine systematics. Acta Paleontol Pol 56(1):61–73

Dias ANC, Chemale F, Candeiro CRA, Lana CC, Guadagnin F, Sales ASW (2020) Unraveling multiple tectonic events and source areas in the intracratonic Bauru Basin through combined zircon geo and thermochronological studies. J South Am Earth Sci 103061. https://doi.org/10.1016/j.jsames.2020.103061

Dias-Brito D et al (2001) The Bauru group: a continental Cretaceous unit in Brazil—Concepts based on micropaleontological, oxygen isotope and stratigraphical data. Revue De Paleobiologie 20(1):245–304

Egerton VM (2011) The Geology, Paleontology and Paleoecology of the Cerro Fortaleza Formation, Patagonia (Argentina), Department of Biology, Drexel University, Philadelphia, p 272

Filippi LS, García RA, Garrido CA (2011) A new titanosaur sauropod dinosaur from the Upper Cretaceous of North Patagonia. Argentina Acta Paleontol Pol 56(3):505–520

Filippi LS, Garrido AC (2008) *Pitekunsaurus macayai* gen. et sp. nov., nuevo titanosaurio (Saurischia, Sauropoda) del Cretácico Superior de la Cuenca Neuquina. Argentina Ameghiniana 45:575–590

Filippi LS, Juárez Valieri RD, Gallina PA, Méndez AH, Gianechini FA, Garrido AC (2016) A new titanosaur specimen with highly derived skull from the Santonian of Northern Patagonia, Argentina. XI Congreso de la Asociación Paleontológica Argentina. Gral. Roca, Río Negro. Ameghiniana 54(4). Suplemento Resúmenes, p 20

Filippi LS, Salgado L, Garrido AC (2019) A new giant basal titanosaur sauropod in the Upper Cretaceous (Coniacian) of the Neuquén Basin, Argentina. Cretaceous Res 100:61–81

França MAG, Marsola JCA, Riff D, Hsiou AS, Langer MC (2016) New lower jaw and teeth referred to *Maxakalisaurus topai* (Titanosauria: Aeolosaurini) and their implications for the phylogeny of titanosaurid sauropods. PeerJ 4:e2054. https://doi.org/10.7717/peerj.2054

Gallina PA, Apesteguía S (2011) Cranial anatomy and phylogenetic position of the titanosaurian sauropod *Bonitasaura salgadoi*. Acta Paleontol Pol 56(1):45–60

Gallina PA, Otero A (2015) Reassessment of *Laplatasaurus araukanicus* (Sauropoda: Titanosauria) from the Upper Cretaceous of Patagonia, Argentina. Ameghiniana 52:487–501

Gallina PA, Canale JI, Carballido JL (2021) The earliest known titanosaur sauropod dinosaur. Ameghiniana 58(1):35–51

González Riga BJ (2003) A new titanosaur (Dinosauria, Sauropoda) from the Upper Cretaceous of Mendoza province. Argentina Ameghiniana 40(2):155–172

González Riga BJ, David LO (2014) A new titanosaur (Dinosauria, Sauropoda) from the Upper Cretaceous (Cerro Lisandro Formation) of Mendoza Province. Argentina Ameghiniana 51(1):3–25

González Riga BJ, Lamanna MC, Otero A, Ortíz David LD, Kellner A, Ibiricu LM (2019) An overview of the appendicular skeletal anatomy of South American titanosaurian sauropods, with definition of a newly recognized clade. An Acad Bras Ciênc 91(2):e20180374. https://doi.org/10.1590/0001-3765201920180374

González Riga BJ, Tomaselli MB (2019) Different trackway patterns in titanosaur sauropods: Analysis of new *Titanopodus* tracks from the Upper Cretaceous of Mendoza, Neuquén Basin, Argentina. Cretaceous Res 93:49–59

González Riga BJ, Mannion PD, Poropat SF, Ortiz David L, Coria JP (2018) Osteology of the Late Cretaceous Argentinean sauropod dinosaur *Mendozasaurus neguyelap*: implications for basal titanosaur relationships. Zool J Linn Soc-Lond. https://doi.org/10.1093/zoolinnean/zlx103

Gorscak E, O'Connor PM (2019) A new African titanosaurian sauropod dinosaur from the middle Cretaceous Galula Formation (Mtuka Member), Rukwa Rift Basin, Southwstern Tanzania. PLoS ONE 14:e0211412. https://doi.org/10.1371/journal.pone.0211412

Hechenleitner EM, Leuzinger L, Martinelli AG, Rocher S, Fiorelli LE, Taborda JRA, Salgado L (2020) Two Late Cretaceous sauropods reveal titanosaurian dispersal across South America. Commun Biol 3:622. https://doi.org/10.1038/s42003-020-01338-w

Heredia S, Salgado L (1999) Posición estratigráfica de los estratos supracretàcicos portadores de dinosaurios en el lago Pellegrini, Patagonia septentrional, Argentina. Ameghiniana 36:229–234

Huene F von (1929) Los Saurisquios y Ornitisquios del Cretácico Argentino. Anales del Museo de La Plata, 2° Serie 3:1–196

Hunt AP, Lockley MG, Lucas, SG, Meyer CA (1994) The global sauropod record. Gaia 10:261–279

Ibiricu LM, Casal GA, Martínez RD, Alvarez BN, Poropat SF (2020) New materials and an overview of Cretaceous vertebrates from the Chubut group of the Golfo San Jorge Basin, central Patagonia, Argentina. L S Am Earth Sci 98. https://doi.org/10.1016/j.jsames.2019.102460

Janensch W (1950) Die Wirbelsäule der von *Brachiosaurus brancai*. Palaeontographica 3(Supp 7):27–92

Juárez Valieri RDJ, Díaz SDR (2013) Assignation of the vertebra CPP 494 to *Trigonosaurus pricei* Campos et al. 2005 (Sauropoda: Titanosauriformes) from the Late Cretaceous of Brazil, with comments on the laminar variation among lithostrotian titanosaurs. Bol Mus Nac Hist Nat Parag 17(1):20–28

Keidel J (1917) Über das patagonische Tafelland und ihre ziehungen zu den geologischen erchein-nugen in den Argentinischen Anden gebiet und Litoral. Zeitschrift Der Deutsche Akademie Wiessenschaft 3(5–6):219–245

Kellner AWA, Azevedo SAK (1999) A new sauropod dinosaur (Titanosauria) from the Late Cretaceous of Brazil In: Tomida Y, Rich TH, Vickers-Rich P (eds) Proceedings of the second gondwanan dinosaur symposium, national science museum monographs, vol 15, pp 111–142

Kellner AWA, Campos DA, Azevedo SAK, Trotta MNF, Henriques DDR, Craik MMT, Silva HP (2006) On a new titanosaur sauropod from the Bauru Group, Late Cretaceous of Brazil. Boletim Do Museu Nacional (geologia) 74:1–31

Kellner AWA, Campos DA, Trotta MNF (2005) Description of a titanosaurid caudal series from the Bauru Group, Late Cretaceous of Brazil. Arq Do Museu Nacional 63(3):529–564

Kellner AWA, Rubilar-Rogers D, Vargas A, Suárez M (2011) A new titanosaur sauropod from the Atacama Desert, Chile. An Acad Bras Ciênc 83(1):211–219

Klein N, Sander M (2008) Ontogenetic stages in the long bone histology of sauropod dinosaurs. Paleobiology 34:247–263

Knoll F, Lautenschlager S, Valentin X, Díez Díaz V, Pereda Suberbiola X, Garcia G (2019) First palaeoneurological study of a sauropod dinosaur from France and its phylogenetic significance. PeerJ 7:e7991

Lacovara KJ, Lamanna MC, Ibiricu LM, Poole JC, Schroeter ER, Ullmann PV, Voegele KK, Boles ZM, Carter AM, Fowler EK, Egerton VM, Moyer AE, Coughenour CL, Schein JP, Harris JD, Novas MRD, FE, (2014) A gigantic, exceptionally complete titanosaurian sauropod dinosaur from southern Patagonia. Argentina Sci Rep 4:6196. https://doi.org/10.1038/srep06196

Leanza AF (1972) Andes Patagónicos australes: Geología Regional Argentina, Academia Nacional de Ciencias:689–706

Leuzinger L, Bernasconi SM, Vennemann T, Luz Z, Vonlanthen P, Ulianov A, Baumgartner-Mora C, Hechenleitner EM, Fiorelli LE, Alasino PH (2021) Life and reproduction of titanosaurians:

Isotopic hallmark of midpalaeolatitude eggshells and its significance for body temperature, diet, and nesting. Chem Geol. https://doi.org/10.1016/j.chemgeo.2021.120452
Lydekker R (1893) Contributions to the study of the fossil vertebrates of Argentina. I, The dinosaurs of Patagonia. Anales Del Museo De La Plata, Paleontología 2:1–14
Machado EB, Avilla LDS, Nava WR, Campos DA, Kellner AWA (2013) A new titanosaur sauropod from the Late Cretaceous of Brazil. Zootaxa 3701(3):301–321
Mannion PD, Otero A (2012) A reappraisal of the Late Cretaceous Argentinean sauropod dinosaur *Argyrosaurus superbus*, with a description of a new titanosaur genus. J Vert Paleo 32:614–638
Mannion PD, Upchurch P (2010) A quantitative analysis of environmental associations in sauropod dinosaurs. Paleobiology 36(2):253–282
Mannion PD, Upchurch P, Carrano MT, Barrett PM (2011) Testing the effect of the rock record on diversity: a multidisciplinary approach to elucidating the generic richness of sauropodomorph dinosaurs through time. Biol Rev 86:157–181
Mannion PD, Upchurch P, Jin X, Zheng W (2019) New information on the Cretaceous sauropod dinosaurs of Zhejiang Province, China: impact on Laurasian titanosauriform phylogeny and biogeography. R. Soc. Open Sci. 6:191057. https://doi.org/10.1098/rsos.191057
Marsh OC (1878) Principal characters of American Jurassic dinosaurs. Am J Sci Pt I Ser 411–416
Martinelli AG, Forasiepi AM (2004) Late Cretaceous Vertebrates from Bajo de Santa Rosa (Allen Formation), Río Negro province, Argentina, with the description of a new sauropod dinosaur (Titanosauridae). Revista Del Museo Argentino De Ciencias Naturales 6:257–305
Martínez RDF, Lamanna MC, Novas FE, Ridgely RC, Casal GA, Martínez JE et al (2016) A basal Lithostrotian Titanosaur (Dinosauria: Sauropoda) with a Complete Skull: implications for the Evolution and Paleobiology of Titanosauria. PLoS ONE 11(4):e0151661. https://doi.org/10.1371/journal.pone.0151661
Meyer CA, Marty D, Thüring B, Thüring S, Belvedere M (2021) The Late Cretaceous dinosaur track record of Bolivia—review and perspective. J S AM Earths Sci. https://doi.org/10.1016/j.jsames.2020.102992
Mezzalira S (1959) Nota preliminar sobre as recentes descobertas paleontológicas no Estado de São Paulo, no período 1958–59. Notas Previas Do Instituto Geográfico e Geológico 2:1–7
Mourier T, Jaillard E, Laubacher G, Noblet C, Pardo A, Sigé B, Taquet P (1986) Découverte de restes dinosauriens et mammalien d´âge Crétacé Superieur á la base des couches rouges du synclinal de Bagua (Andes nord-péruviennes): aspects stratigraphiques, sedimentologiques et paléogeographiques concernant la regression fini-Crétacée. Bulletin De La Société Géologique De France 8(1):171–175
Novas FE (2009) The age of dinosaurs in South America. Indiana University Press, p 480
Novas FE, Agnolin F, Rozadilla S, Aranciaga-Rolando A, Brissón-Eli F, Motta M, Cerroni M, Ezcurra M, Martinelli A, D'Angelo J, Álvarez-Herrera J, Gentil A, Bogan S, Chimento N, García-Marsà J, Lo Coco G, Miquel S, Brito F, Vera E, Loinaze V, Fernandez M, Salgado L (2019) Paleontological discoveries in the Chorrillo Formation (upper Campanian-lower Maastrichtian, Upper Cretaceous), Santa Cruz Province, Patagonia, Argentina. Revista Del Museo Argentino De Ciencias Naturales 21(2):217–293
Novas FE, Salgado L, Calvo JO, Agnolin F (2005) Giant titanosaur (Dinosauria, Sauropoda) from the Late Cretaceous of Patagonia. Revista Del Museo Argentino De Ciencias Naturales, Nueva Serie 7:37–41
Nowinski A (1971) *Nemegtosaurus mongoliensis* n. gen., n. sp. (Sauropoda) from the uppermost Cretaceous of Mongolia. Palaeontol Pol 25:57–81
Oliveira FA, Santucci RM (2020) Palynological analysis of coprolites from the Marília Formation, Bauru Group (Upper Cretaceous), Minas Gerais, Brazil. Cretaceous Res 104545. https://doi.org/10.1016/j.cretres.2020.104545
Otero A (2010) The appendicular skeleton of *Neuquensaurus*, a Late Cretaceous saltasaurine sauropod from Patagonia. Argentina Acta Palaeontologica Polonica 55(3):399–426
Otero A, Vizcaíno SF (2008) Hindlimb musculature and function of *Neuquensaurus australis* (Sauropoda: Titanosauria). Ameghiniana 45(2):333–348

Paulina-Carabajal A, Filippi L, Knoll F (2020) Neuroanatomy of the titanosaurian sauropod *Narambuenatitan palomoi* from the Upper Cretaceous of Patagonia, Argentina. Publicación Electrónica De La Asociación Paleontológica Argentina 20(2):1–9
Porfiri JD, Calvo JO (2010) *Panamericansaurus schroederi* gen. nov. sp. nov. Un nuevo Sauropoda (Titanosauridae-Aeolosaurini) de la Provincia del Neuquén, Cretácico Superior de Patagonia, Argentina. Brazilian Geograph J Geosci Human Res Medium 1:100–115
Powell JE (1986) Revisión de los Titanosáuridos de América del Sur. Universidad Nacional de Tucumán, Tesis Doctoral, p 493
Powell JE (1987) Morfología del esqueleto axial de los dinosaurios titanosauridos (Saurischia, Sauropoda) del Estado de Minas Gerais, Brasil. Anais Do X Congresso Brasileiro De Paleontologia, Rio De Janeiro: Sociedade Brasileira De Paleontologia 1:155–171
Powell JE (1992) Osteología de *Saltasaurus loricatus* (Sauropoda, Titanosauridae) del Cretácico Superior del Noroeste argentino. In: Sanz JL, y Buscalioni AD (eds) Los dinosaurios y su entorno biótico, Actas del Segundo Curso de Paleontología en Cuenca. Instituto 'Juan de Valdes', Ayuntamiento de Cuenca, pp 165–230
Powell JE (2003) Revision of South American titanosaurid dinosaurs: palaeobiological, palaeobiogeographical, and phylogenetic aspects. Records of the Queen Victoria Museum 111:1–173
Prasad V, Strömberg CAE, Alimohammadian H, Sahni A (2005) Dinosaur coprolites and the early evolution of grasses and grazers. Science 310(5751):1177–1180
Rubilar-Rogers D, Gutstein CS (2012) Los titanosaurios de Chile y su contexto filogenético y biogeográfico. Boletín Del Museo Nacional De Historia Natural, Chile 61:55–73
Rubilar-Rogers D, Otero RA, Yury-Yáñez RE, Vargas AO, Gutstein CS (2012) An overview of the dinosaur fossil record from Chile. J S Am Earth Sci 37:242–255
Rubilar-Rogers D, Vargas AO, González Riga B, Soto-Acuña S, Alarcón-Muñoz J, Iriarte-Díaz J, Arévalo C, Gutstein CS (2021) *Arackar licanantay* gen. et sp. nov. a new lithostrotian (Dinosauria, Sauropoda) from the Upper Cretaceous of the Atacama Region, northern Chile. Cretaceous Res 124. https://doi.org/10.1016/j.cretres.2021.104802
Salgado L (1996) *Pellegrinisaurus powelli* nov. gen. et sp. (Sauropoda, Titanosauridae) from the Upper Cretaceous of Lago Pellegrini, northwestern Patagonia. Argentina Ameghiniana 33:355–365
Salgado L, Apesteguía S, Heredia SE (2005) A new specimen of *Neuquensaurus australis*, a Late Cretaceous saltasaurine titanosaur from north Patagonia. J Vert Paleo 25(3):623–634
Salgado L, Azpilicueta C (2000) Un nuevo saltasaurino (Sauropoda, Titanosauridae) de la Provincia de Rio Negro (Formación Allen, Cretácico Superior), Patagonia, Argentina. Ameghiniana 37:259–264
Salgado L, Carvalho IS (2008) *Uberabatitan ribeiroi*, a new titanosaur from the Marília Formation (Bauru Group, Upper Cretaceous), Minas Gerais. Brazil Palaeontology 51(4):881–901. https://doi.org/10.1111/j.1475-4983.2008.00781.x
Salgado L, Coria RA (1993) El género *Aeolosaurus* (Sauropoda, Titanosauridae) en la Formación Allen (Campaniano-Maastrichtiano) de la provincia de Río Negro. Argentina Ameghiniana 2(30):119–128
Salgado L, Coria RA (2009) *Barrosasaurus casamiquelai* gen. et sp. nov., a new titanosaur (Dinosauria, Sauropoda) from the Anacleto Formation (Late Cretaceous: early Campanian) of Sierra Barrosa (Neuquén, Argentina). Zootaxa 2222:1–16
Salgado L, Coria RA, Calvo JO (1997) Evolution of titanosaurid sauropods I: Phylogenetic analysis based on the postcranial evidence. Ameghiniana, Buenos Aires 34(1):3–32
Salgado L, Gallina PA, Paulina-Carabajal A (2015) Redescription of *Bonatitan reigi* (Sauropoda: Titanosauria), from the Campanian-Maastrichtian of the Río Negro Province (Argentina). Hist Biol 27:525–548
Sander PM, Mateus OV, Laven T, Knötschke N (2006) Bone histology indicates insular dwarfism in a new Late Jurassic sauropod dinosaur. Nature 441(7094):739–741

Santucci RM, Arruda Campos AC (2011) A new sauropod (Macronaria, Titanosauria) from the Adamantina Formation, Bauru Group, Upper Cretaceous of Brazil and the phylogenetic relationships of Aeolosaurini. Zootaxa 3085:1–33

Santucci RM, Bertini RJ (2001) Distribuição paleogeográfica e biocronológica dos titanosauros (Saurischia, Sauropoda) do Grupo Bauru, Cretáceo Superior do Sudeste Brasileiro. Revista Brasileira De Geociências 31:307–314

Santucci RM, Bertini RJ (2006a) A new titanosaur from western São Paulo State, Upper Cretaceous Bauru Group, Southeast Brazil. Palaeontology 49:59–66

Santucci RM, Bertini RJ (2006b) A large sauropod titanosaur from Peirópolis, Bauru Group, Brazil. Neues Jahrbuch für Geologie und Paläontologie Monatshefte 2006(6):344–360

Schroeter E, Boles Z, Lacovara K (2011) The histology of a massive titanosaur from Argentina and implications for maximum size. Society of Vertebrate Paleontology, Prog Abst 2011, 189

Sereno PC, Wilson JA, Witmer LM, Whitlock JA, Maga A, Ide O, Rowe TA (2007) Structural extremes in a Cretaceous Dinosaur. PLoS ONE 2(11):e1230. https://doi.org/10.1371/journal.pone.0001230

Sickmann ZT, Schwartz TM, Graham SA (2018) Refining stratigraphy and tectonic history using detrital zircon maximum depositional age: an example from the Cerro Fortaleza Formation, Austral Basin, southern Patagonia. Basin Res 30:708–729

Silva JCG Jr, Marinho TS, Martinelli AG, Langer MC (2019) Osteology and systematics of *Uberabatitan ribeiroi* (Dinosauria; Sauropoda): a Late Cretaceous titanosaur from Minas Gerais. Brazil Zootaxa 4577(3):401–438

Soto M, Perea D, Cambiaso A (2012) First sauropod (Dinosauria: Saurischia) remains from the Guichón Formation, Late Cretaceous of Uruguay. J S Am Earth Sci 33(1):68–79

Souza LM, Santucci RM (2014) Body size evolution in Titanosauriformes (Sauropoda, Macronaria). J Evol Biol 27(9):2001–2012

Tütken T (2011) The diet of sauropod dinosaurs: implications of carbon isotope analysis on teeth, bones, and plants. In: Klein N, Remes K, Gee CT, Sander PM (eds) Biology of the sauropod dinosaurs: understanding the life of giants. Indiana University Press, p 344

Ullmann PV, Lacovara KJ (2016) Appendicular osteology of *Dreadnoughtus schrani*, a giant titanosaurian (Sauropoda, Titanosauria) from the Late Cretaceous of Patagonia, Argentina. J Vert Paleo 36, e1225303

Upchurch P, Barrett PM (2004) Dodson P (2004) Sauropoda. In: Weishampel DB, Dodson P, Osmólska H (eds) The Dinosauria, 2nd edn. University of California Press, Berkeley, pp 259–324

Varela AN, Poiré DG, Martin T, Gerdes A, Goin FJ, Gelfo JN, Hoffmann S (2012) U-Pb zircon constraints on the age of the Cretaceous Mata Amarilla Formation, southern Patagonia, Argentina: its relationship with the evolution of the Austral Basin. Andean Geol 39:359–379

Voegele KK, Lamanna MC, Lacovara KJ (2017) Osteology of the dorsal vertebrae of the giant titanosaurian sauropod dinosaur *Dreadnoughtus schrani* from the Late Cretaceous of Argentina. Acta Paleontol Pol 62(4):667–681

Volkheimer W (1973) Observaciones geológicas en el área de Ingeniero Jacobacci y adyacencias (Provincia de Río Negro). Asociación Geológica Argentina Revista 28:13–44

von Huene F (1927) Short review of the present knowledge of the Sauropoda. Memoirs Queensland Museum 9(1):121–126

Walker CA, Dyke GJ (2010) Euenantiornithin birds from the Late Cretaceous of El Brete (Argentina). Irish J Earth Sci 27:15–62

Wilson JA (2002) Sauropod dinosaur phylogeny: critique and cladistic analysis. Zool J Linn Soc-Lond 136:217–276

Wilson JA (2005) Redescription of the Mongolian sauropod *Nemegtosaurus mongoliensis* Nowinski (Dinosauria: Saurischia) and comments on Late Cretaceous sauropod diversity. J Syst Paleo 3:283–318

Wilson JA, Upchurch P (2003) A revision of *Titanosaurus* Lydekker (Dinosauria—Sauropoda), the first dinosaur genus with a “Gondwanan” distribution. J Sys Paleo 1(3):125–160

Eggs, Nests, and Reproductive Biology of Sauropodomorph Dinosaurs from South America

Mariela Soledad Fernández, Bernat Vila, and Miguel Moreno-Azanza

Abstract The South American sauropodomorph egg record is unrivaled in its richness, with Jurassic and Early and Late Cretaceous eggshell occurrences of up to six different oospecies, included in the oofamilies Megaloolithidae, Fusioolithidae and, probably, Faveoloolithidae, and the oldest putative soft-shelled eggs of the fossil record. In addition, numerous clutches, some nests, exquisitely preserved embryos with skin impressions, and delicate embryological structures such as the egg tooth, and perinatal individuals of stem sauropodomorphs have been reported from South America. Thus, it represents one of the most complete oological and developmental records of an extinct clade and provides a unique opportunity to explore its palaeobiology and palaeoecology. In this chapter, the reader will find a detailed revision of the egg-bearing localities of South America, combined with a critical review of all specimens previously referred to as sauropodomorph taxa and parataxa. Detailed account of the impressive sauropodomorph embryo record is provided. Finally, the different Argentinean nesting grounds are discussed in their paleoenvironmental context, to

M. S. Fernández (✉)
Consejo Nacional de Investigaciones Científicas y Técnicas (CONICET), Buenos Aires, Argentina

Instituto de Investigaciones en Biodiversidad y Medioambiente, Universidad Nacional del Comahue, Quintral 1250, +54 294 4428505 (int 109), Bariloche, (8400), Río Negro Province, Argentina

B. Vila
Institut Català de Paleontologia Miquel Crusafont, Universitat Autònoma de Barcelona, c/Escola Industrial 23, Barcelona 08201, Catalonia, Spain
e-mail: bernat.vila@icp.cat

Museu de la Conca Dellà, CP 25650, Isona i Conca Dellà, (Lleida), Catalonia, Spain

M. Moreno-Azanza
GEOBIOTEC, Department of Earth Sciences, NOVA School of Science and Technology, Campus de Caparica, Caparica 2829-516, Portugal
e-mail: mmazanza@unizar.es

Aragosaurus, Recursos Geológicos y Paleoambientes – IUCA, Pedro Cerbuna 12, 50009 Zaragoza, Spain

Espaço NovaPaleo, Museu de Lourinhã, Rua João Luís de Moura 95, 2530-158 Lourinhã, Portugal

A. Otero et al. (eds.), *South American Sauropodomorph Dinosaurs*, Springer Earth System Sciences, https://doi.org/10.1007/978-3-030-95959-3_11

provide an updated picture of the nesting strategies and reproductive traits of South American sauropodomorphs.

Keywords Megaloolithidae · Faveoloolithidae · Fusioolithidae · Dinosaur embryos · Nesting strategies

1 Introduction

The first non-avian dinosaur eggs and eggshells ever found belong to sauropodomorph dinosaurs (Taquet 2001). In 1846, the French geologist Pierre Philippe Émile Matheron found several eggshell fragments in the foothills of the Pyrenees (Taquet 2001). Years later, Father Jean-Jacques Pouech collected several fragments in the same region that he referred to as a large bird (Pouech 1859). But it was not until 1869 when Matheron published his original findings and attributed them to *Hypselosaurus* (Matheron 1869), then thought to be a marine crocodile, but now regarded as a sauropod titanosaur (Depéret 1900; Buffetaut and Le Loeuff 1994). Despite these early discoveries, the lack of both *in ovo* embryos and gravid females ensured that sauropodomorph reproduction remained a mystery. For almost a century, hundreds of eggs were reported from localities in Europe, South America, and Asia – including the Indian subcontinent (Sigé 1968; Mones 1980; Mikhailov 1991; Vianey-Liaud and López-Martínez 1997; Khosla and Sahni 1995, 2003; Mohabey 1998), and eggshells were parataxonomically ascribed to the oofamilies Megaloolithidae –now Megaloolithidae + Fusioolithidae (Vianey-Liaud et al. 2003; Fernández and Khosla 2015; Khosla and Lucas 2020a, b)—and Faveoloolithidae. In absence of definitive association of *in ovo* embryos, the eggs were related to sauropod dinosaurs on the basis of size and circumstantial association with sauropod osteological remains in the same localities or even formations (e.g., Erben et al. 1979; Mones 1980; Vianey-Liaud et al. 1988).

The fossil record of South America ignited our knowledge of sauropodomorph dinosaur reproduction with the discovery of two "rosetta" localities producing definitive evidence of associated eggs and bone record: the basal-sauropodomorph eggs and post-hatchlings of *Mussaurus patagonicus* Bonaparte and Vince 1979 and the first sauropod *in ovo* embryos (Chiappe et al. 1998). In addition to these keystone findings, the South American record of sauropodomorph reproductive biology includes dozens of localities preserving eggshells, eggs, clutches, nest traces, embryos, and nesting grounds from Argentina, Uruguay, Peru, and Brazil (Table 1). As a whole, it provides exceptional information on how these large reptiles reproduced (Grellet-Tinner et al. 2006). Eggshells provide information on the incubation mode by means of the study of microstructure and water vapor conductance (Seymour 1979; Jackson et al. 2008; Grellet-Tinner et al. 2012); the taphonomic analyses of individual eggs potentially yield information on where the incubation occurred (Mueller-Towe et al. 2002); and the spatial study of complete clutches together with information from nest traces help to elucidate the mode how sauropodomorphs laid the egg mass (Chiappe et al.

Table 1 Main localities with sauropodomorph oological records from South America

Country	Province/Department	Locality	Formation and age	Fossil record	Ootaxonomy	References
Peru	Amazonas	Pongo de Rentema	Fundo El Triunfo Fm.; late Campanian–Maastrichtian	Eggshells	*Fusioolithus baghensis*	Mourier et al. (1988); Vianey-Liaud et al. (1997); Fernández and Khosla (2015)
		Fundo El Triunfo				
		El Pintor				
		Morerilla				
	Puno	Laguna Umayo	Vilquechico Fm.; upper Campanian–lower Maastrichtian			Sigé (1968); Kérourio and Sigé (1984); Vianey-Liaud et al. (1997); Fernández and Khosla (2015)
Brazil	Minas Gerais	Mangabeira	Baurú Group; Upper Cretaceous	Egg	indeterminate, 'titanosaurian eggshells'	Price (1951)
		Peirópolis	Marília Fm.; Maastrichtian	Egg and eggshells	Megaloolithidae; cf. *Fusioolithus* (this study)	Magalhães-Ribeiro (1999a, b, 2000); Grellet-Tinner and Zaher (2007)
		Ponte Alta				
Uruguay	Paysandú	Quebracho	Guichón Fm.; Turonian–Santonian	Eggshells	*Sphaerovum* sp.	Goso and Perea (2004); Soto et al. (2009, 2012); Verde (2012)
		Forestal Caja Bancaria Quarry	Mercedes Fm.; Maastrichtian	Eggshells	*Sphaerovum erbeni*	Alonso-Zarza et al. (2011); Cabrera et al. (2020); Veroslavsky et al. (2019)

(continued)

Table 1 (continued)

Country	Province/Department	Locality	Formation and age	Fossil record	Ootaxonomy	References
		White Quarry	Queguay Fm.; Maastrichtian	Eggshells	*Sphaerovum erbeni*	Veroslavsky et al. (2019)
	Río Negro	Unnamed, locality 11 of Mones (1997)	Mercedes Fm.; Maastrichtian	Eggshells	unknown	Mones (1997)
		Algorta, locality 10 of Mones (1997)	Mercedes Fm.; Maastrichtian	Eggshells	Faveoloolithidae indet	Mones (1997); Faccio (1994)
	Durazno	Unnamed, locality 17 of Mones (1997)	Mercedes Fm.; Maastrichtian	Eggs and eggshells	Faveoloolithidae indet	Faccio and Montaña (1994)
	Flores	Trinidad City's Hippodrome (F11)	Asencio Fm.; Maastrichtian	Eggshells	*Sphaerovum erbeni*	Cabrera et al. (2020); Veroslavsky et al. (2019)
	Soriano	Palmitas	Mercedes Fm; Maastrichtian	Eggs, eggshells	*Sphaerovum erbeni* (paratype Nº 713 MNHN-DP)	Mones (1980, 1997); Faccio et al. (1990)
		Arroyo Laureles	Asencio Fm.; Maastrichtian	Eggshells	Faveoloolithidae indet	Mones (1997)
	Colonia	Arroyo de Tala de Miguelete	Asencio Fm.; Maastrichtian	Eggs	*Sphaerovum erbeni* (holotype Nº 714 MNHN-DP)	Mones (1980)
Argentina	Santa Cruz	Laguna Colorada	El Tranquilo Fm.; Early Jurassic	Eggs and hatchlings	indeterminate	Bonaparte and Vince (1979)

(continued)

Table 1 (continued)

Country	Province/Department	Locality		Formation and age	Fossil record	Ootaxonomy	References
		Titanosaur tibia (locality 2) and Puma Cave (locality 4)		El Chorrillo Fm.; upper Campanian–lower Maastrichtian	Eggshells	*Fusioolithus* sp.	Novas et al. (2019)
	Chubut	Huanimán		Cerro Barcino Fm.; Aptian–Albian	Eggs and eggshells	Megaloolithidae; cf. Fusioolithidae (this study)	Argañaraz et al. (2013)
	Río Negro	Casa de Piedra (Los Alamitos)		Angostura Colorada Fm.; Campanian	Eggshells	'Titanosauridae indet.'	Powell (1987)
		Salitral Moreno		Allen Fm.; Campanian–Maastrichtian	Eggs and eggshells	Faveoloolithidae *Patagoolithus salitralensis*	Frenguelli (1951); Powell (1992); Simón (2006)
		Salitral Ojo de Agua area	Cerro de Guerra		Eggshells	Faveoloolithidae ('*Paquiloolithus rionegrinus*')	Powell (1985, 1987, 1992); Salgado et al. (2007); Fernández and Khosla (2015)
			Arriagada I		Eggs	Faveoloolithidae, *Fusioolithus baghensis*	
			Arriagada II		Eggs	Faveoloolithidae	
			Arriagada III		Eggs and eggshells	Faveoloolithidae, *Megaloolithus jabalpurensis*	

(continued)

Table 1 (continued)

Country	Province/Department	Locality		Formation and age	Fossil record	Ootaxonomy	References
		Salinas de Trapalcó-Salitral de Santa Rosa area	Santa Rosa		Eggs	Faveoloolithidae (‘*Paquiloolithus rionegrinus*’)	Salgado et al. (2007); Coria et al. (2010); Fernández and Khosla (2015)
			Santos I		Eggs	Faveoloolithidae (‘*Paquiloolithus rionegrinus*’)	
			Santos II		Eggs *Fusioolithus berthei*	Faveoloolithidae (‘*Paquiloolithus rionegrinus*’)	
			Santos III		Eggs	Faveoloolithidae (‘*Paquiloolithus rionegrinus*’)	
			Santos IV		Eggshells	Faveoloolithidae (‘*Paquiloolithus rionegrinus*’)	
			Mansilla I		Eggshells	*Megaloolithus jabalpurensis Fusioolithus baghensis*	
			Mansilla II		Eggshells	*Megaloolithus jabalpurensis Fusioolithus baghensis*	
			García I		Eggshells	Faveoloolithidae (‘*Paquiloolithus rionegrinus*’) *Fusioolithus baghensis Fusioolithus berthei*	

(continued)

Table 1 (continued)

Country	Province/Department	Locality		Formation and age	Fossil record	Ootaxonomy	References
			García II		Eggshells	Faveoloolithidae (*'Paquiloolithus rionegrinus'*)	
			Cerro Bonaparte		Eggshells	Faveoloolithidae (*'Paquiloolithus rionegrinus'*) Megaloolithidae	
			Berthe I		Eggshells	Faveoloolithidae (*'Paquiloolithus rionegrinus'*)	
			Berthe II		Eggshells	Faveoloolithidae (*'Paquiloolithus rionegrinus'*) Megaloolithidae	
			Berthe III		Eggshells	Faveoloolithidae (*'Paquiloolithus rionegrinus'*) *Megaloolithus megadermus Fusioolithus baghensis*	
			Berthe IV		Eggs and Eggshells	Faveoloolithidae (*'Paquiloolithus rionegrinus'*) *Megaloolithus jabalpurensis Fusioolithus baghensis*	

(continued)

Table 1 (continued)

Country	Province/Department	Locality		Formation and age	Fossil record	Ootaxonomy	References
			Berthe V		Eggshells	Faveoloolithidae (‘*Paquiloolithus rionegrinus*’) *Megaloolithus cylindricus* *Fusioolithus berthei*	
			Berthe VI		Eggshells	Faveoloolithidae (‘*Paquiloolithus rionegrinus*’) *Megaloolithus cylindricus*	
			Barranca La Laguna		Eggshells	Faveoloolithidae (‘*Paquiloolithus rionegrinus*’)	
			Cerro Tortugas		Eggshells	Faveoloolithidae (‘*Paquiloolithus rionegrinus*’) Megaloolithidae	
			Cerro Laguna Trapalcó		Eggs	Megaloolithidae	
		Yaminué			Eggs, clutches	Faveoloolithidae (‘*Paquiloolithus rionegrinus*’)	Manera de Bianco (1996)
		Gonzalito 1			Eggs and eggshells	Faveoloolithidae	Fernández et al. (2017)
	Neuquén	Neuquén City		Río Colorado Fm.; middle–upper Santonian	Eggs	*Megaloolithus jabalpurensis*	Calvo et al. (1997)

(continued)

Table 1 (continued)

Country	Province/Department	Locality	Formation and age	Fossil record	Ootaxonomy	References
		Auca Mahuevo, Los Barreales	Anacleto Fm; early Campanian	Eggs, clutches, nests	*Fusioolithus baghensis* (levels 1–3) *Megaloolithus patagonicus* (level 4)	Chiappe and Dingus 2001; Garrido (2010); Fernández and Khosla (2015)
	La Pampa	La Pampa	Colorado Fm.; Campanian–Maastrichtian	Eggshells	Faveoloolithidae (‘*Paquiloolithus rionegrinus*’)	Casadío et al. (2002)
	Entre Ríos	Ita-i-cora	Puerto Yeruá Fm.; Late Cretaceous	Egg, eggshells	cf. *Sphaerovum erbeni*	De Valais et al. (2003)
	La Rioja	Pampa de la Virgen and Los Cajones/ Sanagasta	Los Llanos Fm.; Early Cretaceous or Campanian–Maastrichtian	Eggs and eggshells	Faveoloolithidae	Hünicken et al. (2001); Tauber (2007); Grellet-Tinner and Fiorelli (2010)
		Tama		Eggs, clutches	*Megaloolithus* sp. (this study)	Hechenleitner et al. (2016a, b)
		Quebrada de Santo Domingo	Ciénaga del Río Huaco Fm.; Campanian–Maastrichtian	Eggs, clutches	indeterminate, ‘titanosaurian eggshells’	Hechenleitner et al. 2020
	Salta	El Bordo	Yacoraite Fm.; Maastrichtian	Eggs	Fusioolithidae	Capurro et al. (2019)

2004; Sander et al. 2008; Vila et al. 2010; Khosla and Lucas 2020a). In this chapter, we present a summary of all sauropodomorph dinosaur egg evidence found in South America, and we review and discuss the classification, nesting environments, and reproductive biology of this group of extinct dinosaurs.

Institutional Abbreviations CONICET: Consejo Nacional de Investigaciones Científicas y Técnicas; **CRILAR-PV**: Centro Regional de Investigaciones Científicas y Transferencia Tecnológica de La Rioja, colección Palentología de Vertebrados; MCF-PVPH: Museo "Carmen Funes", colección de Paleontología de Vertebrados, paleohistología; **MML-Ph**: Museo Municipal de Lamarque, colección de Paleohistología; **MML-PV**: Museo Municipal de Lamarque, colección de Paleontología de Vertebrados; **MUC-PV**: Museo de la Universidad Nacional del Comahue, Colección Paleontología de Vertebrados.

Other abbreviations PLM: Polarized light microscope; **SEM**: Scanning electron microscope.

2 Sauropodomorph Egg Localities from South America

The South American record of the sauropodomorphs is based on dozens of localities preserving eggshells, eggs, embryos and hatchlings, clutches, nest traces, and nesting grounds from Argentina, Brazil, Peru, and Uruguay (Fig. 1, Table 1).

2.1 Brazil

The first sauropodomorph egg found on the continent dates back to 1951 when Llewellyn Igor Price found an isolated egg in the Upper Cretaceous Baurú Formation (Minas Gerais state) of Brazil and related it to the French sauropod *Hypselosaurus* because of the similarities with the eggs found in France (Price 1951). As *Hypselosaurus* became a *nomen dubium* (Le Loeuff 1993), the egg was reassigned to the family Titanosauridae. Later works of Claudia María Magalhães-Ribeiro in the early 2000s reported two more localities near Uberaba with sauropod eggs and eggshells (Magalhães-Ribeiro 1999a, 2000).

2.2 Uruguay

In Uruguay, eggs and eggshells were originally reported by Álvaro Mones in 1980 at the localities of Arroyo Tala de Miguelete and Palmitas (the Campanian–Maastrichtian Asencio and Mercedes Formations, respectively), where the author described the holotype and paratype of a newly erected ootaxon named *Sphaerovum*

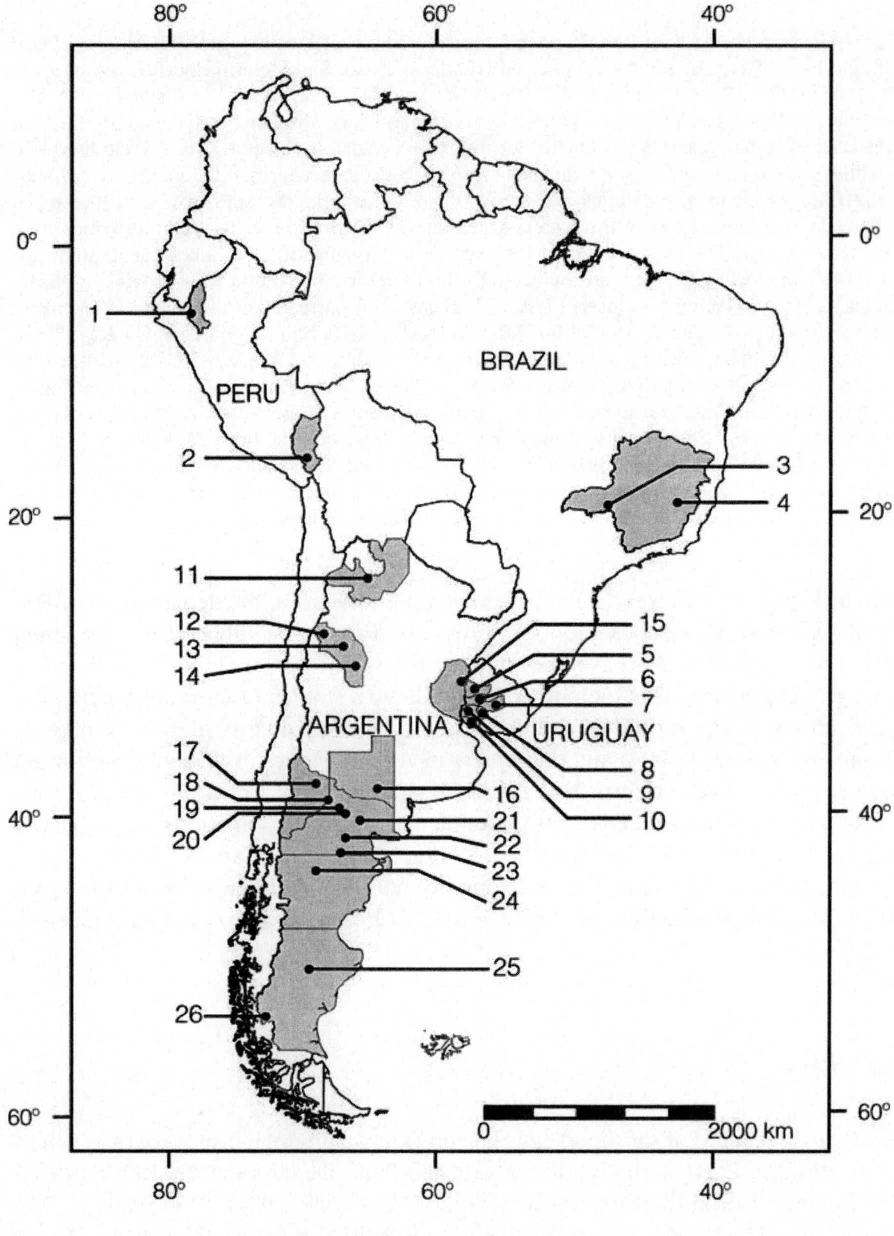
80°
60°
40°
0°
20°
40°
60°
PERU
BRAZIL
ARGENTINA
URUGUAY
1
2
3
4
5
6
7
8
9
10
11
12
13
14
15
16
17
18
19
20
21
22
23
24
25
26
0
2000 km

◂**Fig. 1** Geographic distribution of the main sauropodomorph egg localities in South America. Peru: 1, Bagua basin (Pongo de Rentema, Fundo El Triunfo, El Pintor, and Morerilla localities; Amazonas Department); 2, Laguna Umayo locality (Puno Department). Brazil: 3, Peirópolis and Ponte Alta localities; 4, Mangabeira locality (Minas Gerais Department). Uruguay: 5, Quebracho, Forestal Caja Bancaria Quarry, and White Quarry localities (Paysandú province); 6, Algorta and unnamed localities (Río Negro Province); 7, Unnamed locality (Durazno Province); 8, Trinidad City's Hippodrome (Flores Province); 9, Palmitas and Arroyo Laureles localities (Soriano Province); 10, Arroyo de Tala de Miguelete (Colonia Province). Argentina: 11, El Bordo locality (Salta Province); 12, Quebrada de Santo Domingo locality; 13, Pampa de la Virgen and Los Cajones/Sanagasta localities; 14, Tama locality (La Rioja Province); 15, Ita-i-cora locality (Entre Ríos Province); 16, La Pampa locality (La Pampa Province); 17, Auca Mahuevo-Los Barreales localities; 18, Neuquén city locality (Neuquén Province); 19, Salitral Moreno localities; 20, Salitral Ojo de Agua area (Cerro de Guerra, Arriagada I, Arriagada II, and Arriagada III localities); 21, Salinas de Trapalcó-Salitral de Santa Rosa areas (Santa Rosa, Santos I, Santos II, Santos III, Santos IV, Mansilla I, Mansilla II, García I, García II, Cerro Bonaparte, Berthe I, Berthe II, Berthe III, Berthe IV, Berthe V, Berthe VI, Barranca La Laguna, Cerro Tortugas, and Cerro Laguna Trapalcó localities); 22, Yaminué locality; 23, Casa de Piedra locality (Río Negro Province); 24, Huanimán locality (Chubut Province); 25, Laguna Colorada locality; 26, Titanosaur tibia and Puma Cave localities (Santa Cruz Province). See Table 1 for details. Maps by Free Vector Maps (http://freevectormaps.com)

erbeni (Mones 1980). New material recovered afterward in the departments of Río Negro, Soriano, Durazno, Colonia, Flores, and Paysandú (Table 1) was assigned to this oogenus or to the Faveoloolithidae, an oofamily historically considered to belong to sauropods. This included the middle 90 s studies of Guillermo Faccio in the localities of Algorta and an unnamed locality at Durazno Province (Faccio 1994; Faccio and Montaña 1994) and the review of Álvaro Mones (1997) which reported another locality in Río Negro. Two decades later, the recent works, mostly biostratigraphical studies in the Cretaceous Guichón and Queguay formations, have yielded isolated eggshells at the localities of Quebracho, Forestal Caja Bancaria Quarry, White Quarry, and Trinidad City's Hippodrome (Goso and Perea 2004; Soto et al. 2008, 2012; Alonso-Zarza et al. 2011; Verde 2012; Veroslavsky et al. 2019; Cabrera et al. 2020).

2.3 Peru

The Peruvian record of sauropod eggs became known after the discovery of eggshell material in the Laguna Umayo locality, at the Puno Department, by Bernard Sigé and Philippe Kérourio (Sigé 1968; Kerourio and Sigé 1984). Northwards, field prospects of Thomas Monier and colleagues identified four localities at the Bagua basin (Amazonas Department) which yielded a handful of eggshells (Mounier et al. 1988) that were later studied by Monique Vianey-Liaud (Vianey-Liaud et al. 1997).

2.4 Argentina

In Argentina, evidence of reproductive behavior in sauropodomorphs has been reported in the Provinces of Santa Cruz, Chubut, Río Negro, Neuquén, La Rioja, La Pampa, Entre Ríos and Salta. In Santa Cruz Province the Jurassic deposits of the Laguna Colorada Formation (previously thought to be Late Triassic, Smith et al. 2014) yielded seven hatchling individuals of *Mussaurus patagonicus* that were associated with a pair of eggs and scattered eggshells (Bonaparte and Vince 1979). Four decades later, and in the same province, Fernando Novas and colleagues (2019) found two eggshell localities (Titanosaur tibia-locality 2 and Puma Cave-locality 4) in the upper Campanian–lower Maastrichtian outcrops of El Chorrillo Formation. Northwards, in Chubut Province, Eloísa Argañaraz and colleagues (2013) studied the first eggs and eggshell fragments at the Lower Cretaceous (Aptian–Albian; Cerro Barcino Formation) Huanimán locality. In Río Negro Province, the first putative sauropod egg was collected from the basal part of the "rocanense" bed, a few kilometers from General Roca city (Frenguelli 1951). In the same Province, the early works conducted by Jaime Powell (1985, 1987) in the Allen Formation (middle Campanian–early Maastrichtian) yielded eggshells and isolated eggs from the localities of Casa de Piedra (Los Alamitos), Salitral Ojo de Agua, and Salitral Moreno. A decade later, Teresa Manera de Bianco (1996) reported the egg locality of Yaminué, a set of multiple eggshell-bearing levels from some of which the author described two well-preserved egg clutches. In Allen Formation outcrops, near the first findings of Jaime Powell, Claudia Maria Magalhães-Ribeiro (1997) and María Edith Simón (1999, 2006) studied the egg material from Salitral Moreno and Bajo de Santa Rosa. These works were followed by a field survey in Salinas de Trapalcó and Bajo/Salitral de Santa Rosa by Leonardo Salgado and colleagues who reported further egg-levels and egg-sites with a remarkable record of faveoloolithid and megaloolithid eggshells, eggs, clutches, and nests (Salgado et al. 2007; Coria et al. 2010). Westwards, Jorge Calvo and colleagues described the first oospecies (*Megaloolithus patagonicus*) from Neuquén Province (Calvo et al. 1997), and in November 1997 Luis Chiappe and Lowell Dingus discovered the iconic Auca Mahuevo nesting locality (Chiappe and Dingus 2001) in the Anacleto Formation (early to middle Campanian age; Dingus et al. 2000). In the Auca Mahuevo locality, together with the Los Barreales locality, outcrops with eggs extend over an area of approximately 60 km^2 (Garrido 2010). In the early 2000s new sauropod egg records were found in La Pampa and La Rioja provinces. Silvio Casadío and colleagues studied dinosaur eggshells from the fluvial sedimentary rocks of the Upper Cretaceous Colorado Formation exposed in the south-eastern of La Pampa (Casadío et al. 2002). Mario A. Hünicken and colleagues discovered eggs and clutches in La Rioja province in 2001 after stratigraphic fieldwork at the Los Llanos Formation (Hünicken et al. 2001; Hünicken 2005). Later, Adan Tauber described the eggs and eggshells of the main localities, named Pampa de la Virgen y Los Cajones in the Sanagasta valley (Tauber 2007). At the same sites (re-named as "Sanagasta") Gerard Grellet-Tinner and Lucas E. Fiorelli collected up to 80 clutches of faveoloolithid eggs (Grellet-Tinner and Fiorelli 2010; Fiorelli et al.

2013). Still, in the exposures of the Los Llanos Formation and only 180 km south from the Sanagasta localities, the Tama nesting locality was discovered in 2016 by Martín Hetchenleitner and colleagues. The authors reported a 200 m-long outcrop preserving five distinct egg clutches. More recently, the Quebrada de Santo Domingo yielded titanosaurian egg clutches and eggshells in a site that "is one of the largest nesting sites documented worldwide" and "perhaps is the largest nesting site worldwide" (Hechenleitner et al. 2020, p. 2). Egg clutches were found in three distinct stratigraphic levels that extend over more than three kilometers in the Ciénaga del Río Huaco Formation (Campanian–Maastrichtian). The northernmost province with sauropod egg material is that of Salta, where recently Ivan Capurro and colleagues (2019) reported three subspherical eggs and some fragments of eggshells found in Yacoraite Formation (Maastrichtian).

3 Eggs and Eggshells

All three hard-shelled oofamilies related to sauropodomorphs are prominently recorded in South America (Megaloolithidae, Fusioolithidae, and Faveoloolithidae). In addition, the eggs of the early branching sauropodomorph *Mussaurus patagonicus* have been recently redescribed as soft-shelled instead of mineralized (Norell et al. 2020) although this claim has been disputed (Alleon et al. 2021; but see Wiemann and Briggs 2021 for a reply). We describe each of these oofamilies' records from South America below.

3.1 Megaloolithidae

Megaloolithid eggs are spherical or ellipsoidal, with compactituberculate outer surface ornamentation. They are characterized by isolated, non-bifurcating, spherulitic shell units (spherulites) that usually comprise the whole eggshell thickness, although aborted shell units are common in some oospecies (Mikhailov 1997; Moreno-Azanza et al. 2016). Within Megaloolithidae, oogenera and oospecies are usually distinguished by macroscopic features, such as egg size or eggshell thickness, and some histo- and ultrastructural features, such as abundance and shape of growth lines, and the presence of bifurcating pore channels (Vianey-Liaud and López-Martínez 1997). Three oospecies of this oofamily have been described in South America, uniquely in Argentina.

3.1.1 *Megaloolithus jabalpurensis* Khosla and Sahni (1995)

It was originally reported as *M. patagonicus* from the Rio Colorado Super Group (Santonian–early Campanian) in Neuquén city (Neuquén Province, Argentina; Calvo et al. 1997). Hundreds of eggshell fragments from at least two levels were collected, and although no complete eggs are known, half eggs allowed Calvo et al. (1997) to estimate the egg size. This oospecies has also been reported in the slightly younger Allen Formation in the Salitral Ojo de Agua and Bajos de Santa Rosa sections (Fernández 2013, 2016; Fernández and Khosla 2015; see Table 1 for details in the localities). The eggs of *Megaloolithus jabalpurensis* are medium-sized (140 to 160 *mm* in diameter) and are among the thinnest *Megaloolithus* eggs from South America, with an eggshell thickness of 1.0 to 2.38 mm. This oospecies is characterized by the teardrop-shaped and isolated shell units, where the shell units increase in diameter toward the external surface (Fig. 2a–c, Box 1).

3.1.2 *Megaloolithus cylindricus* Khosla and Sahni (1995)

It is restricted to the Upper Cretaceous of Argentina and has been recorded in the Bajo de Santa Rosa (Allen Formation, late Campanian–early Maastrichtian), in Río Negro Province. The eggs have variable sizes, from 120 to 200 mm in diameter, and present an intermediate thickness between the other two Megaloolithidae oospecies reported in South America, with an eggshell thickness ranging between 1.7 and 3.5 mm. Cylindrical shell units give the oospecies their name, with the shell units having the same diameter through most of the external surface. This translates to an outer surface with more conspicuous compactituberculate ornamentation than other *Megaloolithus* oospecies from South America (Fig. 2d, e, Box 1).

3.1.3 *Megaloolithus megadermus* Mohabey 1998

It is the third *Megaloolithus* oospecies identified in South America, and up to now has only been recorded at a single locality in Argentina (Berthe II, Bajos de Santa Rosa; Fernández 2013; Fernández and Khosla 2015). These oospecies are easily identifiable due to their eggshell thickness, which ranges from 4.0 to 4.8 mm. No complete eggs have been identified of these oospecies in Argentina, but the type material from India shows that eggs can be slightly smaller –130 to 180 mm in diameter—than other Megaloolithidae (Fig. 2f, g, c).

Finally, eggs and clutches were reported at Tama locality (La Rioja Province, Argentina) by Hechenleitner et al. (2016a). The specimens are highly diagenitized, and these authors described the eggshell as "... typical titanosaur eggshell units [..] that consist of acicular calcite crystals radiating from nucleation centers [..] the shape of numerous single, wide, vertical and Y-shaped pore canals that branch outwards ([...] with funnel-shaped apertures" (Hechenleitner et al. 2016a, p. 8). These authors restricted their comparisons to a limited number of putative sauropod eggs and suggested that they are different from those from Auca Mahuevo egg-bearing levels 1 to 3, but strongly resemble those of Auca Mahuevo level 4. According to these authors, the Auca Mahuevo eggshells from egg-level 1 to 3 lack clear limits between units, and have tubocanaliculate pore canals, two characters that differentiate from

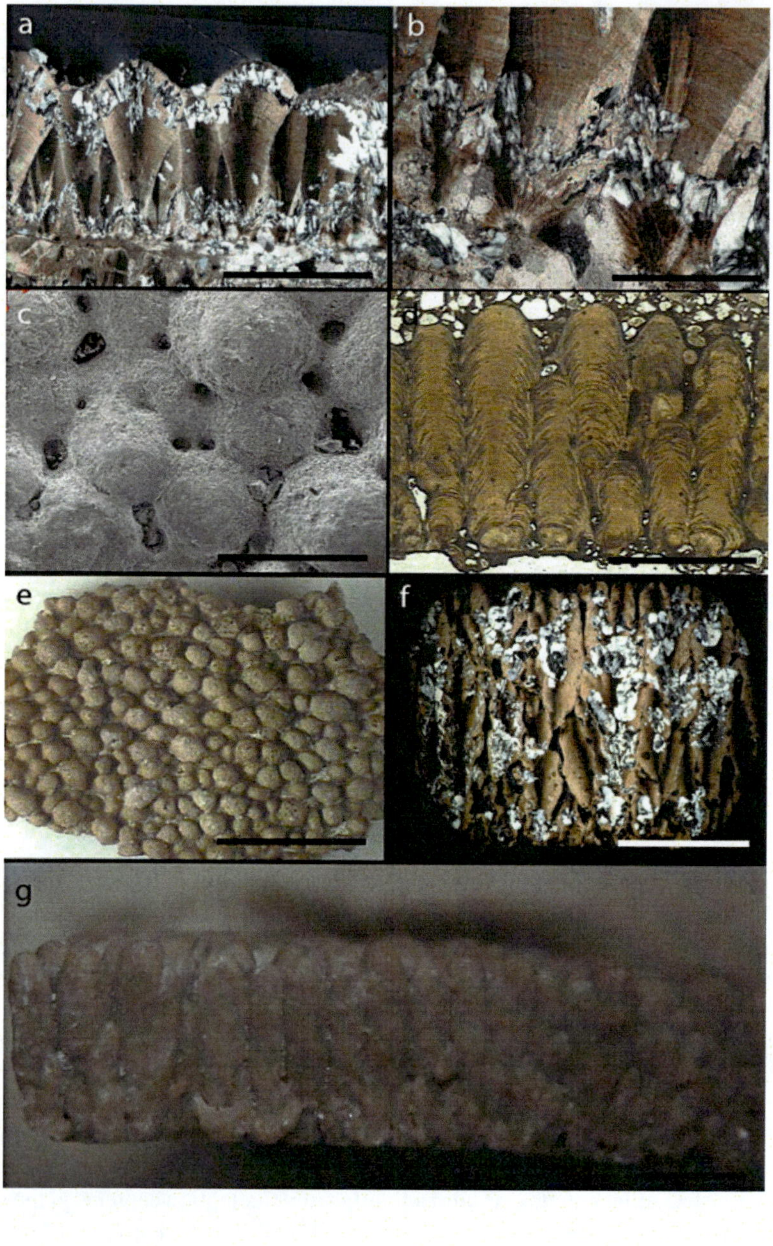
a
b
c
d
e
f
g

◂**Fig. 2** Megaloolithid oospecies from South America. **a, b** *M. jabalpurensis* (MML-PH 1283) collected in Mansilla II, egg level 3, Bajos de Santa, Río Negro. **a** Radial thin section under PLM shows sharply separated shell units, growth lines arched. **b** Detail of shell units with fan-shaped crystals, **c** *M. jabalpurensis* under SEM, picture shows compactituberculate ornamentation, and pore aperture. **d, e** *M. cylindricus*, PLM (MML-PH 1131) collected from Berthe V, egg level 4, Bajos de Santa Rosa, Río Negro, Argentina). **d** Picture shows radial thin section with shell units with straight limits between shell units. Growing lines arched and cylindrical shape of the shell units. **e** External surface shows compactituberculated ornamentation. **f** *M. megadermus* (MML-PH 1143) under common light microscope. Discrete shell units, tall and narrow; lateral margins of spheroliths are straight; shell units show accretion lines. Fan-shaped shell units. Average height/width ratio is 9.6:1. Short basal caps (less than one-tenth of spherolith). **g** *M. megadermus* (MML-PV 914) from Río Negro Province. Radial view under binocular microscope, shows long shell units. Scale bar equals a 1 mm, b 300 μm, c 1 mm, d 2 mm, e 5 mm, f 2 mm, g 2 mm

the Tama material, which has single units without fusion and has bifurcating pore canals. It is noteworthy that the eggshell used by the authors for comparison material (MCF-PVPH-444) was originally reported as coming from the egg-level 3 (Grellet-Tinner et al. 2004), but afterward, the authors reported it from the Auca Mahuevo site without any reference to the precise egg-level (Grellet-Tinner et al. 2006), and finally, they located it from egg-level 4 (Grellet-Tinner et al. 2012). Thus, the materials from Tama need a deeper microstructural and ultrastructural description (including thin sectioning the eggshells) and further comparisons. Currently, these specimens fit better with Megaloolithidae due to their sharply separated shell units, reflected in its external ornamentation with single individualized nodes, with no evidence of fusion of the units.

3.2 *Fusioolithidae*

Fusioolithidae is a mono-oogeneric oofamily recently erected to include oospecies previously belonging to Megaloolithidae, but that differ from true megaloolithids in having the shell units fused in the upper half to upper third of the eggshell (Fernández and Khosla 2015). The fusion of the shell units results in less discrete shell unit contours, growth lines crossing adjacent shell units, and some degree of coalescence in the nodes on the outer surface ornamentation, which combines isolated nodes with vermicular ridges comprising several shell units. This oofamily has a Gondwanan distribution, as evidenced by the reports of eggshells with fused units in South America (Argentina and Peru: Sigé 1968; Kerourio and Sigé 1984; Mourier et al. 1988; Vianey-Liaud and López-Martínez 1997; Chiappe et al. 1998; Magalhães-Ribeiro 1999a, 2000; Simón 1999; Fernández and Khosla 2015), India (Khosla and Sahni 1995; Fernández and Khosla 2015; Kapur and Khosla 2019; Khosla 2019) and probably Africa (Vianey-Liaud and García 2003), but also fusioolithid egg shells have been found in Europe (Spain and France: Vianey-Liaud and López-Martínez 1997; Garcia and Vianey-Liaud 2001; Sellés et al. 2013; Sellés and Vila 2015).

The erection of Fusioolithidae occurred after a long controversy on the Auca Mahuevo oospecies, mostly due to misunderstandings with the egg levels and the provenance of the eggshells, particularly those that have been related with titanosaur embryos (egg-level 3). Originally, Chiappe et al. (1998) assigned the Auca Mahuevo eggshells from egg-level 3 ("embryo quarry") to the Megaloolithidae oofamily, and later Grellet-Tinner et al. (2004) assigned eggs and eggshells presumably from the same level to *Megaloolithus patagonicus*, an oospecies described by Calvo et al. (1997) based on materials from Neuquén. However, important differences exist in the eggshell sample of Grellet-Tinner et al. (2004). The specimens MCF-PVPH 262, 250, 147, 263, and 445 show fused shell units and coalescent nodes (Grellet-Tinner et al. 2004, figs. 2, 3), whereas the eggshell MCF-PVPH-444 have bifurcating Y-shaped pore channels, non-fused units, and non-coalescent and more prominent ornamentation (Grellet-Tinner et al. 2004, Fig. 4). In fact, Grellet-Tinner et al. (2012) indicated that the sample MCF-PVPH-444 was from level 4, and Fernández and Khosla (2015) in their parataxonomical review pointed out that the rest of the samples studied by Chiappe et al. (1998) and Grellet-Tinner et al. (2004) were indistinguishable from *Patagoolithus salitralensis* (now *Fusioolithus baghensis*, see below). *Patagoolithus salitralensis* had been described some years before in the Salitral Moreno area by Simón (2006), who clearly distinguished it from *Megaloolithus patagonicus*. Therefore, significant differences between the eggshells from egg levels 1–3 and egg-level 4 suggest that they cannot be attributed to the same ootaxa, and probably taxa. Particularly, all eggshells from levels 1–3 could be referred to *P. salitralensis* (now *Fusioolithus baghensis*, see below) whereas the sample MCF-PVPH-444 from level 4 fits the diagnosis of *Megaloolithus patagonicus*. This may explain the ongoing controversy between authors working in the same area (see Jackson et al. 2008; Grellet-Tinner et al. 2012) on the presence of both the Y-shaped bifurcating nodes and the horizontal pore system between the *membrana testacea* and the bases of the shell units.

Later, Fernández and Khosla (2015) erected a new oofamily (Fusioolithidae) and oogenus (*Fusioolithus*) based on their interpretations of ultrastructure, microstructure, and macrostructural features and consider *P. salitralensis* as a junior synonym of the *Fusioolithus baghensis*. *Fusioolithus* is diagnosed by both the fusion of shell units and the presence of subhorizontal growth lines crossing the boundaries between adjacent shell units. In addition, the ornamentation shows some coalescent nodes forming short ridges, not present in Megaloolithidae. Bravo and Gaete (2015) objected that the oogenus had been erected with only one character, the fusion of the shell units, and the authors speculated that "other causes such as diagenetic processes (e.g., recrystallization) may also play a role in the fusion of adjacent units" (Bravo and Gaete 2015, p. 8). We agree with Bravo and Gaete (2015) that some variation on the degree of fusion can be observed within the same egg, and some occasional sharp boundaries can be observed, but the dominant condition in Fusioolithidae eggshells is shell units which are indistinguishable in the upper third of the eggshell. Concerning the fused nodes, Bravo and Gaete (2015) argued that this character can be obscured by erosion of the outer surface ornamentation. This can indeed be the case, for instance in the sample figured by Bravo and Gaete (2015). In there, erosion and abrasion are evident and can be easily regarded as a taphonomic alteration. However, the specimens from Auca Mahuevo showing the typical fusioolithid condition are recovered in clutches and sometimes preserved *in ovo* embryos, two pieces of evidence that suggest transport and erosion are highly unlikely to have occurred. Regarding other possible diagenetic alterations that may obscure this condition, Fernández and Matheos (2011) studied the recrystallization mechanisms in Fusioolithidae eggshells and documented them after a complete study on cathodoluminescense of eggshells. In this study, no evidence of further recrystallization that may result in the taphonomic fusion of shell units was found, supporting the hypothesis that this fusion occurs during eggshell growth and is thus a valid anatomical and diagnostic feature. Thus, we consider Fusioolithidae and *Fusioolithus* valid ootaxa, but agree with Bravo and Gaete (2015) that determination of the parataxonomical identity of isolated eggshell fragments may induce errors.

Currently, only one oogenus, *Fusioolithus*, and two oospecies, *Fusioolithus baghensis and Fusioolithus berthei*, have been described in South America. *Fusioolithus baghensis* eggs are spherical (14 to 20 cm in diameter) and their shell thickness ranges between 1 and 1.7 mm. The shell units—or spherulites—are conical and are frequently fused together. These oospecies have been identified in South America, India, and Europe (Table 1, Box 1). *Fusioolithus berthei* is only known on its type locality, Santos II-B, in the Bajo de Santa Rosa (Río Negro Province, Argentina). No complete eggs have been recovered, but associated eggshell fragments are significatively thicker than *Fusioolithus baghensis* (2.45–2.9 mm).

In Argentina, other eggs and eggshells have been assigned to Fusioolithidae. In Santa Cruz and Salta Provinces, the microstructural patterns observed in the eggshells from the "titanosaur tibia" and "Puma cave" localities allow the authors to refer them to *Fusioolithus* oosp. (Novas et al. 2019), and Capurro et al. (2019) assigned the eggs and eggshells from Salta to Fusioolithidae, respectively. Other putative fusioolithid reports come from Argentina and Brazil. In Chubut Province, the eggshell morphology of the Huanimán eggs (Argañaraz et al. 2013) may be referred to Fusioolithidae on the basis of the shape and fusion of the shell units, although they have suffered intense recrystallization, thus hindering a definitive attribution. In addition, a secondary pore canal system, located at the upper third of the eggshell, has been described in these eggshells (Argañaraz et al. 2013), a feature not described in other Fusioolithidae eggs. Thus, we tentatively refer to these materials to ?Fusioolithidae. In Brazil, the eggshells of Péiropolis and Ponte Alta localities (Mina Gerais) show coalescent nodes, a range of diameters, thickness, ultrastructure, partially fused shell units, and pore apertures (Magalhães-Ribeiro 1999a, b, 2000; Grellet-Tinner and Zaher 2007), a set of characters that closely matches the oospecies *Fusioolithus baghensis* (Fernández and Khosla 2015). In both samples, a detailed study is needed to confirm such assignment, and we regard them here as cf. *Fusioolithus*.

3.3 *Faveoloolithidae*

The oofamily Faveoololithidae comprises large subspherical eggs with a filispherulithic eggshell. This oofamily was previously described on the basis of materials from China and Mongolia, to include eggs with smooth or slightly rugose outer surfaces characterized by a complex pore system formed by numerous interconnected pores—multicanaliculate pore system sensu Mikhailov (1991)—that isolate thin spherulitic shell units (filispherulithic morphotype sensu Mikhailov (1991)—(Zhao and Ding 1976; Zhao 1979). The South American "faveoloolithids", share the egg shape, multicanaliculate pore system, and the filispherulithic morphotype, but are significantly larger (150 to 220 mm in diameter), have thicker eggshells (2.1 to 7 mm), and show compactituberculated ornamentation on the outer surfaces. So we choose to follow Simón (2006) in considering them Faveoloolithidae. Probably due to its thickness, this eggshell has a higher resistance to erosion and dissolution (Grellet-Tinner and Fiorelli 2010; Fiorelli et al. 2013), the "faveoloolithid" eggs and eggshells of South America have a high potential of fossilization and are easy to identify. This makes them one of the most common egg remains in the oological fossil record (Fig. 1; Table 1), having been recovered in a number of localities in Argentina (Frenguelli 1951; Powell 1992; Manera de Bianco 1996; Simón 1999, 2006; Tauber 2007; Salgado et al. 2007; Grellet-Tinner and Fiorelli 2010; Fernández 2016) and Uruguay (Alonso-Zarza et al. 2011; Soto et al. 2012; Cabrera et al. 2020). The high porosity of the South American Faveoloolithidae (see Hechenleitner et al. 2016b for a detailed study) results in a high permeability of the eggshell, which has implications for the reproductive biology of the egg layer (because of the high porosity of the eggshells, the egg needs to address a high evapotranspiration; Deeming 2006) and its preservation (as the eggshell is equally permeable to diagenetic fluids during fossilization, South American "faveoloolithid" eggs have often highly recrystallized eggshells, and often show crystal growths inside (Grellet-Tinner and Fiorelli 2010; Grellet-Tinner et al. 2012).

Sphaerovum erbeni was described originally by Mones (1980) based on silicified partial eggs. Later, because of the diagenetically altered characters found in the holotype, Grellet-Tinner et al (2012) argued that it should be considered a *nomen dubium*. Whereas we acknowledge that the holotypic material is diagenetically altered and thus prevents the erection of *Sphaerovum erbeni* as a new ootaxon, we note that the *nomen dubium* condition proposed by Grellet-Tinner et al. (2012) is not formally well-supported as it lacks an in-depth discussion of the characters.

Simón (2006) described isolated eggshell fragments from different localities in Argentina and attributed them to Faveoloolithidae oogenus and oospecies indet. The author addressed the differences between the Argentinean and Asian (Mongolian and Chinese) Faveoloolithidae, and with the Uruguayan *Sphaerovum erbeni*, with particular focus on its slightly greater egg size. Nevertheless, Simón (2006) estimated the egg size from isolated egg fragments, a fact that undoubtedly can entail some degree of error. In her unpublished dissertation (Simón 1999), she also referred the Argentinean eggshells to a newly erected ootaxa "*Paquiloolithus rionegrinus*" and used it in a later conference abstract (Simón 2002). However, "*Paquiloolithus rionegrinus*" is currently unavailable, as its publication did not meet the criteria established in the International Code of Zoological Nomenclature (International Commission for Zoological Nomenclature and Ride, 1999, art. 8.4), and we will avoid its usage pending a future correct erection. Thus, we will refer that Argentinean material to Faveoloolithidae indet. Therefore, a deep revision of the South American Faveoloolithidae is needed, to address its parataxonomic status and several additional controversies associated (see Sect. 4.4).

Box 1 [Parataxonomy of Fossil Eggs From South America]

South America is one of the richest areas with fossil eggshells, eggs, and nests, and six oospecies have been described there. Here, we present a synonymy list for each dinosaur oofamily found in South America.

Oofamily Megaloolithidae Zhao 1975 (emend. 1979).

Diagnosis Zhao (1979, 4) diagnosed the family as follows: "The eggs are large and round, with both transverse and longitudinal diameters over 100 mm. External surface ornamentation as compact, small, and swollen projected striations. Eggshell is composed of circular cones with clearly demarcated boundary lines".

Megaloolithus jabalpurensis Khosla and Sahni 1995, 90–91.

Megaloolithus matleyi (Mohabey 1996, 188–191).

Megaloolithus patagonicus (Calvo, Engelland, Heredia and Salgado 1997, 27–30).

"Tipo 1b" Fernández 2013.

Holotype VPL/KH 250 (Bara Simla Hill) Jabalpur, Madhya Pradesh.

Type Locality Bara Simla Hill (Jabalpur, Madhya Pradesh, India).

Type Horizon and Age Sandy carbonate (¼ limestone) bed; Late Cretaceous, Lameta Formation.

Diagnosis *M. jabalpurensis* eggs are spherical to subspherical in shape with diameter variable from 140 to 160 mm. Eggshells have a discretispherulitic morphotype, and their thickness ranges from 1.0 to 2.38 mm (Fig. 2a, b). External surface shows compactituberculate ornamentation, with subcircular nodes (Fig. 2c). The node diameter ranges from 0.35 to 1 mm (Fig. 2c). Under PLM the shell units are fan-shaped with variable width and shape (Fig. 2a). The lateral margins of shell units are non-parallel (Fig. 2a). The average height/width ratio is 2.45:1. The growth lines are moderately arched upwards and follow the contours of the external nodes. The pores aperture are circular in shape, these oospecies have a tubocanaliculate pore system with straight pore canals. The basal cap is subcircular in shape and smaller (0.1–0.5 mm in diameter) than in *Megaloolithus cylindricus.*

Megaloolithus cylindricus Khosla and Sahni 1995

Megaloolithus rahioliensis (Mohabey 1998, 349).

"Tipo 1 d" by Fernández 2013

Holotype VPL/KH 201 (Chui Hill), Jabalpur, Madhya Pradesh.

Type Locality Chui Hill (Jabalpur, Madhya Pradesh, India).

Type Horizon and Age Sandy carbonate bed, Late Cretaceous, Lameta Formation.

Diagnosis *M. cylindricus* eggs are spherical in shape with diameter variable from 120 to 200 mm, their eggshell thickness ranges from 1.7 to 3.5 mm (Fig. 2c). External surface of the eggshell exhibits compactituberculate ornamentation, most of the nodes appear isolated (Fig. 2d). The nodal diameter ranges between 0.4 and 1.4 mm. The most visible feature is their shell units with a cylindrical shape, the shell units are slender and tall, elongated (Fig. 2c). The lateral margins of spheroliths are straight, parallel, and vertical above mammillae (basal caps) (Fig. 2c). The average height/width ratio is 4:1. The growth lines are highly arched (Fig. 2c). The pore apertures are subcircular and have tubocanaliculate pore canals, where they are long, narrow, and straight. The basal caps are of medium size and subcircular in shape (0.2–0.5 mm in diameter).

Megaloolithus megadermus Mohabey 1998

Hypselosaurus (Kérourio, 1987: 257) (see Mohabey 1998).

"Tipo 1 e" by Fernández 2013.

Holotype OGF-85 Lameta Formation of Dholidhanti and Paori in Dohad area of Panchmahals district.

Type Locality Dholidhanti (Gujarat, India).

Type Horizon and Age Sandy carbonate bed; Late Cretaceous, Lameta Formation.

Diagnosis *M. megadermus* eggs are spherical in shape with variable diameter ranging from 130 to 180 mm; eggshell thickness ranges from 4.0 to 4.80 mm. External surface exhibits compactituberculate ornamentation, densely packed nodes. The shell units are discrete, tall, and narrow, and the lateral margins of shell units are straight (Fig. 2f, g). The average height/width ratio is 9.6:1. The eggshells show a tubocanaliculate pore system, where pore canals are long, straight, and broad. The basal caps are short (less than one-tenth of spherolith).

Oofamily Fusioolithidae Fernández and Khosla 2015.

Diagnosis Eggshells with dinosauroid-spherulitic basic type, tubospherulitic morphotype, tubocanaliculate pore system (Mikhailov 1997). Compactituberculate ornamentation, thin sections show accretion lines crossing the boundaries between shell units, starting on the inner one-third of the eggshell thickness and continuing to the external surface. The shell units are partially fused, under PLM shell units are fan-shaped similar to the eggs of oofamily Megaloolithidae but it is differing in the fused nature of the eggshell units.

Fusioolithus Fernández and Khosla 2015.

Diagnosis Eggs are spherical and large, with diameters ranging 90–200 mm, eggshell thickness ranges from 0.8 to 4.5 mm (Fig. 3). Ornamentation is compactituberculate, external nodes are partially fused, as the nodes are the projections of accretion lines; some nodes are coalescent (Fig. 3b). The oogenus included in this oofamily was previously described as *Megaloolithus*; for more details see Fernández and Khosla (2015). Included oospecies *Fusioolithus baghensis* Fernández and Khosla (2015), *Fusioolithus berthei* Fernández and Khosla (2015).

Fusioolithus baghensis (Khosla and Sahni 1995).

Megaloolithus baghensis (Khosla and Sahni 1995, 91–92).

Mega loolithus pseu (Vianey-Liaud, Hirsch, Sahni and Sigé 1997, 78–81).

Megaloo lithus balasino (Mohabey 1998, 357–358).

Patago olithus salit (Simón 2006, 517, 521–523).

Megal oolithus cf. ***baghensis*** (Sellés, Bravo, Delclòs, Colombo, Martí, Ortega-Blanco, Parellada and Galobart 2013).

" Tipo 1 a" Fernández 2013.

Holotype VPL/KH/551 Bagh cave section, Dhar District, Madhya Pradesh.

Diagnosis The eggs are spherical in shape with diameter ranging from 140 to 200 mm. Eggshell thickness ranges from 1.0 to 1.70 mm (Fig. 3a, b). External surface shows compactituberculate ornamentation, coalesced, and discrete nodes (Fig. 3b). Node diameter ranges from 0.40 to 0.80 mm (Fig. 3b, c). The shell units are short, broad and fan-shaped and distinct, or even partially fused (Fig. 3a). The height/width ratio is 2.32:1. The lateral margins of shell units are straight to conical in radial view (Fig. 3a). The growth lines are moderately arched beneath the nodes and enter into adjacent shell units with concavity while in multimodal shell units growth lines are horizontal to subhorizontal in shape (Fig. 3b). The pore apertures are sub circular to elliptical in shape. Pore canals are tubocanaliculate (Fig. 3c). The pore canals are short, curved, and narrow; swollen-ended, variably spaced basal caps (0.2–0.3 mm in diameter).

Fusioolithus berthei Fernández and Khosla 2015

"Tipo 1 c" (Fernández 2013).

Holotype MML-PH 1269, Bajos de Santa Rosa area, Río Negro province.

Type Horizon and Age Thick succession of sandstones and mudstones, with intercalations of carbonate and evaporitic rocks in its upper part, Allen Formation Campanian—Maastrichtian (Upper Cretaceous).

Diagnosis Eggshell thickness ranges from 2.45 to 2.9 mm (Fig. 3d). External surface shows compactituberculate ornamentation (subcircular nodes) (Fig. 3); average node diameter about 0.66 mm with diameter ranging from 0.3 to 0.7 mm. Fan-shaped units have variable width and shape; lateral margins of shell units are conical and non-parallel, and the eggshell units are partially fused (Fig. 3d). The upper and external parts of the units have a wide difference from the inner parts. The average height/width ratio is 3.27:1; growth lines moderately arched upwards and follow the contour of the external profile; pore apertures are circular to elongate in shape. These eggshells have tubocanaliculate pore canals (pore canals sub-vertical and inclined), subcircular basal caps 0.1–0.5 mm in diameter.

Oofamily Faveoloolithidae Zhao and Ding 1976

Sphaerovum erbeni Mones 1980 *nomen dubium* Grellet-Tinner, Fiorelli, and Salvador (2012)

"Paquiloolithus rionegrinus" Simón 2002. Unavailable name.

Diagnosis Eggshells present a filispherulitic morphotype (Mikhailov 1991), with a multicanaliculated pore system. The external surface shows compactituberculate ornamentation. Spherical eggs range between 18 to 22 cm in diameter. Eggshell thickness ranges between 3.5 to 7.5 mm. The width of the shell units ranges between 0.3 to 1 mm. The diameter of the nodules ranges from 0.3 to 1.1 mm. Shell unit diameter, measured in the internal surface, varies between 0.2 to 1.1 mm. Under PLM fan-shaped extinction patterns are evident.

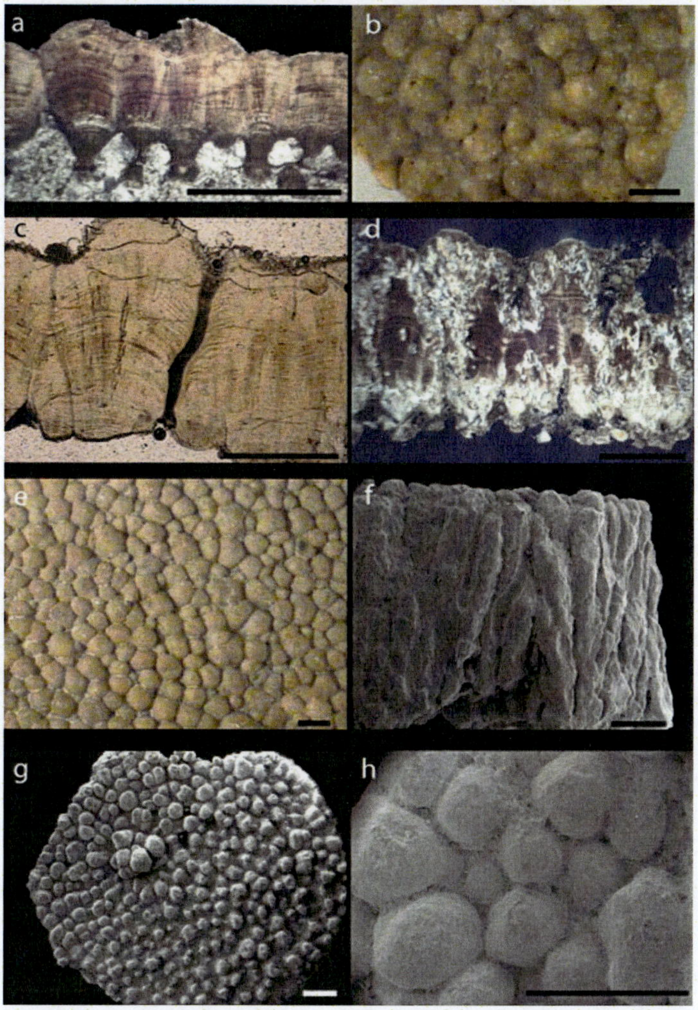

Fig. 3 Fusioolithid ooespecies described from South America. **a–c** *Fusioolithus baghensis*. **a** *F. baghensis* (MML-PH 1133), Mansilla I, egg level 3 Bajos de Santa Rosa, Río Negro Province, Argentina. Radial thin section shows fusion between several shell units. **b** External surface with compactituberculate ornamentation, with coalescent nodes. **c** Thin section showing tubocanalicualte pore canal. **d** *F. berthei* under PLM (MML-PH 1269), Berthe Vegg level 4, Salitral de Santa Rosa, Río Negro, Argentina. Radial thin section shows accretion lines crossing the boundaries of the shell units, and units are partially fused and calcite radial lines are perpendicularly crossing the entire thickness of the shell. **e** External surface with compactituberculate ornamentation. **f** ?Faveoloolithid eggshell MUC-PV 329. Radial view under SEM showing filispherulithic shell units, with multicanaliculate pore canals (Taken from Fernández 2013). **g** External surface showing compactituberculate ornamentation. **h** Detail of nodes on the external surface. Scale bar equals a, b, d–g 1 mm and c 500 μm

3.4 *Did Stem Sauropodomorphs Laid Soft-Shelled Eggs?*

Norell and colleagues (2020) presented mineralogical, organochemical, and ultrastructural evidence for an originally non-biomineralized, soft-shelled nature of exceptionally preserved early sauropodomorph *Mussaurus* eggs. Histology corroborates the organic composition of these soft-shelled dinosaur eggs, revealing a stratified arrangement resembling turtle soft eggshell. Through an ancestral-state reconstruction of composition and ultrastructure, they compare eggshells from *Protoceratops* and *Mussaurus* with those from other diapsids, revealing that the first dinosaur egg was soft-shelled, although this claim has been disputed (Legendre et al. 2020). This conclusion would imply that the calcified, hard-shelled dinosaur egg evolved independently at least three times throughout the Mesozoic era, explaining the bias toward eggshells of derived dinosaurs in the fossil record.

There is an ongoing controversy concerning this and other studies that have used Raman Spectroscopy to detect the presence of fossil biomolecules in fossil specimens (Alleon et al. 2021; Wiemann and Briggs 2021). Critics are mainly concerned with a putative overinterpretation of Raman instrumental errors that produce artifacts (Alleon et al. 2021), an argument disputed by the original authors, who argue that the results of Alleon et al. (2021) are caused by inadequate statistical manipulation of the data (Wiemann and Briggs 2021). At the time of writing this chapter, this controversy is far from resolved, but it is certain that the Argentinian eggs will have a role in solving one of the most interesting scientific disputes in the field of paleoology and paleontology as a whole.

4 Embryos and Hatchlings

Despite a major increase in findings in recent decades, dinosaur embryos and hatchlings are still extremely rare in the fossil record, with a few dozen examples around the world (i.e., Horner and Makela 1979; Norell et al. 1994; Mateus et al. 1997). In the case of sauropodomorphs, identifiable embryos and/or hatchlings have been reported from South Africa (Reisz et al. 2010, 2012), China (Reisz et al. 2013), India (Wilson et al. 2010), and Mongolia (Grellet-Tinner et al. 2011), with other reports of early-stage juvenile individuals in the United States (Britt and Naylor 1994) and Madagascar (Curry-Rogers et al. 2016). Still, the earliest, richest, and most complete record has been recovered in South America, more precisely in Argentina, where three different localities have produced *in ovo* embryos and hatchlings of sauropodomorph dinosaurs, namely the sauropodomorph *Mussaurus patagonicus* (Bonaparte and Vince 1979) post-hatchlings associated with two unhatched eggs recovered in the Lower Jurassic Laguna Colorada Formation. The first confirmed sauropod embryos, represented by at least 40 individuals from different eggs, belonging to a lithostrotian titanosaur sauropod from the Upper Cretaceous Anacleto Formation (Chiappe et al. 1998), and a recently described titanosaurian embryo in *ovo* of unknown provenance within Patagonia, but probably not Auca Mahuevo (Kundrát et al. 2020). This exceptional record has provided unique opportunities to explore sauropodomorph early development and evidence of soft tissue preservation, as well as to unequivocally relate the oofamilies *Megaloolithidae* and *Fusioolithidae* to titanosaurian sauropods.

4.1 The First Steps of the First Sauropodomorphs: *Mussaurus patagonicus* Hatchlings

Bonaparte and Vince (1979) reported seven post-hatchling individuals of a "prosauropod" sauropodomorph for which they erected the taxon *Mussaurus patagonicus* Bonaparte and Vince 1979. All seven specimens were of similar estimated size, about 20 cm from head to tail, being slightly larger than the largest eggs found in the assemblage (long axis of about 6 cm). This size disparity led the authors to suggest that the holotype and associated skeletons here "some weeks old" (and thus interpreted that the hatchlings remained in the nest for a long period after hatching). Nevertheless, previous reports of sauropodomorph adult skeletons from the same formation—originally reported as *Plateosaurus* sp. by Casamiquela (1977)—together with a large collection of juvenile, subadult and adult specimens reported in the last decades (Pol and Powell 2007; Otero and Pol 2013; Otero et al. 2019) allowed further exploration of the growth rates of this taxon. Cerda et al. (2014) calculated the volume of the largest egg and estimated a 63 g body mass for the post-hatchling specimens, which fits well with the 68 to 74 g that Otero et al. (2019) estimated for the holotype and paratype specimens. Osteohistological studies based on juvenile specimens suggest that *Mussaurus* would grow up to reach 23 kg during the first year of life (Cerda et al. 2014). This value has been revised to 8 kg (Otero et al. 2019), but still represents a 100-fold increase in size during the first year of development, thus pointing to the holotype and paratype specimens being much closer to hatchlings than originally proposed by Bonaparte and Vince (1979).

The osteology of the *Mussaurus* post-hatchlings has been studied in depth (Fig. 4a; Pol and Powell 2007; Otero and Pol 2013). These studies revealed a great deal of information not only about the anatomy of this species but about the developmental patterns of sauropodomorph dinosaurs, which has deep implications for their phylogeny. Bonaparte and Vince (1979) already noted the similarities of the post-hatchling *Mussaurus* skull and the adult *Camarasaurus*, suggesting that paedomorphosis had an important role in the evolution of Sauropodomorpha (see also Reisz et al. 2005). Nevertheless, recent studies have shown that, even though some evolutionary traits follow a pattern in which a sauropod like condition in the post-hatchlings is reverted in the adult specimens, other characters show a much more complex scenario with the opposite trend, in which plesiomorphic conditions in the post-hatchlings evolve to a more derived condition during development (Pol and Powell 2007; see also Chap. 12).

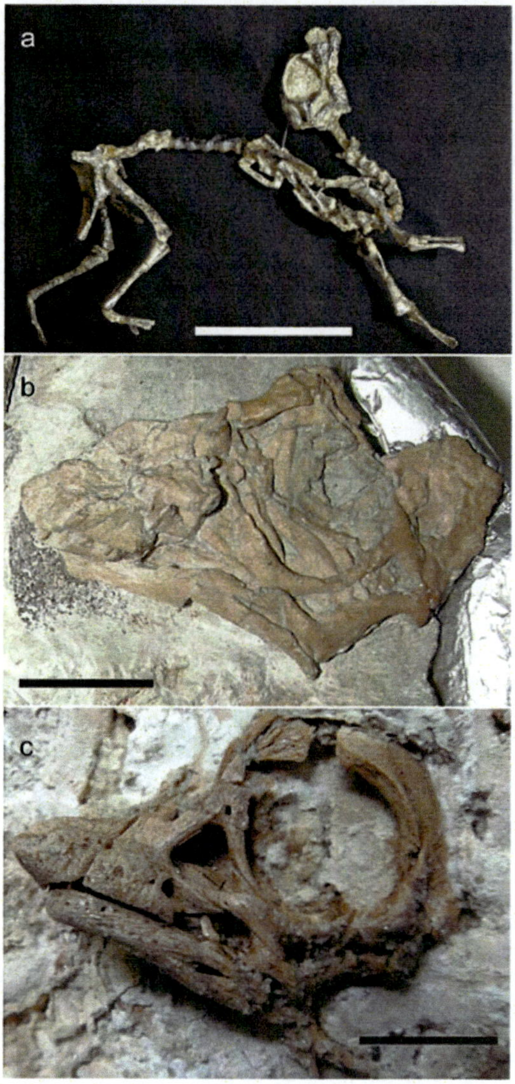

Fig. 4 Sauropodomorph perinates and embryos from South America. **a** PVL 4068, Holotype of *Mussaurus patagonicus*, a several weeks old hatchling preserved as an articulated skeleton. **b** MCF-PVPH-272, Skull of a lithostrotian sauropod embryo from Auca Mahuevo, Neuquén, Argentina, in left lateral view. **c** MCF-PVPH-874, 3-D preserved titanosaurian embryo of unknown procedence within Río Negro, Argentina (Kundrát et al. 2020). Scale bar equals **a** 5 cm, **b** 1 **cm**, and **c** 0.5 cm

4.2 First Look at Sauropod Reproduction: The Auca Mahuevo Embryos

The nesting area of Auca Mahuevo produced the first definitive sauropod embryos (Chiappe et al. 1998) and remains the most prolific outcrop in the world for this kind of fossil (Chiappe et al. 2001). Interestingly, despite eggs and eggshell fragments being found in four different stratigraphic levels at the Auca Mahuevo locality, all embryonic material comes from the egg-level 3 (Chiappe et al. 2003).

In the original publication, Chiappe et al. (1998) reported over a dozen fragmentary embryos, mostly consisting of skull material and skin impressions, found in in situ eggs, with additional eggshell fragments bearing fused embryo fragments. In a subsequent work, Chiappe et al. (2001) reported six additional *in ovo* embryos, this time showing exquisite articulated skulls (Fig. 4a). In contrast, unambiguous postcranial remains have not been reported, with vertebral elements totally absent (Chiappe et al. 2003). The anatomy of the embryonic skull shows diagnostic characteristics of Eusauropoda (Chiappe et al. 1998), which was confirmed with the additional skull material described by Chiappe et al. (2001). The skull specimen shows apomorphic characters that are congruent with a titanosaurian affinity of the eggs, the only clade of sauropods in the latest Cretaceous of South America (Fig. 4b).

The skull of the titanosaurian sauropods is delicate, unlikely to be preserved, and thus relatively unknown, though its record has slightly increased since the description of the Auca Mahuevo embryos (e.g., Zaher et al. 2011; Knoll et al. 2013; Martínez et al. 2016). Remarkable differences can be observed between the embryonic and adult skulls of titanosaurs, suggesting a number of ontogenetic changes. One of the most striking changes is the absence of retracted nares, a shared condition common to all eusauropods that was most likely acquired during development (Chiappe et al. 2001). Noteworthy is the presence of a protruding expansion of the premaxilla, which has been interpreted as an egg tooth, a structure present in crocodiles, birds, turtles, and lepidosaurs that is used to assist eggshell breakage during hatching (García 2007).

The exceptional preservation of the Auca Mahuevo embryos has allowed study of not only the osteology of these sauropods, but also part of their soft tissue, namely the *membrana testacea* of the egg (Chiappe et al. 1998; Grellet-Tinner 2005; Coria and Chiappe 2007). The *membrana testacea* is an organic layer of fibrous material present in the amniote egg, which is attached to the inner surface of the hard eggshell. It serves a structural function and provides a barrier against bacterial invasion (Packard and DeMarci 1995). It is sometimes partially mineralized (Grellet-Tinner 2005), and it was subsequently reported in other eggs and eggshell fragments of the oofamilies Megaloolithidae and Fusioolithidae (Sellés et al. 2013; Bravo and Gaete 2015).

Up to 78 patches of integument were recovered in the Auca Mahuevo egg assemblage, most of them as calcite casts of the original skin, although some negative impressions have been identified (Coria and Chiappe 2007). Sauropod skin impressions are extremely rare (Czerkas 1994) and normally associated with (Fondevilla et al. 2017) or on the lateral surfaces of footprint casts (Romano and Whyte 2012; Paik et al. 2017), thus increasing the relevance of the Auca Mahuevo embryos. Coria and Chiappe (2007) reported very complex integumentary patterns, including round tubercles, large and elongated tubercles, parallel rows of large tubercles, tubercles in rosette-like and flower-like arrangements, and in striate-like rows. These authors highlighted the resemblance of the round tubercles and rosette-like patterns to the arrangement of the osteoderms of *Saltasaurus* (Bonaparte and Powell 1980), a saltasaurine titanosaur of coeval age and distribution as the Auca Mahuevo eggs. Nevertheless, no evidence of osteoderms was observed in the embryos, and this suggests that they could have developed later in ontogeny, during the adult stage. The precise function of titanosaur osteoderms is unknown but there is a certain consensus that they were present in relatively low numbers, therefore defensive function is unlikely (Curry-Rogers et al. 2011). The high blod vessel irrigation and presence of large cavities in some osteoderms suggest they can act as a mineral reservoir, where titanosaurs could store calcium and other minerals that could be later be reworked and used for osteogenesis and oogenesis (Curry-Rogers et al. 2011; Vidal et al. 2017). Most titanosaurs have been found in highly seasonal palaeoenvironments (Curry-Rogers et al. 2011). Thus, osteoderms would represent an adaptation to extreme seasonal changes where resources are limited and would supply minerals for rapid growth and eggshell formation (Vidal et al. 2017).

At the time of their original publication, the Auca Mahuevo embryos were the first *in ovo* embryos associated with an egg presenting a tubospherulithic morphotype. As discussed in Sect. 3.2, the eggshell of the eggs from Auca Mahuevo can be assigned to the oofamily Fusioolithidae, thus providing definitive evidence that fusioolithid eggs were laid by titanosaurian sauropods.

4.3 A 3-D Preserved Sauropodomorph Skull in Ovo

Kundrát and colleagues (2020) reported a pristine, three-dimensionally preserved partial skull of a titanosaur sauropod found inside a megaloolithid partial egg (Fig. 4c). These authors reported that the specimen was illegally smuggled out of Argentina in the first place and that despite having been recovered and now being deposited in a public institution, its exact provenance within the Argentinean Patagonia remains unknown. Nevertheless, the study of the eggshell and sedimentary infilling of the egg shows that it has not undergone the same kind of regional diagenesis that is common to all the eggs from Auca Mahuevo egg-level 3, suggesting a different provenance.

The MCF-PVPH-874 embryo comprises only skull material, again supporting the later ossification of the postcranial skeleton suggested by the specimens from Auca Mahuevo. However, the general preservation of the skull suggests a later stage of development than the Auca Mahuevo embryos, and some key features of the titanosaur skull can be observed, including much more retracted nares than in other Argentinean sauropodomorph embryos (Kundrát et al. 2020). Other striking features observed in this embryo include an anterolateral orientation of the orbit, suggesting binocular vision that is lost during ontogeny, and the lack of occlusion between the premaxilla and dentary, pointing to the presence of an edentulous, corneous horn-like structure, homologous to the egg tooth reported by García (2007). Nevertheless, the inferred position of the *in ovo* embryo leads Kundrát et al. (2020) to discuss the usefulness of this structure as an egg tooth; the authors suggest that this corneous rostral expansion was probably not used during hatching but was retained during early stages of development of the titanosaurian juveniles.

Concerning the eggshell associated with MCF-PVPH-874 embryo, it is much thicker than that of the Auca Mahuevo eggshell and lacks the characteristic fusion of shell units observed in Fusioolithidae eggshells. The spherulithic shell units, tubocanaliculate pore system, and compactituberculate ornamentation allows its classification within Megaloolithidae, indicating that this oofamily was also laid by titanosaurian sauropods.

4.4 *The Lack of Embryos in Faveoloolithid Eggs*

Despite the fact that Faveoloolithidae eggs are widely distributed around South America (see Sect. 3.3 for discussion), no embryonic material associated with these eggs has ever been reported. Since the original works of Mones (1980), these eggs have been attributed to sauropods, and lately to neosauropods or even titanosaurs based on (1) large egg size; (2) rounded shape; (3) clutch arrangement; and (4) presence, sometimes exclusively, of titanosaur remains in the same formation. It is important to note that none of the aforementioned conditions is exclusive to sauropod dinosaurs, with egg size smaller in all definitive sauropod eggs (Chiappe et al. 1998; Grellet-Tinner et al. 2011), rounded shape shared with hadrosaurs and some theropods (Horner and Makela 1979; Mateus et al. 1997), clutch arrangement different to that of confirmed sauropod eggs (Vila et al. 2010); and coeval distribution of eggs and skeletal remains having been proven wrong in the past (Norell et al. 1994, 2020). Even so, some similarities between the eggshell of the Argentinean Faveoloolithidae and some Megaloolithidae eggshells still support a sauropod affinity of these eggs including nucleation centers at their bases, rhombohedric acicular calcite crystals, prominent compactituberculate ornamentation (Grellet-Tinner and Fiorelli 2010), although definitive embryos are needed to test this hypothesis.

Box 2 [Cracking the Eggs: the Tools Used to Understand Eggs, Eggshells, and Embryos]
The exceptional sauropodomorph oological record of South America has resulted in it being at the forefront of new discoveries on egg evolution and embryo development for the last three decades. As a result, the application of new methodologies and analytical techniques has been a constant in the studies of South American eggs.

The first studies of South American eggs were based on macroscopic, naked-eye descriptions using basic stereoscopic microscopes and photographic cameras (Mones 1980). By the end of the 1990s, the use of the Scanning Electron Microscope—used mainly for morphological imaging using secondary electrons—and the petrographic microscope was generalized in egg studies and has been completely adopted by South American researchers (Chiappe et al. 1998). Magalhães-Ribeiro (1999b) pioneered the use of "hot" cathodoluminescence to study possible taphonomic alterations of fossil eggshells, using

Argentinean specimens with a technique further developed by Fernández and Matheos (2011).

In the 2010s, eggshells from Auca Mahuevo were analyzed using Electron Backscatter Diffraction (Trimby and Grellet-Tinner 2011), a technique that allows mapping of crystal orientations of crystallographic materials. This was the second time a fossil eggshell was analyzed with this technique, which has proven to provide deep characterization of eggshell architecture, eggshell growth mechanisms, eggshell systematics, and taphonomy (Moreno-Azanza et al. 2013, 2016; Choi et al. 2018). Eagle et al. (2015) used this tool on eggshells including samples from different levels of Auca Mahuevo to assess the preservation of eggshells before carrying out isotopic analysis.

Compositional studies, both of elemental and isotopic composition, have been carried out on South American specimens using Energy dispersive X-ray (EDX) elemental analyses in a SEM (Schweitzer et al. 2005), and a dual-inlet gas source mass spectrometer (Eagle et al. 2015).

One of the techniques that have had more impact in paleontology during the last decade is the extended use of computed tomography. Micro-CT has been used to characterize the complex pore system of the faveoloolithid eggshells from Sanagasta (Hechenleitner et al. 2016b). On the other hand, propagation phase-contrast synchrotron microtomography has been used to non-destructively obtain 3-D models of *in ovo* embryos (Kundrát et al. 2020).

Finally, eggshells are biominerals, and thus not only their mineralized phase can be addressed, but different techniques have been used to study the presence of fossil organic molecules in the eggshell structure. Schweitzer and colleagues (2005) carried out immunological analysis of organic components extracted from Auca Mahuevo sauropod and compared them with results obtained using modern hen eggshells, confirming that there was molecular similarity between shell-derived immunogens. Furthermore, Norell and colleagues (2020) used Raman Spectroscopy to characterize protein fossilization products (PFPs) in the eggshell of *Mussaurus patagonicus*. This new approach was developed by Wiemann and colleagues (Wiemann et al. 2018a) and was also used to determine the lack of external coloration in the eggs from Auca Mahuevo (Wiemann et al. 2018b), although some critics have been raised concerning these applications (see Sect. 3.4 for discussion).

5 Traces of Sauropodomorph Reproduction

5.1 Clutches and Nests

The *Mussaurus patagonicus* nesting site, the first egg locality in South America providing information on reproductive biology and hatchling anatomy of sauropodomorphs dinosaurs, is a good example of how the association of hatchlings, eggs, and scattered eggshells might provide evidence of nesting activity (Bonaparte and Vince 1979). There, the authors interpreted that babies were still living in the nest at their time of death (Bonaparte and Vince 1979). Four decades after, Norell et al. (2020) concluded, after a detailed examination, that *Mussaurus* eggs were soft (but see Sect. 13.3.4 for discussion) and this implied that, because they were more fragile and prone to water loss, most probably the parents could not brood on them but instead laid them within a moist substrate. Therefore, incubation took place underground, the hatchlings were precocial, and parental care was limited to nest guarding.

In Auca Mahuevo *in ovo* skeletal fossils of titanosaurian embryos were discovered (Chiappe et al. 1998, 2001; Coria and Chiappe 2001; Salgado et al. 2005; García 2007; García and Cerda 2010; Kundrát et al. 2020) alongside complete clutches and well-preserved nest traces (Chiappe et al. 2004) (Fig. 4b). The site consists of four egg levels (referred to as "egg beds" by Chiappe et al. (2000) due to the high concentration of eggs), some of them cropping out extensively and preserving thousands of eggs grouped in clutches (Chiappe et al. 2003, 2004, 2005). Egg levels 1 to 3 are found in the Auca Mahuevo section whereas the fourth is found 25 km southwards, in Los Barreales. The titanosaur eggs and clutches are preferentially found in proximal over bank deposits, on muddy paleosols and pedoturbated channel deposits, which were once buried by muddy sheet-floods (Garrido 2010) that preserved the delicate embryonic skeletons (Chiappe et al. 1998). Maps of the nesting grounds at egg levels 3 and 4 show a very high density of eggs—200 eggs in 25 m^2, in egg-level 3—and clutches per square meter (Chiappe et al. 2000, 2005) and suggest that females nested in herds (Chiappe et al. 2000). Clutches are elongated and contain 20 to nearly 40 eggs stacked one on top of the other in up to three egg levels. However, Jackson and colleagues (2013) realized that the expandable layer clay minerals of the egg beds may have made changes in egg-bearing horizons, and alternating wet and dry seasons may have facilitated pedogenic movement. This pedoturbation would modify or combine eggs from different clutches and different temporal intervals, resulting in time-averaging of nests or nesting horizons. Because of that, according to the authors the resulting pedogenically-formed egg clusters may not accurately portray the egg size and shape, the number of eggs per clutch, the clutch morphology, or the egg distribution produced by individual female sauropods (Jackson et al. 2013). Six subcircular to kidney-shaped depressions (length: 100–140 cm; depth: 10–18 cm), interpreted as nest traces, contain egg clutches preserved in sandstone (Chiappe et al. 2004). The eggs and eggshells were described and analyzed by various authors (Grellet-Tinner et al. 2004, 2006; Jackson et al. 2004, 2008; Grellet-Tinner 2005) and have been recently ascribed to *Fusioolithus baghensis* (Fernández and Khosla 2015).

In the nesting areas of Salitral de Santa Rosa-Salinas de Trapalcó and Salitral Ojo de Agua up to eight egg clutches were identified in the egg-sites of Santos I and II, Berthe IV, and Arriagada I and III (Fig. 5c). The most complete Faveoloolithidae egg clutches are of moderate size, with ten spherical eggs (but larger, 12 egg clutches were reported from an equivalent egg-level; Powell 1992) whereas the fusioolithid (*Fusioolithus baghensis*) clusters contain up to 25 eggs. In all cases the spherical eggs are in close contact with one another and stacked in layers with the deepest eggs in the center of the cluster, indicating that they were originally laid into shallow depressions within the substrate (Fig. 5c; Coria et al. 2010). Occasionally, nest traces consisting of metric and subcircular depressions are preserved. These sedimentary structures with some eggs inside are encircled by an irregular rim of massive sandstone (Fig. 5d). None of the eggs preserve embryos or indicate hatching. Some of the egg levels can be traced laterally to a total extension of 250 km^2, and the vertical distribution of different oospecies indicates that these sauropods shared with other dinosaurs their nesting areas which were occupied differentially throughout time. Interestingly, the different egg producers might have shared the nesting ground, as shown by the co-occurrence of fusioolithid and ?faveoloolithid eggs on the same egg-level (Salgado et al. 2009; Coria et al. 2010).

At the nesting locality of Yaminué the faveoloolithid clutches contain 8–9 spherical eggs in close contact with one another (Fig. 5e; Manera de Bianco 1996), thus producing a grouped arrangement at the same level. Based on this grouped arrangement, the absence of sediment rims around them (and thus, the lack of sedimentary evidence for a nest trace) and the microstructure of the eggshells, Manera de Bianco estimated that the eggs were laid within a hole that the female produced on the substrate, and therefore incubation took place underground.

The first analysis at the Sanagasta nesting site by Mario A. Hünicken (2005) concluded, preliminarily, that the dinosaurs produced a dug hole in the substrate and that "some remains of vegetable stems would have been used in its construction"(Hünicken 2005, p. 80). Later, the works of Grellet-Tinner and Lucas E. Fiorelli (2010) indicated that the most complete clutches contain up to 35 eggs and measure 220 cm in length (although the average clutch size was about 3–12 eggs) with the eggs stacked one over the other in two distinct egg rows. This arrangement confirms more minutely previous inferences (Hünicken 2005) of eggs laid in dug-out holes. Because they were close to active hydrothermal heat sources, the authors suggested that the sauropods were in a symbiotic relationship with the hydrothermal system of the area to take advantage of the heat from the soil, favoring the incubation of the eggs (Grellet-Tinner and Fiorelli 2010). According to the authors, the microstructure of the eggshell (including a well-developed pore canal network and a nodular external surface; Hechenleitner et al. 2016b), as well as its water vapor conductance (Grellet-Tinner et al. 2012), are congruent with this interpretation in which the eggs were laid within a moist sediment, as inferred for hydrothermal region of Sanagasta.

At the Tama locality, most of the five distinct egg clutches are of moderate size except for the largest and best-preserved one which contains up to 21 eggs and

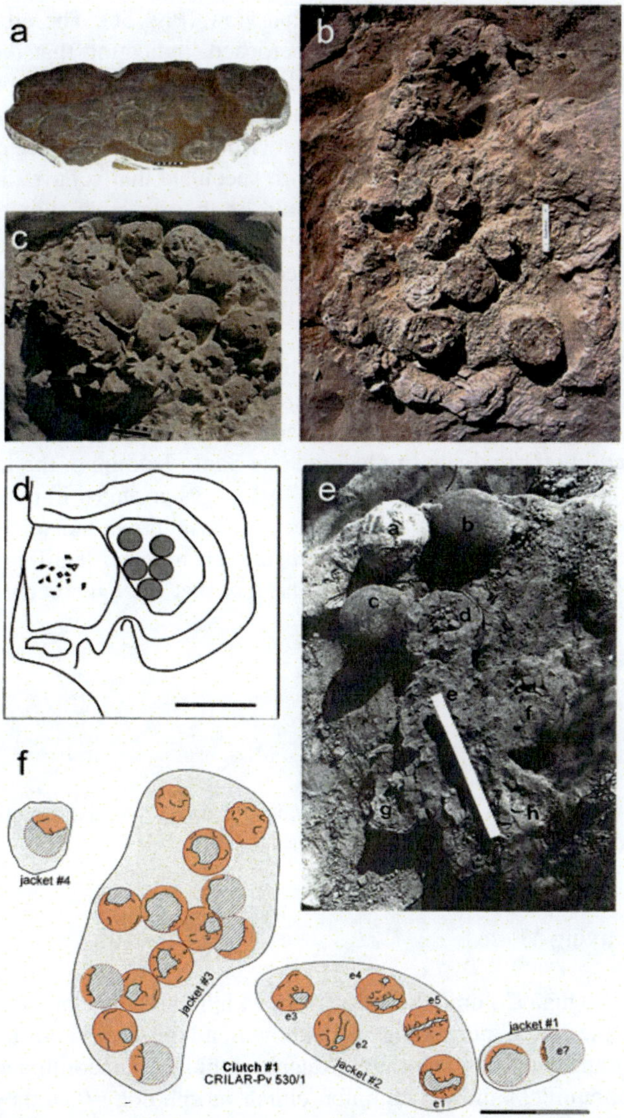

Fig. 5 Nests and clutches. **a** Fusioolithid (*Fusioolithus baghensis*) egg clutch from Auca Mahuevo locality. **b** Titanosaur nest from Auca Mahuevo level 4 (modified from Vila et al. 2010). **c** Fusioolithid (*Fusioolithus baghensis*) nest (MML-PV 41) from Berthe IV locality. **d** ?faveoloolithid egg nest from Arriagada I locality (taken and modified from Salgado et al. 2007). **e** ?faveoloolithid egg clutch from Yaminué locality (taken from Manera de Bianco 1996). **f** Megaloolithid egg clutch (CRILAR-PV 530/1) from Tama locality (taken from Hechenleitner et al. 2016a). Scale bar equals: a, c 10 cm; b 15 cm; e 33 cm; and f 50 cm. Gray circles and letters in e represent individual eggs, respectively

exhibits an elongated-elliptical shape in plan-view (Fig. 5f). The ellipsoidal (but originally spherical in shape) eggs are superposed, indicating that the eggs were laid in excavated holes. Such taphonomic data coupled with the microstructure of the eggshells led the authors to suggest that the Tama titanosaurs were adapted to incubate their eggs underground in a humid environment. The absence of vegetation within or around the clutch led the authors to speculate that solar radiation and/or geothermal activity of the nesting area would be the plausible sources of heat to achieve incubation. The egg locality is also thought to be suggestive of colonial nesting (Hechenleitner et al. 2016a). Detailed analysis of the sedimentary environment of the nesting ground of Tama locality found that the eggs were laid in a poorly developed paleosol (inceptisol, Basilici et al. 2017). These authors interpret the egg-bearing layer as a cumulative paleosol profile, where sauropods dug their nests on the topographic surface, and sediment-covered unsuccessful clutches at a rate of 2 to 10 mm/year, facilitating fossilization (Basilici et al. 2017).

At the recently reported site of Quebrada de Santo Domingo, the largest of the about 20 egg clutches reported contains up to 15 eggs arranged in two superposed rows. The paleoenvironment is interpreted as an alluvial floodplain and although the main authors refer to incubation of eggs within the substrate, no details on taphonomy were provided by the authors, except for the fact that the eggs occur in compact accumulations (Hechenleitner et al. 2020).

5.2 *Nesting Strategies*

With the abovementioned localities and available data, the discussion on nesting strategies focuses on the assessment of several aspects.

5.2.1 Egg-Laying Mode

Except for the Laguna Colorada locality, from which only two eggs were preserved (and therefore the clutch morphology is unknown), all the Cretaceous nesting localities with eggs attributed to titanosaurs (Megaloolithidae, Fusioolithidae, and Faveoloolithidae oofamilies) show a common clutch morphology (García et al. 2015) and therefore, a similar egg-laying mode. In plan-view, the egg clutches typically have an elongated morphology (e.g., Auca Mahuevo, Sanagasta, and Tama localities) and eggs are superposed in rows or egg-layers. In some of the best-preserved clutches (e.g., megaloolithid eggs from clutches at Auca Mahuevo, Chiappe et al. 2000, 2004; ?faveoloolithid eggs from clutch MPCA-501 at Arriagada I in Salitral Ojo de Agua, Salgado et al. 2007; ?faveoloolithid eggs from clutches at Sanagasta, Grellet-Tinner and Fiorelli 2010) authors observe that the deeper eggs are in the center of the accumulation or the greater number of eggs is in the upper row. Thus, we state that all the titanosaur groups that were present in South America during

Cretaceous times excavated a hole in the substrate and laid the eggs in morphologically similar clutches inside. This three-dimensional morphology is shared by various European and Indian megaloolithid clutches (Sahni and Khosla 1994; Sander et al. 2008; Vila et al. 2010; Khosla and Lucas 2020b). Smith and colleagues (2014) performed a taphonomic study in *Mussaurus patagonicus* breeding site at Laguna Colorada, Santa Cruz, the signature for this site is of an attritional accumulation of a stable (non-migratory) population living at this site year-round. They found that the skeletons and egg clusters occur in three distinct horizons within a 3 m-thick sequence of mottled light reddish—brown/olive-gray massive siltstone. Smith and colleagues described the eggs as all of similar dimensions and slightly ovoid in shape. These authors performed a detailed study of compression breakage patterns of complete eggs and stacked shell fragments of hatched eggs, they infer that eggs were left open to incubate. Their interpretation of the spacing of the nests suggests a common breeding ground rather than a colony, and the repeated occurrence of nests and skeletons through the 3 m-thick interval reveals year-on-year breeding at this site (Smith et al. 2014).

5.2.2 Nest Building

Evidence for nest traces understood as physical evidence of nest architecture, consisting of a depression that truncates stratification within the host substrate is extremely rare in the South American sauropodomorph record (Chiappe et al. 2004). *Mussaurus* site yielded so far 25 partial *Mussaurus* skeletons of six different ontogenetic stages, from embryonic to adult, associated with several complete "nests" of unhatched eggs (Bonaparte and Vince 1979; Smith et al. 2014). Smith and colleagues performed a taphonomic study which consists of a series of field observations and X-ray imaging of five complete nests which confirms that 24–27 eggs are arranged in 2 or 3 layers within elongate depressions or trenches with an irregular bottom profile. These authors confirmed that these depressions appear to have been purposely excavated into semi-consolidated loess and as such, they qualify as nest structures. Till this report, the oldest record of nest digging and site fidelity has been from Late Cretaceous titanosaurian sauropods. Smith and colleagues (2014) present new ichnological evidence from Laguna Colorada Formation which strongly indicates that sauropodomorphs of sub-polar Gondwana were practicing herding, nest digging, and breeding site fidelity as far back as the Early Jurassic, at least 100 Ma earlier than previously thought. The nest traces in Auca Mahuevo level 4 and Arriagada I localities share various similarities (size, subcircular morphology, and surrounding rim) and therefore most probably indicate that the titanosaur producers of megaloolithid and ?faveoloolithid eggs share the same nest-building strategy. This would consist of a scratch-digging movement produced by the rear foot of the female (Vila et al. 2010; Fowler and Hall 2011) to produce the subcircular to sub-elliptical to kidney-shaped nest depressions.

5.2.3 Incubation Mode

Important differences exist regarding the incubation mode used by the titanosaurs once the eggs were ovodeposited. On one hand, the titanosaurs that laid fusioolithid egg clutches in the nesting grounds of Auca Mahuevo egg levels 1–3 incubated the eggs in open nests, as indicated by the low water vapor conductance (G_{H2O}) of their eggshells (Jackson et al. 2008). On the other hand, the higher values of water vapor conductance in the megaloolithid (*M. patagonicus*) and ?faveoloolithid eggshells (Grellet-Tinner et al. 2012) indicate that such titanosaurs incubated the eggs underground within a moist substrate. Further, Hechenleitner and colleagues (2015) suggested that in some particular settings, ?faveoloolithid and megaloolithid eggs were incubated with environmental sources of heat. For the sake of untangling controversies, it is worth noting that megaloolithid eggshells of Auca Mahuevo egg-level 4 (sample MCF-PVPH 444 of Grellet-Tinner et al. 2012), have higher G_{H2O} values than those reported for the fusioolithid eggshells of Auca Mahuevo egg-level 3 (Jackson et al. 2008), as they were probably laid by distinct titanosaur clades. Such higher values in the eggs of Auca Mahuevo egg-level 4 are congruent with the high values reported for another megaloolithid (*M. siruguei*) eggs from Europe (Jackson et al. 2008), which were also interpreted as being incubated underground (Vila et al. 2010). Therefore, in spite of the fact that the titanosaurs producing the three ootaxa in South America (Megaloolithidae, Fusioolithidae, and Faveoloolithidae) share the strategy of excavating a dug hole in the substrate to lay the eggs inside, at least one of them (the fusioolithid egg-layers) did not cover the clutch and the eggs were incubated in open nests. Such differences in nesting strategies related to incubation mode probably were the result of differing ecological and/or climatic conditions (Hechenleitner et al. 2015, 2018).

5.2.4 Number of Ovodeposition Episodes Per Season

In at least two of the nesting localities (Sanagasta and Auca Mahuevo) the large clutches (>25 eggs) indicate that the titanosaurs ovodeposited their eggs in mass, as seen in other European localities and modern crocodiles (Deeming 2004). This would imply that ovodeposition occurred once a season. As a rule for most of the clutches reviewed, only with completely preserved accumulations—e.g., Auca Mahuevo clutches (Chiappe et al. 2004, 2005; but see Jackson et al. (2013) for effects of pedogenic processes—a complete clutch 3-D morphology can be defined, and smaller clutches may represent remnants of larger, more complete clutches that were truncated by recent erosion (e.g., Manera de Bianco 1996; Coria et al. 2010); see discussion in Vila et al. 2010). Accordingly, small clutches such as those reported in Tama, Quebrada de Santo Domingo, Yaminué may indicate remnants of larger clutches. The alternative interpretation, in which they represent complete clutches, may indicate that titanosaurs laid multiple clutches per season (Sander et al. 2008).

5.2.5 Nesting Habitat

The South American sauropodomorphs nested in a variety of environmental settings but preferred alluvial or coastal floodplains, particularly in fluvial (Laguna Colorada, Auca Mahuevo, Quebrada de Santo Domingo) or brackish lagoonal and supratidal (Salitral de Santa Rosa-Salinas de Trapalcó and Salitral Ojo de Agua) sub-environments (Salgado et al. 2007, 2009). Other settings correspond to hydrothermal environment (Sanagasta localities; Grellet-Tinner and Fiorelli 2010), paleosols in a semiarid environment (Tama locality; Hechenleitner et al. 2016a) or undescribed in detail (Yaminué locality; Manera de Bianco 1996).

6 Discussion and Conclusions

The fossil record of sauropodomorph reproduction in South America is one of the most complete in the world. With only a single locality reported in the Early Jurassic, and two localities known from the Lower Cretaceous, the bulk of localities are found in the Upper Cretaceous, with hundreds of sauropod eggs and eggshells of three different oofamilies in dozens of localities throughout the continent. This burst in abundance and diversity is more or less coincident with the diversification of titanosaurian sauropods, a globally distributed clade that is particularly diverse in South America (see Chaps. 8, 9, and 10). The Early Jurassic to Lower Cretaceous gap in the record may be partially explained if the currently disputed evidence supporting soft eggshells with low fossilization potential is the plesiomorphic condition for sauropodomorphs, as suggested by the *Mussaurus* eggs (Norell et al. 2020). Nevertheless, little is known of the reproductive strategies of non-titanosaurian sauropods worldwide (D'emic 2012), and South America is not an exception. Further research in the promising geological formations with non-titanosaurian sauropods (see Chaps. 5, 6, and7) may provide new data.

In terms of oological diversity and paleobiogeography, the three distinct oospecies (*M. jabalpurensis*, *M. cylindricus*, and *M. megadermus*) described in South America are also present in India and this reinforces the Gondwanan distribution of these sauropod dinosaurs. Further, *F. baghensis* has the most widespread distribution of all these ootaxa, as it has been reported in Spain, France, India, and South America, and probably also in Africa (Fernández and Khosla 2015; Khosla 2019; Khosla and Lucas 2020a, b). Finally, filispherulitic eggshells here described as Faveoloolithidae represents an endemic ootaxon with no representatives outside of the continent. Asian "true" faveoloolithids are quite different, and thus the South American ?faveoloolithid eggs might indicate an endemic sauropod producer, which inhabited South America during the Late Cretaceous.

Regarding reproductive paleobiology, a common reproductive trait of the South American sauropodomorphs is that all of them laid eggs within a dug hole (García et al. 2015). Based on all the evidence available (from South America and worldwide; Sander et al. 2008; Vila et al. 2010), the deposition of the eggs within a dug hole and the absence of parental care seem to be common traits in the nesting strategy for Sauropodomorpha. Differences occur in the incubation mode between oofamilies, and thus among taxa, as fusioolithid clutches with low values of water vapor conductance are deposited in open nests, and megaloolithid and ?faveoloolithid clutches with high values of water vapor conductance are incubated underground. Taphonomically, the burial of the eggs increases the potential of fossilization, and therefore their abundance in the fossil record, whereas the preservation of open nest traces and eggs is hindered by their subaerial exposure.

Many open questions remain concerning the Argentinean record of fossil eggs. Were the *Mussaurus* eggshells indeed soft? Why were they so different in size? Do Megaloolithidae and Fusioolithidae eggs belong to different clades of titanosaurs that coexisted in the same nesting habitats? Are the Faveoloolithidae eggs true faveoloolithids, or the filispherulithic condition convergent with the eggs from Asia? Is *Sphaerovum* a valid ootaxon? Were they laid by sauropod dinosaurs? These and other questions emphasize that the South American record of sauropodomorph reproduction will have a key role in the development of the discipline of paleoology and dinosaur paleobiology in the future.

Acknowledgements We thank Andrés Batista for his invaluable help in compiling information on Uruguayan localities. Alejandro Otero provided pictures of the holotype of *Mussaurus patagonicus*. Rodolfo Coria allowed the use of pictures of the Auca Mahuevo embryos. Martin Kundrát provided pictures of MCF-PVPH-874. Mainqué Green assisted us with figures. This research was funded by FONCyT PICT 2017-0905; the Fundação para a Ciência e a Tecnologia, project PTDC/CTA-PAL/31656/2017 and GeoBioTec, project (UIDB/04035/2020); by the Spanish Ministry of Science and Innovation, the European Regional Development Fund, the Government of Aragón (Grupo Aragosaurus: Recursos geológicos y Paleoambientes), project CGL2017-85038-P; amd PLEC2021-008203 project, funded by MCIN/AEI/10.13039/501100011033 and by the European Union "NextGenerationEU"/PRTR". M.M-A is supported by a Maria Zambrano Grant, founded by Ministerio de Universidades and by the European Union "NextGenerationEU"/PRTR. B.V. is part of the consolidated research group 2017 SGR 01666 of the Agència de Gestió I Ajuts Universitaris i de Recerca (AGAUR). This research is part of the project I+D+i/PID2020-119811GB-I00 funded by MCIN/ AEI/10.13039/501100011033/. Additional support was also provided by the CERCA Program of the Generalitat de Catalunya. Finally, we thank Dr. Daniel Barta and Dr. Ashu Khosla for their reviews, which represented valuable comments and suggestions to improve the chapter.

References

Alleon J, Montagnac G, Reynard B et al (2021) Pushing Raman spectroscopy over the edge: purported signatures of organic molecules in fossil animals are instrumental artefacts. BioEssays. https://doi.org/10.1002/bies.202000295

Alonso-Zarza AM, Genise JF, Verde M (2011) Sedimentology, diagenesis and ichnology of Cretaceous and Palaeogene calcretes and palustrine carbonates from Uruguay. Sediment Geol 236(1):45–61. https://doi.org/10.1016/j.sedgeo.2010.12.003

Argañaraz E, Grellet-Tinner G, Fiorelli LE et al (2013) Huevos de saurópodos del Aptiano-Albiano, Formación Cerro Barcino (Patagonia, Argentina): un enigma paleoambiental y paleobiológico. Ameghiniana 50(1):33–50. https://doi.org/10.5710/AMGH.9.11.2012.551

Basilici G, Hechenleitner EM, Fiorelli LE et al (2017) Preservation of titanosaur egg clutches in Upper Cretaceous cumulative palaeosols (Los Llanos Formation, La Rioja, Argentina). Paleogeogr Paleoclimatol Paleoecol 482:83–102. https://doi.org/10.1016/j.palaeo.2017.05.034

Bonaparte JF, Powell JE (1980) A continental assemblage of tetrapods from the Upper Cretaceous beds of El Brete, northwestern Argentina (Sauropoda—Coelurosauria—Carnosauria—Aves). Men S Geo F 139:19–28

Bonaparte JF, Vince M (1979) El hallazgo del primer nido de dinosaurios (Saurischia, Prosauropoda), Triásico Superior de Patagonia, Argentina. Ameghiniana 16(1–2):173–182

Bravo AM, Gaete R (2015) Titanosaur eggshells from the Tremp Formation (Upper Cretaceous, Southern Pyrenees, Spain). Hist Biol 27(8):1079–1089. https://doi.org/10.1080/08912963.2014.934231

Britt BB, Naylor BG (1994) An embryonic *Camarasaurus* (Dinosauria, Sauropoda) from the Upper Jurassic Morrison Formation (Dry Mesa Quarry, Colorado). Dinosaur eggs and babies. Cambridge University Press, Cambridge, pp 256–264

Buffetaut E, Le Loeuff J (1994) The discovery of dinosaur eggshells in nineteenth-century France. Dinosaur eggs and babies. Cambridge University Press, Cambridge, pp 31–34

Cabrera F, Martínez S, Verde M (2020) Continental Late Cretaceous gastropod assemblages from Uruguay. Paleoecology, age, and the oldest record of two families and a genus. Hist Biol 32(1):93–103. https://doi.org/10.1080/08912963.2018.1471478

Calvo J, Engelland S, Heredia S et al (1997) First record of dinosaur eggshells (? Sauropoda-Megaloolithidae) from Neuquén, Patagonia, Argentina. Gaia 14:23–32

Capurro IO, Fernández MS, Zacarías G, Apesteguía S, Alonso R (2019) Huevos del tipo fusioolítidos del Cretácico de Salta, Argentina. Reunión de Comunicaciones de la Asociación Paleontológica Argentina R44

Casadío S, Manera T, Parras A et al (2002) Huevos de dinosaurios (Faveoloolithidae) del Cretácico superior de la cuenca del Colorado, província de La Pampa, Argentina. Ameghiniana 39(3):285–293

Casamiquela RM (1977) The presence of the genus *Plateosaurus* (Sauropodomorpha) in the Upper Triassic of the El Tranquilo Formation, Patagonia. In: Abstracts. 4th international Gondwana symposium, pp 30–31

Cerda IA, Pol D, Chinsamy A (2014) Osteohistological insight into the early stages of growth in *Mussaurus patagonicus* (Dinosauria, Sauropodomorpha). Hist Biol 26(1):110–121. https://doi.org/10.1080/08912963.2012.763119

Chiappe LM, Dingus L (2001) Walking on eggs: the astonishing discovery of thousands of dinosaur eggs in the badlands of Patagonia. Scribner, New York

Chiappe LM, Coria RA, Dingus L et al (1998) Sauropod dinosaur embryos from the Late Cretaceous of Patagonia. Nature 396(6708):258–261. https://doi.org/10.1038/24370

Chiappe L, Dingus L, Jackson FD et al (2000) Sauropod eggs and embryos from the Late Cretaceous of Patagonia. First international symposium on dinosaur eggs and babies Isona. Extended abstracts, Spain, pp 23–29

Chiappe LM, Salgado L, Coria RA (2001) Embryonic skulls of titanosaur sauropod dinosaurs. Science 293(5539):2444–2446. https://doi.org/10.1126/science.1063723

Chiappe LM, Coria RA, Jackson FD et al (2003) The Late Cretaceous nesting site of Auca Mahuevo (Patagonia, Argentina): Eggs, nests, and embryos of titanosaurian sauropods. Palaeovertebrata 32(2–4):97–108

Chiappe LM, Schmitt JG, Jackson FD et al (2004) Nest structure for sauropods: sedimentary criteria for recognition of dinosaur nesting traces. Palaios 19(1):89–95

Chiappe LM, Jackson FD, Coria RA et al (2005) Nesting titanosaurs from Auca Mahuevo and adjacent sites. The sauropods University of California Press, Berkeley, pp 285–302

Choi S, Han S, Kim N-H et al (2018) A comparative study of eggshells of Gekkota with morphological, chemical compositional and crystallographic approaches and its evolutionary implications. PLoS ONE 13(6):e0199496. https://doi.org/10.1371/journal.pone.0199496

Coria RA, Chiappe LM (2001) Tooth replacement in a sauropod premaxilla from the Upper Cretaceous of Patagonia, Argentina. Ameghiniana 38:463–466

Coria RA, Chiappe LM (2007) Embryonic skin from Late Cretaceous sauropods (Dinosauria) of Auca Mahuevo, Patagonia, Argentina. J Paleontol 81(6):1528–1532. https://doi.org/10.1666/05-150.1

Coria RA, Salgado L, Chiappe LM (2010) Multiple dinosaur egg-shell occurrence in an Upper Cretaceous nesting site from Patagonia. Ameghiniana 47(1):107–110. https://doi.org/10.5710/AMGH.v47i1.4

Curry-Rogers K, D'Emic M, Vickaryous RR, M, et al (2011) Sauropod dinosaur osteoderms from the Late Cretaceous of Madagascar. Nat Commun 2:564. https://doi.org/10.1038/ncomms1578

Curry-Rogers KC, Whitney M, D'Emic M et al (2016) Precocity in a tiny titanosaur from the Cretaceous of Madagascar. Science 352(6284):450–453. https://doi.org/10.1126/science.aaf1509

Czerkas SA (1994) The history and interpretation of sauropod skin impressions. Gaia 10:173–182

D'Emic MD (2012) The early evolution of titanosauriform sauropod dinosaurs. Zool J Linn Soc 166(3):624–671. https://doi.org/10.1111/j.1096-3642.2012.00853.x

de Valais S, Apesteguía S, Udrizar Sauthier DE (2003) Nuevas evidencias de dinosaurios de la Formación Puerto Yeruá (Cretácico), Província de Entre Ríos, Argentina. Ameghiniana 40(3):507–511

Deeming C (2004) Reptilian incubation: Environment, evolution and behaviour. Nottingham University Press, Nottingham

Deeming DC (2006) Ultrastructural and functional morphology of eggshells supports the idea that dinosaur eggs were incubated buried in a substrate. Paleontology 49(1):171–185. https://doi.org/10.1111/j.1475-4983.2005.00536.x

Depéret C (1900) Sur les dinosauriens des etages de Rognac et de Vitrolles du pied de la Montagne-Noire. C-R Acad Sci 130:637–639

Dingus L, Clarke J, Scott GR et al (2000) Stratigraphy and magnetostratigraphic/faunal constraints for the age of sauropod embryo-bearing rocks in the Neuquén group (Late Cretaceous, Neuquén Province, Argentina). novi 2000(3290):1–11. https://doi.org/10.1206/0003-0082(2000)290<0001:SAMFCF>2.0.CO;2

Eagle RA, Enriquez M, Grellet-Tinner G et al (2015) Isotopic ordering in eggshells reflects body temperatures and suggests differing thermophysiology in two Cretaceous dinosaurs. Nat Commun 6:8296. https://doi.org/10.1038/ncomms9296

Erben HK, Hoefs J, Wedepohl KH (1979) Paleobiological and isotopic studies of eggshells from a declining dinosaur species. Paleobiology 5(04):380–414. https://doi.org/10.1017/S0094837300016900

Faccio G (1994) Dinosaurian eggs from the Upper Cretaceous of Uruguay. Dinosaur eggs and babies. Cambridge University Press, Cambridge, pp 47–55

Faccio G, Montaña J (1994) Registro fósil *in situ* de huevos de dinosaurios del Cretácico Superior de Durazno y la presencia de Faveoloolithidae (Zhao et Ding 1976) en huevos de dinosaurios del Uruguay. Paleociencias Serie Didáctica 2:16–18

Faccio G, Ford I, Gancio F (1990) Primer registro fósil *in situ* de huevos de dinosaurios del Cretácico Superior del Uruguay (Fm. Mercedes). Fac Agron Bol Invest 26:20p.

Fernández MS (2013) Análisis de cáscaras de huevos de dinosaurios de la Formación Allen, Cretácico Superior de Río Negro (Campaniano-Maastrichtiano): Utilidad de los macrocaracteres de interés parataxonómico. Ameghiniana 50(1):79–97. https://doi.org/10.5710/AMGH.14.11.2012.586

Fernández MS, Khosla A (2015) Parataxonomic review of the Upper Cretaceous dinosaur eggshells belonging to the oofamily Megaloolithidae from India and Argentina. Hist Biol 27(2):158–180. https://doi.org/10.1080/08912963.2013.871718

Fernández MS, Matheos SD (2011) Alteraciones en cáscaras de huevos de dinosaurios en el Cretácico Superior de la Província de Río Negro, Argentina. Ameghiniana 48(1):43–52. https://doi.org/10.5710/AMGH.v48i1(272)

Fernández MS (2016) Important contributions of the South American record to the understanding of dinosaur reproduction.In: Khosla A, Lucas SG (eds), Cretaceous period: biotic diversity and biogeography. Bull N M Mus Nat Hist Sci 71:91–105

Fernández MS et al (2017) A new fossil site with eggshells belonging to the oofamily faveoloolithidae from the Allen Formation (Río Negro Province). Rincón de los Sauces, Neuquén. 1ª Reunión de Paleovertebrados de la Cuenca Neuquina, http://rid.unrn.edu.ar/handle/20.500.12049/3466

Fiorelli LE, Grellet-Tinner G, Argañaraz E et al (2013) Tafonomía del sitio de nidificación de neosaurópodos de Sanagasta (La Rioja, Argentina): ejemplo de preservación excepcional en un paleoambiente hidrotermal del Cretácico. Ameghiniana 50(4):389–406

Fondevilla V, Vila B, Oms, et al (2017) Skin impressions of the last European dinosaurs. Geol Mag 154(2):393–398. https://doi.org/10.1017/S0016756816000868

Fowler DW, Hall LE (2011) Scratch-digging sauropods, revisited. Hist Biol 23(1):27–40. https://doi.org/10.1080/08912963.2010.504852

Frenguelli J (1951) Un huevo fósil del Rocanense. Rev Asoc Geológica Argent 6(2):108–112

García RA (2007) An "egg-tooth"–like structure in titanosaurian sauropod embryos. J Vertebr Paleontol 27(1):247–252. https://doi.org/10.1671/0272-4634(2007)27[247:AESITS]2.0.CO;2

García RA, Cerda IA (2010) Dentition and histology in titanosaurian dinosaur embryos from Upper Cretaceous of Patagonia, Argentina. Paleontology 53(2):335–346. https://doi.org/10.1111/j.1475-4983.2010.00937.x

García RA, Salgado L, Fernández MS et al (2015) Paleobiology of titanosaurs: reproduction, development, histology, pneumaticity, locomotion and neuroanatomy from the South American fossil record. Ameghiniana 52(1):29–68. https://doi.org/10.5710/AMGH.16.07.2014.829

Garcia G, Vianey-Liaud M (2001) Nouvelles données sur les coquilles d'œufs de dinosaures Megaloolithidae du Sud du France: systématique et variabilité intraspécifique. C R AcadSci II A 332(3):185–191. https://doi.org/10.1016/S1251-8050(00)91401-0

Garrido AC (2010) Paleoenvironment of the Auca Mahuevo and Los Barreales sauropod nesting-sites (Late Cretaceous, Neuquén Province, Argentina). Ameghiniana 47(1):99–106. https://doi.org/10.5710/AMGH.v47i1.8

Goso C, Perea D (2004) El Cretácico post-basáltico de la Cuenca litoral del río Uruguay: geología y paleontología. In: Veroslavsky G, Ubila M, Martínez S (eds), Cuencas sedimentarias de Uruguay: geología, paleontología y recursos naturales. Mesozoico, Second Edition. DIRAC, Montevideo, Uruguay, pp 143–171

Grellet-Tinner G (2005) Membrana testacea of titanosaurid dinosaur eggs from Auca Mahuevo (Argentina): Implications for exceptional preservation of soft tissue in Lagerstätten. J Vertebr Paleontol 25(1):99–106. https://doi.org/10.1671/0272-4634(2005)025[0099:MTOTDE]2.0.CO;2

Grellet-Tinner G, Fiorelli LE (2010) A new Argentinean nesting site showing neosauropod dinosaur reproduction in a Cretaceous hydrothermal environment. Nat Commun 1:32. https://doi.org/10.1038/ncomms1031

Grellet-Tinner G, Zaher H (2007) Taxonomic identification of the Megaloolithid egg and eggshells from the Cretaceous Bauru Basin (Minas Gerais, Brazil): comparison with the Auca Mahuevo (Argentina) titanosaurid eggs. Papéis Avulsos Zool 47(7):105–112. https://doi.org/10.1590/S0031-10492007000700001

Grellet-Tinner G, Chiappe LM, Coria R (2004) Eggs of titanosaurid sauropods from the Upper Cretaceous of Auca Mahuevo (Argentina). Can J Earth Sci 41(8):949–960. https://doi.org/10.1139/e04-049

Grellet-Tinner G, Chiappe L, Norell M (2006) Dinosaur eggs and nesting behaviors: a paleobiological investigation. Paleogeogr Paleoclimatol Paleoecol 232(2):294–321. https://doi.org/10.1016/j.palaeo.2005.10.029

Grellet-Tinner G, Sim CM, Kim DH et al (2011) Description of the first lithostrotian titanosaur embryo *in ovo* with Neutron characterization and implications for lithostrotian Aptian migration and dispersion. Gondwana Res 20(2–3):621–629. https://doi.org/10.1016/j.gr.2011.02.007

Grellet-Tinner G, Fiorelli LE, Salvador RB (2012) Water vapor conductance of the Lower Cretaceous dinosaurian eggs from Sanagasta, La Rioja, Argentina: paleobiological and paleoecological implications for South American faveoloolithid and megaloolithid eggs. Palaios 27(1):35–47. https://doi.org/10.2110/palo.2011.p11-061r

Hechenleitner EM, Grellet-Tinner G, Fiorelli LE (2015) What do giant titanosaur dinosaurs and modern Australasian megapodes have in common? PeerJ 3:e1341. https://doi.org/10.7717/peerj.1341

Hechenleitner EM, Fiorelli LE, Grellet-Tinner G et al (2016a) A new Upper Cretaceous titanosaur nesting site from La Rioja (NW Argentina), with implications for titanosaur nesting strategies. Paleontology 59(3):433–446. https://doi.org/10.1111/pala.12234

Hechenleitner EM, Grellet-Tinner G, Foley M et al (2016b) Micro-CT scan reveals an unexpected high-volume and interconnected pore network in a Cretaceous Sanagasta dinosaur eggshell. J R Soc Interface 13(116):20160008. https://doi.org/10.1098/rsif.2016.0008

Hechenleitner EM, Jeremías R, Taborda A et al (2018) Biomechanical evidence suggests extensive eggshell thinning during incubation in the Sanagasta titanosaur dinosaurs. Peer J 6:e4971. https://doi.org/10.7717/peerj.4971

Hechenleitner EM, Leuzinger L, Martinelli AG et al (2020) Two Late Cretaceous sauropods reveal titanosaurian dispersal across South America. Commun Biol 3(1):1–13. https://doi.org/10.1038/s42003-020-01338-w

Horner JR, Makela R (1979) Nest of juveniles provides evidence of family structure among dinosaurs. Nature 282(5736):296–298. https://doi.org/10.1038/282296a0

Hünicken MA (2005) La Formación Sanagasta y el Parque Geológico (Cretácico Superior) con nidos y huevos de dinosaurios, Valle de Sanagasta, Provincia de La Rioja, Argentina. INSUGEO Ser Correlación Geológica 19:75–82

Hünicken MA, Tauber AH, Leguizamón RR (2001) Hallazgo de huevos y nidos de dinosaurios, asociados a restos de vegetales silicificados: asignación al Cretácico de las secuencias portadoras aflorantes en Sanagasta, provincia de La Rioja. Ameghiniana 38:R10

Jackson FD, Garrido A, Schmitt JG et al (2004) Abnormal, multilayered titanosaur (Dinosauria: Sauropoda) eggs from *in situ* clutches at the Auca Mahuevo locality, Neuquen Province, Argentina. J Vertebr Paleontol 24(4):913–922. https://doi.org/10.1671/0272-4634(2004)024[0913:AMTDSE]2.0.CO;2

Jackson FD, Varricchio DJ, Jackson RA et al (2008) Comparison of water vapor conductance in a titanosaur egg from the Upper Cretaceous of Argentina and a *Megaloolithus siruguei* egg from Spain. Paleobiology 34(2):229–246. https://doi.org/10.1666/0094-8373(2008)034[0229:COWVCI]2.0.CO;2

Jackson FD, Schmitt JG, Oser SE (2013) Influence of vertisol development on sauropod egg taphonomy and distribution at the Auca Mahuevo locality, Patagonia, Argentina. Paleogeogr Paleoclimatol Paleoecol 386:300–307. https://doi.org/10.1016/j.palaeo.2013.05.031

Kapur VV, Khosla A (2019) Faunal elements from the Deccan volcano-sedimentary sequences of India: a reappraisal of biostratigraphic, palaeoecologic, and palaeobiogeographic aspects. Geol J 54(5):2797–2828

Kerourio P, Sigé B (1984) L'apport des coquilles d'oeufs de dinosaures de Laguna Umayo à l'âge de la Formation Vilquechico (Pérou) et à la compréhension de *Perutherium altiplanense*. Newsl Stratigr 13(3):133–142. https://doi.org/10.1127/nos/13/1984/133

Khosla A, Lucas SG (2020a) Discussion: Oospecies diversity, biomineralization aspects, taphonomical, biostratigraphical, palaeoenvironmental, palaeoecological and palaeobiogeographical inferences of the dinosaur-bearing Lameta Formation of Peninsular India. In: Khosla A, Lucas SG (eds) Late cretaceous dinosaur eggs and eggshells of peninsular India: oospecies diversity and taphonomical, palaeoenvironmental, biostratigraphical and palaeobiogeographical inferences. Springer International Publishing, Cham, pp 207–271

Khosla A, Lucas SG (2020b) Indian Late Cretaceous dinosaur nesting sites and their systematic studies. In: Khosla A, Lucas SG (eds) Late Cretaceous dinosaur eggs and eggshells of Peninsular India: oospecies diversity and taphonomical, palaeoenvironmental, biostratigraphical and palaeobiogeographical inferences. Springer International Publishing, Cham, pp 117–205

Khosla A, Sahni A (1995) Parataxonomic classification of Late Cretaceous dinosaur eggshells from India. J Palaeontol Soc India 40:87–102

Khosla A, Sahni A (2003) Biodiversity during the Deccan volcanic eruptive episode. J Asi Earth Sci 21(8):895–908

Khosla A (2019) Paleobiogeographical inferences of Indian Late Cretaceous vertebrates with special reference to dinosaurs. Hist Biol 1–12. https://doi.org/10.1080/08912963.2019.1702657

Knoll F, Ridgely RC, Ortega F et al (2013) Neurocranial osteology and neuroanatomy of a Late Cretaceous titanosaurian sauropod from Spain (*Ampelosaurus* sp.). PLoS One 8(1):e54991. https://doi.org/10.1371/journal.pone.0054991

Kundrát M, Coria RA, Manning TW et al (2020) Specialized craniofacial anatomy of a titanosaurian embryo from Argentina. Curr Biol 30(21):4263-4269.e2. https://doi.org/10.1016/j.cub.2020.07.091

Smith RMH et al (2014) Ichnology of sauropodomorph nests from Patagonia indicates early Jurassic origin of herd-living and breeding site fidelity. 4° International Paleontological Congress, Mendoza, Argentina

Le Loeuff J (1993) European titanosaurs. Rev Paléobiol 7:105–117

Legendre LJ, Rubilar-Rogers D, Vargas AO et al (2020) The first dinosaur egg remains a mystery. https://doi.org/10.1101/2020.12.10.406678

Magalhães-Ribeiro CM (1997) Descripçāo de caracteres morfológicos e estudo composicional de cascaras de ovo de dinossauros da Formaçao Allen (Cretáceo Superior), do Bajo de Santa rosa, Provincia de Río Negro (Argentina). Master Thesis, Universidad Federal do Río de Janeiro, p 222

Magalhães-Ribeiro CM (1999a) Ovos e cascas de ovos de dinossauros da região de Uberaba, Minas Gerais (Formacão Marília, Bacia Bauru, Cretáceo Superior). PhD Thesis, Universidad Federal do Río de Janeiro, p 212

Magalhães-Ribeiro CM (1999b) Utilizacão de catodoluminescência em cascas de ovos de dinossauros. Ameghiniana 36(4-Resúmenes):15R

Magalhães-Ribeiro CM (2000) Microstructural analysis of dinosaur eggshells from Bauru Basin (Late Cretaceous), Minas Gerais, Brasil. In: First international symposium on dinosaur eggs and babies, Isona, Spain. Extended Abstracts. pp 117–122

Manera de Bianco T (1996) Nueva localidad con nidos y huevos de dinosaurios (Titanosaurldae) del Cretácico Superior, Cerro Blanco, Yaminué, Río Negro, Argentina. Publicación Especial de la Asociación Paleontológica Argentina 4(1)

Martínez RDF, Lamanna MC, Novas FE et al (2016) A basal lithostrotian titanosaur (Dinosauria: Sauropoda) with a complete skull: implications for the evolution and paleobiology of Titanosauria. PLoS ONE 11(4):e0151661. https://doi.org/10.1371/journal.pone.0151661

Mateus I, Mateus H, Antunes MT, Mateus O, Taquet P, Ribeiro V, Manuppella G (1997) Couvée, œufs et embryons d'un Dinosaure Théropode du Jurassique supérieur de Louurinha (Portugal). Annu Rev Earth Planet Sci 325(1):71–78. https://doi.org/10.1016/S1251-8050(97)83275-2

Matheron P (1869) Notice sur les reptiles fossiles des dépôts fluvio-lacustres crétacés du bassin à lignite de Fuveau. B Soc Geol Fr 26(2):781–795

Mikhailov KE (1991) Classification of fossil eggshells of amniotic vertebrates. Acta Palaeontol Pol 36(2):193–238

Mikhailov KE (1997) Fossil and recent eggshell in amniotic vertebrates: fine structure, comparative morphology and classification. Palaeontol Assoc Spec Pap Palaeontol 56:1–80

Mohabey DM (1998) Systematics of Indian Upper Cretaceous dinosaur and chelonian eggshells. J Vertebr Paleontol 18(2):348–362. https://doi.org/10.1080/02724634.1998.10011063

Mohabey DM (1996) A new oospecies, *Megaloolithus matleyi*, from the Lameta Formation (Upper Cretaceous) of Chandrapur district, Maharashtra, India, and general remarks on the palaeoenvironment and nesting behaviour of dinosaurs. Cretaceous Res 17:183–196

Mones Á (1980) Nuevos elementos de la paleoherpetofauna del Uruguay (Crocodylia y Dinosauria). Actas 21 Congreso Argentino de Paleontología y Bioestratigrafía y 1° Congreso Latinoamericano de Paleontología (Buenos Aires, 1978) 1:265–277

Mones Á (1997) Los vertebrados mesozoicos del Uruguay y sus relaciones con los de áreas vecinas. In: Arroyo-Cabrales J, Polaco OJ (eds) Homenaje al Profesor Ticul Álvarez. Instituto Nacional de Antropología e Historia, pp 205–222

Moreno-Azanza M, Mariani E, Bauluz B et al (2013) Growth mechanisms in dinosaur eggshells: an insight from electron backscatter diffraction. J Vertebr Paleontol 33(1):121–130. https://doi.org/10.1080/02724634.2012.710284

Moreno-Azanza M, Bauluz B, Canudo JI et al (2016) Combined use of electron and light microscopy techniques reveals false secondary shell units in Megaloolithidae eggshells. PLoS ONE 11(5):e0153026. https://doi.org/10.1371/journal.pone.0153026

Mourier T, Bengtson P, Bonhomme M et al (1988) The Upper Cretaceous - Lower Tertiary marine to continental transition in the Bagua basin, northern Peru - Paleontology, biostratigraphy, radiometry, correlations. Newsl Stratigr 19(3):143–177. https://doi.org/10.1127/nos/19/1988/143

Mueller-Towe I, Sander PM, Schuller H et al (2002) Hatching and infilling of dinosaur eggs as revealed by computed tomography. Palaeontogr Abt A-Stuttg 267(1/6):119–168

International Commision for Zoological Nomenclature, Ride WDJL (1999) International code of zoological nomenclature. Int Trust Zool Nomenclature

Norell MA, Clark JM, Demberelyin D et al (1994) A theropod dinosaur embryo and the affinities of the Flaming Cliffs dinosaur eggs. Science 266(5186):779–782. https://doi.org/10.1126/science.266.5186.779

Norell MA, Wiemann J, Fabbri M et al (2020) The first dinosaur egg was soft. Nature 583(7816):406–410. https://doi.org/10.1038/s41586-020-2412-8

Novas FE, Agnolin F, Rozadilla S et al (2019) Paleontological discoveries in the Chorrillo Formation (upper Campanian-lower Maastrichtian, Upper Cretaceous), Santa Cruz Province, Patagonia, Argentina. Revista del Museo Argentino de Ciencias Naturales 21:217–293. https://doi.org/10.22179/REVMACN.21.655

Otero A, Pol D (2013) Postcranial anatomy and phylogenetic relationships of *Mussaurus patagonicus* (Dinosauria, Sauropodomorpha). J Vertebr Paleont 33(5):1138–1168. https://doi.org/10.1080/02724634.2013.769444

Otero A, Cuff AR, Allen V et al (2019) Ontogenetic changes in the body plan of the sauropodomorph dinosaur *Mussaurus patagonicus* reveal shifts of locomotor stance during growth. Sci Rep 9(1):7614. https://doi.org/10.1038/s41598-019-44037-1

Packard MJ, DeMarci VG (1995) Shell structure and formationin eggs of oviparous reptiles. Egg Incubation: Its Effects on Embryonic Development in Birds and Reptiles. Cambridge University Press, New York, USA, pp 17–28

Paik IS, Kim HJ, Lee H et al (2017) A large and distinct skin impression on the cast of a sauropod dinosaur footprint from Early Cretaceous floodplain deposits. Korea. Sci Rep 7(1):16339. https://doi.org/10.1038/s41598-017-16576-y

Pol D, Powell JE (2007) Skull anatomy of *Mussaurus patagonicus* (Dinosauria: Sauropodomorpha) from the Late Triassic of Patagonia. Hist Biol 19(1):125–144. https://doi.org/10.1080/08912960601140085

Pouech J-J (1859) Mémoire sur les terrains tertiaires de l'Ariège rapportés à une coupe transversale menée de Fossat à Aillères, passant par le Mas d'Azil et projetée sur le méridien de ce lieu. Bull Soc Geol FR 2(16):381–411

Powell JE (1992) Hallazgo de huevos asignables a dinosaurios titanosáuridos (Saurischia, Sauropoda) de la Provincia de Río Negro, Argentina. Acta Zoológica Lilloana 41:381–389

Powell JE (1985) Hallazgo de nidadas de huevos de dinosaurios (Sauropoda-Titanosauridae) del Cretácico Superior del Salitral Ojo de Agua, provincia de Río Negro. 2° Jornadas Argentinas de Paleontología de Vertebrados

Powell JE (1987) The Late Cretaceous fauna of Los Alamitos, Patagonia, Argentina. VI: The titanosaurids. Revista del Museo Argentino de Ciencias Naturales Bernardino Rivadavia e Instituto Nacional de Investigación de las Ciencias Naturales Paleontología 3(3):147–153

Price LI (1951) Ovo de dinossauro na formação Bauru, do estado de Minas Gerais. Notas Preliminares da Divisão de Geologia de Mineralogia 1–7

Reisz RR, Evans DC, Sues H-D et al (2010) Embryonic skeletal anatomy of the sauropodomorph dinosaur *Massospondylus* from the Lower Jurassic of South Africa. J Vertebr Paleont 30(6):1653–1665. https://doi.org/10.1080/02724634.2010.521604

Reisz RR, Evans DC, Roberts EM et al (2012) Oldest known dinosaurian nesting site and reproductive biology of the Early Jurassic sauropodomorph *Massospondylus*. PNAS 109(7):2428–2433. https://doi.org/10.1073/pnas.1109385109

Reisz RR, Huang TD, Roberts EM et al (2013) Embryology of Early Jurassic dinosaur from China with evidence of preserved organic remains. Nature 496(7444):210–214. https://doi.org/10.1038/nature11978

Romano M, Whyte MA (2012) Information on the foot morphology, pedal skin texture and limb dynamics of sauropods: evidence from the ichnological record of the Middle Jurassic of the Cleveland Basin, Yorkshire, UK. Zubia 30:45

Salgado L, Coria RA, Chiappe LM (2005) Osteology of the sauropod embryos from the Upper Cretaceous of Patagonia. Acta Palaeontol Pol 50(1):79–92

Salgado L, Coria RA, Magalhães Ribeiro CM et al (2007) Upper Cretaceous dinosaur nesting sites of Río Negro (Salitral Ojo de Agua and Salinas de Trapalcó-Salitral de Santa Rosa), northern Patagonia, Argentina. Cretac Res 28(3):392–404. https://doi.org/10.1016/j.cretres.2006.06.007

Salgado L, Magalhães Ribeiro CM, García RA et al (2009) Late Cretaceous Cretaceous Megaloolithid eggs from Salitral de Santa Rosa (Río Negro, Patagonia, Argentina) inferences on the titanosaurian reproductive biology. Ameghiniana 46:605–620. https://doi.org/10.1016/j.cretres.2006.06.007

Sander PM, Peitz C, Jackson FD et al (2008) Upper Cretaceous titanosaur nesting sites and their implications for sauropod dinosaur reproductive biology. Palaeontogr Abt A 69–107. https://doi.org/10.1127/pala/284/2008/69

Schweitzer MH, Chiappe L, Garrido AC et al (2005) Molecular preservation in Late Cretaceous sauropod dinosaur eggshells. Proc R Soc B 272(1565):775–784. https://doi.org/10.1098/rspb.2004.2876

Sellés AG, Vila B (2015) Re-evaluation of the age of some dinosaur localities from the southern Pyrenees by means of megaloolithid oospecies. J Iber Geol 41(1):125–139

Sellés AG, Bravo AM, Delclòs X et al (2013) Dinosaur eggs in the Upper Cretaceous of the Coll de Nargó area, Lleida Province, south-central Pyrenees, Spain: oodiversity, biostratigraphy and their implications. Cretac Res 40:10–20. https://doi.org/10.1016/j.cretres.2012.05.004

Seymour RS (1979) Dinosaur eggs: gas conductance through the shell, water loss during incubation and clutch size. Paleobiology 5(1):1–11

Sigé B (1968) Dents de micromammifères et fragments de coquilles d'oeufs de dinosauriens dans la faune de vértebrés du Crétacé supérieur de Laguna Umayo (Andes peruviennes). Gallica 267:1495

Simón ME (1999) Estudio de fragmentos de cáscaras de huevos de la Formación Allen (Campaniano-Maastrichtiano), provincia de Río Negro. Tesina de Licenciatura, Universidad Nacional de Córdoba, Argentina, Argentina, p 285

Simón ME (2006) Cáscaras de huevos de dinosaurios de la Formación Allen (Campaniano-Maastrichtiano), en Salitral Moreno, provincia de Río Negro, Argentina. Ameghiniana 43(3):513–528

Simón ME (2002) *Paquiloolithus rionegrinus* Simón, 1999, primer oogénero descripto para la Provincia de Rio Negro (Campaniano - Maastrichtiano), Argentina. In: Actas XVI Jornadas Argentinas de Paleontologia de Vertebrados. San Luis, Argentina, p 47

Soto M, Perea D, Cambiaso A (2012) First sauropod (Dinosauria: Saurischia) remains from the Guichón Formation, Late Cretaceous of Uruguay. J South Am Earth Sci 33(1):68–79. https://doi.org/10.1016/j.jsames.2011.08.003

Soto M, Perea D, Veroslasky G et al (2008) New findings of dinosaur remains and considerations on the age of the Guichon Formation. Revista de la Sociedad Uruguaya de Geologia 11–23

Taquet P (2001) Philippe Matheron et Paul Gervais: deux pionniers de la découverte et de l'étude des os et des œufs de dinosaures de Provence (France). Geodiversitas 23(4):611–623

Tauber AH (2007) Primer yacimiento de huevos de dinosaurios (Cretácico Superior) de la provincia de La Rioja, Argentina. Ameghiniana 44(1):11–28

Trimby P, Grellet-Tinner G (2011) The hidden secrets of dinosaur eggs. Infocus 4–21. https://doi.org/10.22443/rms.inf.1.76

Verde M (2012) Icnología de la Formación Asencio y las "Calizas del Queguay." PhD Thesis. Pedeciba Biología. Universidad de la República. Montevideo, p 204

Veroslavsky G, Aubet N, Martínez SA et al (2019) Late Cretaceous stratigraphy of the southeastern Cretaceous Paraná basin ("norte basin" - Uruguay): The Maastrichtian age of the calcretization process. Geociencias 38(2):427–449

Vianey-Liaud M, Garcia G (2003) Diversity among north African dinosaur eggshells. Palaeovertebrata 32:171–188

Vianey-Liaud M, López-Martínez N (1997) Late Cretaceous dinosaur eggshells from the Tremp Basin, southern Pyrenees, Lleida, Spain. J Paleontol 71(6):1157–1171. https://doi.org/10.1017/S002233600003609X

Vianey-Liaud M, Jain SL, Sahni A (1988) Dinosaur eggshells (Saurischia) from the Late Cretaceous Intertrappean and Lameta formations (Deccan, India). J Vert Paleont 7(4):408–424. https://doi.org/10.1080/02724634.1988.10011673

Vianey-Liaud M, Hirsch K, Sahni AS et al (1997) Late Cretaceous Peruvian eggshells and their relationships with laurasian and eastern Gondwanian material. Geobios 30(1):75–90. https://doi.org/10.1016/S0016-6995(97)80259-8

Vianey-Liaud M, Khosla A, Geraldine G (2003) Relationships between European and Indian dinosaur eggs and eggshells of the oofamily Megaloolithidae. J Vert Paleontol 23(3):575–585

Vidal D, Ortega F, Gascó F et al (2017) The internal anatomy of titanosaur osteoderms from the Upper Cretaceous of Spain is compatible with a role in oogenesis. Sci Rep 7(1):42035. https://doi.org/10.1038/srep42035

Vila B, Jackson FD, Fortuny J et al (2010) 3-D Modelling of megaloolithid clutches: insights about nest construction and dinosaur behaviour. PLoS ONE 5(5):e10362. https://doi.org/10.1371/journal.pone.0010362

Wiemann J, Fabbri M, Yang T-R et al (2018a) Fossilization transforms vertebrate hard tissue proteins into N-heterocyclic polymers. Nat Commun 9(1):4741. https://doi.org/10.1038/s41467-018-07013-3

Wiemann J, Yang T-R, Norell MA (2018b) Dinosaur egg colour had a single evolutionary origin. Nature 563(7732):555–558. https://doi.org/10.1038/s41586-018-0646-5

Wiemann J, Briggs DEG (2021) Validation of biosignatures confirms the informative nature of fossil organic Raman spectra. https://doi.org/10.1101/2021.02.07.430162

Wilson JA, Mohabey DM, Peters SE et al (2010) Predation upon hatchling dinosaurs by a new snake from the Late Cretaceous of India. PLoS Biol 8(3):e1000322. https://doi.org/10.1371/journal.pbio.1000322

Zaher H, Pol D, Carvalho AB, Nascimento PM et al (2011) A Complete skull of an Early Cretaceous sauropod and the evolution of advanced titanosaurians. PLoS ONE 6(2):e16663. https://doi.org/10.1371/journal.pone.0016663

Zhao Z, Ding S (1976) Discovery of the dinosaurian eggshells from Alxa, Ningxia and its stratigraphical significance. Vertebrat Palasiatic 14:42–44

Zhao Z K (1975) Microstructure of fossil dinosaur eggs from Nanxiong, Guangdong (I)—A recapitulation of the problems of dinosaur egg taxonomy. Vertebrata PalAsiatica 13(2) (in Chinese).

Zhao Z (1979) Progress in the research of dinosaur eggs. In: Mesozoic and cenozoic red beds of South China selected papers from the "Cretaceous-Tertiary Workshop." Institute of vertebrate paleontology, paleoanthropology & Nanjing Institute of Paleontology Science Press, Nanxiong, Guangdong Province, China, pp 330–340

Body Size Evolution and Locomotion in Sauropodomorpha: What the South American Record Tells Us

Alejandro Otero and John R. Hutchinson

Abstract The transition from early sauropodomorphs to sauropods is of special interest given that a shift from obligatory or facultative bipedalism to an obligatory quadrupedalism is evident. In this chapter, we review and discuss the biological mechanisms underpinning such evolutionary transformations. The discovery of the South American sauropodomorph *Mussaurus patagonicus* has helped elucidate changes in the upper forelimb from bipedality to quadrupedality. Shoulder range of motion studies has shown that *Mussaurus* could not protract its forelimb past vertical, which suggests that quadrupedal locomotion could not have been possible, although it might have been if the elbow was habitually strongly flexed, as has been hypothesized for ornithischian quadrupeds. Yet muscle moment arm studies indicated that *Mussaurus* could not straighten its elbow, suggesting it did not have columnar forelimbs like later, fully quadrupedal sauropods. Quadrupedal locomotion first evolved in the adult forms of the sauropodomorphs closest to Sauropoda (e.g., *Melanorosaurus*). However, it has been suggested that already some early sauropodomorphs (e.g., *Massospondylus*) were quadrupedal during early ontogenetic stages and adopted a bipedal stance, at least facultatively, as adults. The postural shifting that some sauropodomorphs experienced during their ontogeny has important implications for understanding evolutionary processes that caused those shifts. Available ontogenetic series of *Mussaurus* provide additional insight into these evolutionary developmental transitions. The body's center of mass of this species moved from a position in the mid-thorax to a more posterior position close to the pelvis, consistent with a shift from quadrupedalism to bipedalism at a young age. This postural modification could be the product of the relative enlargement of the tail and the reduction of the neck

---

A. Otero (✉)
Consejo Nacional de Investigaciones Científicas y Técnicas (CONICET), Buenos Aires, Argentina
e-mail: aleotero@fcnym.unlp.edu.ar

División Paleontología de Vertebrados, Facultad de Ciencias Naturales y Museo (Anexo Laboratorios), Calle 122 y 60, La Plata (B1900WA), Buenos Aires Province, Argentina

J. R. Hutchinson
Structure and Motion Laboratory, Department of Comparative Biomedical Sciences, The Royal Veterinary College, Hatfield, Hertfordshire, UK
e-mail: JHutchinson@rvc.ac.uk

A. Otero et al. (eds.), *South American Sauropodomorph Dinosaurs*, Springer Earth System Sciences, https://doi.org/10.1007/978-3-030-95959-3_12

during ontogeny, challenging previous studies, which emphasized that that transformation would have been linked to a relative enlargement of the forelimbs. Viewed in a phylogenetic context, the South American sauropodomorph record provides key information regarding the evolution of body size and limb mechanics in this group. An anterior center of mass shift occurred during the evolution of quadrupedalism in the Late Triassic, followed by a more striking anterior shift in Late Jurassic–Cretaceous titanosauriforms, a phenomenon apparently closely linked with locomotion (e.g., weight distribution; reduced athleticism) and environment. As South American titanosaurs included the largest land animals ever, these also inform us about the constraints on terrestrial gigantism and the surprising diversity of giant forms that can exist despite these biomechanical and other constraints.

Keywords Sauropodomorpha · Bipedalism · Quadrupedalism · Gigantism · Evolution

1 Introduction

The transition from early sauropodomorph dinosaurs to eusauropods involved one of the most drastic transformations recorded in the evolutionary history of this group (Galton and Upchurch 2004; Bonnan and Yates 2007; Bates et al. 2016). The evolution of sauropodomorph dinosaurs was characterized by an early radiation of small and bipedal forms (i.e., obligate two-legged stance), such as *Buriolestes*, *Saturnalia*, *Panphagia*, and *Eoraptor*, with a gradual increase in body size (Carrano 2005; Rauhut et al. 2011; Bates et al. 2016; Pol et al. 2021), notably accentuated in Sauropoda, as heavy-weight quadrupedal forms (i.e., obligate four-legged stance).

The lineage of Triassic archosaurs leading to sauropods began as quadrupeds, transitioned to bipedality close to the base of Dinosauria (e.g., Sereno 1991), and then shifted back to quadrupedality close to or at the base of Sauropoda (Wilson and Sereno 1998; Carrano 2005; Apaldetti et al. 2018; McPhee et al. 2018). Although the reversion to quadrupedalism from bipedalism is extremely rare, there are at least four known examples confined to Dinosauria: sauropods themselves, and independently in three branches of ornithischian dinosaurs (ceratopsians, ornithopods, and thyreophorans [Carrano 1998; Brusatte et al. 2010; Maidment and Barrett 2012; VanBuren and Bonnan 2013; Hutson 2014]) (Fig. 1).

It has been proposed that there was mosaic evolution of the morphological features of the forelimb associated with quadrupedalism in the sauropodomorph lineage (Bonnan and Yates 2007). In particular, manus pronation (i.e., the rotation of the manus so that the palm faces posteriorly more than medially) has been proposed as a critical anatomical feature associated with the acquisition of quadrupedal locomotion in different lineages of Dinosauria (e.g., Bonnan 2003; Bonnan and Senter 2007; Bonnan and Yates 2007; Maidment and Barrett 2012; VanBuren and Bonnan 2013; Hutson 2014; Otero et al. 2017). The growing consensus is that increased manus

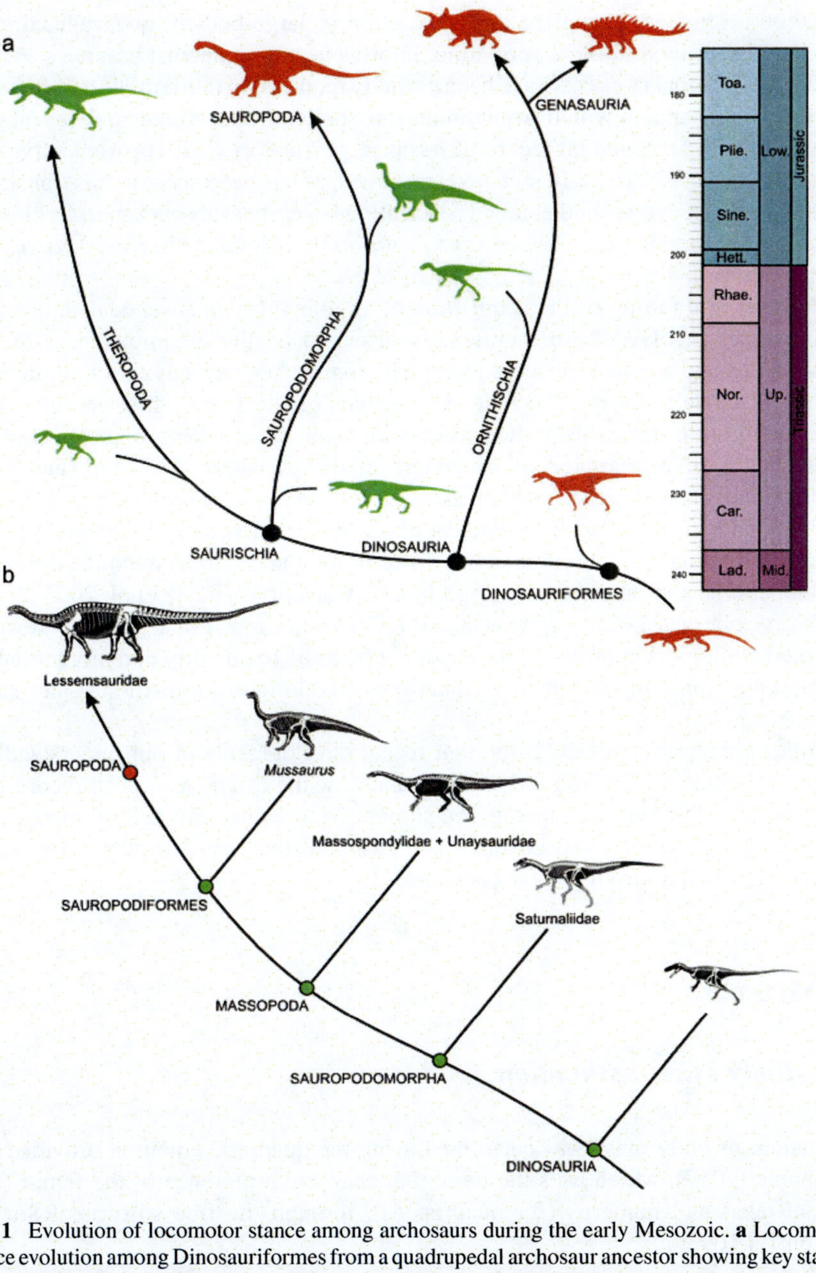

Fig. 1 Evolution of locomotor stance among archosaurs during the early Mesozoic. **a** Locomotor stance evolution among Dinosauriformes from a quadrupedal archosaur ancestor showing key stance shifts. **b** Evolution of stance among Sauropodomorpha depicting most relevant nodes with a South American representative. Red silhouettes and dots represent quadrupedal taxa and node, respectively; green silhouettes and dots represent bipedal taxa and nodes, respectively. Not to scale. (Silhouettes modified from Wilson and Sereno 1998, Martínez and Alcober 2009, Paul 2010, Sereno et al. 2013, Müller et al. 2020, Pol et al. 2021)

pronation originated during the early evolution of large-bodied, quadrupedal, and graviportal sauropodomorphs, providing improved support against gravity.

These evolutionary events have been partially recorded in the Triassic and Jurassic strata of Gondwana, in which Argentinian and South African representatives played a key role, with an abundant record of forms (e.g., Yates et al. 2010; Apaldetti et al. 2018; McPhee et al. 2018). In recent years, new light has been shed on this transition through new discoveries and detailed descriptions of early sauropodomorphs closely related to Sauropoda (e.g., *Lessemsaurus*, *Aardonyx*, *Melanorosaurus*, *Mussaurus*, *Leonerasaurus*).

However, the timing of the acquisition of an obligatory quadrupedal stance and the processes involved were traditionally obscured by the incompleteness of the fossil record and the lack of methodological approaches to analyze it comprehensively. Fortunately, during the first two decades of the twenty-first century, new sauropodomorphs have been described, which have been key to understanding the sequence of appearance of characters in the transition from bipedalism to quadrupedalism, such as *Lessemsaurus*, *Ingentia*, and *Ledumahadi* (Pol and Powell 2007a; Apaldetti et al. 2018; McPhee et al. 2018). Furthermore, the recognition of juvenile individuals of certain taxa has made it possible to understand the developmental patterns and processes involved in such transitions (Reisz et al. 2005, 2010, 2012; Chapelle et al. 2020a; Otero et al. 2017, 2019; Otero and Pol in press). Likewise, new methodological approaches have made it possible to quantify and more reliably interpret the morphological changes that occurred during this transition (Bates et al. 2016).

In this chapter, we review important recent developments in our understanding of the evolution of sauropod locomotion, and how the South American record has contributed to these developments. We summarize the anatomy and biomechanics which characterize these discoveries and examine the key role that body mass variation has had in driving these changes.

2 Methods

2.1 Body Mass Estimation

Estimation of body mass was performed using the quadratic equation provided by Campione (2017), which uses the minimum shaft circumference of the femur (for bipedal taxa) and humerus (for quadrupeds), through the free software RStudio version 1.3.1093.

3 Forelimb Evolution During the Acquisition of Quadrupedalism

The forelimbs of non-sauropod sauropodomorphs have captured the attention of paleontologists because their morphology was likely pivotal to the acquisition of quadrupedalism (Bonnan and Senter 2007; Bonnan and Yates 2007; Mallison 2010a, b; VanBuren and Bonnan 2013; Otero et al. 2017). While the hind limbs are used for terrestrial locomotion (sensu Gatesy and Dial 1996), the biological role of the forelimbs varies, depending on the locomotor pattern of the organism. For example, facultative bipedal vertebrates tend to devote the forelimbs to biological roles other than solely body support or locomotion (e.g., manipulation, digging, display and combat).

The problems associated with the transition from bipedalism to quadrupedalism in Sauropodomorpha received much attention from the beginning of the twenty-first century, in which the functional morphology of the forelimbs has been considered when reconstructing their locomotor habits. The questions of how the forelimbs of early sauropodomorphs were used for functions other than pure locomotion and the functional steps that ultimately produced the derived locomotor mechanisms present in Sauropoda have begun to be answered since then. In addition to the studies performed on taxa such as *Plateosaurus*, *Massospondylus*, and *Melanorosaurus* (Bonnan and Senter 2007; Bonnan and Yates 2007; Mallison 2010b), the discovery of *Mussaurus* has helped elucidate changes in the upper forelimb from bipedality to quadrupedality. The evidence addressing such steps is explored below.

3.1 Evidence from the Shoulder and Elbow Joints

Joint range of motion (ROM) (more broadly termed mobility) analysis can be a useful approach to evaluate the constraints for reconstructing limb poses among fossil tetrapods (Gatesy et al. 2009; Pierce et al. 2012; Manafzadeh et al. 2021), albeit considering the uncertainties regarding the inference of non-preserved articular cartilage (Fujiwara et al. 2010; Holliday et al. 2010; Tsai and Holliday 2014). Analysis of forelimb joint ROMs and mobility performed in the South American early sauropodomorph *Mussaurus patagonicus* tested the likelihood of potential forelimb poses, considering the scapular orientation and elbow flexion (Fig. 2). If the scapula was oriented in an anatomical position of about 55–60° from the horizontal (i.e., posterodorsally), the maximal humeral protraction allowed would not pass vertical (Otero et al. 2017), which concurs with previous reports for other early sauropodomorphs (Bonnan and Senter 2007; Mallison 2010a, b) and theropods (Senter and Robins 2005). This inference partially contradicts the possibility of quadrupedalism as a habitual mode of locomotion in early sauropodomorphs such as adult *Mussaurus*. However, if the elbow was habitually strongly flexed during locomotion, then quadrupedalism might be more achievable, perhaps with shorter

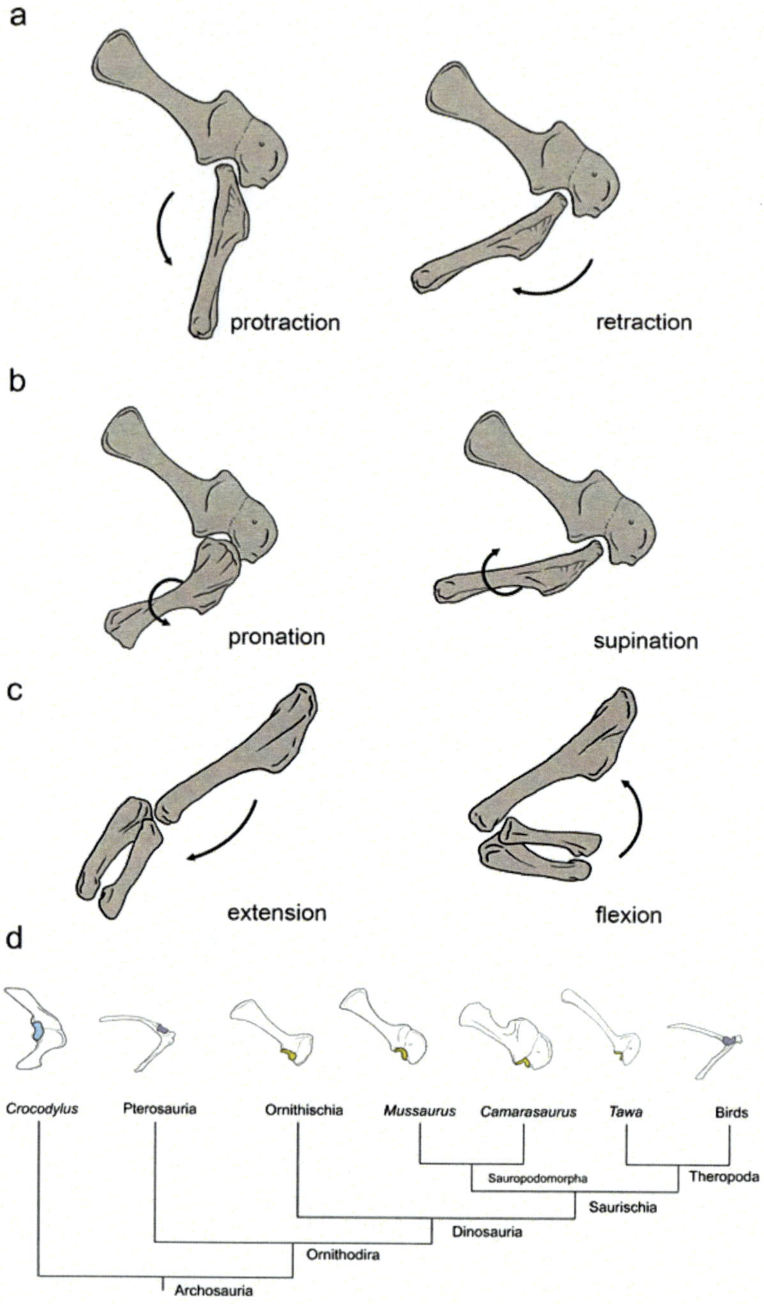
a
protraction
retraction
b
pronation
supination
c
extension
flexion
d
Crocodylus
Pterosauria
Ornithischia
Mussaurus
Camarasaurus
Tawa
Birds
Sauropodomorpha
Theropoda
Saurischia
Dinosauria
Ornithodira
Archosauria

◀**Fig. 2** **a–c** General forelimb joint movements mentioned in the text for *Mussaurus*. **a**, **b** Shoulder joint (scapulocoracoid and humerus). **c** Elbow joint (humerus and radius plus ulna). **d** Orientation of the humeral articulation (glenoid) along the ornithodiran line from the ancestral archosaurian pattern. Reference angles: 0 degrees corresponds to fully humeral protraction (see left sketch in **a**) and full elbow extension (see left sketch in **c**). Same color/tone indicates the same glenohumeral joint orientation. Line drawings modified from *Crocodylus* and generalized bird (Gatesy and Baier 2005); Pterosauria (Witton 2015); Ornithischia (Maidment and Barrett 2011); *Camarasaurus* (Wilson and Sereno 1998) and *Tawa* (Burch 2014)

stride lengths achieved mainly via glenohumeral (i.e., shoulder) movements (Barrett and Maidment 2017). Yet it remains questionable how flexed the limbs of such a large (i.e., 1.4 ton, Otero et al. 2019) animal as an adult *Mussaurus* could have been (e.g., Biewener 1989; Otero et al. 2017), and glenohumeral mobility alone is not indicative of quadrupedal abilities. Moreover, analysis of muscle moment arms or leverage about the elbow in *Mussaurus* showed that peak extensor values were present at ~90° (zero elbow angle corresponds to full extension or a straight elbow joint, whereas larger angles correspond to flexion), meaning that maximal muscle extensor leverage was achieved at a near-maximally flexed elbow (Otero et al. 2017). This mechanical benefit of increased elbow flexion in *Mussaurus* could be speculated to argue against a forelimb with strong specialization for supportive or locomotor functions (Biewener 1989) and is thus inconsistent with habitual quadrupedalism in *Mussaurus*.

The evolution of quadrupedality in sauropods involved modifications to the shoulder joint which allowed the forelimb to be held vertically under the shoulder. In this regard, the proximal shoulder joint surface (the glenoid) was more ventrally oriented, which allowed the humerus to be protracted to a greater degree, even with the same scapular blade orientation (Fig. 2d; see Schwarz et al. 2007). A ventrally oriented glenoid seems to have allowed sauropods to protract their humerus anterior to the vertical, facilitating glenohumeral flexion (i.e., joint movement parallel to the long axis of the glenoid) during quadrupedal locomotion, as proposed for some eusauropods (Wilhite 2003; Klinkhamer et al. 2019).

3.2 *Manus Pronation: Evidence From the Radius and Ulna*

The evolution of forearm posture and rotational ability has played a key role in locomotor stance among tetrapods in general (Hutson 2014; Hutson and Hutson 2015, 2017; VanBuren and Bonnan 2013) and in sauropodomorphs in particular (Bonnan 2003; Bonnan and Yates 2007; Otero et al. 2017). While bipedal dinosaurs had forelimbs modified for grasping, in which the palms faced each other so that food could be clasped between the hands (Sereno 1997; Butler et al. 2007; Maidment and Barrett 2011), most quadrupedal dinosaurs probably must have faced their palms to the substrate; or posteriorly; in order to bear their weight (Bonnan 2003; Bonnan and Yates 2007; Otero et al. 2017, but see Wright 1999 for contrasting evidence for iguanodontid ornithischians). This pronation requires the wrist or forearm to rotate about its long axis. It is important to keep in mind that supination vs. pronation is not a binary state but a continuum of postures, requiring a nuanced consideration of the issue.

Active pronation was defined as the muscle-driven ability to rotate the manus around its longitudinal axis, from pronation to supination, by any kind of rearrangements of the antebrachial bones (Otero et al. 2017; but see Hutson and Hutson 2015 for an alternative definition). Increased pronation of the manus should have aided the forelimbs to generate anteroposteriorly directed propulsive or braking forces that roughly paralleled the actions of the pes in the parasagittal plane (Bonnan 2003). Active pronation could facilitate facultative quadrupedalism, whereas passive pronation implies a manus fixed into pronation, with no clear ability to supinate, leading to obligate quadrupedalism.

It has been suggested that the "U"-shape of the sauropod manus resulted from forearm pronation, not from weight-bearing adaptations, as traditionally thought (Wilson and Sereno 1998; Bonnan 2003). Specifically, the development of the anterolateral process of the ulna in forms close to Sauropoda (e.g., *Melanorosaurus*) modified the position of the radius with respect to the latter. This development positioned the radius anteriorly to the ulna in its proximal portion and slightly medially to it in its distal portion, generating some pronation of the hand; thusly, a semi-pronated manus was concluded to have evolved at least in sauropodomorphs close to Sauropoda, at the base of the so-called "quadrupedal clade" (Bonnan and Yates 2007; Yates et al. 2010). Such a clade retained other clearly "prosauropod-like" forelimb features (e.g., well-developed deltopectoral crest of the humerus, an arched metacarpus, three manus claws, and a medially divergent pollex), lacking a "U"-shaped manus (Fig. 3a). This indicates a potential decoupling of manus shape and quadrupedalism. Nonetheless, quadrupedalism in Sauropodomorpha likely originated well before the rise of sauropods (McPhee et al. 2018). The evolution of the anterolateral process of the ulna and the concomitant modification of the position of the radius are features that were already present in primitive sauropods (e.g., *Vulcanodon*, Cooper 1984). Therefore, the "U" morphology of the manus, typical of Eusauropoda, would not have been temporally related to the change in orientation of the radius with respect to the ulna, being separate evolutionary modifications (Bonnan and Yates 2007). In contrast, Hutson (2014) and Hutson and Hutson (2015) proposed that the evolution of the anterolateral process of the ulna was a specialization to immobilize the proximal radioulnar joint. These ideas are not mutually exclusive; radioulnar immobility may have been associated with forearm pronation as well.

The morphology of the radius is an important determinant of pronation capabilities, such as the presence of radial shaft curvature (allowing the radius to cross the ulna) and a rounded proximal articular face (permitting the radius to rotate around the proximal end of the ulna during active pronation), a condition typical of extant mammals (VanBuren and Bonnan 2013; Hutson and Hutson 2017). Nonetheless, the absence of radial shaft curvature and presence of a mediolaterally expanded radial head may have precluded active manus pronation in most dinosaurs

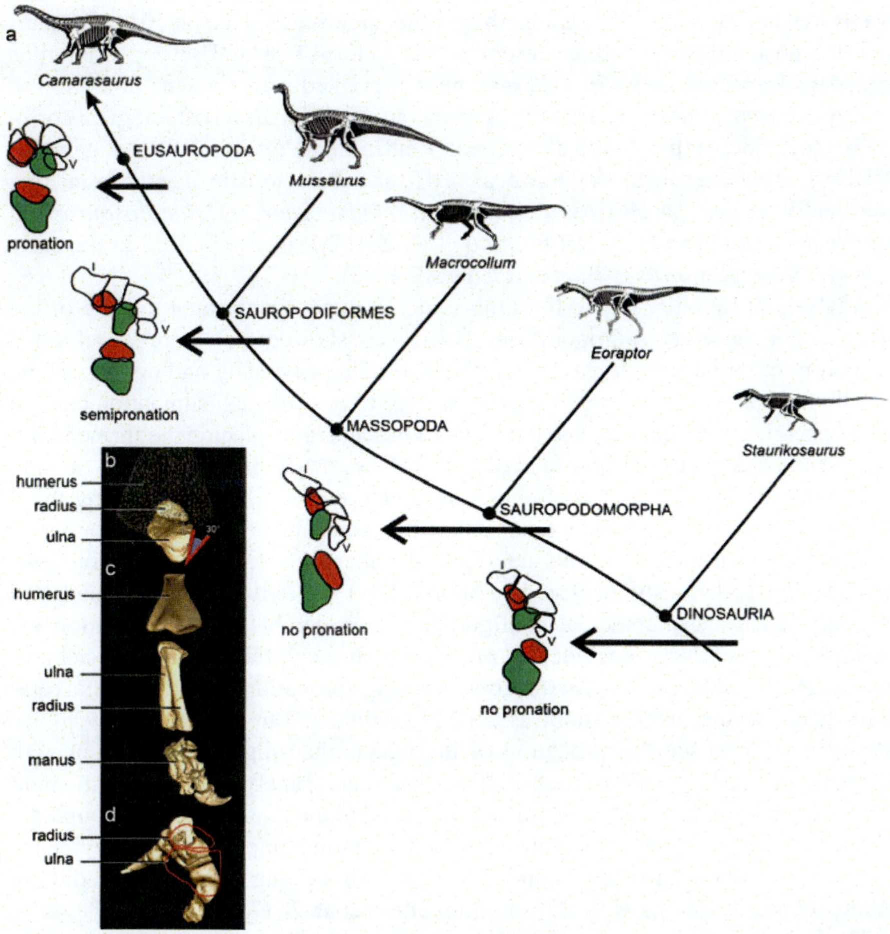

Fig. 3 Evolution of manus in Sauropodomorpha. **a** Relative position of antebrachial bones in relation to the manus among Sauropodomorpha. Green bones represent the ulna (proximal outline below and distal outline above); red bones represent the radius (proximal outline below and distal outline above); the manus is represented by the proximal outline of the metacarpus. Semipronation is represented by *Mussaurus*. **b–d** Semi-pronated (i.e., 30° of medial rotation/pronation from **a**) poses depicting the relationships among antebrachial bones in *Mussaurus*' right forelimb, as follows: in **b** the radius and ulna in proximal view and their relationship with the humerus (meshed texture); in **c** relationships of distal humerus, antebrachium and manus in anterior view; in **d** manus in proximal view with silhouettes of the outlines of radius and ulna (in red) in proximal view. (Silhouettes in a modified from Wilson and Sereno 1998, Paul 2010, Sereno et al. 2013, Müller et al. 2020, Pol et al. 2021; b modified from Otero et al. 2017)

(VanBuren and Bonnan 2013) and perhaps other archosaurs (Hutson 2014). In addition, the posterodistal tubercle present on the radius of several sauropodomorphs across the transition to Sauropoda may have prevented active manus pronation in sauropodomorph dinosaurs. This tubercle was suggested to be an osteological correlate of the radioulnar ligament's attachment (Remes 2008; Yates et al. 2010; Otero and Pol 2013; McPhee et al. 2014; Otero et al. 2015) and characteristic of early-branching sauropodiforms such as *Mussaurus*, *Aardonyx*, *Sefapanosaurus*, *Melanorosaurus* and *Antetonitrus* (McPhee et al. 2014; Otero et al. 2015). It is also present in the early sauropod *Tazoudasaurus* (Allain and Aquesbi 2008).

Analysis of mobility of the forelimb joints in three-dimensional models of the South American early sauropodiform *Mussaurus* showed that there was limited possibility of radial movement against the ulna both proximally and distally. Here, the elliptical proximal surface of the radius precluded pronation/supination and the distal tubercle would have locked the distal radius and ulna, placing the former anterior to the latter. Furthermore, the radius of *Mussaurus* is rather straight, making radial crossing around the ulna impossible (Otero et al. 2017). Considering these constraints, the most likely way to articulate the radius and ulna in an anatomically plausible way was with the radius anterior to the ulna proximally, and slightly medially distally, as previously suggested by Bonnan (2003). Otero et al. (2017) concluded that, with such an antebrachial configuration, considerable manus pronation via radioulnar rotation was precluded in *Mussaurus*. Instead, the only way to achieve some degree of pronation in *Mussaurus* was through pronation (internal/medial rotation) of the whole antebrachium as a single unit (e.g., around the elbow joint), although far from the full pronation of the manus that might be consistent with permanently quadrupedal locomotion (Fig. 3b). Finally, analysis of muscle moment arms performed in *Mussaurus* estimated that most of the muscles acting around the elbow joint were consistently supinators, which is another important line of evidence against full, active manus pronation in this taxon, suggesting that this taxon did not habitually use its forelimbs for locomotion (Otero et al. 2017).

The conclusions presented for *Mussaurus* are similar to those for the Early Jurassic massopodans *Plateosaurus* and *Massospondyus*, which had poor abilities for quadrupedal locomotion (thus favoring bipedalism) based on the restricted ROMs of their distal forelimb joints and the morphology of their radius and ulna (i.e., straight radius, not crossing the ulna), which may have precluded active or passive pronation (Bonnan and Senter 2007; Mallison 2010a,b). More ambiguous findings were obtained for the possibly close relative from Brazil, *Unaysaurus* (Vargas-Peixoto et al. 2015). These conclusions on joint ROMs deserve re-examination using innovations in semi-automated and more objectively comparative mobility analysis (e.g., Manafzadeh et al. 2021).

As pointed out above, the evolution of a pronated manus has been postulated to have begun at least prior to the rise of sauropods, at the origin of the quadrupedal sauropodiform clade (i.e., *Melanorosaurus*, Bonnan and Yates 2007; Yates et al. 2010). *Aardonyx*, a distant outgroup to Sauropoda + *Melanorosaurus* outside the quadrupedal clade, was proposed to have had some features that preceded quadrupedal locomotion in sauropodomorphs, such as an incipient anterolateral process of the ulna and a rather straight femoral shaft (Yates et al. 2010), but the question of how much earlier this evolution began has remained unresolved.

Considering the evidence provided so far, passive manus pronation was not present at the base of Sauropodiformes (sensu Sereno 2007), but instead much closer to the origin of Sauropoda than previously thought (see also Yates et al. 2010). However, some capacity for active pronation (while seemingly unlikely) cannot be ruled out in *Mussaurus* and presumably some other sauropodiforms, as a potential intermediate state in this transformational series of forelimb function. One alternative is that quadrupedalism did not merely evolve once in the sauropodomorph lineage, but rather that mosaic evolution in early Sauropodiformes resulted in some taxa tending to use quadrupedalism more often than others did. Ultimately, reconstruction of the origin and perhaps stepwise acquisition, of manus pronation in Sauropodomorpha will depend upon further analyses using not only qualitative, descriptive approaches but also quantitative, explicitly three-dimensional methods such as the one performed for *Mussaurus*, and further such innovations.

In summary, then, the origin of quadrupedalism in Sauropoda would be linked not only to manus pronation, which should have occurred very close to the node Sauropoda. This quadrupedalism was also enabled by shifting forelimb morphology as a whole, allowing larger extension/flexion excursions of the glenohumeral joint and a more columnar forelimb posture.

4 Postural Shifts During Ontogeny

Reisz et al. (2005) suggested that some early-branching sauropodomorphs (e.g., *Massospondylus*) may have been quadrupedal during early ontogenetic stages and adopted a bipedal posture, at least facultatively, in adult stages. They proposed that the quadrupedal stance in sauropods evolved through the heterochronic mechanism of paedomorphosis, taking into account that *Massospondylus* embryos show proportions in the skeleton more characteristic of obligate quadrupedalism (e.g., proportionally large skull and forelimbs, long and horizontal neck, small caudal vertebrae with poorly developed transverse processes and haemal arches). In contrast, the proportions of the adult bones of *Massospondylus* are more characteristic of, at least, facultative bipedalism (Reisz et al. 2005). In addition, Reisz et al. (2012) suggested that some *Otozoum* footprints found near nests and bones of young *Massospondylus* might be quadrupedal trackways of juvenile *Massospondylus*, with pronated manus, which supports the idea of quadrupedalism at early juvenile stages.

Available ontogenetic stages and computational models of CoM position provide novel information regarding the life history of *Mussaurus patagonicus* and a postural shift along ontogeny which reveals a complex evolutionary scenario. In *Mussaurus*, the relative shortening of the forelimbs (and unlikelihood of quadrupedal function in adults; Otero et al. 2017) along with the progressive reduction of the relative size of the skull and neck strongly correlate with a postural shift throughout ontogeny (Otero et al. 2019; see below), as previously proposed for *Massospondylus* (Reisz et al. 2005) and *Psittacosaurus* (Zhao et al. 2013). If viewed in a phylogenetic context, the postural shifting that sauropodomorphs (such as *Mussaurus*) seem to have undergone along their ontogeny has important implications for understanding evolutionary processes that generated it.

As Otero et al. (2019) estimated, the main segmental influences on shifting the body CoM position during *Mussaurus* ontogeny were more linked to a relative increase in the tail size, compared to the neck/head, rather than a relative reduction of the pectoral appendages, compared to the hind limbs. It was inferred that the posterior shift of the CoM from the hatchling toward adulthood stage involved a shift from quadrupedal to bipedal stance, considering that the CoM of hatchlings was placed too far forward (about halfway along the trunk) relative to the short hind limbs (Fig. 4). In addition, the relative position of the CoM of the *Mussaurus* hatchling corresponded to the CoM estimated for several quadrupedal sauropods, showing a similar position to that of *Camarasaurus* (Bates et al. 2016) and far from any bipeds, supporting a quadrupedal stance or, at least, the inability to maintain a bipedal stance in *Mussaurus* early ontogenetic stages.

The pattern estimated across ontogeny in *Mussaurus* reveals that at least some early sauropodomorphs experienced a posterior shifting of CoM within their life history. Although previous studies based on limb proportions and footprints suggested that *Massospondylus* also experienced a locomotor shift during ontogeny (Reisz et al. 2005, 2012), new analyses of long bone dimensions and inner ear

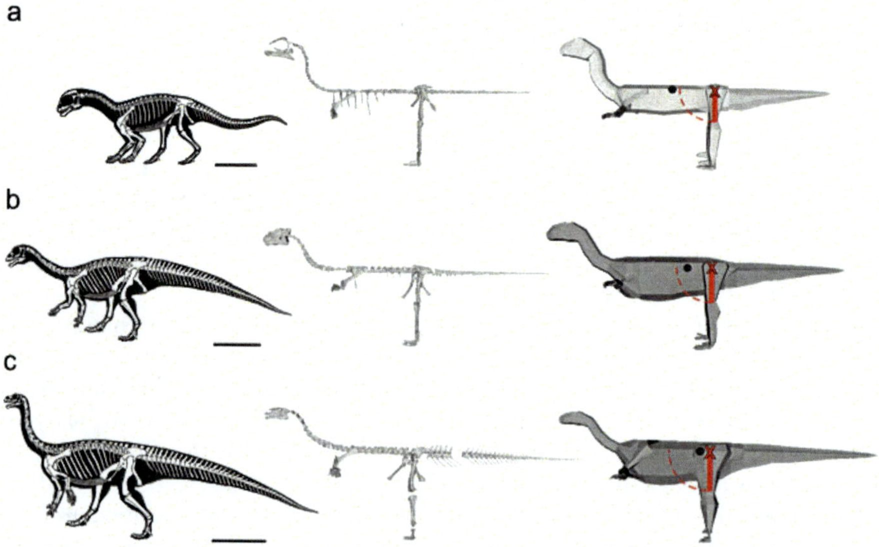

Fig. 4 *Mussaurus* silhouettes, 3D skeletons, and spline-based digital models. **a** Hatchling. **b** Yearling. **c** Adult. Estimated centers of mass (CoM) are denoted with a black dot relative to femur length from the acetabula (red "X"). Bipedal static stability is possible where the CoM is within one femur length (red line) of the acetabula (dashed line), i.e., in the yearling and adult models. Scale bars equal 5 cm (a), 15 (b) and 100 cm (c). (modified from Otero et al. 2019)

morphology did not support such a locomotor shift (Neenan et al. 2019; Chapelle et al. 2020a). Other studies have questioned the ability of inner ear morphology to infer head, neck, or body orientation in tetrapods (Taylor et al. 2009; Marugán-Lobón et al. 2013; Benoit et al. 2020), but further analysis of this problem and data on ontogenetic changes of inner ear morphology in *Mussaurus* would be valuable. Consequently, and considering the available evidence to date, it is not possible to elucidate if the locomotor shift estimated for *Mussaurus* corresponded to a trend (i.e., an ontogenetic shift common to early sauropodomorphs) or a unique case among early sauropodomorphs. Nau et al. (2020) found tantalizing evidence from a ~40 kg juvenile skeleton of *Plateosaurus* that similar ontogenetic changes in body shape might have occurred in this taxon, too, adding support to the inference that the pattern in *Mussaurus* is a trend for early Sauropodomorpha. If it was indeed a trend, it contrasts with the available evidence in sauropods indicating a single stance (quadrupedal) across all known ontogenetic stages (Gilmore 1925; Lehman and Coulson 2002; Ikejiri et al. 2005; Schwarz et al. 2007), although this lack of an ontogenetic shift in Sauropoda might be a derived character shared by all members of the clade.

Despite the evidence suggesting a quadrupedal to bipedal ontogenetic shift in *Mussaurus* that appears as inverse to the bipedal to quadrupedal shift in the phylogeny of Sauropodomorpha, the skeletons of adult sauropods and juvenile early sauropodomorphs are built in different ways. For example, the enlarged (reconstructed) tail of adult *Mussaurus* had the dominant influence on moving the CoM posteriorly in the late ontogeny, constituting a previously unreported key factor (Otero et al. 2019; see also Allen et al. 2013; Bates et al. 2016). Thus, although previous studies emphasized the influence of hind limb/forelimb length proportions in dinosaurian stance (Reisz et al. 2005; Zhao et al. 2013), the study performed in *Mussaurus* indicates that the relative development of tail (and head/neck) size was more influential in constraining/determining the locomotor stance in this clade, in both ontogenetic and phylogenetic changes. These changes were reinforced by a follow-up comparative morphospace analysis by Bishop et al. (2020), combining CoM position, body mass, and limb lengths to estimate locomotor stance, which recovered the same results assigning hatchling *Mussaurus* as quadrupeds and adults as bipeds. One important finding in that paper was that juvenile crocodylians were sometimes misassigned as bipeds due to their unusual body dimensions (as in McPhee et al. 2018; Chapelle et al. 2020a), so assessments on ontogenetic shifts of locomotor stance need multiple lines of congruent evidence (e.g., Reisz et al. 2012).

5 Heterochrony

Heterochrony is the phenomenon of changes that affect the appearance or rate of development of ancestral characters (de Beer 1937) and such changes are related to morphology and size variation (Alberch et al. 1979; McNamara 1982, 1986). Heterochrony is important because it may direct morphological variation of a species along particular pathways (Ede 1978; Alberch 1980; Gould 1980; Levinton and

Simon 1980). These changes occur along with an ontogenetic series and are described by different patterns whose morphological expressions are paedomorphosis (i.e., retention of ancestral juvenile characters in the descendant adult) and peramorphosis (i.e., ancestral adult morphology in a descendant). Considering these definitions, evidence shows that ancestral forms across the sauropodomorph lineage (the non-paedomorphic form or "apaedomorph", in this case, represented by *Mussaurus* and putatively *Massospondylus*) passed through two possible locomotor stages along with their life history: quadrupedal—bipedal (regardless of the fact that more morphological ontogenetic stages actually existed, based on external anatomy and bone microstructure [Pol and Powell 2007b; Cerda et al. 2013; Otero and Pol in press]). It is important to note that in the case of *Mussaurus'* neonates, the evidence suggests that these individuals were not capable of bipedal locomotion (Otero et al. 2019), which constrains quadrupedalism as a plausible stance for this ontogenetic stage. These locomotor stages correspond to stage 1 (neonate) and 3 (adult) of the ontogenetic series (Otero et al. 2019) (Fig. 5a). Regarding stage 2, which corresponds to the one-year-old juveniles, locomotor stance is ambiguous because of uncertainty in the model estimates, although at least a facultatively bipedal stance is inferred based on the proximity of the CoM to the acetabula (Otero et al. 2019). Available evidence in eusauropods (i.e., the descendant) shows that there is only one postural stage for that clade, quadrupedal, most probably present in all ontogenetic stages, considering that juvenile specimens yet known for this group show limb proportions and other aspects of morphology that suggest obligate quadrupedalism (Gilmore 1925; Lehman and Coulson 2002; Ikejiri et al. 2005; Schwarz et al. 2007). This means that, in terms of postural changes, the descendant species passed through fewer morphological stages during growth. If we consider a morphofunctionally intermediate species between the ancestral form (*Mussaurus*) and the descendant (any eusauropod), for example, *Melanorosaurus* or *Aardonyx*, they are traditionally considered as habitual or obligate quadrupeds (Bonnan and Yates 2007; Yates et al. 2010). In spite of this, the manus of these species still retained some degree of functionality for non-locomotor purposes, including an offset and mobile pollex with some grasping ability (Yates and Kitching 2003; Bonnan and Yates 2007), suggesting facultative bipedalism (Carrano 2005).

The trend along the sauropodomorph lineage is bipedal—quadrupedal stance (Bates et al. 2016), and if the facultatively bipedal stance of the "transitional" forms (e.g., *Aardonyx, Melanorosaurus*) is considered as equivalent of the possibly facultatively bipedal stance of the *Mussaurus* juveniles of stage 2, then the postural change along the sauropodomorph lineage shows the opposite pathway to the ontogenetic development of the earliest species (i.e., *Mussaurus*, at least). The sequence of adult morphologies showing such a temporal morphological gradient constitutes a phylogenetic trend named a "paedomorphocline" (McNamara 1982) and this particular heterochronic pattern identified in a phylogenetic trend (i.e., directional morphological change between more than two species) then would constitute a "paedomorphoclade" (sensu Long and McNamara 1982) (Fig. 5). Among dinosaurs, the ceratopsian lineage, mostly obligate quadrupeds, seems to show the same heterochronic pattern, in which the early ontogenetic stages of the early ceratopsian *Psittacosaurus* present

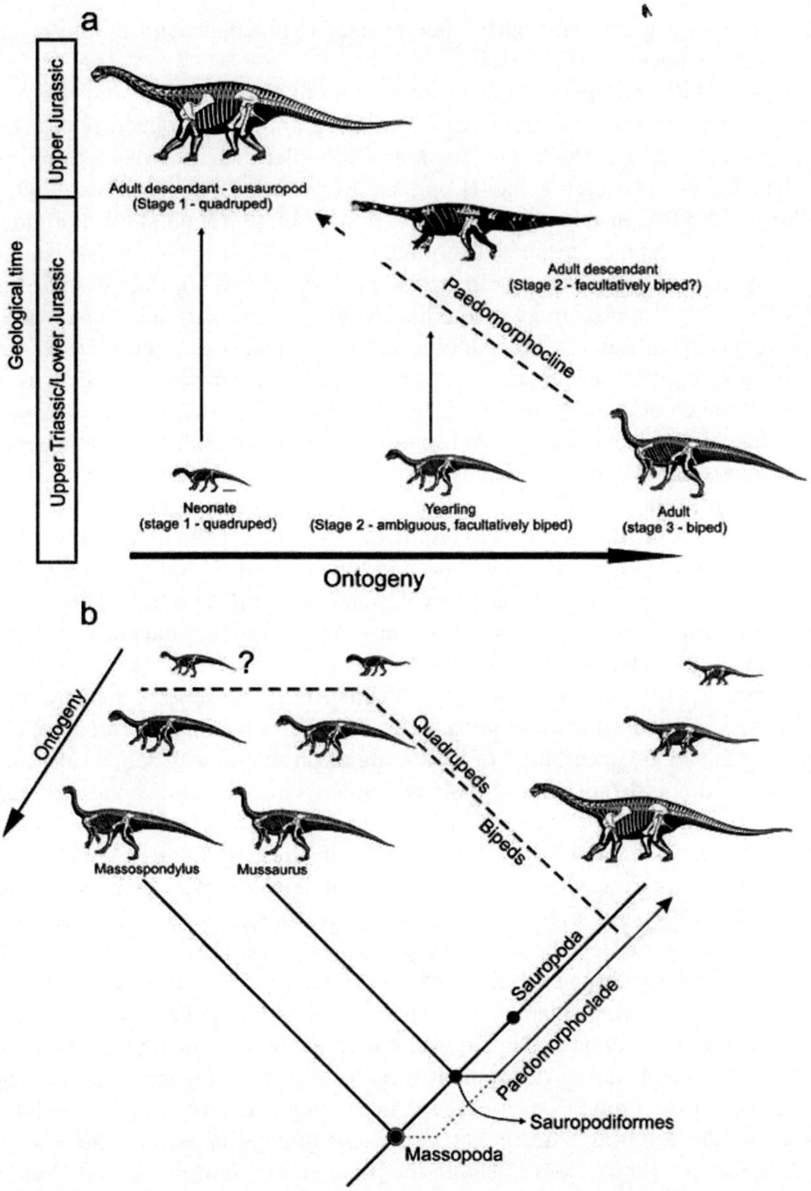

Fig. 5 Heterochrony and phylogenetic trends during the early evolution of Sauropodomorpha. **a** discontinuous morphological gradient of progressively more paedomorphic sauropodomorph species through time (paedomorphocline). Stages 1 to 3 represent the ontogenetic stages for each stance type (unambiguous quadruped, ambiguous quadruped, unambiguous biped). **b** Simplified phylogeny of sauropodomorphs depicting the postural trend along the massopodan lineage, in which the postural change shows the opposite pathway to the ontogenetic development of some species (i.e., *Mussaurus*). This particular heterochrony identified in a phylogenetic trend then would constitute a paedomorphoclade

evidence of obligate quadrupedalism, but reversed to bipedalism at adulthood (Zhao et al. 2013; Chappelle et al. 2020a).

Paedomorphosis and peramorphosis have numerous examples among dinosaurs, mostly focused on specific structures of the skeleton, (i.e., isolated bones, Long and McNamara 1997), which tend to show decoupled heterochrony among them and mixed patterns acting together (Long and McNamara 1997; Salgado 1999; Pol and Powell 2007b). In this regard, major features of sauropod skull morphology could be paedomorphic, originated through some paedomorphic mechanism such as neoteny, as first proposed by Bonaparte and Vince (1979), due to some similarities in the skull morphology of the holotype of *Mussaurus* and *Camarasaurus* (e.g., short premaxilla and maxilla, dorsoventrally high antorbital fenestra, expanded mandibular symphysis, and enlarged teeth). Other characters, however, seem to have an opposite developmental peramorphic trend, showing a plesiomorphic condition in the post-hatchlings and a more derived condition in the sub-adult skulls (e.g., relative enlargement of the descending process of the postorbital, the increase in the posterior extension of the posterodorsal process of the prefrontal along the orbital rim, among others). A similar situation of conflicting developmental trends was recently noted in the ontogeny of derived neosauropods (Salgado et al. 2005), suggesting that heterochronic patterns underlying the evolution of sauropodomorph skulls may be more complex than previously thought (Pol and Powell 2007b; Chapelle et al. 2020b; Kundrát et al. 2020; Fabbri et al. 2021).

Assigning a particular heterochronic mechanism (e.g., progenesis, neoteny, hypermorphosis, acceleration) in dinosaurs is not an easy task due to the scarcity of ontogenetic series and the uncertainty of the timing of onset maturity and, as mentioned above, such patterns do not act in an isolated fashion, but combined (Long and McNamara 1997). In the particular case of sauropods, it has been postulated that acceleration was the most suitable heterochronic mechanism underlying the phylogenetic size increase in this group, based on the evidence of rapid growth from bone histology (Sander et al. 2004). As originally defined, acceleration implies an increase in the rate of morphological development during ontogeny, meaning that the descendant (sauropod) will pass through the adult stage of the ancestor (early sauropodomorph) during its ontogeny (McNamara 1986). The problem with assigning acceleration as an explanation of sauropod gigantism is that such an increase in the rate of morphological development during descendant (i.e., sauropod) ontogeny does not imply any increase in the growth rate (i.e., size increasing). That is why in accelerated descendants the resulting adult is not larger, but often smaller, than the ancestral adult (McNamara 1986). Conversely, in the particular case of the sauropodomorph lineage and considering the general proportions of the limbs, the postural shifting and the general trend of size increase toward Sauropoda, the heterochronic pattern that better describes this scenario should be neoteny, as originally proposed by Bonaparte and Vince (1979) (see also Reisz et al. 2005). However, together sexual maturity and growth rate will ultimately determine the final size of the descendant. For example, if the onset of sexual maturity in the descendant is delayed (thus extending the juvenile phase) and/or the growth rate is increased, the result would be a larger size. Sauropods seem to have reached the age of sexual maturity in the second decade of life, before full growth (Sander 2000; Sander and Tückmantel 2003). Osteohistological data from other sauropods (i.e., *Janenschia, Apatosaurus*) show that final body size was reached between 25–30 years. Although the information of sexual maturity in early sauropodomorphs is less available, previous studies have shown that *Plateosaurus engelhardti* could have reached sexual maturity at about 8 years (Klein and Sander 2008), and a similar age was corroborated for *Mussaurus* (Cerda et al. 2014). If this is the common pattern for early sauropodomorphs, then the delay of sexual maturity was fulfilled in sauropods relative to their ancestors, as predicted by neoteny.

Nonetheless, could sauropods have reached such gigantic sizes only by a delay of the onset of maturity? A distinctive change in the growth dynamics occurred during the evolution of Sauropodomorpha. Early sauropodomorphs were characterized by cyclical growth, with a combination of fibro-lamellar bone and regularly spaced lines of arrested growth (LAGs), whereas in juvenile sauropods growth was rapid, evidenced by uninterrupted, highly vascularized bone tissue, denoting that sauropods grew considerably faster than early sauropodomorphs (Sander 2000; Sander et al. 2004; Klein and Sander 2008) and such a fast growth pattern was already present in early sauropods *Volkheimeria* and *Patagosaurus* (Cerda et al. 2017). In members of Lessemsauridae (e.g., *Lessemsaurus* and *Ingentia*), interestingly, periodic LAGs and annuli interrupt the deposition of highly vascularized fibrolamellar bone with reticular vascular organization, suggesting that although cyclical growth occurred, growth was accelerated during the fast-growing periods. Hence, while *Lessemsaurus* and *Ingentia* retain the plesiomorphic pattern of cyclical growth, during the periods of rapid growth (even at sub-adult stages) they grew at rates equivalent to those of the later sauropods (Cerda et al. 2017; Apaldetti et al. 2018). This is perhaps not unexpected since lessemsaurids are considered to have had key transitional characteristics relative to the origin of Sauropoda, showing both derived and plesiomorphic anatomical features and a considerably larger body size than the early sauropodomorph average (e.g., McPhee et al. 2018). Thus, an accelerated early growth rate combined with a deceleration of the rate of morphological change in sauropods could be the factors framing the pattern of neoteny which ultimately favored sauropod gigantism.

6 Evolution of Gigantism in Sauropodomorpha and the Role of the South American Record

Body size evolution in Sauropodomorpha involved drastic anatomical transformations from relatively small (<1 kg) bipedal (or facultatively bipedal) ancestors from the Triassic period into large-bodied (>30 ton) obligate quadrupeds from the Jurassic–Cretaceous. In this regard, CoM position played a key role in determining bipedal versus quadrupedal stance and thus boosting size increase among Sauropodomorpha. Thus, a posterior shift in CoM in Middle Triassic dinosauriforms (e.g., *Marasuchus*), associated with the evolution of bipedalism in various dinosaur lineages, including Carnian sauropodomorphs (e.g., *Eoraptor*, *Buriolestes*), was reversed in Late Triassic Lessemsaurid sauropodomorphs (*Ingentia*, *Lessemsaurus*, *Antetonitrus*), which surpassed 10 metric tons and constitute the first undisputed clade of quadrupedal sauropodomorphs (Apaldetti et al. 2018; McPhee et al. 2018). A more striking anterior shift in CoM took place in Late Jurassic–Cretaceous titanosauriforms, which included the largest sauropods (Bates et al. 2016; see Table 1).

Table 1 Estimates of body mass for South American sauropodomorphs from the literature. Specimen/species vary. See Campione and Evans (2020) for more estimates and discussion

Taxon	Lineage	Body mass (kg)	Method	Reference
Eoraptor	ES	1.9	PVM	Paul (2010)
Eoraptor	ES	17.3	BCS	McPhee et al. (2018)
Adeopapposaurus	ES	39.6	BCS	Benson et al. (2018)
Buriolestes	ES	4.5	DVM	Müller et al. (2020)
Bagualosaurus	ES	23.5	BCS	This contribution
Chromogisaurus	ES	13.1	BCS	Benson et al. (2018)
Pampadromaeus	ES	8.5	BCS	Benson et al. (2018)
Panphagia	ES	12.5	BCS	Benson et al. (2018)
Macrocollum	ES	84	BCS	This contribution
Riojasaurus	ES	2233	BCS	Benson et al. (2018)
Saturnalia	ES	10.6	BCS	Benson et al. (2018)
Mussaurus (adult)	SF	1418	DVM	Otero et al. (2019)
Lessemsaurus	SA	8500	BCS	Apaldetti et al. (2018)
Volkheimeria	SA	1780	BCS	Benson et al. (2018)
Patagosaurus	EU	8584	DVM	Bates et al. (2016)
Patagosaurus	EU	24,385	BCS	Benson et al. (2018)
Tehuelchesaurus	MA	31,433	BCS	Benson et al. (2018)
Amargasaurus	DI	3304	DVM	Bates et al. (2016)
Amargasaurus	DI	10,195	BCS	Benson et al. (2018)
Brachytrachelopan	DI	16,164	BCS	Benson et al. (2018)
Cathartesaura	DI	15,759	BCS	Benson et al. (2018)
Comahuesaurus	DI	12,331	BCS	Benson et al. (2018)
Limaysaurus	DI	11,688	BCS	Benson et al. (2018)
Rayososaurus	DI	9353	BCS	Benson et al. (2018)
Amargatitanis	TF	11,406	BCS	Benson et al. (2018)
Chubutisaurus	TF	29,174	BCS	Benson et al. (2018)
Ligabuesaurus	TF	20,435	BCS	Benson et al. (2018)
Aeolosaurus	TI	17,180	BCS	Benson et al. (2018)

(continued)

Table 1 (continued)

Taxon	Lineage	Body mass (kg)	Method	Reference
Antarctosaurus	TI	20,851	BCS	Benson et al. (2018)
Argentinosaurus	TI	94,844	BCS	Benson et al. (2018)
Argentinosaurus	TI	83,230	DVM	Sellers et al. (2013)
Argentinosaurus	TI	70,000	PVM	Paul (2019)
Argyrosaurus	TI	41,937	BCS	Benson et al. (2018)
Atacamatitan	TI	4726	BCS	Benson et al. (2018)
Bonatitan	TI	1016	BCS	Benson et al. (2018)
Bravasaurus	TI	2890	BCS	Hechenleitner et al. (2020)
Dreadnoughtus	TI	26,912	DVM	Bates et al. (2016)
Dreadnoughtus	TI	59,400	BCS	McPhee et al. (2018)
Elaltitan	TI	35,371	BCS	Benson et al. (2018)
Epachthosaurus	TI	10,415	BCS	Benson et al. (2018)
Futalognkosaurus	TI	38,139	BCS	Benson et al. (2018)
Maxakalisaurus	TI	7097	BCS	Benson et al. (2018)
Neuquensaurus	TI	1464	DVM	Bates et al. (2016)
Neuquensaurus	TI	6118	BCS	Benson et al. (2018)
Panamericansaurus	TI	11,258	BCS	Benson et al. (2018)
Patagotitan	TI	57,000	BCS	Otero et al. (2020)
Patagotitan	TI	52,000	PVM	Paul (2019)
Petrobrasaurus	TI	154,877	BCS	Benson et al. (2018)
Rinconsaurus	TI	4080	BCS	Benson et al. (2018)
Rocasaurus	TI	4652	BCS	Benson et al. (2018)
Saltasaurus	TI	5770	BCS	Benson et al. (2018)
Traukutitan	TI	38,129	BCS	Benson et al. (2018)

Lineage: ES = early Sauropodomorpha; SF = Sauropodiformes; SA = early Sauropoda; EU = Eusauropoda; MA = Macronaria; DI = Diplocoidea; TF = Titanosauriformes (non-titanosaurian); TI = Titanosauria. Methods: BCS = bone circumference scaling; DVM = digital volumetric modeling; PVM = physical volumetric modeling. *Eoraptor*, *Patagosaurus*, *Argentinosaurus*, *Dreadnoughtus*, *Neuquensaurus* and *Patagotitan* have multiple estimates provided to emphasize areas of controversy or inter-specimen/species variation

At the same time, the evolution of gigantism in this clade was driven by crucial changes in several aspects of their biology, including faster population recovery via retention of the plesiomorphic oviparous mode of reproduction, strong pneumatization of the axial skeleton resulting from the evolution of an avian-style respiratory system, and extremely long necks, which allowed more efficient food uptake by covering a much larger feeding envelope, among other factors (Sander et al. 2011). As corroborated for living large herbivores and suggested for sauropods, abundance of giant herbivores is limited mainly by food availability (Owen-Smith and Mills

2008; Sander et al. 2011). The increase of the flora worldwide during the Jurassic period, particularly with the diversification and establishment of coniferous forests (Stewart and Rothwell 1993; Leslie et al. 2012), seems to have facilitated the diversification of the gigantic sauropods. The South American titanosaurian phylogenetic and ecological diversification also coincided with a global period of major environmental changes that include a warmer climate (Huber et al. 2002; Klages et al. 2020), and the onset of dominance of angiosperms in continental environments (Upchurch and Dilcher 1990; Herendeen et al. 2017). This is not surprising, considering that similar correlations have been identified for other archosaurian clades in the Southern Hemisphere, such as theropods and crocodyliforms (Novas et al. 2013; Pol et al. 2014). In particular, South America provides the richest record of giant sauropods worldwide, concentrating the heaviest known taxa in Patagonia, Argentina. This record of giant taxa persists regardless of the mass estimation method used (i.e., scaling equation or volumetric reconstruction; Campione and Evans 2020; Bates et al. 2016; Paul 2019; and references therein; also Table 1), such as *Patagotitan* and *Dreadnoughtus* (Lacovara et al. 2014; Carballido et al. 2017; Campione and Evans 2020), not to mention other forms whose body mass estimation is not very reliable using current long bone measurement methods, like *Argentinosaurus*, *Puertasaurus*, *Notocolossus* and Zapala's titanosaur (Bonaparte and Coria 1993; Novas et al. 2005; González Riga et al. 2016; Otero et al. 2021).

Considering the quadratic scaling equation method (Campione 2017), in which femoral and humeral circumferences are needed, *Patagotitan* yields an estimated mean body mass of 55–57 tons, exceeding by almost 15% the mass estimate obtained for the possibly sub-adult *Dreadnoughtus*, and 20% more than the estimated body mass of several large Jurassic sauropods using the same equation (e.g., *Giraffatitan*, *Apatosaurus*) (Campione and Evans 2020; Otero et al. 2020; Table 1; Fig. 6). Direct comparisons using overlapping bone elements (i.e., anterior dorsal vertebrae) allowed estimations of the area of these vertebrae for *Argentinosaurus*, *Puertasaurus*, and *Notocolossus*, yielding values that are approximately 10% smaller for these taxa in comparison with *Patagotitan* (Carballido et al. 2017). It is important to note that previous mass estimations provided for *Argentinosaurus* using linear regressions and scaling equations (Mazzetta et al. 2004; Benson et al. 2014; Campione and Evans 2020; Table 1) should be considered with caution since this taxon lacks a preserved humerus and the currently preserved femur is extremely fragmentary and mostly reconstructed with plaster. Hence, body mass estimations based on long bone measurement or volumetric modeling cannot be confidently considered for this taxon (Table 1). On the other hand, the Zapala's titanosaur currently lacks long bones as well as dorsal vertebrae; however, measurements made on the scapula and pelvic bones retrieved body mass estimates roughly 10 to 20% larger relative to *Patagotitan* (Otero et al. 2021). These estimates place the former as one of the putatively largest individual animals that ever walked on land, with the caveat that estimations based on long bone circumference or volumetric modeling are needed.

Nonetheless, the locomotion of giant sauropods surely was slow (limited by tissue strength vs. square-cube scaling principles of area-mass; e.g., Biewener 1989), as estimated by Sellers et al. (2013). This conclusion is firmly consistent with South

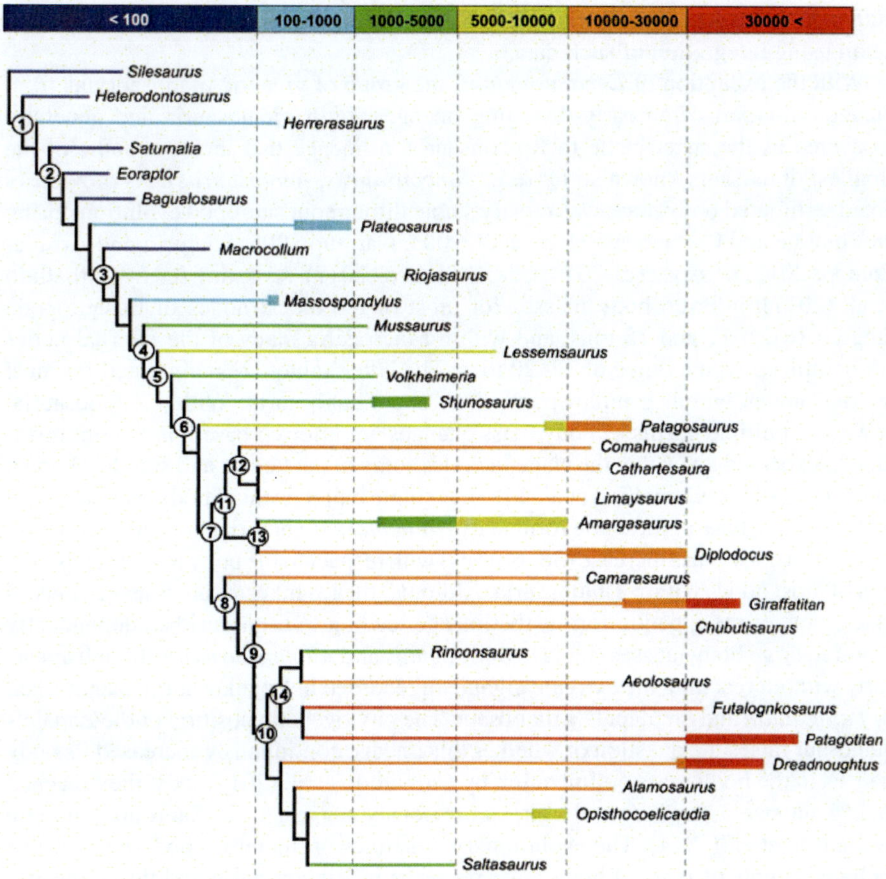

Fig. 6 Simplified phylogeny showing evolution of body mass (in kilograms) across South American sauropodomorphs, with some non-South American taxa for comparison. Long bars denote taxa with multiple body mass estimates. Nodes: 1 = Dinosauria; 2 = Sauropodomorpha; 3 = Massopoda; 4 = Sauropodiformes; 5 = Sauropoda; 6 = Eusauropoda; 7 = Neosauropoda; 8 = Macronaria; 9 = Somphospondily; 10 = Titanosauria; 11 = Diplodocoidea; 12 = Rebbachisauridae; 13 = Flagellicaudata; 14 = Colossosauria. (phylogeny based on Pol et al. 2021; Mannion et al. 2019; Otero et al. 2021)

American fossil trackways of titanosaurs at <<2 ms^{-1} walking speeds (e.g., González Riga 2011; González Riga and Tomaselli 2019), although claims of multiple gaits in titanosaurs should be viewed warily (see Stevens et al. 2016). Similar biomechanical analyses using more complete skeletons across titanosaur phylogeny, such as *Neuquensaurus* and *Epachthosaurus* from Patagonia (Otero and Vizcaíno 2008; Ibiricu et al. 2018), could reveal interesting insights about locomotor diversity within this clade. For example, the wide-gauge limb posture and its evolution revealed by the unusual appendicular morphologies (Wilson and Carrano 1999) and fossil trackways (e.g., *Titanopodus* in Argentina: González Riga 2011; González Riga and Calvo

2009; González Riga and Tomaselli 2019) of titanosaurs could be analyzed in more detail using integration of such data.

With the exception of *Dreadnoughtus*, the group of gigantic titanosaurians from Patagonia is part of an early-diverging lineage within Titanosauria and generally recovered as the sister clade to Rinconsauria, a lineage that includes some of the smallest titanosaurs known so far (e.g., *Rinconsaurus*, *Aeolosaurus*). As previously noted, estimated body mass was mostly stable throughout sauropod evolution during the Jurassic and Cretaceous (Hone et al. 2005; Carrano 2005; Santucci and Arruda-Campos 2012; Benson et al. 2014; Carballido et al. 2017). As optimized by Carballido et al. (2017), inferred body masses for most of the oldest nodes of Eusauropoda range between 12 and 15 tons, and within Macronaria many of the internal nodes are optimized with a range of 12–20 tons. The latter values are maintained for most internal nodes within somphospondylians and Titanosauria. Within Titanosauria, however, multiple events of body mass changes are inferred based on the optimization, as expected given that the clade includes some of the largest and smallest known sauropod species (Carballido et al. 2017). A single major event of body mass increase is found at the base of the clade formed by *Notocolossus* and Lognkosauria, implying a threefold body mass increase (38–60 tons) with respect to the body mass reconstruction of most nodes within Titanosauria. Although the Lognkosaurian lineage concentrates most of the gigantic forms, the lineage leading to saltasaurids independently acquired large body masses (i.e., *Dreadnoughtus* and *Alamosaurus*), although apparently without reaching the extreme gigantism recorded in Lognkosauria, exemplified by *Patagotitan* and the Zapala's titanosaur. The phylogenetic position of the Zapala's titanosaur raises the question of whether this lineage continuously increased its body size, as in the body mass optimization by Carballido et al. (2017), or if they reached their giant size as an ancestral trait, as previously suggested for Sauropodomorpha by Benson et al. (2014). The evolutionary dynamics of titanosaur body masses also included events of marked body mass decrease in Rinconsauria and three lineages within the saltasaurid clade (Bonaparte and Powell 1980; Salgado and Azpilicueta 2000; Company 2011; Otero 2010; Stein et al. 2010).

The stratigraphic age of *Patagotitan* fossils indicates that the radiation of the lognkosaurian lineage started in the late Early Cretaceous (Albian), temporally coincident with the diversification of the other major clade of titanosaurs, the lithostrotians (Gomani 2005; Zaher et al. 2011; Wilson et al. 2016). With the discovery of the Zapala's titanosaur, the record of >30-ton titanosaur sauropods in Patagonia is virtually continuous from the Albian to the latest Cretaceous, covering a temporal range of about 20 Myr (Fig. 5). At the beginning of the Late Cretaceous in South America, there was remarkable co-existence of extremely large titanosaurs (e.g., *Argentinosaurus* and the Zapala's titanosaur) together with medium-sized titanosaurs (e.g., *Andesaurus* and *Choconsaurus*) and small-sized rebbachisaurids (e.g., *Limaysaurus* and *Cathartesaura*). These size differences could explain the existence of such sauropod diversity in the Neuquén Basin during the Late Cretaceous in terms of niche partitioning, which has already been suggested on the basis of the co-occurrence of several titanosaur genera (Barrett and Upchurch 2005). Niche diversification in sauropod taxa has traditionally been inferred from differences in dental and cranial anatomy, neck length and inferred posture, estimated adult body mass, and dietary preferences (Bakker 1978; Stevens and Parrish 1999; Christian and Dzemski 2011; Tütken 2011; Upchurch and Barrett 2000; Barrett and Upchurch 2005; Barrett 2014; Button et al. 2014, 2017). Most of these traits are related directly or indirectly to possible browsing height, resulting in a marked vertical division of plant resources among different sauropod lineages. Putative niche partitioning in Late Cretaceous sauropods from Patagonia could have been related to adult body mass, which would affect browsing height, nutrient assimilation rates, migration capacity, and population sizes, among other factors (Barrett and Upchurch 2005). The body size differences among the sauropods known from the Candeleros and Huincul formations, coupled with the morphological disparity between rebbachisaurids and titanosaurs, could therefore be indicative of niche partitioning during the Late Cretaceous of Patagonia that would best be addressed by means of deeper analyses of the paleoecologies of those taxa.

7 Perspectives and Future Work

The beginning of the twenty-first century has involved exploring novel methodologies for the study of sauropodomorph functional morphology and biomechanics of locomotion in a quantitative fashion (e.g., Mallison 2010a, b; Otero et al. 2017; Klinkhamer et al. 2018, 2019). Although such studies have generally focused on specific taxa, the results allow us to begin to understand broader trends in locomotion among Sauropodomorpha, for example, shifting of locomotor stance during the transition from early bipedal forms to quadrupedal sauropods (Bates et al. 2016; Otero et al. 2017, 2019). In addition, improvements in the methodologies for estimating body mass facilitated their usage by researchers, which is reflected in the novel hypotheses of body mass evolution among different dinosaurian lineages that have been published since early 2000s (e.g., Benson et al. 2014, 2018; Campione and Evans 2012, 2020; Bates et al. 2016; Carballido et al. 2017).

Many important discoveries of early sauropodomorphs, including very well-preserved skeletons, have been made recently in Brazil, including *Bagualosaurus* (Pretto et al. 2018), *Macrocollum* (Müller et al. 2018), and *Buriolestes* (Cabreira et al. 2016; Müller et al. 2020). As most or all of these taxa seem to have been obligate bipeds, they do not fundamentally alter the polarity of evolution of quadrupedalism, but studies of their functional morphology and biomechanics would still be highly valuable in understanding how early bipeds functioned and what the roles of the forelimbs were in diverse taxa and disparate morphologies/paleoecologies.

Acknowledgements We thank the following funding sources: Agencia Nacional de Promoción Científica y Tecnológica (PICT 2015-0504 and 2019-00456 to AO) and an ERC Horizon 2020 Advanced Investigator Grant (695517 to JRH). We would like to thank reviews by F. Pretto and S. Maidment, which greatly improved this work.

References

Alberch P, Gould SJ, Oster GF, Wake DB (1979) Size and shape in ontogeny and phylogeny. Paleobiology 5:296–317

Alberch P (1980) Ontogenesis and morphological diversification. Am Zool 20:653–667

Allen VA, Bates KT, Li Z, Hutchinson JR (2013) Linking the evolution of body shape and locomotor biomechanics in bird-line archosaurs. Nature 497:104–107

Allain R, Aquesbi N (2008) Anatomy and phylogenetic relationships of *Tazoudasaurus naimi* (Dinosauria, Sauropoda) from the late Early Jurassic of Morocco. Geodiversitas 30:345–424

Apaldetti CG, Martínez RN, Cerda IA, Pol D, Alcober OA (2018) An early trend towards gigantism in Triassic sauropodomorph dinosaurs. Nature Ecol Evol 2:1227–1232

Bakker RT (1978) Dinosaur feeding behavior and the origin of flowering plants. Nature 274:661–663

Barrett PM, Upchurch P (2005) Sauropod diversity through time: possible macroevolutionary and palaeoecological implications. In: Curry Rogers KA, Wilson J (eds) The Sauropods: evolution and paleobiology. University of California Press, Berkeley, pp 125–156

Barrett PM (2014) Paleobiology of herbivorous dinosaurs. Annu Rev Earth Planet Sci 42:207–230

Barrett PM, Maidment SCR (2017) The evolution of ornithischian quadrupedality. J Iber Geol 43:363–377

Bates KT, Mannion PD, Falkingham PL, Brussatte SL, Hutchinson JR, Otero A, Sellers WI, Sullivan C, Stevens KA, Allen V (2016) Temporal and phylogenetic evolution of the sauropod dinosaur body plan. Royal Soc Open Sci 3:150636

Benoit J, Legendre LJ, Farke AA, Neenan JM, Mennecart B, Costeur L, Merigeaud S, Manger PR (2020) A test of the lateral semicircular canal correlation to head posture, diet and other biological traits in "ungulate" mammals. Sci Rep 10:19602

Benson RBJ, Campione NE, Carrano MT, Mannion PD, Sullivan C, Upchurch P, Evans DC (2014) Rates of dinosaur body mass evolution indicate 170 million years of sustained ecological innovation on the avian stem lineage. PLoS Biol 12:e1001853

Benson RBJ, Hunt G, Carrano MT, Campione N (2018) Cope's rule and the adaptive landscape of dinosaur body size evolution. Palaeontology 61(1):13–48

Biewener AA (1989) Scaling body support in mammals: limb posture and muscle mechanics. Science 245:45–48

Bishop PJ, Bates KT, Allen VR, Henderson DM, Randau M, Hutchinson JR (2020) Relationships of mass properties and body proportions to locomotor habit in terrestrial Archosauria. Palaeobiology 46:550–568

Bonaparte JF, Vince M (1979) El hallazgo del primer nido de Dinosaurios Triásicos (Saurischia, Prosauropoda), Triásico Superior de Patagonia, Argentina. Ameghiniana 16:173–182

Bonaparte JF, Powell JE (1980) A continental assemblage of tetrapods from the Upper Cretaceous beds of El Brete, North-Western Argentina (Sauropoda-Coelurosauria-Carnosauria-Aves). Mémories De La Sociéte Géologique De France, Nouvelle Série 139:19–28

Bonaparte JF, Coria RA (1993) Un nuevo y gigantesco sauropodo titanosaurio de la Formación Rio Limay (Albiano-Cenomaniano) de la Provincia del Neuquén, Argentina. Ameghiniana 30:271–282

Bonnan MF (2003) The evolution of manus shape in sauropod dinosaurs: implications for functional morphology, forelimb orientation, and phylogeny. J Vertebr Paleontol 23:595–613

Bonnan MF, Senter P (2007) Were the basal sauropodomorph dinosaurs *Plateosaurus* and *Massospondylus* habitual quadrupeds? Spec Pap Palaeontol 77:139–155

Bonnan MF, Yates A (2007) A new description of the forelimb of the basal sauropodomorph *Melanorosaurus*: implications for the evolution of pronation, manus shape, and quadrupedalism in sauropod dinosaurs. Spec Pap Palaeontol 77:157–168

Brusatte SL, Nesbitt SJ, Irmis RB, Butler RJ, Benton MJ, Norell MA (2010) The origin and early radiation of dinosaurs. Earth Sci Rev 101:68–100

Butler RJ, Smith RMH, Norman DB (2007) A primitive ornithischian dinosaur from the Late Triassic of South Africa, and the early evolution and diversification of Ornithischia. Proc R Soc B 274:2041–2046

Button DJ, Rayfield EJ, Barrett PM (2014) Cranial biomechanics underpins high sauropod diversity in resource-poor environments. Proc R Soc B 281:20142114

Button DJ, Barrett PM, Rayfield EJ (2017) Craniodental functional evolution in sauropodomorph dinosaurs. Paleobiology 43:435–462

Cabreira SF, Kellner AWA, da-Silva SD, da Silva LR, Bronzati M, Marsola JCA, Müller RT, Bittencourt JS, Batista BJA, Raugust T, Carrilho R, Brodt A, Langer MC (2016) A unique late Triassic dinosauromorph assemblage reveals dinosaur ancestral anatomy and diet. Curr Biol 26:390–395

Campione NE (2017) Extrapolating body masses in large terrestrial vertebrates. Paleobiology 43:693–699

Campione N, Evans D (2012) A universal scaling relationship between body mass and proximal limb bone dimensions in quadrupedal terrestrial tetrapods. BMC Biol 10:60

Campione N, Evans D (2020) The accuracy and precision of body mass estimation in non-avian dinosaurs. Biol Rev 95:1759–1797

Carballido JL, Pol D, Otero A, Cerda IA, Salgado L, Garrido AC, Ramezani J, Cúneo NR, Krause JM (2017) A new giant titanosaur sheds light on body mass evolution among sauropod dinosaurs. Proc R Soc B 284:20171219

Carrano MT (1998) Locomotion in non-avian dinosaurs: integrating data from hind limb kinematics, in vivo strains, and bone morphology. Paleobiology 24:450–469

Carrano MT (2005) The evolution of Sauropod locomotion: Morphological diversity of a secondarily quadrupedal radiation. In: Curry Rogers KA, Wilson JA (eds) The Sauropods: evolution and paleobiology. University of California Press, Berkeley, pp 229–249

Cerda IA, Pol D, Chinsamy A (2013) Osteohistological insights into the early stages of growth in *Mussaurus patagonicus* (dinosaurian, sauropodomorpha). Historical Biol: An Int J Paleobiol 26:110–121

Cerda IA, Chinsamy A, Pol D (2014) Unusual endosteally formed bone tissue in a Patagonian basal sauropodomorph dinosaur. Anat Rec 297:1385–1391

Cerda IA, Chinsamy A, Pol D, Apaldetti C, Otero A, Powell JE, Martínez RN (2017) Novel insight into the origin of the growth dynamics of sauropod dinosaurs. PLoS ONE 12:e0179707

Chapelle KEJ, Benson RBJ, Stiegler J, Otero A, Zhao Q, Choiniere JN (2020a) A quantitative method for inferring locomotory shifts in amniotes during ontogeny, its application to dinosaurs and its bearing on the evolution of posture. Palaeontology 63:229–242

Chapelle KEJ, Fernandez V, Choiniere JN (2020b) Conserved in-ovo cranial ossification sequences of extant saurians allow estimation of embryonic dinosaur developmental stages. Sci Rep 10:4224

Christian A, Dzemski G (2011) Neck posture in sauropods. In: Klein N, Remes K, Gee CTv Sander PM, (eds) Biology of the Sauropod dinosaurs: understanding the life of giants. Indiana University Press, Bloomington, pp 249–260

Company J (2011) Bone histology of the titanosaur *Lirainosaurus astibiae* (Dinosauria: Sauropoda) from the Latest Cretaceous of Spain. Naturwissenschaften 98:67–78

Cooper MR (1984) A reassesment of *Vulcanodon karibaensis* Raath (Dinosauria: Saurischia) and the origin of the Sauropoda. Palaeontol Afr 25:203–231

de Beer GS (1937) The development of the vertebrate skull. University of Chicago Press, Chicago

Ede DA (1978) An introduction to developmental biology. Wiley, New York

Fabbri M, Navalón G, Koch NM, Hanson M, Petermann H, Bhullar BA (2021) A shift in ontogenetic timing produced the unique sauropod skull. Evolution. https://doi.org/10.1111/evo.14190

Fujiwara SI, Taru H, Suzuki D (2010) Shape of articular surface of crocodilian (Archosauria) elbow joints and its relevance to sauropsids. J Morphol 271:883–896

Galton PM, Upchurch P (2004) Prosauropoda. In: Weishampel DB, Dodson P, Osmólska H (eds) The Dinosauria, 2nd edn. University of California Press, Berkeley, pp 232–258

Gatesy SM, Dial KP (1996) Locomotor modules and the evolution of avian flight. Evolution 50:331–340
Gatesy SM, Bäker M, Hutchinson JR (2009) Constraint-based exclusion of limb poses for reconstructing theropod dinosaur locomotion. J Vertebr Paleontol 29:535–544
Gilmore CW (1925) A nearly complete articulated skeleton of *Camarasaurus*, a saurischian dinosaur from the Dinosaur National monument, Utah. Memoirs of the Carnegie Museum 10:347–384
Gomani EM (2005) Sauropod dinosaurs from the Early Cretaceous of Malawi Africa. Paleontol Electron 27A:37
González Riga BJ (2011) Speeds and stance of titanosaur sauropods: analysis of *Titanopodus* tracks from the Late Cretaceous of Mendoza, Argentina. An Acad Bras Ciênc 83(1):279–290
González Riga BJ, Calvo JO (2009) A new wide-gauge sauropod track site from the Late Cretaceous of Mendoza, Neuquén basin Argentina. Palaeontology 52(3):631–640
González Riga BJ, Lamanna MC, Ortiz David LD, Calvo JO, Coria JP (2016) A gigantic new dinosaur from Argentina and the evolution of the sauropod hind foot. Sci Rep 6:19165
González Riga BJ, Tomaselli MB (2019) Different trackway patterns in titanosaur sauropods: analysis of new *Titanopodus* tracks from the Upper Cretaceous of Mendoza, Neuquén Basin, Argentina. Cretac Res 93:49–59
Gould SJ (1980) The promise of paleobiology as a nomothetic, evolutionary discipline. Paleobiology 6:96–118
Hechenleitner EM, Leuzinger L, Martinelli AG, Rocher S, Fiorelli LE, Taborda JR, Salgado L (2020) Two late Cretaceous sauropods reveal titanosaurian dispersal across South America. Commun Biol 3(1):1–13
Herendeen PS, Friis EM, Pedersen KR, Crane PR (2017) Palaeobotanical redux: revisiting the age of the angiosperms. Nature Plants 3:17015
Holliday CM, Ridgely RC, Sedlmayr JC, Witmer LM (2010) Cartilaginous epiphyses in extant archosaurs and their implications for reconstructing limb function in dinosaurs. PLOS ONE 5(9):e13120
Hone DWW, Keesey TM, Pisani D, Purvis A (2005) Macroevolutionary trends in the Dinosauria: cope's rule. J Evol Biol 18:587–595
Huber BT, Norris RD, MacLeod KG (2002) Deep-sea paleotemperature record of extreme warmth during the Cretaceous. Geology 30:123–126
Hutson JD (2014) Quadrupedal dinosaurs did not evolve fully pronated forearms: new evidence from the Ulna. Acta Palaeontol Pol 60:599–610
Hutson JD, Hutson KN (2015) An examination of forearm bone mobility in *Alligator mississippiensis* (Daudin, 1802) and *Struthio camelus* Linnaeus, 1758 reveals that *Archaeopteryx* and dromaeosaurs shared an adaptation for gliding and/or flapping. Geodiversitas 37:325–344
Hutson JD, Hutson KN (2017) An investigation of the locomotor function of therian forearm pronation provides renewed support for an arboreal, chameleon-like evolutionary stage. J Mamm Evol 24:159–177
Ibiricu LM, Martínez RD, Casal GA (2018) The pelvic and hind limb myology of the basal titanosaur Epachthosaurus sciuttoi (Sauropoda: Titanosauria). Hist Biol 32:773–788
Ikejiri T, Tidwell V, Trexler DL (2005) New adult specimens of *Camarasaurus lentus* highlight ontogenetic variation within species. In: Tidwell V, Carpenter K (eds) Thunder-Lizards: the Sauropodomorph dinosaurs. Indiana University Press, Bloomington, pp 154–179
Klages JP, Salzmann U, Bickert T, Hillenbrand CD, Gohl K, Kuhn G, Bohaty SM, Titschack J, Müller J, Frederichs T, Bauersachs T, Ehrmann W, van de Flierdt T, Simões Pereira P, Larter RD, Lohmann G, Niezgodzki I, Uenzelmann-Neben G, Zundel M, Spiegel C, Mark C, Chew D, Francis JE, Nehrke G, Schwarz F, Smith JA, Freudenthal T, Esper O, Pälike H, Ronge TA, Dziadek R, the Science Team of Expedition PS104 (2020) Temperate rainforests near the South Pole during peak Cretaceous warmth. Nature 580:81–86
Klein N, Sander PM (2008) Ontogenetic stages in the long bone histology of sauropod dinosaurs. Paleobiology 34:247–263

Klinkhamer AD, Mallison H, Poropat SF, Sinapius GHK, Wroe S (2018) Three-dimensional musculoskeletal modeling of the sauropodomorph hind limb: the effect of postural change on muscle leverage. Anat Rec 301:2145–2163

Klinkhamer AD, Mallison H, Poropat SF, Sloan T, Wroe S (2019) Comparative three-dimensional moment arm analysis of the sauropod forelimb: implications for the transition to a wide-gauge stance in titanosaurs. Anat Rec 302:794–817

Kundrát M, Coria RA, Manning TW, Snitting D, Chiappe LM, Nudds J, Ahlberg PE (2020) Specialized craniofacial anatomy of a titanosaurian embryo from Argentina. Curr Biol 30:4263–4269

Lacovara KJ, Lamanna MC, Ibiricu LM, Poole JC, Schroeter ER, Ullmann PV, Voegele KK, Boles ZM, Carter AM, Fowler EK, Egerton VM, Moyer AE, Coughenour CL, Schein JP, Harris JD, Martınez RD, Novas FE (2014) A gigantic, exceptionally complete titanosaurian sauropod dinosaur from Southern Patagonia Argentina. Sci Rep 4:6196

Lehman TM, Coulson AB (2002) A juvenile specimen of the sauropod dinosaur *Alamosaurus sanjuanensis* from the Upper Cretaceous of Big Bend National Park Texas. J Paleont 76:156–172

Leslie AB, Beaulieu JM, Rai HS, Crane PR, Donoghue MJ, Mathews S (2012) Hemisphere-scale differences in conifer evolutionary dynamics. PNAS 109:16217–16221

Levinton JS, Simon CM (1980) A critique of the punctuated equilibria model and implications for the detection of speciation in the fossil record. Syst Zool 29:130–142

Long JA, McNamara KJ (1997) Heterochrony: the key to dinosaur evolution. Dinofest lnternational, Arizona, pp 113–123

Maidment SCR, Barrett PM (2011) The locomotor musculature of basal ornithischian dinosaurs. J Vertebr Paleontol 31:1265–1291

Maidment SCR, Barrett PM (2012) Does morphological convergence imply functional similarity? A test using the evolution of quadrupedalism in ornithischian dinosaurs. Proc Royal Soc B: Biol Sci 279:3765–3771

Mallison H (2010a) The digital *Plateosaurus* I: body mass, mass distribution, and posture assessed using CAD and CAE on a digitally mounted complete skeleton. Palaeontol Electron 13:1–26

Mallison H (2010b) The digital *Plateosaurus* II: an assessment of the range of motion of the limbs and vertebral column and of previous reconstructions using a digital skeletal mount. Acta Palaeontol Pol 55:433–458

Manafzadeh AR, Kambic RE, Gatesy SM (2021) A new role for joint mobility in reconstructing vertebrate locomotor evolution. Proc Natl Acad Sci 118: e2023513118

Mannion PD, Upchurch P, Schwarz D, Wings O (2019) Taxonomic affinities of the putative titanosaurs from the Late Jurassic Tendaguru formation of Tanzania: phylogenetic and biogeographic implications for eusauropod dinosaur evolution. Zool J Linn Soc 185:784–909

Martínez RN, Alcober OA (2009) A Basal Sauropodomorph (Dinosauria: aurischia) from the Ischigualasto Formation (Triassic, Carnian) and the Early Evolution of Sauropodomorpha. PLoS ONE 4:e4397

Marugán-Lobón J, Chiappe LM, Farke AA (2013) The variability of inner ear orientation in saurischian dinosaurs: testing the use of semicircular canals as a reference system for comparative anatomy. PeerJ 1:e124

Mazzetta GV, Christiansen P, Fariña R (2004) Giants and Bizarres: body size of some southern south American Cretaceous dinosaurs. Hist Biol 16:71–83

McNamara KJ (1982) Heterochrony and phylogenetic trends. Paleobiology 8:130–142

McNamara KJ (1986) A guide to the nomenclature of heterochrony. J Paleontol 60:4–13

McPhee BW, Yates AM, Choiniere JN, Abdala F (2014) The complete anatomy and phylogenetic relationships of *Antetonitrus ingenipes* (Sauropodiformes, Dinosauria): implications for the origins of Sauropoda. Zool J Linn Soc 171:151–205

McPhee BW, Benson RBJ, Botha-Brink J, Bordy EM, Choiniere JN (2018) A giant dinosaur from the earliest Jurassic of South Africa and the transition to quadrupedality in early sauropodomorphs. Curr Biol 28:3143–3151

Müller RT, Langer MC, Bronzati M, Pacheco CP, Cabreira SF, Dias-da-Silva D (2018) Early evolution of sauropodomorphs: anatomy and phylogenetic relationships of a remarkably well preserved dinosaur from the Upper Triassic of southern Brazil. Zool J Linn Soc 184:1187–1248

Müller RT, Garcia MS, Silva SD (2020). Evidências da origem e ascensão dos dinosauros sauropodomorfos preservadas em leitos fossilíferos do Triássico do Sul do Brasil. Terræ Didatica 16:e020013

Nau D, Lallensack JN, Bachmann U, Sander PM (2020) Postcranial osteology of the first early-stage juvenile skeleton of Plateosaurus trossingensis from the Norian of Frick, Switzerland. Acta Palaeontol Pol 65:679–708

Neenan JM, Chapelle KEJ, Fernandez V, Choiniere JN (2019) Ontogeny of the *Massospondylus* labyrinth: implications for locomotory shifts in a basal sauropodomorph dinosaur. Palaeontology 62:255–265

Novas FE, Salgado L, Calvo JO, Agnolín F (2005) Giant titanosaur (Dinosauria, Sauropoda) from the Late Cretaceous of Patagonia. Revisto Del Museo Argentino De Ciencias Naturales 7:37–41

Novas FE, Agnolín FL, Ezcurra MD, Porfiri J, Canale JI (2013) Evolution of the carnivorous dinosaurs during the Cretaceous: the evidence from Patagonia. Cretac Res 45:174–215

Otero A (2010) The appendicular skeleton of *Neuquensaurus*, a Late Cretaceous saltasaurine sauropod from Patagonia, Argentina. Acta Palaeontol Pol 55:299–326

Otero A, Vizcaíno SF (2008) Hindlimb musculature and function of *Neuquensaurus australis* (Sauropoda: Titanosauria). Ameghiniana 45:333–348

Otero A, Pol D (2013) Postcranial anatomy and phylogenetic relationships of *Mussaurus patagonicus* (Dinosauria, Sauropodomorpha). J Vertebr Paleontol 33:1138–1168

Otero A, Krupandan E, Pol D, Chinsamy A, Choiniere J (2015) A new basal sauropodiform from South Africa and the phylogenetic relationships of basal sauropodomorphs. Zool J Linn Soc 174:589–634

Otero A, Allen V, Pol D, Hutchinson JR (2017) Forelimb muscle and joint actions in Archosauria: insights from *Crocodylus johnstoni* (Pseudosuchia) and *Mussaurus patagonicus* (Sauropodomorpha). PeerJ 5:e3976

Otero A, Cuff A, Allen V, Sumner-Rooney PD, Hutchinson JR (2019) Ontogenetic changes in the body plan of the sauropodomorph dinosaur Mussaurus patagonicus reveal shifts of locomotor stance during growth. Sci Rep 9:7614

Otero A, Carballido JL, Pérez Moreno A (2020) The apendicular osteology of *Patagotitan mayorum* (Sauropoda, Titanosauria). J Vertebr Paleontol 40:e1793158

Otero A, Carballido, JL, Salgado, L, Canudo, I, Garrido A (2021) Report of a giant titanosaur sauropod from the Upper Cretaceous of Neuquen Province, Argentina. Cretaceous Res 122:104754

Otero A, Pol D (In Press) Ontogenetic changes in the postcranial skeleton of *Mussaurus patagonicus* (Dinosauria, Sauropodomorpha) and phylogenetic relationships of basal sauropodomorphs. J Syst Palaeontol. https://doi.org/10.1080/14772019.2022.2039311

Owen-Smith N, Mills MGL (2008) Shifting prey selection generates contrasting herbivore dynamics within a large-mammal predator–prey web. Ecology 89:1120–1133

Paul G (2010) The princeton field guide to dinosaurs. Princeton University Press, Princeton

Paul G (2019) Determining the largest known land animal: a critical comparison of differing methods for restoring the volume and mass of extinct animals. Ann Carnegie Museum 85:335–358

Pierce SE, Clack JA, Hutchinson JR (2012) Three-dimensional limb joint mobility in the early tetrapod *Ichthyostega*. Nature 486:523–526

Pol D, Powell JE (2007a) New information on *Lessemsaurus sauropoides* (Dinosauria: Sauropodomorpha) from the Upper Triassic of Argentina. Spec Pap Palaeontol 77:223–243

Pol D, Powell JE (2007b) Skull anatomy of *Mussaurus patagonicus* (Dinosauria: Sauropodomorpha) from the Late Triassic of Patagonia. Hist Biol 19:125–144

Pol D, Nascimento PM, Carvalho AB, Riccomini C, Pires-Domingues RA, Zaher H (2014) A new notosuchian from the Late Cretaceous of Brazil and the phylogeny of advanced notosuchians. PLoS ONE 9:e93105

Pol D, Otero A, Apaldetti C, Martínez (2021) Triassic sauropodomorph dinosaurs from South America: the origin and diversification of dinosaur dominated herbivorous faunas. J South Am Earth Sci 107: 103145

Pretto FA, Langer MC, Schultz CL (2018) A new dinosaur (Saurischia: Sauropodomorpha) from the Late Triassic of Brazil provides insights on the evolution of sauropodomorph body plan. Zool J Linn Soc 185:388–416

Rauhut OWM, Fechner R, Remes K, Reis K (2011) How to get big in the Mesozoic: the evolution of the sauropodomorph body plan. In: Klein N, Remes K, Gee CT, Sander PM (eds) Biology of the Sauropod dinosaurs: understanding the life of giants. Indiana University Press, Bloomington and Indianapolis, pp 119–149

Reisz RR, Scott D, Sues HD, Evans DC, Raath MA (2005) Embryos of an early prosauropod dinosaur and their evolutionary significance. Science 309:761–764

Reisz RR, Evans DC, Sues HD, Scott D (2010) Embryonic skeletal anatomy of the sauropodomorph dinosaur *Massospondylus* from the Lower Jurassic of South Africa. J Vert Paleontol 30:1653–1665

Reisz RR, Evans DC, Roberts EM, Sues HD, Yates AM (2012) Oldest known dinosaurian nesting site and reproductive biology of the Early Jurassic sauropodomorph Massospondylus. Proc Natl Acad Sci 109:2428–2433

Remes K (2008) Evolution of the pectoral girdle and forelimb in Sauropodomorpha (Dinosauria, Saurischia): Osteology, myology and function. Unpublished D. Phil, Thesis, Universität München

Salgado L (1999) The macroevolution of Diplodocimorpha (Dinosauria; Sauropoda): a developmental model. Ameghiniana 36:203–216

Salgado L, Azpilicueta C (2000) Un nuevo saltasaurino (Sauropoda, Titanosauridae) de la provincia de Río Negro (Formación Allen, Cretácico Superior), Patagonia, Argentina, Ameghiniana 37:259–264

Salgado L, Coria RA, Chiappe LM (2005) Osteology of the sauropod embryos from the Upper Cretaceous of Patagonia. Acta Palaeontol Pol 50:79–92

Sander PM (2000) Long bone histology of the Tendaguru sauropods: implications for growth and biology. Paleobiology 26:466–488

Sander PM, Tückmantel C (2003) Bone lamina thickness, bone apposition rates, and age estimates in sauropod humeri and femora. Paläontol Z 76:161–172

Sander PM, Klein N, Buffetaut E, Cuny G, Suteethorn V, Le Loeuff J (2004) Adaptive radiation in sauropod dinosaurs: Bone histology indicates rapid evolution of giant body size through acceleration. Org Divers Evol 4:165–173

Sander PM, Christian A, Clauss M, Fechner R, Gee CT, Griebeler EM, Gunja HCH, Hummel J, Mallison H, Perry SF, Preuschoft H, Rauhut OWM, Remes K, Tütken T, Wings O, Witzel U (2011) Biology of the sauropod dinosaurs: the evolution of gigantism. Biol Rev 86:117–155

Santucci RM, De Arruda-Campos AC (2012) A new sauropod (Macronaria, Titanosauria) from the Adamantina Formation, Bauru Group, Upper Cretaceous of Brazil and the phylogenetic relationships of Aeolosaurini. Zootaxa 3085:1–33

Schwarz D, Frey E, Meyer CA (2007) Novel reconstruction of the orientation of the pectoral girdle in Sauropods. Anat Rec 290:32–47

Sellers WI, Margetts L, Coria RA, Manning PL (2013) March of the Titans: the locomotor capabilities of Sauropod dinosaurs. PLoS ONE 8:e78733

Senter P, Robins JH (2005) Range of motion in the forelimb of the theropod dinosaur *Acrocanthosaurus atokensis*, and implications for predatody behaviour. Zool J London 266:307–318

Sereno PC (1991) Basal archosaurs: phylogenetic relationships and functional implications. Soc Vertebr Paleontol Memoir 2:53

Sereno PC (1997) The origin and evolution of dinosaurs. Annu Rev Earth Planet Sci 25:435–489

Sereno PC (2007) Basal Sauropodomorpha: historical and recent phylogenetic hypotheses, with comments on *Ammosaurus major* (Marsh, 1889). Spec Pap Palaeontol 77:261–289

Sereno PC, Martínez RN, Alcober OA (2013) Osteology of *Eoraptor lunensis* (Dinosauria, Sauropodomorpha). Soc Vertebr Paleontol Memoir 12:83–179

Stein K, Csiki Z, Curry Rogers K, Weishampel DB, Redelstorff R, Carballido JL, Sander PM, Stanley SM (2010) Small body size and extreme cortical bone remodeling indicate phyletic dwarfism in *Magyarosaurus dacus* (Sauropoda: Titanosauria). Proc Natl Acad Sci 107:9258–9263

Stevens KA, Parrish JM (1999) Neck posture and feeding habits of two jurassic sauropod dinosaurs. Science 284:798–800

Stevens KA, Ernst S, Marty D (2016) Uncertainty and ambiguity in the interpretation of sauropod trackways (Chapter 13). In: Falkingham PL, Marty D, Richter A (eds) Dinosaur tracks—The next steps. Indiana University Press, Bloomington, Indiana, USA, pp 227–243

Stewart WN, Rothwell GW (1993) Paleobotany and the evolution of plants, 2nd edn. Cambridge University Press, Cambridge, UK

Taylor MP, Wedel MJ, Naish D (2009) Head and neck posture in sauropod dinosaurs inferred from extant animals. Acta Palaeontol Pol 54:213–220

Tsai HP, Holliday CM (2014) Articular soft tissue anatomy of the archosaur hip joint: structural homology and functional implications. J Morphol 276:601–630

Tütken T (2011) The diet of sauropod dinosaurs: implications of carbon isotope analysis on teeth, bones, and plants. In: Klein N, Remes K, Gee CT, Sander PM (eds) Biology of the Sauropod dinosaurs: understanding the life of giants. Indiana University Press, Bloomington, pp 57–82

Upchurch GR Jr, Dilcher DL (1990) Cenomanian angiosperm leaf megafossils, Dakota formation, Rose Creek locality, Jefferson County, southeastern Nebraska. US Geol Surv Bull 1950:1–55

Upchurch P, Barrett PM (2000) The evolution of sauropod feeding mechanisms. In: Sues HD (ed) Evolution of herbivory in terrestrial vertebrates. Cambridge University Press, Cambridge, pp 79–122

VanBuren CS, Bonnan M (2013) Forearm posture and mobility in quadrupedal dinosaurs. PLOS ONE 8:e74842

Vargas-Peixoto D, Da-Rosa AAS, Franca MAG (2015) Functional and biomechanic aspects of the scapular girdle and forelimbs of *Unaysaurus tolentinoi* Leal et al., 2004 (Saurischia: Sauropodomorpha). J South Am Earth Sci 61:129–133

Wilhite R (2003) Biomechanical reconstruction of the appendicular skeleton in Three North American Jurassic Sauropods. Unpublished D. Phil, Thesis, Louisiana State University

Wilson JA, Sereno PC (1998) Early evolution and higher-level phylogeny of sauropod dinosaurs. Soc Vertebr Paleontol Memoir 5:1–68

Wilson JA, Carrano MT (1999) Titanosaurs and the origin of "wide-gauge" trackways: a biomechanical and systematic perspective on sauropod locomotion. Paleobiology 25:252–267

Wilson JA, Pol D, Carvalho AB, Zaher H (2016) The skull of the titanosaur *Tapuiasaurus macedoi* (Dinosauria: Sauropoda), a basal titanosaur from the Lower Cretaceous of Brazil. Zool J Linn Soc 178:611–662

Wright JL (1999) Ichnological evidence for the use of the forelimb in iguanodontid locomotion. Spec Pap Palaeontol 60:209–219

Yates AM, Kitching JW (2003) The earliest known sauropod dinosaur and the first steps towards sauropod locomotion. Proc R Soc Lond B 270:1753–1758

Yates AM, Bonnan MF, Neveling J, Chinsamy A, Blackbeard MG (2010) A new transitional sauropodomorph dinosaur from the Early Jurassic of South Africa and the evolution of sauropod feeding and quadrupedalism. Proc R Soc Lond B 277:787–794

Zaher H, Pol D, Carvalho AB, Nascimento PM, Riccomini C, Larson P, Juarez-Valieri R, Pires-Domingues R, da Silva Jr JN, de Almeida Campos D (2011) A complete skull of an early Cretaceous sauropod and the evolution of advanced titanosaurians. PLoS ONE 6:e16663

Zhao Q, Benton MJ, Sullivan C, Sander PM, Xu X (2013) Histology and postural change during the growth of the ceratopsian dinosaur *Psittacosaurus lujiatunensis*. Nature Commun 4:2079

South American Sauropodomorphs: What Their Bone Histology Has Revealed to Us

Ignacio A. Cerda

Abstract Given that bone microstructure is a very important source of paleobiological information, several paleohistological studies have been conducted on sauropodomorph dinosaurs, possibly making this clade the most studied histologically. Despite these paleohistological studies on sauropodomorph dinosaurs from South America are relatively scarce in comparison with other regions of the world, significant progress on this matter (i.e. paleohistology of South American sauropodomorphs) has been made during the last decade. Following an order from rather specific to more generalized issues, the most important advances are related to the origin of particular skeletal elements (i.e. osteoderms, sacral supraspinous ossifications and extremely elongated cervical ribs), the growth patterns of basal sauropodomorphs and the variation on sauropod growth dynamics and its relationship with gigantism. Regarding the origin of osteoderms and extremely elongated cervical ribs, these structures have been formed by metaplastic ossification of dermal and tendinous tissues, respectively. Their histological characterization has been helpful to discover that the alleged osteoderms of *Agustinia ligabuei* were actually dorsal and cervical ribs. The long bone histology of basal sauropodomorphs has revealed that the cyclical growth pattern assumed for this group actually showed some degree of variation. Finally, a clade of basal sauropods (i.e. lessemsaurids) was characterized by a cyclical growth pattern, not previously reported for other sauropods, combined with episodes of highly accelerated growth rates. The gigantic body sizes obtained by lessemsaurids were therefore reached through a growth strategy different from that developed by eusauropods (i.e. rapid and continuous growth).

Keywords Paleohistology · Sauropodomorpha · Growth rate · Osteoderms · Life history

---

I. A. Cerda (✉)
Consejo Nacional de Investigaciones Científicas y Técnicas (CONICET), Río Negro, Argentina

Instituto de Investigación en Paleobiología y Geología (IIPG), Universidad Nacional de Río Negro (UNRN), Río Negro, Argentina

Museo 'Carlos Ameghino', Belgrano 1700, Paraje Pichi Ruca (Predio Marabunta), Cipolletti (8300), Río Negro, Argentina

A. Otero et al. (eds.), *South American Sauropodomorph Dinosaurs*, Springer Earth System Sciences, https://doi.org/10.1007/978-3-030-95959-3_13

1 Introduction

The study of the bone microstructure has become a valuable source of paleobiological information for extinct vertebrates, particularly in sauropodomorph dinosaurs (e.g. Reid 1996; Sander 2000; Chinsamy-Turan 2005; Erickson 2014). Such special attention is mostly due to the fact that this group of non-avian dinosaurs includes the largest terrestrial vertebrates that have ever existed. Paleohistological studies of sauropodomorph dinosaurs have mostly provided insights about traits related with: growth rate and growth dynamics, age for sexual and somatic maturity attainment, osseous pathologies, bone histogenesis and soft tissue reconstruction, among others (e.g. Currey 1962; Chinsamy 1993; Curry 1999; Sander 2000; Sander et al. 2004; Sander and Klein 2005; Klein et al. 2012a, b; Griebeler et al. 2013; Hofmann et al. 2014; Curry Rogers and Kulik 2018; González et al. 2020). Most of these contributions have been conducted on taxa from Europe, North America and Africa, being the South American forms relatively less studied. This is rather surprising considering the high diversity and abundance of sauropodomorph remains in this extensive landmass. The main reason for this is mostly because, contrary to other areas of the vertebrate paleontology, paleohistology in South America has begun to become an important discipline just during the last decade. Despite paleohistological studies on sauropodomorph dinosaurs from South America being relatively scarce in comparison with other regions of the world, significant progress on this matter has been made during the last several years.

In this chapter, I provide a general review of the most important contributions to the sauropodomorph paleobiology based on paleohistological studies on South American taxa. Given the numerous paleobiological traits that can be assessed from bone histology, I have divided the chapter into different topics, from specific to more generalized issues. I have tried to avoid the incorporation of data from congress presentations and dissertations. Contributions that include histological analyses of few elements with the purpose of obtaining information about the relative age or ontogenetic stage (e.g. Pol et al. 2011; Carballido et al. 2011; Lacovara et al. 2014; Ghilardi et al. 2016; Ibiricu et al. 2017a, b; Pretto et al. 2016) are also omitted. The main goal of this contribution is to provide a general view of the advances in paleohistological studies of South American sauropodomorph dinosaurs, showing how they have contributed to the general knowledge of sauropodomorph biology.

Institutional Abbreviations **MCF-PVPH**: Museo Municipal 'Carmen Funes' de Plaza Huincul, Vertebrate Paleontology collection, Plaza Huincul, Neuquén, Argentina; **MCS-Pv**: Museo de Cinco Saltos, Vertebrate Paleontology collection, Cinco Saltos, Rio Negro, Argentina; **MDT-Pv**: Museo 'Desiderio Torres', Vertebrate Paleontology collection, Sarmiento, Chubut, Argentina; **MPCA-Pv**: Museo Provincial 'Carlos Ameghino', Vertebrate Paleontology collection, Cipolletti, Río Negro, Argentina; **MPM-PV**: Vertebrate Paleontology collection of the Museo Regional Provincial 'Padre Molina', Río Gallegos, Santa Cruz, Argentina; **OMNH**: Oklahoma Museum of Natural History, Norman, Oklahoma, United States of America, **PVL**: Instituto 'Miguel Lillo', Vertebrate Paleontology collection, San Miguel del

Tucumán, Tucumán, Argentina; **UNPSJB-PV**: Universidad Nacional de la Patagonia 'San Juan Bosco', Vertebrate Paleontology collection, Comodoro Rivadavia, Chubut, Argentina.

2 Osteoderms Origin and Function

Dermal ossifications (i.e. osteoderms, dermal plates) represent the most commonly identified and documented element of the tetrapod integumentary skeleton (Vickaryous and Sire 2009). These skeletal structures have been defined as mineralized organs entrenched within the dermis (Francillon-Vieillot et al. 1990; Vickaryous and Hall 2008) and are widely distributed among different groups of tetrapods, including non-avian dinosaurs (Moss 1969; Hill 2005; Sire et al. 2009; Vickaryous and Sire 2009). Among sauropodomorph dinosaurs, however, osteoderms have only been documented in derived sauropods (i.e. Titanosauria) (e.g. Powell 1980; Le Loeuff et al. 1994; Dodson et al. 1998; Salgado 2003; D'Emic et al. 2009; Curry Rogers et al. 2011; Vidal et al. 2014; Zurriaguz 2017; Fronimos 2021). In spite of detailed morphological descriptions of titanosaur osteoderms (Huene 1929; Powell 1980, 2003; Dodson et al. 1998; Salgado 2003), particularly from South America, their bone microanatomy and histology is relatively unknown.

Paleohistological studies on titanosaur osteoderms, which were mostly conducted on South American taxa, have focused on two main traits, namely, their ossification mechanism (i.e. histogenesis) and possible functions. Although fossil materials do not directly record the process of ossification, the structure of the resulting primary bone (i.e. the tissue first formed in the element) and comparisons with extant groups can provide clues as to how it was produced (Cerda and Powell 2010). The hypothesis most frequently invoked for osteoderm development in extant and fossil tetrapods is metaplastic ossification, a process in which a preexisting non-osseous and fully developed tissue is transformed into bone (Haines and Mohuiddin 1968). In fossil groups, metaplastic tissue has been identified in osteoderms by the presence of interwoven bundles of mineralized collagen fibers formally named 'structural fibers' (Scheyer and Sander 2004, 2007; Main et al. 2005; Witzmann and Soler-Gijón 2008). A second possible mode of osteoderm formation is intramembranous ossification, in which the newly formed osseous tissue displaces the preformed soft tissue structures instead of incorporating them. This kind of ossification has been proposed on the basis of the absence of mineralized structural fibers (e.g. Scheyer and Sander 2009; Cerda and Desojo 2011; Buchwitz et al. 2012; Cerda et al. 2018). A combination of both types of mechanism has also been proposed in some instances (e.g. Reid 1996; Ponce et al. 2017; Cerda et al. 2018). The first attempt to infer the ossification mechanism of titanosaur osteoderms was made by Salgado (2003), who analyzed the bone microstructure of an isolated element collected from sediments of the Allen Formation (Campanian–Maastrichtian) in the Salitral Moreno locality (Río Negro Province, Argentina). The abundance of mineralized fibers in the primary

bone tissue of the element, interpreted by the author as Sharpey's fibers (i.e. mineralized ingrowing fibrillary processes from adjacent soft tissues; Francillon-Vieillot et al. 1990; Peterman and Sander 2013), was considered as evidence for metaplastic ossification. Later, Cerda and Powell (2010) carried out a comprehensive histological study of the osteoderms (a single dermal plate and several millimeter-sized ossicles) of *Saltasaurus loricatus*, a relatively a small-bodied taxon recovered from Upper Cretaceous (?late Campanian–Maastrichtian) sediments of the Lecho Formation at the locality of El Brete (south of Salta Province, Argentina) (Fig. 1a, b). The predominance of structural fibers in both the dermal plate and ossicles was interpreted

◀**Fig. 1** Bone microstructure of osteoderms and hyperelongated cervical ribs in sauropods. **a** Large osteoderm (i.e. dermal plate) of *Saltasaurus loricatus* (PVL 4017 – 113) in superficial view. **b** Dermal ossicles of *Saltasaurus loricatus* (PVL 4017 – 208) in deep view. **c** Detail of the *Saltasaurus loricatus* dermal plate histology showing a thin outer cortex of primary bone formed by irregularly oriented bundles of mineralized structural fibers. Secondary remodeling is profuse. **d** Complete section a *Saltasaurus loricatus* dermal ossicle. The same is entirely formed by mineralized structural fibers, which exhibits a higher degree of spatial organization in comparison with the dermal plate. **e** Detail of the structural fibers in the ossicles. **f** Natural breakage of Titanosauria indet. MPCA-Pv 67 osteoderm. The distinct system of internal neurovascular canals has been enlightened in blue. **g** Articulated cervical vertebrate 5–8 of the basal titanosauriform *Sauroposeidon proteles* (OMNH 53062) showing hyperelongated cervical ribs. Note the formation of 'bundles' of at least three ribs. Modified from Wedel et al. (2000). **h** Cervical rib cross section of Titanosauria indet. MCS-Pv uncatalogued specimen. **i** Detailed view of secondary osteons and unremodeled primary bone tissue in transverse section (box inset in h). The matrix of primary bone is entirely formed by mineralized fibers of a tendinous tissue. A detailed view of the fibers, which exhibits a distinct rhomboidal shape, is showed in the lower left corner. **j** Longitudinal section of the same sample. Despite the high degree of secondary remodeling, primary bone tissue is preserved in the outer cortex. Pictures have taken under normal transmitted light (**h**), cross-polarized light (**i, j**) and cross-polarized light with a lambda filter (**c**, **d** and **e**). Abbreviations: cr: cervical rib; de: deep side; Hc: Haversian canal; lsf: longitudinally sectioned structural fibers; pb: primary bone tissue; rc: resorption cavity; rl: resorption line; so: secondary osteon; su: superficial side; tsf: transversally sectioned structural fibers; vs: vascular space

as evidence of metaplastic ossification of the osteoderms (Fig. 1c, d). Contrary to the dermal plates, which exhibit a rather disorganized arrangement of the bundles of structural fibers, these bundles show a highly organized pattern in the ossicles (Fig. 1d, e). In this regard, dermal ossicles of *Saltasaurus* are formed by at least three orthogonal systems of fiber bundles: one vertical and two horizontals. Such an arrangement resembles those described for the *stratum compactum* of the dermis (Scheyer et al. 2007; Vickaryous and Hall 2008; Sire et al. 2009), indicating that it is in the portion of the integument in which mineralization occurs (at least in the ossicles). It is interesting to note that the histological structure of the *Saltasaurus* ossicles is identical to that reported for the same elements in the ankylosaur *Antarctopelta oliveroi* (de Ricqlès et al. 2001a; Cerda et al. 2019). Such a strong resemblance has been interpreted as an example of deep homology, where a common ossification process occurs in the same integumental layer of two non-related groups and generates non-homologous structures (Cerda and Powell 2010). Whereas the osteoderms of titanosaurs and ankylosaurs are not homologous structures, the capability of the dermis to produce them actually is homologous (Zylberberg and Castanet 1985; Hill 2006; Vickaryous and Hall 2008; Vickaryous and Sire 2009). The interpretation of a metaplastic origin of the osteoderms has been supported in a more recent study (Cerda et al. 2015a), in which histological data from four osteoderms of indeterminate titanosaurs from the Upper Cretaceous of the Northern Patagonia were provided.

One of the most intriguing questions regarding titanosaur osteoderms is their function. Possible functions include passive and/or active defense, thermoregulation, mineral reservoir and display and/or species recognition, among others (Powell 1980; Sanz and Buscalioni 1987; Le Loeuff et al. 1994; Salgado 2003; Cerda and

Powell 2010; Curry Rogers et al. 2011; Marihno and Iori 2011; Cerda et al. 2015a; Vidal et al. 2017). Some of these hypotheses originated and/or were tested using microanatomical and histological data. Taking into account the information provided by the South American taxa from the Upper Cretaceous of Northern Patagonia, Cerda et al. (2015a) considered that the mineral-storage function is actually supported by the presence of several large vascular spaces that cross the osteoderms (Fig. 1f). This vascular network probably allowed rapid mobilization of the mineral reserves from the osteoderms to other parts of the body (Curry Rogers et al. 2011). Large internal cavities as those described in other titanosaur osteoderms (e.g. Curry Rogers et al. 2011; Vidal et al. 2017) were not reported in South American forms. The role of passive defense has been supported for the dermal ossicles of *Saltasaurus* (Cerda and Powell 2010). The tight arrangement of these elements (Fig. 1b), which were surrounded by the thick collagenous fiber bundles of the unmineralized portions of the dermis, may have provided a high resistance to the integument (Cerda and Powell 2010; Cerda et al. 2015a). Therefore, different types of osteoderms can serve for different functions, which also could vary during ontogeny and between sexes (Salgado 2003; Marinho and Iori 2011; Cerda et al. 2015a). For example, Marinho and Iori (2011) proposed that a possible defensive function of osteoderms in titanosaurs was most effective in juveniles that could have had osteoderms proportionally larger and composing a cuirass with the osteoderms closer to each other than in adults, thus providing some protection against predators. Also, some authors have considered that the ornamentation in the bulb of some kind of osteoderms suggests the presence of a keratin sheath, which might imply sexual display (Vidal et al. 2017).

Although the studies performed on titanosaur osteoderms from South America have provided relevant paleobiological information, there are several aspects that are still unexplored. For example, their potential for age estimation (i.e. skeletochronology) has not been evaluated to date. Osteoderms of other groups of archosaurs (e.g. crocodylians) have been particularly important for skeletochronology, because they preserve cyclical growth marks (i.e. lines of arrested growth and annuli), which are formed annually (e.g. Hutton 1986; Tucker 1997; Erickson and Brochu 1999; Erickson et al. 2003). The potential of any bony structure to be used for skeletochronological analysis is inversely proportional to the degree of internal bone resorption that occurs. Although cyclical growth marks are actually formed during the titanosaur osteoderm histogenesis, the extensive secondary remodeling usually precludes its use for age estimations. Nevertheless, Cerda and Powell (2010) proposed that dermal ossicles, which are entirely formed by primary bone tissue, and hence preserve a complete growth record, have a strong potential for skeletochronological studies.

3 Hyperelongated Cervical Rib Origins and Tendinous Ossifications

Hyperelongated cervical ribs (i.e. cervical ribs in which the shaft extends backward to two or more vertebrae) have been reported in several sauropod dinosaurs, including basal titanosauriforms as *Giraffatitan brancai* and *Sauroposeidon proteles* (Janensch 1950; Wedel et al. 2000), derived lithostrotians as *Uberabatitan riberoi* and *Bonitasaura salgadoi* (Salgado and Carvalho 2008; Gallina and Apesteguía 2015) and mamenchisaurids (Young and Zhao 1972). Their particular size, shape and arrangement in the neck (forming thick 'bundles' of superposing elements) have raised questions about their origins (Fig. 1g). Specifically, the main question is related to determining whether these structures originated through endochondral and periosteal ossification (as dorsal ribs) or, conversely, if they correspond with tendinous structures formed by metaplasia (i.e. ossified tendons) (Cerda 2009). To test these hypotheses, I analyzed the bone microstructure of a hyperelongated cervical rib from an indeterminate titanosaur collected in sediments of the Plottier Formation (late Coniacian) from the El Anfiteatro locality (Río Negro Province, Argentina) (Cerda 2009). As expected from bone tissue originating from metaplastic ossification of a tendinous structure, the primary bone tissue of the cervical rib shaft is formed by coarse bundles of mineralized collagenous fibers, which are oriented parallel to the rib main axis (Fig. 1h–j). These observations and interpretations were later supported by other authors (Gallina 2012; Klein et al. 2012b; Díez Díaz et al. 2018). Furthermore, Klein et al. (2012b) determined that the tendinous origin of the cervical rib shaft was not restricted to hyperelongated elements, but also to short ribs. It is interesting to note that hyperelongated cervical ribs are always strongly remodeled, indicating a rapid replacement of the primary, metaplastic bone tissue with secondary osteons (e.g. Cerda 2009; Gallina 2012). Such a feature could be related to different biomechanical features of the bone tissues, which may strongly affect the function of the cervical ribs. Future studies that combine both histological data with finite element analyses will surely provide insights on this regard.

Tendinous ossifications have also been reported in structures other than cervical ribs. Martínez et al. (2016) mention the presence of a thin (diameter of around 3 mm), cable-like structure, located ventrolateral to the right side of the cervical vertebrae and ribs of *Sarmientosaurus musacchioi*, a basal lithostrotian from the Upper Cretaceous (Cenomanian–Turonian) Lower Member of the Bajo Barreal Formation of southern Chubut Province in central Patagonia, Argentina. This enigmatic structure, which originated posterior to the skull and coursed along multiple vertebrae, was sectioned for histological analysis. Although the microstructure of this element is mostly composed of secondary bone, the authors mentioned the presence of longitudinally oriented mineralized fibers. As in the cervical rib, the thin cervical rod of *Sarmientosaurus* was also interpreted as an ossified tendon by Martínez et al. (2016). Similar structures have not apparently been reported in other sauropodomorph dinosaurs. This structure may correspond to a pathological ossification originating from a natural process of aging.

4 The Elastic Supraneural Ligament of the Titanosauriforms

Apart from being able to discern the origin of well-identified anatomical structures such as osteoderms and hyperelongated cervical ribs, bone histology can provide information on the nature of particular and 'enigmatic' elements. The studies identifying the origin of the sacral 'supraspinous rod' recorded in some titanosauriforms is a good example of this. The 'supraspinous rod' consists of an elongated rod that runs along the apices of the neural spines of the sacral vertebrae (Fig. 2a). The same structure has been identified in at least 11 titanosauriform specimens of several taxa, including *Tastavinsaurus sanzi*, *Epachthosaurus sciuttoi*, *Atsinganosaurus velauciensis*, *Malawisaurus dixeyi*, *Huabeisaurus allocotus*, *Huanghetitan ruyangesis*, *Huanghetitan liujianxensis* and three indeterminate titanosaurs (see review in Cerda et al. 2015b). Based solely on the gross morphology, the sacral supraspinous rod has been considered alternatively as an ossified tendon (Campos et al. 1999; Giménez et al. 2008; García et al. 2010), an ossified ligament (Powell 1987, 2003) or as calcified cartilage (Bonaparte 1996). The first attempt to determine the origin of this structure by histological analysis was performed by Giménez et al. (2008). They studied thin sections from the supraspinous rod of *Epachthosaurus sciuttoi* from the Upper Cretaceous of Patagonia, which consists entirely of dense secondary (Haversian) bone tissue. The absence of primary bone tissue in the sample examined by Giménez et al. (2008) prevented any inferences being made about the origin of the supraspinous rod. Later, Cerda et al. (2015b) examined new samples from the same *Epachthosaurus sciuttoi* specimen analyzed by Giménez et al. (2008) and from an indeterminate lithostrotian titanosaur from the Upper Cretaceous of Patagonia. Although the microstructure of these elements are mostly formed by dense Haversian bone, remains of primary bone tissue revealed the presence of two types of fiber-like structures, which are coarse and longitudinally oriented (i.e. parallel to the main axis of the element) (Fig. 2b–e). The histological features were interpreted as evidence for a ligamentous origin of the sacral supraspinous rod; more specifically, the authors infer that this structure originated from a non-pathological ossification of the elastic supraspinous ligament, possibly induced by the continuous tensile forces applied to the element. This interpretation has been supported by a more recent study (Woodruff et al. 2016).

5 Osteoderms, Ribs of Anything Else?

Most of the contributions reviewed in the previous sections of this chapter have shown the importance of the comparative histological analyses for the determination of the origin of skeletal structures. Indeed, such determination allows us to histologically distinguish these structures (at least from their primary bone tissue) from others. Therefore, histological information can be useful for identifying specific skeletal

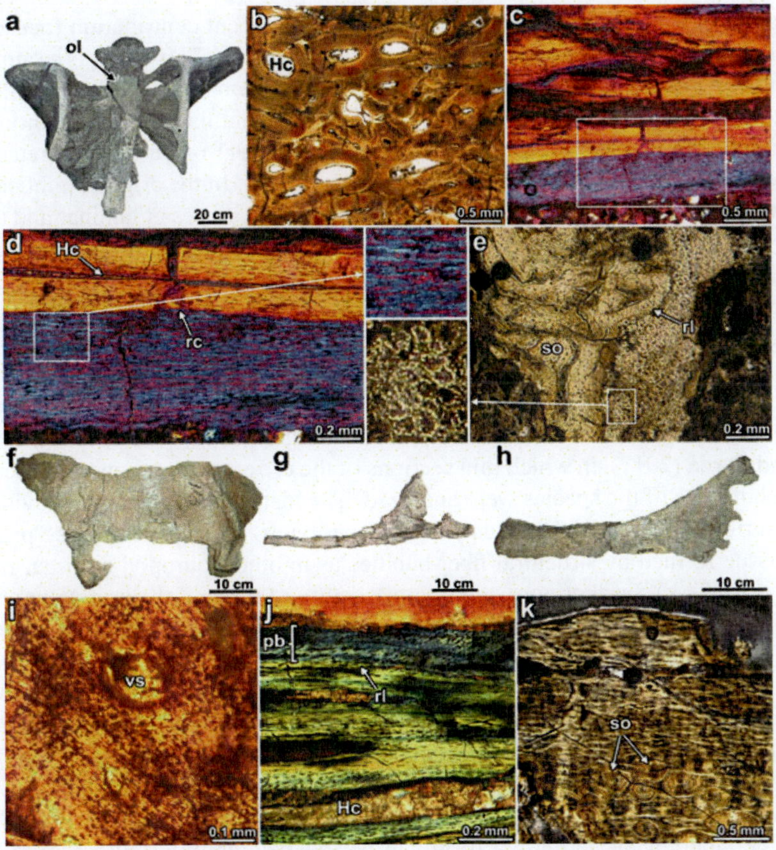

Fig. 2 Bone histology of mineralized elastic ligament (**a–e**) and putative *Agustinia ligabuei* osteoderms (**f–k**). **a** Sacrum of *Epachthosaurus sciuttoi* UNPSJB-PV 920 in dorsal view showing a distinct ossified elastic ligament over the sacral vertebrae spines. **b–e** Histological features of Lithostrotia indet. MDT-Pv 4 mineralized elastic ligament. **b** Detail of a cross section showing dense Haversian bone, which predominates in the sample. **c, d** General view and detail of a longitudinal section showing Haversian tissue (upper portion of the pictures) and remains of primary bone tissue (below). A detailed view of the primary bone is shown in the upper right corner. The primary bone is avascular and composed of abundant mineralized collagenous fibers (blue structures) which are intercalated with other structures, interpreted as mineralized elastic fibers (fuchsia structures). **e** General view and detail of the same element in cross section. A detailed view of the primary bone is shown in the lower left corner. Note the irregular shape of the transversely sectioned fibers. **f–h** Representative set of *Agustinia ligabuei* putative osteoderms. **f** MCF-PVPH-110/08 (morphotype 2). **g** MCF-PVPH-110/10 (morphotype 3). **h** MCF-PVPH-110/03 (morphotype 4). **i** Detailed view of the primary bone tissue of MCF-PVPH-110/10, which is entirely composed by Sharpey's fibers. Compare with the structural fibers of *Saltasaurus loricatus* osteoderms showed in the Fig. 1c–e. **j** Detailed view of MCF-PVPH-110/10 longitudinal section. The primary bone is preserved in the outer cortex. Note the strong resemblance with the longitudinal section of hyperelongated cervical rib showed in Fig. 1j. **k** Cross section of MCF-PVPH-110/03. Bundles of structural fibers, typical of sauropod osteoderms, are also absent in this structure. Pictures have taken under normal transmitted light (**b**, **e** and **i**), cross-polarized light (**k**) and cross-polarized light with a lambda filter (**c**, **d** and **j**). Abbreviations: Hc: Haversian canal; ol: ossified supraspinous ligament: pb: primary bone tissue; rl: resorption line; so: secondary osteon; vs: vascular space

structures. This premise was taken into account in a recent contribution focused on the nature of the putative osteoderms of *Agustinia ligabuei*. This taxon is a sauropod dinosaur of uncertain phylogenetic affinities whose remains have been collected from the Cerro Leon locality (Neuquén Province, Argentina), from the continental outcrops of the upper member (Cullin Grande Member) of the Early Cretaceous Lohan Cura Formation (early Albian) (Bonaparte 1999). In the original publication, Bonaparte (1999) mentioned the presence of four morphotypes of laminar and transversely elongated putative osteoderms (Fig. 2f–h). The morphology of these structures, however, strongly differs from that reported for other sauropods (e.g. lithostrotian titanosaurs, Salgado 2003; D'Emic et al. 2009; Cerda et al. 2015a). Furthermore, some authors suggested that the putative osteoderms of *Agustinia* correspond with other, non-dermal skeletal structures such as cervical and dorsal ribs (D'Emic et al. 2009; Mannion et al. 2013). The hypothesis about the presence of osteoderms in *Agustinia* was tested in a comparative paleohistological study performed by Bellardini and Cerda (2017), in which thin sections of the different morphotypes described for the holotype of the species were analyzed. The histological evidence provided in that contribution revealed that none of the morphotypes of *Agustinia* shows a primary bone tissue formed by structural fiber bundles as in other sauropod dinosaur osteoderms (Fig. 2i–k). Furthermore, the microstructure of three of the original morphotypes is more comparable with dorsal and cervical ribs than any other kind of dermal or bony element. The histological analysis of the putative osteoderms of *Agustinia* is of particular interest for several reasons. First, it allowed for the reassessment of the osteological composition of the *Agustinia* holotype. Second, since the presence of osteoderms is a character relevant among sauropod dinosaurs, their supposed presence in *Agustinia* biased the phylogenetical relationships of this taxon. Finally, the study of Bellardini and Cerda (2017) nicely illustrates how a particular hypothesis about the anatomy of an extinct form can be independently tested using its bone microstructure.

6 Soft Tissue Reconstruction

Bone histology can provide insights about the soft tissue associated with different skeletal elements. For example, muscle insertions can alter the bone microstructure of the attachment site, generating modifications in the vascularization pattern and leaving distinct histological structures known as Sharpey' fibers. A soft tissue of particular interest in sauropodomorph dinosaurs which, nevertheless, has been poorly studied from histological approaches, correspond with the thin epithelium of the pneumatic diverticula derived from the lung and air sacs recorded and inferred in ornithodiran archosaurs (e.g. Britt 1997; Wedel 2005; O'Connor 2006; Cerda et al. 2012). The pneumatic features of the postcranial skeleton of sauropodomorph dinosaurs and other ornithodirans have been mainly assessed from both external and internal anatomical features (e.g. Janensch 1947; Britt 1997; Schwarz et al. 2007; Wedel 2003, 2009; Yates et al. 2012; Zurriaguz and Cerda 2017; Apaldetti et al.

2018). The causal relationship between pneumatic diverticula and osseous structures has been just recently analyzed by Lambertz et al. (2018) using bone histology. These authors reported the presence of thin Sharpey's fibers associated with the diverticula epithelium and proposed the term 'pneumosteon' for this tissue. Later, Aureliano et al. (2020) investigated the presence of pneumosteonal tissue in the South American titanosaur *Uberabatitan ribeiroi*, from the Maastrichtian (Marília Formation) of Minas Gerais, Brazil. The authors reported the presence of this bone tissue in the septa that bound the internal pneumatic cavities (i.e. 'camellae') of a cervical vertebra and mentioned that the absence of this bone tissue in some parts of the element is related to diagenetic alteration of the specimen. As much as is known, the study of Aureliano et al. (2020) corresponds with the only report of pneumosteonal tissue in a titanosauriform sauropod and provides additional evidence for the intraskeletal invasion of pneumatic diverticula from the respiratory system.

The relationship between distinct superficial marks on the bones surfaces and the presence of particular soft tissues has also been explored in a giant South American form: *Patagotitan mayorum*. Carballido et al. (2017) used histological examination to infer whether the heavily striated external texture observed in the transverse process and the tip of the neural spines of the anterior caudal vertebrae had a histological signature of a strong ligament/tendinous attachment. Their analysis revealed that, effectively, the primary bone tissue preserved in the outer cortex of both regions was avascular and entirely formed by densely grouped bundles of Sharpey's fibers. The authors inferred that these extrinsic fibers correspond to the presence of an interspinous or supraspinous ligament (vertebral tip) and a strong muscle attachment (transverse process). Although not explored by the authors, the histological information obtained from osseous scars derived from muscle attachment can provide important insights about different paleobiological issues. For example, the spatial orientation of Sharpey's fibers can inform about the main orientation of the tensile forces (Petermann and Sander 2013).

7 Osseous Pathologies

Diseases and traumas suffered by organisms usually leave distinct traces on bones that allow identifying different types of pathologies in living and extinct organisms (e.g. Rothschild and Martin 2006; Rothschild 2009). Although the determination of the pathologies is made from a characterization of internal and/or external macroscopic features of the osseous elements, histological analyses have recently been performed in paleopathological studies. Regarding South American forms, *Mussaurus patagonicus* from the Early Jurassic (Laguna Colorada Formation, Sinemurian) of southern Patagonia (Laguna Colorada locality, Santa Cruz Province, Argentina), is, to date, the only early branching sauropodomorph for which osseous pathologies have been determined from its bone histology (Cerda et al. 2014a). An abnormal tissue formed by a highly vascularized matrix characterized by a noticeable abundance and density of osteocyte lacunae was recorded in the medullary cavity and within several erosion

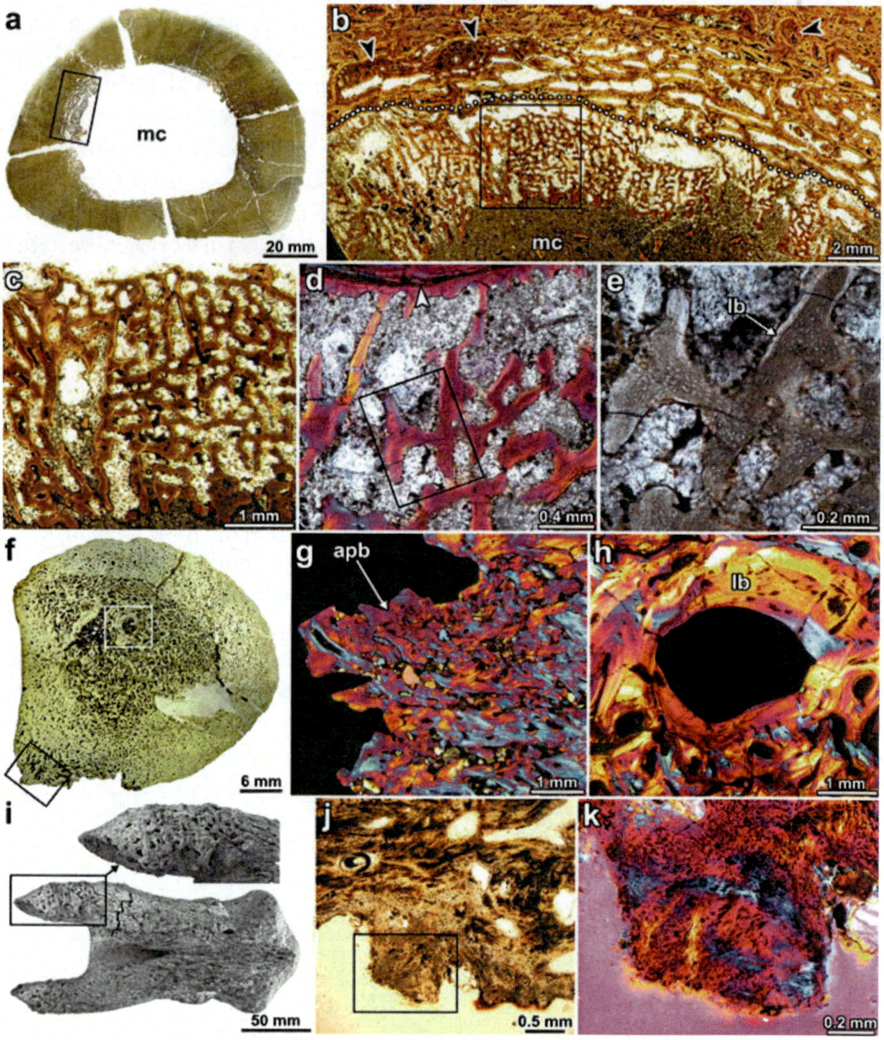

cavities of the inner cortex of an adult femur of an adult individual of this species (Fig. 3a–e). This abnormal tissue was interpreted as the result of a retrovirus-induced disease similar to the avian osteopetrosis produced in extant birds. I must note, nevertheless, that this pathological origin has been more recently challenged by Prondvai (2017), who interpret the abnormal tissue in *Mussaurus* as corresponding to medullary bone (i.e. ephemeral tissue that forms before ovulation in the medullary cavities of female bird bones and acts as a calcium source for egg shelling, Taylor and Moore 1953; Bonucci and Gherardi 1975; Schraer and Hunter 1985).

◂**Fig. 3** Pathological tissues in sauropodomorph dinosaurs. **a–e** Femur of *Mussaurus patagonicus* MLP 61-III-20–22. **f–h** Metatarsal of indeterminate titanosaur PVL 4017–127. **i–k** Caudal vertebra of indeterminate titanosaur MCS-PV 183/8. **a** Complete transverse midshaft section. **b** General view of the pathological tissue formed around the medullary cavity (box inset in **a**). The boundary between the pathological tissue and the normal bone is signed with a dotted line. Resorption cavities filled with pathological bone (black arrowheads) are also observed in the perimedullary region. **c** Detail of the pathological bone tissue (box inset in **b**). Note the high degree of vascularization. **d** Transition between pathological and normal bone tissue (white arrowhead). **e** Detail of the pathological bone tissue (box inset in d). Note the profuse density of osteocyte lacunae in the matrix and the layer of lamellar bone tissue around the vascular spaces. **f** Complete transverse section of the metatarsal. Note the rather compacted general appearance of the bone, even in the cancellous region. The external cortex exhibits a shallow excavation in portion of the sample (left bottom margin), which also exhibits a high degree of porosity. **g** Detail of the external cortex (black box inset in **f**) showing a highly fibrous primary bone tissue which has been profusely remodeled. **h** Internal cavity in the medullary region (white box inset in **f**), which is coated a thick layer of lamellar bone tissue. **i** General aspect of caudal vertebrae and detail of the right prezygapophysis (both in dorsal view). Note the irregular surface texture of the prezygapophysis. **j** General view of the external cortex at the pathological region of the bone, in this case, a prezygapophysis longitudinally sectioned. **k** Detail of the bone pathological bone tissue (box inset in j). Note the variable orientation of the fiber bundle and the absence of vascular spaces. In **a**, **g** and **h** the sediment that contacts the pathological bone has been digitally erased to enhance contrast. Pictures have been taken under normal transmitted light (a–c, e, f and j) and cross-polarized light with a lambda filter (d, g, h and k). Abbreviations: apb: abnormal periosteal bone; lb: lamellar bone; mc: medullary cavity

Among South American sauropod dinosaurs, histological information about osseous pathologies has only been recorded in titanosaurs. The first report of a possible osseous pathology was reported by Gallina (2012), who analyzed it both anatomical and histologically. This author described an abnormal tissue associated with an osseous outgrowth in a prezygapophysis from a caudal vertebra of the holotype of the listhostrotian *Bonitasaura salgadoi*, which was collected in sediments of the Bajo de la Carpa Formation (Santonian), Río Negro Province, Argentina. The abnormal tissue, interpreted as pathological, was described as poorly vascularized fibrolamellar bone different from the primary bone formed in other parts of the same element (Gallina 2012). However, an explanation regarding the possible origin of the pathological tissue was not provided by the author. The study on *Bonitasaura* holotype pathologies was later expanded by González et al. (2017). In the same study, the authors not only reanalyzed the pathological prezygapophysis, but also examined the microstructure of a large outgrowth of ovoid appearance located on the posterior side of the femoral midshaft. The abnormal periosteal tissue formed in the prezygapophysis was determined to be an infection identified by the presence of reactive new bone associated with a local widening of the subperiosteal margin and a sinus of drainage. In the case of the bony overgrowth of the femur, the abnormal structure is formed by well-vascularized fibrolamellar tissue with mostly radially oriented vascular canals, which gives a general spiculated microstructural pattern to the bone tissue. Such a structure was identified as an osteoblastic tumor.

Other studies that include microstructural analyses of pathological bones have been conducted on at least three undetermined titanosaurs (Barbosa et al. 2016; Chinsamy et al. 2016; García et al. 2017). Barbosa et al. (2016) described both the external and internal morphology as well as the histology of an isolated small bony button-shaped protuberance at the anterior articular surface of an indeterminate titanosaur from the Late Cretaceous (Adamantina Formation, Turonian–Santonian) of south-eastern Brazil. The authors describe the protuberance as having 'a micro-trabecular pattern, contrasted with that of normal trabecule' and 'The component bone was well-vascularized with copious osteocyte lacunae' (*sic* Barbosa et al. 2016, p 14). These proposed histological features, alongside the anatomical macroscopic characteristics of the anomalous structure, were considered as evidence for an osteoma (i.e. a benign, slow growing bone-forming tumor neoplasm). Another study was conducted by Chinsamy et al. (2016) on three different elements (an osteoderm, a metatarsal and a posterior caudal vertebra) from undetermined titanosaurs (possibly *Saltasaurus loricatus*) from the Upper Cretaceous of Northern Argentina. These elements showed the presence of well-vascularized endosteally formed bone tissue. Whereas the nature of the vascularized endosteal tissue in both the osteoderm and caudal vertebra cannot be determined, the authors inferred that the tissue had a pathological origin in the metatarsal (Fig. 3f–h). This bone also preserved an unusual excavation in the outer cortex. The primary bone tissue associated with this structure is composed of coarse bundles of mineralized fibers. A high degree of remodeling was reported in that area, forming a distinct cancellous tissue. The identification of the osseous pathology in the metatarsal, however, was not assessed by the authors. The last contribution on this topic was presented by García et al. (2017), who analyzed a distinct pathology in a caudal series from an indeterminate titanosaur collected in sediments of the Anacleto Formation (Campanian, Late Cretaceous), at the Cinco Saltos locality (Río Negro Province, Argentina). Among different anomalies, both the centra and neural arches of the vertebrae showed irregular 'microbubbly' texture of bone surfaces produced by periosteal reactive bone (Fig. 3i). Histological analysis of the periosteal reactive bone revealed that it is composed of an avascular and highly fibrous bone matrix, in which the intrinsic fibers are organized into thick bundles oriented in different directions (Fig. 3j, k). Based mostly on the macroscopic data, García et al. (2017) inferred that the abnormalities recorded in the specimen are consistent with an osteomyelitis (i.e. infection of the bone).

The review of the paleohistological studies related to osseous pathologies in South American sauropodomorphs reveals some important issues that deserve discussion. First, perhaps the most obvious is related with the scarce number of taxa analyzed to date (*Mussaurus patagonicus*, *Bonitasaura salgadoi* and three indeterminate titanosaurs). Second, the kind of osseous pathology can hardly ever be inferred from its bone histology. In this regard, whereas the paleopathologies reported by Chinsamy et al. (2016) could not be determined, those reported by García et al. (2017) and González et al. (2017) were mainly inferred by external and internal morphology. This situation is possibly derived from the fact that, contrary to other groups (e.g. Schultz 2003), the histological features of several types of osseous pathologies in sauropodomorph (and other) non-avian dinosaurs are still poorly characterized. Furthermore, some particular tissues (e.g. reactive periosteal fibrolamellar bone with radial vascularization) can be originated from very different pathologies. Third, in some cases the histological characterization of the pathological tissues is problematic. The study of Barbosa et al. (2016) provides a good example of this. An important issue is related to the histological sections shown by the authors (Barbosa et al. 2016, Fig. 3f, g), which are too thick for a standard histological section, precluding the identification of fine histological features. Irrespective of this, the figures appear to show secondary cancellous bone, which does not look particularly different from other, 'normal' cancellous tissues. Furthermore, the use of terms such as 'micro-trabecular pattern' makes it more difficult to understand the histological descriptions provided by the authors. The characterization of the secondary, rather than primary bone tissue, also appears to be problematic for evaluating the origin of osseous pathology. As a final comment, contrary to other paleobiological issues that can be assessed from paleohistological approaches, the informative value of osseous microstructure appears to be very limited for paleopathological studies, at least for the moment. Note that only in a few cases (e.g. Cerda et al. 2014a) has the determination of specific pathologies been derived solely from histological data.

8 Dental Attachment

Features such as tooth replacement and tooth implantation and attachment are particularly well studied in vertebrates because of their importance in the systematics and evolutionary relationships among various lineages (Bertin et al. 2018; Reisz et al. 2020). From these studies, tooth attachment to the tooth-bearing bone can be distinguished from teeth that are fused to the jaws by hard tissue (ankylosis) and those that are anchored by a periodontal ligament (gomphosis) (LeBlanc and Reisz 2013). The histological study of dental attachment tissues in non-avian dinosaurs and other extinct vertebrates has only lately been studied in depth. In this regard, important contributions in the field have been mostly on ornithischian dinosaurs (e.g. Erickson et al. 2015; LeBlanc et al. 2016, 2017; Bramble et al. 2017; Chen et al. 2018) and, to a lesser degree, theropods (e.g. Fong et al. 2016; LeBlanc et al. 2017). In spite of several studies providing information regarding sauropodomorph teeth microstructure (e.g. Sereno et al. 2007; D'Emic et al. 2013; Reisz et al. 2020), the complete set of dental attachment tissues (i.e. root cementum, alveolar bone and periodontal ligament) has only been described in a fragmentary dentary from an undetermined titanosaur from the Upper Cretaceous (Anacleto Formation) of Patagonia (Río Negro Province) (García and Cerda 2010a; García and Zurriaguz 2016) (Fig. 4a). As in other amniotes with gomphosis, the indeterminate titanosaur exhibits three tooth attachment tissues: root cementum, alveolar bone and periodontal ligament (Fig. 4b–d). The cementum consists of a very thin layer that covers the root area of the tooth. The alveolar bone occupies the inner walls of the tooth socket and is formed by a highly vascularized, woven fibered bone tissue. This tissue contains the extrinsic fibers (i.e.

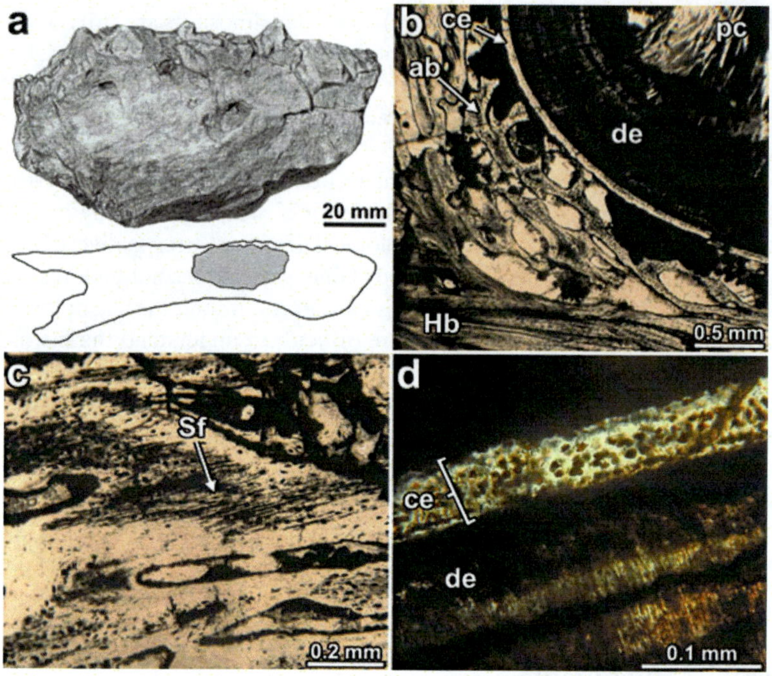

Fig. 4 Indeterminate Titanosauria indet. MCS-Pv 061 dental attachment tissues. **a** Sampled fragment of right dentary in lateral view. The scheme below indicates its relative position. Illustration made by Tech. Prebitero Pacheco. **b** Longitudinal section at the level of the root tooth. The image includes part of the root tooth with cementum, alveolar bone formed by woven fibered bone and jaw bone formed by dense Haversian tissue. **c** Detail of the alveolar bone showing Sharpey's fibers that correspond with the periodontal ligament. **d** Detail of cellular cementum at the root tooth. Pictures have taken under normal transmitted light (b and c) and cross-polarized light (d). Abbreviations: ab: alveolar bone; ce: cementum; de: dentine; Hb: Haversian bone; pc: pulp cavity; Sf: Sharpey's fibers

Sharpey's fibers) that correspond to the periodontal ligament, which anchor the tooth to the alveolus. García and Cerda (2010a) interpreted that the fine cancellous appearance of the alveolar bone and the presence of resorption lines were related to a high rate of tooth replacement in titanosaurs. Also, the absence of Sharpey's fibers in the cementum and its noticeable thinness was considered by García and Zurriaguz (2016) as evidence of a weak attachment of the tooth to the tooth-bearing element. The contributions of García and Cerda (2010a) provide the first direct evidence of gomphosis in a sauropodomorph dinosaur. It is interesting to note that, despite some differences (e.g. relative thickness of the cementum), no particular histological modifications were found in the attachment tissues of the indeterminate titanosaur in comparison with other amniotes. Such invariability reinforces the idea of conservatism in the identity of the tooth attachment tissues among amniotes (LeBlanc and Reisz 2013).

9 Early Growth

As sauropodomorph dinosaurs include the largest terrestrial vertebrates that ever existed, their life history and growth dynamics have historically received particular attention. The knowledge of these biological traits is mostly derived from paleohistological studies on long bones usually from incomplete growth series (e.g. Chinsamy 1993; Curry 1999; Klein and Sander 2007, 2008; Klein et al. 2009, 2012a; Stein et al. 2010; Curry Rogers and Kulik 2018). Most of these studies, however, do not include specimens that represent early ontogenetic stages (i.e. embryonic, hatching, early juveniles), so our knowledge about the sauropodomorph early growth is largely limited. Despite these limitations, some significant contributions focusing on South American taxa have been published during the last several years.

To date, the only South American early branching sauropodomorph for which hatching and early juvenile individuals have been recorded is for *Mussaurus patagonicus* (Fig. 5a–f). The relative abundance of articulated skeletons of hatching individuals described by Bonaparte and Vince (1979) encouraged a histological analysis of this taxon. Unfortunately, the first sections that were obtained revealed a strong alteration of the bone tissues (personal observation), which dampened further studies on these remains. Despite the general bad preservation, a longitudinal section of the distal femur epiphysis revealed a thick layer of calcified epiphyseal cartilage perforated by prominent cartilage canals (Fig. 5b). These features have also been reported in other dinosaurs, including birds, but are absent in basal archosaurs. The presence of cartilage canals indicates an acceleration of longitudinal bone growth in *Mussaurus* and other taxa that have them (Horner et al. 2001; Ricqlès et al. 2001b; Weishampel et al. 2008).

Much more informative results were obtained from four early juveniles described by Cerda et al. (2014b). These individuals are part of an important monospecific accumulation of partially articulated remains of at least 14 specimens, which are significantly larger than the post-hatchling specimens (post-hatchling femur length, 30 mm; largest juvenile femur length, 120 mm) (Pol and Powell 2007; Otero et al. 2019; Otero and Pol in press). The sampled individuals exhibited a series of histological features (i.e. higher porosity of the cortical bone, predominance of woven fibered matrix, primary osteons absent or just poorly developed, thick layer of epiphyseal calcified cartilage perforated by prominent cartilage canals) that revealed a higher early ontogenetic growth rate in this taxon (Fig. 5c–f). Furthermore, a comparison with a specimen tentatively assigned to the South African massopondylid, *Massospondylus carinatus* (BP/1/5253), indicates a higher growth rate in the South American taxon. In this regard, while an individual of *Mussaurus* would have reached a body mass of 23 *kg* in less than a year, *Massospondylus* would have needed about 2 years to reach an almost equivalent body mass (Cerda et al. 2014b). Given that maximum growth is positively correlated with adult sizes (Case 1978), this variation in growth strategies in *Massospondylus* and *Mussaurus* are expected given the larger adult body size of *Mussaurus* (Cerda et al. 2014b).

Fig. 5 Histological features of sauropodomorph dinosaurs at early ontogenetic stages. **a** Life restoration of three different ontogenetic stages of *Mussaurus patagonicus* (from left to right, adult, early juvenile and hatching). Illustration made by Jorge Blanco. **b–f** Bone histology of hatching (**b**) and early juvenile specimens (**c–f**). **b** Longitudinally oriented section of the femur distal end (PVL 4068). A thick layer of calcified epiphyseal cartilage, which is performed by prominent canals, is observed below the articular surface. **c** Detail of cartilage canals in a longitudinally oriented section of a tibia distal end (MPM-PV 1813/12). The cartilage canals are coated by a thin layer of lamellar bone. **d, e** General view (**d**) and detail (**e**, box inset in **d**) of the humerus cortex (MPM-PV 1813/2). The osseous matrix is entirely formed by woven fibered bone tissue. **f** Dorsal rib in cross section (MPM-PV 1813/8). The cortex is entirely formed by azonal, fibrolamellar bone tissue. **g** Life restoration of a titanosaur embryo. Illustration made by Tech. Jorge González. **h** Complete section of an undetermined long bone (MCF-PVPH-808). The same is formed by highly vascularized woven fibered bone. **i** Detail of the cortical bone (box inset in h). A Kastschenko's line, signaling the starting of the perichondral ossification, lines the perimedullary border. In **h** and **i**, the resin that surrounds the bone has been digitally erased to enhance contrast. Pictures have taken under normal transmitted light (c and d), cross-polarized light (b and e) and cross-polarized light with a lambda filter (f, h and i). Abbreviations: ca: cartilage canal; cc: calcified cartilage; Kl: Kastschenko's line; mc: medullary cavity; lb: lamellar bone

Regarding more derived sauropodomorph dinosaurs (i.e. Sauropoda), histological insights on the earliest phases of growth are only known for titanosaurs from the Upper Cretaceous of Argentina (García and Cerda 2010b; González et al. 2020). The Anacleto Formation of Argentina yielded embryonic titanosaurs preserved in eggs (Fig. 5g). Microstructural analysis of cranial and limb bones of these embryonic specimens revealed a high degree of cortical porosity and dense osteocyte lacunae in a context of woven bone consistent with a rapid growth rate in early development (Fig. 5h, i). A more recent review of the same sections has revealed the presence of a Kastchenko's line around the medullary cavity (González et al. 2020). The Kastchenko's line signals the position that perichondral ossification starts (Francillon-Vieillot et al. 1990) and can be used to estimate the size of the cartilage template of long bones. The presence of a Kastchenko's line also reinforces the hypothesis that these titanosaur embryos died at an advanced developmental stage but not close to hatching (Chinsamy-Turan 2005; García and Cerda 2010b).

Apart from the re-examination of osseous histology of embryonic titanosaurs, González et al. (2020) studied two femora of indeterminate juvenile titanosaurs from the Upper Cretaceous Bajo de la Carpa and Allen formations of Argentina (Neuquén and Río Negro Provinces, respectively). Both femora showed predominantly highly vascularized fibrolamellar bone tissue, interpreted as evidence of a fast growth rate. Despite the early stage of development of the specimens, secondary remodeling was reported in one of them, reinforcing previous ideas about a higher secondary remodeling rate in titanosaurs (e.g. Company 2011; García et al. 2015). The presence of a line of arrested growth in the other sample was considered as evidence for a cyclically interrupted growth strategy in the individual. Such a strategy departs from that reported for most sauropod dinosaurs, which are characterized by uninterrupted growth through most of ontogeny (Sander et al. 2004, 2011).

9.1 *Evolution and Diversity of Sauropodomorph Growth Strategies*

Based on the presence and distribution of cyclical growth marks and primary bone typology, previous studies based on long bone histology (Sander et al. 2004, 2011) have proposed that sauropodomorph dinosaurs exhibit mainly two basic growth patterns. A cyclical and comparatively slow growth has been proposed for early diverging sauropodomorphs on the basis of its relative predominance of parallel fibered bone and the presence of cyclical growth marks throughout the entire cortex (Sander et al. 2004, 2011). Conversely, more derived forms (i.e. Sauropoda) exhibits mostly uninterrupted, highly vascularized fibrolamellar bone tissue, which has been interpreted as evidence of a sustained rapid growth during most of their ontogeny, which becomes cyclical only in late stages (Sander et al. 2004, 2011). This 'continuous' growth pattern was considered as a key evolutionary innovation that, along with other anatomical and physiological modifications, allows the gigantism attainment in Sauropoda. Recent contributions focusing on the long bone histology of basal sauropodomorphs and basal sauropods from South America have revealed some distinct variations on these basic growth strategies. First, *Lessemsaurus sauropoides* and *Ingentia prima*, both members of Lessemsauridae, a clade of Late Triassic–Early Jurassic early sauropods (see McPhee et al. 2018 for a different interpretation about the phylogenetic position of the clade), exhibit evidence of a 'third' pattern, which combines both cyclical and fast growth dynamics (Cerda et al. 2017; Apaldetti et al. 2018). Second, early branching sauropodomorphs can exhibit the 'continuous' growth strategy, which was originally proposed as being unique to sauropod dinosaurs (Cerda et al. 2017). This last interpretation was based on the long bone histology of *Mussaurus* (Cerda et al. 2017). Furthermore, a more recent study reveals that *Mussaurus* is polymorphic with regard to the cyclical/continuous growth strategy (Cerda et al. in press).

Besides their cyclical growth strategy, lessemsaurids exhibit another particular feature, which is considered key to reaching gigantic size in this lineage. The long bone histology of *Lessemsaurus* and *Ingentia* reveals that the cyclical growth pattern is accompanied by the presence of highly vascularized fibrolamellar bone with reticular vascular organization, indicative of accelerated growth rates during active growth intervals (Cerda et al. 2017; Apaldetti et al. 2018). Relatively high growth rates are also proposed from the exceptional osteogenesis rates reported in lessemsaurids. In this regard, following Sander and Tückmantel (2003), lessemsaurid apposition rates were obtained by Apaldetti et al. (2018) dividing the thickness of a growth cycle by the number of days per year, using the value of 365 days per year. The obtained results revealed that lessemsaurids exhibit the highest rates for bone apposition among Sauropoda (20 and 37 microns per day in *Lessemsaurus* femora and 19.1 microns per day in *Ingentia* humerus). Apposition rates in other sauropods never reach more than 12 microns per day (Sander and Tückmantel 2003). Finally, the reticular and plexiform arrangement of the vascular spaces is another important difference between the lessemsaurids and the rest of the sauropods, which always exhibit a laminar pattern (Sander and Tückmantel 2003; Klein and Sander 2008; Cerda et al. 2017; Apaldetti et al. 2018). These histological features have also been reported in other lessemsaurids such as *Antetonitrus ingenipes*, from the Early Jurassic of South Africa (Krupandan et al. 2018). As lessemsaurids can be considered giants themselves, reaching 10 tons or even more (Apaldetti et al. 2018; McPhee et al. 2018), osteohistological information indicates that a sustained, uninterrupted growth is not a strict prerequisite for gigantism, but that an increase in appositional growth rates might be.

9.2 Future Perspectives and Concluding Remarks

Paleohistological studies on sauropodomorph dinosaurs and other vertebrates from South America have become important only in recent years. Despite this, some contributions have provided relevant information on different aspects of sauropodomorph paleobiology, being in several cases the first (and often the only) to be conducted in this group (e.g. Salgado 2003; Cerda 2009; Cerda et al. 2015b; García and Cerda 2010a, b).

The paleohistological studies on South American sauropodomorphs have increased our general knowledge about their paleobiology in at least two main fundamental aspects. First, they have allowed for the determination of the origin of particular anatomical structures (e.g. osteoderms, ossified supraneural elastic ligament). Note that while in some cases the histological data have provided clues about the ossification mechanism related to the origin and development of a well identified structure (e.g. osteoderms), in others it has allowed for the identification of the structure itself (e.g. ossified elastic ligament). Second, these studies have enhanced our knowledge regarding the growth strategies of early diverging sauropodomorphs, mostly in those taxa close to the origin of Sauropoda. In this regard, the evidence for both cyclical and fast growth dynamics in lessemsaurids and the inferred polymorphism in the growth strategy of *Mussaurus* revealed a more complex scenario in the evolution of life history traits in sauropodomorph dinosaurs.

The knowledge about other paleobiological issues that have been poorly studied to date (e.g. correlation between histological structures and soft tissues, dental implantation) can actually be enhanced by taking into account the considerable abundance and diversity of sauropodomorph remains in South America. On the other hand, paleopathological inferences based mainly or solely on paleohistological data appear to be highly problematic because of the still poorly understood relationship between specific histological structures and osseous diseases.

Despite the current advances performed in South American sauropodomorphs and other dinosaurs, the enormous potential of bone histology for paleobiological inferences is just beginning to be explored. It is evident that there is still much work to do, beginning with increased sampling. In this regard, although more than seventy valid species of sauropodomorph dinosaurs have been recognized in South America, histological data has only been provided for a much lower number. This is particularly noticeable for titanosaurs, a highly abundant and diverse lineage for which paleohistological studies have been conducted on only a few specimens, sometimes from indeterminate taxa. It is worth noting that many of the published contributions have included only a brief histological description from a single individual or a single fragmentary bone. Also, paleohistological information from other groups that are poorly represented in the South American fossil record (e.g. basal Macronaria, Rebbachisauridae) is too scarce or nonexistent. Although the bone histology of at least seven taxa has been analyzed, *Mussaurus patagonicus* is currently the best-known basal sauropodomorph in this regard.

Fortunately, the number of research projects focused on sauropodomorph paleohistology and the paleobiological implications has increased during the last several years. In this regard, there is much work in progress, including the analysis of long bone histology of dicraeosaurids, rebbachisaurids and titanosaurs and dental tissue implantation of basal sauropodomorphs, among others. The information obtained from the long bone histology of dicraeosaurid and rebbachisaurids sauropods will provide insights about the growth dynamics of these lineages, allowing us to test if they exhibit the typical sauropod growth strategy (Sander et al. 2004, 2011). Regarding titanosaurs, it will be important to determine if the new data supports the hypothesis about increased secondary remodeling in this lineage (e.g. Company 2011). Finally, the further study of dental tissue implantation will contribute to determining the degree of variation in this character, bringing valuable new data to both systematics and paleobiology. Besides the classical histological techniques employed for all the paleohistological studies reviewed here, other, non-destructive methodologies (e.g. micro-computed tomography scanning) have revealed to be strongly informative (e.g. Sanchez et al. 2014). Their future application will dramatically expand the availability of rare and/or very important specimens, for which destructive sampling is not possible. Taken into account all these considerations, future of the paleohistological research in South American sauropodomorph dinosaurs, and hence its contribution toward general sauropodomorph paleobiology, is strongly promising.

Acknowledgements I would like to express my sincere gratitude to the editors for their invitation to participate in this book. Several of the contributions reviewed here have been possible because of the curators of many institutions of Argentina, who always understood the importance of paleohistological studies and have provided permission for sampling specimens under their care. I'm also grateful to my dear colleagues, who have entrusted me to carry on several projects focused on sauropodomorph histology. Pictures shown in Fig. 2 f–h courtesy of Flavio Bellardini. I want to dedicate this work to Leonardo Salgado, my former advisor, who introduced me to the wonderful field of paleohistology. Sci-hub library and Wikipaleo group shared valuable publications for this research. Part of my research was developed with the financial support of Agencia Nacional de Promoción Científica y Tecnológica (PICT-2011-1181; 2015-1021). Jennifer Botha and Leonardo Salgado are kindly acknowledged for their comments on an earlier draft of the manuscript.

References

Apaldetti C, Martínez RN, Cerda IA, Pol D, Alcober O (2018) An early trend towards gigantism in Triassic sauropodomorph dinosaurs. Nat Ecol Evol 2:1227–1232. https://doi.org/10.1038/s41559-018-0599-y

Aureliano T, Ghilardi AM, Silva-Junior JCG, Martinelli AG, Ribeiro LCB, Marinho T, Fernandes MA, Ricardi-Branco F, Sander PM (2020) Influence of taphonomy on histological evidence for vertebral pneumaticity in an Upper Cretaceous titanosaur from South America. Cretac Res 108:104337. https://doi.org/10.1016/j.cretres.2019.104337

Barbosa FHS, Costa Pereira PVLG, Bergqvist LP, Rothschild BM (2016) Multiple neoplasms in a single sauropod dinosaur from the upper Cretaceous of Brazil. Cretac Res 62:13–17. https://doi.org/10.1016/j.cretres.2016.01.010

Bertin TJC, Thivichon-Prince B, LeBlanc ARH, Caldwell MW, Viriot L (2018) Current perspectives on tooth Implantation, attachment, and replacement in Amniota. Front Physiol 9:1630. https://doi.org/10.3389/fphys.2018.01630

Bellardini F, Cerda IA (2017) New histological data sheds light on the nature of the "dermal armor" of the enigmatic sauropod dinosaur *Agustinia ligabuei* Bonaparte, 1999. Sci Nat (naturwissenschaften) 104:1. https://doi.org/10.1007/s00114-016-1423-7

Bonaparte JF (1996) Cretaceous tetrapods of Argentina. Münch Geowiss Abh A 30:73–130

Bonaparte JF (1999) An armoured sauropod from the Aptian of northern Patagonia, Argentina. Natl Mus Nat Sci Monogr 15:1–12

Bonaparte JF, Vince M (1979) El hallazgo del primer nido de Dinosaurios Triásicos (Saurischia, Prosauropoda), Triásico Superior de Patagonia, Argentina. Ameghiniana 16:173–182

Bonucci E, Gherardi G (1975) Histochemical and electron microscope investigations on medullary bone. Cell Tissue Res 163:81–97

Bramble K, LeBlanc ARH, Lamoureux DO, Wosik M, Currie PJ (2017) Histological evidence for a dynamic dental battery in hadrosaurid dinosaurs. Sci Rep 7:15787. https://doi.org/10.1038/s41598-017-16056-3

Britt BB (1997) Postcranial pneumaticity. In: Currie PD, Padian K (eds) The encyclopedia of dinosaurs. New York Academic Press, New York, pp 590–593

Buchwitz M, Witzmann F, Voigt S. Golubev V (2012) Osteoderm microstructure indicates the presence of a crocodylian-like trunk bracing system in a group of armoured basal tetrapods. Acta Zool 93:260–280

Campos D, Kellner AWA (1999) On some sauropod (Titanosauridae) pelves from the continental Cretaceous of Brazil. Natl Mus Nat Sci Monogr 15:143–166

Carballido JL, Pol D, Cerda IA, Salgado L (2011) The osteology of *Chubutisaurus insignis* del Corro, 1975 (Dinosauria, Neosauropoda) from the "middle" Cretaceous of Central Patagonia, Argentina. J Vertebr Paleontol 31:93–110

Carballido JL, Pol D, Otero A, Cerda IA, Salgado L, Garrido AC, Ramezani J, Cúneo NR, Krause MJ (2017) An assemblage of giant dinosaurs from the mid-Cretaceous of Patagonia. Proc R Soc B 284:20171219. https://doi.org/10.1098/rspb.2017.1219

Case TJ (1978) On the evolution and adaptive significance of postnatal growth rates in the terrestrial vertebrates. Q Rev Biol 53:243–282

Cerda IA (2009) Consideraciones sobre la histogénesis de las costillas cervicales en los dinosaurios saurópodos. Ameghiniana 46:193–198

Cerda IA, Desojo JB (2011) Dermal armour histology of Aetosaurinae aetosaurs (Archosauria: Pseudosuchia) from the Upper Triassic of Argentina and Brazil. Lethaia 44:417–428

Cerda IA, Powell JE (2010) Dermal armor histology of *Saltasaurus loricatus*, an Upper Cretaceous sauropod dinosaur from Northwest Argentina. Acta Palaeontol Pol 55:389–398

Cerda IA, Salgado L, Powell JE (2012) Extreme postcranial pneumaticity in sauropod dinosaurs from South America. Paläontol Z 86:441–449

Cerda IA, Chinsamy A, Pol D (2014a) Unusual endosteally formed bone tissue in a Patagonian basal sauropodomorph dinosaur. Anat Rec 297:1385–1391

Cerda IA, Pol D, Chinsamy A (2014b) Osteohistological insight into the early stages of growth in *Mussaurus patagonicus* (Dinosauria, Sauropodomorpha). Hist Biol 26:110–121

Cerda IA, García RA, Powell JE, López O (2015a) Morphology, microanatomy and histology of titanosaur (Dinosauria, Sauropoda) osteoderms from the Upper Cretaceous of Patagonia. J Vertebr Paleontol 35:1. https://doi.org/10.1080/02724634.2014.905791

Cerda IA, Casal G, Martínez R, Ibiricu L (2015b) Histological evidence for a supraspinous ligament in sauropod dinosaur. R Soc Open Sci 2:150369. https://doi.org/10.1098/rsos.150369

Cerda IA, Chinsamy A, Pol D, Apaldetti C, Otero A, Powell JE, Martínez RN (2017) Novel insight into the origin of the growth dynamics of sauropod dinosaurs. PLoS ONE 12:e0179707. https://doi.org/10.1371/journal.pone.0179707

Cerda IA, Desojo JB, Scheyer TM (2018) Novel data on aetosaur (Archosauria, Pseudosuchia) osteoderm microanatomy and histology: palaeobiological implications. Palaeontology 61:721–745

Cerda IA, Gasparini Z, Coria RA, Salgado L, Reguero M, Ponce D, Gonzalez R, Jannello JM, Moly J (2019) Paleobiological inferences for the Antarctic dinosaur *Antarctopelta oliveroi* (Ornithischia: Ankylosauria) based on bone histology of the holotype. Cretac Res 103:104171. https://doi.org/10.1016/j.cretres.2019.07.001

Cerda IA, Pol D, Otero A. Chinsamy A (in press) Palaeobiology of the early sauropodomorph *Mussaurus patagonicus* inferred from its long bone histology. Palaeontology

Chinsamy A (1993) Bone histology and growth trajectory of the prosauropod dinosaur *Massospondylus carinatus* (Owen). Mod Geol 18:319–329

Chinsamy-Turan A (2005) The microstructure of dinosaur bone. Johns Hopkins University Press, Baltimore

Chinsamy A, Cerda IA, Powell JE (2016) Unusual bone tissue in an armored sauropod dinosaur: reproductive or pathological? Sci Rep 6:24858. https://doi.org/10.1038/srep24858

Chen J, LeBlanc ARH, Jin L, Huang T, Reisz RR (2018) Tooth development, histology, and enamel microstructure in *Changchunsaurus parvus*: implications for dental evolution in ornithopod dinosaurs. PLoS ONE 13:e0205206. https://doi.org/10.1371/journal.pone.0205206

Company J (2011) Bone histology of the titanosaur *Lirainosaurus astibiae* (Dinosauria: Sauropoda) from the Latest Cretaceous of Spain. Naturwissenschaften 98:67–78

Currey JD (1962) The histology of the bone of a prosauropod dinosaur. Palaeontology 5:238–246

Curry KA (1999) Ontogenetic histology of *Apatosaurus* (Dinosauria: Sauropoda): new insights on growth rates and longevity. J Vert Paleontol 19:654–665

Curry Rogers K, D'Emic MD, Rogers R, Vickaryous M, Cagan A (2011) Sauropod dinosaur osteoderms from the Late Cretaceous of Madagascar. Nat Commun 564. https://doi.org/10.1038/ncomms1578

Curry Rogers K, Kulik Z (2018) Osteohistology of *Rapetosaurus krausei* (Sauropoda: Titanosauria) from the Upper Cretaceous of Madagascar. J Vertebr Paleontol 38:1–24. https://doi.org/10.1080/02724634.2018.1493689

Díez Díaz V, Garcia G, Pereda-Suberbiola X, Jentgen-Ceschino B, Stein K, Godefroit P, Valentin X (2018) The titanosaurian dinosaur *Atsinganosaurus velauciensis* (Sauropoda) from the Upper Cretaceous of southern France: New material, phylogenetic affinities, and palaeobiogeographical implications. Cretac Res 91:429–456

Dodson P, Krause DW, Forster CA, Sampson SD, Ravoavy F (1998) Titanosaurid (Sauropoda) osteoderms from the Late Cretaceous of Madagascar. J Vertebr Paleontol 18:563–568

D'Emic MD, Wilson JA, Chatterjee S (2009) The titanosaur (Dinosauria: Sauropoda) osteoderm record: review and first definitive specimen from India. J Vertebr Paleontol 29:165–177. https://doi.org/10.1671/039.029.0131

D'Emic MD, Whitlock JA, Smith KM, Fisher DC, Wilson JA (2013) Evolution of high tooth replacement rates in sauropod dinosaurs. PLoS ONE 8:e69235. https://doi.org/10.1371/journal.pone.0069235

Erickson GM (2014) On dinosaur growth. Annu Rev Earth Pl Sc 42:675–697

Erickson GM, Brochu CM (1999) How the 'terror crocodile' grew so big. Nature 398:205–206

Erickson GM, de Ricqlès A, de Buffrénil V, Molnar RE, Bayless MK (2003) Vermiform bones and the evolution of gigantism in *Megalania*—how a reptilian fox became a lion. J Vertebr Paleontol 23:966–970

Erickson GM, Sidebottom MA, Kay DI, Turner KT, Ip N, Norell MA, Sawyer WG, Krick BA (2015) Wear biomechanics in the slicing dentition of the giant horned dinosaur *Triceratops*. Sci Adv 1:e1500055. https://doi.org/10.1126/sciadv.1500055

Fong RKM, LeBlanc ARH, BermanDS RRR (2016) Dental histology of *Coelophysis bauri* and the evolution of tooth attachment tissues in early dinosaurs: dinosaur dental histology. J Morphol 277:916–924. https://doi.org/10.1002/jmor.20545

Francillon−Vieillot H, Buffrénil V de, Castanet J, Geraudie J, Meunier FJ, Sire JY, Zylberberg L, Ricqlès A de (1990) Microstructure and mineralization of vertebrate skeletal tissues. In: Carter JG (ed) Skeletal Biomineralization: patterns, processes and evolutionary trends. Van Nostrand Reinhold, New York, pp 471–548

Fronimos JA (2021) Morphology and neurovascular anatomy of a titanosaurs (Dinosauria, Sauropoda) osteoderm from the upper Cretaceous of Big Bend National Park, Texas. Cretac Res 120:104670. https://doi.org/10.1016/j.cretres.2020.104670

Gallina PA (2012) Histología del titanosaurio *Bonitasaura salgadoi* (Dinosauria: Sauropoda) del Cretácico superior de Patagonia. Ameghiniana 49:289–302

Gallina PA, Apesteguía S (2015). Postcranial anatomy of *Bonitasaura salgadoi* (Sauropoda, Titanosauria) from the Late Cretaceous of Patagonia. J Vertebr Paleontol 35:1e22. https://doi.org/10.1080/02724634.2014.924957

García G, Amico S, Fournier F, Thouand E, Valentín X (2010) A new titanosaur genus (Dinosauria, Sauropoda) from the Late Cretaceous of southern France and its paleobiogeographic implications. Bull Soc Geol Fr 181:269–277. https://doi.org/10.2113/gssgfbull.181.3.269

García RA, Cerda IA (2010a) Dentición de los titanosaurios del Cretácico Superior de la provincia de Río Negro, Argentina: aspectos morfológicos, reemplazo e inserción. Ameghiniana 47:45–60

García RA, Cerda IA (2010b) Dentition and histology in titanosaurian dinosaur embryos. Palaeontology 53:335–346

García RA, Salgado L, Fernández MS, Cerda IA, Paulina Carabajal A, Otero A, Coria RA, Fiorelli LE (2015) Paleobiology of titanosaurs: reproduction, development, histology, pneumaticity, locomotion and neuroanatomy from the South American fossil record. Ameghiniana 52:29–68

García RA, Cerda IA, Heller M, Rothschild BM, Zurriaguz VL (2017) The first evidence of osteomyelitis in adult sauropod (Dinosauria: Saurischia). Lethaia 50:227–236. https://doi.org/10.1111/let.12189

García RA, Zurriaguz V (2016) Histology of teeth and tooth attachment in titanosaurs (Dinosauria; Sauropoda). Cretac Res 57:248–256. https://doi.org/10.1016/j.cretres.2015.09.006

Ghilardi AM, Aureliano T, Duque RRC, Fernandes MA, Barreto AMF, Chinsamy A (2016) A new titanosaur from the Lower Cretaceous of Brazil. Cretac Res 67:16–24. https://doi.org/10.1016/j.cretres.2016.07.001

Giménez O, Salgado L, Cerda IA (2008) Osteohistología de la viga supraneural de *Epachthosaurus sciuttoi* (Cretácico Tardío del Chubut). Natur Patag 4:1–14

González R, Cerda IA, Filippi LS, Salgado L (2020) Early growth dynamics of titanosaur sauropods inferred from bone histology. Palaeogeogr Palaeoclimatol Palaeoecol 537:109404. https://doi.org/10.1016/j.palaeo.2019.109404

González R, Gallina PA, Cerda IA (2017) Multiple paleopathologies in the dinosaur *Bonitasaura salgadoi* (Sauropoda: Titanosauria) from the upper Cretaceous of Patagonia, Argentina. Cretac Res 79:159–170

Griebeler EM, Klein N, Sander PM (2013) Aging, maturation and growth of sauropodomorph dinosaurs as deduced from growth curves using long bone histological data: an assessment of methodological constraints and solutions. PLoS ONE 8(6):e67012. https://doi.org/10.1371/journal.pone.0067012

Haines RW, Mohuiddin A (1968) Metaplastic bone. J Anat 103:527–538

Hill RV (2005) Integration of morphological data sets for phylogenetic analysis of Amniota: the importance of integumentary characters and increased taxonomic sampling. Syst Biol 54:530–547

Hill RV (2006) Comparative anatomy and histology of xenarthran osteoderms. J Morphol 267:1441–1460

Hofmann R, Stein K, Sander PM (2014) Constraints on the lamina density of laminar bone architecture of large-bodied dinosaurs and mammals. Acta Palaeontol Pol 59:287–294

Horner JR, Padian K, de Ricqlès A (2001) Comparative osteohistology of some embryonic and perinatal archosaurs: developmental and behavioral implications. Paleobiology 27:39–58

Huene F von (1929) Los saurisquios y ornitisquios del Cretáceo Argentino. Anales del Museo de La Plata (Series 2) 3:1–196

Hutton JM (1986) Age determination of living Nile crocodiles from the cortical stratification of bone. Copeia 2:332–341. https://doi.org/10.2307/1444994

Ibiricu LM, Casal GA, Martínez RD, Luna M, Svoboda EG, Cerda IA (2017a) New materials of lithostrotian titanosaurs (Dinosauria: Sauropoda) from the late Cretaceous of central Patagonia. Cretac Res 72:1–15

Ibiricu LM, Lamanna M, Martínez RD, Casal GA, Cerda IA, Martínez G, Salgado L (2017b) A novel form of postcranial skeletal pneumaticity in a sauropod dinosaur: implications for the paleobiology of Rebbachisauridae. Acta Palaeontol Pol 62:221–236

Janensch W (1947) Pneumatizitat bei Wirbeln von sauropoden und anderen saurischien. Palaeontographica 3:1–25

Janensch W (1950) Die Wirbelsaule von *Brachiosaurus brancai*. Palaeontographica (Suppl. 7) 3:27–93

Klein N, Sander PM (2007) Bone histology and growth of the prosauropod *Plateosaurus engelhardti* Meyer, 1837 from the Norian bonebeds of Trossingen (Germany) and Frick (Switzerland). Spec Pap Palaeontol 77:169–206

Klein N, Sander PM (2008) Ontogenetic stages in the long bone histology of sauropod dinosaurs. Paleobiology 34:247–263

Klein N, Sander PM, Suteethorn V (2009) Bone histology and its implications for the life history and growth of the Early Cretaceous titanosaur *Phuwiangosaurus sirindhornae*. Geol Soc Spec Publ 315:217–228

Klein N, Sander PM, Stein K, Le Loeuff J, Carballido JL, Buffetaut E (2012a) Modified laminar bone in *Ampelosaurus atacis* and other titanosaurs (Sauropoda): Implications for Life history and physiology. PLoS ONE 7:e36907. https://doi.org/10.1371/journal.pone.0036907

Klein N, Christian A, Sander PM (2012b) Histology shows that elongated neck ribs in sauropod dinosaurs are ossified tendons. Biol Lett 8:1032–1035. https://doi.org/10.1098/rsbl.2012.0778

Krupandan E, Chinsamy A, Pol D (2018) The long bone histology of the sauropodomorph, *Antetonitrus ingenipes*. Anat Rec 301:1506–1518. https://doi.org/10.1002/ar.23898

Lacovara KJ, Lamanna MC, Ibiricu LM, Poole JC, Schroeter ER, Ullmann PV (2014) A gigantic, exceptionally complete titanosaurian sauropod dinosaur from Southern Patagonia Argentina. Sci Rep 4:6196. https://doi.org/10.1038/srep06196

Lambertz M, Bertozzo F, Sander PM (2018) Bone histological correlates for air sacs and their implications for understanding the origin of the dinosaurian respiratory system. Biol Lett 14:20170514. https://doi.org/10.1098/rsbl.2017.0514

LeBlanc ARH, Reisz RR, Evans DC, Bailleul AM (2016) Ontogeny reveals function and evolution of the hadrosaurid dinosaur dental battery. BMC Evol Biol 16:152. https://doi.org/10.1186/s12862-016-0721-1

LeBlanc ARH, Brink KS, Cullen TM, Reisz RR (2017) Evolutionary implications of tooth attachment versus tooth implantation: a case study using dinosaur, crocodilian, and mammal teeth. J Vertebr Paleontol 37:e1354006. https://doi.org/10.1080/02724634.2017.1354006

LeBlanc ARH, Reisz RR (2013) Periodontal ligament, cementum, and alveolar bone in the oldest herbivorous tetrapods, and their evolutionary significance. PLoS ONE 8:e74697. https://doi.org/10.1371/journal.pone.0074697

Le Loeuff J, Buffetaut E, Cavin L, Martin M, Martin V, Tong H (1994) An armoured titanosaurid sauropod from the Late Cretaceous of Southern France and the occurrence of osteoderms in the Titanosauridae. Gaia 10:155–159

Main RP, de Ricqlès A, Horner JR, Padian K (2005) The evolution and function of thyreophoran dinosaur scutes: implications for plate function in stegosaurs. Paleobiology 31:291–314

Mannion PD, Upchurch P, Barnes RN, Mateus O (2013) Osteology of the late Jurassic Portuguese sauropod dinosaur *Lusotitan atalaiensis* (Macronaria) and the evolutionary history of basal titanosauriforms. Zool J Linnean Soc 168:98–206. https://doi.org/10.1111/zoj.12029

Marinho TS, Iori FV (2011) A large titanosaur (Dinosauria, Sauropoda) osteoderm with possible bite marks from Ibirá, São Paulo State, Brazil. In: Carvalho I de S, Srivastava NK, Strochschoen Jr O, Lana CC (Eds) Paleontologia: Cenários de Vida, Volume 4. Editora Interciência, Rio de Janeiro, pp 339–349

Martínez RDF, Lamanna MC, Novas FE, Ridgely RC, Casal GA, Martínez JE, Vita JR, Witmer LM (2016) A basal lithostrotian titanosaur (Dinosauria: Sauropoda) with a complete skull: implications for the evolution and paleobiology of Titanosauria. PLoS ONE 11:e0151661. https://doi.org/10.1371/journal.pone.0151661

McPhee BW, Benson RBJ, Botha Brink J, Bordy EM, Choiniere JN (2018) A giant dinosaur from the Earliest Jurassic of South Africa and the transition to quadrupedality in early sauropodomorphs. Curr Biol 28:3143-3151.e7. https://doi.org/10.1016/j.cub.2018.07.063

Moss ML (1969) Comparative histology of dermal sclerifications in reptiles. Acta Anat 73:510–533. https://doi.org/10.1159/000143315

O'Connor PM (2006) Postcranial pneumaticity: an evaluation of soft-tissue influences on the postcranial skeleton and the reconstruction of pulmonary anatomy in archosaurs. J Morphol 267:1199–1226. https://doi.org/10.1002/jmor.10470

Otero A, Cuff A, Allen V, Summer-Rooney L, Hutchinson PD, J R, (2019) Ontogenetic changes in the body plan of the sauropodomorph dinosaur *Mussaurus patagonicus* reveal shifts of locomotor stance during growth. Sci Rep 9:7614. https://doi.org/10.1038/s41598-019-44037-1

Otero A, Pol D (in press) Ontogenetic changes in the postcranial skeleton of *Mussaurus patagonicus* (Dinosauria, Sauropodomorpha) and phylogenetic relationships of basal sauropodomorphs. Syst. Paleo

Petermann H, Sander PM (2013) Histological evidence for muscle insertion in extant amniote femora: implications for muscle reconstruction in fossils. J Anat 222:419–436

Pol D, Garrido A, Cerda IA (2011) A new sauropodomorph dinosaur from the Early Jurassic of Patagonia and the origin and evolution of the sauropod-type sacrum. PlosOne 6:e14572. https://doi.org/10.1371/journal.pone.0014572

Pol D, Powell JE (2007) Skull anatomy of *Mussaurus patagonicus* (Dinosauria: Sauropodomorpha) from the Late Triassic of Patagonia. Hist Biol 19:125–144

Ponce D, Cerda IA, Desojo JB (2017) The osteoderm microstructure of *Doswellia kaltenbachi, Vancleavea campi*, and *Chanaresuchus bonapartei* and its implications for palaeobiology of stem archosaurs. Acta Palaeontol Pol 62:819–831

Powell JE (1980) Sobre la presencia de armadura dérmica en algunos dinosaurios titanosauridos. Acta Geol Lilloana 15:41–47

Powell JE (1987) Morfología del esqueleto axial de los dinosaurios titanosauridos (Saurischia, Sauropoda) del Estado de Minas Gerais Brasil. An. X Congresso Brasileiro Paleontol 10:155–159

Powell JE (2003) Revision of South American titanosaurid dinosaurs: palaeobiological, palaeobiogeographical and phylogenetic aspects. Rec Queen Vic Mus 111:1–173

Pretto FA, Veiga FH, Langer MC, Schultz CL (2016) A juvenile sauropodomorph tibia from the 'Botucarai Hill', Late Triassic of Southern Brazil. Rev Bras Paleontol 19:407–414

Prondvai E (2017) Medullary bone in fossils: function, evolution and significance in growth curve reconstructions of extinct vertebrates. J Evol Biol 30:440–460

Reid REH (1996) Bone histology of the Cleveland-Lloyd dinosaurs and of dinosaurs in general. Part I: introduction to bone tissues. Bringham Young Univ Geol Stud 41:25–72

Reisz RR, LeBlanc ARH, Maddin HC, Dudgeon TW, Scott D, Huang T, Chen J, Chen C-M, Zhong S (2020) Early Jurassic dinosaur fetal dental development and its significance for the evolution of sauropod dentition. Nat Commun 11:2240. https://doi.org/10.1038/s41467-020-16045-7

de Ricqlès A, Pereda Suberbiola X, Gasparini Z, Olivero E (2001a) Histology of the dermal ossifications in an ankylosaurian dinosaur from the Late Cretaceous of Antartica. Asoc Paleontol Argentina, Public Esp 7:171–174

de Ricqlès A, Mateus O, Antunes MT, Taquet T (2001b) Histomorphogenesis of embryos of Upper Jurassic from Lourinha (Portugal). C R Acad Sci Series IIA-Earth Planet Sci 332:647–656

Rothschild BM (2009) Scientifically rigorous reptile and amphibian osseous pathology: lessons for forensic herpetology from comparative and paleopathology. Appl Herpetol 6:47–79. https://doi.org/10.1163/157075409X413842

Rothschild BM, Martin LD (2006) Skeletal impact of disease. New Mexico Museum of Natural History, Albuquerque

Salgado L (2003) Considerations on the bony plates assigned to titanosaurs. Ameghiniana 40:441–456

Salgado L, Carvalho IS (2008) *Uberabatitan ribeiroi*, a new titanosaurs from the Marília Formation (Buaru Group, Upper Cretaceus), Minas Gerais, Brazil. Palaeontology 51:881–901

Sanchez S, Tafforeau P, Ahlberg PE (2014) The humerus of *Eusthenopteron*: a puzzling organization presaging the establishment of tetrapod limb bone marrow. Proc R Soc B 281: 20140299. https://doi.org/10.1098/rspb.2014.0299

Sander PM (2000) Long bone histology of the Tendaguru sauropods: implications for growth and biology. Paleobiology 26:466–488

Sander PM, Klein N (2005) Developmental plasticity in the life history of a prosauropod dinosaur. Science 310:1800–1802

Sander PM, Klein N, Buffetaut E, Cuny G, Suteethorn V, Le Loeuff J (2004) Adaptive radiation in sauropod dinosaurs: bone histology indicates rapid evolution of giant body size through acceleration. Org Divers Evol 4:165–173

Sander PM, Klein N, Stein K, Wings O (2011) Sauropod bone histology and implications for sauropod biology. In: Klein N, Remes K, Gee CT, Sander PM (eds) Biology of the Sauropod dinosaurs: understanding the life of giants. Indiana University Press, Bloomington, pp 276–302

Sander PM, Tückmantel C (2003) Bone lamina thickness, bone apposition rates, and age estimates in sauropod humeri and femora. Paläontol Z 76:161–172

Sanz JL, Buscalioni AD (1987) New evidence of armored dinosaurs in the upper Cretaceous of Spain. In: Currie PM, Koster EH (eds) 4th symposium of Mesozoic terrestrial ecosystems, Drumheller, Alberta, Short Papers. Royal Tyrrell Museum, Paleontology, Drumheller, Alberta, pp 199–204

Scheyer TM, Sander PM (2004) Histology of ankylosaur osteoderms: implications for systematics and function. J Vertebr Paleontol 20:874–893

Scheyer TM, Sander PM (2007) Shell bone histology indicates terrestrial palaeoecology of basal turtles. Proc R Soc B 274:1885–1893

Scheyer TM, Sander PM (2009) Bone microstructures and mode of skeletogenesis in osteoderms of three pareiasaur taxa from the Permian of South Africa. J Evol Biol 22:1153–1162

Scheyer TM, Sander PM, Joyce WG, Böhme W, Witzel U (2007) A plywood structure in the shell of fossil and living soft-shelled turtles (Trionychidae) and its evolutionary implications. Org Divers Evol 7:136–144

Schraer H, Hunter SJ (1985) The development of medullary bone: a model for osteogenenesis. Comp Biochem Physiol 82:13–17

Schultz M (2003) Light microscopic analysis in skeletal paleopathology. In: Ortner DJ (ed) Identification of pathological conditions in human skeletal remains, 2nd edn. Academic Press, London, pp 73–109

Schwarz D, Frey ED, Meyer C (2007) Pneumaticity and soft tissue reconstructions in the neck of diplodocid and dicraeosaurid sauropods. Acta Palaeontol Pol 52:167–1188

Sereno PC, Wilson JA, Witmer M, Whitlock JA, Maga A, Ide O, Rowe TA (2007) Structural extremes in a Cretaceous dinosaur. PLoS ONE 2:e1230. https://doi.org/10.1371/journal.pone.0001230

Sire J-Y, Donoghue PCJ, Vickaryous MK (2009) Origin and evolution of the integumentary skeleton in non-tetrapod vertebrates. J Anat 214:409–440

Stein K, Csiki Z, Curry Rogers K, Weishampel DB, Redelstorff R, Sander PM (2010) Small body size and extreme cortical bone remodeling indicate phyletic dwarfism in *Magyarosaurus dacus* (Sauropoda: Titanosauria). PNAS 107:9258–9263

Taylor TG, Moore JH (1953) Avian medullary bone. Nature 172:504–505

Tucker AD (1997) Validation of skeletochronology to determine age of freshwater crocodiles (*Crocodylus johnstoni*). Mar Freshw Res 48:343–351

Vickaryous MK, Hall BK (2008) Development of the dermal skeleton in *Alligator mississippiensis* (Archosauria, Crocodylia) with comments on the homology of osteoderms. J Morphol 269:398–422. https://doi.org/10.1002/jmor.10575

Vickaryous MK, Sire JY (2009) The integumentary skeleton of tetrapods: origin, evolution, and development. J Anat 214:441–464. https://doi.org/10.1111/j.1469-7580.2008.01043.x

Vidal D, Ortega F, Sanz JL (2014) Titanosaur osteoderms from the Upper Cretaceous of Lo Hueco (Spain) and their implications on the armor of Laurasian titanosaurs. PLoS ONE 9:e102488. https://doi.org/10.1371/journal.pone.0102488

Vidal D, Ortega F, Gascó F, Serrano-Martínez A, Sanz JL (2017) The internal anatomy of titanosaur osteoderms from the Upper Cretaceous of Spain is compatible with a role in oogenesis. Sci Rep 7:42035. https://doi.org/10.1038/srep42035

Wedel MJ (2003) Vertebral pneumaticity, air sacs, and the physiology of sauropod dinosaurs. Paleobiology 29:243–255. https://doi.org/10.1666/0094-8373(2003)029\0243:VPASAT[2.0.CO;2

Wedel MJ (2005) Postcranial skeletal pneumaticity in sauropods and its implications for mass estimates. In: Wilson JA, Curry-Rogers K (eds) The Sauropods: evolution and paleobiology. University of California Press, Berkeley, pp 201–228
Wedel MJ (2009) Evidence for bird-like air-sacs in saurischian dinosaurs. J Exp Zool A 311:611–628. https://doi.org/10.1002/jez.513
Wedel MJ, Cifelli RL, Sanders RK (2000) Osteology, paleobiology, and relationships of the sauropod dinosaur *Sauroposeidon*. Acta Palaeontol Pol 45:343–388
Weishampel DB, Fastovsky DE, Watabe M, Varricchio D, Jackson F, Tsogtbaatar K, Barsbold R (2008) New oviraptorid embryos from Bugin-Tsav, Nemegt Formation (Upper Cretaceous), Mongolia with insights into their habitat and growth. J Vertebr Paleontol 28:1110–1119
Witzmann F, Soler-Gijón R (2008) The bone histology of osteoderms in temnospondyl amphibians and in the chroniosuchian *Bystrowiella*. Acta Zool 89:1–19
Woodruff DC, Atwood NJ, Madill A (2016) The structural preservation of a titanosaurid (Dinosauria: Sauropoda) vertebral ligament. Cretac Res 60:253–266
Yates AM, Wedel MJ, Bonnan MF (2012) The early evolution of postcranial skeletal pneumaticity in sauropodomorph dinosaurs. Acta Palaeontol Pol 57:85–100
Young CC, Zhao X-J (1972) *Mamenchissaurus hochuanensis* sp. nov. Inst Vert Paleontol Paleoanthropol Monogr 8:1–30
Zylberberg L, Castanet J (1985) New data on the structure and growth of the osteoderms in the reptile *Anguis fragilis* L. (Anguidae, Squamata). J Morphol 186:327–342
Zurriaguz V (2017) New record of titanosaurian (Dinosauria: Sauropodomorpha) osteoderms from the Upper Cretaceous of North Patagonia. Cretac Res 74:175–180. https://doi.org/10.1016/j.cretres.2017.02.015
Zurriaguz V, Cerda IA (2017) Caudal pneumaticity in derived titanosaurs (Dinosauria: Sauropoda). Cretac Res 73:14–24

Sauropod Ichnology: Overview and New Research Lines from a South American Perspective

Jorge Orlando Calvo, Bernardo J. González Riga, Sebastián Apesteguía, and María Belén Tomaselli

Abstract Forty years ago, L. Branisa and G. Leonardi discovered the first sauropodomorph tracks in South America during expeditions to Toro Toro (Bolivia). Since then, numerous findings, mainly in Argentina, Bolivia, and Brazil, have increased the record. The first research lines mainly covered morphological description, ichnotaxonomic identification, and behavioral analyzes (e.g., gregariousness and speed). Some Cretaceous tracksites allowed the description of three new ichnotaxa: *Sauropodichnus giganteus* (Calvo in Ameghiniana 28:241–258, 1991), *Titanopodus mendozensis* (González Riga and Calvo in Palaeontology 52:631–640, 2009), and *Calorckosauripus lazari* (Meyer et al. in Ann Soc Geol Pol 88:223–241, 2018), the former two corresponding to Argentina, and the latter to Bolivia. A new research line named 'ichnology and comparative anatomy' has become relevant in the last years linking the skeletal information to the ichnological record, thus providing an integral interpretation of the set. This kind of approach allowed making more accurate inferences about paleoecological aspects, including limb posture, gauge, gait, speed, and size diversity recorded. In sum, this chapter aims to provide an

J. O. Calvo (✉)
Departamento de Geología y Petróleo, Grupo de Transferencia Proyecto Dino, Facultad de Ingeniería, Universidad Nacional del Comahue, Bs. As. 1400 (8300), Neuquén, Argentina

Parque Geo-Paleontológico Proyecto Dino, Universidad Nacional del Comahue, Ruta Prov. 51, km 65, lago Barreales, Neuquén, Argentina

Departamento de Geología, Facultad de Ciencias Exactas y Naturales, Universidad Nacional de la Pampa, Av. Uruguay 151 (6300), Santa Rosa, La Pampa, Argentina

B. J. González Riga · S. Apesteguía · M. B. Tomaselli
Consejo Nacional de Investigaciones Científicas y Técnicas (CONICET), Buenos Aires, Argentina
e-mail: briga@mendozaconicet.gob.ar

B. J. González Riga · M. B. Tomaselli
Laboratorio y Museo de Dinosaurios, Facultad de Ciencias Exactas y Naturales, Universidad Nacional de Cuyo. Padre Contreras 1300, Parque Gral. San Martín (5500), Mendoza, Argentina

Instituto Interdisciplinario de Ciencias Básicas, CONICET-UNCUYO, Mendoza, Argentina

S. Apesteguía
Área de Paleontología, Fundación de Historia Natural 'Félix de Azara', Universidad Maimónides, Hidalgo 775 (1405), Buenos Aires, Argentina

A. Otero et al. (eds.), *South American Sauropodomorph Dinosaurs*, Springer Earth System Sciences, https://doi.org/10.1007/978-3-030-95959-3_14

overview of the South American sauropodomorph ichnotaxonomy, the history of the discoveries, and the results of new research lines in development.

Keywords Ichnology · Sauropod · South America · Sauropodomorph tracks

1 Introduction

South America is characterized by having an abundant dinosaur fossil record that includes bones, tracks, skin marks, and eggs. This heterogeneous source of data provides abundant evidence about their life history. The first paleontologist dedicated to the study of dinosaur tracks in South America was Dr. Giussepe Leonardi, who spent most of his life to record dinosaur tracks and tracksites, thus producing the first *Glossary and Manual of tetrapod footprint paleoichnology* (Leonardi 1987). Later on, Leonardi (1994) published his *Atlas of South America Tetrapod footprints*, including all the sites and localities. After him, many researchers were incorporated to explore and study dinosaur tracks all over the continent. During the first two decades of the twenty-first century, the dinosaur track record has been improved with hundreds of dinosaur sites from Argentina, Chile, Perú, Bolivia, Uruguay, and Brazil. The morphological diversity of dinosaur tracks is high and includes both minute and large forms, made by bipedal and quadrupedal animals, herbivorous and carnivorous forms that lived in floodplain deposits, coasts of lakes, seas and rivers, deserts, among other paleoenvironments. Sauropods tracks are displayed in a wide range of environments across South America (Fig. 1), including sites like Cal Orck'o where 12.093 single imprints were counted, and the longest sauropod trackway of the world, with 380 m. However, only three ichnotaxa have been described: *Sauropodichnus giganteus* (Calvo 1991), *Titanopodus mendozensis* (González Riga and Calvo 2009), both from Argentina, and *Calorckosauripus lazari* Meyer et al. (2018), from Bolivia. In the last 30 years, Argentina has given many sauropod tracks; most of them from the Neuquén Basin in northern Patagonia. The Neuquén Basin covers part of Neuquén, Río Negro, La Pampa, and Mendoza Provinces and has provided all the sauropod tracks discovered in Argentina, except for three tracksites located at San Juan, Jujuy, and Salta Provinces (Argentine Northwest Basin). The most important areas are located on the coast of the Ezequiel Ramos Mexía Lake, at Neuquén and Río Negro Provinces (Albian–Cenomanian) of Candeleros Formation (Calvo 1991, 1999; Ortiz et al. 2008; Candia Halupczok et al. 2017; Calvo and Rivera 2018); at Paso Córdoba locality, Río Negro Province, Campanian of Allen Formation (Calvo and Ortiz 2011); at Agua del Choique, Mendoza Province, early Campanian Anacleto Formation (González Riga et al. 2015) and late Campanian–early Maastrichtian Loncoche Formation (González Riga and Calvo 2009; González Riga 2011; González Riga and Tomaselli 2019). Recently, new findings were found over the coast of the Barreales Lake, Neuquén Province, at the Campanian of Bajo de la Carpa Formation (Calvo et al. 2011) and Coniacian-Santonian of Portezuelo and Plottier formations. In this

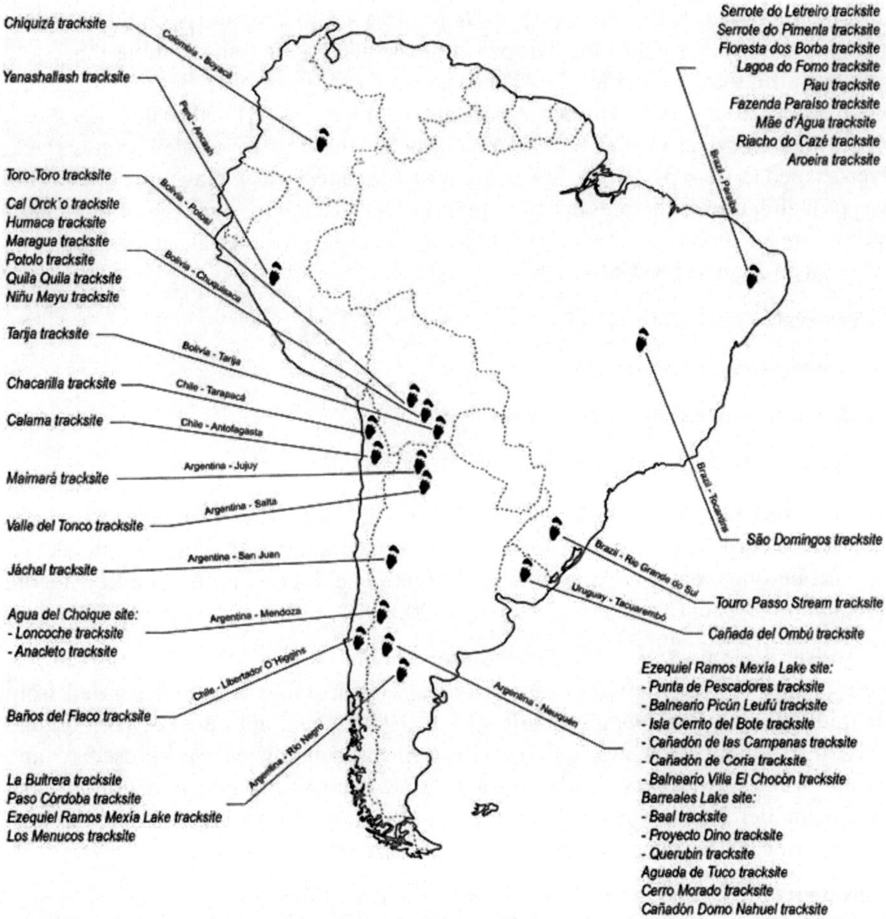

Fig. 1 Record of sauropod tracksites in South America (modified from Tomaselli et al. 2021)

contribution, we will describe the main sauropod dinosaur track and tracksites found in South America up to date, including main paleobiological aspects related to them.

2 Systematic Paleontology

Sauropodomorpha von Huene 1932

Sauropodomorpha indet. Bianco and Calvo 1999

Locality and Age Los Menucos (40°48′53″ S; 68°08′35″ W), Río Negro Province (Fig. 1). Vera Formation. Triassic.

Comments Manus track subcircular in shape (150 × 120 mm) and pes track subtriangular in shape (410 × 300 mm). The pes impressions present four striations that could belong to the toes print (Fig. 2a). These are not well-preserved, but they resemble those observed in *Tetrasauropus unguiferus* (Rainforth 2003). Other nine tracks are bigger and elliptical (1,000 × 600 mm). Finally, there is an isolated manus track, liver-shaped (510 × 310 mm). It was not possible to recognize claw impressions but was probably due to the preservation quality. The shape, size, and age of the manus and pes impressions indicate that these tracks were left by a basal sauropodomorph (Manera de Bianco and Calvo 1999).

Sauropoda Marsh 1878

Sauropodichnus Calvo 1991, 1999

Sauropodichnus giganteus Calvo 1991, 1999

Holotype MUCPv-145-146.

Locality and Age Península Nueva and Cerrito del Bote island, Neuquén Province, Candeleros Formation, Neuquén Group, Albian–Cenomanian. Other tracksites are on the surrounding area of Ezequiel Ramos Mexía Lake, Picún Leufú-Villa El Chocón, Neuquén Province; (Fig. 1) (Calvo et al. 1990, 2000; Calvo and Rivera 2018).

Diagnosis Wide trackway of a huge sauropod (Fig. 2c, d). The pace angulation is more than 60°. The internal border of left and right tracks is well separated from the midline as in *Brontopodus* (Farlow et al. 1989). Pes tracks are between 60 and 100 cm in length (Fig. 2e); the shape is subtriangular when well-preserved and circular when bad-preserved. Manus tracks are wider (40 cm) than long (25 cm), but always smaller than the pes tracks (Fig. 2f). Manus tracks are crescent-shaped with the posterior border concave and anterior border convex (Calvo 1991, 1999).

Comments This ichnospecies was the first evidence of sauropod tracks discovered in Argentina. Calvo (1989, 1991, 1999) recognized the existence of better impressions of sauropod tracks at Cerrito del Bote and El Chocón localities (Mazzetta and Calvo 2004). A complete record of all these tracksites was made by Calvo and Rivera (2018). At the trackways, the footprints are separated from the midline; for that reason, they were classified as a wide-gauge sauropod (Calvo 1991), but a new interpretation indicates that they are medium-gauge trackways (Heredia et al. 2019). Calvo (1999) interpreted that all these tracks and trackways could have been left by the sauropod *Andesaurus delgadoi* according to the wide-gauge of *Sauropodichnus giganteus* and the lack of phalanges and claws on its forefoot leaving liver-shaped manus track morphology. *Sauropodichnus giganteus* validity was confronted by Lockley et al. (1994a) and Krapovickas et al (2020).

Sauropoda Marsh 1878

Iguanodonichnus Casamiquela and Fasola 1968

Iguanodonichnus frenkii Casamiquela and Fasola 1968 (sauropod tracks sensu Moreno and Benton 2005)

Fig. 2 Sauropod tracks and trackways from Río Negro and Neuquén Provinces (Argentina). **a** Early sauropodomorph pes and manus track from Los Menucos tracksite. Arrows show toe impressions (Manera de Bianco and Calvo 1999). **b** Early sauropodomorph prints (arrow) and a trackway of Triassic mammal (Manera de Bianco and Calvo 1999). **c** *Sauropodichnus giganteus* trackway at Punta de Pescadores, Picún Leufú (MUCPv-145, Holophototype) (Calvo 1991). **d** *Sauropodichnus giganteus* trackway at Punta de Pescadores, Picún Leufú (MUCPv-146, Hypophototype) (Calvo 1991). **e** Sauropod pes track from Balneario Picún Leufú; 100 cm in diameter, 20 cm depth. **f** Manus print of *Sauropodichnus* from Balneario Villa El Chocón (Calvo and Rivera 2018)

Locality and Age Termas del Flaco, San Fernando, Región del Libertador Bernardo O'Higgins, Chile. Baños del Flaco Formation (Tithonian, Upper Jurassic).

Comments Originally, the Termas del Flaco footprints were attributed to iguanodontids (Casamiquela and Fasola 1968). The tracks were later redescribed by Moreno and Benton (2005), who attributed them to sauropods on the basis of the following features: step angles average less than 110°; pes prints intersect the trackway midline; pes prints are longer than wide, with the long axis rotated outward; the claw impression of digit I is prominent and directed forward; and claws on digits II, III, and IV are strongly reduced. The authors mentioned that the main features of *I. frenkii* are similar to the ichnogenus *Parabrontopodus* (Lockley et al. 1994a). *I. frenkii* is attributed to a basal sauropod, diplodocoid or basal macronarian, which are all narrow-gauge sauropods.

Titanopodus González Riga and Calvo 2009

Titanopodus mendozensis González Riga and Calvo 2009

Holotype Original manus and pes tracks in situ, trackway AC-1, Agua del Choique geological section (Fig. 3e).

Reference Material Trackway AC-4 and AC-3 from the same tracksite.

Locality and Age Malargüe Department, Mendoza Province, Argentina. Loncoche Formation (late Campanian–early Maastrichtian), Tracksite Lo-1, Agua del Choique geological section.

Revised Diagnosis Sauropod tracks of medium size (pes length about 46–50 cm) characterized by the following unique combination of characters: a wide-gauge trackway [manus and pes trackway ratios (MTR and PTR) of 24–22.9% and 25.5–24%, respectively]; manus tracks without claw impressions and strongly rotated outward (25–48°); manus with asymmetrical crescent shape and acuminate or thinner external border; pes tracks with subtriangular to subcircular shape, longer than wide, and rotated slight negative (outward); heteropody of manus-pes of 1:3; outer limits of trackway defined, in some cases, by manus tracks; manus-pes distance about 1.7 times pes footprint length; stride length around 4.9 times pes track length; step angle 92–110° (González Riga and Calvo 2009; González Riga and Tomaselli 2019).

Comments This ichnospecies was the first ichnotaxon described and published in South America from the late Campanian–early Maastrichtian and is a good example of wide-gauge trackways produced by derived titanosaurs. In relation to the diagnosis, *Titanopodus* is characterized by a unique association of characteristics and not just one. So far, no tracks from another site have been assigned to *Titanopodus*. Up to now, two unnamed manus tracks show an asymmetrical shape, similar to *Titanopodus*. However, this shape is not present in other manus tracks here; moreover, the trackways do not exhibit other ichnological features of *Titanopodus*. *Titanopodus* is associated with a narrow-gauged trackway and assigned to large-sized rebbachisaurids, despite being considered largely extinct by that time (Leanza et al. 2004). *Titanopodus*' trackmakers were specimens of 12.2 and 13.3 m length.

Fig. 3 Photographs of sauropod tracks and trackways from Neuquén and Mendoza Provinces (Argentina). **a** Querubin tracksite, Bajo de la Carpa Formation (Coniacian) at Lago Barreales, Neuquén at level III. **b** Querubin tracksite at Lago Barreales at level I. **c** Titanosaurid sauropod manus and pes tracks from Querubin tracksite at level III. **d** Paso Córdoba tracksite, Allen Formation, Río Negro Province (Calvo and Ortiz 2011). **e** *Titanopodus mendozensis* (González Riga and Calvo 2009) from the late Campanian–early Maastrichtian, Loncoche Formation, Agua del Choique, Mendoza. Holotype's trackway and detail of the manus track with their characteristic acuminate external border. **f** Manus and pes tracks of the morphotype 1 (González Riga et al. 2015) from the early Campanian Anacleto Formation, Agua del Choique, Mendoza. Note the claw impressions of the pes track

The gleno-acetabular distance calculated is 2.25 and 2.45 m, respectively, following the formula of González Riga (2011). The author interpreted that these speeds and some taphonomical features like very deep tracks (10–25 cm), prominent rims, and some distorted and elongated shapes indicate the capacity of derived titanosaurs for walking effectively over a very wet and slippery substrate. González Riga and Tomaselli (2019) interpreted that these conditions did not allow the preservation of foot digit impressions; furthermore, these authors recognized two different gait patterns (see Sect. 10.2).

Calorckosauripus Meyer et al. 2018

Calorckosauripus lazari Meyer et al. 2018

Comments This ichnospecies is characterized by coupled oval-shaped manus and large pes imprints (pes length: 70 cm), with a slender track shape, an outward position of the manus and a narrow-gauge trackway configuration with an estimated gleno-acetabular distance of 2.8–3.2 m). Manus impressions are circular, rather blunt, with the presence of marked impressions of digit I and V (sensu Meyer et al. 2018). These tracks have an anterior–posterior lengthening with a broad anterior part. Pes prints are oval. An interesting matter here is that when the specimen walks at low speed, the manus seems 'primitive' or in crescent shape as a consequence of the pes pushing a load of mud to cover the center of the manus imprint with sediments (e.g., Fig. 4a, from Toro Toro), but when a change of speed exposes the full manus imprint, this is rounded to slightly angled as in most derived Maastrichtian titanosaurs.

3 Sauropod Tracks from Argentina

3.1 Neuquén Province

Except for Cerro Morado, Cañadón Domo Nahuel and Aguada de Tuco tracksites, the rest of the sauropod tracksites are around the Ezequiel Ramos Mexía and Barreales lakes. The tracks from Cerro Morado and Cañadón Domo Nahuel tracksites are Lower Cretaceous in age and belong to the Bajada del Agrio Group (Zavala and Ponce 2011), while the rest of the tracks are 'mid' and Upper Cretaceous in age and belong to the Neuquén and Malargüe Groups.

3.1.1 Cerro Morado Tracksite

Canudo et al. (2017) described a level with a large number of subcircular tracks, assigned to sauropods, preserved as undertracks and concave epirreliefs in sandstone of about 8 cm thickness. The tracks belong to the Rayoso Formation, late Aptian–Albian in age. A trackway composed of four consecutive tracks has been recognized. Apparently, the rest of tracks could belong to other trackways, but it has not been

Fig. 4 Sauropod tracksites from Bolivia, Brazil and Perú. **a** Titanosaurian trackway in Toro Toro, Bolivia, showing how the mud displaced by the pes enters in the manus track center. **b** Titanosaurian trackway reaching the base of the Cal Orck'o cliff, Bolivia. **c** Titanosaurian trackway in Toro Toro with well-preserved mud rims. **d** Manus track with the liver-shaped morphology at Floresta dos Borba tracksite, Brazil. **e** Sauropod manus and pes tracks at Floresta dos Borba tracksite, Brazil (Calvo et al. 2002). **f** Lower Cretaceous sauropod tracks on a wall at 4.840 m above the sea level, Yanashallash tracksite, Perú (Vildoso et al. 2011b)

possible to recognize them due to the superposition of tracks and the small size of the outcrop. Pes tracks are subtriangular in shape (average length and width of 41 cm and 27.5 cm, respectively). Manus tracks have an oval shape (7–11 cm length and 11–15 cm width) and are partially obliterated by the overlapping of the pes.

3.1.2 Cañadón Domo Nahuel Tracksite

This tracksite, also described by Canudo et al. (2017), shows a sauropod trackway composed by three pes tracks and two manus tracks preserved as concave epirreliefs. Pes tracks are subtriangular in shape (20–24 cm length and 17–23 cm width), with a narrow heel turned toward the midline, while manus tracks are suboval and large (14–16 cm length and 17–19 cm width). This outcrop also belongs to the Rayoso Formation (late Aptian–Albian).

3.1.3 Aguada de Tuco Tracksite

One sauropod tracksite has been found at this locality. It is located on the west of Neuquén Province at the locality of Aguada de Tuco (37°26′43″ S, 69°54′02″ W) (Fig. 1). The trackway consists of five manus-pes track set of a narrow-gauge trackway (Heredia et al. 2019). The pes track shape is not uniform, varying from rhomboidal to subcircular or to subrectangular in shape. The posterior margin of the track, narrower than the anterior, is subtriangular to subcircular and lacks digit impressions. The pes imprints are longer than wide (average length and width of 98 cm and 82 cm, respectively) and the major axis is almost parallel to the trackway midline (Heredia et al. 2019). It was related by the authors to Rebbachisauridae.

3.1.4 Ezequiel Ramos Mexía Lake Site

The area around the Ezequiel Ramos Mexía Lake corresponds to numerous and important tracksites, one of the richest on the Neuquén Basin. In fact, on the coast of this lake were recognized the first sauropod tracks studied in Argentina (Calvo 1989). The first description on sauropod tracks was done on the southern part of the lake at Picún Leufú area (Calvo 1991). After that, more materials were found at Cerrito del Bote (39°31′35″ S, 69°08′29″ W), Punta de Pescadores (39°31′44″ S, 69°09′12″ W), Cañadón de Coria (39°16′00″ S, 68°50′55″ W), and Balneario Villa El Chocón (39°16′20″ S, 68°49′29″ W) in the Neuquén Province (Calvo et al. 1990, Calvo and Salgado 1995; Mazzetta and Calvo 2004; Calvo 2007). A review of all dinosaur tracks on the east coast of E.R. Mexía was done by Calvo and Rivera (2018). The only sauropod ichnospecies described for this locality is *Sauropodichnus giganteus* (Calvo 1991, 1999) (see Sect. 2).

Punta de Pescadores Tracksite

This is the most important tracksite of the Ezequiel Ramos Mexía Lake because all the studied holotype tracks (theropods, ornithopods, and sauropods) were described here (Calvo 1991, 1999). Two main tracksites were identified as *Sauropodichnus* (Museo Universidad Nacional del Comahue, Neuquén, Argentina, MUCPv-145-146). Moreover, new findings of isolated tracks and a couple of tracks are widespread on the coast of the lake. Pes tracks are 80 cm in diameter and manus tracks around 50 cm (Fig. 2c). Some isolated tracks are liver-shaped with 70 cm width and 38 cm long (Calvo and Rivera 2018).

Balneario Picún Leufú Tracksite

On this site, many sauropod tracks were found (Calvo and Rivera 2018). The best-preserved one is 100 cm in diameter and 20 cm depth (Fig. 2e). No phalanx impressions are present. One of the trackways is 220 cm wide, indicating a wide-gauged sauropod, and the distance between manus and pes is 33 cm. The pace is 165 cm. Pes impressions are subtriangular or rounded, depending on the preservation.

Isla Cerrito del Bote Tracksite

One trackway with 12 tracks and 6 m in length was recognized (Calvo 1999; Mazzetta and Calvo 2004). The separation between tracks is 60 cm. The pace angulation for the manus tracks is between 75° and 95° and for the pes tracks is between 100° and 120°. The external width is 140 cm. The pace length is 2.1 m (Mazzetta and Calvo 2004). The pes impressions are bigger than those of the manus, and they have a subtriangular shape, 70 cm in length, and 60 cm in width. The manus impressions are liver-shaped, and they are 25 cm in length and 40 cm in width.

Cañadón de las Campanas Tracksite

Many badly preserved sauropod tracks were recognized (Krapovickas 2010). The manus prints are rounded or liver-shaped without digit impressions. According to the morphology, they were assigned to *Sauropodichnus* (Calvo and Rivera 2018) but Krapovickas et al. (2020) assigned them to *Brontopodus patagonicus* and cf. *Brontopodus*, all of them recorded in playa lake and wet interdune deposits (Krapovickas 2010). Some specimens recorded by one of the authors (S.A.) reach 100 cm in pes diameter.

Cañadón de Coria Tracksite

Many isolated and badly preserved tracks are on this creek. Undetailed, rounded pes print remains were found on the creek. Manus print is liver-shaped with a size between 40 and 80 cm width (Calvo and Rivera 2018).

Balneario Villa El Chocón Tracksite

Nine trackways have been recognized (Calvo and Rivera 2018). Some of them are represented by one or two tracks. Manus tracks are liver-shaped (Fig. 2f) similar to those described for *Sauropodichnus giganteus* (Calvo 1999). Pes tracks are subtriangular.

3.1.5 Barreales Lake Site

On the east and northeast coast of the Barreales Lake, seven dinosaur tracksites were discovered in the last ten years; however, only one was briefly published (Calvo et al. 2011) (Fig. 3a–c). These tracksites were found at different formations and both theropod and sauropod as well as ornithopod tracks have been recognized. Sauropod tracks were identified at Proyecto Dino tracksite (Plottier Formation), Querubin tracksite (Bajo de la Carpa Formation), and Baal tracksite (Portezuelo Formation), but these are not well-preserved.

Baal Tracksite

The site shows some sauropod tracks in very bad condition, just as holes on the sedimentary rocks, and few data could be obtained. The tracks are on the Northeast coast of Barreales Lake, 92 km northwest of Neuquén city (38°27′04″ S, 68°44′40″ W). These rocks belong to the middle section of the Portezuelo Formation. The paleoenvironment is interpreted as an anastomosed river where all tracks were left near a floodplains deposit (Calvo et al. 2015).

Proyecto Dino Tracksite

The dinosaur site is located 70 m west of the buildings of the Proyecto Dino Geo-paleontological Park (38°27′09″ S, 68°43′22″ W), 90 km northwest of Neuquén city. These rocks belong to the Plottier Formation. The original tracks are in a red and green claystone, and they were filled with gray medium-size sandstone. Tracks are not well-preserved. They are visible in lateral and bottom views because the platform forms a balcony. Twelve dinosaur tracks were identified; six of them are visible in lateral view and the other six from the bottom view. They are rounded in shape and some deeper than others without impression of toes and general morphology (Calvo et al. 2021). Some of the shallow tracks might belong to manus prints and the deeper ones to pes prints. They are separated 30 cm apart. The displacement of the clay on the border of the tracks is well-marked on deeper tracks.

Querubin Tracksite

The site is placed on the East coast of Barreales Lake, 80 km northwest of Neuquén city (38°31′56″ S, 68°40′48″ W). All tracks are on a gray and green, medium to fine sandstones belonging to the Santonian age of lower levels of Bajo de la Carpa Formation (Calvo et al. 2016). Sauropod tracks are abundant in this site, and they were left by sauropods of different sizes, probably from the same ichnotaxonomic group (Fig. 3a, b). The site is composed by four platforms with hundreds of dinosaur tracks. In some places, it is possible to detect some small theropod and probably ornithopod tracks. A brief description of the site was made by Calvo et al. (2011). Sauropod trackways are difficult to identify because there are overprinting tracks indicating a movement of many sauropods in the area. Manus tracks are liver-shaped, smaller in size than pes tracks (Fig. 3c). In general, it is not possible to see data on phalanx and claw impressions on both manus and pes prints (Calvo et al. 2021). In one of the trackways, the tracks are separated from the midline, like in the other wide-gauged trackways, camarasaurids, brachiosaurids, and titanosaurids (Farlow 1992; Lockley et al. 1994a). Pes prints have subtriangular shape similar to *Sauropodichnus* from Albian–Cenomanian of Neuquén Province (Calvo 1999). Manus tracks are liver-shaped; they have pointed and convex internal borders different from those strongly convex of *Sauropodichnus* or to those more acuminate presents in *Titanopodus* (González Riga and Calvo 2009).

3.2 *Río Negro Province*

3.2.1 Los Menucos Tracksite

The early branching sauropodomorph tracks are scarce in South America; however, some in situ tracks were identified at Río Negro Province in Patagonia (Fig. 2a, b), Argentina (Manera de Bianco and Calvo 1999; Calvo 2007). Likewise, sauropodomorph dinosaur manus and pes tracks were also found (see Sect. 2).

3.2.2 Ezequiel Ramos Mexía Lake Tracksites

In spite of the abundance of sauropod tracks at the Neuquén Province coast, only two dinosaur tracksites were recorded on the southeast coast of the E.R. Mexía at Río Negro Province, the Salas tracksite briefly mentioned by Apesteguía (2005: Fig. 15.3a), Apesteguía et al. (2010: Fig. 6), and de Valais et al. (2020: Fig. 3d, e), about 10 km southeast of the Villa El Chocón town, and the Bahía Saihueque tracksite (39°32′44″ S, 68°58′55″ W) about 25 km south of the Villa El Chocón town (Ortiz et al. 2008; Novas 2009; de Valais et al. 2020: Fig. 3f), both in surfaces of the Candeleros Formation. The Salas tracksite shows a heavy sauropod trampling (de Valais et al. 2020) with details of a highly heteropodic manus-pes print set, related by Apesteguía et al. (2010) to rebbachisaurid sauropods. There are several well-preserved sauropod trackways at Bahía Saihueque tracksite, and they are associated with pterosaur tracks. Manus and pes prints have toe impressions. The axis of each track is directed outside from the midline at 45°. The midline trackway is tangential to all tracks, and they do not cut the midline as in other narrow-gauged sauropods as diplodocimorphs. The manus tracks are similar to *Brontopodus*, although in some tracks the anterior border has a prominence different to that taxon. The pes impression shows digit I–IV with the same size; digit V is reduced and laterally directed (Ortiz et al. 2008). These sauropod tracks are different from those of *Sauropodichnus*, but they were probably left by a wide-gauge sauropod as a basal titanosaurian.

3.2.3 La Buitrera Tracksite

The La Buitrera Paleontological Area is a fossiliferous site located on the East coast of Ezequiel Ramos Mexía Lake 80 km northwest of Neuquén city (38°31′56″ S, 68°40′48″ W). Tracks in the area are found in dune slipfaces, wet interdunes, and sandsheet deposits of the Kokorkom Desert, preserved in the Candeleros Formation (Candia Halupczok et al. 2017). They are preserved mostly in cross-section but some others in plain view. Those tracks in cross-bedded structures bear a distinctive rim on the down-slope side of the footprint. The tracks on sand sheets and wet interdune deposits look wider and deeper, and mostly symmetrical. Tracks found in plain view correspond to an ornithischian trackmaker that has affinities with *Limayichnus major* (Calvo 1991). Sauropod tracks were also identified in the area but remain unpublished. They show digit impressions as well as scaly skin marks.

3.2.4 Paso Córdoba Tracksite

In Paso Córdoba locality, crossing the Río Negro River, several sauropod tracks were found and briefly described (Fig. 3d) (Calvo and Ortiz 2011). The impressions were left at the bottom of the Allen Formation (Maastrichtian), Malargüe Group. The tracksite is 6 × 10 m, and it is covered with many sauropod tracks, probably left by a herd (39°07′28″ S, 67°36′55″ W). Manus tracks are semilunar in shape, and the pes impressions are rounded. It is not possible to see finger or claw impressions.

3.3 Mendoza Province

3.3.1 Agua del Choique Site

Agua del Choique site comprises two tracksites and is located in the Late Cretaceous strata of Mendoza Province (Fig. 1). It was discovered by B. González Riga and has *ca.* 400 sauropod tracks and some tridactyl theropod footprints (González Riga and Calvo 2006, 2009). Two different ichnocoenoses are recorded, one in the Anacleto Formation (early Campanian) and the other in the Loncoche Formation (late Campanian–early Maastrichtian).

Anacleto Tracksite

In the Anacleto Formation, titanosaurian tracks are preserved in facies of poorly drained floodplain deposits with ephemeral channels and tabular sandstone bodies deposited during flash flood episodes (González Riga et al. 2011, 2013). An exceptionally preserved pes track with three claw impressions laterally directed has been studied (González Riga et al. 2015). It is comparable with articulated and complete pedes of an unnamed titanosaur discovered in the same Province, Plottier Formation, Neuquén Group (Fig. 3f). During the editing of this chapter, a new ichnotaxon named

Teratopodus malarguensis (Tomaselli et al. 2022) was erected from new studies made in this tracksite.

Loncoche Tracksite

In the Loncoche Formation, the ichnocoenosis is dominated by titanosaur tracks assigned to *Titanopodus mendozensis* González Riga and Calvo (2009) that correspond to wide-gauge trackways (Fig. 3e). The *Titanopodus* trackmakers were titanosaurs that walked in very humid substrates of marine marginal environment of the first Atlantic transgression of the region. Theropod tracks were also preserved in the same level (Ortiz David et al. 2012; González Riga et al. 2015).

3.4 San Juan Province

3.4.1 Jáchal Tracksite

Sauropod tracks have been discovered in the type locality of Ciénaga del Río Huaco Formation (Maastrichtian), in the Jáchal department (30°13′58″ S, 68°37′12″ W), located north of the San Juan Province (Fig. 1). The manus tracks are kidney-shaped with a significant circular rim. The pes tracks are almost circular in shape; some of them have four anterior incisions that may correspond to short and thick digit marks. There is no evidence of claws or heel prints. The tracks are in the transitional environment levels between the Middle Member (fluvial environment) and the Upper Member (lacustrine environment), in a layer of medium to thick sandstone, somewhat conglomerate, with the presence of salts, corresponding to a typical sandy bar of river channel (Contreras et al. 2010).

3.5 Jujuy Province

3.5.1 Maimará Tracksite

This site is located on the eastern side of the Quebrada de Humahuaca (23°37′30.92″ S, 65°23′56.07″ W), Jujuy Province (Cónsole-Gonella et al. 2017) (Fig. 1). There, the deposits of the Yacoraite Formation correspond to a lagoon-type littoral system. The tracks may have been preserved in the presence of water; therefore, both manus and pes impressions lack clear digital or claw impressions. The footprints are longer than wide, with an average width of 42 cm, and an average length of 71 cm. Pes imprints with subtriangular posterior edge, located posteriorly to the manus imprints. The manus imprints have no clear details and are about half the size of the footprints. They are semicircular with a major axis of 33 cm rotated outwards relative to the trackways midline. The trackways are moderately wide-gauge. There is variation in the size of manus-pes tracks, probably due to slight or incipient sediment collapse and not to morphological differences among trackmakers.

3.6 Salta Province

3.6.1 Valle del Tonco Tracksite

A new sauropod trackway was described in the Yacoraite Formation (Maastrichtian–Danian), in the Quebrada de El Candado locality, located southwest of the city of Salta (Fig. 1). The trackway consists of twelve manus-pes sets tracks preserved as convex hyporelief (natural casts). The pes tracks are longer than wide (averages of length and width: 67 and 42 cm, respectively) and have a general rhomboidal shape, with a subtriangular posterior edge, and lack of lateral notch. They are located posteriorly with respect to the manus tracks. Some pes tracks show the digit imprints laterally located. The manus tracks are subrounded to rectangular shaped, with at least two short and posteriorly oriented digit marks. They display about half of the size of the pes tracks (averages of length and width are 21 and 32 cm, respectively). The trackway-bearing surface belongs to an inverted section from the uppermost Cretaceous Caliza Amblayo Member, the lower unit of the Yacoraite Formation in that area. This formation is composed mainly of gray carbonates with varied depositional textures that are largely dolomitic; in some of the proximal sections, calcareous sandstones and pelites are the major component (Díaz-Martínez et al. 2018).

4 Sauropod Tracks from Bolivia

4.1 Toro Toro Tracksite

The most famous dinosaur track locality in Bolivia is in Toro Toro, at the Potosi department (e.g., Apesteguía et al. 2011). Hundreds of tracks were left by theropods, sauropods and ornithischians (Leonardi 1994). Some sauropod and theropod trackways are almost parallel, showing the pattern expected for either theropods chasing sauropods or environmental features such as walking on the coast of a lake or a river. Toro Toro sauropods are very homopodal (1:1). The gauge in the Toro Toro sauropods is about 31% (Fig. 4a, c) (Meyer et al. 2021).

4.2 Cal Orck'o Tracksite

Another important site in Bolivia is the Cal Orck'o tracksite, situated in the quarry of FANCESA (*Fábrica Nacional de Cementos*). It bears nine levels with dinosaur tracks situated in the middle part of the El Molino Formation (Maastrichtian) at the shore of a big lake or seaway where no less than ten dinosaur types were recognized. Sauropod tracks are abundant, and Meyer et al. (2021) declares to have recognized two forms (Fig. 4b). An unnamed morphotype (morphotype E) is considered by Meyer et al. (2021) as a rather indistinct morphotype (but rare as declared by authors) of derived titanosaur, with more rounded and axially compressed pes imprints (pes length: 50 cm), and more rounded manus. The most relevant species is another sauropod morphotype (F) considered by its authors as a basal titanosaurid sauropod after the presence of a horseshoe manus impression with marked digits I and V. However, the pace changes reveal a more rounded manus impression. Cal Orck'o sauropods bear a low heteropody (1:2). The gauge in Cal Orck'o sauropods is about 36.24%. The only ichnospecies described is *Calorckosauripus lazari* Meyer et al. (2018).

4.3 Quila Quila, Humaca and Niñu Mayu Tracksites

A series of sites in the outer rim of the Maragua Syncline captured the attention of several ichnologists since their discovery by the local fossil association SOCIUPA (*Sociedad Universitaria de Paleontología*) by 1995 and documented by Mr. David Keremba Mamani. Particularly, the Humaca tracksite (Ore Pajlla) published by Lockley et al. (2002) (19°04′40.59″ S, 65°28′0.45″ W) represents a vast majority of titanosaur tracks imprinted in gray rocks which authors attribute to the Chaunaca Formation. Meyer et al. (2021) attribute these tracks to *Calorckosauripus* for showing impressions of digits I and V (crescent shape), but this could easily be an artifact of pes adding sediments to the central part of a rounded manus track. In Niñu Mayu, there are several trackways, one of them isolated, belonging to a large titanosaur (pes length: 70 cm) with low heteropody. The Kimsa Saruska site, close to Niñu Mayu, shows three very complete huge tracks of a heteropodic titanosaur with triangular pedes (narrow heels) provided of five digits clearly marked. Manus are here rounded and much smaller. Humaca sauropods show a moderate heteropody (1:2–1:3), and their purported juvenile tracks (after Lockley et al. 2002) are mid-gauged (pes trackway ratio of 45.30%). They are completely different from the wide-gauged *Titanopodus* from Argentina (manus trackway ratio and pes trackway ratio of 18–22 and 26–31%).

4.4 *Potolo Tracksite*

This site is located on the hill right above Potolo town. Trackways are parallel and produced by two middle-sized sauropods, with low heteropody, as common in derived titanosaurs, especially young specimens, about 5 m in total length. Both trackways are composed of 48 and 13 tracks each. Every pes is 30 cm width and 40 cm in length. Manual tracks are similar but inverted, being 40 cm width and 30 cm in length. The pace is 67 cm. Additionally, there are four tracks of a much larger, 12 m specimen which overprinted some of the smaller tracks (S.A. pers. observ.).

4.5 *Tarija Tracksite*

A brief outcrop was located in 2019 close to the Tambo River, in the Entre Ríos Municipality, 74 km from Tarija city (21°29′58.6″ S, 64°13′12.25″ W). The site shows three levels, and in the third, there are several trackways of large adult sauropods (pes length: 100 cm) and many more of juveniles (pes length: 22–44 cm), with manus deformed by the pes imprint (Méndez Torrez 2019).

5 Sauropod Tracks from Perú

Dinosaur tracks and tracksites in Peru are very scarce; however, on the Conococha–Yanacancha route, many important dinosaur tracksites belonging to theropod, ornithopods, and sauropod tracks are on display in many almost vertical walls (Vildoso et al. 2011a). The Yanashallash locality is the only one with sauropod tracks (Vildoso et al. 2011b).

5.1 *Yanashallash Tracksite*

This is the only locality in Peru with recorded sauropod tracks (Vildoso et al. 2011b). Tracks were left on the Cretaceous rocks of the Pariatambo Formation (early to middle Albian) at Yanashallash locality, 70 km SE of Huaraz city (Fig. 1). The main outcrop includes more than 25 tracks. At least one trackway was recognized, that includes two pes and two manus prints (Fig. 4f). The pes tracks are oval-shaped with the longest axis anteromedially directed. The pes length is 58 cm while the width is 55 cm. They show four toe impressions with rounded ends. The manus track area is oval in shape, and wider (51 cm) than longer (39 cm). The pes pace angulation is 110°, and the manus pace angulation is 120–125°. The internal border of the manus and pes tracks is closer to the contacts with the midline, showing that this is a medium-gauged sauropod. According to the track morphology and trackway evidence, the sauropod that walked on this shallow carbonate environment close to the Pacific coast was probably a diplodocoid. This sauropod tracksite is placed today on a vertical wall at 4.840 m above the sea level, being the highest dinosaur tracksite for southern continents.

6 Sauropod Tracks from Chile

Most of the dinosaur tracksites from Chile are located in the north of the country; mainly on Chacarilla and Calama localities (Fig. 1) (Blanco et al. 2000; Moreno et al. 2004). These dinosaur tracks are lower Cretaceous outcrops and belong to sauropod and theropod trackmakers. The most important tracksite with sauropod tracks is located in Central Chile (Baños del Flaco tracksite), 100 km from the city of San Fernando (Moreno and Pino 2002; Moreno and Benton 2005).

6.1 Baños Del Flaco Tracksite

On the offspring of Baños del Flaco, an ichnocoenosis Ornithopod-Theropod-Sauropod was found and recognized the ornithopod ichnoespecies *Camptosaurichnus* and the diplodocimorph sauropod *Parabrontopodus* (Lockley et al. 1994a). The age of the locality is Upper Jurassic (Tithonian). The paleoenvironment inferred for the track bed corresponds to a shallow lagoonal carbonate continental platform and tropical–subtropical, approximately 18 °C (Moreno 2000; Moreno and Pino 2002). An overview of the dinosaur tracks from Chile shows tracksites indicating an assemblage of sauropod and theropod tracks (Rubilar et al. 2005). The Baños del Flaco sauropod ichnites correspond to the genus *Parabrontopodus*, typically related to diplodocids because tracks are closer to the midline. The sauropod was moved at 2.2 km/h toward N 40 W (Moreno and Benton 2005).

6.2 Chacarilla Tracksite

This site placed on the Chacarilla Formation at the Chacarilla Creek has been dated as Upper Jurassic–Lower Cretaceous in age (Blanco et al. 2000). The sauropod tracks of this site have been shallowly studied, just to mention the presence. Trackmakers were left with both narrow-gauged and wide-gauged sauropods (Moreno 2001; Moreno et al. 2004). The environment was a meandering river.

6.3 Calama Tracksite

This tracksite includes theropod and sauropod footprints. The site is placed west of Calama city on red beds of the Estación Member, San Salvador Formation (Upper Jurassic). The sauropod material has not been described, but they were assigned to narrow-gauged sauropods (Moreno et al. 2004).

7 Sauropod Tracks from Brazil

The ichnological sites from Brazil are abundant, and some of them have thousands of dinosaur tracks. Dinosaur tracks are in Jurassic rocks of San Pablo, Paraná, Santa Catarina, and Rio Grande do Sul states that cover an area of 1.300.000 km^2 (Leonardi 1994; Dentzien-Dias and Bertoni-Machado 2005). Most of the sites are not well-preserved, but eighth ichnospecies were described. On the northeast of Brazil, mainly on the Ceará and Paraíba states, the ichnological record belongs to the lower Cretaceous, and many ichnospecies were described such as the iguanodontids *Caririchnium* from Serrote do pimenta; *Sousaichnium* from Passagem das Pedras, and the theropod *Moraesichnium* found on the last locality. Sauropod tracks have been recorded at Floresta dos Borba (Calvo et al. 2002; Leonardi and dos Santos 2004), Touro Passo Stream (Dentzien-Dias and Bertoni-Machado 2005), São Domingos, Serrote do Letreiro, Serrote do Pimenta, Aroeira (Leonardi 1994; Leonardi and dos Santos 2004), Lagoa do Forno, Riacho do Cazé, Piau, Mae d'Agua, and Fazenda Paraíso tracksites (Leonardi and dos Santos 2004) (Fig. 1). On the other hand, at Engheno Novo and Piau-Cai Zara, some tracks could belong to sauropods, but there are some doubts (Leonardi 1994).

7.1 Floresta dos Borba Tracksite

The sauropod tracks from Floresta dos Borba, Sousa Basin, come from Lower Cretaceous outcrops of Antenor Navarro Formation (Calvo et al. 2002). Two pes and two manus tracks were found (Fig. 4e). One of the suboval pes has 0.90 × 0.75 cm, whereas the other one has an external dimension of 1–1.20 m and the internal is 0.80–0.60 m (Fig. 4d); these differences are due to partial destruction after they were left. All sauropod pes are associated with suboval manus tracks of 0.30 by 0.20 m. The manus sizes are too small, indicating a putative taphonomic influence. In that area, another manus track was found; it has liver shape, the width is 70 cm and the length is 45 cm. No finger impression has been recorded.

7.2 *Touro Passo Stream Tracksite*

The sauropod dinosaur tracks found near Touro Passo Stream, in Rosário do Sul city at the southwest region of Rio Grande do Sul State were common at the Guará Formation. It is a trackway consisting of three rounded footprints with diameters around 45 cm; the depth of each footprint is around 25 cm. Tracks can be seen both on the surface as in section, in an interdune facies (Dentzien-Dias and Bertoni-Machado 2005). The size and shape of the footprints lead us to attribute them to a middle-sized sauropod.

7.3 *São Domingos Tracksite*

The sauropod tracksite from São Domingos is located in the Tocantins River at the Tocantins state. These tracks were left on a coarse sandstone in a floodplain deposit of the Lower Cretaceous Corda Formation (05°43′24″ S, 47°31′10″ W). Six trackways of large dimension were recognized but not too much information was given, but some tracks have claw impressions (Leonardi 1994).

7.4 *Serrote do Letreiro Tracksite*

These tracks were found on the Lagoa farm at Paraíba State (06°41′48″ S, 38°18′32″ W). Outcrops belong to sandstone of the Lower Cretaceous Antenor Navarro Formation. The sauropod tracks are not well-preserved, but some claws are seen. They are big and inferred as left by a sauropod herd (Leonardi 1994).

7.5 *Serrote do Pimenta Tracksite*

These tracks were found on the Estreito farm at the Paraiba State (06°43′40″ S and 38°12′15″ W). The material comes from the Antenor Navarro Formation (Lower Cretaceous). Many trackways are in an area of 190 × 140 m and up to three layers. In one place, there are eight sub-parallel trackways of a heavy quadruped animal (Leonardi 1994). The pedes are elliptical with a diameter of 60–80 cm without morphological details.

7.6 *Aroeira Tracksite*

This dinosaur site is located near Morada Nova at the Paraiba State (06°41′44″ S, 38°22′06″ W). The tracks are in sandstone intercalated with siltstone of the Antenor Navarro Formation. One trackway with three large tracks and no morphological details are inferred to be left by sauropods (Leonardi 1994).

7.7 *Lagoa do Forno Tracksite*

It is situated at 7.1 km SE of the center of Sousa (06°48′06″ S, 38°11′ W) at Paraíba State. At least six poor-quality sauropod tracks and four theropod footprints were found. The carrier formation is uncertain, although it is inferred to be the Sousa or Piranhas Formation (Leonardi and dos Santos 2004).

7.8 *Riacho do Cazé Tracksite*

At 4.2 km NNE of Sousa (06°43′17″ S, 38°13′30″ W), some poor-quality sauropod and theropod footprints were recorded in the Antenor Navarro Formation (Leonardi and dos Santos 2004).

7.9 *Piau Tracksite*

Piau Farm includes at least 25 horizons with a vertebrate ichnocoenoses of almost 200 individual dinosaurian trackways and isolated footprints. Tracks were left by many vertebrate kinds including sauropods with large, rounded subtracks. Seven of the trackways probably belong to a herd. The best-preserved sauropod trackway of Piau tracksite includes four manus-pes sets, with very wide and high displacement rims. The pes tracks are oval (~78 cm long and ~46 wide) and do not have morphological details. The manus tracks have a horseshoe shape and are almost completely infilled by the anterior part of the displacement rim of the pes track. The trackway is narrow-gauged and belongs to the *Parabrontopodus* type. While the footprints do not overlap the midline, they are very close to it. The animal was heading in a westerly direction (Leonardi and dos Santos 2004).

7.10 *Mae d'Agua Tracksite*

Some footprints have been discovered in this locality placed at 7 km SSE of Sousa. The tracks are in the Piranhas Formation and correspond to sauropods, theropods, and to a long-heeled ornithopod (Leonardi 1994; Leonardi and dos Santos 2004).

7.11 *Fazenda Paraíso Tracksite*

The tracksite belongs to the Piranhas Formation and is located at the municipality of Sousa (06°48′31″ S, 38°09′30″ W). Although theropod tracks predominate at the site, poorly preserved tracks attributed to sauropods have been recorded (Leonardi and dos Santos 2004).

8 Sauropod Tracks from Uruguay

8.1 *Cañada del Ombú Tracksite*

The first sauropod tracks from Uruguay were found in the sandstones of the Tacuarembó Formation (Late Jurassic–?earliest Cretaceous) in the Northern Basin of Uruguay (southern part of the Paraná Basin) in the Cañada del Ombú tracksite, on the southern margin of Route 26, near the city of Tacuarembó (Fig. 1). They are interpreted as interdune deposits. Two pes-only narrow-gauged trackways plus some isolated tracks were described by Mesa and Perea (2015). Trackway A is 12 m with 16 tracks of 54 × 47 cm. The average oblique pace is 94.9 cm while the average stride is 138.7 cm. Trackway B is much larger in all sense, being a trackway of 14.5 m with 14 tracks of 109.7 × 103 cm. The average oblique pace is 137.5 cm while the average stride is 198.3 cm.

9 Sauropod Tracks from Colombia

9.1 *Chiquizá Tracksite*

Colombia records only one site with sauropod tracks, located 2 km northwest of Chiquizá locality (Boyacá Department) (Fig. 1). The tracks were found in levels of quartz sandstones, at the top Member of the Arcabuco Formation (Tithonian–Berriasian), interpreted as fluvial plain facies. The best-preserved trackway is composed of eight tracks, which show four claw marks (Moreno-Sánchez et al. 2011).

10 Paleobiological Approaches

In the last two decades, the study of dinosaur tracks has incorporated new perspectives changing the paradigm of traditional paleoichnology (Fig. 5). Instead of analyzing the traces almost independently of their producers, or with minimal reference to them, the aim here is to understand paleoichnology as a branch of paleontology as a whole, and to link it to paleobiology. In this way, paleoichnological and paleoenvironmental studies are related to data on anatomy, functional morphology, and taphonomy in order to know the speed, gaits, and herding behavior.

10.1 Speed Analysis

The first method for estimating the speed of dinosaurs was proposed by Alexander (1976), who obtained a relationship between speed, stride length, and body size from living animals and supported his analysis using the hypothesis of physical similarity (see Duncan 1953 for further details). In order to establish comparisons between animals of different sizes, Alexander used the Froude number as a non-dimensional parameter, where inertia and gravity interact. Applied to terrestrial locomotion, the Froude number becomes v^2/gh, where v is the velocity, g the acceleration of free fall, and h is the height of the hip from the ground. According to this hypothesis, the movements of animals with geometrically similar shapes, but different in size, will be dynamically similar only when they move with equal Froude number. This method had suffered some revision from the same author (Alexander 2006) and others. First, some taphonomic processes related with the substrate conditions affect the track and stride lengths measurements, data key for further analyses (e.g., Allen 1997; Gatesy et al. 1999; Nadon 2001; Manning 2004; Milàn and Bromley 2006; Marty et al. 2009; Falkingham et al. 2011). Second, some anatomical aspects were analyzed by Thulborn (1989, 1990), who points out that the relationship between the height at the hip joint and the footprint length (h/L) varies in systematic fashion among dinosaur taxa, and that this h/L ratio certainly changes during ontogeny, on account of the allometric growth that prevails in terrestrial vertebrates. However, Alexander's method continues being valid in trackways speed estimations (e.g., Moratalla et al. 1988; Lockley et al. 1994b; Castanera et al. 2011). In an attempt to make it more precise, several authors have made modifications to the original equation (Thulborn 1982, 1989, 1990; Thulborn and Wade 1984; Lockley 1986; Alexander 1991; Christiansen 1998; Henderson 2003; González Riga 2011). The track record indicates that sauropods usually progressed at slow speeds, with estimates from trackways ranging from about 2 to 7 km/h (Thulborn 1990; Christiansen 1997; Lockley and Meyer 2000; Mazzetta and Blanco 2001). Alexander (1976) pointed out that extant terrestrial mammals change from a walk to a run or trot when the ratio of stride length to hip height (s/h) reaches a value of about 2.0, and suggested that the same was probably true for dinosaurs. In subsequent studies of dinosaur locomotion (Thulborn

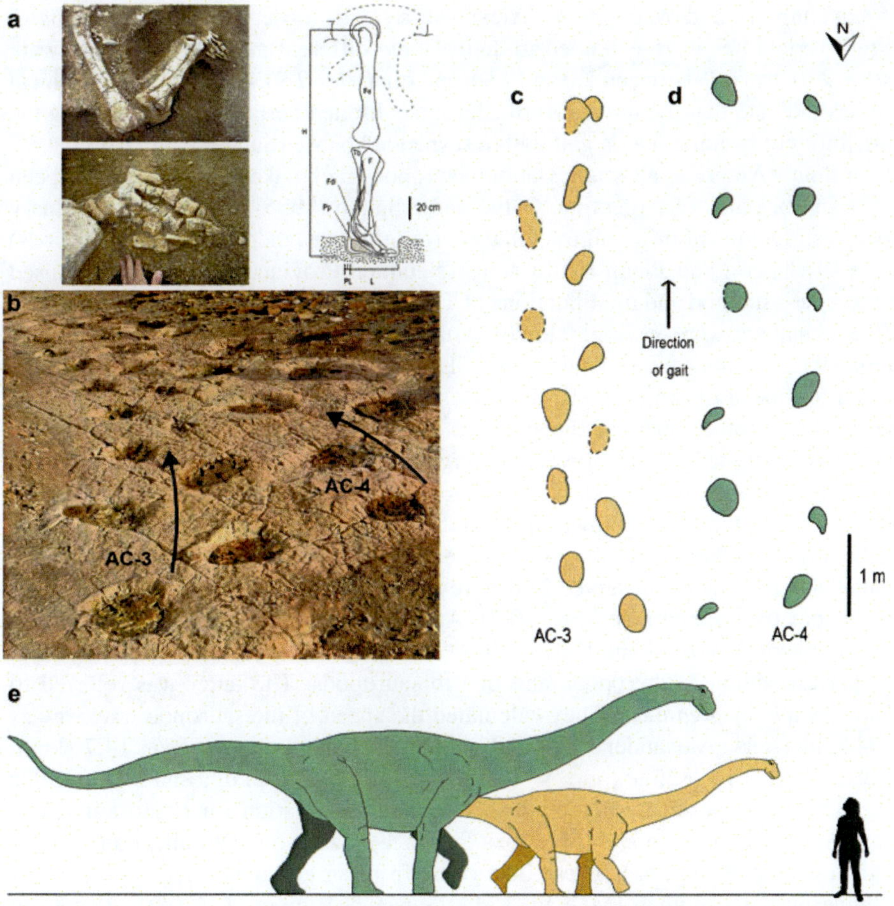

Fig. 5 Paleobiological approaches. **a** Speed estimations based on tracks and anatomical studies. Photographs of the articulated hind limb of the titanosaur MUCPv-1533 from La Invernada (Neuquén Province, Argentina), and schematic reconstruction of the limb and pes impression in lateral view. This analysis indicates that hip height can be estimated as 4.586 times the length of the pes track in derived titanosaurian taxa (modified from . **b–e** Study of speed, gaits and gregarious behavior in *Titanopodus mendozensis* trackways from the Loncoche Formation, Mendoza Province, Argentina. **b** Photograph of the AC-3 and AC-4 parallel trackways; the arrows indicate the direction of gait towards the South (175°). **c** Segment of the AC-3 trackway, showing partial or complete overlap of the manus and pes tracks; the internal and external width of the trackway, as well as the size of the identifiable manus and pes tracks, are smaller than those of the AC-4 trackway. **d** Segment of the AC-4 trackway, showing the alternation of manus and pes tracks. c and d are at scale. e Sketch of two titanosaurian trackmakers walking together at ~4.2 km/h in an amble gait (smaller individual, a juvenile or subadult, trackway AC-3) and an asynchronous gait (bigger individual, an adult, trackway AC-4). Trackmakers are to scale according to the estimated gleno-acetabular distances. Woman's size: 1.70 m (**b**, **c**, **d** and **e**, modified from González Riga 2011), González Riga and Tomaselli 2019)

1982; Thulborn and Wade 1984), Alexander's (1976) observations on the gaits of extant vertebrates were extended to define three different gaits in dinosaurs: walk ($s/h < 2.0$), trot (s/h between 2.0 and 2.9), and run ($s/h > 2.9$). Clearly, biomechanical studies indicate that the huge size of sauropods limited certain gaits, excluding the possibility of running, i.e., a gait with a suspended phase (Sander et al. 2011).

In South America, all studies about sauropod trackmaker's velocity have been made in tracksites of Argentina. Calvo and Salgado (1995) presented preliminary speed estimations of different trackmakers (ornithopods, theropods, and sauropods) from El Chocón and Picún Leufú areas (Neuquén Province). These authors used Alexander's method and modifications of Thulborn (1989), and the data showed that all trackmakers were walking slowly, with s/h values less than one (sensu Thulborn 1982; Thulborn and Wade 1984). Later, Mazzetta and Blanco (2001) estimated the speed of ornithopods, theropods, and sauropods trackways of Peninsula Nueva site, another place of Picún Leufú area. They calculated the speed for two trackways assigned to *Sauropodichnus giganteus* (MUCPv-145 and MUCPv-146) applying Alexander's method. Neither trackway showed manus tracks. The speeds obtained were remarkably low; all the trackmakers were walking ($s/h < 2.0$). The authors raised several hypotheses with the aim of explaining this situation such as manus and pes tracks preserved in different levels (true tracks and undertracks) or manus-pes overlapped (see Sect. 10.2). At 15 km east of Picún Leufú, in Cerrito del Bote islet, Mazzetta and Calvo (2004) estimated the speed of seven trackways, six corresponding to theropods and one to sauropods. The latter was referred to *Sauropodichnus giganteus*. They calculated the speed of the sauropod trackway as 3.24 km/h (sensu Alexander 1976) and estimated the maximum speed as 14.76 km/h [sensu Thulborn 1982; i.e., $v_{\max} = (6.12\ h)^{0.5}$ for graviportal dinosaurs]. González Riga (2011) proposed a refinement to Alexander's equation for titanosaur tracks from Agua del Choique site (Mendoza Province), based on anatomical and ichnological evidence (Fig. 5a). These data came from an articulated left hind limb of a titanosaurian specimen (MUCPv-1533, González Riga et al. 2008) collected in strata that are regarded as correlative to those that have yielded *Titanopodus* tracks (Allen Formation, Neuquén Basin). This analysis indicates that hip height (h) can be estimated as 4.586 times the length of the pes track (L) for the La Invernada specimen, a derived lithostrotian titanosaur that lived toward the end of the Cretaceous. With this datum, the speed of two wide-gauge *Titanopodus* trackways was calculated (AC-1 and AC-4 trackways, Figs. 3e and 5b, d). This study together with the field mapping indicates that tracks were produced by small-sized titanosaurs (hip height of 211–229 cm) that walked at 4.7–4.9 km/h toward the south and southwest following, in part, a sinuous pathway. These speeds and some taphonomic features of tracks (prominent rims, distorted elongated shapes) indicate the capacity of derived titanosaurs for walking effectively over a very wet and slippery substrate. The González Riga's estimation that considers $h = 4.586\,L$ has been used by different researchers (e.g., Vila et al. 2013; Castanera et al. 2014; Xing et al. 2014, 2015, 2019, 2021; Ghilardi et al. 2016; Boudchiche et al. 2017; Stettner et al. 2018). Later, González Riga and Tomaselli (2019) estimated the speed of a new and problematic *Titanopodus* trackmaker (corresponding to the AC-3 trackway) (Fig. 5b, c).

Compared with the trackways analyzed by González Riga (2011), AC-3 trackway has certain peculiarities that make it difficult to accurately estimate the length of the pes to calculate *h*, because most of pes tracks overprint the manus ones (Fig. 5c). In fact, most of the AC-3 tracks are enlarged and distorted. As an alternative way to calculate *h*, the authors used the *GAD*/*h* ratio (i.e., the gleno-acetabular distance in function of the height of the hip joint). Analyzing two titanosaur species represented by one articulated specimen each: *Opisthocoelicaudia* (Borsuk-Bialynicka 1977) and *Epachthosaurus* (Martínez et al. 2004), they obtained a ratio *GAD*/*h* of 1.07. Thereby, *h* was calculated in 1.35 *m* for the AC-3 trackmaker. To test the estimation of *h* by the method of González Riga (2011), the authors used the length of one well-preserved pes track. The result was $h = 1.44$ m, which is similar to the obtained by *GAD*/*h* ratio. In conclusion, the speed of the AC-3 trackmaker was estimated at 3.46 and 3.73 *km*/h, according to 4.586 *L* and *GAD*/*h*, respectively.

10.2 Gaits

Gait studies are very complex because they need comparison with living animals, and sauropods were unique as land vertebrates. In these studies, elephants are used as analogous models (e.g., Hutchinson et al. 2003, 2006; Schmitt et al. 2006). Following the study's case of only-pedes tracks of *Sauropodichnus* in the Peninsula Nueva site, Mazzetta and Blanco (2001) proposed different interpretations as (1) the sauropod trackmakers were using some sort of unconventional progression, like that adopted by horses proceeding quadrupedally on a slippery surface, or (2) they adopted a bipedal stance while moving in a very slow walking gait. Evidently, these interpretations are different from each other for the same fact and are not supported by anatomical data in the paper. On the other hand, González Riga and Tomaselli (2019) presented the case of *Titanopodus* tracks in Agua del Choique site. The authors identified two different trackway patterns: one of them characterized by alternating manus and pes tracks (AC-1 and AC-4 trackways, Figs. 3e and 5b, d, respectively), and another trackway that shows some elongated and distorted tracks interpreted as manus and pes overlapped (AC-3 trackway, Fig. 5b, c). This overprinting interpretation is supported by the presence of clear juxtaposed or partially overlapped manus and pes prints in some parts of the trackway (e.g., manus-pes set of the Fig. 5b upper part). Considering these situations, two different hypotheses were raised. The pattern 1 is congruent with an asynchronous gait produced by large individuals, while the pattern 2 is assigned to an amble gait produced by a smaller individual (a juvenile or sub-adult trackmaker) (Fig. 5e). The different types of gait can be analyzed through the ichnological estimation of the gleno-acetabular distance (*GAD*). Three main formulas have been proposed for tetrapods (Leonardi 1987; Farlow et al. 1989). By applying the different formulas, both hypotheses could be corroborated. The *GAD* was calculated at 2.26, 2.45, and 1.46 *m* for AC-1, AC-4, and AC-3 trackmakers, respectively. The trackmaker AC-3 not only has a smaller *GAD* (1.46 *m*, equal to *S* according to the formula for an amble gait), but also smaller tracks and a narrower trackway width. This study of new *Titanopodus* trackways in the Upper Cretaceous strata of Argentina shows two patterns produced by the same trackmakers of different size and probably, belonging to two ontogenetic stages: the small individual walked using an amble gait, whereas the large ones using an asynchronous gait (Fig. 5e). A similar case is observed in living elephants that show asynchronous gait at slow speed and an amble gait at high speed. According to Leonardi (1987), the asynchronous gait is produced when the feet of the same diagonal limb pair are not synchronized in their movement; the animal rests constantly during its progression on three supports. Choosing a section with three successive manus-pes sets, the *GAD* can be estimated on the midline as the distance between the reference point of the more advanced manus track and the intersection of the midline with the reference point of the line that unit the two pes tracks of the other two sets. Observation shows that the glenoid articulation is vertical to one of the anterior autopodia, while the acetabular is found upon the middle point of the line that unites the hind-feet (Leonardi 1987). The formula to calculate the gleno-acetabular distance is summarized in:

$$GAD = {}^{3}\!/_{4}S + m - pD$$

In contrast, the amble gait corresponds to a rarer type of walking gait present in vertebrates of long legs like camels and elephants. The animal shifts forward the two legs on the same side (lateral support) at the same time, while the other two legs sustain the body and thrust it ahead. Really, in the actual ambling gait, a slight phase-displacement exists in the movements of the legs of the same lateral limb pair: the foot shifts with little advancement in relation to the hand (Leonardi 1987). The *GAD* can be estimated on the midline, as the distance between the intersection point of the line that unit left and right manus tracks and the intersection of the line that unit left and right pes tracks. Pes tracks are not necessarily those that follow immediately the manus tracks. The lines that unit both manus tracks and the one that unit both pes tracks are more or less parallel. The formula to calculate the gleno-acetabular distance is expressed as:

$$GAD = S + m - pD$$

10.3 Herding Behavior

As Ostrom (1972) well defined, gregariousness is more than just multiple occurrences or close habitation. Gregarious behavior implies group activity and requires a relatively high degree of organization. The gregarious animals are usually aware of the presence and behavior of others of their own kind, and their individual behavior is stimulated by the actions of the group, at least in part. Therefore, the term 'gregarious' involves much more complex biological aspects than simple group displacement (Senar 1994). According to several studies, sauropods are considered animals with gregarious behavior, supported by both the skeletal and the ichnological records. The gregarious behavior has been analyzed through taphonomy studies in bone accumulations (Coria 1994; Winkler et al. 2000; Bandyopadhyay et al. 2002; González Riga and Astini 2007; Carballido et al. 2017; González Riga et al. 2018), in nests and nesting sites (Chiappe et al. 2004; Chiappe and Coria 2013), and in tracks and trackway associations. The latter constitutes one of the best evidences of herding behavior in sauropods (e.g., Bird 1944; Ostrom 1985; Lockley et al. 1986, 1994b; Pittman and Gillette 1989; Barnes and Lockley 1994; Pittman and Lockley 1994; Schulp and Brokx 1999; Lockley and Meyer 2000; Day et al. 2004; Myers and Fiorillo 2009; Castanera et al. 2011, 2012, 2014; Plotnick 2012). The association of different trackways can provide information about herd speed, direction, size and structure (relative position of juveniles and adults during their movement), and their composition (juvenile–adult relationship) (Lockley 1986; Myers and Fiorillo 2009). In South America, herding has been proposed in Brazil (Leonardi 1989; Carvalho 2000), Bolivia (Leonardi 1984, 1989; Lockley et al. 2002), and Argentina (González Riga and Ortiz David 2011; Tomaselli 2014). In 1989, Leonardi described some tracksites localities according to their distribution by behavior. The author mentions three localities of Brazil that can be consider with sauropod 'social' organization: (1) São Domingos (Itaguatins, Goiás), with seven sauropod trackways, mostly subparallel; (2) Serrote do Pimenta (Souza, Paraíba), with six parallel trackways which go forward in a front of about 20 m indicating a probable gregarious behavior of a herbivore herd; and (3) Piau (Souza, Paraíba), with seven parallel sauropod trackways, advancing in a narrow front of about 18 m indicating that the trackmakers could be walking side by side. On the other hand, Carvalho (2000) described an ichnocoenosis of twenty isolated tracks and four short trackways in Serrote do Letreiro (Souza, Paraíba), mainly produced by sauropods. The sauropod tracks and trackways are in a main north–northwest to south–southeast direction. The author suggests a herding behavior among sauropod trackmakers, composed by sixteen individuals. Three of the largest tracksites of South America correspond to Bolivia, with the Cal Orck'o, Toro Toro, and Humaca localities. In Toro Toro tracksite, Leonardi (1984, 1989) described eight parallel trackways of large sauropods (six adults and two juveniles), which walked together in a front of about 200 m. This record represented the first documentation of probable social behavior among Late Cretaceous sauropods (presumably derived lithostrotian titanosaurs) in South America (Lockley et al. 2002). On the other hand, in Humaca tracksite, Lockley et al. (2002) described a social group of at least 11 sub-adult titanosaurs moving in west–southwest direction. The trackmakers were traveling close together, as the total spacing perpendicular to the direction of travel is only 20 m (average spacing about 2 m). Cal Orck'o does not represent herding behavior except for the two parallel trackways, which suggest two individuals walking in succession or at least following a definite pathway (Lockley et al. 2002; Fig. 5). In Argentina, González Riga and Ortiz David (2011) and Tomaselli (2014) have proposed a social behavior for the ichnotaxon *Titanopodus mendozensis* of Loncoche tracksite. At least four individuals (one of them juvenile or sub-adult) walked together in a north–south direction (between 174° and 208°). Two of the trackways, one belonging to an adult and the other to a juvenile or sub-adult, are parallel and very close to each other (~0.6–2 m of distance) (Fig. 5b). This trackway also describes similar moves along 22.4 m (total length of the juvenile or sub-adult trackway). The authors bring together several criteria proposed by different authors (e.g., Bird 1944; Ostrom 1972; Lockley 1991; Lockley and Matsukawa 1999; Barco et al. 2006) to support a herding behavior among the *Titanopodus* trackmakers: (1) Stratigraphic level. The tracks must be in the same track-bearing surface to be considered simultaneous. The depth of the tracks and the development of rims should be similar (although these can be conditioned by variations in humidity, porosity, compaction, among others). (2) Direction of travel. The trackways association must be arranged to represent a set of tracks that are oriented in the same direction of travel. The trackways should be parallel or sub-parallel and maintain a distance among them, consistent with the sizes of the trackmakers. (3) Velocity of displacement. The estimation of speed at which the individuals in a group moved should be similar, whether they are adults or juveniles. Paleoichnological and anatomical studies reveal a great potential of data to know paleobiological aspects. This new line of work, which includes functional anatomy and computational models, opens up new perspectives to better understand the diverse fauna of South American sauropodomorphs.

11 Conclusions and Perspectives

The paleoichnological study of vertebrates has been improved in recent years. Whereas at the beginning, most studies were focused on describing track and trackways, now the evidence points to understanding behavior, paleoenvironment, and paleoecology. South America is very rich in dinosaur tracks, mainly theropods and ornithopods; therefore, this summary of all the sauropodomorph fossil tracksites is of great value. The only recognized early branching sauropodomorph tracks come from Patagonia, Argentina, and the best sauropod trackways come from Neuquén Basin and Bolivia. Gregarism in sauropods is very common, and it is rare to find just one track or trackway, so if that occurs, it is because the outcrops were covered or eroded. The study of sauropod velocity based on trackways shows that they mainly moved slowly. The study of diversity of sauropod tracks and trackways in South America is growing with new specialists that also include the study of invertebrate traces. This overview of sauropodomorph tracks has the objective to display the record of them that up to now are present in Argentina, Uruguay, Brazil, Bolivia, Chile, and Perú.

Acknowledgements We are grateful to the editors Dr. Alejandro Otero, Dr. José Luis Carballido and Dr. Diego Pol for including our work in this volume. We also appreciate the valuable suggestions made by the reviewers for improving this manuscript with their technical suggestions. We thank Dr. Leonardo Ortiz David for his valuable contribution in editing the photos and designing the figures of this chapter. We also thank the technician Juan Mansilla and geologist Federico Alvarez for fieldwork and his help in the preparation of tracks and molds. We thank all the people, companies and university students. We also thanks especially to José Brillo, Teresa Galaz Bravo, Gladys Quiroga, Alejandra Molina, Elsa del Valle Luna, Nestor Hernán Figueroa, Cynthia Rivera, Laura Avila, Gastón Pantoja, Nicolas Fuentes, Mabel Muñoz and Elda Quinteros who help continuously to maintain the Geo-Paleontological Park Proyecto Dino. Thanks are due to Dra. Silvina de Valais, Dr. Juan Canale and the Prefecture of Villa El Chocón. We also thank the team of Laboratorio y Museo de Dinosaurios, Lic. Juan Pedro Coria, Lic. Claudio Mercado, Tec. Germán Sánchez Tiviroli, Tec. Mauricio Guerra and volunteer students, for their constant contributions and commitment. Our investigations are funded by the following projects: [04/I-231-259] UNComahue, M069 SIIP-UNCUYO 2019, M06/M112 SIIP-UNCUYO 2019, M085 SIIP-UNCUYO 2019, and CONICET PIP.

References

Alexander RMCN (1976) Estimates of speeds of dinosaurs. Nature 261:129–130.https://doi.org/10.1038/261129a0

Alexander RMCN (1991) Doubts and assumptions in dinosaur mechanics. Interdiscipl Sci Rev 16(2):175–181. https://doi.org/10.1179/isr.1991.16.2.175

Alexander RM (2006) Dinosaur biomechanics. P R Soc B 273(1596):1849–1855. https://doi.org/10.1098/rspb.2006.3532

Allen JR (1997) Subfossil mammalian tracks (Flandrian) in the Severn Estuary, SW Britain: mechanics of formation, preservation and distribution. Philos T Roy Soc B 352(1352):481–518. https://doi.org/10.1098/rstb.1997.0035

Apesteguía S (2005) Evolution of the Titanosaur metacarpus. In: Tidwell V, Carpenter K (eds) Thunder lizards. Indiana University Press, Bloomington, pp 321–345

Apesteguía S, Gallina PA, Haluza A (2010) Not just a pretty face: anatomical peculiarities in the postcranium of rebbachisaurids (Sauropoda: Diplodocoidea). Hist Biol 22:165–174

Apesteguía S, De Valais S, Cordero GR, Ramírez OM (2011) New ichnological record from the late Campanian Toro Toro Formation at Toro Toro, Potosí (Bolivia): first probable dromaeosaurid tracks from South America. Ameghiniana 48(4):662–667

Bandyopadhyay S, Roy-Chowdhury TK, Sengupta DP (2002) Taphonomy of some Gondwana vertebrate assemblages of India. Sediment Geol 147:219–245

Barco JL, Canudo JI, Ruiz-Omeñaca JI (2006) New data on *Therangospodus oncalensis* from the Berriasian Fuentesalvo tracksite (Villar del Río, Soria, Spain): an example of gregarious behaviour in theropod dinosaurs. Ichnos 13:237–248

Barnes FA, Lockley MG (1994) Trackway evidence for social sauropods from the Morrison Formation, Eastern Utah (USA). Gaia 10:37–42

Bird RT (1944) Did Brontosaurus ever walks on land? Nat Hist 53:60–67

Blanco N, Tomlinson A, Moreno K, Rubilar D (2000) Importancia estratigráfica de icnitas de dinosaurios en la Formación Chacarilla (Jurásico Superior—Cretácico Inferior), I Región, Chile. En: Resúmenes del IX Congreso Geológico Chileno, Servicio Nacional de Geología y Minería, Puerto Varas, Chile, 31 de Julio–4 de Agosto de, p 5

Borsuk-Bialynicka M (1977) A new camarasaurid sauropod *Opisthocoelicaudia skarzynskii* gen. n., sp. n. from the Upper Cretaceous of Mongolia. Palaeontol Pol 37(5):5–64

Boudchiche L, Masrour M, Boutakiout M, Hamzaoui L, Gascón JH, Pérez-Lorente F (2017) Jarrate Lbel: new upper cretaceous continental site in Morocco. A probable ornithischian non-onithopod trackway and three amble gait titanosauriform trackways. Geogaceta 61:19–22

Calvo JO (1989) Nuevos hallazgos de huellas de dinosaurios en el Albiano–Cenomaniano de la localidad de Picún Leufú. Pcia. del Neuquén, Patagonia, Argentina. In: Resúmenes del VI Jornadas Argentinas de Paleontología de Vertebrados. San Juan, Argentina, pp 68–70

Calvo JO (1991) Huellas de dinosaurios en la Formación Río Limay (Albiano–Cenomaniano) Picún Leufú. Provincia del Neuquén, Argentina. (Ornithischia-Saurischia: Saurópoda-Terópoda). Ameghiniana 28(3–4):241–258

Calvo JO (1999) Dinosaurs and other vertebrates of the Ezequiel Ramos Mexía área, Neuquén - Patagonia Argentina. In: Tomida Y, Rich TH, Vickers-Rich P (eds) Proceedings of the second Gondwanan Dinosaur symposium, National Science Museum Monographs. 2nd symposium of Dinosaurs from Gondwana, Tokio, Japan, pp 13–45
Calvo JO (2007) Mesozoic reptilian tracks. In: Gasparini Z, Coria RA, Salgado L (eds) Patagonian Mesozoic reptiles. Indiana University Press, pp 314–334
Calvo JO, Salgado L (1995) Huellas de Dinosaurios en la localidad de El Chocón, Neuquén. Actas de la II Reunión Argentina de Icnología de Vertebrados. Universidad Nacional de San Juan, San Juan, Argentina, pp 23–26
Calvo JO, Ortiz RA (2011) Nuevos registros de Dinosauria en la Formación Allen, (Cretácico Superior) en el Área Natural Protegida Paso Córdoba. General Roca, Río Negro. In: Actas del IV Congreso Latinoamericano de Paleontología de Vertebrados, San Juan, Argentina, CD Rom
Calvo JO, Rivera C (2018) Huellas de dinosaurios en la costa oeste del embalse Ezequiel Ramos Mexía y alrededores (Cretácico Superior, Provincia de Neuquén, República Argentina). B Soc Geol Mex 70(2):449–496
Calvo JO, Coria R, Salgado L (1990) Nuevas localidades con huellas de dinosaurios del Miembro Candeleros (Albiano-Cenomaniano) (Formación Río Limay, Grupo Neuquén) Provincia del Neuquén. In: Actas de las VII Jornadas Argentinas de Paleontología de Vertebrados, Ameghiniana 26(3–4):241
Calvo JO, Moreno K, Rubilar D (2000) First record of Lower Cretaceous dinosaur tracks (Sauropoda-Theropoda) in Río Negro Province, Patagonia, Argentina. In: Abstracts of the 31st International Geological Congress, Río de Janeiro, Brasil. Vol. 427
Calvo JO, González Riga B, Mansilla JE (2015) Nuevos hallazgos de vertebrados en la Formación Portezuelo, Cretácico Superior, Neuquén, Argentina. In: Actas de las IX Jornadas de Ciencias de la Tierra, Córdoba, Argentina, p 10
Calvo JO, Avila L, Engelland S (2016) Icnofacies características de la Formación Bajo de la Carpa, Grupo Neuquén, Patagonia, Argentina. In: Actas del IX Congreso Latinoamericano de Paleontología, Lima, Perú, p 82
Calvo JO, Campos HBN, Kellner AWA, Saraiva FAA (2002) Floresta dos Borba, a new locality with dinosaur tracks in the Sousa Basin, Paraiba. Brazil Arq Mus Nac 60(4):277–284
Calvo JO, Bellardini F, Porfiri JD, Navarrete G (2011) Primer hallazgo de huellas de saurópodos en el Cretácico superior de la Costa Norte del Lago Barreales, Neuquén. Argentina. In: Actas del XVIII Congreso Geológico Argentino, Neuquén, Argentina, CD Rom
Calvo JO, Rivera C, Avila L (2021) An overview of upper cretaceous dinosaur tracks and other traces fossils from the Barreales lake area, Neuquén province, Patagonia, Argentina. J S Am Earth Sci 110:03375. https://doi.org/10.1016/j.jsames.2021.103375
Canudo JI, Castanera D, Garrido A, Salgado L, Carballido JL (2017) Icnitas de dinosaurios saurópodos en la Formación Rayoso (Cuenca Neuquina, Albiense, Argentina). Soc Geol Esp 61:43–46
Candia Halupczok DJ, Sánchez ML, Veiga G, Apesteguía S (2017) Dinosaur tracks in the Kokorkom desert, Candeleros Formation (Cenomanian, Upper Cretaceous), Patagonia Argentina: Implications for deformation structures in dune fields. Cretaceous Res 83:194–206
Carballido JL, Pol D, Otero A, Cerda IA, Salgado L, Garrido AC, Ramezani J, Cúneo NR, Krause JM (2017) A new giant titanosaur sheds light on body mass evolution among sauropod dinosaurs. P R SOC B 284(1860):20171219. https://doi.org/10.1098/rspb.2017.1219
Carvalho IS (2000) Huellas de saurópodos de la Formación Antenor Navarro (Cretácico Temprano de la Cuenca de Sousa), Serrote do Letreiro, Paraíba, Brasil. Ameghiniana 37:353–362
Casamiquela RM, Fasola A (1968) Sobre pisadas de dinosaurios del Cretácico Inferior de Colchagua (Chile). Chile Universidad, Departamento de Geología. Publicaciones 30:24
Castanera D, Barco JL, Díaz-Martínez I, Gascón JH, Pérez-Lorente F, Canudo JI (2011) New evidence of a herd of titanosauriform sauropods from the lower Berriasian of the Iberian range (Spain). Palaeogeogr Palaeoclim Palaeoecol 310(3–4):227–237

Castanera D, Vila B, Razzolini N, Dos Santos VF, Pascual C, Canudo JI (2012) Gregarious behaviour inferred from sauropod footprints in the Iberian Peninsula. In: Paleodiversity and Paleoecology of Iberian Ecosystems, Valencia, Spain, Sot de Chera, pp 61–62

Castanera D, Vila B, Razzolini N, Dos Santos VF, Pascual C, Canudo JI (2014) Sauropod trackways of the Iberian Peninsula: palaeoetological and palaeoenvironmental implications. J Iber Geol 40(1):49–59

Chiappe LM, Coria RA (2013) Auca Mahuevo, un extraordinario sitio de nidificación de dinosaurios saurópodos del Cretácico Tardío, Neuquén. Argentina Ameghiniana 41(4):591–596

Chiappe LM, Schmitt JG, Jackson FD, Garrido A, Dingus L, Grellet-Tinner G (2004) Nest structure for sauropods: sedimentary criteria for recognition of dinosaur nesting traces. Palaios 19(1):89–95

Christiansen P (1997) Locomotion in sauropod dinosaurs. GAIA 14:45–75

Christiansen P (1998) Strength indicator values of theropod long bones, with comments on limb proportions and cursorial potential. GAIA 15:241–255

Cónsole-Gonella C, Silvina de Valais R, Marquillas A, Sánchez MC (2017) The Maastrichtian-Danian Maimará tracksite (Yacoraite Formation, Salta Group), Quebrada de Humahuaca, Argentina: environments and ichnofacies implications. Palaeogeogr Palaeoclim Palaeoecol 468:327–350. https://doi.org/10.1016/j.palaeo.2016.11.008

Contreras VH, León LI, Ripoll Y, Jordán M (2010) Trazas fósiles continentales del Cretácico Tardío en la Precordillera de San Juan, Argentina. In: Actas de las Jornadas de Ciencia y Técnica, Universidad Nacional de San Juan, San Juan, Argentina, CD ROM Trabajos Científicos, pp 1–7

Coria RA (1994) On a monospecific assemblage of sauropod dinosaurs from Patagonia: implications for gregarious behavior. GAIA 10:209–213

Day JJ, Norman DB, Gale AS, Upchurch P, Powell HP (2004) A middle Jurassic dinosaur trackway site from Oxfordshire UK. Palaeontology 47(2):319–348

Dentzien-Dias PC, Bertoni-Machado C (2005) New discoveries of dinosaurs footprints from late Jurassic (?) Guará formation, southern Brazil. In: Actas de las XXI Jornadas Argentinas de Paleontología de Vertebrados, Plaza Huincul, Neuquén, pp 15

De Valais S, Díaz-Martínez I, Citton P, Cónsole-Gonella C (2020) Vertebrate tracks of the Río Negro Province, Patagonia, Argentina: Stratigraphy, palaeobiology and environmental contexts. Rev Asoc Geol Argent 77(3):402–426

Díaz-Martínez I, de Valais S, Cónsole-Gonella C (2018) New sauropod tracks from the Yacoraite Formation (Maastrichtian–Danian), Valle del Tonco tracksite, Salta, northwestern Argentina. J Iber Geol 44(1):113–127

Duncan WJ (1953) Physical similarity and dimensional analysis: an elementary treatise. Arnold, London, p 156

Falkingham PL, Bates KT, Margetts L, Manning PL (2011) The 'Goldilocks' effect: preservation bias in vertebrate track assemblages. J R Soc Interface 8(61):1142–1154. https://doi.org/10.1098/rsif.2010.0634

Farlow JO (1992) Sauropod tracks and trackmakers integrating the ichnological and skeletal records. Zubia 10:89–138

Farlow JO, Pittman JG, Hawthorne JM (1989) *Brontopodus birdi*, Lower Cretaceous sauropod footprints from the US Gulf coastal plain. In: Gillette DD, Lockley MG (eds) Dinosaur tracks and traces. Cambridge University Press, Cambridge, pp 371–394

Gatesy SM, Middleton KM, Jenkins FA, Shubin NH (1999) Three-dimensional preservation of foot movements in Triassic theropod dinosaurs. Nature 399:141–144. https://doi.org/10.1038/20167

Ghilardi AM, Aureliano T, Duque RR, Fernandes MA, Barreto AM, Chinsamy A (2016) A new titanosaur from the lower cretaceous of Brazil. CretacOus Res 67:16–24. https://doi.org/10.1016/j.cretres.2016.07.001

González Riga BJ (2011) Speeds and stance of titanosaur sauropods: analysis of Titanopodus tracks from the Late Cretaceous of Mendoza Argentina. An Acad Bras Cienc 83(1):279–290

González Riga BJ, Calvo JO (2006) Primer estudio sobre huellas de dinosaurios en la provincia de Mendoza, Argentina. In: Resúmenes del IX° Congreso Argentino de Paleontología y Bioestratigrafía. Universidad Nacional de Córdoba, Córdoba, p 284

González Riga BJ, Astini R (2007) Fossil preservation of large titanosaur sauropods in overbank fluvial facies: a case study in the Cretaceous of Argentina. J S Am Earth Sci 23:290–303

González Riga BJ, Calvo JO (2009) A new wide-gauge sauropod track site from the Late Cretaceous of Mendoza, Neuquén Basin Argentina. Palaeontology 52(3):631–640

González Riga BJ, Ortiz David L (2011) Yacimiento de las icnitas *Titanopodus mendozensis*, Cretácico Superior de Mendoza (Argentina): una síntesis. In: Resúmenes del 7° Encuentro del international center for earth sciences (ICES) E-ICES 7, Malargüe, Mendoza, Argentina, pp 1–14

González Riga BJ, Tomaselli MB (2019) Different trackway patterns in titanosaur sauropods: analysis of new *Titanopodus* tracks from the upper cretaceous of Mendoza, Neuquén Basin, Argentina. Cretaceous Res 93:49–59

González Riga BJ, Calvo JO, Porfiri JD (2008) An articulated titanosaur from Patagonia (Argentina): new evidence of neosauropod pedal evolution. Paleoworld 17:33–40

González Riga BJ, Ordoñez M, Nasif D, Croce C, Crack J (2011) First record of dinosaur tracks from the Anacleto Formation in the Mendoza Province, Neuquén Basin, Argentina. In: Resúmenes de la Reunión Anual de comunicaciones de la Asociación Paleontológica Argentina, Luján, Buenos Aires, Argentina, pp 35–36

González Riga BJ, Ortiz David LD, Gutiérrez P, Martínez L, Pinto L, Resa L, Sánchez Tiviroli G, Pascual M, Pena V, Poblete S (2013) Huellas de saurópodos titanosaurios en la Formación Anacleto, Cretácico Superior de Malargüe, Mendoza, Argentina. In: Resúmenes de las XXVII Jornadas Argentinas de Paleontología de Vertebrados, La Rioja, vol 50, no (4), Argentina. Ameghiniana, p 51

González Riga BJ, Ortiz David LD, Tomaselli MB, dos Anjos Candeiro CR, Coria JP, Prámparo M (2015) Sauropod and theropod dinosaur tracks from the upper cretaceous of Mendoza (Argentina): trackmakers and anatomical evidences. J S Am Earth Sci 61:134–141

González Riga BJ, Mannion PD, Poropat SF, Ortiz David LD, Coria JP (2018) Osteology of the Late Cretaceous Argentinean sauropod dinosaur *Mendozasaurus neguyelap*: implications for basal titanosaur relationships. Zool J Linn Soc-Lond 184(1):136–181

Henderson D (2003) Footprints, trackways, and hip heights of bipedal dinosaurs—testing hip height predictions with computer models. Ichnos 10(2–4):99–114. https://doi.org/10.1080/10420940390257914

Heredia AM, Pazos PJ, Fernández DE, Díaz-Martínez I, Comerio M (2019) A new narrow-gauge sauropod trackway from the Cenomanian Candeleros Formation, northern Patagonia, Argentina. Cretaceous Res 96:70–82

Hutchinson JR, Famini D, Lair R, Kram R (2003) Are fast-moving elephants really running? Nature 422(6931):493–494

Hutchinson JR, Schwerda D, Famini DJ, Dale RH, Fischer MS, Kram R (2006) The locomotor kinematics of Asian and African elephants: changes with speed and size. J Exp Biol 209(19):3812–3827

Krapovickas V (2010) El rol de las trazas fósiles de tetrápodos en los modelos de icnofacies continentales en ambientes de climas áridos –semiáridos: Buenos Aires, Argentina. Ph.D. Dissertation, Universidad de Buenos Aires (Buenos Aires, Argentina), p 358

Krapovickas V, Garrido AC, Canale JI (2020) Dinosaur tracks at Cañadón de las Campanas and Villa El Chocón area (Late Cretaceous, Candeleros Formation), Patagonia Argentina. Rev Asoc Geol Arg 77(3):447–462

Leanza HA, Apesteguía S, Novas FE, De la Fuente MS (2004) Cretaceous terrestrial beds from the Neuquén basin (Argentina) and their tetrapod assemblages. Cretaceous Res 25(1):1–96

Leonardi G (1984) Le impronte fossili di dinosauri. In: Bonaparte JF, Colbert EH, Currie PJ, De Ricqles AJ, Kielan-Jaworowska Z, Leonardi G, Morello F, Taquet P (eds) Sulle orme dei dinosauri. Erizzo (Esplorazioni e ricerche IX), Italia, pp 161–186

Leonardi G (1987) Glossary and manual of tetrapod footprint palaeoichnology. In: Departamento Nacional de Produção Mineral, Brasil, p 75

Leonardi G (1989) Inventory and statistics of the South American dinosaurian ichnofauna and its paleobiological interpretation. In: Gillette DD, Lockley MG (eds) Dinosaur tracks and traces. Cambridge University Press, Cambridge, pp 165–178

Leonardi G (1994) Annotated atlas of South America Tetrapod footprints (Devonian to Holocene) with appendix on Mexico and Central America: Brasilia, Brasil. Ministério de Minas e Energia, Companhia de Pesquisa de Recursos Minerais, Geological Services of Brazil, p 247

Leonardi G, dos Santos MFCF (2004) New dinosaur tracksites from the Sousa Lower Cretaceous basin (Paraíba, Brasil). Studi Trentini Di Scienze Naturali, Acta Geologica 81:5–21

Lockley MG (1986) The paleobiological and paleoenvironmental importance of dinosaur footprints. Palaios 1(1):37–47

Lockley MG (1991) Tracking dinosaurs: a new look at an ancient world. Cambridge University Press, Cambridge, p 238

Lockley MG, Matsukawa M (1999) Some observations on trackway evidence for gregarious behavior among small bipedal dinosaurs. Palaeogeogr Palaeoclimatol Palaeoecol 150(1–2):25–31

Lockley MG, Meyer C (2000) Dinosaur tracks and other fossil footprints of Europe. Columbia University Press, New York, p 323

Lockley MG, Houck KJ, Prince NK (1986) North America's largest dinosaur trackway site: Implications for Morrison Formation paleoecology. Geol Soc Am Bull 97(10):1163–1176

Lockley MG, Farlow JO, Meyer CA (1994a) *Brontopodus* and *Parabrontopodus* ichnogen. nov. and the significance of wide -and narrow-gauge sauropod trackways. Gaia 10:135–145

Lockley MG, Meyer CA, Santos VF (1994b) Trackway evidence for a herd or juvenile sauropods from the late Jurassic of Portugal. Gaia 10:27–35

Lockley MG, Schulp AS, Meyer CA, Leonardi G, Mamani DK (2002) Titanosaurid trackways from the upper cretaceous of Bolivia: evidence for large manus, wide-gauge locomotion and gregarious behaviour. Cretaceous Res 23(3):383–400

Manera de Bianco T, Calvo JO (1999) Hallazgo de huellas de gran tamaño en el Triásico de Los Menucos, provincia de Río Negro, Argentina. In: Resúmenes de las XV Jornadas Argentinas de Paleontología de Vertebrados, La Plata, vol 36, no 4, Argentina, Ameghiniana, p 14R

Manning PL (2004) A new approach to the analysis and interpretation of tracks: examples from the dinosauria. Geol Soc Spec Publ 228(1):93–123

Marsh OC (1878) Principal characters of American Jurassic dinosaurs, Part I. Am J Sci (series 3) 16:411–416

Martínez RD, Giménez O, Rodríguez J, Luna M, Lamanna MC (2004) An articulated specimen of the basal titanosaurian (Dinosauria: Sauropoda) *Epachthosaurus sciuttoi* from the early Late Cretaceous Bajo Barreal Formation of Chubut province, Argentina. J Vertebr Paleontol 24(1):107–120. https://doi.org/10.1671/9.1

Marty D, Strasser A, Meyer CA (2009) Formation and taphonomy of human footprints in microbial mats of present-day tidal-flat environments: implications for the study of fossil footprints. Ichnos 16(1–2):127–142. https://doi.org/10.1080/10420940802471027

Mazzetta GV, Blanco RE (2001) Speeds of dinosaurs from the Albian-Cenomanian of Patagonia and sauropod stance and gait. Acta Palaeontol Pol 46(2):235–246

Mazzeta GV, Calvo JO (2004) On the locomotory implications of the dinosaur ichnofauna at Ezequiel Ramos Mexía lake, Neuquén, Patagonia, Argentina. In: Abstracts of the First International Congress on Ichnology, Trelew, Chubut, Argentina. Ichnia Special Issue, p 54

Méndez Torrez G (2019) Descubrimiento de huellas fósiles de dinosaurios en Tarija. Informe Técnico de la Secretaría Departamental de Energía e Hidrocarburos del Gobierno Autónomo del Departamento de Tarija (Unpublished), p 8

Mesa V, Perea D (2015) First record of theropod and ornithopod tracks and detailed description of sauropod trackways from the Tacuarembó Formation (Late Jurassic-? Early Cretaceous) of Uruguay. Ichnos 22(2):109–121

Meyer CA, Marty D, Belvedere M (2018) Titanosaur trackways from the Late Cretaceous El Molino Formation of Bolivia (Cal Orck'o, Sucre). Ann Soc Geol Pol 88:223–241

Meyer CA, Marty D, Thüring B, Belvedere M (2021) The Late Cretaceous dinosaur track record of Bolivia—Review and perspective. J South Am Earth Sci 106:102992. https://doi.org/10.1016/j.jsames.2020.102992
Milàn J, Bromley RG (2006) True tracks, undertracks and eroded tracks, experimental work with tetrapod tracks in laboratory and field. Palaeogeogr Palaeoclim Palaeoecol 231(3–4):253–264. https://doi.org/10.1016/j.palaeo.2004.12.022
Moratalla JJ, Sanz JL, Jiménez S (1988) Nueva evidencia icnológica de Dinosaurios en el Cretácico inferior de La Rioja (España). Estud Geol 44(1–2):119–131
Moreno K (2000) Huellas de dinosauros (ornithopoda-theropoda-sauropoda) de la Formación Baños del Flaco, VI Región, Chile; Paloeambiente y Paleoetología. Tesis de Licenciatura, Universidad Austral de Chile (Valdivia, Chile), p 50
Moreno K (2001) Icnocenosis e icnofacies de la Formación Chacarilla (Jurásico Superior – Cretácico Inferior), Norte de Chile. In: Actas de la IV Reunión Argentina de Icnología y II Reunión de Icnología del Mercosur, Universidad Nacional de Tucumán, Tucumán, Argentina, p 59
Moreno K, Pino M (2002) Dinosaur footprints in theBanos del Flaco Formation (Tithonian–Upper Jurassic), VI Region, Chile: paleoethology and paleoenvironment. Rev Geol Chile 29:151–165
Moreno K, Benton MJ (2005) Occurrence of sauropod dinosaur tracks in the Upper Jurassic of Chile (redescription of *Iguanodonichnus frenki*). J S Am Earth Sci 20(3):253–257
Moreno K, Blanco N, Tomlinson A (2004) New dinosaur footprints from the Upper Jurassic of northern Chile. Ameghiniana 41(4):535–544
Moreno-Sánchez M, Cruz AG, Tapias JG (2011) Reporte de huellas de dinosaurios en el santuario de fauna y flora de Iguaque, en cercanías de Chíquiza (Boyacá, Colombia). Bol Geol 33(2):107–118
Myers TS, Fiorillo AR (2009) Evidence for gregarious behaviour and age segregation in sauropods. Palaeogeogr Palaeoclim Palaeoecol 274:96–104
Nadon GC (2001) The impact of sedimentology on vertebrate track studies. In: Tanke DH, Carpenter K (eds) Mesozoic vertebrate life. Indiana University Press, Indiana, pp 395–407
Novas FE (2009) The age of dinosaurs in South America. Indiana University Press, Bloomington, IN, EEUU, p 437
Ortiz David LD, González Riga BJ, Prámparo M, Porfiri J, González V (2012) Primer registro de huella de terópodo (Dinosauria, Theropoda) para la Provincia de Mendoza, Formación Loncoche (Campaniano tardío–Maastrichtiano temprano, Cretácico Superior, Malargüe, Argentina). In: Resúmenes del 8° Encuentro del International Center for Earth Sciences (ICES) E-ICES 8, Mar del Plata, Argentina, p 62
Ortiz R, Novas E, Calvo JO (2008) Nuevos hallazgos de huellas de saurópodos en la provincia de Río Negro, Argentina. In: Resúmenes del III Congreso Latinoamericano de Paleontología de Vertebrados, Neuquén, Argentina, p 188
Ostrom JH (1972) Were some dinosaurs gregarious? Palaeogeogr Palaeoclim Palaeoecol 11:287
Ostrom JH (1985) Social and unsocial behavior in dinosaurs. Field Museum Nat Hist Bull 55(9):10–21
Pittman JG, Gillette DD (1989) The Briar Site: a new sauropod dinosaur tracksite in Lower Cretaceous beds of Arkansas, USA. In: Gillette DD, Lockley MG (eds) Dinosaur tracks and traces. Cambridge University Press, Cambridge, pp 313–332
Pittman JG, Lockley MG (1994) A review of sauropod dinosaur tracksites of the Gulf of Mexico Basin. Gaia 10:95–108
Plotnick RE (2012) Behavioral biology of trace fossils. Paleobiology 38(3):459–473
Rainforth EC (2003) Revision and Re-evaluation of the early Jurassic dinosaurian ichnogenus *Otozoum*. Palaeontology 46(4):803–838
Rubilar-Rogers D, Otero RA, Yury-Yañez RE, Vargas A, Gutstein CS (2012) An overview of the dinosaur fossil record from Chile. J S Am Earth Sci 37:242–255
Sander PM, Christian A, Clauss M, Fechner R, Gee CT, Griebeler EM, Gunga HC, Hummel J, Mallison H, Perry SF, Preuschoft H, Rauhut OWM, Remes K, Tütken T, Wings O, Witzel U (2011) Biology of the sauropod dinosaurs: the evolution of gigantism. Biol Rev 86(1):117–155. https://doi.org/10.1111/j.1469-185X.2010.00137.x

Schmitt D, Cartmill M, Griffin TM, Hanna JB, Lemelin P (2006) Adaptive value of ambling gaits in primates and other mammals. J Exp Biol 209(11):2042–2049

Schulp AS, Brokx WA (1999) Maastrichtian sauropod footprints from the Fumanya site, Berguedà. Spain. Ichnos 6(4):239–250

Senar JC (1994) Vivir y convivir: la vida en grupos sociales. In: Carranza J (ed) Etología: introducción a la ciencia del comportamiento. Universidad de Extremadura, pp 205–232

Stettner B, Persons WS IV, Currie PJ (2018) A giant sauropod footprint from the Nemegt Formation (Upper Cretaceous) of Mongolia. Palaeogeogr Palaeoclim Palaeoecol 494:168–172

Thulborn RA (1982) Speeds and gaits of dinosaurs. Palaeogeogr Palaeoclim Palaeoecol 38(3–4):227–256

Thulborn RA (1989) The gaits of dinosaurs. In: Gillette DD, Lockley MG (eds) Dinosaur tracks and traces. Cambridge University Press, Cambridge, pp 39–50

Thulborn RA (1990) Dinosaur tracks. Chapman and Hall, London, UK, p 410

Thulborn RA, Wade M (1984) Dinosaur trackways in the Winton formation (mid-Cretaceous) of Queensland. Memoirs of the Qld Museum 21(2):413–517

Tomaselli MB (2014) Desplazamiento en manada de saurópodos titanosaurios: Estudio de las icnitas *Titanopodus*, Cretácico Superior de Mendoza, Argentina. Tesis de Licenciatura, Universidad del Aconcagua (Mendoza, Argentina), p 70

Tomaselli MB, Ortiz David LD, González Riga BJ (2021) Paleoenvironmental contexts of South American Cretaceous sauropod tracks. J S Am Earth Sci 110:103393. https://doi.org/10.1016/j.jsames.2021.103393

Tomaselli MB, Ortiz David LD, González Riga BJ, Coria JP, Mercado CR, Guerra M, Sánchez Tiviroli G (2022) New titanosaurian sauropod tracks with exceptionally well-preserved claw impressions from the Upper Cretaceous of Argentina. Cretaceous Res 129:104990

Vila B, Oms O, Galobart À, Bates KT, Egerton VM, Manning PL (2013) Dynamic similarity in titanosaur sauropods: ichnological evidence from the Fumanya dinosaur tracksite (southern Pyrenees). PLoS One 8(2):e57408

Vildoso C, Calvo JO, Porfiri J (2011a) New dinosaur tracksite from Perú (Lower Cretaceous). In: Resúmenes de la Reunión Anual de Comunicaciones de la Asociación Paleontológica Argentina (RCAPA), Luján, Buenos Aires, vol 48, no 4, Argentina, Ameghiniana, p R27

Vildoso C, Calvo JO, Porfiri JD (2011b) The highest sauropod tracksite in the World, Lower Cretaceous, Peru. In: Resúmenes del IV Congreso Latinoamericano de Paleontología de Vertebrados, San Juan, Argentina, CD Rom #173

Von Huene F (1932) Die fossile Reptil-Ordnung Saurischia, ihre Entwicklung und Geschichte. Monographien Geologie und Paläontologie 4:1–361

Winkler DA, Gomani EM, Jacobs LL (2000) Comparative taphonomy of an Early Cretaceous Sauropod Quarry, Malawi, Africa. Paleontol Soc Korea Spec Publ 4:99–114

Xing LD, Lockley MG, Jiang Z, Klein H, Persons IV WS, Zhang P, Wan X (2021) Variation in sauropod trackway pattern from the Tuchengzi Formation (Jurassic-Cretaceous boundary) in western Liaoning, China. Cretaceous Res 118:104662.https://doi.org/10.1016/j.cretres.2020.104662

Xing LD, Lockley MG, Klein H, Qiu L, Chou C, Wang D, Persons IV WS (2019) A Lower Cretaceous dinosaur track assemblage from the Taoqihe Formation in Central Heilongjiang, China. Palaeogeogr Palaeoclim Palaeoecol 532:109275.https://doi.org/10.1016/j.palaeo.2019.109275

Xing LD, Lockley MG, Zhang J, Klein H, Marty D, Peng G, Ye Y, McCrea RT, Persons WS IV, Xu T (2015) The longest theropod trackway from East Asia, and a diverse sauropod-, theropod-, and ornithopod-track assemblage from the Lower Cretaceous Jiaguan Formation, southwest China. Cretaceous Res 56:345–362. https://doi.org/10.1016/j.cretres.2015.05.008

Xing LD, Lockley MG, Zhang J, Klein H, Persons IV WS, Dai H (2014) Diverse sauropod-, theropod-, and ornithopod-track assemblages and a new ichnotaxon *Siamopodus xui* ichnosp. nov. from the Feitianshan Formation, Lower Cretaceous of Sichuan Province, southwest China. Palaeogeogr Palaeoclimatol Palaeoecol 414:79–97. https://doi.org/10.1016/j.palaeo.2014.08.011
Zavala C, Ponce JJ (2011) La Formación Rayoso (Cretácico Temprano) en la Cuenca Neuquina. In: Resúmenes del XVIII Congreso Geológico Argentino, Geología y Recursos Naturales de la Provincia del Neuquén, Argentina, pp 205–222

Taphonomy: Overview and New Perspectives Related to the Paleobiology of Giants

Bernardo J. González Riga, Gabriel A. Casal, Anthony R. Fiorillo, and Leonardo D. Ortiz David

Abstract Most taphonomy studies of South American sauropodomorphs have addressed extrinsic factors such as sedimentary environments, bone dispersal, and mineralogical processes that occurred during fossil diagenesis. These studies provide important data on the taphonomic modes which are associated with bone accumulations in different paleoenvironmental contexts. However, these analyses have generally not considered intrinsic factors like the shape, size, and structural integrity of the skeletal elements, variables that can produce some taphonomic bias. Sauropodomorphs include dinosaurs of highly varied sizes, ranging from small (less than 8 m long) to remarkably giant forms (around 30 m long). In the largest sauropods, such as the huge titanosaurs, very incomplete skeletons are commonly found and most notably skull and articulated pedes rarely are preserved. We focus here on some intrinsic anatomical factors as they relate to articulation in some key parts of the skeletons. Further, this study suggests that the preservation of fragile portions of sauropodomorph skeletons was possible only under specific combinations of sedimentological and biological processes.

Keyword Taphonomy · Sauropoda · Extrinsic factors · Taphonomic modes · South America

---

B. J. González Riga (✉) · L. D. Ortiz David
Laboratorio y Museo de Dinosaurios, Facultad de Ciencias Exactas y Naturales, Universidad Nacional de Cuyo, Parque Gral. San Martín, Padre Contreras 1300, 5500 Mendoza Province, Argentina
e-mail: briga@mendoza-conicet.gob.ar

L. D. Ortiz David
e-mail: lortiz@mendoza-conicet.gob.ar

Instituto Interdisciplinario de Ciencias Básicas, CONICET-UNCUYO, Mendoza, Argentina

G. A. Casal
Laboratorio de Paleontología de Vertebrados, Facultad de Ciencias Naturales y Ciencias de La Salud, Universidad Nacional de la Patagonia San Juan Bosco, km 4 (9000) Comodoro Rivadavia, Chubut Province, Argentina

A. R. Fiorillo
Huffington Department of Earth Sciences, Southern Methodist University, Dallas, TX 75275, USA
e-mail: fiorillo@mail.smu.edu

A. Otero et al. (eds.), *South American Sauropodomorph Dinosaurs*, Springer Earth System Sciences, https://doi.org/10.1007/978-3-030-95959-3_15

1 Introduction

Taphonomy is the study of all biotic and abiotic factors that influence the preservation of organismal remains after death (Behrensmeyer and Kidwell 1985; Behrensmeyer et al. 2000). Fiorillo and Eberth (2004) succinctly point out that 'Taphonomic factors remove or modify information about living organisms and assemblages (e.g., soft tissue decomposition, bone dispersal) and therefore create a biased picture of their biology and environmental and ecological associations.' Dinosaur taphonomy, then, is a multidisciplinary science with diverse aims and applications (Lyman 1994).

In many regions of the world. taphonomy studies have had a relatively important multidisciplinary development. These studies typically include sedimentologists and paleontologists combining their expertise to understand the origin of fossil assemblages (Behrensmeyer and Kidwell 1985; Fastovsky et al. 1997; Eberth et al. 2001; Rogers et al. 2001; Fiorillo and Eberth 2004; Moore and Norman 2009; Csiki et al. 2010; Orr et al. 2016; Botfalvai et al. 2017).

In South America, taphonomic studies have increased in recent years, incorporating new methods and techniques, but they fall short with respect to the growing and numerous discoveries of new species of sauropodomorphs. This is particularly unfortunate given that the last few decades have seen a tremendous growth in discoveries of new taxa of sauropods (Martínez et al. 2016; González Riga et al. 2016, 2019; Carballido et al. 2017). Further, given the enormous body size of many of these dinosaurs, in this chapter we explore how sauropods can provide unique insights into taphonomic processes that smaller-bodied vertebrates, including other dinosaurian clades, cannot offer. This is particularly evident in the examples of large, massive animals with skeletons 30 m long that cannot be covered by most fluvial sedimentary processes. This report will provide an overview that highlights the insights gained by studying such an unusual group of dinosaurs.

In South America, most taphonomic studies of Triassic and Jurassic dinosaurs are focused on whole vertebrate assemblages, including some paleoecological aspects related to the increase of abundance, diversity, and body size of sauropodomorphs from the Late Triassic onwards (Martínez et al. 2011, 2013). Some taphonomic reviews focus on taphonomic modes (sensu Behrensmeyer 1988; Behrensmeyer and Hook 1992) and sedimentary environments (Colombi et al. 2012, 2017; Otero et al. 2019a).

The Cretaceous fossil record from this continent includes 71 known and valid sauropod species, most of them titanosaurians. Recent studies on sauropod occurrences have included information about the preservational processes (i.e., biostratinomy, fossil diagenesis), and in some cases, the characterization of taphonomic modes, such as in the case of *Mendozasaurus neguyelap* (González Riga et al. 2007), *Bonitasaura salgadoi* (Pérez et al. 2009), *Aeolosaurus colhuehuapensis* (Casal et al. 2014a), and *Pilmatueia faundezi* (Pino et al. 2021). Additionally, important taphonomic studies of exceptionally preserved nest, eggs, and embryos have been made in Auca Mahuevo in northern Argentinean Patagonia (Chiappe et al. 1998, 2005; Salgado et al. 2005; García et al. 2010) as well as in Sanagasta in northwestern of

Argentina, shedding insights on both paleoenvironmental aspects and the gregarious behavior of sauropods (Fiorelli et al. 2012, 2013).

The aim of this chapter is to bring forward primary taphonomic observations that focus specifically on South American Cretaceous sauropodomorphs, including the analysis of some intrinsic and extrinsic factors related to accumulation of large skeletal elements. Moreover, a new nomenclatural code is proposed in the analysis of taphonomic modes, to facilitate their description and understanding with respect to sauropods, and other relatively large no-avian dinosaurs. Lastly, it is important to point out that for the purposes of this study, as most of the case studies we cite are coincident with the now decades long growing appreciation of the value of taphonomic practices to an excavation, we make the assumption that all available bones were excavated, and removal of skeletal elements was not selective.

Institutional Abbreviations MUCPv: Museo de Geología y Paleontología Universidad Nacional del Comahue, Neuquén, Argentina; **UNCUYO-LD**: Universidad Nacional de Cuyo, Instituto de Ciencias Básicas, Laboratorio de Dinosaurios, Mendoza, Argentina; **MDT-PV**: Museo Desiderio Torres, Paleontología de Vertebrados, Sarmiento, Chubut, Argentina.

2 Remarks on Primary Taphonomic Studies

A brief review of the most significant detailed taphonomic studies available to date is presented here. In the Triassic of South America, taphonomic studies have focused more on the analysis of the entire fossil vertebrate associations and not on a specific taxon. Ischigualasto-Villa Unión is a world-famous sedimentary basin that includes the most diverse dinosaur faunas from the Late Triassic of South America recognized thus far (Bonaparte 1972, 1999a, b; Martínez and Alcober 2009; Ezcurra and Apaldetti 2011), including one of the oldest-known dinosaur assemblages (Rogers et al. 1993; Martínez et al. 2011). In particular, in the Ischigualasto Formation sauropodomorphs are represented by *Panphagia protos* (Martínez and Alcober 2009) *Eoraptor lunensis* (Sereno et al. 1993), and *Chromogisaurus novasi* (Ezcurra 2010) that come from the *Scaphonyx-Exaeretodon-Herrerasaurus* biozone (sensu Martínez et al. 2011). In the Ischigualasto Formation, Colombi et al. (2012) described the environmental settings of different types of preservation in vertebrates, including specific taphonomic modes. In their study, floodplain deposits, which are divided into proximal and distal components, preserve most fossils (approximately 88%). The quality of fossil preservation is highly variable, from isolated and weathered specimens to complete and well-preserved skeletons. In contrast, coarser-grained deposits of ancient fluvial channels preserve just under 12% of the fossil occurrences, including some well-preserved cynodont skulls. Vertebrate fossils are exceptionally rare in abandoned channel facies of the Ischigualasto Formation. Moreover, the fluvial architecture and paleosols of this formation provided a framework to reconstruct the paleoenvironmental evolution of this basin during the Late Triassic using continental sequence stratigraphy (Colombi et al. 2017).

Sauropodomorphs increased in size and abundance toward the end of the Triassic in the overlying Los Colorados Formation (Martínez et al. 2013). This formation has yielded a remarkably high diversity of basal sauropodomorphs such as *Riojasaurus incertus* (Bonaparte 1972), *Coloradisaurus brevis* (Bonaparte 1978), and *Lessemsaurus sauropoides* (Bonaparte 1999a; Pol and Powell 2007). In their preliminary taphonomic study, Otero et al. (2019a) described aspects of fossil diagenesis and the main taphonomic features present in sauropodomorph assemblage. Their study showed that the most abundant bones (i.e., appendicular elements and vertebrae) are generally found disarticulated. A different faunal assemblage was recently discovered in the Quebrada del Barro Formation, Marayes-El Carrizal Basin, Argentina (Martínez et al. 2015) which comprises numerous vertebrate specimens of Cynodontia, Testudinata, Sphenodontia, Pseudosuchia, Pterosauria, and Dinosauromorpha. In particular, sauropodomorph specimens are represented by partial skeletons without skulls.

Mussaurus patagonicus (Bonaparte and Vince 1979) is known from the Laguna Colorada Formation (El Tranquilo Group, Norian, Late Triassic of Patagonia). The remains of this taxon were found along with eggshells, and the individuals were interpreted to be just a few days old (Bonaparte and Vince 1979). Subsequently, additional sauropodomorph bones discovered at the fossil site were attributed to adults of *Mussaurus* (Otero and Pol 2013; Otero et al. 2019b). Additional fieldwork resulted in the discovery of new and abundant materials from Laguna Colorada. The remains here, which include juvenile individuals, subadults, and adults of *Mussaurus*, as well as their nesting sites, are an extraordinary discovery as it helped advance the understanding of the paleobiology of sauropodomorphs (Smith et al. 2014).

In the Jurassic of South America, there is little taphonomic information on sauropods. One available study corresponds to *Patagosaurus fariasi*, from the early Middle Jurassic Cañadón Asfalto Formation, Patagonia (Bonaparte and Vince 1979; Holwerda and Pol 2018). This discovery received some attention due to the preservation of several specimens, which was interpreted as evidence of gregarious behavior by Coria (1994).

In the Cretaceous of South America, diverse sauropod lineages have been discovered, and most of the taphonomic work has focused on Argentinean titanosaurian taxa. *Mendozasaurus neguyelap* was discovered in the middle–upper Coniacian Sierra Barrosa Formation, in Mendoza Province (González Riga 2003, 2005). The anatomy of this taxon has been extensively studied (González Riga et al. 2018), helping to define two new titanosaurian clades: Lognkosauria (Calvo et al. 2007a) and Colossosauria (González Riga et al. 2019). Approximately 200 bones and bone fragments, which represent four individuals, were recovered from the Arroyo Seco site. Most remains were assigned to *Mendozasaurus neguyelap*, a taxon represented by an associated set of elements consisting of 22 articulated caudal vertebrae and several others disarticulated bones that include cervical and dorsal vertebrae, appendicular elements, and large osteoderms corresponding to two adult individuals (~18 m long). Another set of disarticulated bones belonging to a larger specimen (~25–27 m long) of the same taxon were also identified. In the same locality, other fragmentary and indeterminate titanosaurian specimens and a small theropod referred to Maniraptora indet. were recovered. These remains represent a specific taphonomic mode, 'overbank bone assemblage' (sensu González Riga and Astini 2007). This taphonomic mode is characterized by disarticulated to partially articulated accumulations of bones (some of them are oriented) that have been exposed to weathering prior to burial within a proximal overbank deposit of an ancient meandering fluvial system. Within this setting, the dynamics of crevasse splay deposition trapped these sets of bones (González Riga and Astini 2007). At the same site, a few meters above the *Mendozasaurus* level, a new ichnospecies named *Cubiculum levis* was discovered and described (Pirrone et al. 2014). It is a bioerosion trace preserved in a dinosaur bone and assigned to dermestid beetles.

The fossil remains of *Bonitasaura salgadoi* were discovered in the Santonian-Campanian Bajo de la Carpa Formation, at La Bonita site in the Río Negro Province (Apesteguía 2004). Taphonomic and sedimentary analyses concluded that the dinosaur died close to a river margin and was rapidly incorporated into the fluvial sediments that buried the bones in successive events (Pérez et al. 2009). The spatial distribution of bones and their state of association suggest a short transport distance from the source area. The presence of two articulated caudal series with an opisthotonous articulation pattern suggests that the animal was exposed long enough allowing the onset of rigor mortis and desiccation. The degree of disarticulation also points to some amount of weathering of the skeleton before being buried. However, the many examples of exceptional periostium preservation suggest a very brief window of time for subaerial exposure. The bones are preserved in massive sandstones (facies E after Pérez et al. 2009) located over channel sands indicating a relatively rapid sedimentary event.

The bones of *Aeolosaurus colhuehuapensis* were collected in the Coniacian-Maastrichtian Lago Colhué Huapi Formation, at Lago Colhué Huapi (Chubut Province, Argentina, Casal et al. 2007). The fossil materials include twenty-one caudal vertebrae and seven hemaphophyses, which were found articulated in facies corresponding to overbank deposits from multi-channel river systems of moderate to high sinuosity. The taphonomic analyses of *Aeolosaurus colhuemhuapensis* remains presented by Casal et al. (2014a) are the first detailed taphonomic case study of a dinosaur preserved in proximal floodplain facies for the Chubut Group in Patagonia. The presence of slight longitudinal striation, scarce exfoliation, the degree of articulation, and dorsal bowing of the caudal series (opisthotonos articulation) suggests a short subaerial exposure time and rapid burial, with little fluvial transport. This type of bone concentration may correspond to the taphonomic mode 'overbank bone assemblages' proposed by González Riga and Astini (2007).

Notocolossus gonzalezparejasi was discovered in the upper Coniacian-lower Santonian Plottier Formation (Mendoza Province, Argentina, González Riga et al. 2016). *Notocolossus* is one of the largest known dinosaurs ever found (Paul, 2019) and is represented by two specimens: a large and disarticulated individual, and a small and partially articulated one. Previteria (2017, 2019) described the taphonomy of the Cerro Guillermo sites, including that of *Notocolossus*, and recognized three taphonomic modes in fluvial systems where isolated bones commonly occurred in channel deposits, and the associated or articulated bones were preserved in distal floodplain deposits (especially well-drained ones). In the case of *Malarguesaurus florenciae* (González Riga et al. 2009a, b), which was discovered in the same area, their disarticulated bones are preserved in massive red mudstones that are interpreted as a relatively well-drained floodplain facies overlaying to crevasse splay deposits in a fluvial meandering system. The presence of eolian sandstones at the Cerro Colorado section, laterally correlated and very close to the fossil site of *Malarguesaurus* as well as the features of the association of facies suggests the development of sub-arid episodes within the floodplain deposits (González Riga et al. 2009a, b; Previteria 2017).

Pino et al. (2021) described the biostratinomic processes associated with an accumulation of bones from the early Valanginian Mulichinco Formation (Neuquén Province, Argentina). In the lower levels of the formation, lag deposits contain fragmentary bones of an indeterminate ornithopod. At the middle and upper levels of the formation, the channel filling contains a multispecific accumulation of bones tentatively referred to dicraeosaurid sauropods, with large and relatively more complete specimens, including a partial articulated skeleton without significant transport. In this last case, we interpret the skeletal integrity was likely due the presence of soft tissue being present during transport.

Sarmientosaurus musacchioi (Bajo Barreal Formation, Chubut Province, Argentina) was discovered by Martínez et al. (2016) and includes a complete skull with all the teeth located in their corresponding alveoli, the atlas, axis and other cervical vertebrae, cervical ribs, and part of a cervical tendon. The fossils were found articulated in such a way that the vertebrae lay on the right side while the skull was disposed ventrally. The weathering surfaces of the remains show evidence of a short time of subaerial exposure and then a rapid burial. The skeletal remains were entombed in coarse to medium green tuffaceous sandstones, with vitreous shreds and clayey intraclasts, and abundant matrix (>40%), interpreted as a hyper-concentrated flow corresponding to an overflow lobe located on the proximal floodplain (Casal and Nillni 2020). Recent erosion cut a deep channel after the seventh cervical, likely destroying any additional skeletal elements. The bones of *Sarmientosaurus* do not show tooth marks or other evidence of bioerosion; however, during the excavation an abelisaurid tooth was discovered near the occipital region of the skull (Martínez et al. 2016).

Detailed diagenetic studies on Cretaceous sauropods were made using thin sections, X-ray diffractometry, scanning electron microscope, and geochemistry to analyze the diagenetic processes of fracturing, plastic deformation, and permineralization events. This line of research is important; however, it is beyond the scope of this chapter. Examples of these analyzes are focused on sauropod remains from the provinces of Mendoza (González Riga and Astini 2007; Previteria 2017, 2019) and Chubut (Casal et al. 2017a, 2019, Casal and Nillni 2020). Some studies allowed paleoclimatic interpretations; for instance, permineralization of vascular channels with abundant hematite is frequent in well-drained distal floodplains and in arid or semiarid climates, since seasonal fluctuations in the water table favor the release by hydrolysis of Fe^{+2} from phyllosilicates that make up the rocks and their precipitation in the vascular channels (Pereda-Suberbiola et al. 2000; Pfretzschner 2004; Casal et al. 2017a).

Other taphonomic studies were carried out on extraordinary sauropod nesting sites and their associated egg remains discovered in Argentina. Chiappe et al. (1998) reported large numbers of sauropod eggs that provided the first unequivocal embryonic remains of sauropod dinosaurs. The eggs come from the early Campanian Anacleto Formation from Auca Mahuevo (Neuquén Province, Argentina). Chiappe et al. (1998) described at least four different levels of eggs, two of which (levels 3 and 4) have lateral continuity for several kilometers. Clutch sizes vary from 15 to almost 40 eggs stacked without spatial ordering (Chiappe and Coria 2004) and were deposited in depressions excavated in the substrate (Garrido 2010). Titanosaur nests consist of shallow depressions, which were excavated in the ground and rimmed by sediment removed from the excavation (Vila et al. 2010). In Auca Mahuevo, this structure was documented only in those few cases where the nests had been dug in sandy fluvial channels and later covered by mud (Chiappe et al. 2004). Eggs were deposited by titanosaurs in a low-gradient alluvial plain environment. However, in a few cases, these eggs are found in sandstones interpreted as deposits of abandoned channels or crevasse splay (Dingus et al. 2000).

The Sanagasta nesting site located in La Rioja Province (Argentina) is another important area for taphonomic studies as it shed insights on the reproductive behavior of titanosaurs, suggesting colonial behavior and site fidelity with possible phylopatry (Grellet-Tinner and Fiorelli 2010). Geologic, paleontologic, and taphonomic evidence (Fiorelli et al. 2012, 2013) suggests that the Sanagasta titanosaurs chose a hydrothermal area as nesting ground, an opportunistic environment-dependent reproduction relationship with a geothermally active paleoenvironment. Grellet-Tinner et al. (2012) suggested that a biological adaptation to this unique environment may have been that the outer eggshell surfaces of the eggs in the hydrothermal substrate were thickened, then thinned by dissolution and acidification, a process that likely had a great effect on the embryo physiology and development and, therefore, incubation time. The presence of extremely thick eggshells in eggs laid in geothermal environments represents a natural reproductive adaptation to resist chemical dissolution in this extreme environment by buffering external acidic hydrothermal fluids.

In some taxonomic and phylogenetic studies, general taphonomic remarks have been included such as studies focused on the rebbachisaurid *Katepensaurus goicoecheai* (Casal 2015; Casal et al. 2017a,b), the titanosauriform *Malarguesaurus florenciae* (González Riga et al. 2009a, b), and the titanosaurians *Epachthosaurus sciuttoi* (Rodríguez 1993), *Futalognkosaurus dukei* (Calvo et al. 2007a, b), *Drusilasaura deseadensis* (Navarrete et al. 2011), *Aeolosaurus maximus* (Santucci and Arruda-Campos 2011), La Invernada specimen (González Riga et al. 2008), and in the monotaxic association of the giant *Patagotitan* (Carballido et al. 2017).

Similarly, taphonomic comments have been made in some marine-marginal deposits related to the Late Cretaceous transgression of the Atlantic Sea over Patagonia. In these cases, the sauropod record comprises only teeth and small bone fragments preserved as lag deposits in channels of tidal-dominated deltas associated with several freshwater turtles, theropods, plesiosaurs, and fishes (Parras et al. 1998; González Riga 1999; Martinelli and Forasiepi 2004; Previtera and González Riga 2008). These mixed assemblages are preserved in the same late Campanian-early Maastrichtian formations that have exceptionally preserved sauropod trackways (González Riga and Calvo 2009; González Riga and Tomaselli 2019; Tomaselli et al. 2021).

3 Analysis of Intrinsic Factors

The preservation of sauropodomorphs, as in other vertebrates, is conditioned by both intrinsic (biotic) and extrinsic (abiotic and biotic) factors that are closely related to each other. Intrinsic factors can be thought of as driven in large part by animal behavior or related to the structural integrity of individual bones of skeletons, while extrinsic factors cross over to physical aspects of the environmental hazards such as droughts, wildfires, and flooding. In general, each group of dinosaurs (and vertebrates in general) has more robust and more fragile skeletal elements, which indicates that intrinsic factors also determine which elements are more frequently preserved and which elements are extremely rare. Considering that some sauropods achieved volumes and lengths that exceeded any other terrestrial vertebrate, like *Notocolossus, Patagotitan,* and *Argentinosaurus* (González Riga et al. 2016; Carballido et al. 2017; Paul 2019), the taphonomic processes for their preservation must be considered as potentially unique and analyzed carefully. Specifically, sedimentary extrinsic factors are essential to understand different accumulation processes that could bury skeletons of huge animals.

Among intrinsic factors which are potentially relevant for the fossilization of sauropod remains are: (1) body size, (2) general skeletal plan (3) weak point of disarticulation, (4) anatomical structural fragility, (5) shape and size of each bone element, (6) ontogeny, (7) paleopathology, (8) behavior including aspects of locomotion, gregariousness, feeding and diet, and reproduction. It has long been recognized that there is bias for the preservation of larger animal skeletons (Behrensmeyer et al. 1979; Behrensmeyer and Dechant Boaz 1980; Arribas and Palmqvist 1998; Germonpré 2003; White and Diedrich 2012), though this phenomenon may not be directly applicable to dinosaur skeletons (Dodson 1971). In addition, while skeletons have their own taphonomic attributes (Rogers et al. 2007) sauropod skeletal elements have a very broad range in size as well as other unique attributes that provide their own taphonomic biases (see below). Given that dinosaur surficial bone texture shows changes with ontogeny (e.g., Ryan et al. 2001), ontogeny's role as an intrinsic factor may contribute to the truism that sauropod juveniles are quite rare. Lastly, aspects of behavior such as age segregation have been shown for at least some sauropods (e.g., Myers and Fiorillo 2009), and such behavioral characters likely played a taphonomic role in the formation of the sauropod fossil record. From these parameters, in South America only some aspects of the second and sixth have been studied in taphonomic and ichnologic analyses (Coria 1994; González Riga 2011; Fiorelli et al. 2013; García et al. 2014; González Riga et al. 2015; González Riga and Tomaselli 2019). In this chapter, the first four parameters are discussed in some detail.

3.1 Body Size

Sauropodomorphs include dinosaurs of very varied sizes, ranging from small (less than 10 m long) to giant forms (around 35 m long) (Wilson 2005; Paul 2019). A detailed study of the South American record allows us to define small, medium, large, and giant-sized sauropods. These general body size class categories proposed here are admittedly arbitrary but serve to better understand the extreme variation of body size. These classes are recognized based on the length of the most frequent appendicular bones in the record, the humerus (H) and the femur (F). For taxa that do not have preserved these bones, other axial or appendicular elements were included for size estimates. It is worth noting that the total body length (L) is based on relatively complete titanosaurian (*Futalognkosaurus, Dreadnoughtus, Patagotitan* and *Epachthosaurus*) and diplodocoid specimens (*Amargasaurus*). These four categories are:

(1) Small body size (H < 80 cm; F < 110 cm, L < 12 m) which includes: *Amazonsaurus maranhensis* (Carvalho et al. 2013); *Tapuiasaurus macedoi* (Zaher et al. 2011); *Laplatasaurus araukanicus* (Huene 1929); *Amargasaurus cazaui* (Salgado and Bonaparte 1991); *Amargatitanis macni* (Apesteguía 2007); *Rinconsaurus caudamirus* (Calvo and González Riga 2003); *Muyelensaurus pecheni* (Calvo et al. 2007a, b, c); *Sarmientosaurus musacchioi* (Martínez et al. 2016); *Saltasaurus loricatus* (Bonaparte and Powell 1980); *Rocasaurus muniozi* (Salgado and Azpilicueta 2000); *Brasilotitan nemophagus* (Machado et al. 2013) *Overosaurus paradasorum* (Coria et al. 2013); *Trigonosaurus pricei* (Campos et al. 2005); *Pilmatueia faundezi* (Coria et al. 2019); *Gondwanatitan faustoi* (Kellner and Azevedo 1999); *Yamanasaurus lojaensis* (Apesteguía et al. 2019); *Neuquensaurus australis* (Lydekker 1893; Powell 1986); *Bravasaurus arrierosorum.*

   Hechenleitner et al. (2020); *Bonatitan reigi* (Martinelli and Forasiepi 2004); *Arackar licanantay* (Rubilar-Rogers et al. 2021); *Leikupal laticauda* (Gallina et al. 2014); *Bajadasaurus pronuspinax* (Gallina et al. 2019).

(2) Medium body size (H = 80–130 cm; F = 110–155 cm, L 12–22 m) which includes: *Rayososaurus agrioencis* (Bonaparte 1996, 1997); *Comahuesaurus windhauseni* (Carballido et al. 2012); *Bonitasaura salgadoi* (Apesteguía 2004); *Panamericansaurus schroederi* (Calvo and Porfiri 2010); *Narambuenatitan palomoi* (Filippi et al. 2011); *Epachthosaurus sciuttoi* (Powell 1990); *Pitekunsaurus macayai* (Filippi and Garrido 2008); *Ligabuesaurus leanzai* (Bonaparte et al. 2006); *Kaijutitan maui* (Filippi et al. 2019); *Elaltitan lilloi* (Mannion and Otero 2012); *Drusilasaura deseadensis* (Navarrete et al. 2011);

Aeolosaurus maximus (Santucci and Arruda-Campos 2011); *Katepensaurus goicoecheai* (Ibiricu et al. 2013); *Choconsaurus baileywillisi* (Simón et al. 2017); *Zapalasaurus bonapartei* (Salgado et al. 2006); *Agustinia ligabuei.*

Bonaparte (1999b); *Nullotitan glaciaris* (Novas et al. 2019); *Atacamatitan chilensis* (Kellner et al. 2011); *Punatitan coughlini* (Hechenleitner et al. 2020); *Limaysaurus tessonei* (Calvo and Salgado 1995; Salgado et al. 2004); *Quetecsaurus rusconii* (González Riga and Ortiz David 2014); *Barrosasaurus casamiquelai* (Salgado and Coria 2009); *Maxakalisaurus topai* (Kellner et al. 2006); *Adamantisaurus mezzalirai* (Santucci and Bertini 2006); *Baurutitan britoi* (Kellner et al. 2005); *Aeolosaurus rionegrinus rionegrinus* (Powell 1987); *Aeolosaurus colhuehuapensis* (Casal et al. 2007); *Andesaurus delgadoi* (Calvo and Bonaparte 1991); *Triunfosaurus leonardii* (Carvalho et al. 2017); *Pellegrinisaurus powelli* (Salgado 1996); *Cathartesaura anaerobica* (Gallina and Apesteguía 2005); *Padillasaurus leivaensis* (Carballido et al. 2015); *Baalsaurus mansillai* (Calvo and González Riga 2018); *Nopcsaspondylus alarconensis* (Apesteguía 2007).

(3) Large body size (H $\geq$ 130–155 cm; F $\geq$ 155–200 cm, L 22–28 m) which includes: *Mendozasaurus neguyelap* (González Riga 2003; González Riga et al. 2019); *Chubutisaurus insignis* (Del Corro 1975); *Malarguesaurus florenciae* (González Riga et al. 2008); *Traukutitan eocaudata* (Juárez Valieri and Calvo 2011); *Argyrosaurus superbus.*

Lydekker (1893); *Austroposeidon magnificus* (Bandeira et al. 2016); *Uberabatitan ribeiroi* (Salgado and Carvalho 2008).

(4) Giant body size (H > 155; F > 2,000; L > 28 m) which includes: *Futalognkosaurus dukei* (Calvo et al. 2007a, b, c); *Dreadnoughtus schrani* (Lacovara et al. 2014); *Argentinosaurus huinculensis* (Bonaparte and Coria 1993); *Notocolossus gonzalezparejasi* (González Riga et al. 2016); *Patagotitan mayorum* (Carballido et al. 2017); *Puertasaurus reuili* (Novas et al. 2005); *Antarctosaurus wichmanianus* (Huene 1929).

Traditionally, taphonomic observations predict that animals of large body size may potentially be overrepresented in the fossil record (Behrensmeyer et al. 1979). In the case of non-avian theropods, Cashmore et al. (2020) recover no significant relationship between the body size of theropod taxa and their skeletal completeness, even when Lagerstätten taxa are removed. In contrast, in the case of sauropods, the probability of preservation of complete individual declines with larger size (Dodson 1971). Thus, complete sauropod skeletons are exceedingly rare. For example, in 2004, from 100 valid genera, only 21 are known from 90% or more of the entire skeleton (Upchurch et al. 2004). This record produces missing data on sauropod taxonomy and systematics (McIntosh 1990), and this problem, like on other continents that have produced sauropodomorphs, is also present in South America.

Body size is the most obvious intrinsic factor, and it is related to two taphonomic features: 'completeness' of individuals and 'frequencies of skeletal parts,' the last one described in the next paragraphs of Sect. 3.2. Here we describe 'skeletal completeness' as a percentage in relation to the total number of elements expected in a complete skeleton, averaged over the number of individuals found at a quarry site (sensu Voorhies 1969; Behrensmeyer and Dechant Boaz 1980; Fiorillo 1988; Lyman 1994). Other concepts defined by Mannion and Upchurch (2010) were named 'skeletal completeness metric' (SCM) and are useful in other approaches. One of them, SCM1, is 'the completeness (expressed as a percentage) of the most complete specimen known for that taxon,' and the SCM2 'quantifies how much of the skeleton is known for a given taxon as a whole; that is, it utilizes all known individuals of that taxon.'

3.2 General Skeletal Plan

Here we define 'the general skeletal plan' (GSP) as an intrinsic factor related to the morphology of the skeleton and the number of bones in each sector. It can be analyzed by different taphonomic attributes, among them, the completeness of the individuals and the frequency of the skeletal parts.

The general skeletal plan in sauropods comprises the following structure: a long neck with ~13–15 cervical vertebrae, ~10–13 dorsal vertebrae, ~5–6 sacral vertebrae, and an extremely long tail with ~35 caudal vertebrae, and columnar limb elements with reduced manual and pedal bones being represented by two scapulae, two sternal plates, two humeri, two ulnae, two radius, five metacarpals, variable numbers of manual phalanges, two ilia, two pubes, two ischia, two femora, two tibiae, two fibulae, one astragalus, a variable numbers of pedal phalanges, and in some cases, osteoderms (McIntosh 1990; Upchurch 1998; Wilson and Sereno 1998; Wilson 2002, 2005; Upchurch et al. 2004; González Riga et al. 2008, 2016; Nair and Salisbury 2012; Mannion et al. 2013). In this general plan, the skull is very small in relation to the total length of the body and comprises ~40 elements, which are mostly fused (Zaher et al. 2011; Martínez et al. 2016).

The taphonomic feature 'frequencies of skeletal parts' (sensu Lyman 1994: 223) is one of the most obvious and visible properties of a faunal assemblage. To analyze the data, here we use the criterion of the presence or absence of different portions of the skeleton. In this way, we do not consider the total number of bones, instead of the presence of a sector of the skeleton. For example, if an individual had preserved caudal vertebrae, it is indicated by '1' in 'caudals,' independently that the number of caudals recovered (Table 1). The elements that present the greatest occurrence in South American sauropods are the axial bones, mainly the caudal vertebrae. This is the case of most titanosaurian taxa, like *Andesaurus, Baurutitan, Bonitasaura, Dreadnoughtus, Epachthosaurus, Gondwanatitan, Mendozasaurus, Overosaurus, Saltasaurus, Rinconsaurus,* and *Pellegrinisaurus.*

The anterior and posterior appendicular elements have similar representation, with the femora and humeri as the most abundant. At the opposite extreme, the skeletal elements with the least potential for preservation are the bones of the pedes, manus, and the skull (Fig. 1a).

Table 1 Frequencies of skeletal parts of sauropods in South American records

Bones		Number	Articulated
Cranial remains		24	
Skull	Almost complete skull	3	3
	Braincase	7	–
	Dentary	7	–
	Isolated teeth	12	–
	Other elements isolated	14	–
Axial remains		65	–
Axial skeleton	Cervical	41	5
	Dorsal	50	8
	Sacral	26	7
	Caudal	56	22
	Ribs (cervical 16, dorsal 29. Sacral 4)	49	6
	Chevrons	25	5
Anterior appendicular remains		42	–
Anterior appendicular skeleton	Scapula	25	4
	Coracoid	15	4
	Sternal plate	18	–
	Humerus	36	4
	Ulna	21	4
	Radius	21	4
	Metacarpals	22	3
Posterior appendicular remains		53	–
Posterior appendicular skeleton	Ilium	22	5
	Ischium	26	5
	Pubis	24	5
	Femur	37	3
	Tibia	23	6
	Fíbula	23	6
	Astragalus	14	4
	Metatarsals	13	5
	Phalanges	9	5
Other elements	Osteoderms	3	–
Total taxa of sauropods from the South American Cretaceous		71	

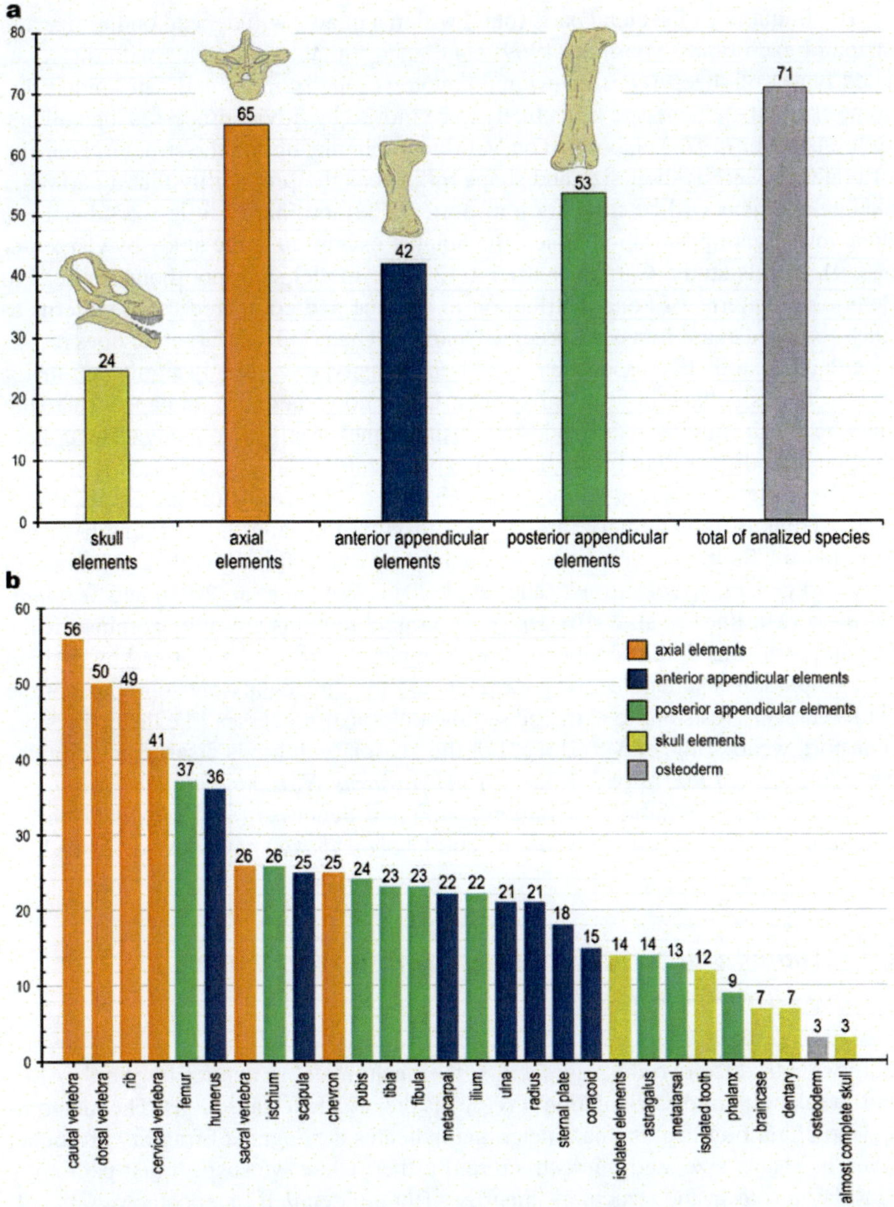

Fig. 1 Frequencies of skeletal elements of the 71 currently recognized South American Cretaceous taxa. For further explanation see the text and Table 1. **a** Axial and posterior appendicular elements are most preserved bones elements. **b** Detail of the frequency of skeletal parts in the South American Cretaceous sauropod record. In orange (axial elements), in green (posterior appendicular elements), in blue (anterior appendicular elements), and in yellow (cranial elements)

The frequency of skeletal parts found within a quarry will depend on the size and shape of each bone. Voorhies (1969) conducted flume experiments with disarticulated bones of sheep (*Ovis aries*) and coyote (*Canis latrans*) and determined that some skeletal elements are more likely to be moved by fluvial processes than others. Behrensmeyer (1975) elaborated on Voorhies' scheme, noting the structural density of bones as well as their size and shape influences the probability that a particular skeletal element will be fluvially transported. Extraordinarily little work has been done investigating sauropod bones, the notable exception is the study by Carpenter (2020). In this study, Carpenter used 1/12 scale model sauropod bone replicas to determine patterns of bone distribution in a flume and compared those patterns to ones observed at the Jurassic Carnegie Quarry in Utah (USA). Carpenter does recognize that his study dismisses the role of bone density, a seemingly significant factor in how bones behave in a fluvial system. Up to now, there are no papers referring specifically to full-size sauropod bones using bone's replicas to analyze accurately the potential dispersion of each bone in experiments.

It has often been said through the history of the study of sauropods that a preserved skull is extremely rare in their fossil record and this remains true. For example, in South America, almost complete, nearly fused skulls are known in only two genera, *Tapuisaurus* (Zaher et al. 2011; Wilson et al. 2016) and *Sarmientosaurus* (Martínez et al. 2016). However, cranial remains are more common if the elements are considered separately. For example, the braincase (formed by ossified basioccipital, basisphenoid, and parasphenoid) and the dentary are solid parts of the skull relatively resistant to extrinsic sedimentological processes like the dispersion, transport, weathering, and crushing. For this reason, numerous disarticulated braincases (e.g., *Bonatitan, Muyelensaurus, Pitekunsaurus, Narambuenatitan, Kaijutitan, Amargasaurus, Limaysaurus, Bajadasaurus*) and dentaries (e.g., *Maxakalisaurus, Baalsaurus, Bonitasaura, Brasilotitan,* and *Choconsaurus*) have been described.

3.3 *Anatomical Structural Fragility and Weak Point of Disarticulation*

Here we introduce two intrinsic factors: the 'anatomical structural fragility' (ASF) and 'weak point of disarticulation' (WPD) as they relate to sauropods. The 'anatomical structural fragility' is a parameter that indicates the degree of fragility of a bone, given its shape, size, and microstructure. Further, while we suspect that pathology can play a role in the structural integrity of bone, detailed paleopathology is still an emerging field with much still to be learned, we focus here on non-pathologic aspects defining 'anatomical structural fragility' as a means of providing a predictive framework for future studies.

The concept of anatomical structural fragility is linked to the 'structural density' sensu Lyman (1996: 237). Lithostatic or confining pressure is the most frequent process that deforms bones plastically or breaks them, both in sauropods (e.g.,

González Riga and Astini 2007; Casal et al. 2014a; Müller et al. 2018) and other vertebrates like huge pterosaurs (Ortiz David 2019). Indirectly, the skeletal frequency graphs (Fig. 2b) indicate that sauropod skulls are skeletal elements with high structural fragility, being the first to break, disarticulate, and disperse. We recognize that

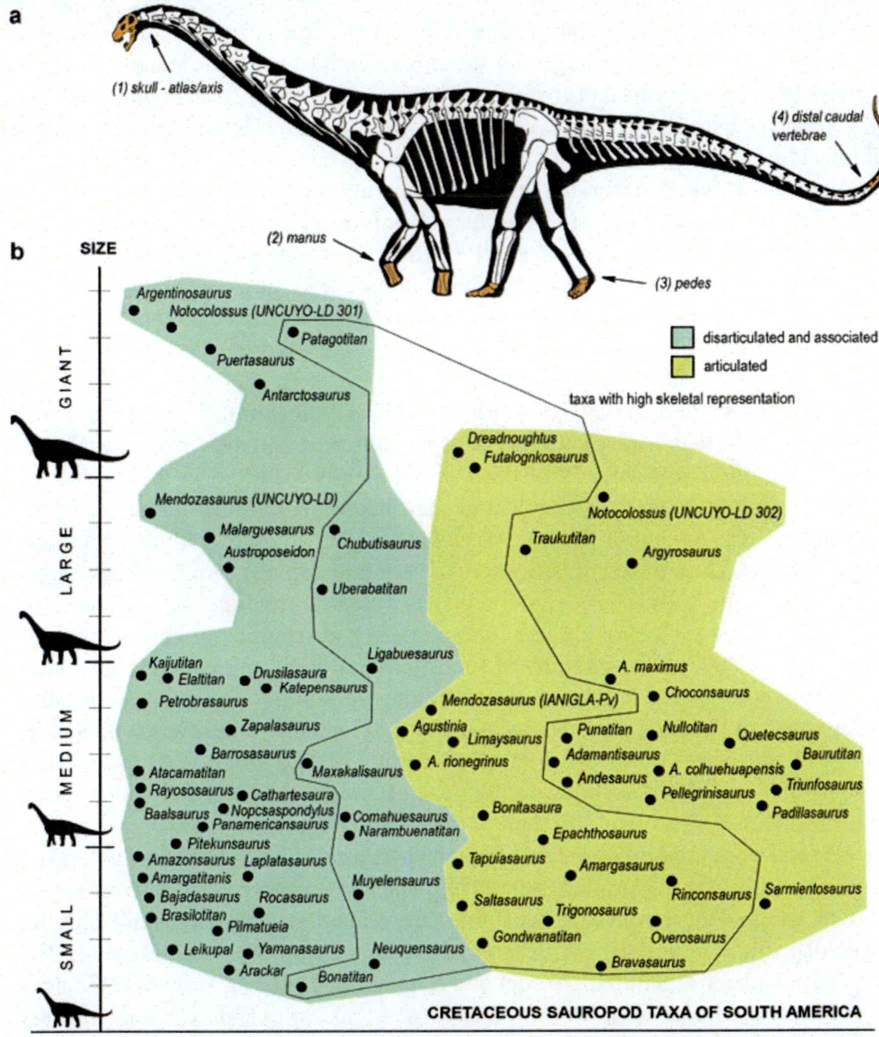

Fig. 2 **a** Main point of disarticulation recognized in South American Cretaceous sauropods: articulation atlas/axis to skull, articulation of manus, articulations of pes, and articulation of distal caudal vertebrae. **b** Body size class categories in relation to taphonomic preservation. Three types of preservation are represented here: (1) partially articulated specimens (yellow), (2) well-represented disarticulated or partially articulated specimens (black line), and (3) disarticulated specimens (green)

to define this parameter by a scale more useful for subsequent workers, more detailed studies are needed, which are beyond the scope of this chapter.

The 'weak point of disarticulation' is a parameter that addresses the fact that different joints of a skeleton have different morphologies, structures, and sizes both in appendicular and in axial bones. During the disarticulation process of a tetrapod carcass, there are weak points of disarticulation. The points are defined here as the joints of a skeleton that first disarticulate due to some form of taphonomic process. These joints produce the first stage of dismemberment in a skeleton, and the dispersal of certain skeletal elements that are relatively rare in the fossil record. For the purposes of this report, the available sauropod fossil record is used for defining this parameter and indicating the main points of disarticulation.

In the case of South American sauropods, mainly represented by titanosaurians, and to a lesser extent by diplocoids, four main points of early disarticulation are recognized: (1) articulation atlas/axis to skull, (2) articulation of manus (3) articulations of pes, and (4) articulation of distal caudal vertebrae (Fig. 2a).

These main points of disarticulation are areas where the muscular and ligaments patterns are scarce. Thus, the skeletal elements are separated in an early stage of the carcass degradation and dispersion processes.

The articulation of the atlas-axis-skull complex is typical of the anatomical structure of sauropods, which have extremely small skulls (up to 40–60 cm long) for necks of 3–9 m length. The weakness of this joint is clearly reflected in the absence of skulls in most South American Cretaceous sauropods. Neck elongation in sauropods involved a rearrangement of the bony structure to gain strength with the minimum weight increase. Thus, the complex system of vertebral laminae evolved as an adaptation to mitigate the body mass requirements, as well as correlating with the development of a complex system of air sacs (Otero 2018) and specific muscle morphology (Schwarz et al. 2007). The anatomy of sauropods that links an extremely long neck with an exceedingly small head is unique among vertebrates and is one of the five 'evolutionary cascade hypotheses' that explain the giantism of these animals (Sander et al. 2011, 2013). The obvious selective advantages of a small head are the low moments of force that it bestows on the neck (Taylor and Wedel 2013; Preuschoft and Klein 2013). The skeletal structure of the neck in huge sauropods comprises very long cervical vertebrae up to 1.2 m long which contrasts with the 0.5 m long skull (Carballido et al. 2017; Martínez et al. 2016). The neck comprises a relatively strong structure where long ossified cervical ribs are related with the hypaxial muscles that belong to the *m. longus colli* group based on the homology with birds (Klein et al. 2012; Cobley et al. 2013; Wedel and Sander 2002, 2011). However, the neck progressively decreases in size toward the first cervicals, atlas and axis, and with very small structures articulated with the occipital condyle of the basicranium forming a relatively weak articulation. These anatomical remarks explain the 'weak point of disarticulation' in the base of the skull.

Another interesting 'weak point of disarticulation' is the tibia/fibula with the astragalus/pes. For example, complete pedes are extremely scarce in the fossil record of titanosaurians, and indeed, in all groups of sauropods (Bonnan 2005; González Riga et al. 2016). Intrinsic factors related to the large body dimensions of these dinosaurs coupled with the relatively small size of their pedal elements evidently led to the early disarticulation, and therefore loss, of these comparatively small skeletal elements during the biostratinomic stage of necrokinesis (González Riga et al. 2009a, b). If we analyze the fossil record, of ca. 90 titanosaurian species recognized at present (Carballido et al. 2017), only three are known from complete and articulated pedes: *Opisthocoelicaudia skarzynskii* from Mongolia (Borsuk-Bialynicka 1977), and *Epachthosaurus sciuttoi* (Martínez et al. 2004) and *Notocolossus gonzalezparejasi* (González Riga et al. 2016) from Argentina. Two additional titanosaurs with complete, articulated hind feet may be added to this group: the Agua de Padrillo taxon (UNCUYO-LD 313, González Riga et al. 2015), which preserves both complete pedes, and the La Invernada taxon (González Riga et al. 2008), which preserves the complete left fore- and hind limbs (presently under study). Pedal elements are known for many other titanosaurs, but none of these preserve the pes completely and in articulation, an unfortunate aspect of sauropod taphonomy as the finding of articulated pedes would allow for the analysis of additional paleobiological aspects focused on locomotion, stance, and gaits (e.g., González Riga 2011; González Riga and Tomaselli 2019). Further, such elements have a role in phylogenetic studies where a progressive reduction in both the number and length of the pedal phalanges has been documented (Bonnan 2005; González Riga et al. 2016). For example, *Epachthosaurus* has a pedal phalangeal formula of 2-2-3-2-0 (Martínez et al. 2004), and an even more reduced formula of 2-2-2-2-0 occurs in the Padrillo (UNCUYO-LD 313) and Invernada (MUCPv-1533) taxa, *Notocolossus* (González Riga et al. 2016), and probably in *Mendozasaurus* (González Riga et al. 2018). In the Titanosauria, two primary skeletal morphotypes termed 'long-footed' and 'short-footed' were recently described (González Riga et al. 2016, 2019). In sum, taphonomic analyses of articulated pedes, some of them in anatomical position, are relevant to understanding their anatomical diversity and evolutionary trends.

The examination of the four sauropodomorph body size class categories in relation to degree of articulation is represented in Fig. 1b. Three broad types of preservation are recognized here: (a) disarticulated specimens, (b) specimens with some articulated parts, and (c) articulated or disarticulated specimens represented by numerous skeletal elements. From Fig. 2b, some general conclusions can be drawn.

First, numerous taxa are represented by disarticulated skeletal elements. In the record is frequent the presence of isolated or associated but disarticulated elements (Fig. 3a–b).

Second, the number of well-preserved specimens with some articulated portions of the skeleton is well-represented in the small and medium body size class (e.g., *Epachthosaurus, Trigonosaurus,* La Invernada Taxon; Fig. 3c) and decreases in large and giant taxa. In other words, the degree of articulation decreases with body size.

Third, in large taxa, most of them are represented by disarticulated bone elements; however, two new unnamed large titanosaurian individuals are preserved partially articulated. They are the specimen MDT-PV 4 discovered in the Lago Colhué Huapi Formation (Chubut) and specimen UNCUYO-LD 303 from the Plottier Formation (Mendoza) (Fig. 3d–f). In some cases, individuals with some articulated portions of the skeleton correspond to subadult specimens. This is the case of *Notocolossus,*

Fig. 3 Different types of preservation in South American sauropods. **a** Fragment of humerus with high degree of weathering and abrasion from the Barreales Formation (Chubut). **b** Caudal vertebrae of *Malarguesaurus* from the Los Bastos Formation (Mendoza) preserved as disarticulated but associated individual. **c** Partially articulated titanosaurian skeleton of La Invernada specimen MUCPv-1533 from the Allen Formation (Neuquén). **d** Articulated titanosaurian specimen MDT-PV 4 from the Lago Colhué Huapi Formation (Chubut). Next to the fossil, Dr. L. Ibiricu. **e–f** A new and recently discovered titanosaurian taxon from the Plottier Formation (Mendoza). This specimen (UNCUYO-LD 304) preserves a partially articulated skeleton, including a complete cervical series of 8 **m** long. **g** Excavation quarry of a giant titanosaurian *Patagotitan* from the Cerro Barcino Formation (Chubut). It is one of the biggest dinosaurs of the world and is represented by a monospecific assemblage of several disarticulated specimens

which is represented by a subadult individual (UNCUYO-LD 302) that preserved a part of the caudal sequence and a complete foot, while the adult and giant specimen (UNCUYO-LD 301) is completely disarticulated. For this reason, both specimens are plotted in two different body size classes. Similarly, in *Mendozasaurus* the medium-sized holotype specimen (IANIGLA-PV 065) has an articulated caudal series, but a large specimen (UNCUYO-LD 356) is poorly represented by some disarticulated skeletal elements.

Fourth, most individuals of giant titanosaurians (*Argentinosaurus, Notocolossus, Puertasaurus,* and *Antarctosaurus)* are poorly represented, suggesting that the sedimentary processes of fluvial environments do not fully encase and preserve an articulated skeleton that is more than 30 m long before other factors (i.e., scavenging or bone weathering) degrade the skeleton. However, exceptionally examples that have at least 40–50% of the skeleton are *Futalognkosaurus* (Calvo et al. 2007a, b), *Dreadnoughtus* (Lacovara et al. 2014; Ullmann and Lacovara 2016) and a new and unnamed specimen recently published (MOZ PV 1221, Otero et al. 2021). In the record studied, monotaxic or multitaxic associations are frequent in small to large taxa (e.g., *Muyelensaurus, Mendozasaurus*), being an exceptional case in giant taxa like *Patagotitan* (Fig. 3g).

4 Analysis of Extrinsic Factors and Taphonomic Modes

Sedimentological and taphonomic analyses are essential components of paleoecology; they provide information on the environmental context where organisms lived and were buried as well as the potential interactions between biotic and abiotic components of ancient ecosystems (Behrensmeyer and Hook 1992).

Given the general incompleteness of sauropod skeletons, it seems that the large bodies of sauropods required specific sedimentological processes to preserve them. In the case of extrinsic factors, sedimentary environments comprise a set of processes that can act separately or interact with each other, processes such as weathering, sedimentation rate, selection, dispersal, abrasion, and bone transport. At fine-scale resolution, driven by local environmental conditions, each skeleton or set of skeletons in each river, lake, or delta system can produce a somewhat unique taphonomic story. However, despite the differences in individual cases, it is possible to recognize standardized taphonomic modes, based on the main sedimentological contexts and their taphonomic attributes, as described below.

4.1 Taphonomic Modes

To understand taphonomic processes, it is necessary to review key definitions. Up to now, two concepts are important: taphofacies and taphonomic mode. Taphofacies is a suite of rock characterized by combinations of preservational features of the contained fossils, defined based on the consistency of those features (Brett and Speyer 1990: 258). A taphonomic mode is a recurring pattern of preservation of organic remains in a particular sedimentary context, accompanied by characteristic taphonomic features (Behrensmeyer 1988: 183). Here the focus is on taphonomic modes; they are useful for describing patterns of preservation and their associated paleoenvironmental and paleoclimatic contexts. In fluvial environments, most of the bone accumulations of large vertebrates contain a significant number of incomplete skeletons and fragmented bones (e.g., Fiorillo 1988, 1991; Rogers 1990; Coombs and Coombs 1997; Cladera et al. 2004; Orr et al. 2016). In the case of sauropods, complete skeletons are decidedly uncommon, and well-preserved skulls, pedes, and manus are extremely rare (González Riga et al. 2016). Sauropod fossils in South America are preserved in fluvial and lacustrine sub-environments (Tomaselli et al. 2021). However, the issue of preferential burial within specific sub-environments is unclear. Here the South American record is used to investigate with high resolution such environments.

Tables 2 and 3 provide a proposed new code of letters for describing fluvial environments and the main taphonomic attributes that can be applied to fossil assemblages. The proliferation of names and descriptions of similar taphonomic modes can be avoided using this code. Obviously, this code is likely to be modified and extended to additional sedimentary environments as new discoveries are made. The practical idea is that this coding serves to quickly understand where the bones have been preserved, their taphonomic history, and the possible habitat. Each of these sub-environments presents specific conditions of hydrodynamic transport, sedimentation

Table 2 Main taphonomic modes in the fluvial system including a new and easy code for its description. Sedimentological context is modified from Miall (1996)

Taphonomic mode	Sedimentological context
CHL-BA Channel-lag bone assemblage	Lag's deposits in channels
LA-BA Lateral accretion bone assemblage	Lateral-accretion deposits
ACH-BA Abandoned channel bone assemblage	Abandoned channel (oxbow lake) deposits
LE-BA Levee bone assemblage	Levee deposits
O-BA Overbank bone assemblage	Overbank lobules (crevasse splay)
OCH-BA Overbank channel bone assemblage	Channels of overbank deposits (crevasse splay)
LS-BA Laminated sand bone assemblage	Laminated sand sheets produced by flash floods deposits
WDF-BA Well-drained floodplain bone assemblage	Well-drained floodplain
VDF-BA Variable drained floodplain bone assemblage	Variable drained floodplain
PDF-BA Poorly drained floodplain bone assemblage	Poorly drained floodplain and swamps

Table 3 Taphonomic parameters used in the new taphonomic code proposed here. Taxonomic composition after (Fiorillo and Eberth 2004), and weathering modified from (Fiorillo 1988); completeness after the 'skeletal completeness' described in this chapter

Taphonomic parameters

Taxonomic composition	Skeletal articulation	Completeness	Weathering
M1. Monotaxic site (single individual)	A1. Articulated	C1. Complete	W1. Absent or incipient. bone surface shows no sign of cracking of flaking
M2. Monotaxic site (multiple individuals of the same taxon)	A2. Mostly articulated (more than 75%)	C2. Almost complete (more than75%)	W2. Low bone surface show cracking, usually parallel to the fibrous of the bone. Cracking confined to the outermost layers of bones
M3. Multitaxa site (multiple individuals of different taxa)	A3. Partially articulated (75–25%)	C3. Partially represented (75–25%)	W3. Moderate bone surface shows flaking, as well as craking, on outer surface. cracking has started to penetrate into the bone cavities
	A4. Poorly articulated (less than 25%)	C4. Poorly represented (less than 25%)	W4. High extreme outermost layers are gone, fibrous texture present. Most of the cracks penetrate into the bone cavities
	A5. Disarticulated	C5. Isolated and/or fragmentary bones	W5. Extremely high Large splinters present, bone can loss the morphology

rate and time of subaerial exposure, aspects that strongly condition the preservation of sauropod bones. Table 4 is an application of this coding as it shows several examples of preservation, using best-studied South American sauropodomorphs.

In South America, the sedimentological context of sauropod record indicates that these dinosaurs mainly inhabited fluvial and sometimes lacustrine environments, although the ichnological record is particularly abundant in marine-marginal facies related with the late Campanian–Maastrichtian Atlantic epicontinental transgression (Tomaselli et al. 2021). Up to now, no well-preserved sauropod skeletons have been found in these marine-marginal facies, and this stands in contrast to other taxa (e.g., articulated freshwater turtles, (González Riga 1999; de la Fuente et al. 2017).

Table 4 Taphonomic modes of relevant of sauropodomorph fossil assemblages of South America using coding provided in Tables 2 and 3

Taphonomic modes	Code	Taxonomic composition	Taphonomic features			Fm	Ref
			Skeletal articulation	Completeness	Bone modification		
Overbank bone assemblage	O-BA M3/A3-A5/C3-C4/W2-3	M3 Multitaxa site *Mendozasaurus neguyelap*, Titanosauria indet., Theropoda indet	A3 Partially articulated for *Mendozasaurus* holotype A5. Disarticulated in other specimens of *Mendozasaurus*, Titanosauria indet. and Theropoda	C3 Partially represented for *Mendozasaurus* holotype C4 Poorly represented for other specimens	W2-3 Low to moderate weathering	Sierra Barrosa Fm	González Riga and Astini (2007)
Overbank bone assemblage	O-BA M1/A1/C4/W2	M1 Monotaxic site (single individual) *Aeolosaurus colhuehuapensis*	A1 Articulated Composed by articulated caudal vertebrae	C4 Poorly represented	W2 Low weathering	Lago Colhué Huapi Fm	Casal et al. (2014a, 2019), Casal (2015)
Overbank bone assemblage	O-BA M1/A1/C4/W1	M1 Monotaxic site (single individual) *Sarmientosaurus musacchioi*	Skull and articulated anterior cervical vertebrae	C4 Poorly represented	W1 Incipient weathering	Bajo Barreal Fm	Martínez et al. (2016) Casal and Nillni, (2020); Casal (pers. observ.)
Variable drained floodplain bone assemblage	VDF-BA M1/A3/C3/W2	Sites 1 and 2 M1 Monotaxic site (single individual each one) *Notocolossus González parejasi,* UNCUYO-LD 302 and 301	A3 Site 1, Partial skeleton articulated that include a complete pes and an anterior-middle caudal vertebrae A5 Site 2, Disarticulated bones	C4 Poorly represented	W2 Low weathering	Plottier Fm	González Riga et al. (2016), Previtera (2019)

(continued)

Table 4 (continued)

Taphonomic modes	Code	Taxonomic composition	Taphonomic features			Fm	Ref
			Skeletal articulation	Completeness	Bone modification		
Poorly drained floodplain bone assemblage	PDF-BA M1/A2/C2/W1	M1 Monotaxic site (single individual) Padrillo specimen A, new titanosaurian taxon, UNCUYO-LD 304	A2 Mostly articulated skeleton from cervical to caudal vertebrae, and disarticulated appendicular bones	C2 Almost complete	W1 Incipient weathering	Plottier Fm	González Riga et al. (2012, 2013)
Poorly drained floodplain bone assemblage	PDF-BA M1/A3/C4/W1	M1 Monotaxic site (single individual) Padrillo specimen B, new titanosaurian taxon, UNCUYO-LD 313	A3 Partially articulated, represented by both complete pedes and some disarticulated axial bones	C4 Poorly represented	W1 Incipient weathering	Plottier Fm	González Riga et al. (2015)
Well drained floodplain bone assemblage	WDF-BA M1/A5/C4/W3	M1 Monotaxic site (single individual) *Katepensaurus goicoecheai*	A5. Disarticulated, composed by cervical, dorsal and caudal vertebrae, ribs	C4 Poorly represented	W3 Moderate weathering	Bajo Barreal Fm	Ibiricu et al. (2013, 2015), Casal et al. (2017a, b)

(continued)

Table 4 (continued)

Taphonomic modes	Code	Taxonomic composition	Taphonomic features			Fm	Ref
			Skeletal articulation	Completeness	Bone modification		
Well drained floodplain bone assemblage	WDF-BA M1/A4/C4/W3-4	M1 Monotaxic site (single individual) *Drusilasaura deseadensis*	A5. Disarticulated, composed by dorsa, sacral and caudal vertebrae and appendicular bones	C4 Poorly represented	W4 Moderate to high weathering	Bajo Barreal Fm	Navarrete et al. (2011); Casal (pers. observ.)
Levee bone assemblage	LE-BA M1/A3/C3/W1-2	M1 Monotaxic site (single individual) MDT-PV 4, unamend titanosaurian taxon	A3 Partially articulated, represented by axial and appendicular bones	C3 Partially represented	W1 Incipient to low weathering	Lago Colhué Huapi Fm	Casal et al. (2010, 2019); Casal (2015)
Laminated sand bone assemblage	LS-BA M1/A1/C2/W1	M1 Monotaxic site (single individual) *Epachthosaurus sciuttoi*	A1 Articulated. Composed by a skeleton lacking the skull, neck, some anterior dorsal an distastal caudal vertebrae	C2 Almost complete (more than75%)	W1 Incipient weathering	Bajo Barreal Fm	Rodríguez (1993), Martínez et al. (2004); Casal (pers. observ.)

Fm: Geological Formation, Ref: References

4.2 Weathering and Taphonomic Modes in Selected Cases

Weathering is one of the key extrinsic factors in taphonomy. It increases with the time exposed to the atmospheric agents like diurnal and seasonal temperature changes, wetting and drying, freezing and thawing, and exposure to UV rays (Behrensmeyer 1978; Lyman and Fox 1989; Fiorillo 1995). These processes produce cracking parallel to fiber structure, flaking of outer surface (exfoliation), and splinters on the external surfaces of skeletal elements. The effects of weathering vary according to intrinsic attributes such as the size, shape, and density of the bones and the ontogenetic state of individuals (Behrensmeyer 1978; Muñoz et al. 2008; Gutiérrez et al. 2018).

Much of the knowledge we have about weathering comes from studies that sought to understand the relationship between environmental variables and their impacts on organic remains, through current observations in different taxa and climatic regions (Cruz 2015). Several reference weathering scales have been proposed, based on actuotaphonomic data, which reflect the increase in bone deterioration and destruction as a function of exposure time under certain climatic and environmental conditions (Behrensmeyer 1978; Fiorillo 1988; Alcalá 1994; Lyman 1994). Tappen (1994) proposed that weathering varies between habitats, showing that the destruction of a bone would be slower in heavily vegetated areas such as a forest, where vegetation plays an important role in mitigating the effects of atmospheric agents; and conversely, weathering would be more intense in arid to semiarid zones devoid of vegetation. Most, but not all, weathering processes stop with the burial of the remains. Some bone alteration can continue in the shallow soil horizons (Behrensmeyer 1978; Todisco and Monchot 2008) though the additional alteration tends to be associated with etching from plant roots. The cessation of bone weathering with limited cover is important in fluvial sub-environments with low sedimentation rates.

Furthermore, for good preservation of nearly complete fossil skeletons it is important that the carcasses experienced little mobilization so that the mechanical action of transport did not disarticulate and fragment the bone elements. In fluvial environments, these two conditions, rapid burial and short transport distance, are closely related to the sedimentary supply rate, the type of flow that moves a carcass, and characteristics of the bone remains. The rapid burial and minimal transport allow a higher probability of articulation and skeletal completeness being retained. Also, the possibility that predators and scavengers intervene in the fragmentation and disarticulation of a skeleton has a certain dependence on the sedimentary environment and climate. For the former, if the sedimentation rate is high, the carcasses are quickly buried.

In a fluvial system, where sauropodomorphs have been preserved, the most active sedimentation area is associated with the channel where the higher hydraulic energy of transport and a higher sedimentary rate are present. Within active channels, usually skeletons are transported, increasing the likelihood of disarticulation as well as abrasion and fragmentation of the bone elements. This occurrence is commonly observed in deltaic channels where isolated bones are preserved with intraclasts in lag deposits (e.g., Loncoche Formation, Mendoza Province, Argentina, González Riga 1999). A similar deposit is observed in crevasse-channel deposits of meandering systems of the *Mendozasaurus*' quarry (Sierra Barrosa Formation, Mendoza Province, Argentina, González Riga and Astini 2007; Fig. 5). This example corresponds to the taphonomic mode named 'overbank channel bone assemblage' (OCH-BA, Table 4).

In fluvial systems, moving more distally from the active channel, the possibility of fragmentation by weathering increases, because the remains are exposed for a longer time in the surface. These conditions are accentuated in the distal floodplain deposits characterized by low sedimentation rate with little to no degree of bone mobilization (Cladera et al. 2004; Casal et al. 2014b). There, sedimentation occurs mainly by settling of fine particles or sporadic overflow that reaches distal areas far from the channel. Examples of remains preserved in this sub-environment are *Katepensaurus* (Casal et al. 2017a, b), *Malarguesaurs* (González Riga et al. 2009a, b; Previtera 2017), and *Notocolossus* (González Riga et al. 2016). In these sub-environments, we can differentiate 'well-drained,' 'variable drained,' and 'poorly drained floodplain bone assemblages' (WDF-BA, VDA-BA, and PDF-BA, respectively, Tables 2 and 4). These taphonomic modes are related to the position of the water table. If the water table is relatively low, lesser vegetation and frequent oxidative conditions in seasonal and semiarid climates are present. In contrast, if the table is relatively high, the substrate is partially saturated and a greater development of vegetation under humid climatic condition is present (Retallack 1988; Ghosh et al. 2006).

The poorly drained floodplain deposits are characterized here by saturated substrates during significant portions of the annual cycle and frequently form ephemeral swamps. In these facies, sauropod specimens are preserved with a low degree of weathering, such is the case of Agua del Padrillo, a new and unnamed titanosaurian from Plottier Formation, Mendoza Province, Argentina (UNCUYO-LD 313, González Riga et al. 2015). It preserves the two complete and articulated pedes but in one of them, the tibia and fibula are cut at level of the exposed portion, which is interpreted as the burial surface, having weathered and destroyed at their proximal end. Another specimen discovered in the same area and rock unit is represented by almost all articulated axial skeleton from the cervical to the caudal vertebrae, being one of the most complete titanosaurian taxa recorded in South America (UNCUYO-LD 304, González Riga et al. 2012, 2013). These two cases are included in a taphonomic mode named 'poorly drained floodplain bone assemblage' (PDF-BA after Table 4) (Fig. 4).

The variable drained floodplain deposits with different substrate saturation conditions can preserve articulated to completely disarticulate skeletal elements. With respect to a *Notocolossus*, for example, a small individual with some articulated portions of the skeleton, and a large individual totally disarticulated have been recorded (taphonomic mode VDF-BA; Table 4; Fig. 5a–c). The small individual was preserved with the foot in anatomical position associated with a portion of the caudal sequence articulated. This fully articulated foot suggests that an individual walked on a saturated muddy substrate and became trapped. After death, the lower limb was buried and preserved intact while most of the carcass was exposed and destroyed. Similar differential preservation in the skeleton of the same individual

Fig. 4 Different types of preservation in sauropod assemblages from South America. Taphonomic modes after Tables 2, 3 and 4. Exceptionally preserved new titanosaurian taxa recently discovered in the Plottier Formation (Mendoza). **a** Articulated specimen of a new taxon next to B. González Riga. This fossil specimen (UNCUYO-LD 303) preserves almost all articulated axial skeleton from the cervical to the caudal vertebrae and is one of the most complete titanosaurian taxa recorded in South America (taphonomic mode PDF-BA). **b** Technicians and L. Ortiz David working in taphomomic mapping of the titanosaurian articulated specimen. It was done using a Maptek I-Site 8,800 laser scanner and two Topcon GR3 geodetic GPS receivers. **c** Three-dimensional model referenced spatially by geographic coordinates that reproduce the orientation of the bones and their size, with an accuracy of ±1 cm. This is obtained by the methodology explained in the figure. **d** Complete and articulated pes (UNCUYO-LD 313) preserved in anatomical position (scale bar 10 cm). This finding suggested that titanosaurians had their feet buried in saturated muddy floodplains (taphonomic mode PDF-BA). **e** Detail of the articulated caudal vertebrae of the specimen UNCUYO-LD 303, scale bar 20 cm

Fig. 5 Different types of preservation in sauropod assemblages from South America. Taphonomic modes after Tables 2, 3 and 4. **a, b** Articulated foot in anatomical position of *Notocolossus* preserved in the Plottier Formation, Late Cretaceous of Mendoza. It is associated with the taphonomic mode 'variable drained floodplain bone assemblage' (VDF-BA), scale bar 20 cm). **c** *Notocolossus*, one of the largest dinosaur known in the world (life restoration of B. González Riga), Scale bar 20 cm. **d** Skull of *Sarmientosaurus* discovered in Bajo Barreal Formation, Patagonia, and assigned to taphonomic mode 'overbank bone assemblage' (O-BA). It is one of the two titanosaurian skulls know in South America, scale bar 10 cm. **e** Computed tomography-based digital visualizations of the skull of *Sarmientosaurus* (Martínez et al. 2016). **f** Humerus of *Mendozasausrus* preserved in the taphonomic mode 'overbank bone asseambleage' (O-BA), Mendoza, scale bar 30 cm. **g** Overbank facies association from the Arroyo Seco section (*Mendozasaurus*' quarry). Ripple lamination and trough cross-bedding of crevasse channels overlaying massive silty shales of the floodplain deposits, scale bar 1 m

has been observed in extant animals (Abler 1985) and has been invoked to explain the preservation patterns observed for some dinosaurs (e.g., Eberth et al. 2010). Such taphonomic modeling has not been proposed for sauropods.

Some sub-environments close to the channel (crevasse splay and levee) have a high potential for preserving fossil remains. There, rapid burial is combined with minimal carcass transport and exposure to weathering. Deposition takes place in a timely manner during floods that overflow the main channel (Willis and Behrensmeyer 1995; Therrien 2005; González Riga and Astini 2007; Csiki et al. 2010; Casal et al. 2014b, 2017b). Therefore, the possibility of finding more complete and articulated skeletons with a low degree of weathering increases. The dynamics of the flooding of rivers cause an overflow onto the floodplain when the water rises above the levee of the channel. These floods can be controlled by tectonic processes but on a small scale, are frequently controlled by climatic factors (periods of high seasonal runoff). In semiarid or arid climates these flood events are sporadic, subject to torrential rains. When overflows expand rapidly in the floodplain, they lose energy and carrying capacity. They deposit large volumes of sediment quickly, causing little transport of sauropod skeletons that may be present on the floodplain (Fiorillo and May 1996; Lovelace 2006; Myers and Storrs 2007).

Examples of sauropods found in relatively well-preserved and partially articulated in overbank (crevasse splay) deposits are those of *Mendozasaurus neguyelap* (González Riga and Astini 2007), *Sarmientosaurus musacchioi* (Martínez et al. 2016; Casal and Nillni 2020), and *Aeolosaurus colhuehuapensis* (Casal et al. 2014a, 2019; Casal and Nillni 2020). These cases correspond to the taphonomic mode 'named overbank bone assemblage' (O-BA after Table 4; Fig. 5).

An example of the taphonomic mode 'levee bone assemblage' (LE-BA, Table 4) is recorded in the site of the unnamed titanosaurian MDT-PV 4 (25 m long) from Lago Colhué Huapi Formation (Chubut Province, Argentina). It comprises a large part of an articulated skeleton with a low degree of weathering (Casal et al. 2010; Casal 2015).

Another interesting case is that of *Epachthosaurus sciuttoi* a small size basal titanosaurian found completely articulated and discovered in the Bajo Barreal Formation (Chubut Province, Argentina). It corresponds to the taphonomic mode 'laminated sand bone assemblage' (Table 4), and unfortunately, the skull and cervical vertebrae are missing due to modern erosion. *Epachthosaurus* is preserved in laminated sand sheets produced by flash floods deposits (Rodríguez 1993; Martínez et al. 2004).

It should be pointed out, however, that although the presence of articulated remains may be indicative of a rapid burial, the record of articulated skeletal elements does not necessarily indicate little or no transport. There are examples of articulated skeletal elements mobilized by carcass flotation (Casal 2015; Pino et al. 2021).

The degree of disarticulation, transport, dispersion, and abrasion of a skeleton depends on the characteristics of the flow and the form of transport (e.g., bottom load vs. suspension load), as well as the shape, density, size, weight, and degree of articulation. Low viscosity and turbulent flows favor roll and drag abrasion, impact breaks, and dispersion, while laminar and dense flows are less abrasive and, in some cases, can mobilize bones without bone-to-bone contact, avoiding breakage. Therefore, overbank environments with their high sedimentary load, favor the preservation of fossils.

To sum up, sedimentological data actively intervene in the preservation of bones and are a key source of information for understanding the origin of the sauropod skeletal accumulations. Hydrodynamic conditions in low viscous, high-velocity flows have a higher ability to select the skeletal remains and generate accumulations of bones of similar size, shape, or density. In contrast, hydraulic conditions of hyper-concentrated flows in overbank facies, such as high sedimentary load, low to moderate energy, and high viscosity, cause short transport distances and poorly sorted accumulations of bones.

5 Conclusions

In this analysis, the exploration of a complex relationship between intrinsic and extrinsic factors gives us new interpretations on the preservation of sauropodomorphs. In South American Cretaceous sauropods, mainly represented by titanosauriforms and to a lesser extent by dicraeosaurids, rebbachisaurids, and one possible diplodocid, four main points of disarticulation are recognized: articulation atlas-axis-skull, articulation of pedes, articulations of manus, and articulation of distal caudal vertebrae.

The skeletal portions best preserved in South American sauropodomorphs are axial and appendicular elements. The preferential preservation of humeri and femora allows the recognition of four body size classes: small, medium, large, and giant, based on the length of these long bones. In contrast, articulated pedes and manus elements, and the skull, are very rare in the fossil record, a pattern consistent with what is observed in sauropodomorphs found in other continents.

Small- and medium-sized sauropod taxa (up to 22 m long) are the best preserved including articulated skeletons (*Epachthosaurus, Amargasaurus*). In contrast, giant titanosaurians, with humerus larger than 1.55 m and femur greater than 2.0 m (like is the case of *Argentinosaurus, Notocolossus* large individual-, *Puertasaurus* and *Antarctosaurus*) are represented by disarticulated or very incomplete skeletons.

In South America, sauropods inhabited fluvial environments and associated small lakes and swamps. Better preserved specimens are the result of: (a) rapid burial and (b) short transport distances and the related sedimentary supply rate, as well as flow type that moves the specimen, and (c) the characteristic of the bone remains. Thus, the preserved sauropod record shows that the most favorable sub-environments for the preservation of skeletons partially or fully articulated correspond to overbank, levee, and flash floods facies. Moreover, some new specimens are preserved in wet mudstones of variable and poorly drained floodplain facies. The close relationship of sedimentary paleoenvironment with taphonomic processes allows differentiating sub-environments that have better potential for fossil preservation and the predictive power to allow workers to focus explorations in their search for new fossils. And here, a new code of names and letters is proposed to better describe the taphonomic modes in fluvial systems, avoiding long and repeated descriptions. Finally, the inclusion of sedimentological data as variables that actively intervene in the preservation of bones constitutes a key source of information to understand the origin of these sauropod bone accumulations.

Acknowledgements We are grateful to the editors A. Otero, J. L. Carballido and D. Pol, for inviting us to participate in this book. The following institutions and projects supported our researches: authorities and colleagues of the Universidad Nacional de Cuyo and Facultad de Ciencias Exactas y Naturales (Mendoza, Argentina), and Instituto Argentino de Nivología, Glaciología y Ciencias Ambientales (IANIGLA), and projects of Consejo Nacional de Investigaciones Científicas y Técnicas de Argentina PIP0695 (to B.J. González Riga), Universidad Nacional de Cuyo (Mendoza, Argentina) SIIP-UNCUYO 725 2019-06/M112 (to B.J. González Riga), SIIP-UNCUYO 2019-M085 (to L.D. Ortiz David, and SIIP-UNCUYO 2019-M069 (to J.P. Coria), Universidad Nacional de la Patagonia San Juan Bosco Ciunpat PI 1089; PI 1249, and PI 1409 (to G. Casal), and ANCYT PICT 201-0459 (to L. Ibiricu and G. Casal). J. L. Carballido. J. O. Calvo, P. Gallina and M. Luna gently provided photographs.

References

Abler WL (1985) Skulls in fossil material: one mechanism contributing to their rarity. J Paleontol 249–250

Alcalá L (1994) Macromamíferos Neógenos de la fosa de Alfambra-Teruel. Instituto de Estudios Turolenses y Museo Nacional Ciencias Naturales, Madrid, p 554

Apesteguía S (2004) *Bonitasaura salgadoi* gen. et sp. nov.: a beaked sauropod from the Late Cretaceous of Patagonia. Naturwissenschaften 91(10):493–497

Apesteguía S (2007) The sauropod diversity of the La Amarga Formation (Barremian), Neuquén (Argentina). Gondwana Res 12(4):533–546

Apesteguía S, Soto J, Gallina P, Granda J, Guaman G (2019) The first dinosaur remains from the Cretaceous of Ecuador. Cretac Res 108:104345. https://doi.org/10.1016/j.cretres.2019.104345

Arribas A, Palmqvist P (1998) Taphonomy and palaeoecology of an assemblage of large mammals: hyaenid activity in the Lower Pleistocene site at Venta Micena (Orce, Guadix-Baza, Granada, Spain). Geobios 31(3):3–47

Bandeira KL, Medeiros Simbras F, Batista Machado E, de Almeida Campos D, Oliveira GR, Kellner AWA (2016) A new giant Titanosauria (Dinosauria: Sauropoda) from the Late Cretaceous Bauru Group, Brazil. PLoS One 11(10):e0163373

Behrensmeyer AK (1975) The Taphonomy and paleoecology of Plio-Pleistocene vertebrate assemblages of Lake Rudolf, Kenya. Bull Mus Comp Zool 146:473–578

Behrensmeyer AK (1978) Taphonomic and ecologic information from bone weathering. Paleobiology 4:150–162

Behrensmeyer AK (1988) Vertebrate preservation in fluvial channels. Palaeogeogr Palaeoclim Palaeoecol 63:183–189

Behrensmeyer AK, Dechant Boaz DE (1980) The recent bones of Amboseli National Park, Kenya, in relation to East African paleoecology. In: Behrensmeyer AK, Hill AP (eds) Fossils in the making: vertebrate taphonomy and paleoecology, pp 72–92

Behrensmeyer AK, Kidwell SM (1985) Taphonomy's contributions to paleobiology. Paleobiology 11:105–119

Behrensmeyer AK, Hook RW (1992) Paleoenviromental contexts and taphonomic modes. In: Behrensmeyer AK, Damuth J, Di Michele W, Potts R, Sues H-D, Wings S (eds) Terrestrial

ecosystems through time: evolutionary paleoecology of terrestrial plants and animals. University of Chicago Press, Chicago, pp 15–136
Behrensmeyer AK, Western KD, Boaz D (1979) New perspectives in vertebrate paleoecology from a recent bone assemblage. Paleobiology 5:12–21
Behrensmeyer AK, Kidwell SM, Gastaldo RA (2000) Taphonomy and paleobiology. Paleobiology 26(4):103–144
Bonaparte JF (1972) Los tetrápodos del sector superior de la Formación Los Colorados, La Rioja, Argentina (Triásico superior). Opera Lilloana 22:1–183
Bonaparte JF (1978) El mesozoico de América del Sur y sus tetrápodos (geología, reptiles). Opera Lilloana 26:1–596
Bonaparte JF (1996) Cretaceous tetrapods of Argentina. In: Arratia G (ed) Contribution of Sourthern South America to Vertebrate Paleontology, Münchner Geowissenschaftliche Abhandlungen (A), München 30:73–130
Bonaparte JF (1997) Rayososaurus agrioensis Bonaparte 1995. Ameghiniana 34:116
Bonaparte JF (1999a) Evolución de las vértebras presacras en Sauropodomorpha. Ameghiniana 36:115–187
Bonaparte JF (1999b) An armoured sauropod from the Aptian of northern Patagonia, Argentina. In: Proceedings of the second Gondwanan Dinosaur symposium, vol 15. National Science Museum Monographs, pp 1–12
Bonaparte JF, Vince M (1979) El hallazgo del primer nido de dinosaurios triásicos (Saurischia, Prosauropoda), Triásico Superior de Patagonia, Argentina. Ameghiniana 16:173–182
Bonaparte JF, Powell JE (1980) A continental assemblage of tetrapods from the Upper Cretaceous beds of El Brete, northwestern Argentina (Sauropoda-Coelurosauria-Carnosauria-Aves). Mémoires de la Société géologique de France, Serie 139:19–28
Bonaparte JF, Coria RA (1993) Un nuevo y gigantesco saurópodo titanosaurio de la Formación Río Limay (Albiano-Cenomaniano) de la Provincia del Neuquén, Argentina. Ameghiniana 30:271–282
Bonaparte JF, González Riga BJ, Apesteguía S (2006) *Ligabuesaurus leanzai* gen. et sp. nov. (Dinosauria, Sauropoda), a new titanosaur from the Lohan Cura Formation (Aptian, lower Cretaceous) of Neuquén, Patagonia, Argentina. Cretac Res 27(3):364–376
Bonnan M (2005) Pes anatomy in sauropod dinosaurs: implications for functional morphology, evolution, and phylogeny. In: Tidwell V, Carpenter K (eds) Thunder-Lizards: the sauropodomorph dinosaurs. Indiana University Press, Bloomington, pp 346–380
Borsuk-Bialynicka M (1977) A new camarasaurid sauropod *Opisthocoelicaudia skarzynskii*, gen. n., sp. n. from the Upper Cretaceous of Mongolia. Palaeont Pol 37:1–64
Botfalvai G, Csiki-Sava Z, Grigorescu D, Vasile S (2017) Taphonomical and palaeoecological investigation of the Late Cretaceous (Maastrichtian) Tuştea vertebrate assemblage (Romania; Haţeg Basin)—insights into a unique dinosaur nesting locality. Palaeogeogr Palaeoclim Palaeoecol 468:228–262
Brett CE, Speyer SE (1990) Taphofacies. In: Briggs DEG, Crowther RR (eds) Palaeobiology: a synthesis. Blackwell, Oxford, pp 258–263
Calvo JO, Bonaparte JF (1991) Andesaurus delgadoi gen. et. sp. nov. (Saurischia-Sauropoda), dinosaurio Titanosauridae de la Formación Río Limay (Albiano-Cenomaniano), Neuquén Argentina. Ameghiniana 28:303–310
Calvo JO, Salgado L (1995) *Rebbachisaurus tessonei* sp. nov. A new Sauropoda from the Albian-Cenomanian of Argentina; new evidence on the origin of the Diplodocidae. Gaia 11:13–33
Calvo JO, González Riga BJ (2003) *Rinconsaurus caudamirus* gen. et sp. nov., a new titanosaurid (Dinosauria, Sauropoda) from the Late Cretaceous of Patagonia, Argentina. Revista Geológica de Chile 30(2):333–353
Calvo JO, Porfiri JD (2010) *Panamericansaurus schroederi* gen. nov. sp. nov. un nuevo Sauropoda (Titanosauridae-Aeolosaurini) de la Provincia del Neuquén, Cretácico Superior de Patagonia, Argentina. Braz Geogr J: Geosci Humlties Res Mediu 1:100–115

Calvo JO, González Riga BJ (2018) *Baalsaurus mansillai* gen. et sp. nov. a new titanosaurian sauropod (Late Cretaceous) from Neuquén, Patagonia, Argentina. Anais da Academia Brasileira de Ciências 91(2):e20180661. https://doi.org/10.1590/0001-3765201820180661

Calvo JO, González Riga BJ, Porfiri JD (2007a) A new titanosaur sauropod from the Late Cretaceous of Neuquén, Patagonia, Argentina. Arquivos do Museu Nacional 65(4): 485–504

Calvo JO, Porfiri JD, González Riga BJ, Kellner AWA (2007b) A new terrestrial ecosystem from Gondwana with the description of a new sauropod dinosaur. An Acad Bras Ciênc 79(3):529–541. https://doi.org/10.1590/50001-37652007000300013

Calvo JO, Porfiri JD, González Riga BJ, Kellner AWA (2007b) Anatomy of *Futalognkosaurus dukei* Calvo, Porfiri, González Riga and Kellner, 2007c (Dinosauria, Titanosauridae) from the Neuquén Group (Late Cretaceous), Patagonia, Argentina. Arquivos do Museu Nacional 65(4):511–526

Campos DDA, Kellner AWA, Bertini RJ, Santucci RM (2005) On a titanosaurid (Dinosauria, Sauropoda) vertebral column from the Bauru group, Late Cretaceous of Brazil. Arquivos do Museu Nacional 63(3):565–593

Carballido JL, Salgado L, Pol D, Canudo JI, Garrido A (2012) A new basal rebbachisaurid (Sauropoda, Diplodocoidea) from the Early Cretaceous of the Neuquén Basin: evolution and biogeography of the group. Hist Biol 24:631–654

Carballido JL, Pol D, Parra Ruge ML, Padilla Bernal S, Páramo-Fonseca ME, Etayo-Serna F (2015) A new Early Cretaceous brachiosaurid (Dinosauria, Neosauropoda) from northwestern Gondwana (Villa de Leiva, Colombia). J Vertebr Paleontol 35(5):e980505

Carballido JL, Pol D, Otero A, Cerda IA, Salgado L, Garrido AC, Ramezani J, Cúneo NR, Krause JM (2017) A new giant titanosaur sheds light on body mass evolution among sauropod dinosaurs. Proc R Soc B: Biol Sci 284(1860):20171219

Carpenter K (2020) Use of scaled dinosaur bones in taphonomic water flume experiments. Sci Nat 107:15. https://doi.org/10.1007/s00114-020-01673-2

Carvalho IS, Santos Avilla L, Salgado L (2003) *Amazonsaurus maranhensis* gen. et sp. nov. (Sauropoda, Diplodocoidea) from the Lower Cretaceous (Aptian-Albian) of Brazil. Cretac Res 24:697–713

Carvalho IS, Salgado L, Lindoso RM, de Araújo-Júnior HI, Nogueira FCC, Soares JÁ (2017) A new basal titanosaur (Dinosauria, Sauropoda) from the Lower Cretaceous of Brazil. J S Am Earth Sci 75:74–84

Casal GA (2015) Paleontología, tafonomía y paleoambientes en yacimientos con dinosaurios del Cretácico Superior, Patagonia central, Argentina. (Unpublished Ph.D. tesis). Universidad Nacional de la Patagonia San Juan Bosco. Comodoro Rivadavia. Chubut, Argentina, p 274

Casal GA, Nillni AM (2020) Mineralogía y geoquímica de huesos de dinosaurios del Cretácico Superior del Grupo Chubut Argentina. Rev Soc Geol Esp 33(1):11–26

Casal GA, Martínez RD, Luna M, Sciutto JC, Lamanna MC (2007) *Aeolosaurus colhuehuapensis* sp. nov. (Saurópoda, Titanosauria) de la Formación Bajo Barreal, Cretácico Superior de Argentina. Revista Brasileira de Paleontologia 10:53–62

Casal G, Martínez R, Luna M, Ivany E, Ibiricu LM (2010) Un nuevo titanosaurio articulado de la Formación Bajo Barreal, Cretácico Superior tardío del Sur de Chubut. Asp ambient y tafonómicos. Ameghiniana 47(4):6R

Casal GA, Ibiricu LM, Allard JO, Martínez RD, Luna M, González Riga BJ (2014a) Tafonomía del titanosaurio *Aeolosaurus colhuehuapensis*, Cretácico Superior, Patagonia central, Argentina: un ejemplo de preservación en facies continentales fluviales. Rev Mex Cienc Geol 31:163–173

Casal GA, Foix N, Allard JO, González Riga BJ, Ibiricu LM (2014b) Controles tafonómicos y sedimentológicos en la preservación de dinosaurios en la Fm. Bajo Barreal, Cretácico Superior, Patagonia Argentina. Actas XIX Cong Geol Arg. Córdoba, Argentina, S7-3, pp 777–778

Casal GA, Foix N, Ibiricu LM, Martínez RD, Luna M (2017a) Tafonomía de un saurópodo (Dinosauria) de la Formación Bajo Barreal (Cretácico Superior). Análisis de atributos bioestratinómicos y sedimentológicos. Libro de Resúmenes V Jornadas de las Ciencias de la Tierra "Dr. Eduardo Musacchio" (De Sosa Tomas, Casal, Castro eds.), UNPSJB. Comodoro Rivadavia, Chubut, pp 31–33

Casal GA, Nillni AM, Valle MN, González Sbovoda E, Tiedemann C (2017b) Permineralización de restos de dinosaurios preservados en depósitos de desbordamiento de la Formación Bajo Barreal (Cretácico Superior), Patagonia central, Argentina. Rev Mex Cien Geol 34:12–24
Casal GA, Nillni AM, Valle MN, González Svoboda E, Tiedemann C, Ciapparelli H, Ibiricu LM, Luiz MM (2019) Fosildiagénesis en restos de dinosaurios preservados en depósitos fluviales de la Formación Lago Colhué Huapi (Cretácico Superior). Cuenca del Golfo San Jorge, Argentina. Andean Geol 46(3):670–692
Cashmore DD, Mannion PD, Upchurch P, Butler RJ (2020) Ten more years of discovery: revisiting the quality of the sauropodomorph dinosaur fossil record. Palaeontology 63(6):951–978
Chiappe LM, Coria R (2004) Auca Mahuevo, un extraordinario sitio de nidificación de dinosaurios saurópodos del Cretácico Tardío, Neuquén Argentina. Ameghiniana 41:591
Chiappe LM, Coria RA, Dingus L, Jackson F, Chinsamy A, Fox M (1998) Sauropod dinosaur embryos from the Late Cretaceous of Patagonia. Nature 396:258–261
Chiappe LM, Schmitt JG, Jackson FD, Garrido A, Dingus L, Grellet-Tinner G (2004) Nest structure for sauropods: sedimentary criteria for recognition of dinosaur nesting traces. Palaios 19(1):89–95
Chiappe LM, Jackson F, Coria RA, Dingus L (2005) Nesting titanosaurs from Auca Mahuevo and adjacent sites: understanding sauropod reproductive behavior and embryonic development. In: Curry Rogers KA, Wilson JA (eds) The sauropods: evolution and paleobiology. University of California Press, Berkeley, pp 285–302
Cladera G, Limarino C, Alonso MS, Rauhut O (2004) Controles estratigráficos en la preservación de restos de vertebrados en la Formación Cerro Barcino (Cenomaniano), Provincia del Chubut. AAS Revista 11:1–17
Cobley M, Rayfield E, Barrett PM (2013) Intervertebral flexibility of the ostrich neck: implications for estimating sauropod neck flexibility. PLoS ONE 8:e72187
Colombi CE, Rogers RR, Alcober OA (2012) Vertebrate taphonomy of the Ischigualasto Formation. J Vertebr Paleontol 32(1):31–50
Colombi CE, Limarino CO, Alcober OA (2017) Allogenic controls on the fluvial architecture and fossil preservation of the Upper Triassic Ischigualasto Formation NW Argentina. Sediment Geol 362:1–16. https://doi.org/10.1016/j.sedgeo.2017.10.003
Coombs MC, Coombs WP (1997) Analysis of the geology, fauna, and taphonomy of Morava Ranch Quarry, early Miocene of northwest Nebraska. Palaios 12(2):165–187
Coria RA (1994) On a monospecific assemblage of sauropod dinosaurs from Patagonia: implications for gregarious behavior. Gaia 10:209–213
Coria RA, Filippi LS, Chiappe LM, Garcia R, Arcucci AB (2013) *Overosaurus paradasorum* gen. et sp. nov., a new sauropod dinosaur (Titanosauria: Lithostrotia) from the Late Cretaceous of Neuquén, Patagonia, Argentina. Zootaxa 3683(4):357–376
Coria RA, Windholz GJ, Ortega F, Currie PJ (2019) A new dicraeosaurid sauropod from the Lower Cretaceous (Mulichinco Formation, Valanginian, Neuquén Basin) of Argentina. Cretac Res 93:33–48
Cruz I (2015) Las investigaciones sobre preservación de huesos de aves y mamíferos grandes en Patagonia (Argentina). Archaeofauna 24:209–224
Csiki Z, Grigorescu D, Codrea V, Therrien F (2010) Taphonomic modes in the Maastrichtian continental deposits of the Haţeg Basin, Romania, palaeoecological and palaeobiological inferences. Palaeogeogr Palaeoclim Palaeoecol 293:375–390. https://doi.org/10.1016/j.palaeo.2009.10.013
de la Fuente M, Maniel I, Jannello M, Sterli J, Garrido A, Garcia R, Salgado L, Canudo J, Bolatti R (2017) Unusual shell anatomy and osteohistology in a new Late Cretaceous panchelid turtle from northwestern Patagonia, Argentina. Acta Palaeontol Pol 62. https://doi.org/10.4202/app.00340.2017
Del Corro G (1975) Un nuevo saurópodo del Cretácico Superior, Chubutisaurus insignis gen. et sp. nov. (Saurischia, Chubutisauridae, nov.) del Cretácico Superior (Chubutiano), Chubut, Argentina. En: 1° Congreso Argentino de Paleontología y Bioestratigrafía. Actas 2:229–240

Dingus L, Clarke J, Scott GR, Swisher C, Chiappe LM, Coria RA (2000) Stratigraphy and magnetostratigraphic/faunal constraints for the age of sauropod embryo-bearing rocks in the Neuquén Group (late Cretaceous, Neuquén Province, Argentina). Am Mus Novit 3290:11

Dodson P (1971) Sedimentology and taphonomy of the Oldman Formation (Campanian), Dinosaur Provincial Park, Alberta (Canada). Palaeogeogr Palaeoclim Palaeoecol 10:21–74

Eberth DA, Brinkman DB, Chen PJ, Yuan FT, Wu SZ, Gang L, Cheng XS (2001) Sequence stratigraphy, paleoclimate patterns and vertebrate fossil preservation in Jurassic-Cretaceous strata of the Junggar basin, Xinjiang Autonomous Region, People's Republic China. Can J Earth Sci 38:1627–1644

Eberth DA, Xing XU, Clark JM (2010) Dinosaur death pits from the Jurassic of China. Palaios 25:112–125

Ezcurra MD (2010) A new early dinosaur (Saurischia: Sauropodomorpha) from the Late Triassic of Argentina: a reassessment of dinosaur origin and phylogeny. J Syst Palaeontol 8:371–425

Ezcurra MD, Apaldetti C (2011) A robust sauropodomorph specimen from the Upper Triassic of Argentina and insights on the diversity of the Los Colorados Formation. Proc Geologist's Assoc 123:155–164

Fastovsky DED, Badamgarav H, Ishimoto M, Watabe M, Weishampel DB (1997) The paleoenvironments of Tugrikin-shireh (Gobi Desert, Mongolia) and aspects of the taphonomy and paleoecology of Protoceratops (Dinosauria: Ornithischia). Palaios 12:59–70

Filippi LS, Garrido AC (2008) *Pitekunsaurus macayai* gen. et sp. nov., nuevo titanosaurio (Saurischia, Sauropoda) del Cretácico Superior de la Cuenca Neuquina, Argentina. Ameghiniana 45(3):575–590

Filippi LS, García RA, Garrido AC (2011) A new titanosaur sauropod dinosaur from the Upper Cretaceous of North Patagonia Argentina. Acta Palaeontol Pol 56(3):505–520

Filippi LS, Salgado L, Garrido AC (2019) A new giant basal titanosaur sauropod in the Upper Cretaceous (Coniacian) of the Neuquen Basin, Argentina. Cretac Res 100:61–81

Fiorelli LE, Grellet-Tinner G, Alasino PH, Argañaraz E (2012) The geology and palaeoecology of the newly discovered Cretaceous neosauropod hydrothermal nesting site in Sanagasta (Los Llanos Formation), La Rioja, northwest Argentina. Cretac Res 35:94–117

Fiorelli LE, Grellet-Tinner G, Argañaraz E, Salgado L (2013) Tafonomía del sitio de nidificación de neosaurópodos de Sanagasta (La Rioja, Argentina): ejemplo de preservación excepcional en un paleoambiente hidrotermal del Cretácico. Ameghiniana 50(4):389–406

Fiorillo AR (1988) Taphonomy of Hazard Homestead Quarry (Ogallala Group), Hitchcock County, Nebraska. Contribut Geol University of Wyoming 26:57–97

Fiorillo AR (1991) Taphonomy and depositional setting of Careless Creek Quarry (Judith River Formation), Wheatland County, Montana, USA. Palaeogeogr Palaeoclim Palaeoecol 81(3–4):281–311

Fiorillo AR (1995) The possible influence of low temperature on bone weathering in Curecanti National recreation area, southwest Colorado. Curr Res Pleistocene 12:69–71

Fiorillo AR, May CL (1996) Preliminary report on the taphonomy and depositional setting of a new dinosaur locality in the Morrison Formation (Brushy Basin Member) of Curecanti National Recreation Area, Colorado. Bull MusM North Ariz 60:555–561

Fiorillo AR, Eberth DA (2004) Dinosaur taphonomy. In: Weishampel DB, Dodson P, Osmólska H (eds). University of California Press, Berkeley, CA, pp 607–613

Gallina PA, Apesteguía S (2005) *Cathartesaura anaerobica* gen. et sp. nov., a new rebbachisaurid (Dinosauria, Sauropoda) from the Huincul Formation (Upper Cretaceous), Río Negro, Argentina. Revista del Museo Argentino de Ciencias Naturales 7(2):153–166

Gallina PA, Apesteguía S, Haluza A, Canale JI (2014) A diplodocid sauropod survivor from the Early Cretaceous of South America. PLoS ONE 9(5):e97128. https://doi.org/10.1371/journal.pone.0097128

Gallina PA, Apesteguía S, Canale JI, Haluza A (2019) A new long-spined dinosaur from Patagonia sheds light on sauropod defense system. Sci Rep 9(1):1–10

García RA, Salgado L, Coria RA, Chiappe LM (2010) Osteología embrionaria de saurópodos titanosaurios de Neuquén (Argentina): aspectos ontogenéticos y evolutivos. Ameghiniana 47:409–430. https://doi.org/10.5710/AMGH.v47i4.4

García R, Salgado L, Fernández M, Cerda I, Paulina Carabajal A, Otero A, Coria R, Fiorelli L (2014) Paleobiology of titanosaurs: reproduction, development, histology, pneumaticity, locomotion and neuroanatomy from the South American fossil record. Ameghiniana 52:29–68. https://doi.org/10.5710/AMGH.16.07.2014.829

Garrido AC (2010) Estratigrafía del Grupo Neuquén, Cretácico Superior de la Cuenca Neuquina (Argentina): nueva propuesta de ordenamiento litoestratigráfico. Buenos Aires Rev Mus Argent CiencS Nat 12(2):121–177

Germonpré M (2003) Mammoth taphonomy of two fluvial sites from the Flemish Valley, Belgium. In: Reumer JWF, De Vos J, Mol D (eds) Advances in mammoth research (Proceedings of the second international mammoth conference, Rotterdam), DEINSEA 9, pp 171–183. ISSN 0923-9308

Ghosh P, Sarkar S, Maulik P (2006) Sedimentology of a muddy alluvial deposit: Triassic Denwa Formation, India. Sediment Geol 191:3–36

González Riga BJ (1999) Hallazgo de vertebrados fósiles en la Formación Loncoche, Cretácico Superior de la provincia de Mendoza, Argentina. Ameghiniana 36:401–410

González Riga BJ (2003) A new titanosaur (Dinosauria, Sauropoda) from the Upper Cretaceous of Mendoza Province, Argentina. Ameghiniana 40:155–172

González Riga BJ (2005) Nuevos restos fósiles de *Mendozasaurus neguyelap* (Sauropoda: Titanosauria) del Cretácico Tardío de Mendoza, Argentina. Ameghiniana 42:535–538

González Riga BJ (2011) Speeds and stance of titanosaur sauropods: analysis of *Titanopodus* tracks from the Late Cretaceous of Mendoza Argentina. An Acad Bras Ciênc 83(1):279–290

González Riga BJ, Astini R (2007) Fossil preservation of large titanosaur sauropods in overbank fluvial facies: a case study in the Cretaceous of Argentina. J S Am Earth Sci 23:290–303

González Riga BJ, Calvo JO (2009) A new wide-gauge Sauropod track site from the Late Cretaceous of Mendoza, Neuquén Basin Argentina. Palaeontology 52(3):631–640

González Riga BJ, Ortiz David LD (2014) A new titanosaur (Dinosauria, Sauropoda) from the Upper Cretaceous (Cerro Lisandro Formation) of Mendoza Province, Argentina. Ameghiniana 51:3–25

González Riga BJ, Calvo JO, Previtera E (2007) Taphonomic analyses of titanosaur sauropods from the Neuquén Basin (Argentina) and their implications in systematic studies. Libro de Resúmenes IV Jornadas Internacionales sobre Paleontología de Dinosaurios y su entorno, Salas de los Infantes, Burgos, España, pp 73–76

González Riga BJ, Calvo JO, Porfiri J (2008) An articulated titanosaur from Patagonia (Argentina): new evidence of neosauropod pedal evolution. Palaeoworld 17:33–40

González Riga BJ, Calvo JO, Previtera E (2009a) Taphonomic analyses of titanosaur sauropods from the Neuquén Basin (Argentina) and their implications in systematic studies. In: Colectivo Arqueológico y Paleontológico de Salas, CAS (eds) IV Jornadas Internacionales sobre Paleontología de Dinosaurios y su entorno. Burgos, España, pp 235–241

González Riga BJ, Previtera E, Pirrone CA (2009b) Malarguesaurus florenciae gen. et sp. nov., a new titanosauriform (Dinosauria, Sauropoda) from the Upper Cretaceous of Mendoza Argentina. Cretac Res 30:135–148

González Riga BJ, Porfiri JD, Calvo JO, Dos Santos D, Ortiz David LD, Coelho F (2012) Hallazgo de titanosaurios parcialmente articulados en el sur de Mendoza (Argentina): contexto estratigráfico. Resúmenes XXVI Jornadas Argentinas de Paleontología de Vertebrados (Buenos Aires, 21–23 de mayo)

González Riga BJ, Ortiz David LD, Londero S, Calvo JO, Porfiri JD, Dos Santos D (2013) Hallazgo de un gigantesco titanosaurio parcialmente articulado del Cretácico Tardío de Mendoza, Argentina. In: Abstract Book 1° Brazilian Dinosaur Symposium. Ituitaba, Minas Gerais, Brasil, p 35

González Riga BJ, Ortiz David LD, Tomaselli MB, Candeiro C, Coria JP, Prámparo M (2015) Sauropod and theropod dinosaur tracks from the Upper Cretaceous of Mendoza (Argentina): trackmakers and anatomical evidences. J S Am Earth Sci 61:134–141

González Riga BJ, Lamanna MC, Ortiz David LD, Calvo JO, Coria JP (2016) A gigantic new dinosaur from Argentina and the evolution of the sauropod hind foot. Scient Rep 6:19165

González Riga BJ, Mannion P, Poropat S, Ortiz David LD, Coria JP (2018) Osteology of the Late Cretaceous Argentinean sauropod dinosaur *Mendozasaurus neguyelap*: implications for basal titanosaur relationships. Zool J Lin Soc 184:136–181

González Riga BJ, Lamanna MC, Otero A, Ortiz David LD, Kellner AWA, Ibiricu LM (2019) An overview of the appendicular skeletal anatomy of South American titanosaurian sauropods, with definition of a newly recognized clade. An Acad Bras Ciênc 91(2):e20180374

González Riga BJ, Tomaselli MB (2019) Different trackway patterns in titanosaur sauropods: analysis of new *Titanopodus* tracks from the Upper Cretaceous of Mendoza, Neuquén Basin, Argentina. Cret Res 93:49–59

Grellet-Tinner G, Fiorelli LE (2010) A new Argentinean nesting site showing neosauropod dinosaur reproduction in a Cretaceous hydrothermal environment. Nat Commun 1:32. https://doi.org/10.1038/ncomms1031

Grellet-Tinner G, Fiorelli LE, Salvador RB (2012) Water vapor conductance of the Lower Cretaceous dinosaurian eggs from Sanagasta, La Rioja, Argentina—Paleobiological and paleoecological implications for South American faveoloolithid and megaloolithid eggs. Palaios 27:35–47

Gutiérrez MA, Rafuse DJ, Alvarez MC, Massigoge A, González ME, Scheifler NA, Kaufmann CA (2018) Ten years of actualistic taphonomic research in the Pampas region of Argentina: contributions to regional archaeology. Quat Int 492:40–52. https://doi.org/10.1016/j.quaint.2017.09.025

Hechenleitner M, Leuzinger L, Martinelli A, Rocher S, Fiorelli L, Taborda J, Salgado L (2020) Two Late Cretaceous sauropods reveal titanosaurian dispersal across South America. Commun Biol 3(1):Article number 622. https://doi.org/10.1038/s42003-020-01338-w

Holwerda FM, Pol D (2018) Phylogenetic analysis of Gondwanan basal eusauropods from the early-middle Jurassic of Patagonia Argentina. Span J Palaeontol 33(2):289–298

Huene F (1929) Los saurisquios y ornitisquios del Cretáceo Argentino. Anales del Museo de La Plata 3(2):1–194

Ibiricu LM, Casal GA, Martínez RD, Lamanna MC, Salgado L (2013) *Katepensaurus goicocheai* gen. et sp. nov., a Late Cretaceous rebbachisaurid (Sauropoda, Diplodocoidea) from central Patagonia, Argentina. J Vertebr Paleontol 33(6):1351–1366

Ibiricu LM, Casal GA, Martínez RD, Lamanna MC, Luna M, Salgado L (2015) New material of *Katepensaurus goicoecheai* (Sauropoda: Diplodocoidea) and its significance for the morphology and evolution of Rebbachisauridae. Ameghiniana 52:430–446. https://doi.org/10.5710/AMGH.24.04.2015.2830

Juárez Valieri RD, Calvo JO (2011) Revision of MUCPv 204, a Senonian Basal Titanosaur from Northern Patagonia. In: Calvo JO, González Riga BJ, Porfiri JD, Dos Santos D (eds) Paleontología y Dinosaurios desde América Latina, pp 143–152

Kellner AWA, Azevedo SAK (1999) A new sauropod dinosaur (Titanosauria) from the Late Cretaceous of Brazil. In: Tomida Y, Rich TH, Vickers-Rich P (eds) Proceedings of the Gondwanan Dinosaur symposium, vol 15. National Science Museum Monograph, pp 111–142

Kellner AWA, Campos DA, Trotta MN (2005) Description of a titanosaurid caudal series from the Bauru Group, Late Cretaceous of Brazil. Arquivos do Museu Nacional 63(3):529–564

Kellner AWA, Campos D, De Azevedo S, Trotta M, Henriques D, Craik M, De Paula SH (2006) On a new titanosaur sauropod from the Bauru Group, Late Cretaceous of Brazil. Bol Mus Nac Nova Série 74:1–31

Kellner AWA, Rubilar-Rogers D, Vargas A, Suárez M (2011) A new titanosaur sauropod from the Atacama Desert, Chile. An Acad Bras Ciênc 83:211–219

Klein N, Christian A, Sander PM (2012) Histology shows that elongated neck ribs in sauropod dinosaurs are ossified tendons. Biol Lett 8:1032–1035

Lacovara KJ, Lamanna MC, Ibiricu LM, Poole JC, Schroeter ER, Ullmann PV, Voegele KK, Boles ZM, Carter AM, Fowler EK, Egerton VM, Moyer AE, Coughenour CL, Schein JP, Harris JD, Martínez RD, Novas FE (2014) A gigantic, exceptionally complete titanosaurian sauropod dinosaur from Southern Patagonia Argentina. Sci Rep 4:6196. https://doi.org/10.1038/srep06196

Lovelace DM (2006) An Upper Jurassic Morrison Formation fire-induced debris flow: taphonomy and paleoenvironment of a sauropod (Sauropoda: *Supersaurus vivianae*) locality, east-central Wyoming. In: Foster JR, Lucas SG (eds) Paleontology and geology of the upper Jurassic Morrison Formation, vol 36. New Mexico Museum of Natural History and Science Bulletin, pp 47–56

Lydekker R (1893) The Dinosaurs of Patagonia. An Del Mus Plata 2:1–14

Lyman RL (1994) Vertebrate taphonomy. Cambridge University Press, Cambridge, p 524

Lyman RL (1996) Applied zooarchaeology: the relevance of fauna analysis to wildlife management. World Archaeol 28(1):110–125

Lyman RL, Fox GL (1989) A critical evaluation of bone weathering as an indication of bone assemblage formation. J Archaeol Sci 16(3):293–317. https://doi.org/10.1016/0305-4403(89)90007-1

Machado E, Avilla L, Nava W, Campos D, Kellner AWA (2013) A new titanosaur sauropod from the Late Cretaceous of Brazil. Zootaxa 3701:301–321

Mannion PD, Upchurch P (2010) Completeness metrics and the quality of the sauropodomorph fossil record through geological and historical time. Paleobiology 36(2):283–302

Mannion PD, Otero A (2012) A reappraisal of the Late Cretaceous Argentinean sauropod dinosaur *Argyrosaurus superbus*, with a description of a new titanosaur genus. J Vertebr Paleontol 32:614–638

Mannion PD, Upchurch P, Barnes R, Mateus O (2013) Osteology of the Late Jurassic Portuguese sauropod dinosaur *Lusotitan atalaiensis* (Macronaria) and the evolutionary history of basal titanosauriforms. Zool J Lin Soc 168:98–206

Martinelli AG, Forasiepi AM (2004) Late Cretaceous vertebrates from Bajo de Santa Rosa (Allen Formation), Río Negro Province, Argentina, with the description of a new sauropod dinosaur (Titanosauridae). Rev Mus Argent Cienc Nat 6:57–305

Martínez RN, Alcober OA (2009) A basal sauropodomorph (Dinosauria: Saurischia) from the Ischigualasto Formation (Triassic, Carnian) and the Early evolution of Sauropodomorpha. PLoS ONE 4:1–12

Martínez RD, Giménez O, Rodríguez J, Luna M, Lamanna MC (2004) An articulated specimen of the basal Titanosaurian (Dinosauria: Sauropoda) *Epachthosaurus sciuttoi* from the Early Late Cretaceous Bajo Barreal Formation of Chubut Province, Argentina. J Vertebr Paleontol 24:107–120

Martínez RN, Sereno PC, Alcober OA, Colombi CE, Renne PR, Montañez IP, Currie BS (2011) A basal dinosaur from the dawn of the dinosaur era in Southwestern Pangaea. Science 331:206–210

Martínez RN, Apaldetti C, Alcober OA, Colombi C, Sereno PC, Fernandez E, Santi Malnis P, Correa G, Abelín D (2013) Vertebrate succession in the Ischigualasto Formation. In: Sereno PC (ed) Basal sauropodomorphs and the vertebrate fossil record of the Ischigualasto Formation (Late Triassic: Carnian–Norian) of Argentina, vol 12. Society of Vertebrate Paleontology Memoir, pp 10–30

Martínez R, Apaldetti C, Correa G, Colombi C, Fernandez E, Santi Malnis P, Pradeiro A, Abelin D, Benegas G, Aguilar-Cameo A, Alcober O (2015) A new late triassic vertebrate assemblage from northwestern argentina. Ameghiniana 52. https://doi.org/10.5710/AMGH.27.04.2015.2889

Martínez RD, Lamanna MC, Novas FE, Ridgely RC, Casal GA, Martínez JE, Vita JR, Witmer LM (2016) A basal lithostrotian titanosaur (Dinosauria: Sauropoda) with a complete skull: implications for the evolution and paleobiology of Titanosauria. PLoS One 11:e0151661

McIntosh JS (1990) Sauropoda. In: Weishampel DB, Dodson P, Osmólska H (eds) The Dinosauria. University of California Press, Berkeley, pp 345–401

Miall AD (1996) The geology of fluvial deposits. Springer, Berlin, p 582

Moore JR, Norman DB (2009) Quantitatively evaluating the sources of taphonomic biasing of skeletal element abundances in fossil assemblages. Palaios 24:591–602. https://doi.org/10.2110/palo.2008.p08-135r

Müller RT, Silva Garcia M, Stock Da-Rosad AA, Dias-da-Silva S (2018) Under pressure: effect of sedimentary compression on the iliac morphology of early sauropodomorphs. J S Am Earth Sci 88:345–351

Muñoz AS, Mondini M, Durán V, Gasco A (2008) Los pumas (Puma concolor) como agentes tafonómicos. Análisis actualístico de un sitio de matanza en los Andes de Mendoza Argentina. Geobios 41:123–131. https://doi.org/10.1016/j.geobios.2006.11.010

Myers TS, Storrs GW (2007) Taphonomy of the mother's day quarry, upper jurassic morrison formation, South-Central Montana, USA. Palaios 22:651–666

Myers TS, Fiorillo AR (2009) Evidence for gregarious behavior and age segregation in sauropod dinosaurs. Palaeogeogr Palaeoclim Palaeoecol 274(1–2):96–104

Nair J, Salisbury S (2012) New Anatomical Information on *Rhoetosaurus brownei* Longman, 1926, a Gravisaurian Sauropodomorph Dinosaur from the Middle Jurassic of Queensland, Australia. J Vertebr Paleontol 32:369–394. https://doi.org/10.1080/02724634.2012.622324

Navarrete C, Casal GA, Martínez RD (2011) *Drusilasaura deseadensis* gen. et sp. nov., un nuevo titanosaurio (Dinosauria-Sauropoda) de la Formación Bajo Barreal, Cretácico Superior del Norte de Santa Cruz, Argentina. Rev Bras de Paleontol 14(1):1–14. https://doi.org/10.4072/rbp.2011.1.01

Novas FE, Salgado L, Calvo JO, Agnolin F (2005) Giant titanosaur (Dinosauria, Sauropoda) from the Late Cretaceous of Patagonia. Revista del Museo Argentino de Ciencias Nat 7:37–41

Novas F, Agnolin F, Rozadilla S, Aranciaga-Rolando A, Brissón-Eli F, Motta M, Cerroni M, Ezcurra M, Martinelli A, D'Angelo J, Álvarez-Herrera J, Gentil A, Bogan S, Chimento N, García-Marsà J, Lo Coco G, Miquel S, Brito F, Vera E, Loinaze V, Fernandez M, Salgado L (2019) Paleontological discoveries in the Chorrillo Formation (upper Campanian–lower Maastrichtian, Upper Cretaceous), Santa Cruz Province, Patagonia, Argentina. Revista del Museo Argentino de Ciencias Nat 21(2):217–293

Orr PJ, Adler LB, Beardmore SR, Furrer H, McNamara ME, Peñalver-Mollá E, Redelstorff R (2016) "Stick 'n' peel": Explaining unusual patterns of disarticulation and loss of completeness in fossil vertebrates. Palaeogeogr Palaeoclim Palaeoecol 457:380–388

Ortiz David LD (2019) Análisis anatómico y filogenético de un pterosaurio gigante de Cuenca Neuquina, Mendoza, Argentina. (Unpubl. Ph.D. tesis). Universidad Nacional de Cuyo—ProBiol. Mendoza, Argentina, p 448

Otero A (2018) Forelimb musculature and osteological correlates in Sauropodomorpha (Dinosauria, Saurischia). PLoS One 13(7):e0198988. https://doi.org/10.1371/journal.pone.0198988

Otero A, Pol D (2013) Postcranial anatomy and phylogenetic relationships of *Mussaurus patagonicus* (Dinosauria, Sauropodomorpha). J Vertebr Paleontol 33:1138–1168

Otero A, Perez, LM, Gaetano, L, Leardi, JM, Apaldeti, C, Krapovickas, V, Alonso Muruaga, PJ (2019a). Nuevos aportes al conocimiento de la fauna de tetrápodos de la Formación Los Colorados (Triásico Superior, Argentina): aspectos tafonómicos y paleobiogeográficos. In: II Simposio Internacional de Paleontología del Perú, pp 112–115

Otero A, Cuff AR, Allen V, Sumner-Rooney L, Pol D, Hutchinson JR (2019b) Ontogenetic changes in the body plan of the sauropodomorph dinosaur *Mussaurus patagonicus* reveal shifts of locomotor stance during growth. Sci Rep 9:7614

Otero A, Carballido JL, Salgado L, Canudo JI, Garrido A (2021) Report of a giant titanosaur sauropod from the Upper Cretaceous of Neuquen Province, Argentina. Cretac Res 122:104754

Parras A, Casadio S, Pires M (1998) Secuencias depositacionales del Grupo Malargüe (límite Cretácico–Paleógeno), sur de la Provincia de Mendoza, Argentina. In: Casadío S (ed) Paleógeno de América del sur y de la Península Antártica. Pub Esp Asoc Paleontol Arg 5:181–192

Paul GS (2019) Determining the largest known land animal: A critical comparison of differing methods for restoring the volume and mass of extinct animals (PDF). Ann Carnegie Mus 85(4):335–358. https://doi.org/10.2992/007.085.0403

Pereda-Suberbiola X, Astibia H, Murelaga X, Elorza JJ, Gómez-Alday JJ (2000) Taphonomy of the Late Cretaceous dinosaur-bearing beds of the Laño Quarry (Iberian Peninsula). Palaeogeogr Palaeoclim Palaeoecol 157(3–4):247–275

Pérez LM, Otero A, Apesteguía S, Gallina P (2009) Estratigrafía y análisis tafonómico de *Bonitasaura salgadoi* Apesteguía, en el sitio La Bonita (Cretácico Superior, Río Negro, Argentina). Rev Mus Argent Cienc Nat 11:39–48

Pfretzschner HU (2004) Fossilization of Haversian bone in aquatic environments. CR Palevol 3(6–7):605–616

Pino DA, Coria RA, Díaz-Martínez I, Tunik MA (2021) An accumulation of dinosaur remains in fluvial deposits of Mulichinco Formation (lower Valanginian, Neuquén Basin), Patagonia, Argentina: Taphonomic and paleoenvironmental inferences. J S Am Earth Sci 105:102979

Pirrone C, Buatois L, González Riga BJ (2014) A new ichnospecies of *Cubiculum* from Upper Cretaceous dinosaur bones in Western Argentina. Ichnos 21(4):251–260 https://doi.org/10.1080/10420940.2014.958225

Pol D, Powell JE (2007) New information on *Lessemsaurus sauropoides* (Dinosauria: Sauropodomorpha) from the Late Triassic of Argentina. In: Barrett PM, Batten DJ (eds) Evolution and paleobiology of early sauropodomorph dinosaurs. Special Papers in Palaeontology, vol 77, pp 223–243

Powell JE (1986) Revisión de los titanosauridos de América del Sur. Ph.D. thesis. Universidad Nacional de Tucumán, Argentina. Inédito

Powell JE (1987) The Late Cretaceous Fauna from Los Alamitos, Patagonia, Argentina. The titanosaurids. Rev Del Mus Argent Cienc Nat 3:147–153

Powell JE (1990) *Epachthosaurus sciuttoi* (gen. et sp. nov.) un dinosaurio saurópodo del Cretácico de Patagonia (provincia de Chubut, Argentina). Congreso Argentino de Paleontología y Bioestratigrafía, Actas 5:123–128

Preuschoft H, Klein N (2013) Torsion and bending in the neck and tail of sauropod dinosaurs and the function of cervical ribs: insights from functional morphology and biomechanics. PLoS One 8(10):e78574

Previteria E (2017) Bone microstructure and diagenesis of saurischian dinosaurs from the Upper Cretaceous (Neuquén Group) Argentina. Andean Geol 44(1):39–58

Previtera E (2019) Taphonomic analysis of saurischian dinosaurs from the plottier formation (Upper cretaceous), Mendoza, Argentina. Andean Geol 46:345–367. https://doi.org/10.5027/andgeov46n2-3161

Previteria E, González Riga BJ (2008) Vertebrados cretácicos de la Formación Loncoche en Calmu-Co, Mendoza, Argentina. Ameghiniana 45:349–359

Retallack GJ (1988) Field recognition of paleosoils. In: Reinhardt J, Sigleo WR (eds) Paleosols and weathering through geologic time: principles and applications. Geol Soc Am Spec, vol 216, pp 1–20

Rodríguez JFR (1993) La depositación de las areniscas verdes (Formación Bajo Barreal, Cretácico Tardío) y sus implicancias tafonómicas. In: Actas 2° Congreso Geológico Argentino y 12° Congreso de Exploración de Hidrocarburos, vol 1. Mendoza, Argentina, pp 194–199

Rogers RR (1990) Taphonomy of three dinosaur bone beds in the Upper Cretaceous Two Medicine Formation of northwestern Montana; evidence for drought-related mortality. Palaios 5(5):394–413

Rogers RR, Swisher CC III, Sereno PC, Forster CA, Monetta AM (1993) The Ischigualasto tetrapod assemblage (Late Triassic) and 40Ar/39Ar calibration of dinosaur origins. Science 260:794–797

Rogers RR, Arcucci AB, Abdala F, Sereno PC, Forster CA, May CL (2001) Paleoenvironment and taphonomy of the chañares formation tetrapod assemblage (middle triassic), northwestern argentina: spectacular preservation in volcanogenic concretions. Palaios 16:461–481

Rogers RR, Eberth DA, Fiorillo AR (2007) Bonebeds: genesis, analysis, and paleobiological significance. University of Chicago Press

Rubilar-Rogers D, Vargas AO, González Riga BJ, Soto-Acuña S, Alarcón-Muñoz J, Iriarte-Díaz J, Arévalo C, Gutstein CS (2021) *Arackar licanantay* gen. et sp. nov. a new lithostrotian (Dinosauria,

Sauropoda) from the Upper Cretaceous of the Atacama Region, northern Chile. Cretac Res 124:104802. https://doi.org/10.1016/j.cretres.2021.104802

Ryan MJ, Russell AP, Eberth DA, Currie PJ (2001) The taphonomy of a *Centrosaurus* (Ornithischia: Certopsidae) bone bed from the Dinosaur Park Formation (Upper Campanian), Alberta, Canada, with comments on cranial ontogeny. Palaios 16(5):482–506

Salgado L (1996) Pellegrinisaurus powelli nov. gen. et sp. (Sauropoda, Titanosauridae) from the Upper Cretaceous of Lago Pellegrini, North-western Patagonia Argentina. Ameghiniana 33:355–365

Salgado L, Bonaparte JF (1991) Un nuevo saurópodo Dicraeosauridae, *Amargasaurus cazaui* gen. et sp. nov. de la Formación La Amarga, Neocomiano de la Provincia del Neuquén Argentina. Ameghiniana 28:333–346

Salgado L, Azpilicueta C (2000) Un nuevo saltasaurino (Sauropoda, Titanosauridae) de la provincia de Río Negro (Formación Allen, Cretácico Superior), Patagonia, Argentina. Ameghiniana 37:259–264

Salgado L, Carvalho IS (2008) *Uberabatitan ribeiroi*, a new titanosaur from the Marília Formation (Bauru Group, Upper Cretaceous), Minas Gerais, Brazil. Palaeontology 51:881–901

Salgado L, Coria RA (2009) *Barrosasaurus casamiquelai* gen. et sp. nov., a new titanosaur (Dinosauria, Sauropoda) from the Anacleto Formation (Late Cretaceous: early Campanian) of Sierra Barrosa (Neuquén, Argentina). Zootaxa 2222:1–16

Salgado L, Garrido A, Cocca S, Cocca JR (2004) Lower Cretaceous rebbachisaurid sauropods from Cerro Aguada del León (Lohan Cura Formation), Neuquén Province, northwestern Patagonia, Argentina. J Vertebr Paleontol 24:903–912

Salgado L, Coria RA, Chiappe LM (2005) Osteology of the sauropod embryos from the Upper Cretacoeus of Patagonia. Acta Palaeontol Pol 50:79–92

Salgado L, de Souza CI, Garrido AC (2006) *Zapalasaurus bonapartei*, un nuevo dinosaurio saurópodo de la Formación La Amarga (Cretácico Inferior), noroeste de Patagonia, Provincia de Neuquén, Argentina. Géobios 39:695–707

Sander PM, Klein N, Stein K, Wings O (2011) Sauropod bone histology and its implications for sauropod biology. In: Klein N, Remes K, Gee CT, Sander PM (eds) Biology of the sauropod dinosaurs: understanding the life of giants, pp 276–302

Sander PM (2013) An evolutionary cascade model for sauropod dinosaur gigantism—Overview, update and tests. PLoS One 8(10):e78573. https://doi.org/10.1371/journal.pone.0078573

Santucci RM, Bertini RJ (2006) A new titatanosaur from western São Paulo state, Upper Cretaceous Bauru Group, southeast Brazil. Palaeontology 49:59–66

Santucci RM, de Arruda-Campos AC (2011) A new sauropod (Macronaria, Titanosauria) from the Adamantina Formation, Bauru Group, Upper Cretaceous of Brazil and the phylogenetic relationships of Aeolosaurini. Zootaxa 3085:1–33

Schwarz D, Frey E, Meyer CA (2007) Novel reconstruction of the orientation of the pectoral girdle in sauropods. Anat Rec 290:32–47

Sereno PC, Forster CA, Rogers RR, Monetta AM (1993) Primitive dinosaur skeleton from Argentina and the early evolution of the Dinosauria. Nature 361:64–66

Simón E, Salgado L, Calvo JO (2017) New titanosaur sauropod from the Upper Cretaceous of Patagonia, Neuquén Province Argentina. Ameghiniana 55(1):1–29

Smith RMH, Marsicano CA, Pol D, Mancuso AC (2014) Ichnology of sauropodomorph nests from Patagonia indicates Early Jurassic origin of herd-living and herding site fidelity. In: Actas 4th international paleontological congress. Mendoza, Argentina, p 227

Tappen M (1994) Bone weathering in the tropical rain forest. J Archaeol Sci 21:667–673

Taylor MP, Wedel MJ (2013) Why sauropods had long necks; and why giraffes have short necks. PeerJ 1:e36. https://doi.org/10.7717/peerj.36

Therrien F (2005) Palaeoenvironments of the latest Cretaceous (Maastrichtian) dinosaurs of Romania: insights from fluvial deposits and paleosols of the Transylvanian and Hateg basins. Palaeogeogr Palaeoclim Palaeoecol 218:15–56

Todisco D, Monchot H (2008) Bone weathering in a periglacial environment: the Tayara site (KbFk-7), Qikirtaq Island, Nunavik (Canada). Arctic 61:87–101

Tomaselli MB, Ortiz David LD, González Riga BJ (2021) Remarks on paleoenvironmental Cretaceous contexts of South American sauropod tracks. J S Am Earth Sci 110:103393

Ullmann PV, Lacovara KJ (2016) Appendicular osteology of *Dreadnoughtus schrani*, a giant titanosaurian (Sauropoda, Titanosauria) from the Upper Cretaceous of Patagonia, Argentina. J Vertebr Paleontol 36(6):e1225303

Upchurch P (1998) The phylogenetic relationships of sauropod dinosaurs. Zool J Linn Soc 124:43–103

Upchurch P, Barrett P, Dodson P (2004) Sauropoda. In: Weishampel DB, Dodson P, Osmólska H (eds) The Dinosauria, 2nd edn. University of California Press, Berkeley, pp 259–324

Vila B, Galobart A, Oms O, Poza B, Bravo AM (2010) Assessing the nesting strategies of Late Cretaceous titanosaurs: 3-D clutch geometry from a new megaloolithid eggsite. Lethaia 43(2):197–208

Voorhies MR (1969) Taphonomy and population dynamics of the early Pliocene vertebrate fauna, Knox County, Nebraska. Geol Soc Am Spec Pap 1:1–69

Wedel MJ, Sanders RK (2002) Osteological correlates of cervical musculature in Aves and Sauropoda (Dinosauria: Saurischia), with comments on the cervical ribs of *Apatosaurus*. PaleoBios 22:1–6

White PA, Diedrich CG (2012) Taphonomy story of a modern African elephant *Loxodonta africana* carcass on a lakeshore in Zambia (Africa). Quatern Int 276:287–296

Willis BJ, Behrensmeyer AK (1995) Fluvial systems in the Siwalik Miocene and Wyoming Eocene. Palaeogeogr Palaeoclim Palaeoecol 115(1–4):13–35

Wilson JA (2002) Sauropod dinosaur phylogeny: critique and cladistic analysis. Zool J Linn Soc 136:217–276

Wilson JA (2005) Redescription of the Mongolian sauropod *Nemegtosaurus mongoliensis* Nowinski (Dinosauria: Saurischia) and comments on Late Cretaceous Sauropod diversity. J Syst Palaeontol 3:283–318

Wilson JA, Sereno PC (1998) Early evolution and higher-level phylogeny of sauropod dinosaurs. J Vertebr Paleontol 18(2):1–68

Wilson JA, Pol D, Carvalho AB, Zaher H (2016) The skull of the titanosaur *Tapuiasaurus macedoi* (Dinosauria: Sauropoda), a basal titanosaur from the Lower Cretaceous of Brazil. Zool J Linn Soc 178(3):611–662. https://doi.org/10.1111/zoj.12420

Zaher H, Pol D, Carvalho AB, Nascimento PM, Riccomini C, Larson P, Juárez-Valieri R, Pires-Domingues R, da Silva Jr, de Almeida Campos NJD (2011) A complete skull of an Early Cretaceous Sauropod and the evolution of advanced Titanosaurians. PLoS One 6(2):e16663